ISBN 978-3-642-45877-4 ISBN 978-3-642-45875-0 (eBook)
DOI 10.1007/978-3-642-45875-0

Ursprünglich erschienen bei Springer-Verlag oHG, Berlin . Göttingen . Heidelberg 1957
Softcover reprint of the hardcover 1st edition 1957

HANDBUCH DER PHYSIK

HERAUSGEGEBEN VON

S. FLÜGGE

BAND XL

KERNREAKTIONEN I

MIT 280 FIGUREN

SPRINGER-VERLAG BERLIN HEIDELBERG GMBH

1957

ENCYCLOPEDIA OF PHYSICS

EDITED BY

S. FLÜGGE

VOLUME XL

NUCLEAR REACTIONS I

WITH 280 FIGURES

SPRINGER-VERLAG BERLIN HEIDELBERG GMBH

1957

Contents.

Nuclear Reactions, Levels, and Spectra of Light Nuclei.

By

W. E. BURCHAM.

With 102 Figures.

Introduction.

The study of the nuclear reactions of the light nuclei is not at present one of the basic investigations of nuclear physics. Already in 1935 attention was being concentrated on the simpler scattering problems, which would lead to some knowledge of the forces between nucleons. At present, emphasis is placed increasingly on attempts to interpret the inter-nucleon force in terms of a quantised field theory in which π-mesons are the quanta. The behaviour of the meson-nucleon system and the role played by new unstable particles in field theory have therefore become the central questions of nuclear physics. At first sight it might appear that the interactions of relatively complex systems of nucleons, moving with non-relativistic velocities, could contribute little to an answer to these questions. This is however not so, at least in so far as the force between nucleons is concerned. Although important properties of this force can be inferred from the two-body system and from the knowledge that nuclear forces show saturation, some properties, such as charge independence, do seem to be best exhibited in the existence of nuclear charge multiplets. The existence of non-central couplings is perhaps basically a matter for high energy experiments, but such properties are also revealed in nuclear level splittings and in the sequence of single particle states of the nuclear shell model, and the possibility of many-body forces can only be investigated through a detailed knowledge of many-body systems. These reasons, apart from the general interest of the subject as a field for experimental endeavour, justify the effort still being expended in investigating the behaviour of light nuclei, and in tabulating their excited states.

Both the theoretical and the experimental study of the excited states of nuclei have been much helped by progress in the understanding of atomic structure. Although the nucleus and the atom are held together by forces of a fundamentally different character, there has been no reason to doubt that quantum mechanics provides a satisfactory description of most of the properties of both these systems. For this reason there are notable resemblances between atoms with a size of 10^{-8} cm and nuclei, whose extent is 10^5 times smaller. The boundary conditions for the two systems both lead to a set of discrete energy levels above the ground state, in which atoms or nuclei may exist for a finite time. Transitions of the atom or nucleus between two of its quantised energy levels may be accompanied by emission or absorption of radiation, and the radiation field is coupled to the multipole moment of the radiating system in the same way, despite the great disparity in size. It is true that the difference in origin of the fields of electric and nuclear force leads to corrections to such properties as magnetic moments which are small for electrons and large for protons, but the general similarity between more complex atoms and nuclei as radiating systems exists.

The periodic classification of the elements, based on the quantum theory of the atom and the PAULI exclusion principle has clearly demonstrated the existence of a shell structure in atoms. Although the relatively simple atomic-type of structure with electrons moving round an attracting centre does not exist in nuclei it has recently become clear that many nuclear properties do exhibit that marked periodicity with atomic weight which suggests such a structure. The reconciliation of these discoveries with the earlier ideas of BOHR, which envisaged the nucleus as a closely packed system of strongly interacting particles has been a major concern of theoretical nuclear physics since 1950. It is a matter to which the detailed experimental study of nuclear states can contribute.

The experimental methods used in the study of atomic states are only infrequently of direct use in nuclear physics owing to the different scale of energies involved, but the general collision theory developed to describe the bombardment of atoms by electrons is of basic importance in nuclear dynamics. The classification of such collision processes into elastic and inelastic types is also directly relevant. Since however the apparatus of nuclear physics is generally more complex and often different in conception from the equipment used in atomic spectroscopy, and since it often influences very markedly the type of information which may be sought, the proper appreciation of experimental results demands an account of the methods by which they have been obtained. This is given in this article in Part C.

In Part A an attempt is made to pick out those general properties of inter-nucleon forces which may be expected to influence the structure of complex nuclei in a way that is susceptible of experimental study. A brief account is given of the theoretical approach to an interpretation of discrete level schemes and of the consequences of particular coupling schemes for angular momentum vectors in a complex system. Part B reviews the properties of the excited states of a complex nucleus with particular reference to their formation and decay in the course of a nuclear reaction. Part C is devoted to the experimental methods which have been most successful in revealing the properties of nuclear states. In Part D the experimental results obtained from nuclear reactions (up to July 1955) are reviewed according to reaction type. Only the main results are given, with emphasis on those which may be interpreted theoretically. Fuller details are given in [*32*] and [*33*]. Nuclear energy level systems and reaction constants for particular levels are reviewed according to mass number in Part E. The review extends to the end of the nuclear *d*-shell at mass number $A = 40$.

Similar treatments of the general subject of the excited states of light nuclei, including both experimental and theoretical aspects, are given in [*1*] to [*4*].

A. Nuclear level schemes for the light nuclei.

1. The interaction between nucleons[1]. Detailed calculations of nuclear structure and in particular the estimation of the energies of excited states, must be based on knowledge of the force between nucleons in a nucleus. Since no experiment gives directly the properties of this force, it is customary to make certain definite assumptions and to see how far experimental results can be so explained[2]. The first assumptions, against which there is no evidence, are that the

[1] Details on this problem are given in HULTHÉN and SUGAWARA's article, Vol. XXXIX of this Encyclopedia.

[2] R. E. PEIERLS: Sixth RUTHERFORD Lecture: The atomic nucleus and its constituents. Proc. Phys. Soc. Lond. A **66**, 313 (1953).

force is of a *two-body type and attractive*, that it is *static*, i.e. that no terms depending on nucleon velocities enter into the potential energy, and that it is mainly of a *central* character, so that the potential energy can be written

$$V_{12} = J(r_{12}) \tag{1.1}$$

where r_{12} is the distance between nucleons 1 and 2. The main evidence as a result of which these assumptions may be modified and supplemented comes from a study of the two body problem and from observation of regularities in nuclear structure. In the first place it is known from scattering experiments that the force is of short range; in contrast with the long range COULOMB forces familiar in the atom, the force between nucleons vanishes if r_{12} is greater than a few times 10^{-13} cm. Since the force is assumed to be attractive[1] all the nucleons of a complex nucleus might be expected to come so close together that the binding energy of a nucleon to a nucleus would increase with the number of particles. This is not so, since the binding energy and nuclear density seem to be nearly independent of mass number. The generally accepted explanation of this phenomenon of saturation is that the force between two nucleons is partly of an *exchange* character, and that attraction only arises when the pair of interacting particles can exchange position, or spin, or both. Since the PAULI principle limits the number of particles with which exchanges can take place, such forces will show saturation. The exchange nature of the force also provides an explanation of the angular distribution observed in the scattering of high energy neutrons by protons.

The exchange properties are included in the nucleon-nucleon interaction by writing

$$V_{12} = J(r_{12})\,(W + M P_x + B P_\sigma + H P_\tau) \tag{1.2}$$

where the operators P_x, P_σ, P_τ interchange respectively position, spin and charge co-ordinates in nucleon wave functions on which V_{12} may operate. The coefficients W, M, B, H (WIGNER, MAJORANA, BARTLETT, HEISENBERG) determine the precise nature of the force; a frequently used set of values, chosen to ensure saturation, is given in [*8*], p. 234.

The inclusion of the spin exchange terms B and H in the two-body interaction is sufficient to interpret the difference in energy between the ^3S and ^1S states of the deuteron, which is clear evidence of the spin dependence of the neutron-proton force. This property is of considerable importance in determining the spin of the ground state of more complex nuclei, particularly of mass number $4n+2$. The general interaction (1.2) is *charge-independent* in the sense that it postulates the same interaction between all pairs of nucleons in corresponding states. In such states it must be possible for a neutron to be changed into a proton or vice versa without violating the PAULI principle. It follows that the ^1S state is possible for the (pp) (nn) and (np) systems, while the ^3S state is only possible for the (np) interaction. It is customary to label states of motion of the two nucleon system in which all three pairs of nucleons may take part as states of *isotopic spin* $T=1$, and those describing only the neutron-proton system as states of *isotopic spin* $T=0$. This rather formal classification[2] has been found useful in comparing the energy levels of isobaric nuclei. The charge independence of nuclear forces receives considerable direct support from low energy nucleon scattering experiments, according to which the singlet intrinsic range

[1] There is evidence from high energy experiments that the internucleon force becomes repulsive at very small distances.

[2] The term *isobaric spin* is also frequently used.

for both the (np) and (pp) scattering appear to be about the same. Charge independence follows very readily from the symmetrical meson theory of nuclear forces. Deviations of nuclear behaviour from the consequences of strict charge independence can probably be ascribed to the effect of the charge dependent COULOMB forces.

The central potential V_{12} is known to be only a first approximation because of the existence of an electric quadrupole moment in the deuteron. This is most easily explained by a tensor interaction according to which the force between nucleons depends on the relative orientation of the spin and orbital vectors. Such a force may considerably modify the wave functions obtained with a pure central potential and the consequences for nuclear theory in general have not yet been fully worked out. It may be found that the phenomenological spin-orbit interaction

$$H = a(\boldsymbol{l}\cdot\boldsymbol{s}) \tag{1.3}$$

which is found necessary to explain doublet splitting in nuclear spectra, and possibly also the ${}^1S - {}^3S$ splitting in the deuteron are consequences of the tensor force.

Any general theory of nuclear structure suitable for comparison with experiment should include the properties of nuclear forces so far discussed in successive approximations. The first general discussion along these lines of the level systems of complex nuclei was given by WIGNER and his collaborators.

2. The WIGNER approximation. *α) Classification of states.* In the general classification of nuclear states proposed by WIGNER ([*8*], p. 199, and [*9*]) it assumed that nuclear forces are independent of both charge and spin. The coefficients H and B in Eq. (1.2) are put equal to zero, so that the exchange part of the force is of the MAJORANA type calling for exchange of spatial positions of nucleons. The quantum numbers used to specify a nuclear state are based on conservation laws. The most general are (cf. [*8*], p. 199).

(a) J, the total angular momentum quantum number, related to the total angular momentum vector $\boldsymbol{J}$ by the equation

$$|\boldsymbol{J}| = \sqrt{J(J+1)}\,\hbar\,. \tag{2.1}$$

(b) π, the parity, which must be ± 1, i.e. even or odd, because of the invariance of nuclear structure with respect to reflection of the wave function through the origin of coordinates.

(c) T_z the third component of total isotopic spin which is defined as

$$T_z = \tfrac{1}{2}\sum_i \tau_z^i = \tfrac{1}{2}(Z-N) = \tfrac{1}{2}A - N \tag{2.2}$$

and measures the excess of protons over neutrons in the nucleus. Z is the atomic number, A the mass number and N the neutron number. τ_z is the third component of the isotopic variable $\boldsymbol{\tau}$ which is used, following HEISENBERG, to label the charge state of a nucleon. Here $\tau_z = 1$ is taken to mean a proton and $\tau_z = -1$ a neutron.

These quantum numbers are related directly to the properties of space and time and to the conservation of charge, and have definite values for a nuclear state $(J, \pm, T_z)$. Other quantum numbers which are valid under special circumstances are:

(d) T, the total isotopic spin quantum number which may be defined if nuclear forces are charge independent, and is related to a formal vector $\boldsymbol{T} = \tfrac{1}{2}\sum \boldsymbol{\tau}^i$ of total isotopic spin by the equation

$$|\boldsymbol{T}| = \sqrt{T(T+1)}\,. \tag{2.3}$$

For a given T, T_z has $2T+1$ values each corresponding to the same binding energy and defining a set of isobaric nuclei of different charge in each of which this particular energy state may be identified. This group of nuclear states is a *charge-multiplet*.

(e) L and S, the quantum numbers for total orbital and total intrinsic angular momentum. The corresponding vectors add to make the total angular momentum J of a nuclear state:

$$\boldsymbol{J} = \boldsymbol{L} + \boldsymbol{S}. \tag{2.4}$$

If the interaction between spin and orbital motion may be neglected, L and S are good quantum numbers and J may have $2S+1$ values, between $|L-S|$ and $L+S$. These states form a *spin multiplet*. The total spin S is compounded from the spins of individual nucleons by the summation $\boldsymbol{S} = \frac{\hbar}{2}\sum_i \sigma^i$ and L is compounded from the orbital vectors. These orbital vectors are conveniently obtained from a simple shell model for a given number of nucleons and the directions of the spin vectors are then determined by application of the PAULI principle.

In the WIGNER theory, the assumption of charge and spin independence of the nuclear forces implies that the states of charge and spin multiplets do not differ in energy. A group of degenerate nuclear states, including perhaps several charge and spin multiplets is known as a *supermultiplet*. The states of a supermultiplet have all the same orbital angular momentum and parity, since these are related to the spatial part of the wave function. The charge and spin multiplets are inter-related through the exclusion principle, and together determine the symmetry of the spatial part of the wave function with respect to exchange of nucleons.

The symmetry properties of the spatial wave functions of a WIGNER supermultiplet cannot be simply described in the many-body case, because the function may be symmetric for spatial exchange of some pairs of particles and antisymmetric for exchange of other pairs. The supermultiplet may be characterised by four numbers $\lambda_1, \lambda_2, \lambda_3, \lambda_4$, with $\sum_i \lambda_i = A$ the mass number, representing respectively the number of neutrons and protons with spin "up" and spin "down". An equivalent representation is in terms of the numbers

$$\left.\begin{aligned} P &= \tfrac{1}{2}(\lambda_1 - \lambda_2 + \lambda_3 - \lambda_4), \\ P' &= \tfrac{1}{2}(\lambda_1 + \lambda_2 - \lambda_3 - \lambda_4), \\ P'' &= \tfrac{1}{2}(\lambda_1 - \lambda_2 - \lambda_3 + \lambda_4) \end{aligned}\right\} \tag{2.5}$$

which are respectively the maximum values of S_z, T_z, and $Y_z\left(=\frac{1}{2}\sum_i \sigma_z^i \tau_z^i\right)$ in the supermultiplet.

The individual states of a given supermultiplet (P, P', P'') are thus classified by writing down the permitted spin-charge multiplets (S, T) or (T, S) and combining with these the orbital momentum allowed for the component particles; the individual multiplet symbol is then ${}^{(2T+1)(2S+1)}L$.

The states of n equivalent particles with orbital angular momentum l may be classified by group theoretical methods. The technique and results for the s, p, d and f shells are reviewed by FLOWERS [*11*].

β) Calculation of binding energies. In the WIGNER approximation the binding energy of a nuclear state for a given nuclear interaction depends on the symmetry properties of the spatial part of the wave function Ψ of the supermultiplet to which the (degenerate) state belongs. The potential energy $\int \Psi^* V \Psi dv$ for s

and p shell nuclei may be expressed in the form

$$pL + qK \tag{2.6}$$

where

$$\left.\begin{aligned} L &= \iint |\psi_1(v_1)|^2 V_{12} |\psi_2(v_2)|^2 \, dv_1 dv_2 \\ K &= \iint \psi_1^*(v_2)\, \psi_2^*(v_1)\, V_{12} \psi_1(v_1)\, \psi_2(v_2)\, dv_1 dv_2 \end{aligned}\right\} \tag{2.7}$$

and

are the direct and exchange integrals taken for pairs of nucleons and $\psi_1\psi_2$ are single particle wave functions. The ratio of p to q in the final interaction energy is determined by the exchange nature of the potential V_{12}, and by the number of interacting particles and their spatial states. The evaluation of the integral for the two-body problem is discussed simply by INGLIS [10].

In numerical calculations a particular form for the radial dependence of V_{12} must be assumed. This is usually taken to be either a YUKAWA potential

$$V_c \frac{e^{-r_{12}/r_0}}{r_{12}/r_0} \tag{2.8}$$

or a GAUSS potential

$$V_c e^{-r_{12}^2/r_0^2}, \tag{2.9}$$

and the depth of the central potential V_c is either taken from two-body data or used as a variable parameter. The single particle wave functions are usually taken to be those of an oscillator potential for convenience and the extent of these wave functions, determined by the width of the potential well, introduces a further parameter b into the calculation. For p-shell nuclei the ratio of r_0, the range of the forces to b, the extent of the wave function, determines the ratio of L to K. For long range forces $L = V_c$ and $K = 0$; for reasonable values of r_0/b [10]

$$L = 6K. \tag{2.10}$$

The radial integrals for d-shell nuclei are more numerous and the procedures adopted are discussed by ELLIOTT and FLOWERS [45].

Calculations of total nuclear binding energy based on quasi-atomic models in which nucleons are assumed to move in definite orbital states are not particularly satisfactory [9]. The general supermultiplet theory does however make some useful predictions, in particular that the ground states of light nuclei should belong to supermultiplets in which the wave functions have maximum spatial symmetry. This is because of the strong component of MAJORANA (space exchange) force assumed in the basic interaction. Moreover the quasi-atomic calculations may be expected to give approximately the correct order of the different multiplets $^{(2T+1)(2S+1)}L$ within a given supermultiplet, in terms of the radial integrals.

The WIGNER scheme leads naturally to the RUSSELL-SAUNDERS or LS system of coupling for spin and orbital angular momenta in which $\boldsymbol{L}$ and $\boldsymbol{S}$ are combined to give $\boldsymbol{J}$, and in this system the separation of the lowest multiplets is given by FEENBERG and WIGNER [9] and by FEENBERG and PHILLIPS [9]. The results of the latter authors are used in this article, following INGLIS [10], for comparison with experimental data. They show that in LS coupling the ground states of even-even nuclei should be ^1S states and those of odd nuclei should be doublets with $L = l$, the orbital quantum number of the ground state configuration. The predictions of the LS coupling scheme for the spins of odd-odd nuclei are discussed by FLOWERS [11]. For $A = 2Z$ the ground state is ^3S if the forces are of the type known in the deuteron, but if $A \neq 2Z$ the lowest states are $^3P_{0,1,2}$ in the p-shell and $^3F_{2,3,4}$ in the d-shell. The order of the different J-values is determined by the spin-orbit forces.

γ) Resolution of the (ST) multiplets. The introduction of charge-dependent and spin-dependent forces resolves the (ST) multiplets into individual states. The degeneracy with respect to T_z is removed by COULOMB forces and the $(2T+1)$ states with $|T_z| \leq T$ represent isobaric nuclei in each of which there exists a state of given spin and parity and with a binding energy which differs from that of the state in the adjacent isobar by a COULOMB energy. Thus in the odd mirror nuclei of the p-shell the lowest supermultiplet always contains the (S, T) value $(\frac{1}{2}, \frac{1}{2})$ and the energy difference between the ground states of the two odd mirror nuclei with $T_z = \pm\frac{1}{2}$ is directly predictable from the COULOMB energy (Sect. 56). In the nuclei of mass $4n+2$ the lowest supermultiplet is $(P, P', P'') = (1, 0, 0)$ which contains the (S, T) combinations $(1, 0)$ and $(0, 1)$, so that there should be three nuclei, with $T_z = \pm 1$ and 0, in which a state of even parity and spin zero is found with a predictable difference in energy from that in the adjacent isobar. The verification of the existence of these charge multiplets is strong evidence for the charge independence of nuclear forces, which is one of the basic assumptions of the theory of WIGNER.

For given T, the introduction of HEISENBERG-BARTLETT or tensor forces separates out states of different spin S. As is known from the deuteron, the state of higher spin lies lower. The degeneracy with respect to S_z is removed by the introduction of forces coupling spin to orbital motion, leading to $2S+1$ states of different J for $L > 0$.

The separation of states of different T_z implies that one of a set of isobaric nuclei is stable and the others unstable against beta decay. The WIGNER theory predicts superallowed decay between the states of a given supermultiplet because no change of spatial wave function is needed. This is found (a) for the positron decay of odd mirror nuclei; (b) for transitions between the low states of nuclei with mass number $4n+2$ in which both $T=1$ and $T=0$ states are found in the $(1, 0, 0)$ supermultiplet.

Any charge independent theory would predict superallowed transitions between the states of a charge multiplet, but the fact that they are also found between the states of a supermultiplet (see Shap. E II and E III) is strong evidence for the WIGNER classification. The observation of such superallowed transitions, which are discussed by BOLSTERLI and FEENBERG [*46*], sometimes permits identification of analogous states in isobaric nuclei.

3. The nuclear shell model; coupling schemes[1]. In discussing the orbital momenta of the WIGNER multiplets a shell model for the individual nucleons has been assumed. The possibility of predicting nuclear properties from the model of a single particle moving in a central field of force has been discussed since the discovery of the neutron in 1932, but seemed inconsistent with the strong interaction postulated in the compound nucleus model of BOHR. By 1948 however it had become clear that many nuclear properties showed discontinuities at the neutron or proton numbers 2, 8, 20, 28, 50, 82 and 126, and this led MAYER and HAXEL, JENSEN and SUESS to propose the *single particle shell model.* The historical development of this model is reviewed in [*11*] and [*12*]. The important assumptions are that:

(a) each nucleon moves in an average field due to the remaining nucleons of the nucleus,

(b) the average potential may be approximately represented as that of an isotropic harmonic oscillator,

[1] The nuclear shell model is treated in detail by LANE and ELLIOTT in vol. XXXIX of this Encyclopedia. See also R. J. BLIN-STOYLE: Rev. Mod. Phys. **28**, 75 (1956) for a recent discussion of nuclear models.

(c) a strong spin-orbit interaction lowers the doublet state of a single particle of the higher j; i.e. $j=l+\frac{1}{2}$ is a more stable state than $j=l-\frac{1}{2}$. The spin-orbit coupling is required to increase with l.

With these assumptions it is possible to explain the "magic" numbers. The single particle level schemes for N and $Z<50$ are shown in Fig. 1; the model is essentially one of (jj) coupling and the single particle states are characterised by the symbol l_j. The building up of a sequence of nuclei is then a matter of filling each level with $2(2j+1)$ nucleons. A complete allocation of nucleons to shell model states is given by KLINKENBERG [13]. The ground states of odd nuclei are single particle states according to this model, and further single particle states may in some cases be reached by exciting a single nucleon.

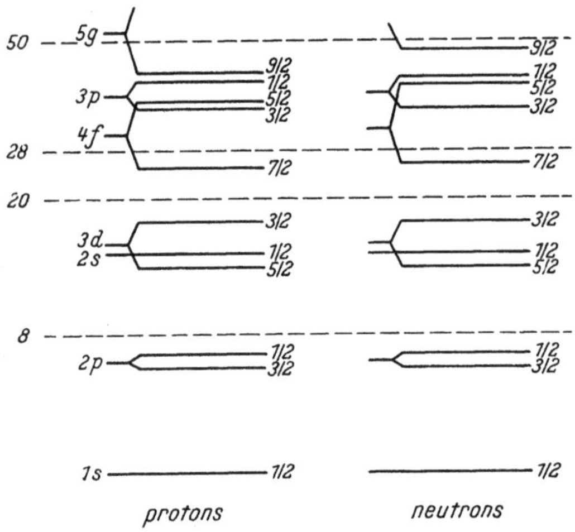

Fig. 1. Single particle levels for $N, Z<50$. [11], [13].

The single particle shell model gives a good account of many properties of the ground state such as spin, magnetic moment and electric quadrupole moment for odd mass and even-even nuclei. In particular even-even nuclei should have spin zero and even parity in their ground state and an odd configuration $(l_j)^n$ should have spin j and parity $(-1)^l$. There are however sufficient important deviations from the predictions of the single particle model to make it evident that this model is incomplete. Thus there is the well known spin anomaly for Na^{23} (predicted $\frac{5}{2}$, observed $\frac{3}{2}$) and the fact that for many nuclei the magnetic moments fall between the single particle values corresponding to $j=l\pm\frac{1}{2}$. For this reason it has been necessary to develop the single particle model into the *individual particle shell model*[1], in which all nucleons outside closed (jj) shells are considered to exist in equivalent single particle orbits. Thus the nucleus Li^7 is treated as a case of $p^3_{3/2}$ rather than as a single $p_{3/2}$ nucleus. The classification of the states of the configuration j^n in jj coupling according to the independent particle model has been given by FLOWERS[2] for $j=\frac{3}{2}, \frac{5}{2}, \frac{7}{2}$. Total angular momentum J and total isotopic spin T remain good quantum numbers, but L and S are no longer significant.

The order of energy levels in the jj coupling model is found by making calculations with a potential of the type (2.8) or (2.9) and with oscillator wave functions. With these assumptions EDMONDS and FLOWERS, KURATH and TALMI [14] have calculated the ordering of the levels of the configurations j^n with $j=\frac{3}{2}, \frac{5}{2}$ or $\frac{7}{2}$. This order is dependent on the ratio of the range of the forces to the extent of the wave function as in the LS calculation and this ratio may be chosen to give agreement with observed experimental characteristics such as ground state spin. For zero range forces it is found that a configuration containing an even number of neutrons or of protons has $J=0$ while a configuration containing an odd number of particles of one kind has $J=j$ in the ground state.

[1] Often described as the *independent particle model*.

[2] B. H. FLOWERS: Proc. Roy. Soc. Lond., Ser. A **212**, 248 (1952).

The energy associated with the spin-orbit force is written in the form (1.3) for a single nucleon with orbital and spin angular momenta $\boldsymbol{l}$ and $\mathbf{s}$, and the total spin-orbit energy of a configuration is a sum of such terms. Extreme jj coupling arises when a is very large compared with the exchange integral K (2.7) which measures the extent to which orbital wave functions overlap. In extreme LS coupling K is large compared with a. The introduction of a small spin orbit coupling term splits the states of the (LS) multiplets and as the ratio a/K increases the pattern of energy levels changes to that predicted in the jj scheme. The level order in many p-shell nuclei seems to be well described by an intermediate coupling scheme (cf. [10] and Part E).

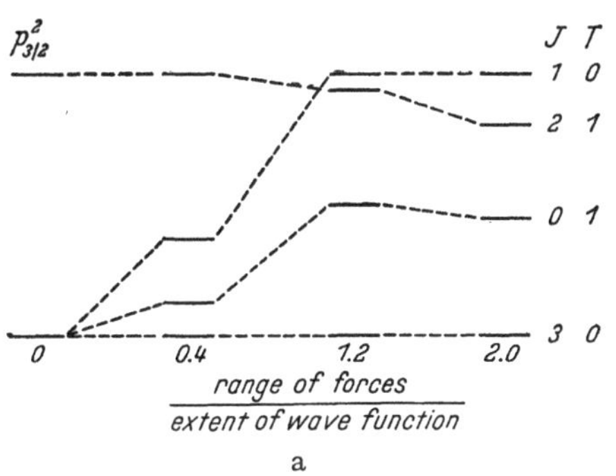

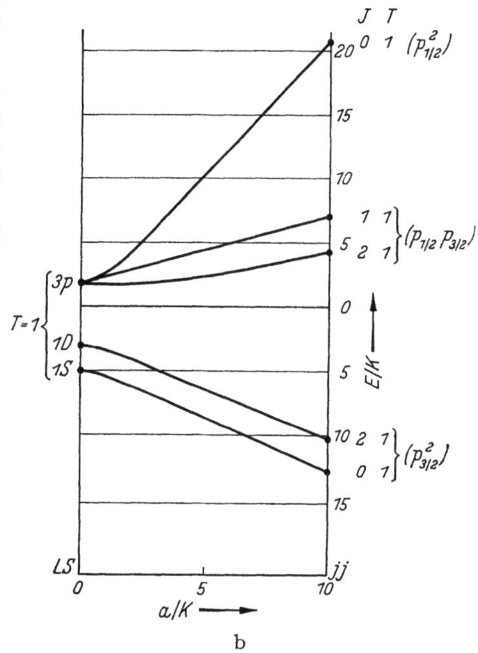

Fig. 2. (a) States of the nucleon configuration $p^2_{3/2}$ in jj coupling (Edmonds and Flowers [14]). (b) Transition from LS to jj coupling for $T=1$ states of two nucleons [10].

The states of two $p_{3/2}$ particles as a function of the ratio of range of forces to extent of wave function is shown in Fig. 2a, and the transition from LS to jj coupling for two neutrons or two protons $(T=1)$ is shown in Fig. 2b. In most of the intermediate coupling curves of reference [10] only the jj and LS asymptotes are accurately known. Detailed calculations of the transition curves have been given for p-shell nuclei by Crosbie[1] and by Kurath[2] and for certain p and d shell nuclei by Tauber and Wu[3], who investigate the effect of varying both the type and range of the central force.

The wave functions obtained from LS, jj and intermediate coupling calculations enable predictions of spin, magnetic moment, electric quadrupole moment, radiative transition probability and reduced width to be made. The coupling scheme may then be deduced by comparison with experiment (e.g. $C^{13}N^{13}$, Sect. 61).

The individual particle model describes in the first place the states of a given configuration $(l_j)^n$. If one of the nucleons of this configuration is excited to another single particle state, the configuration $(l_j)^{n-1}\,(l'_{j'})$ is obtained, with a higher energy. The wave function of a particular state of given (J, T) may contain contributions from several different configurations, which are mixed in by the particular type of inter-nucleon force assumed. This configurational interaction is particularly

[1] E. A. Crosbie: Phys. Rev. **90**, 138 (1953).
[2] D. Kurath: Phys. Rev. **101**, 216 (1956).
[3] G. E. Tauber and Ta-You Wu: Phys. Rev. **93**, 295 (1954).

important for instance where single particle states lie close together, as in the $2s-1d$ shell for mass number near 20. Near magic numbers the configurational interaction may be expected to be rather small compared with the energy difference between different configurations, but at the middle of a shell the mixing may become very complicated, particularly for the heavier nuclei. This has led to the development of models in which more emphasis is placed on collective aspects of nuclear motion.

4. Collective models and the nuclear level spectrum. Many of the low-lying bound states of light nuclei cannot be adequately described by the individual particle shell model because of lack of agreement in excitation energy, spin, parity or other properties such as electric quadrupole and radiative transition moments. It is possible to obtain some experimental features by postulating configurations in which a particle is removed from a closed shell. The effect of the closed shells has also been introduced in a very successful way by A. BOHR[1] who considers the coupling of the motion of a single particle in an orbit to the nuclear core responsible for the potential field in which the single particle moves. In this model the potential field is no longer static or spherically symmetrical, but becomes deformed by the influence of the single particle motion[2]. The energy of such a system may be calculated from the rotation of the nuclear deformation in terms of an effective moment of inertia I. For even-even nuclei the excited states have spins 2, 4, ..., with even parity and the energies are

$$E_J = \frac{\hbar^2}{2I} J(J+1) \tag{4.1}$$

while for odd nuclei with spin $\frac{3}{2}$ the rotational sequence is

$$E_J = \frac{\hbar^2}{2I} \left(J(J+1) - J_0(J_0+1)\right) \tag{4.2}$$

where J_0 is the spin of the ground state and $J = J_0+1$, J_0+2, ...

The model appears to work best for mass numbers near the middle of shells, where there are several particles available for coupling with the motion of the core. It predicts a rotational fine structure of energy levels on a scale small compared with the single particle level spacings, and it also predicts large quadrupole moments and $E2$ transition probabilities which are large compared with the single particle values. On the other hand single particle gamma ray transitions may be reduced in probability in this model because of the complexity of the motion. These features have been verified experimentally, particularly in the COULOMB excitation of low lying states (Sect. 22) of the heavier elements, and it is almost certain that many features of the level spectra of light elements may also be explained by this model. In particular, it has been shown by SCHARFF-GOLDHABER[3] that the spin of the first excited state of even-even nuclei is frequently 2^+; this is the result expected from the collective model, although it is also predicted by the extreme (jj) coupling model for an even configuration of like nucleons. Collective features are known in the COULOMB excitation cross section for the low levels of F^{19} and in the properties of the levels of Al^{25}. Further

---

[1] A. BOHR: Kgl. danske Vid. Selsk., mat-fys. Medd. **26**, No. 14 (1952). — A. BOHR and B. MOTTELSON: Kgl. danske Vid. Selsk., mat-fys. Medd. **27**, No. 16 (1953).

[2] Details in the article of S. MOSZKOWSKI in vol. XXXIX, of this Encyclopedia.

[3] G. SCHARFF-GOLDHABER: Phys. Rev. **90**, 587 (1953). See also P. STÄHELIN and P. PREISWERK: Nuovo Cim. **10**, 1219 (1953).

studies of transition probabilities in light nuclei will probably further establish at least the partial validity of the BOHR-MOTTELSON model[1].

Among collective models which have been developed for special purposes is the *alpha particle model*. If it is assumed that the four-structure already apparent in the supermultiplet theory of WIGNER has some permanence in an actual nucleus, the methods of molecular physics may be applied to deduce a level sequence (WHEELER[2]). The symmetry of the system imposes special restrictions on the level spins of the lower excited states. The energies of these states are obtained from a decomposition of the motion into rotation and vibration. The treatment is especially suitable for nuclei of mass number $4n$, but it is not restricted to these[3] and the detailed application of the model to a number of nuclei is discussed in references [*7*] and [*10*].

5. Virtual levels. The bound levels, of which a fairly complete description can be given, form only a small part of the complete nuclear level scheme. Because of the short range of nuclear forces the spacing of levels in a nucleus is much more nearly uniform than in an atom, and there is no great difference between the levels just below and just above the binding energy of a given particle in the nucleus. This is a consequence of the change of wave number of an incident or emergent particle as it crosses the nuclear surface, which leads to boundary conditions for the internal wave function which are not greatly affected by the possibility of nucleon escape until the excitation energy greatly exceeds the internal well depth. Levels unstable to particle emission are known as *virtual levels* and may be examined as compound states in nuclear reactions. Although it is only occasionally that theoretical predictions of the properties of virtual levels can be made, the study of nuclear reactions provides a great deal of information about their properties and particularly about their width and separation.

The earliest theories of nuclear reactions considered the interaction of an incident particle with a potential well. This model predicts (a) a predominance of scattering over absorption (because the nucleon spends very little time near the nucleus), and (b) broad well-separated resonance levels corresponding to states of the particle in the potential well. These are in fact the single particle states of the nuclear shell model (Sect. 3) and for a given angular momentum the separation of these states is perhaps 10 MeV; the width is about 1 MeV. Experimentally it was found that slow neutron reactions in fairly heavy elements showed narrow closely spaced resonances of the order of 0.1 eV width and perhaps 10 eV spacing and that radiative capture rather than scattering was the main interaction. The N. BOHR[4] theory of the compound nucleus offered an interpretation of these facts by discarding the idea of interaction with a potential well and by concentrating on the strong coupling between the incident particle and the many particles inside a nucleus. In a many body problem of this type the process of sharing the energy of the incident particle with the other particles of the system takes a time long compared with the time of transit of a particle through a potential well. The resonances are therefore narrow and numerous; since they are now the quantum states of a

[1] Since this article was written a great deal of evidence for the validity of a collective description of the levels of light nuclei has accumulated. It seems likely that the individual particle model, modified by configurational interaction, and the collective model with coupling to single particle motion, are alternative modes of description.

[2] J. A. WHEELER: Phys. Rev. **52**, 1083 (1938).

[3] L. R. HAFSTAD and E. TELLER: Phys. Rev. **54**, 681 (1938). — R. R. HAEFNER: Rev. Mod. Phys. **23**, 228 (1951).

[4] N. BOHR: Nature, Lond. **137**, 344 (1936). — N. BOHR and F. KALCKAR: Kgl. danske Vid. Selsk., mat-fys. Medd. **14**, No. 10 (1937).

many body system scattering becomes less probable than radiative capture, as found experimentally.

The assumptions of N. BOHR led to a theory of nuclear reactions which was extraordinarily successful and which diverted attention from the fact that the many regularities of nuclear properties demanded some kind of shell model interpretation in which distinct orbits for the incident particle could still be assumed. The need for extending the N. BOHR model was also very clearly shown when the results of the scattering of fast neutrons (100 keV to 3 MeV) by nuclei (BARSCHALL[1]) became available. According to the BOHR picture, experiments of moderate resolution should be incapable of distinguishing the narrow, closely spaced resonances of the compound nucleus. The total cross section as a function of energy for this case has been calculated by FESHBACH and WEISSKOPF[2] and shows a monotonic decrease with energy to the asymptotic value $2\pi R^2$ for a scattering nucleus of radius R. Both the total and differential cross section however, plotted as a function of E, the bombarding energy and A, the target mass number, show "giant" resonances. These occur at very nearly the mass numbers, for a given energy, which would be predicted by the simple potential well theory and correspond to $s, p, d \ldots$ resonances in the quantity kR where k is the wave number of the particle inside the nucleus. On the other hand the lower energy neutron phenomena, with high resolution, do show the narrow resonances of the N. BOHR theory.

These facts led FESHBACH, PORTER and WEISSKOPF[3], following earlier suggestions, to propose in some detail the *cloudy crystal-ball* model of the nucleus in which an incident nucleon moves in a complex potential field:

$$V = -[V_0 + i V_1].$$

In this model, which is also known as the *optical model* following FERNBACH *et al.*[4], (who used it to describe high energy reactions) the complex potential $i V_1$ is supposed to represent compound system formation, in the sense of the N. BOHR or *uniform*, model. It represents any processes which are not described by the single particle interaction with the potential V_0. This model reproduces many of the experimental features of medium energy reactions. An important advance towards an interpretation of the potentials of the optical model in terms of two-nucleon interaction has recently been made by BRUECKNER[5].

A different approach to the problem of combining single particle and collective aspects of nuclear motion has been made by LANE, THOMAS and WIGNER[6]. In their model (the *intermediate model*) it is supposed that each single particle state is replaced by a group of levels of which the single particle state is the parent in the sense that the wave functions of these states contain contributions from that of the single particle state. In the BOHR model this process is carried so far that no trace of the single particle states is left, but in the intermediate model the mixing is important only over localised regions of the level spectrum near the single particle states.

Both these models are useful in describing the general trend of nuclear reactions over a wide range of mass numbers and bombarding energies. The reactions of the light nuclei are however highly individual and require more particular treatment.

[1] H. H. BARSCHALL: Phys. Rev. **86**, 431 (1952).
[2] H. FESHBACH and V. F. WEISSKOPF: Phys. Rev. **76**, 1550 (1949).
[3] H. FESHBACH, C. E. PORTER and V. F. WEISSKOPF: Phys. Rev. **96**, 448 (1954).
[4] S. FERNBACH, R. SERBER and T. B. TAYLOR: Phys. Rev. **75**, 1352 (1949).
[5] References are given in the paper by K. A. BRUECKNER, R. J. EDEN and N. C. FRANCIS: Phys. Rev. **100**, 891 (1956), which discusses the application of the theory to neutron reactions.
[6] A. M. LANE, R. G. THOMAS and E. P. WIGNER: Phys. Rev. **98**, 693 (1955).

B. General features of nuclear reactions.

6. Types of reaction. In this article mainly reactions of the general type

$$X + a \rightarrow Y + b + Q \tag{6.1}$$

are considered, in which there are only two product particles, and an energy release Q is observed. This reaction is usually abbreviated to $X(ab)Y$. In some cases the nucleus Y will be formed in an unstable state and will emit a further radiation. If the unstable state of Y is reasonably well defined the reaction may still be analysed in detail, but if the lifetime of Y is so short that the transition is really to a three body system

$$X + a \rightarrow Y' + b + c \tag{6.2}$$

it is more difficult to extract useful information[1]. Reactions such as (6.2) do however throw light on the statistical aspects of fairly heavy nuclei and are a limiting case of the spallation type of reaction observed with high energy particles[2]. The competition between processes of the type (xn), $(x\,2n)$ $(x\,3n)\ldots$ has been described by many authors in terms of evaporation of nucleons from a compound state and can be used to determine nuclear level densities (Sect. 8). The probability per unit time of a nuclear reaction such as (6.1) is determined by the general formula[3]

$$w = \frac{2\pi}{\hbar} |H|^2 \frac{dn}{dE} \tag{6.3}$$

where H is the matrix element causing the transition, and dn/dE is the number of final states of the system per unit energy range. This expression is related to the cross section $\sigma(ab)$ for the reaction by the formula

$$w = n\, v_a\, \sigma(ab) \tag{6.4}$$

where n is the number of nuclei X per cm³ and v_a is the velocity of a relative to X. The detailed mechanism of the reaction is contained in H and the following main types of reaction may be distinguished.

α) *Direct transitions*, in which the matrix element is of the form $\int \psi_f^* U \psi_i \, d\tau$ with U an interaction energy and ψ_i and ψ_f the initial and final wave functions. In processes of this type, in which stripping and pick-up reactions (Sect. 51) are included, the product particle has typically a high energy and an angular distribution with forward peak. There is no complete amalgamation of the particle a with the nucleus X but only perhaps some local interaction with the surface. The process is complete in the time, of the order of 10^{-22} sec., necessary for the particle a to traverse a nuclear diameter.

β) *Compound nuclear processes*, in which according to the theory of N. Bohr, the incident particle is assimilated by the target nucleus and enters into the motion of a many-body system. The lifetime of this compound system may be as much as 10^{-15} sec, and is influenced by the number of decay possibilities or *reaction channels* available at the given energy of excitation. The reaction $X(ab)Y$ proceeding through a compound nucleus C is written

$$X + a \rightarrow C \rightarrow Y + b \tag{6.5}$$

[1] See K. M. Watson: Phys. Rev. **88**, 1163 (1952) for a discussion of the effect of final state interactions on reaction cross sections.

[2] This type of reaction is discussed in the article by Wattenberg, vol. XL of this Encyclopedia.

[3] See, e.g., L. I. Schiff: Quantum Mechanics, p. 193. New York: McGraw Hill 1949.

and the matrix element H in (6.3) has the form[1]

$$\frac{H_{\mathrm{XC}} H_{\mathrm{CY}}}{E_{\mathrm{C}} - E_{\mathrm{X}}}$$

where H_{XC} and H_{CY} are matrix elements describing the two stages of the process, namely formation and decay of the compound nucleus C. In such processes, because of the sharing of energy among the many nucleons of the system, the emission of low energy particles is favoured if barrier effects are disregarded and angular distributions are more nearly isotropic than in the case of direct interactions.

It is commonly assumed in discussions of the N. BOHR theory of reactions that the cross section $\sigma(ab)$ for a nuclear reaction may be written in the form

$$\sigma(ab) = \sigma_c(a) \frac{\Gamma_b}{\Gamma} \tag{6.6}$$

where $\sigma_c(a)$ is the cross section for the formation of the compound nucleus from the system $\mathrm{X} + a$ and Γ_b is proportional to the probability of decay of the compound nucleus into the system $\mathrm{Y} + b$. As there are usually several modes of decay we have

$$\Gamma = \sum_b \Gamma_b \tag{6.7}$$

and if we write $\Gamma = \hbar/\tau$, where τ is the mean life of the compound state, then the quantities Γ are measured in energy units and, from their role in resonance reactions, are known as widths. The form (6.6) for the reaction cross section assumes that the decay of a compound nucleus is independent of its mode of formation. This is true if C is a well-defined state, well separated from other states, but if several quantum states are excited in varying proportions at the energy considered the independence assumption is almost certainly wrong because of the unknown phases of the component wave functions. The assumption regains some validity when there is such a large amount of mixing that a random distribution of phases may be assumed. Whether or not a compound state is well-defined, depends on the relative magnitude of the width Γ and the level spacing D, and it is convenient to consider separately two cases:

(a) $\Gamma \gg D$ in which no features of individual levels may be distinguished in the formation cross section $\sigma_c(a)$ and the reaction may generally be discussed by statistical methods (*continuum theory*).

(b) $\Gamma \ll D$ in which the reaction cross section shows well separated resonances. This type of reaction is the more important for the spectroscopy of the light nuclei; the evaluation of the matrix element leads to the *dispersion theory* of nuclear reactions.

The total cross section for interaction of particle a with the nucleus X is greater than $\sigma_c(a)$ because of elastic scattering of a by the potential fields surrounding X. Some elastic scattering is also obtained by re-emission of a particle a from the compound nucleus, with partial width Γ_a, which is included in Γ. It is therefore also useful to write

$$\sigma_{\mathrm{total}} = \sigma_{\mathrm{el}} + \sigma_{\mathrm{inel}} \tag{6.8}$$

where the two terms describe elastic and inelastic processes separately. These cross sections are related by the following conservation theorems according to the wave mechanical theory of scattering for a relative angular momentum $l\hbar$

[1] L. I. SCHIFF: Quantum Mechanics, p 196. New York: McGraw Hill 1949.

(see [7], p. 317)

$$\left.\begin{aligned}\sigma^l_{\text{total}} &\leq (2l+1)\,4\pi\lambda\!\!\!^-{}^2,\\ \sigma^l_{\text{el}} &\leq (2l+1)\,4\pi\lambda\!\!\!^-{}^2,\\ \sigma^l_{\text{inel}} &\leq (2l+1)\,\pi\lambda\!\!\!^-{}^2\end{aligned}\right\} \tag{6.9}$$

and when σ^l_{inel} has its maximum value σ^l_{el} has the same value so that σ^l_{total} is then $2(2l+1)\pi\lambda\!\!\!^-{}^2$. These relations are useful in setting an upper limit to the possible angular momentum l occurring in a reaction.

7. Penetration of potential barriers. In all types of nuclear reaction the incident particle must pass through, or into, a region of varying potential $V(r)$ before the reaction can take place. In general, the wave function of the incident particle is modified by the field, and the amplitude of this function at the nucleus differs from its asymptotic value. This leads to a penetrability factor in the reaction cross section. In addition, the incident particle having penetrated the external barrier suffers a change of wave number inside the nucleus and this in itself leads to some reflection at the nuclear surface. These factors are discussed in full in [7], p. 358, and only the main results are summarised here.

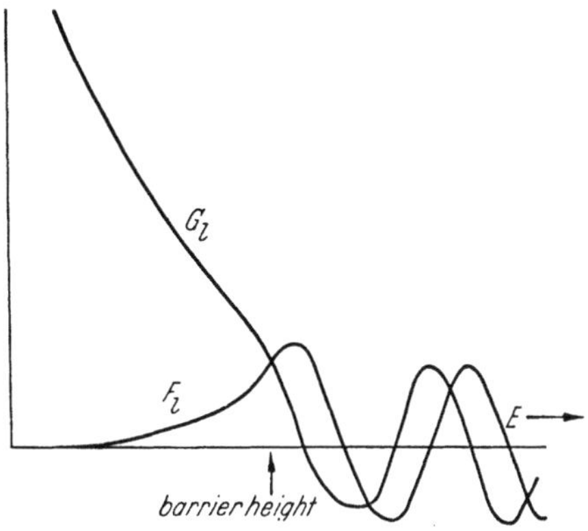

Fig. 3. The functions F_l, G_l of $\varrho = kr$ evaluated at a fixed radius as a function of energy $E = \frac{k^2\hbar^2}{2M}$ (schematic).

The radial wave equation for a particle moving with orbital angular momentum l in a Coulomb field is

$$\frac{d^2(r\psi_l)}{dr^2} + \left[k^2 - \frac{l(l+1)}{r^2} - \frac{2M}{\hbar^2}\frac{ZZ'e^2}{r}\right](r\psi_l) = 0 \tag{7.1}$$

where M is the reduced mass of the system and k the wave number at infinity. This equation has two independent solutions; in practice the two used are the only regular solution F_l having asymptotic form

$$F_l \sim \sin\left(\varrho - \tfrac{1}{2}l\pi - \eta\log 2\varrho + \sigma_l\right) \tag{7.2}$$

and the special irregular solution G_l with asymptotic form

$$G_l \sim \cos\left(\varrho - \tfrac{1}{2}l\pi - \eta\log 2\varrho + \sigma_l\right) \tag{7.3}$$

where

$$\varrho = kr, \qquad \eta = \frac{ZZ'e^2}{\hbar v}$$

and

$$\sigma_l = \arg\Gamma(l+1+i\eta) \qquad (\text{Coulomb phase shift}) \tag{7.4}$$

F_l vanishes at $r=0$, and G_l is the solution that is $\pi/2$ out of phase with F_l at $r=\infty$; the general behaviour of the functions F_l and G_l is shown in Fig. 3. Eq. (7.1) shows that $r\psi_l$ has a turning point for an energy corresponding to the top of the barrier E_B i.e. when

$$\frac{\hbar^2k^2}{2M} = E_B = \frac{ZZ'e^2}{R} + \frac{\hbar^2}{2M}\frac{l(l+1)}{R^2} \tag{7.5}$$

and the solutions F_l and G_l cease to be oscillatory at this point. R is the nuclear radius.

If the wave number of the incident particle in the nucleus (before amalgamation to form a compound state) is K then the barrier penetration coefficient T_l may be written:

(a) *For neutrons* for which there is only the centrifugal barrier, ($\eta = 0, \sigma_l = 0$),

$$T_l = \frac{4k}{K} v_l \tag{7.6}$$

if $k \ll K$, where

$$v_l = \left[\frac{1}{F_l^2 + G_l^2}\right]_{\varrho = kR}. \tag{7.7}$$

Exact values for v_l can be given; they are

$$\left.\begin{aligned} v_0 &= 1, \\ v_1 &= \varrho^2/1 + \varrho^2, \\ v_2 &= \varrho^4/9 + 3\varrho^2 + \varrho^4, \\ &\ldots\ldots\ldots \end{aligned}\right\} \tag{7.8}$$

with $\varrho = kR$. The expression for T_l is still only approximate and a more accurate value will be found in [7].

(b) *For charged particles* no simple approximate expressions can be given except at very low energies when

$$T_0 \approx \frac{A}{E} \exp\left(-\frac{2\pi Z Z' e^2}{\hbar v}\right), \tag{7.9}$$

the familiar GAMOW formula. For higher energies, proton penetration factors are given in graphical form by CHRISTY and LATTER[1]. More extensive information on COULOMB wave functions is given in the tables prepared by BLOCH *et al.*[2], who also discuss their application to nuclear calculations[3].

The radius used[4] in penetrability calculations is not strictly the nuclear radius $R (= r_0 A^{1/3})$, regarded as a property of a particular nucleus and used, for instance, in simple calculations of COULOMB energy (Sect. 56). Following CHRISTY and LATTER the radius generally used is[5]

$$a = \frac{1}{2} \frac{e^2}{mc^2} (A_1^{1/3} + A_2^{1/3}) = 1 \cdot 4 \times 10^{-13} (A_1^{1/3} + A_2^{1/3}) \text{ cm}, \tag{7.10}$$

where A_1 and A_2 are the mass numbers of the interacting nuclei. This conventional interaction radius is probably too large, but is used throughout this article.

The barrier height (COULOMB plus centrifugal) is then given by Eq. (7.5); the centrifugal barrier is of major importance only for very light nuclei. In this article it will be assumed that an appropriate value for the barrier penetrability for charged particles is given by v_l, as in Eq. (7.7). This gives the probability of reaching the nuclear surface; the effect of change of wave number will be included in nuclear factors to be discussed later.

---

[1] R. F. CHRISTY and R. LATTER: Rev. Mod. Phys. **20**, 185 (1948). See also J. L. JOHNSON and H. M. JONES: Phys. Rev. **93**, 1286 (1954) for a correction to the GAMOW formula.

[2] I. BLOCH, M. H. HULL, A. A. BROYLES, W. G. BOURICIUS, B. E. FREEMAN and G BREIT Rev. Mod. Phys. **23**, 147 (1951).

[3] See also the article by BREIT and coworkers in this Encyclopedia, vol. XLI.

[4] A special article by D. HILL in vol. XXXIX of this Encyclopedia deals with nuclear radii. See also J. M. C. SCOTT: Phil. Mag. **45**, 441 (1954).

[5] In this report the symbol a is used (i) for the spin-orbit coupling strength [in which connection it nearly always occurs as part of the intermediate coupling parameter a/K (Sect. 3)]; (ii) as the interaction radius in penetrability calculations and reduced widths $\hbar^2/(Ma)$ (Sect. 9); and (iii) as the bombarding particle in the reaction X (ab) Y (Sect. 6). These meanings are sufficiently distinct to avoid confusion.

8. Statistical theory of nuclear reactions. A relatively simple picture of a nuclear reaction, and particularly of the energy dependence of the cross section, can be given if resonance phenomena are disregarded. This may be justified if the resolution is such as to average over many resonances as in the neutron scattering work of BARSCHALL described in Sect. 5 or if reactions at high energies are considered, for then the total width Γ is expected to exceed the level spacing D. In the theory of cross sections given by FESHBACH and WEISSKOPF[1] it is assumed that the compound nucleus is able to decay into many channels and that the probability of decay by re-emission of a particle into the entrance channel, i.e. by elastic scattering, is small. The wave function of the incident particle inside the nucleus is then entirely a converging wave e^{-iKr} where K is the wave-number of the incident particle in the nucleus before formation of the compound state. The cross section for formation of the compound nucleus is found to be, in the case of particles incident with wave number k and angular momentum $l\hbar$

$$\sigma_c^l = \pi \lambda\!\!\!^{-2} (2l+1)\, T_l \tag{8.1}$$

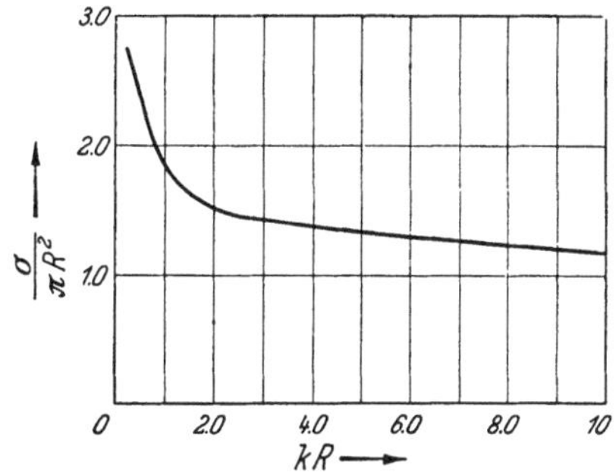

Fig. 4. Energy variation of reaction cross section for neutrons of energy E incident on a nucleus of radius R (in 10^{-13} cm), averaged over resonances (FESHBACH and WEISSKOPF). $k = 0.22\, E^{\frac{1}{2}}$ MeV.

where T_l is the barrier penetrability. Only particles with l less than the value $R/\lambda\!\!\!^-$ are expected to interact strongly with a nucleus of radius R. In the limiting case of very high energy particles, high orbital momenta may be involved and the total cross section for compound nucleus formation is then

$$\sigma_c = \sum_l \sigma_c^l = \pi \lambda\!\!\!^{-2} \sum_0^{R/\lambda\!\!\!^-} (2l+1) = \pi R^2 \tag{8.2}$$

assuming $T_l = 1$, which is approximately true, at least for neutrons. In the high energy limit elastic scattering contributes a further πR^2 so that the total cross section has the limiting value $2\pi R^2$. FESHBACH and WEISSKOPF suggest that the formula

$$\sigma_c = \pi (R + \lambda\!\!\!^-)^2\, T \tag{8.3}$$

gives a better account of observed reaction cross sections. A typical energy variation of reaction cross section for neutrons predicted by this non-resonant theory is shown in Fig. 4. The cross sections for the formation of the compound nucleus by charged particles have been calculated by SHAPIRO[2]; his results show an energy variation which is determined essentially by the barrier penetration factors and agrees in general with experimentally determined curves for the $(p\,n)$ reaction.

The decay of the compound nucleus is assumed in the statistical theory to be independent of the mode of formation of the compound state. The probabilities of decay into different product systems are calculated by analogy with the evaporation of a liquid drop heated to a uniform temperature. The level system of the residual nucleus Y will be excited with a probability, for a given level, dependent on the energy available. Arguments based on the theorem of detailed balancing show that if $\omega(\varepsilon)$ $(=1/D)$ is the level density at excitation ε in the

[1] H. FESHBACH and V. F. WEISSKOPF: Phys. Rev. **76**, 1550 (1949).
[2] M. M. SHAPIRO: Phys. Rev. **90**, 171 (1953).

nucleus Y, then the energy spectrum of the particles emitted is given by

$$I(\varepsilon)\,d\varepsilon = C\,\varepsilon\,\sigma_c(\varepsilon)\,\omega(\varepsilon_{b\,\max} - \varepsilon)\,d\varepsilon \tag{8.4}$$

where $\varepsilon_{b\,\max} - \varepsilon$ is the kinetic energy available to particle b.

It is important to know nuclear level densities in order to check the sum rules of resonance theories (Sect. 9). A summary of available information is given by VOGT [18]. Fig. 5 is extracted from this reference and shows the level density at 7 MeV excitation as a function of mass number for $A < 40$. The values plotted are obtained from summaries of reaction data. Relative values are also obtainable from observations of the energy spectrum of the particles evaporated in reactions to which the statistical theory can be applied; a plot of $\frac{I(\varepsilon)}{\varepsilon\,\sigma_c}$ against ε determines the relative level density (8.4). The total density of nuclear levels, and the density of levels having a given angular momentum, have been calculated by a statistical method based on the shell model by BLOCH[1].—The statistical theory is of importance in the discussion of reactions leading to, or proceeding through, highly excited states. It is also useful in the theory of photodisintegration, but for a detailed discussion of discrete nuclear states a resonance theory is required.

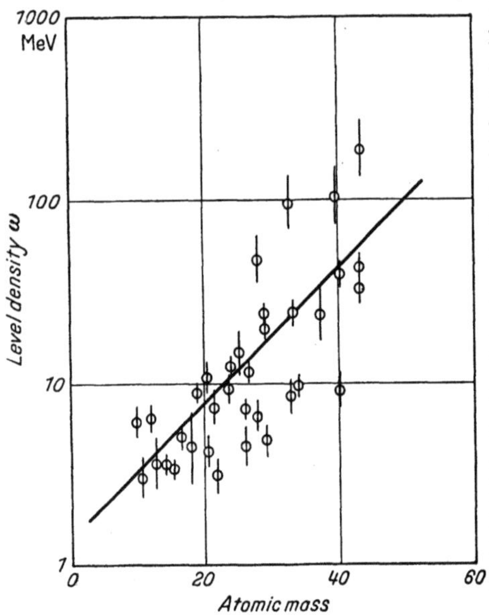

Fig. 5. Number of levels per MeV at 7 MeV excitation in nuclei of $A < 40$ (VOGT [18]).

9. Resonance theory of nuclear reactions. α) *General*[2]. The variation of cross section of a nuclear reaction with energy given by σ_c, Eq. (8.3) and shown in Fig. 4 is a monotonic function, in contrast with the known behaviour of many nuclear reactions which exhibit sharp resonances. The theory however assumes that the incident particle is absorbed by the target nucleus X and that coherent re-emission of the incident particle is relatively unimportant because the energy is so high that many alternative processes can take place. For incident particles of low energy there may be only a few reaction channels possible and elastic scattering with compound nucleus formation becomes a relatively important process. In order to allow a significant amount of elastic scattering, it is necessary to impose certain conditions on the internal and external wave functions at the nuclear boundary, as emphasized by FESHBACH *et al.*[3] (see also [7], p. 379). Fig. 6 shows that if the internal and external wave functions (for an incident

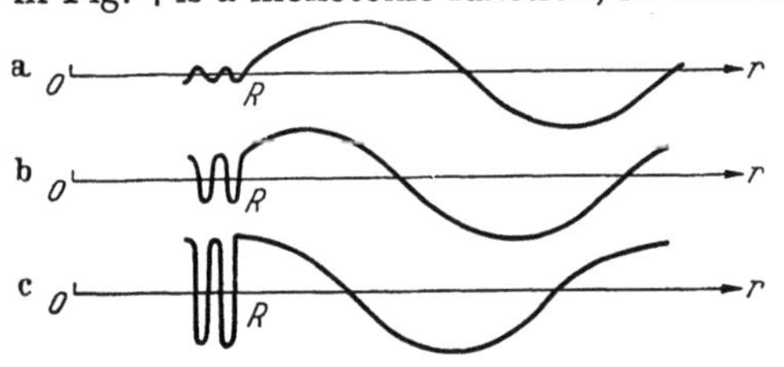

Fig. 6 a—c. Neutron wave functions at the nuclear surface as function of distance r from the centre of the nucleus (a) between resonances; (b) near resonance; (c) at resonance (BLATT and WEISSKOPF [7], p. 382).

[1] C. BLOCH: Phys. Rev. **93**, 1094 (1954).
[2] Electromagnetic transitions are discussed in Sects. 19 to 24.
[3] H. FESHBACH, D. C. PEASLEE and V. F. WEISSKOPF: Phys. Rev. **71**, 145 (1947).

neutron) are made to join smoothly at the nuclear boundary, the amplitude of the former will in general be small because of the change of wave number. If however the slope of the internal wave function is zero at the nuclear surface the two wave functions may be joined with similar amplitudes. This is a resonance, and the occurrence of such resonances is determined by the behaviour at the nuclear boundary of the function

$$f(E) = R\left[\frac{d\psi/dr}{\psi}\right]_{r=R}. \tag{9.1}$$

Zero values of this function define a set of resonance energies of the compound nucleus; the large amplitude of the internal wave function means a long lifetime for the compound state.

The spacing of these resonances is specified by D, and their total width, which is related to their mean life, by Γ. The quantities Γ and D are not independent because the complicated nuclear motion in any particular state can be described by a wave function constructed by superimposing a number of wavefunctions belonging to states of energy $E_n = E_0 + nD$, if the spacing is assumed uniform (see [7], p. 387). This gives a function of the form

$$e^{-\frac{iE_0 t}{\hbar}} \sum a_n \varphi_n e^{-\frac{inDt}{\hbar}} \tag{9.2}$$

which shows that the nuclear motion repeats a given configuration after a time

$$\frac{h}{D}. \tag{9.3}$$

Now the occurrence of well defined resonances implies that the width Γ is $\ll D$ i.e. that the lifetime τ of a virtual state is long compared with the time h/D in which the nuclear motion repeats. This is because the barrier effects at the nuclear boundary in the resonance region reduce the probability of emission of a particle. It then follows that if T is the barrier penetration [given by Eq. (7.6)]

$$\tau = \frac{h}{D\,T} \tag{9.4}$$

and

$$\Gamma = \frac{\hbar}{\tau} = \frac{D\,T}{2\pi} = \frac{2k}{K} v_l \frac{D}{\pi} \tag{9.5}$$

where v_l is defined by (7.7). Strictly, this equation refers to a given process of decay of the virtual state i.e. emission of a given particle b; if many such processes are possible the resonance condition $\Gamma \ll D$ may no longer be valid. Strictly also D relates to levels of given spin and parity.

It is customary to write (9.5) in the form

$$\Gamma_b = 2k\, v_l \gamma_b^2 \tag{9.6}$$

where

$$\gamma_b^2 = \frac{D}{K\pi} \tag{9.7}$$

is defined as the *reduced width* for the emission of particle b. From (9.7) it follows that γ_b^2 has the dimensions energy × length (MeV-cm) and that the ratio γ_b^2/D should be independent of energy. Observed values of γ_b^2/D are tabulated by Vogt [18]. The reduced width γ^2 is an important parameter of nuclear levels, closely related to the character of the nuclear motion. If for instance this motion is represented by that of a particle in a potential well of radius a, the level spacing would be $D \sim \pi^2 \hbar^2/M a^2$ and the internal wave number would be $K \sim \pi/a$ so that in this case $\gamma^2 \sim \hbar^2/M a$. Extraction of γ^2 from experimental results is thus a direct method of testing nuclear models.

β) Cross section formula (no spin). The variation with energy of the cross section for the reaction X(ab)Y in the resonance region may be calculated, as indicated in Eq. (6.3) by time dependent perturbation theory[1]; in this theory, of which a simple example is given in [*16*], the experimental widths appear as the squares of matrix elements such as H_{XC}, H_{CY}. An alternative treatment sets up a stationary state problem in which a flux of particles a inwards towards X to form the compound system C is balanced by an outward flow of particles b into reaction channels and of elastically scattered particles a back into the entrance channel. In such theories[2], which are fully discussed by BREIT and coworkers in Vol. XLI of this Encyclopedia and also in [*6*], [*7*], [*19*] resonances are determined, as indicated in the previous section, by the matching of external and internal wave functions on a surface of radius $r = a$ [Eq. (7.10)]. The external wave functions are calculable from the known centrifugal and COULOMB barriers. The internal wave function depends on the choice of interaction radius a and of boundary conditions, and the resonance energies E_λ obtained from the boundary conditions also depend on a. The observed resonance energies E_{res} cannot of course depend on choice of radius and are related to the resonances E_λ of the compound state by the introduction into the theory of a *level shift* Δ_λ. The probability of the appearance of the pair of particles Y + b at the nuclear surface during the lifetime of the compound state is measured by the *reduced width* γ_λ^2.

For a *reaction* between spinless particles in which only one resonance level of the compound state is important the total cross section may be written in the amiliar BREIT-WIGNER form

$$\sigma(a\,b) = \pi\,\lambda\!\!\!^{-2}\,(2l+1)\,\frac{\Gamma_{\lambda a}\,\Gamma_{\lambda b}}{(E-E_{res})^2+\frac{1}{4}\Gamma_\lambda^2} \tag{9.8}$$

with

$$\Gamma_{\lambda a} = 2 k_a v_{l a} \gamma_{\lambda a}^2 \tag{9.9}$$

and

$$\sum_a \frac{2 M a\,\gamma_{\lambda a}^2}{3\hbar^2} \sim 1\,. \tag{9.10}$$

The sum rule (9.10), given by TEICHMANN and WIGNER, may be used to place an upper limit on a single reduced width $\gamma_{\lambda a}^2$. Only if the wave function for the internal region can be fully expressed in a single particle form should $\gamma_{\lambda a}^2$ be as large as $3\hbar^2/2Ma$. The calculation of γ^2 from Γ by use of formula (9.9) is not exact because it neglects the variation of Δ_λ $(= E_{res} - E_\lambda)$ with energy over the resonance. The correction for this effect, which is small except when γ^2 is a large fraction of the single particle width, is given by THOMAS[3] and in [*19*].

Formula (9.8) must be modified to describe *elastic scattering* because of interference between the waves from the compound nucleus and those scattered by the potential fields.

The result for spinless particles is

$$\sigma^l(a\,a) = (2l+1)\,4\pi\,\lambda\!\!\!^{-2}_a \left| \frac{\frac{1}{2}\Gamma_{\lambda a}}{E - E_{res} + \dfrac{i\Gamma_\lambda}{2}} + \frac{A^l_{pot}}{2i} \right|^2 \tag{9.11}$$

---

[1] G. BREIT and E. P. WIGNER: Phys. Rev. **49**, 519 (1936). — S. FLÜGGE: Z. Naturforsch. **3a**, 97 (1948).

[2] H. FESHBACH, D. C. PEASLEE and V. F. WEISSKOPF: Phys. Rev. **71**, 145 (1947). — E. P. WIGNER and L. EISENBUD: Phys. Rev. **72**, 29 (1947). — T. TEICHMANN and E. P. WIGNER: Phys. Rev. **87**, 123 (1952).

[3] R. G. THOMAS: Phys. Rev. **81**, 148 (1951).

where the term A^l_{pot} refers to potential scattering and is given by ([7], p. 334)

$$A^l_{\text{pot}} = e^{-2i\delta_l} - 1 = -2i\, e^{-i\delta_l} \sin \delta_l. \qquad (9.12)$$

The phase shifts for neutrons are approximately

$$\left.\begin{aligned} \delta_0 &= -k a, \\ \delta_1 &= -k a + \tfrac{1}{2}\pi - \operatorname{arc\,cot} k a, \\ \delta_2 &= -k a + \pi - \operatorname{arc\,cot} \frac{(k a)^2 - 3}{3 k a}, \\ &\ldots\ldots\ldots \end{aligned}\right\} \quad \text{for} \quad k a \gg 1 \qquad (9.13)$$

and these values are also approximately valid for high energy charged particles. For low energy neutrons only the s-phase shift is important and is still

$$\delta_0 = -k a \qquad (9.14)$$

For low energy charged particles δ is the COULOMB phase shift (σ_l, Sect. 16).

The scattering cross section has a maximum value $4\pi\lambda\!\!\!^{-2}(2l+1)$ at resonance; between resonances, if $ka \ll 1$, it is approximately equal to $4\pi a^2$, as for a hard sphere of radius a. Elastic scattering between resonances is usually described in this way; it may also be considered as due to contributions from many distant resonances.

Formulae (9.8) and (9.11) are valid and sharp resonances in cross section are observed, only when $\Gamma < D$. This is likely, according to (9.9), if k or v_l are small i.e. for slow particles or for low barrier penetrability. Special selection rules may also reduce Γ, particularly in the reactions of light nuclei.

γ) Introduction of spin. If the intrinsic spins of the nucleus X and particle a are j_1 and j_2 (in units of $\hbar$), the *channel spin* of the initial system may take any value between $j_1 - j_2$ and $j_1 + j_2$, giving $(2j_1+1)(2j_2+1)$ states. An unpolarised initial system is described by an incoherent mixture of all possible channel spins; each of these may combine with a particular orbital momentum to form angular momentum J of the compound state. For incident orbital momentum zero J is equal to the channel spin and the probability of forming J from the randomly oriented j_1, j_2 of an unpolarised system is

$$g(J) = \frac{2J+1}{(2j_1+1)(2j_2+1)}. \qquad (9.15)$$

This factor then multiplies (9.8) and the *reaction cross section* for incident particles with $l = 0$ becomes:

$$\sigma(a\,b) - \pi\lambda\!\!\!^{-2} g(J) \frac{\Gamma_{\lambda a}\Gamma_{\lambda b}}{(E - E_{\text{res}})^2 + \frac{1}{4}\Gamma_\lambda^2}. \qquad (9.16)$$

For *elastic scattering* the channels with channel spin not equal to J also contribute to the potential scattering and (9.11) must be replaced by

$$\sigma(a\,a) = 4\pi\lambda\!\!\!^{-2} g(J) \left| \frac{\frac{1}{2}\Gamma_a}{E - E_{\text{res}} + \frac{i\Gamma_\lambda}{2}} + \frac{A^0_{\text{pot}}}{2i} \right|^2 + [1 - g(J)]\,\pi\lambda\!\!\!^{-2}_a\, |A^0_{\text{pot}}|^2. \qquad (9.17)$$

The generalisation of these formulae to arbitrary l-values is given in [7], p. 426; the statistical factor is still $g(J)$. Complete expressions for the differential cross section are given by BLATT and BIEDENHARN [28].

Eq. (9.16) gives a relation between the cross sections for inverse processes passing through the same compound state. If the wave lengths $\lambda\!\!\!^{-}_a$ and $\lambda\!\!\!^{-}_b$ refer to the centre of mass system

$$\frac{\sigma(a\,b)}{\sigma(b\,a)} = \frac{g_a(J)}{g_b(J)}\,\frac{\lambda_a^2}{\lambda_b^2}\,. \tag{9.18}$$

If the incident particle is of low energy the reaction cross section (9.16) has the asymptotic forms (see [7], p. 394)

$$\left.\begin{aligned} \sigma(a\,b) &= \text{const}\; E_a^{-\frac{1}{2}} \quad \text{(neutrons)}\,,\\ \sigma(a\,b) &= \frac{\text{const}}{E_a}\exp\left(-\frac{2\pi Z Z' e^2}{\hbar v_a}\right) \quad \text{(charged particles)}\,. \end{aligned}\right\} \tag{9.19}$$

If the outgoing particle is slow the asymptotic forms are

$$\left.\begin{aligned} \sigma(a\,b) &= \text{const}\; E_b^{\frac{1}{2}} \quad \text{(neutrons)}\,,\\ \sigma(a\,b) &= \text{const}\exp\left(-\frac{2\pi Z Z' e^2}{\hbar v_b}\right) \quad \text{(charged particles)} \end{aligned}\right\} \tag{9.20}$$

and for elastic scattering of slow neutrons

$$\sigma(a\,a) = \text{const}\,. \tag{9.21}$$

δ) Conservation theorems. In any nuclear reaction total angular momentum J and its resolved part J_z along an axis must be conserved. In special cases total orbital and intrinsic spin momenta L and S and their resolved parts may be conserved, but the existence of forces coupling spin to orbit will mix states of different L to form a given J. Parity (π) is also conserved in nuclear reactions; the parity of orbital wave functions is $(-1)^l$.

The operator for total isotopic spin T and that for its resolved part T_z commute with the nuclear Hamiltonian for charge independent forces. Both T and T_z should be conserved in a nuclear reaction if this may be regarded as a small perturbation produced by a charge independent interaction. The evidence for conservation of total isotopic spin in nuclear reactions has been summarised in several articles[1]. Isotopic spin conservation may fail, as discussed by LANE and THOMAS [19], when a nuclear reaction passes through a long lived state in which the COULOMB perturbation has time to mix wave functions of different T.

The quantities J, T, π and perhaps also L and S for well defined nuclear states thus impose selection rules on nuclear reactions. These quantities and the associated nucleon configuration may in many cases be determined by measurements of angular distributions or transition probabilities, particulary in resonance reactions. Many examples of this procedure are discussed in Part D.

10. Reduced widths and nuclear models. Reduced widths extracted from observed widths by application of Eq. (9.9) are usually found to be very much less than the single particle value $\frac{\hbar^2}{M a}$, except in obviously simple cases such as the transition $He^5 \to He^4 + n$. This can be explained by invoking the independent particle model. Since the reduced width measures the probability of a nucleus dissociating into a certain pair of particles, it must be possible to express its amplitude as the integral of an internal wave function over the nuclear surface. This wave function must be appropriate to the necessary channel spin, orbital momentum and nature of the dissociating particles. If the compound nucleus, of mass A, has n equivalent particles outside a closed shell the reduced width

[1] D. H. WILKINSON: Nature, Lond. **172**, 576 (1953). — R. E. MARSHAK: J. Phys. Radium **16**, 325 (1955). — W. E. BURCHAM: Progress in Nuclear Physics, Vol. 4. London: Pergamon Press 1955.

amplitude will contain contributions from several states of the configuration. These states may conveniently be specified by products of wave functions for $(n-1)$ nucleons and single particle wave functions. A wave function for $(n-1)$ nucleons represents a "parent state" in the nucleus of mass $(A-1)$. The single particle may be coupled to the parent state according to the LS, jj or intermediate scheme, e.g.

$$(LS) \quad l_0 + [l^{n-1}]_{T_p S_p L_p J_p} \rightarrow [l^{n-1} l_0]_{TSLJ}, \tag{10.1}$$

$$(jj) \quad j_0 + [j^{n-1}]_{J_p T_p} \quad \rightarrow [j^{n-1} j_0]_{JT} \tag{10.2}$$

in which suffix p refers to the parent state and 0 to the nucleon. The coefficients appearing in the final wave function, and hence the value of the reduced width, depend on the coupling scheme, and have been given explicity for equivalent nucleons of the p-shell. Since the individual terms of the total wave function need not all have the same phase, the reduced width will usually be much less than the single particle value. If on the other hand the nuclear motion is highly co-ordinated, the reduced width may exceed this value; this is so in the collective models discussed in Sect. 4.

The detailed theory of reduced width expansions is given by LANE and THOMAS [*19*], who give formulae from which the p-shell reduced widths for nucleon, alpha particle and deuteron processes may be calculated. They show that experimental results generally favour a form of intermediate coupling. The importance of the concept of parentage of nuclear states has been emphasised by LANE and WILKINSON[1].

C. Experimental methods for studying energy levels.

I. General techniques.

11. Introduction. The energy release Q in reactions of the type

$$X + a \rightarrow Y + b \tag{11.1}$$

was first determined by BLACKETT[2] and by CHADWICK, CONSTABLE and POLLARD[3] for the (αp) reaction, in experiments following the discovery of artifical transmutation by RUTHERFORD[4]. Excitation energies of the residual nucleus Y were obtained in these and later experiments as differences between the various Q-values observed and the greatest of these values for a particular reaction. Early experiments[5] also established that reactions like (11.1) would proceed most readily for certain definite energies of the incident particle a, now known as resonance energies. In all this work natural sources of alpha particles were used and the detecting instruments were usually scintillation screens with visual observation or simple ionisation chambers or counters. The largely increased precision of similar but recent measurements arises in approximately equal measure from the use of particle accelerators, and improved methods of detection. Energy releases in typical nuclear reactions of the light elements are now known with errors as small as one part in 2000.

---

[1] A. M. LANE and D. H. WILKINSON: Phys. Rev. **97**, 1199 (1955). See also A. M. LANE: Phys. Rev. **92**, 839 (1953).

[2] P. M. S. BLACKETT: Proc. Roy. Soc. Lond. **107**, 349 (1925).

[3] J. CHADWICK, J. E. R. CONSTABLE and E. C. POLLARD: Proc. Roy. Soc. Lond., Ser. A **130**, 463 (1930).

[4] E. RUTHERFORD: Phil. Mag. **37**, 538 (1919).

[5] E. POSE: Phys. Z. **30**, 780 (1929).

The experimental techniques described in Sect. 12 to 15 have been mainly used to obtain excitation energies of both bound and virtual states and cross sections for nuclear reactions as a function of energy. From such observations total and partial widths (Γ, Γ_b) of resonance levels may be obtained and in some cases the spin of the level. Elastic scattering processes are treated separately in Sect. 16 to 18 because interference with potential scattering presents special features. Radiative processes are also discussed separately in Sect. 19 to 24 although the bombardment techniques are the same as those used for inducing charged particle emission. The information on spin, parity and coupling mechanisms obtainable[1] from angular distributions and correlation experiments is discussed in Sect. 25 to 27.

12. Particle accelerators and neutron sources. *α) Charged particles.* The cascade generator, used by COCKCROFT and WALTON[2], is a convenient source of accelerating potentials up to about 1.5 MeV. Both voltage ripple and generator impedance increase as the number of voltage doubling stages increases, but smoothing circuits can be used to remove most of the ripple. Pressure-insulated cascade generators using metal rectifiers have been made and may extend the range of usefulness of this type of accelerator. The voltage developed by the cascade generator is applied to an accelerating tube which carries some form of ion source capable of yielding good currents of protons, deuterons or alpha particles. Target currents of about 100 to 1000 μA after acceleration are possible and the homogeneity can be improved, after full use has been made of high tension smoothing circuits, by magnetic analysis. A typical installation for 1.5 MeV is described by MILEIKOWSKY[3]; this type of equipment has the great advantage of accessibility; but the voltage obtainable in a room of moderate size is a limiting factor.

The most compact and versatile source of unidirectional voltage for ion acceleration is the electrostatic generator of VAN DE GRAAFF[4]. This type of machine has been extensively developed since 1931 and in its latest pressurised form can supply a current of several microamperes at a voltage up to 8 MeV. By accelerating negative ions and then converting them to positive ions in a foil it is in principle possible to double the ion energy available from such an accelerator. The voltage developed can be electronically stabilised to about 0.1% and a beam of greater homogeneity can if necessary be obtained by magnetic or electrostatic analysis (FOWLER *et al.*[5], WARREN *et al.*[6]). In this way an energy resolution of 0.01% is possible with beam intensities of about 1 μA per keV of width, and such facilities permit the direct observation of the width of many virtual levels. Progress in the development of electrostatic generators is reviewed by FORTESCUE[7].

It seems likely that problems of insulation will limit electrostatic generators to voltages below 10×10^6 volts. Particles of higher energy can be obtained from the cyclotron, which since its original conception by LAWRENCE, has become one of the most powerful instruments of nuclear physics. The general design of cyclotrons has recently been reviewed by PICKAVANCE[8] and by FREMLIN and GOODEN[9]; fixed-frequency machines are limited to an energy of about 25 MeV for deuterons, but synchrocyclotrons, in which frequency modulation is used,

[1] See also Vols. XLIV and XLV of this Encyclopedia.
[2] J. D. COCKCROFT and E. T. S. WALTON: Proc. Roy. Soc. Lond., Ser. A **136**, 619 (1932).
[3] C. MILEIKOWSKY and R. T. PAULI: Ark. Fysik **4**, 287 (1952).
[4] R. J. VAN DE GRAAFF: Phys. Rev. **38**, 1919 (1931). See also Phys. Rev. **43**, 149 (1933).
[5] W. A. FOWLER, C. C. LAURITSEN and T. LAURITSEN: Rev. Sci. Instrum. **18**, 818 (1947).
[6] R. E. WARREN, J. L. POWELL and R. G. HERB: Rev. Sci. Instrum. **18**, 559 (1947).
[7] R. L. FORTESCUE: Progress in Nuclear Physics, vol. 1. London: Pergamon Press 1950.
[8] T. G. PICKAVANCE: Progress in Nuclear Physics, vol. 1. London: Pergamon Press 1950.
[9] J. H. FREMLIN and J. S. GOODEN: Rep. Progr. Phys. **13**, 295 (1950).

are now operating at energies up to 660 MeV. Only the smaller synchrocyclotrons have been useful for precise nuclear reaction studies. Several fixed-frequency machines have been equipped with deflecting and focussing magnets by means of which a well collimated ion beam of 1 μA homogeneous to ± 10 keV can be obtained as much as 20 feet from the machine (BENDER *et al.*[1]). Such installations lack the easy variability of energy which is characteristic of electrostatic generators, but a number of cyclotrons are now being designed in which this limitation is removed. Another instrument for a fixed energy of which some use has recently been made is the proton linear accelerator; the best known is the 32 MeV machine constructed by ALVAREZ[2].

All heavy particle accelerators have been much improved by advances in ion source design, and in particular by the development of the *RF* (THONEMANN[3]) and *PIG* (GOW and FOSTER[4]) type sources which provide an intense beam of singly charged ions with small voltage spread[5]. Special techniques (BITTNER[6]) have resulted in the availability of sources of He^{++} ions.

β) Neutrons. Most of the early work on neutron induced reactions used the radon-beryllium source which gives a nearly continuous neutron spectrum. Homogeneous sources of neutrons can however be obtained from charged particle reactions and a discussion of suitable reactions is given by HANSON *et al.*[7]. The most useful sources have been the reactions

$$Li^7 + H^1 \to Be^7 + n - 1.647\ \text{MeV}, \tag{12.1}$$

$$H^3 + H^1 \to He^3 + n - 0.764\ \text{MeV}, \tag{12.2}$$

$$H^3 + H^2 \to He^4 + n + 17.6\ \text{MeV}, \tag{12.3}$$

$$H^2 + H^2 \to He^3 + n + 3.28\ \text{MeV}. \tag{12.4}$$

Of these, the first two are especially suitable for use with high energy electrostatic accelerators. If thin targets are used, the homogeneity of the neutron beam at a given angle depends mainly on the homogeneity of the incident protons and may be as good as 5 keV; the neutron energy is continuously variable from a low value of perhaps 20 keV determined by the kinetics of the reaction, to the maximum permitted by the accelerator. Until recently the maximum neutron energy available has been about 2 MeV and for this reason reaction (12.4) has also been much used, since the neutron energy in this case varies considerably with angle, and neutrons of energies between about 2.5 and 7 MeV could be obtained with existing accelerators. The development of 10 MeV electrostatic machines will extend the useful range of all these reactions. Homogeneous high energy (14 MeV) neutrons are very conveniently supplied by the reaction (12.3) which has a maximum cross section for deuterons of about 200 keV energy. Such a neutron source can be a very compact and simple installation and it is in use in many laboratories. Neutrons of energy below 10 keV may also be obtained from nuclear reactors equipped with "fast choppers" and time-of-flight circuits,

---

[1] R. S. BENDER, E. M. REILLEY, A. J. ALLEN, R. ELY, J. S. ARTHUR and H. J. HAUSMAN: Rev. Sci. Instrum. **23**, 542 (1952).

[2] L. W. ALVAREZ: Phys. Rev. **70**, 799 (1946); see also U.S. Atomic Energy Commission Document AECU 120, or University of California Radiation Laboratory Report UCRL 236.

[3] P. C. THONEMANN: Progress in Nuclear Physics, Vol. 3. London: Pergamon Press 1954.

[4] J. D. GOW and J. S. FOSTER: Rev. Sci. Instrum. **24**, 606 (1953).

[5] For details see D. KAMKE in Vol. XXXIII of this Encyclopedia.

[6] J. W. BITTNER: Rev. Sci. Instrum. **25**, 1058 (1954).

[7] A. O. HANSON, R. F. TASCHEK and J. H. WILLIAMS: Rev. Mod. Phys. **21**, 635 (1949).

or from pulsed linear electron accelerators acting as photoneutron sources[1]. Intense beams of thermal neutrons are of course available directly from reactors. Low energy neutrons produce only a few charged particle transmutations, but are extremely important for an accurate study of virtual levels just above the neutron binding energy and of the resulting cascade of radiative transitions (Sect. 40).

Sources of gamma radiation for nuclear transmutation experiments are mentioned in Sect. 23.

13. Detection and analysis of reaction products. The determination of the energy changes (Q) in transmutation and inelastic scattering reactions is one of the most important methods of locating excited states. The Q-values are derived from the observed energies of particle groups by simple formulae (given for instance in [*1*]) which correct for the bombarding energy and angle of observation and effectively convert the results to the centre-of-mass system. The calculated energy of capture radiation and of the widths of resonance levels must also allow for centre-of-mass motion. In angular distribution experiments cross sections to be compared with theory must be referred to the centre-of-mass system. Conversion formulae for angle and solid angle are given by SCHIFF[2]. Frequently the same experiment gives both energy releases and resonance energies; the techniques of detection may be the same for each, and are described in this section. The detection of gamma radiation produced in nuclear reactions is discussed in Sect. 14.

α) Particle groups from transmutations. The energy of a homogeneous group of charged particles can be found by measuring the range of the group in a standard material, and using known range-energy relations. The energy can also be measured directly by finding the ionisation produced by stopping a single particle in a proportional counter or ionisation chamber, or the light produced by stopping the particle in a scintillating material. Electrostatic analysis of a group of particles gives a direct measure of the energy per unit charge and magnetic analysis gives momentum per unit charge.

The methods based on range measurement are limited by straggling to an accuracy of the order of 1% in energy. They can however be adapted to discriminate between particles of different charge because of the different variation of specific ionisation of protons, deuterons, tritons and alpha particles near the end of their range. The performance of a triple proportional counter used for range measurements and particle discrimination in nuclear reaction studies is described by HOLT and MARSHAM[3]. A similar counter employing the differential technique has been used by REYNOLDS[4] for detecting alpha-particle groups in the presence of protons of much greater range. Nuclear emulsions discriminate easily between singly and doubly charged particles, but only with difficulty between particles of the same charge and different mass. They have been used for energy measurements by LATTES *et al.*[5] and by many others; details of the technique are described by ROTBLAT[6], whose observations of charged particle groups from oxygen bombarded by 7.7 MeV deuterons are shown in Fig. 7. The plotting of differential range curves point by point is a tedious and rather ineffi-

[1] A. W. MERRISON and E. R. WIBLIN: Nature, Lond. **167**, 346 (1951).
[2] L. I. SCHIFF: Quantum Mechanics, p. 96. New York: McGraw Hill 1949.
[3] J. R. HOLT and T. N. MARSHAM: Proc. Phys. Soc. Lond. A **66**, 249 (1953).
[4] J. B. REYNOLDS: Phys. Rev. **98**, 1289 (1955).
[5] C. M. G. LATTES, P. H. FOWLER and P. CUER: Proc. Phys. Soc. Lond. A **59**, 883 (1947).
[6] J. ROTBLAT: Progress in Nuclear Physics, Vol. 1. London: Pergamon Press 1950. See also Y. GOLDSCHMIDT-CLERMONT: Ann. Rev. Nucl. Sci. **3** (1953).

cient method of finding an energy spectrum. It has however been found that the light output of inorganic scintillating crystals such as thallium-activated sodium iodide is a linear function of incident particle energy for protons and deuterons of energies up to 20 MeV and for alpha particles of energies above about 15 MeV (EBY and JENTSCHKE[1]). A scintillation spectrometer consisting of a sodium iodide crystal fixed to the face of a photomultiplier, with associated pulse amplifying and analysing equipment, can be used to record the energy spectrum of a distribution of charged particles. The 100-channel pulse analyser designed by HUTCHINSON and SCARROTT[2] has proved a very powerful instrument for this purpose. The sodium iodide spectrometer does not discriminate between different types of particle, but ASCHENBRENNER[3] has added this feature by allowing the incident particles to pass through a thin organic scintillator, in which their rate of energy loss is measured, before they reach the sodium iodide crystal. The resolution of scintillation spectrometers is limited by the statistical fluctuation of electron emission in the photomultiplier and is therefore energy dependent; the width of a homogeneous group of collimated alpha particles of energy about 5 MeV may be about 5%. Better resolution is obtained by stopping the particles in the gas of a proportional counter rather than in a crystal, but the stopping power available in a convenient size is then much reduced.

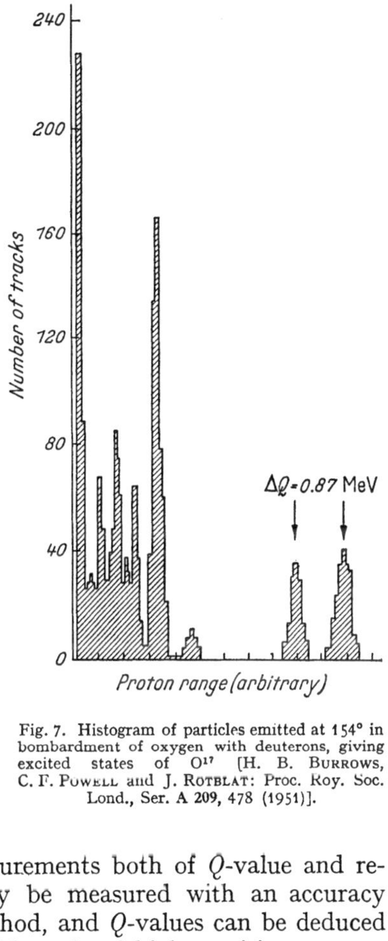

Fig. 7. Histogram of particles emitted at 154° in bombardment of oxygen with deuterons, giving excited states of O^{17} [H. B. BURROWS, C. F. POWELL and J. ROTBLAT: Proc. Roy. Soc. Lond., Ser. A **209**, 478 (1951)].

The methods of detection so far described are at present mainly used for selection of certain particle groups, on which other observations, such as angular distributions, are to be made. Precise measurements of energy release have been made mainly with 180° magnetic spectrometers of which the instruments of BUECHNER *et al.*[4] and of COLLINS *et al.*[5], are good examples. Since the geometrical arrangement of such instruments is simple they may be used for accurate absolute measurements both of Q-value and reaction cross-section. The deflecting field may be measured with an accuracy of 1 part in 10^5 by the nuclear resonance method, and Q-values can be deduced with an accuracy of 1 part in 2000. The solid angle of high precision spectrometers is low and for some purposes, such as determination of angular distributions, magnets with specially shaped fields are used, e.g. the double focussing magnet of SNYDER *et al.*[6] and the 90° second-order focussing magnet of

[1] F. S. EBY and W. K. JENTSCHKE: Phys. Rev. **96**, 911 (1954).
[2] G. W. HUTCHINSON and G. G. SCARROTT: Phil. Mag. **42**, 792 (1951).
[3] F. A. ASCHENBRENNER: Phys. Rev. **98**, 657 (1955).
[4] W. W. BUECHNER, E. N. STRAIT, C. G. STERGIOPOULOS and A. SPERDUTO: Phys. Rev. **74**, 1569 (1948).
[5] E. R. COLLINS, C. D. MCKENZIE and C. A. RAMM: Proc. Roy. Soc. Lond. A **216**, 219 (1953).
[6] C. W. SNYDER, S. RUBIN, W. A. FOWLER and C. C. LAURITSEN: Rev. Sci. Instrum. **21**, 852 (1950).

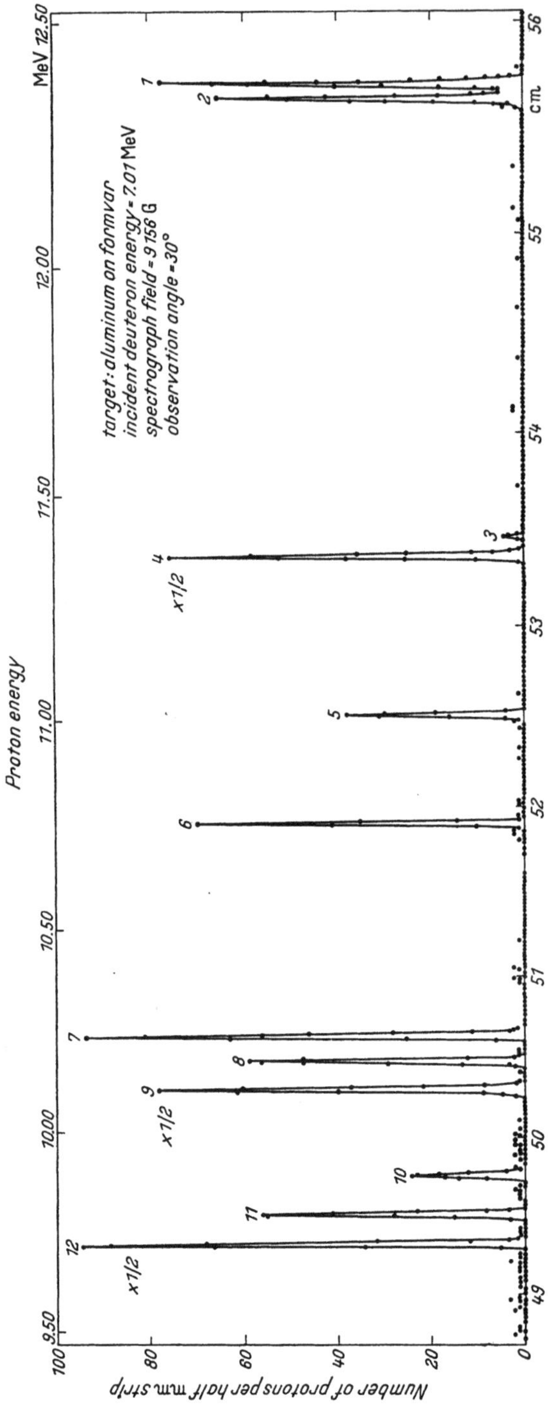

Fig. 8. Proton groups from Al^{27} $(d p)$ reaction observed by high resolution magnetic spectroscopy. This is part of a spectrum in which more than 100 groups were found (BUECHNER).

BUECHNER[1], which focusses particles of a range of energies along a hyperbola. These spectrometers are not strictly absolute instruments, but may easily be calibrated with natural alpha particle sources and have a resolution 50 to 100 times greater than most other methods of detection. The particles deflected in magnetic spectrometers are usually recorded in photographic emulsions, although counters have been used and are more convenient for excitation function work. KHROMCHENKO[2] detected magnetically resolved particles in a photographic plate and scanned the resulting spectra with a microphotometer. The general theory of magnetic spectrometers is discussed by JUDD[3]. Electrostatic analysers, such as that of BROWNE *et al.*[4], are capable of high precision and have large acceptance angles but are limited to rather low energies. The precision of possible measurements of nuclear reaction energies is well demonstrated by the work of BUECHNER[1] on the Al^{27} $(d p)$ Al^{28} reaction. The incident deuteron energy was 7.0 MeV and a magnetic spectrometer with a nuclear emulsion detector was used for recording an extensive proton spectrum in one exposure. The large number of proton groups found is shown in Fig. 8; the excellent resolution is due to the combination of a stable electrostatic generator and magnetic analysis. The Q-value for the ground state of Al^{28} is (5.494 ± 0.010) MeV and energy differences between proton groups

[1] W. W. BUECHNER, M. MAZARI and A. SPERDUTO: Phys. Rev. **101**, 188 (1956).
[2] L. M. KHROMCHENKO: Dokl. Akad. Nauk. SSSR. **93**, 451 (1953).
[3] D. L. JUDD: Rev. Sci. Instrum. **21**, 213 (1950).
[4] C. P. BROWNE, D. S. CRAIG and R. M. WILLIAMSON: Rev. Sci. Instrum. **22**, 952 (1951).

are known even more accurately than this. The observed proton groups determine the excited states of Al^{28}.

The reactions studied by these methods have usually been of the non-resonant type and the bombarding energy is generally chosen either to be as high as possible, to obtain the maximum number of groups, or for convenience in analysing the results. The combination of a stable high tension generator and a magnetic

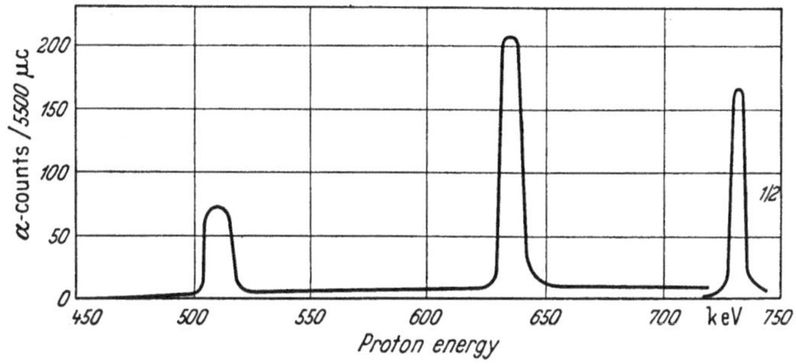

Fig. 9. Excitation curve for Al^{27} ($p\alpha$) Mg^{24} (RUTHERGLEN and SMITH).

analyser has also been much used for the study of virtual levels. The interest in this case is primarily the determination of resonance energies, but Q-values for the ground state and possibly a few excited states of the residual nucleus have sometimes been obtained at the same time. The results of RUTHERGLEN and SMITH[1] on the reaction Al^{27} ($p\alpha$) Mg^{24} are shown in Fig. 9. The sharp levels, unstable to particle emission, at an excitation of about 13 MeV in the compound nucleus Si^{28} are found from this work. Knowledge of the constants of the magnetic analyser enables absolute cross sections and partial widths for the several processes to be derived.

Energy releases in reactions of type (11.1) in which particle b is a neutron cannot be measured as accurately as in the case of charged particles, except (a) in the case of threshold reactions, and (b) when particle Y is a light charged particle whose energy can be measured by the methods already discussed e.g. $H^2(d, n)\, He^3$. Usually the energies of neutron groups have been found with an accuracy of 50 to 100 keV by observing proton recoils in an expansion chamber or in a nuclear emulsion; the groups shown in Fig. 10 were obtained by GREEN *et al.*[2] in work on the $C^{13}(dn)\, N^{14}$ reaction, in which Q-values of (5.41 ± 0.06); (3.07 ± 0.04); (1.39 ± 0.02) MeV and others were obtained. Endothermic reactions such as most (pn) transmutations [and (dn) reactions to excited states] give thresholds for neutron production as a function of incident particle energy. The Q-value is therefore determined by the accuracy with which (a) the response of a neutron detector, such as a boron-trifluoride filled ionisation chamber can be extrapolated to threshold, and (b) the incident energy can be measured.

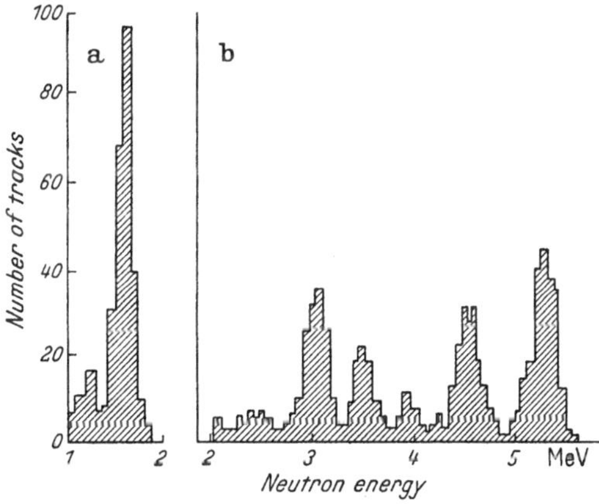

Fig. 10 a and b. Neutron spectrum from C^{13} (dn) reaction (GREEN *et al.*). Sects. (a) and (b) represent different areas scanned.

[1] J. G. RUTHERGLEN and R. D. SMITH: Proc. Phys. Soc. Lond. A **66**, 800 (1953).
[2] L. L. GREEN, J. P. SCANLON and J. C. WILLMOTT: Proc. Phys. Soc. Lond. A **68**, 386 (1955).

Thresholds for slow neutron production in (dn) reactions have been observed by BONNER and COOK[1]; the ratio of the counting rate in a slow neutron detector to that in a fast neutron counter increases sharply when the incident deuteron energy becomes sufficient to excite a new level of the residual nucleus [see for example Be^9 (dn) Sect. 50].

HERB *et al.*[2] have measured the threshold for the $Li^7(pn)Be^7$ reaction as 1.882 MeV with an accuracy of 0.1%, and similar measurements for a number of resonant reactions establish a "high voltage scale" for the measurement and inter-comparison of nuclear reaction energies.

Recoil ionisation chambers and proportional counters can be used for neutron detection but are useless for analysis of a complex neutron spectrum. HORNYAK[3] has described a simple and convenient fast neutron detector consisting of a uniform dispersion of zinc sulphide grains in a hydrogenous plastic. Proton recoils produced in the plastic cause light scintillations in the phosphor and the light pulses are detected by a photomultiplier. The detector is relatively insensitive to gamma radiation and has an efficiency roughly proportional to neutron energy. An ingenious counting method for determining the range of the recoil proton from a hydrogeneous radiator has been described by HOLT and LITHERLAND[4]..

β) Particle groups from inelastic scattering. Inelastic scattering is a particular case of a transmutation in which the incident particle loses energy to the bombarded nucleus as a result of some type of interaction and a particle of the same kind as the incident particle but with reduced energy is observed after the reaction. Alternatively radiation emitted from the excited states of the residual nucleus may be detected. The importance of this type of reaction is that it gives information about bound levels of stable nuclei which cannot always be investigated conveniently by transmutation reactions. The techniques used in inelastic scattering work are the same as those used for transmutation studies, but since inelastic scattering reactions necessarily absorb energy, the most useful work has come from accelerators able to cover a wide energy range. The energy levels of Al^{27} have been studied by BROWNE *et al.*[5] using an electrostatic generator with a maximum energy of 8.4 MeV and a high resolution magnetic analyser, to detect scattered protons; their results, some of which are shown in Fig. 11, give the levels up to an excitation of 6 MeV with an accuracy of ± 6 keV. Similar work has been carried out with deuterons and alpha particles. The cyclotron, has been used for similar work with protons of energies up to about 10 MeV (REILLEY *et al.*[6]), with deuterons up to about 10 MeV (GREENLEES[7]) and with alpha particles up to about 20 MeV (MCMINN *et al.*[8]), but the results are less precise, even when a magnetic spectrometer is used for analysing the scattered particles. Resolution of a few keV is necessary for a satisfactory study even of the lowest levels of odd mass nuclei such as Al^{27}.

Resonance energies of the compound state in a nuclear reaction may be detected by observation of inelastic scattering, usually of protons or alpha particles. Electrostatic generators and magnetic analysis techniques are again

[1] T. W. BONNER and C. F. COOK: Phys. Rev. **96**, 122 (1954).

[2] R. G. HERB, S. C. SNOWDON and O. SALA: Phys. Rev. **75**, 246 (1949). See also Phys. Rev. **94**, 947 (1954).

[3] W. F. HORNYAK: Rev. Sci. Instrum. **23**, 264 (1952).

[4] J. R. HOLT and A. E. LITHERLAND: Rev. Sci. Instrum. **25**, 298 (1954).

[5] C. P. BROWNE, S. F. ZIMMERMAN and W. W. BUECHNER: Phys. Rev. **96**, 725 (1954).

[6] E. M. REILLEY, A. J. ALLEN, J. S. ARTHUR, R. S. BENDER, R. L. ELY and H. J. HAUSMAN: Phys. Rev. **86**, 857 (1952).

[7] G. W. GREENLEES: Proc. Phys. Soc. Lond. A **68**, 98 (1955).

[8] W. O. MCMINN, M. B. SAMPSON and V. K. RASMUSSEN: Phys. Rev. **84**, 963 (1951).

particularly suited for this work (SHOEMAKER *et al.*[1]). It is important in studies of resonance reactions to obtain partial widths for a given level for all processes, including inelastic scattering.

The spectra of protons of incident energy about 20 MeV scattered inelastically from moderately heavy elements lose the group structure shown with lower energy protons and lighter elements and have to be interpreted by continuum theory. The experiments (Sect. 38) then lead to an estimation of relative level densities of the target nuclei. The excitation of low-lying levels of beryllium by 190 MeV electrons has also been observed (McINTYRE *et al.*[2]).

γ) Reactions produced by neutrons may be studied by the expansion chamber or ionisation chamber or proportional counter. If a continuous distribution of neutron energies is used, groups may be found which require for their interpretation both excited states of the residual nucleus and resonance energies of the

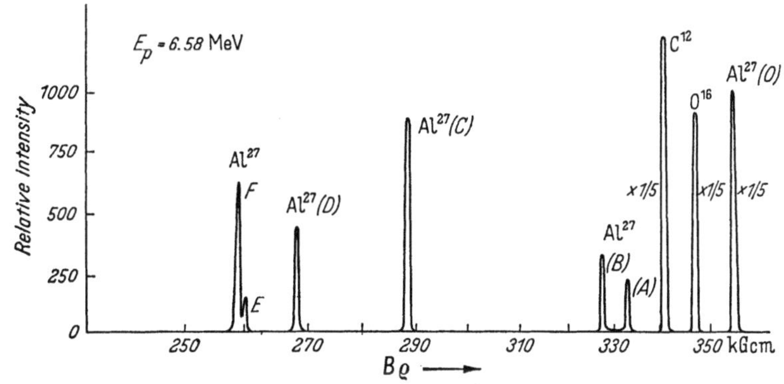

Fig. 11. Proton groups from scattering by Al^{27} target including groups scattered by C^{12} and O^{16} contamination (BROWNE *et al.*). Actual experimental points omitted.

compound state. It is however desirable if possible to separate these effects by using homogeneous neutrons of variable energy, as in the work of JOHNSON and BARSCHALL[3] on the neutron induced reactions of nitrogen. Their results, given in part in Fig. 35, show the precision obtainable by the use of the Li^7 (pn) neutron source, and also clearly demonstrate the resonance energies for proton emission. The neutrons were allowed to react with nitrogen in a cylindrical ionisation chamber and the disintegration energies were obtained from the pulse size distribution observed; the resonance energies were obtained by recording the yield of a particular group as the neutron energy was varied. Several other investigators (e.g. HURLIMAN and HUBER[4]) have used ionisation chambers and pulse analyser techniques for Q-value measurements.

LILLIE[5] used an expansion chamber for studying the disintegrations of oxygen and nitrogen by 14 MeV neutrons; this technique is well suited to a study of such reactions and gives detailed information about each event. These advantages are shared by the nuclear emulsion, although the number of elements which may be covered by both these techniques is rather limited. The exothermic reactions $N^{14}(np)C^{14}$, $B^{10}(n\alpha)Li^7$ and $Li^6(nt)He^4$ have been studied by both methods,

---

[1] F. C. SHOEMAKER, J. E. FAULKNER, G. M. B. BOURICIUS and F. P. MOORING: Phys. Rev. **83**, 1011 (1951).
[2] J. A. McINTYRE, B. HAHN and R. HOFSTADTER: Phys. Rev. **94**, 1084 (1954).
[3] C. H. JOHNSON and H. H. BARSCHALL: Phys. Rev. **80**, 818 (1950).
[4] T. HURLIMAN and P. HUBER: Helv. phys. Acta **28**, 33 (1955).
[5] A. B. LILLIE: Phys. Rev. **87**, 716 (1952).

and the last has been developed as a method of neutron energy determination (KEEPIN and ROBERTS[1]).

Cross sections for (np), $(n\alpha)$ and $(n2n)$ reactions with 14 MeV neutrons have been obtained by PAUL and CLARKE[2] by absolute measurement of the activity of the residual nucleus. Such work is useful in checking the predictions of the continuum theory of reactions for the case of competing reactions. In this and many other measurements with neutrons, an absolute measure of neutron flux is required. This is usually obtained from a "long counter" of the type designed by HANSON and MCKIBBEN[3] in which a long paraffin wax tube surrounds a proportional counter containing boron trifluoride. This counter is relatively insensitive to changes of neutron energy and may be calibrated with a Ra—Be neutron standard, which has in turn been calibrated by an integration method. Alternatively the activity observed may be compared with that induced in a standard material (e.g. $Cu^{63}(n2n)Cu^{62}$) or with the number of recoils produced from a hydrogenous radiator.

Inelastic scattering of neutrons has recently received much attention because of its convenience for investigating reaction thresholds. The variation of yield of a reaction near threshold may be difficult to establish if there is an outgoing charged particle because of barrier effects, but with neutrons these are small and detailed comparison of the variation of yield may be made with theory.

14. Detection and analysis of energy of gamma radiation. Many excellent reviews and books [*20*], [*21*] now deal fully with this subject and only the methods chiefly used for the study of energy levels of light nuclei will be listed here. Further details will be found in the articles in Vol. XLV of this Encyclopedia.

Transitions between the low lying levels of nuclei give rise to photons with energies between about 10 keV and perhaps 4 MeV, while capture radiation gives some transitions with energies up to 12 to 13 MeV. Absorption methods are useless for analysing complex transitions of this kind and the most generally useful method is scintillation spectrometry. Sodium iodide crystals activated with thallium can be grown in very large sizes, but even a moderate crystal of the order of $1'' \times 1'' \times 1''$ has sufficient stopping power to absorb the electrons produced by 10 MeV gamma radiation. The sensitivity of these detectors is very high in comparison with that of spectrometers relying on conversion in thin foils, and the resolution is adequate for most purposes. The size of the pulse produced in a scintillation crystal by a photon, and observed by a photomultiplier and linear amplifier is closely proportional to the energy of the electron produced in the primary event. The response of NaI (Tl) crystals to gamma radiation is discussed by GARLICK[4] and the factors contributing to line width are fully analysed by BREITENBERGER[5] who also gives references to papers which describe the measurement of absolute counting efficiency of the crystal for gamma-rays. Fig. 12a shows a typical spectrum of high energy gamma radiation obtained with a sodium iodide scintillation spectrometer. The pulses are recorded and analysed electronically or photographically[6] so that the whole spectrum is obtained in one run. Provision is usually made in such pulse analysers for

[1] G. R. KEEPIN and J. H. ROBERTS: Rev. Sci. Instrum. **21**, 163 (1950).

[2] E. B. PAUL and R. L. CLARKE: Canad. J. Phys. **31**, 267 (1953).

[3] A. O. HANSON and J. L. MCKIBBEN: Phys. Rev. **72**, 673 (1947).

[4] G. F. J. GARLICK: Progress in Nuclear Physics, Vol. 2. London: Pergamon Press 1952.

[5] E. BREITENBERGER: Progress in Nuclear Physics, Vol. 4. London: Pergamon Press 1955.

[6] The pulse analyser described by G. W. HUTCHINSON and G. G. SCARROTT: Phil. Mag. **42**, 792 (1951) or the gray-wedge analyser of W. BERNSTEIN, R. L. CHASE and A. W. SCHARDT: Rev. Sci. Instrum. **24**, 437 (1953) are frequently used.

recording the spectrum of a particular line in coincidence with some other radiation, so that genetic relationships in a cascade may be studied.

The resolution of NaI (Tl) spectrometers depends on the energy of the radiation. At energies of about 10 MeV when the response in the crystal is mainly due to pair production, the resolution is about 10%. A slight difficulty in interpretation arises from the presence of subsidiary peaks due to escape of one or both of the pair annihilation quanta. At lower energies the pair lines are also found but the COMPTON distribution becomes more important and the pair peaks are superimposed on this. At energies below the pair threshold there are pulses from the COMPTON recoil electrons and from photoelectrons and as the quantum energy decreases the photoelectric peak becomes more important but the resolution decreases, as $\sqrt{E}$, until it is only of the order of 20 to 30% for 50 keV quanta. Despite the relatively poor resolution, the sodium iodide spectrometer has such enormous advantages in sensitivity owing to the high density of the crystals, that it is almost universally used now, except where resolution is of cardinal importance.

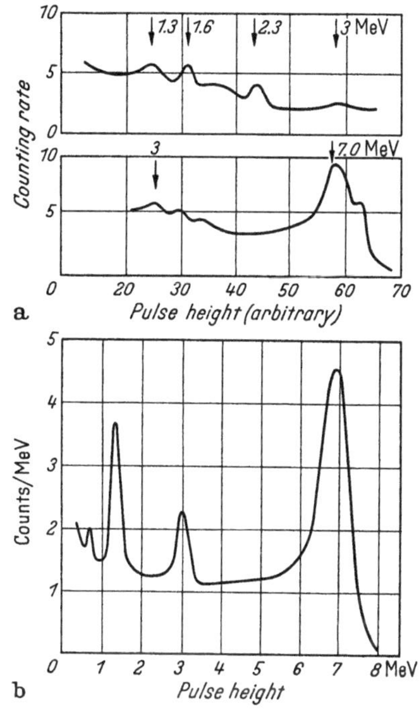

Fig. 12 a and b. Pulse height distribution of gamma radiation from C^{13} $(p\gamma)$ N^{14} reaction. (a) Single crystal [H. H. WOODBURY, R. B. DAY and A. V. TOLLESTRUP: Phys. Rev. 92. 1199 (1953)]. (b) Three-crystal [B. HIRD, C. WHITEHEAD, J. BUTLER and C. H. COLLIE: Phys. Rev. 96, 702 (1954)].

The ambiguity due to the occurrence of multiple peaks in NaI (Tl) spectrometry for energies above the pair threshold can be removed, at the expense of sensitivity, by the use of a three-crystal spectrometer. In this instrument pulses from the centre crystal are only accepted when there is simultaneously a pulse due to an annihilation quantum in each of two side crystals. The difference between single- and three-crystal spectra is shown in Fig. 12. For energies below the pair threshold improved resolution can be obtained by using a two-crystal coincidence spectrometer to measure the pulse height due to COMPTON recoil electrons (HOFSTADTER and MCINTYRE[1]).

For accurate measurements of high energy quanta, if intensity is sufficient, a magnetic pair spectrometer may be used. The instruments used by WALKER and MCDANIEL[2] for a study of $(p\gamma)$ capture reactions and by KINSEY [36] for $(n\gamma)$ capture reactions are typical and give a resolution of about 5% independent of energy, although the sensitivity falls off rapidly as quantum energy decreases. Fig. 13 shows the spectrum of gamma radiation in the $F^{19}(p\alpha\gamma)O^{16}$ reaction studied by WALKER and MCDANIEL[2]. Rather better resolution can be obtained by magnetic spectrographs used to analyse COMPTON electron distributions from thin radiators (HORNYAK *et al.*[3]) but a high intensity of primary radiation is required. If this is available a precision of about 0.2% in energy may be obtained which is not much worse than the accuracy of Q-value determinations by heavy particle spectrographs. The connection between the gamma transitions and the states

[1] R. HOFSTADTER and J. A. MCINTYRE: Phys. Rev. **78**, 619 (1950).
[2] R. L. WALKER and B. D. MCDANIEL: Phys. Rev. **74**, 315 (1948).
[3] W. F. HORNYAK, T. LAURITSEN and V. K. RASMUSSEN: Phys. Rev. **76**, 731 (1949).

indicated by heavy particle emission may then be established with fair certainty. Confirmation is often obtained with the lower resolution, but higher solid angle scintillation spectrometer used in coincidence with a heavy particle detector.

The magnetic spectrograph has been much used for studying internal conversion electrons corresponding to radiative transitions in heavy nuclei because of its high resolution. Internal conversion is less frequently encountered in the transitions of light nuclei, but low energy radiation, including X-radiation, is often emitted, and in the energy range from 1 keV up to 100 keV is best studied by the proportional counter developed for this purpose by Curran and others [*21*]. The resolution of this instrument depends on the gas amplification and can be as good as 5% at an energy of 50 keV. The study of the K and L capture radiation of A^{37} by Pontecorvo *et al.*[1] is a good example of the power of this method.

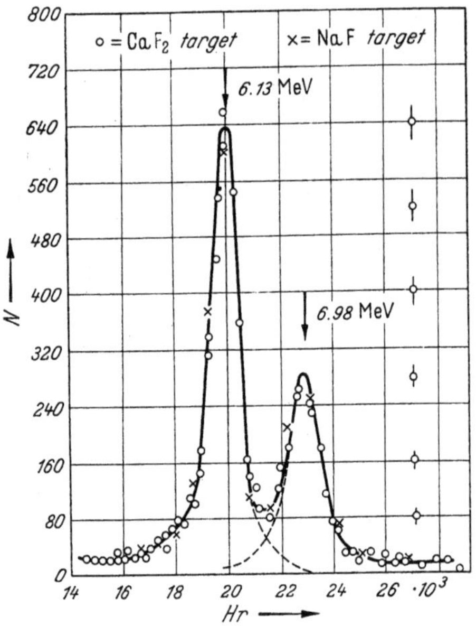

Fig. 13. Measurement of energy of gamma radiation from F^{19} $(p\alpha\gamma)$ O^{16} reaction by pair spectrometer (Walker and McDaniel).

In some experiments, particularly of the coincidence type, in which time correlation of successive radiations is important rather than energy resolution, organic scintillators are used. Anthracene and other similar crystals have a response or rise time of the order of 10^{-8} to 10^{-9} sec and enable coincidence resolving times to be so far reduced that high disintegration rates can be used. Time-of-flight methods using fast efficient scintillators such as solutions of terphenyl in benzene have been developed in several laboratories (James and Treacy[2]) for both neutron and gamma-ray detection.

By application of the two methods of high resolution and coincidence spectrometry thereis now very definite and satisfactory correlation of radiative transitions between bo und levels with the groups of particles emitted in artificial transmutation. In a resonant reaction the observed transitions include gamma ray lines originating from resonant states as well as bound states. The former may be distinguished from the latter by the fact that their energy varies with the resonance energy. The differences between the energies of a set of gamma ray lines originating in a virtual state can be used to deduce the excitation of the lower bound levels to which the transitions proceed.

15. Beta-spectroscopy. Many light nuclei formed by transmutation processes, including radiative capture, decay into stable nuclei, either in the ground state or in an excited state, by the emission of beta-particles. The particular aspects of the theory of beta-decay (see Vol. XLII of this Encyclopedia) which are of significance for the spectroscopy of the light nuclei arise from the dependence of transition probabilities on the relative symmetry properties of the initial and final nuclei. The selection rules for spin and parity changes in beta-transitions are given in Table 1.

---

[1] B. Pontecorvo, D. H. W. Kirkwood and G. C. Hanna: Phys. Rev. **75**, 982 (1949).
[2] D. B. James and P. B. Treacy: Proc. Phys. Soc. Lond. A **64**, 847 (1951).

The FERMI and GAMOW-TELLER rules correspond respectively to the emission of an electron-neutrino pair in a singlet or triplet state. Convincing evidence has accumulated that transitions of the type $0 \to 0$, *No* and $0 \to 1$, *No* both occur with high probability, so that the general beta-interaction may contain both scalar and tensor terms.

The forbidden beta-transitions are those in which the electron and neutrino take away orbital momentum. The difference in probability is due to the appearance of a factor of the form R/λ_β in the matrix element in analogy with the higher multipole terms in radiation theory. The empirical evidence on beta-emitters enables them to be classified according to their comparative half-life or ft-value, where t is the half-life and f is an integral of the energy spectrum. Determination of the ft-value entails observation of the upper limit of the beta-spectrum and use of the graphs of f-values given by FEENBERG and TRIGG[1]. From the ft-values the probable change of spin and parity can be estimated; FEENBERG and TRIGG show

Table 1. *Selection rules for beta-decay.*

Type of transition	FERMI rules		GAMOW-TELLER rules	
	spin ch.	parity ch.	spin ch.	parity ch.
Allowed . . .	0	No	0, ±1 except 0 → 0	No
1st Forbidden	0, ±1 except 0 → 0	Yes	0, ±1, ±2	Yes
2nd Forbidden	±1, ±2	No	0 → 0, ±2, ±3	No

the grouping of observed transitions into superallowed (Sect. 2), allowed, and forbidden classes. In analysis of forbidden spectra the detailed spectral shape must be observed, using generally a magnetic spectrometer, or for low energy transitions a proportional counter. In certain cases of anomalous ft-values the effect of additional selection rules such as the separate conservation of L and S may be seen. The actual transition probabilities may be compared with predictions from particular models such as the independent particle model with jj, LS, or intermediate coupling.

The techniques of beta spectroscopy are discussed in [*21*] and in Vols. XXXIII and XLII of this Encyclopedia. A simple beta-spectrum shows a straight line "FERMI" plot. Complexity is shown by deviations from the straight line plot and may be verified by beta-gamma coincidence measurements. Measurements on the gamma radiation can be made by the methods discussed in the preceding section, particularly those appropriate for low energy radiation. In a few special cases beta-spectra have been examined by absorbing the radiation in an anthracene crystal and plotting a pulse size distribution (LAUBITZ[2]).

II. Elastic scattering.

16. General. The analysis of experiments on the elastic scattering of nuclei is complicated by the existence of COULOMB scattering, with which at least part of the nuclear scattering is coherent. The formulae (9.11) and (9.17) give the total cross section for elastic scattering near an isolated resonance of the compound nucleus. These were obtained by adding terms containing the potential scattering

[1] E. FEENBERG and G. L. TRIGG: Rev. Mod. Phys. **22**, 399 (1950). See also: Tables for Analysis of Beta Spectra. Nat. Bur. Stand., Appl. Math. Ser. No. 13 (1952) and [*21*].

[2] M. J. LAUBITZ: Proc. Phys. Soc. Lond. A **68**, 1033 (1955).

amplitude A_{pot} to the general reaction formula (9.8). It is however useful to consider elastic scattering from first principles using the method of partial waves. General treatments of the problem have been given by FLÜGGE[1] and by BLATT and BIEDENHARN [28].

The differential cross section for elastic scattering of a charged particle by a nucleus is given in terms of the asymptotic phase shifts δ_l of the partial wave of angular momentum $l\hbar$ by the formula (SCHIFF[2])

$$\sigma(\vartheta) = \left| f_c(\vartheta) + \bar{\lambda} \sum_{l=0}^{\infty} (2l+1)\, e^{(i\delta_l + i\alpha_l)} \sin\delta_l P_l(\cos\vartheta) \right|^2 \qquad (16.1)$$

where ϑ is the centre-of-mass angle of scattering, $f_c(\vartheta)$ is the amplitude for scattering in a pure COULOMB field and

$$\begin{aligned} \alpha_l &= 2(\sigma_l - \sigma_0) & (16.2)\\ &= 2\{\arg\Gamma(l+1+i\eta) - \arg\Gamma(1+i\eta)\} \\ &= 2\sum_{n=1}^{l} \operatorname{arc\,tan}\frac{\eta}{n} \quad \text{for} \quad l>0 & (16.3) \end{aligned}$$

and $\alpha_0 = 0$[3]. The symbols σ, η are defined in Sect. 7.

The phase shift δ_l represents the refraction of the incident partial wave of angular momentum l by the non-COULOMB field; if all δ_l are zero (16.1) reduces to the RUTHERFORD scattering formula. If the non-COULOMB field is due to a nucleus of radius R, and if $\bar{\lambda} \ll R$, then the phase shifts are important up to an l-value given by

$$l\bar{\lambda} = R \qquad (16.4)$$

according to a simple classical argument, and must be determined by fitting the external wave function to the internal wave function at the nuclear surface in such a way that $\left(\frac{1}{\psi_l}\frac{d\psi_l}{dr}\right)$, Eq. (9.1), is continuous. In the limiting case of no internal wave function, the scattering is that from a hard sphere of radius R and the phase shifts are those given by (9.13). If however the amplitude of the internal wave function is large for the particular incident energy and l-value, there is resonant scattering, and it is useful to split the phase shift δ_l into two parts (ADAIR[4])

$$\delta_l = \varphi_l - \beta_l \qquad (16.5)$$

where

$$\tan\varphi_l = -\left[\frac{F_l}{G_l}\right]_{r=R} \qquad (16.6)$$

and

$$\tan\beta_l = \frac{\frac{1}{2}\Gamma_\lambda}{E - E_{\text{res}}}. \qquad (16.7)$$

The quantities F_l, G_l, Γ_λ, E_{res} are defined in Sect. 9. The separation into hard sphere scattering (phase shift φ_l) and nuclear resonant scattering (phase shift β_l) is a mathematically convenient way of distinguishing between potential scattering

[1] S. FLÜGGE: Ergebn. exakt. Naturw. **25**, 476 (1951).

[2] L. I. SCHIFF: Quantum Mechanics, p. 120. New York: McGraw Hill 1949.

[3] The equivalent expressions

$$e^{i\alpha_l} = \prod_1^l \frac{l+i\eta}{l-i\eta}, \qquad e^{i\alpha_0} = 1$$

are found in many papers.

[4] R. K. ADAIR: Phys. Rev. **81**, 310 (1951); **86**, 155 (1952).

and compound nucleus scattering, but it is somewhat artificial physically since potential scattering may well be due to the superposition of many resonant contributions from distant levels. In practice the hard sphere contribution is evaluated at an interaction radius a chosen to give the observed ratio between resonant and potential scattering. Choice of an interaction radius determines the reduced widths and level shifts of dispersion theory (Sect. 9). The energy at which maximum yield is observed is not necessarily E_{res} because of interference with the potential scattering term. The maximum yield of resonant scattering, according to (9.11) cannot exceed $4\pi\lambda\!\!\!^{-2}(2l+1)$. As the incident energy varies through the resonance, the nuclear phase shift β_l varies from nearly 0 to nearly π in an energy range of a few half-widths, while the potential phase shift φ_l varies much more slowly. The observed scattering at a given angle then exhibits a characteristic pattern.

The possibility of the interference effect being experimentally observable depends on the ratio of the width of the resonance to the homogeneity of the beam of incident particles, but with stable electrostatic generators these can be of the same order. The interference effect was well demonstrated by BENDER[1] *et al.* who studied the variation of elastic scattering of protons with energy near one of the gamma-ray resonances known from the $Al^{27}(p\gamma)Si^{28}$ reaction. Fig. 14 shows the variation of gamma ray intensity and of the number of scattered protons per unit solid angle at 135°. The form of the curves agrees with that predicted by the single level formula, and gives, with the assumption of zero orbital momentum for the incident protons, a natural width of about 100 volts.

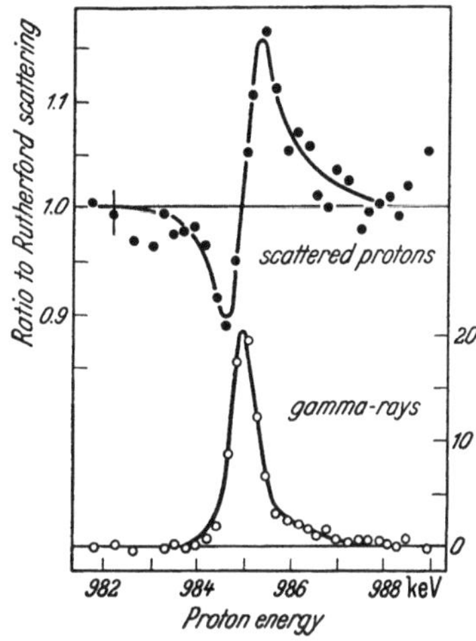

Fig. 14. Resonant scattering of protons by Al^{27} (BENDER *et al.*).

This brief discussion disregards the existence of spin-orbit forces which result in different phase shifts for waves with different orientations of spin with respect to orbital momentum.

17. Elastic scattering of neutrons. The absence of COULOMB scattering simplifies formula (16.1) in which $f_c(\vartheta)$ and α_l may be put equal to zero. The differential cross section for scattering near a resonance formed by the partial wave of orbital momentum $l\hbar$ reduces to

$$\sigma(\vartheta) = \lambda\!\!\!^{-2}\,|(2l+1)\,e^{i\delta_l}\sin\delta_l P_l(\cos\vartheta)|^2 \tag{17.1}$$

and this may be written, using (16.5) as

$$\sigma(\vartheta) = \lambda\!\!\!^{-2}\,|(2l+1)\,(e^{-i\beta_l}\sin\beta_l - e^{-i\varphi_l}\sin\varphi_l)\,P_l(\cos\vartheta)|^2. \tag{17.2}$$

In the case of neutron scattering, a total cross section may be defined, since there is no small angle COULOMB term and this is

$$\sigma = 2\pi\int_0^\pi \sigma(\vartheta)\sin\vartheta\,d\vartheta, \tag{17.3}$$

$$= 4\pi\lambda\!\!\!^{-2}(2l+1)\left|\frac{\frac{1}{2}\Gamma_n}{E - E_{\mathrm{res}} + \frac{i}{2}\Gamma_n} - e^{-i\varphi_l}\sin\varphi_l\right|^2 \tag{17.4}$$

[1] R. S. BENDER, F. C. SHOEMAKER, S. G. KAUFMANN and G. M. B. BOURICIUS: Phys. Rev. **76**, 273 (1949).

using the customary normalisation of $P_l(\cos\vartheta)$. This must be further modified to take account of the potential scattering of partial waves with l-values which do not contribute to the resonance. The final result (ADAIR[1]) is, if neutron scattering is the only process possible

$$\left.\begin{aligned}\sigma &= (2l+1)\, g'\, 4\pi \lambda\!\!\!^{-2} \left| \frac{\frac{1}{2}\Gamma_n}{E - E_{\rm res} + \frac{i}{2}\Gamma_n} - e^{-i\varphi_l}\sin\varphi_l \right|^2 + \\ &+ (2l+1)(1-g')\, 4\pi \lambda\!\!\!^{-2} \sin^2\varphi_l + \sum_{l' \neq l} (2l'+1)\, 4\pi \lambda\!\!\!^{-2} \sin^2\varphi_{l'}\end{aligned}\right\} \tag{17.5}$$

where the spin of the resonance is J and

$$g'(J) = \frac{g(J)}{2l+1} = \frac{2J+1}{(2j_1+1)(2j_2+1)(2l+1)} \tag{17.6}$$

j_1 and j_2 $(=\frac{1}{2})$ being the spins of the target nucleus and neutron respectively.

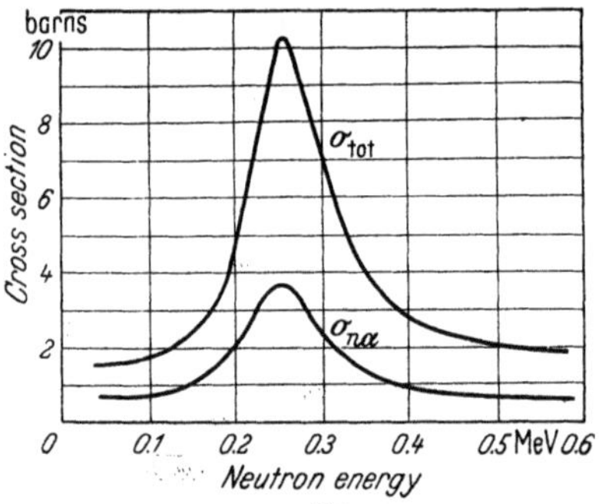

Fig. 15. Total cross section of Li^6 for neutrons (JOHNSON *et al.*). The lower curve gives the cross section for the Li^6 $(n\alpha)$ H^3 reaction.

The cross section has maximum and minimum values at energies $E_{\rm res} - \frac{1}{2}\Gamma_n \tan\varphi_l$ and $E_{\rm res} + \frac{1}{2}\Gamma_n \cot\varphi_l$. The difference between these cross section values is (ADAIR[1])

$$\left.\begin{aligned}&\sigma_J(\max) - \sigma_J(\min) \\ &\quad = 4\pi \lambda\!\!\!^{-2} \frac{2J+1}{(2j_1+1)(2j_2+1)}\end{aligned}\right\} \tag{17.7}$$

and the depth of the dip is

$$4\pi \lambda\!\!\!^{-2} \frac{2J+1}{(2j_1+1)(2j_2+1)} \sin^2\varphi_l. \tag{17.8}$$

These relations determine the spin J of the compound state, the hard sphere phase shift φ_l and the incident orbital momentum forming the compound state. For neutrons the phase shifts φ_l are given by formula (9.13) evaluated at the interaction radius. Thus, for an s-wave resonance $\varphi_0 = -ka$ and as the neutron energy increases towards the resonance, Eq. (17.4) shows that a dip precedes the peak.

If reactions are energetically possible the resonant scattering amplitude becomes

$$\frac{\frac{1}{2}\Gamma_n}{E - E_{\rm res} + \frac{i}{2}\Gamma} \tag{17.9}$$

where $\Gamma = \Gamma_n + \Gamma_p$. The scattering anomaly, Eq. (17.8), then depends on Γ_n/Γ and if $\Gamma_n \ll \Gamma$, so that the total resonant cross section is equal to the reaction cross section, the anomaly may not be observable in a given experiment. The peak in the total cross section is displaced from that in the reaction cross section by the interference effect.

Measurement of total cross sections is usually carried out by a transmission method in which the attenuation of a neutron flux by a known thickness of material is observed. Fig. 15 shows the results obtained by JOHNSON *et al.*[2] for Li^6. Analysis of the total cross section gives $J = \frac{5}{2}$ for the compound state in Li^7 so that the incident l-value must be between 1 and 4. The observed width suggests that $l = 1$. Differential cross section measurements (WILLARD *et al.*[3])

[1] R. K. ADAIR: Rev. Mod. Phys. **22**, 249 (1950).
[2] C. H. JOHNSON, H. B. WILLARD and J. K. BAIR: Phys. Rev. **96**, 985 (1954).
[3] H. B. WILLARD, J. K. BAIR and J. D. KINGTON: Phys. Rev. **98**, 669 (1955).

remove some of the ambiguities in assignment of l-values from total cross section measurements.

18. Elastic scattering of charged particles. *α) Channel spin zero.* In the case of the scattering of spinless particles by nuclei without spin the differential cross section is given directly by (16.1). Inspection of this equation shows that scattering anomalies due to the partial wave of order l, corresponding to a resonant state of spin $J = l$ and parity $(-1)^l$, are not observed at the angle at which $P_l(\cos\vartheta) = 0$. For this reason excitation curves for elastic scattering are usually taken at centre-of-mass angles 90° $(P_l(\cos\vartheta) = 0$ for l odd), 125° $(P_2(\cos\vartheta) = 0)$, 140° $(P_3(\cos\vartheta) = 0)$ and as near as possible to 180° where $P_l(\cos\vartheta)$ is a maximum for all l-values. The direct interpretation of scattering resonances by this method in this simple case is shown by the work of CAMERON[1] on the scattering of alpha particles by oxygen (Fig. 16b). The Fig. 16c and d show the potential scattering phase shifts φ_l and the penetrability amplitudes used in obtaining reduced widths from observed widths in this experiment. The levels of the compound nucleus Ne^{20} shown have the spin values 0^+ (2.5 and 3.09 MeV), 2^+ (3.39 and 3.89 MeV) and 3^- (3.045 MeV). This means of identification of l-values is always possible in the scattering of spinless particles, for which spin-orbit coupling does not occur. Absolute cross sections are obtained in such work by comparison with RUTHERFORD scattering.

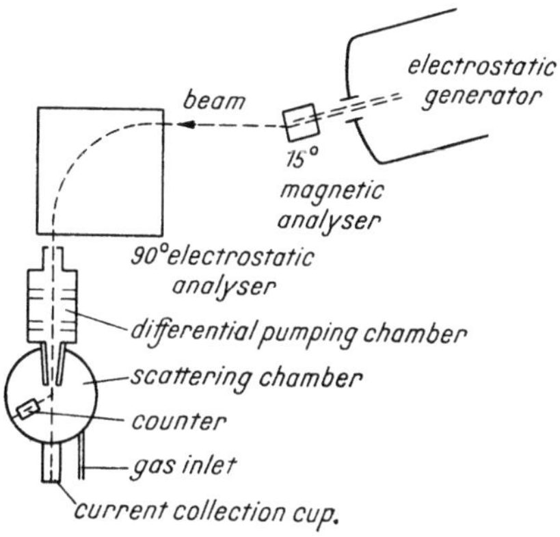

Fig. 16 a—d. Scattering of alpha particles by oxygen (CAMERON). (a) Experimental apparatus.

Fig. 16a shows the apparatus used in this work and in many similar elastic scattering experiments in the Wisconsin laboratory.

β) Channel spin ½. In the scattering of protons or neutrons by nuclei of zero spin it is necessary to separate the incident partial wave of order l into two sub-waves which form compound states with $J = l \pm \frac{1}{2}$. Then l is no longer a good quantum number and the spin-orbit forces may induce transitions with a change of the orientation of spin with respect to orbital momentum, i.e. with a change in the z-component, of channel spin. The scattering with change of channel spin component ("spin flip" scattering) is incoherent with the COULOMB scattering and if δ_l^+ and δ_l^- are the phase shifts for the two sub-waves of the partial wave of order l, then the differential cross section is found[2] to be

$$\sigma(\vartheta) = \lambda\!\!\!^{-2}\{|A|^2 + |B|^2\} \tag{18.1}$$

where

$$\left.\begin{aligned} A = f_c(\vartheta) + \sum_0^\infty (l+1)\, P_l(\cos\vartheta)\, e^{i\alpha_l + i\delta_l^+} \sin\delta_l^+ + \\ + \sum_0^\infty l\, P_l(\cos\vartheta)\, e^{i\alpha_l + i\delta_l^-} \sin\delta_l^-, \end{aligned}\right\} \tag{18.2}$$

$$B = \sin\vartheta \sum_1^\infty P_l'(\cos\vartheta)\, e^{i\alpha_l}\,[\sin\delta_l^-\, e^{i\delta_l^-} - \sin\delta_l^+\, e^{i\delta_l^+}] \tag{18.3}$$

[1] J. R. CAMERON: Phys. Rev. **90**, 839 (1953).
[2] C. L. CRITCHFIELD and D. C. DODDER: Phys. Rev. **76**, 602 (1949).

with

$$P_l'(\cos\vartheta) = \frac{d}{d(\cos\vartheta)}\,[P_l(\cos\vartheta)]\,. \tag{18.4}$$

The term B, which represents the incoherent scattering due to spin-orbit forces does not exist for $l=0$ and is small at backward angles because of the $\sin\vartheta$ term.

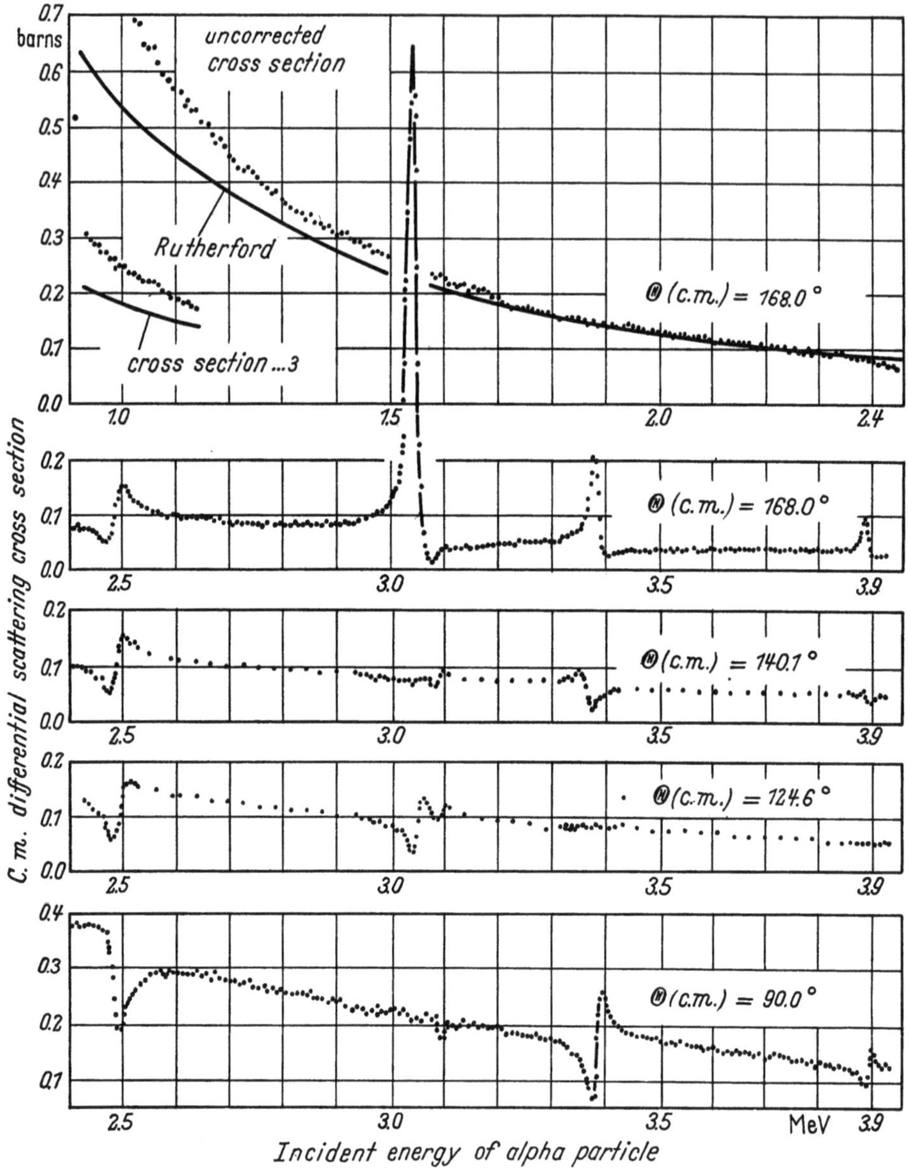

Fig. 16b. Results.

LAUBENSTEIN and LAUBENSTEIN[1] give a geometric representation of the amplitude and phase of each partial wave which is useful in the analysis of experimental results. As the incident energy passes through a resonance the l-wave scattering vector moves once round a circle of diameter $(J+\frac{1}{2})\,P_l\,(\cos\vartheta)$ and the resultant

[1] R. A. LAUBENSTEIN and M. J. W. LAUBENSTEIN: Phys. Rev. **84**, 18 (1951).

of this vector and that representing RUTHERFORD and potential scattering gives the form of the observed resonance. A useful result which may be simply deduced from this treatment is that for a given angle of observation the maximum and minimum values of differential cross section near a resonance J are related by

$$\sqrt{\sigma_{\max}(\vartheta)} \pm \sqrt{\sigma_{\min}(\vartheta)} = \lambda\!\!\!^{-} (J + \tfrac{1}{2})\, P_l(\cos\vartheta). \tag{18.5}$$

This is analogous to the relation (17.7) for total neutron cross sections and enables J to be estimated. This graphical method has been applied to most of the results of the Wisconsin laboratory for the elastic scattering of charged particles; in particular to the elastic scattering of protons by C^{12}, O^{16} and Mg^{24}. The absolute cross sections in experiments such as these are

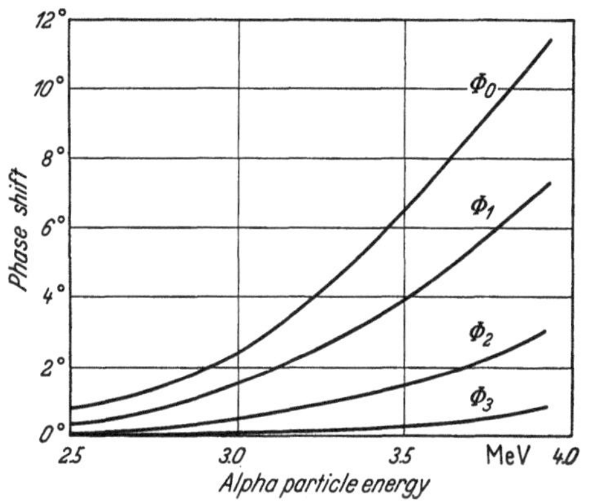

Fig. 16c. Potential scattering phase shifts $(-\varphi_l)$.

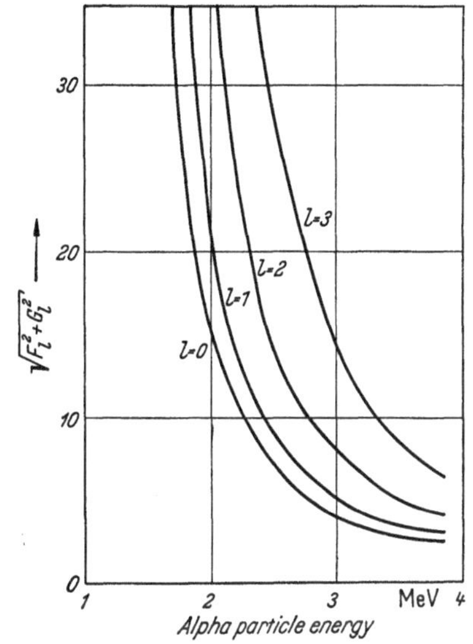

Fig. 16d. Penetrability amplitudes.

obtained by observation of RUTHERFORD scattering at low energies away from resonances. The existence of spin-orbit coupling implies different cross sections for the scattering of the partial waves corresponding to angular momentum $l \pm \frac{1}{2}$ near a resonance. This leads to partial polarisation of the scattered particles.

γ) *Higher values of channel spin. Competing reactions.* The formulae become very complex because of the number of possible transitions to states with different z-components of channel spin. If reactions are energetically possible the ratio of reaction width to total width also appears as a further nuclear parameter. Examples of the application of the general theory of BLATT and BIEDENHARN [28] to particular cases are given by BARANGER[1] ($F^{19} + p$) and LAURITSEN[2] *et al.* ($He^4 + H^2$).

III. Radiative transitions.

19. Introduction; lifetime formulae. All excited states of nuclei may decay by electromagnetic transitions (radiation, pair production) and in the case of bound levels these transitions are (excepting for beta-decay) the only means by which the nucleus may lose its excess energy. Radiative processes from virtual levels have to compete with the re-emission of particles, and owing to the smallness of the coupling between the nuclear charge distribution and the electromagnetic field, the probability of particle break-up is usually much larger. The smallness of the

[1] E. U. BARANGER: Phys. Rev. **99**, 145 (1955).
[2] T. LAURITSEN, T. HUUS and S. G. NILSSON: Phys. Rev. **92**, 1501 (1953).

electromagnetic coupling has however the advantage that a semi-classical treatment based on nuclear charge distributions is satisfactory for the estimation of radiative transition probabilities in nuclear reactions. In particular it has been possible to set up energy-lifetime relations which although based on simple and inadequate assumptions, have proved extremely useful in classifying experimental information.

Radiation may be treated formally in the same way as particles in the sense that for both a plane wave may be decomposed into partial waves with different orbital momenta or l-values. A radiative transition carries away a certain number of units of orbital momentum, just as does a particle emitted from a nuclear state, and for the same reason. There is the difference that the parity of the radiation field is not determined solely by the orbital momentum of the photon as in the case of a particle. The expressions for the electric and magnetic fields $\boldsymbol{E}$ and $\boldsymbol{H}$ radiated by a classical charge distribution are given in [7], p. 796. From these it can be seen that quantum numbers l, m, related in the classical picture to the order of the multipole moment, are necessary to describe the radiation. The angular momentum carried away by one quantum is $l\hbar$ with z-component $m\hbar$. Multipole fields of order $l=0$ vanish; the corresponding transition (monopole) is an internal conversion or pair production process.

Table 2.
Parity selection rule in radiative transitions.

	$\Delta J = 0, 1, 2 \ldots$	
	Electric multipoles	Magnetic multipoles
l even	no parity change	parity change
l odd	parity change	no parity change

The selection rules for a radiative transition between two states of a nucleus are based on the conservation of angular momentum and are

$$|J' - J| \leq l \leq J' + J \tag{19.1}$$

and

$$J'_z - J_z = m \tag{19.2}$$

with photon emission between $J' = J = 0$ forbidden. The transition probability is proportional to a matrix element of the form

$$H \sim \int \psi_f^* \, r^l \, \psi_i \, d\tau \tag{19.3}$$

for a multipole radiation of order l. In order that this shall not vanish the parity must change between ψ_f and ψ_i for "odd" radiation and must remain the same for "even" radiation. This gives the selection rules shown in Table 2.

A further selection rule which seems to have some validity for radiative processes prescribes the change of total isotopic spin T in emission or absorption. The rules given by RADICATI[1] and by GELL-MANN and TELEGDI[2] are that

(a) For all radiative processes $\Delta T = 0$ or 1.

(b) Electric dipole radiation in self-conjugate nuclei ($T_z = 0$) is forbidden unless $\Delta T = \pm 1$.

These rules are not absolute in the sense of the parity selection rule, since they are based on the assumption of charge independence. Rule (b) in particular is weakened in many cases by COULOMB perturbation which mixes wave functions of different T for a given nuclear state. Other causes of isotopic spin impurity are discussed by MACDONALD[3]; the experimental evidence for radiative transitions is reviewed by WILKINSON[4].

[1] L. A. RADICATI: Phys. Rev. **87**, 521 (1952). — Proc. Phys. Soc. Lond. A **66**, 139 (1953); **67**, 39 (1954).
[2] M. GELL-MANN and V. L. TELEGDI: Phys. Rev. **91**, 169 (1953).
[3] W. M. MACDONALD: Phys. Rev. **98**, 60 (1955).
[4] D. H. WILKINSON: Nature, Lond. **172**, 580 (1953). See also [*24*].

The *multipolarity* of a transition $J' \to J$ is given by Eq. (19.1). The fields of electric 2^l-pole and magnetic 2^{l-1}-pole radiation may both be derived from the time variation of the l-th moment of a charge distribution. Although the magnetic radiation is weaker than the associated electric radiation by a factor of the order $(v/c)^2$ in intensity, the two types of transition have the same parity and show similar behaviour in internal conversion and pair production phenomena which depend on the form of the field near the origin of the radiation. It is therefore convenient to describe both electric 2^l-pole and magnetic 2^{l-1}-pole transitions as of *order* $\Lambda = l$.

The probability of emission of radiation can be estimated from classical electrodynamics if the radiating moments are given. Thus for a moment of order eR^l leading to electric 2^l-pole radiation of energy $E = \hbar\omega = k\hbar c$ the transition probability per unit time is

$$T(l) = \frac{e^2 k}{\hbar} \frac{(kR)^{2l}}{[(l-1)!]^2} \tag{19.4}$$

which already shows useful features, namely that the radiative widths for 2^l-pole radiation are proportional to E^{2l+1} and that since kR is usually $\ll 1$, the emission probability decreases rapidly as multipolarity increases. A similar quantum mechanical formula has been given by WEISSKOPF[1] who takes account of the contribution to magnetic radiation from intrinsic spins. The matrix elements are calculated for a single particle moving in a potential field and are summed over all nucleons, but this procedure ignores collective motion (which shortens the lifetime), relativistic effects and exchange currents. WEISSKOPF gives for the transition probabilities per unit time for electric and magnetic radiation the formulae:

$$\left.\begin{aligned} T_E(l) &= \frac{2(l+1)}{l[(2l+1)!!]^2}\left(\frac{3}{l+3}\right)^2 \frac{e^2}{\hbar c}(kR)^{2l}\omega \\ &= \frac{4.4(l+1)}{l[(2l+1)!!]^2}\left(\frac{3}{l+3}\right)^2\left(\frac{\hbar\omega}{197\text{ MeV}}\right)^{2l+1} \times (R\text{ in }10^{-13}\text{ cm})^{2l}\cdot 10^{21}\text{ sec}^{-1} \end{aligned}\right\} \tag{19.5}$$

and

$$T_M(l) = 10\left(\frac{\hbar}{McR}\right)^2 T_E(l) \tag{19.6}$$

where

$$(2l+1)!! \equiv 1\cdot 3\cdot 5\ldots(2l+1).$$

If the half-life for the transition is $t_{1/2}$ then

$$T(l) = \frac{0.693}{t_{1/2}}. \tag{19.7}$$

These formulae are expected to be only approximately accurate, but they have been surprisingly successful. A nomogram for calculation of the lifetimes has been given by MONTALBETTI[2]. The lifetimes may easily be converted into radiative widths and simple formulae relating width to energy for a given multipolarity may be obtained. For dipole radiation from a nucleus of mass number A and radius $1.5\cdot 10^{-13} A^{1/3}$ cm:

$$\Gamma_\gamma(E) = 0.11\, E_\gamma^3 A^{2/3}, \tag{19.8}$$

$$\Gamma_\gamma(M) = 0.021\, E_\gamma^3 \tag{19.9}$$

where Γ_γ is measured in eV and E_γ in MeV.

[1] V. F. WEISSKOPF: Phys. Rev. **83**, 1073 (1951).
[2] R. MONTALBETTI: Canad. J. Phys. **30**, 660 (1952).

The WEISSKOPF calculation strictly gives the lifetime for a transition of a single particle from a state of angular momentum J to a state of angular momentum $\frac{1}{2}$. If the initial and final states have other spin values, for the same spin change, a statistical factor enters the lifetime formula to give an average over initial states and a sum over final states. These factors for electric dipole transitions are listed by HUGHES[1]; they are small in comparison with the uncertainty of the theoretical formulae but should be used in making comparisons between observed widths.

The experimental information on radiative widths in the light elements has been reviewed by WILKINSON [*24*] who lists a quantity $|M|^2$, equal to the observed radiative width divided by the value expected according to the WEISSKOPF formulae (19.5) and (19.6). For $E1$ transitions, $|M|^2$ is distributed about a value of about 0.03, and for $M1$ transitions about 0.15. Similar conclusions were reached by KINSEY and BARTHOLOMEW[2] in a survey of radiation widths for heavier nuclei. These ratios of observed to single particle properties suggest the interpretation used to explain the similar ratios found for nucleon reduced widths (Sect. 10). LANE and RADICATI[3] show in detail how the independent particle model predicts values for $|M|^2$ using the concept of parent states. In LS coupling the radiative transition is

$$l^{n-1}\, l_0(T\,S\,L,\,J) \rightarrow l^{n-1}\, l_1(T'\,S'\,L',\,J') \tag{19.10}$$

and in jj coupling:

$$j^{n-1}\, j_0(J\;T) \rightarrow j^{n-1}\, j_1(J'\;T') \tag{19.11}$$

where the $(n-1)$ nucleons of the parent state are in equivalent shell model orbits l, or j and the single particle makes the change $l_0 \rightarrow l_1$, or $j_0 \rightarrow j_1$. The strength of the transition is then determined by the extent to which the initial and final states derive from the same parent state. Specific calculations of the quantity $|M|^2$ are given by LANE and THOMAS [*19*], and WILKINSON [*24*] shows that the general distribution of $|M|^2$ values for $E1$ transitions follows from a systematic permutation of the quantum numbers in the general formula.

In the following Sects. 20 to 24 a brief account is given of the experimental methods used in the study of radiative transitions, with particular reference to the determination of radiative widths Γ_γ or lifetimes, and of multipolarity. If these quantities are known the quantum numbers and possibly the configuration of the corresponding states may be inferred. The angular correlation of successive radiations is an important source of information and is separately treated (Sect. 25 to 27). Many of the detection techniques have been discussed in earlier sections, particularly Sect. 14.

20. Radiative capture. The main characteristic of the proton capture process

$$\mathrm{X} + p \rightarrow \mathrm{Y}^* \rightarrow \mathrm{Y} + \gamma \tag{20.1}$$

is the occurrence of the sharp resonances in the intensity of gamma radiation as a function of proton energy. These may be accompanied by similar resonances in the yield of competing reactions, and will always be accompanied by anomalies in the elastic scattering of protons by the nucleus X, although experimental resolution may not always permit their observation. The resonances are displayed experimentally as steps in the excitation function with a thick target or as sharp

[1] D. J. HUGHES: Phys. Rev. **94**, 740 (1954).
[2] B. B. KINSEY and G. A. BARTHOLOMEW: Phys. Rev. **93**, 1260 (1954).
[3] A. M. LANE and L. A. RADICATI: Proc. Phys. Soc. Lond. A **67**, 167 (1954).

peaks with a thin target. The natural width of the peaks is often determined by the competition of elastic scattering of the incident protons, but at low energies the proton width may be reduced by the COULOMB barrier so that the radiation width determines the true width of the resonance. The observed resonance width is then given by the experimental resolution, which unfortunately is still many times greater than radiation widths even for the most homogeneous beams available and the thinnest possible targets. Even when these factors are brought under satisfactory control there remains the thermal motion of the target atoms which contributes a DOPPLER broadening to the observed resonance.

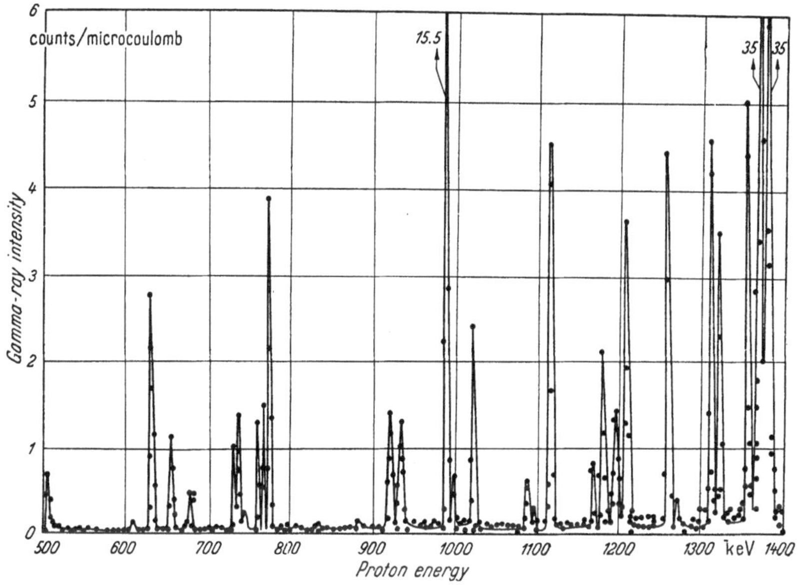

Fig. 17. Radiative capture of protons in Al^{27} (BROSTRÖM *et al.*).

The type of excitation function obtained by bombarding Al^{27} with protons is shown in Fig. 17, taken from the work of BROSTRÖM *et al.*[1]. The gamma radiation was detected by a shielded GEIGER counter near the aluminium target of an electrostatic generator. The sharp peaks determine resonance levels of the compound nucleus Si^{28} and show clearly how the level spacing has diminished from the values of a few hundred keV found for the lightest elements. The widths shown are experimental. The accuracy of determination of resonance energies and width has continuously improved; thus HUNT[2] was able, by using an electrostatically resolved proton beam to obtain for the first resonance in the bombardment of Li^7 by protons a resonant energy of (441.5 ± 0.5) keV and a width Γ of (12.2 ± 0.5) keV. This is the true width of the resonance level, determined in this case by proton re-emission.

The shape of $(p\gamma)$ resonance curves for isolated levels is well described by the single-level formula (9.16) which reduces for the case of a $(p\gamma)$ reaction at resonance with competition only from elastic scattering to the form

$$\sigma_R = 4\pi \lambda\!\!\!^{-2} g \frac{\Gamma_p \Gamma_\gamma}{\Gamma^2} \tag{20.2}$$

[1] K. J. BROSTRÖM, T. HUUS and R. TANGEN: Phys. Rev. **71**, 661 (1947).
[2] S. E. HUNT: Proc. Phys. Soc. Lond. A **65**, 982 (1952).

where $\Gamma = \Gamma_p + \Gamma_\gamma$ is the observed width of the resonance and Γ_p, Γ_γ are the partial widths for proton emission and radiation. g is the statistical factor defined in Sect. 9 and is generally of order unity. In the case that the experimental resolution (including target thickness) is $\ll \Gamma$, a direct determination of σ_R at the peak of the resonance curve can be made by observing the gamma ray yield per incident proton. Knowledge of the angular distribution of the gamma radiation and of the number of atoms per cm² of the target is necessary. If however the resonance width is comparable with the target thickness then the cross section σ_R must be obtained from an integration of the single level formula (FOWLER *et al.*[1]). The yield per proton at the peak is

$$Y_{\max}(\xi) = \frac{\sigma_R \Gamma}{\varepsilon} \arctan \frac{\xi}{\Gamma} \tag{20.3}$$

where ε is the stopping cross section for the incident particle per disintegrable nucleus and ξ is the energy loss in the target. In the limiting case of a thick target

$$Y_{\max}(\infty) = \frac{\pi}{2} \frac{\sigma_R \Gamma}{\varepsilon} = \frac{\lambda^2}{2} g \frac{\Gamma_p \Gamma_\gamma}{\Gamma}. \tag{20.4}$$

The apparent width for a target of thickness ξ is

$$\Gamma' = \sqrt{\Gamma^2 + \xi^2}, \tag{20.5}$$

and the area under a thick target curve is

$$A(\xi) = \int Y \, dE = \frac{\pi}{2} \frac{\sigma_R \Gamma}{\varepsilon} \xi = \xi \, Y_{\max}(\infty), \tag{20.6}$$

a result which is independent of homogeneity of beam, and which is useful for determining target thicknesses in resonance experiments by comparing thick and thin target yields. The stopping cross sections of most targets can be obtained from the data contained in [*22*].

From the cross section σ_R the quantity

$$g \frac{\Gamma_p \Gamma_\gamma}{\Gamma^2} \tag{20.7}$$

is found, which gives either $g\Gamma_p$ or $g\Gamma_\gamma$ according as Γ_p is very much less than or very much greater than Γ_γ. Since the initial spins are usually known, the experiment gives $(2J+1)\,\Gamma_p$ or $(2J+1)\,\Gamma_\gamma$. When $(2J+1)\,\Gamma_\gamma$ is found, the total width may give Γ_p and the l-value necessary to give a reasonable reduced width γ_p^2 can then be estimated, not always without ambiguity. The radiation width may be compared with the theoretical estimates (19.5) and (19.6) to determine multipolarity.

If Γ_p and Γ_γ are of comparable size, then the observed cross section determines the quantity

$$g \cdot \frac{\Gamma_p (\Gamma - \Gamma_p)}{\Gamma^2} = g\, x (1 - x) \tag{20.8}$$

where $x = \Gamma_p / \Gamma$. Two values of the widths are then found and the most reasonable combination must be selected. Observation of an elastic scattering anomaly at the resonance gives x directly. In some reactions alpha particle emission may compete with proton emission and elastic scattering. This reduces the resonant yield of gamma radiation in about the proportion in which it increases the total width, and may even effectively suppress the radiative transition.

---

[1] W. A. FOWLER, C. C. LAURITSEN and T. LAURITSEN: Rev. Mod. Phys. **20**, 236 (1948). (This article surveys techniques for determining gamma ray energies and intensities.)

An alternative method of studying $(p\gamma)$ resonance curves, which is sometimes useful, is based on observation of positron activity of the residual nucleus, e.g. in the reactions C^{12} $(p\gamma)$ N^{13} and Mg^{24} $(p\gamma)$ Al^{25}. Absolute thick target yields may be more easily obtained by this method, and it is also useful for weak resonances since the efficiency of positron detection is high.

Not all proton capture processes lead to resonant states and capture gamma radiation is observed in some reactions at all proton energies. This non-resonant radiation may be interpreted as due to overlapping of the tails of many remote levels, or, more probably, as a direct radiative process, inverse to photo-disintegration. A discussion of this type of radiation in the $Li^7(p\gamma)$ Be^8 reaction has been given by WILKINSON[1].

Radiative capture of deuterons would be expected to occur with cross sections of 10^{-28} to 10^{-32} cm² since the inverse reaction, photodisintegration with the emission of deuterons, is well known. The only known example however, is the $He^3(d\gamma)$ reaction. Upper limits for the $H^2(d\gamma)$, $He^4(d\gamma)$ and $O^{16}(d\gamma)$ reactions have been set by SINCLAIR[2] and for the $Be^9(d\gamma)$ and $C^{12}(d\gamma)$ reactions by ALLAN and SARMA[3].

There is good astrophysical evidence for the occurrence of $(\alpha\gamma)$ reactions and they have been observed in Li^6, Li^7 and N^{14}. ALLAN and SARMA[3] set an upper limit for the cross section of the $C^{12}(\alpha\gamma)$ reaction. The $(\alpha\gamma)$ reaction should be somewhat more difficult to observe than the $(p\gamma)$ reaction at comparable energies because of the increased rate of energy loss in the target and because of the smaller value of $\lambda\!\!\!^{-2}$, Eq. (20.2).

The radiative capture of neutrons is the main interaction of slow neutrons with heavy nuclei. The cross sections of neutron processes analogous to proton capture in the light elements are low and the resonance phenomena are usually observed in total cross sections. No certain identification of high energy neutron capture radiation has been made, owing largely to the low intensity of fast neutron beams of controllable energy. Fast neutron capture resonance may be observed by detecting the beta activity of the product nucleus (see Sect. 41, F^{19} and Al^{27}). The gamma radiation resulting from slow neutron capture in many nuclei has been extensively studied and is discussed in Sect. 40. The resonances resemble those found in proton capture processes and are well described by the single level formula.

21. Determination of internal conversion coefficients. The transfer of energy of nuclear excitation directly to an atomic electron is a process alternative to radiative emission. This process of internal conversion furnishes an important method of determination of multipolarity for radiative transitions in the heavy elements, since the ratio of internal conversion to gamma emission depends on the field strengths near the radiating nucleus. Internal conversion has been observed in light nuclei [see $O^{16}(dp)$] but the conversion coefficients decrease with Z and also with increase of quantum energy, so that determinations of multipolarity by this method are difficult. This is not so for an alternative form of internal conversion in which an electron-positron pair is emitted instead of a photon[4]. This pair internal conversion increases in probability with transition energy and is nearly independent of atomic number. The angular correlation between the positive and negative electrons provides a rather sensitive test of multipolarity. The method has been developed by DEVONS and GOLDRING [25]

[1] D. H. WILKINSON: Phil. Mag. **45**, 259 (1954).
[2] R. M. SINCLAIR: Phys. Rev. **93**, 1082 (1954).
[3] H. R. ALLAN and N. SARMA: Proc. Phys. Soc. Lond. A **68**, 535 (1955).
[4] M. E. ROSE: Phys. Rev. **76**, 678 (1949).

and applied to a number of transitions. In these experiments the electron and positron are detected by GEIGER counter telescopes in coincidence; it is necessary to take precautions against the production of external pairs by reducing material surrounding the target to a minimum. Fig. 18 shows the correlation found for the $Li^7(p\gamma)$ 17+14 MeV radiation and the theoretical correlation curves for orders (Sect. 19) $\Lambda=1$ and 2. The discrimination in favour of $\Lambda=2$ is very satisfactory. The absolute yield of internal pairs is about 10^{-3} pairs per quantum for a transition of energy 6 MeV in a nucleus of $Z=10$, and is dependent on multipolarity. Absolute yields are difficult to measure but have been obtained for Mg^{24} by BLOOM[1] using a magnetic spectrometer. Transitions with $\Delta J=0$ (No) and in particular the transition $0^+ \rightarrow 0^+$, for which single quantum radiation is absolutely forbidden, can proceed by internal conversion because electron wave functions can penetrate into the nucleus. The pair correlation for this transition in O^{16} has been measured by DEVONS *et al.* [25].

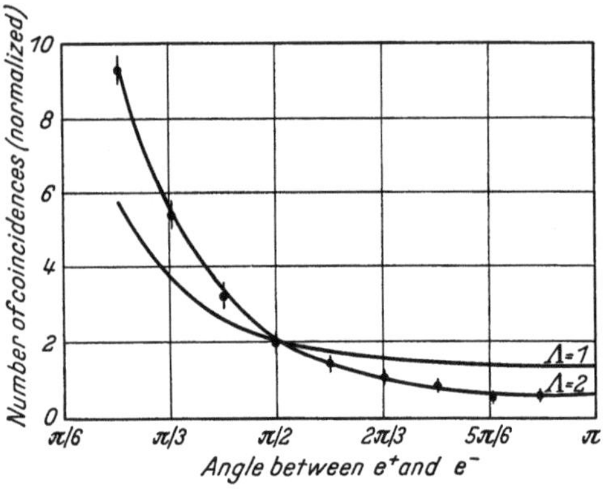

Fig. 18. Angular correlation of internal pairs in Li^7+p reaction (DEVONS and GOLDRING [25]).

22. COULOMB excitation. The possibility of the excitation of nuclear states by the electric field of a bombarding particle has been appreciated for some time. It was suggested in 1938 by WEISSKOPF[2] but lacked experimental confirmation for many years because in early experiments excitation by compound nucleus formation could not be excluded and also because the characteristic radiations of nuclei were poorly known. The possibility that electric excitation could be responsible for the large cross sections observed in some (dp) reactions with heavy nuclei was also explored (HUBY and NEWNS[3]) but this type of reaction is complicated by the occurrence of stripping phenomena (Sect. 51). A convincing demonstration of the existence of COULOMB excitation was made in 1952 by HUUS and ZUPANČIČ[4]; they used a beam of 1 MeV protons and succeeded in exciting a well known gamma ray transition between the ground state and first excited state of tantalum. Under the conditions of their experiment, penetration of the potential barrier was negligible so that compound nucleus formation could not contribute. The identity of the radiative transition was established by sodium iodide spectrometry. Since their work COULOMB excitation has been observed throughout the periodic system, with many different types of bombarding particle, and has become one of the most important methods of nuclear excitation. It has been outstandingly successful in the study of the low rotational states of medium and heavy nuclei which are predicted by the collective motion model of BOHR and MOTTELSON (Sect. 4) and which result in large $E2$ radiative widths. The corresponding probabilities for COULOMB excitation are also large and a great many $E2$ transitions have been induced by this method. For light nuclei COULOMB excitation is perhaps less useful because of the greater probability of barrier penetration but it has been used in the study of a few nuclei with very low-lying states such as F^{19}.

[1] S. D. BLOOM: Phys. Rev. **88**, 312 (1952).
[2] V. F. WEISSKOPF: Phys. Rev. **53**, 1018 (1938).
[3] R. HUBY and H. C. NEWNS: Proc. Phys. Soc. Lond. A **64**, 619 (1951).
[4] T. HUUS and Č. ZUPANČIČ: Kgl. danske Vid. Selsk., mat.-fys. Medd. **28**, 1 (1953).

The theory of COULOMB excitation is given by many authors, in particular TER-MARTIROSYAN and ALDER and WINTHER[1]. The interaction potential between the bombarding particle and the nucleus is taken to be

$$V = \sum_{p=1}^{Z'} \frac{Z e^2}{|r - r_p|} \tag{22.1}$$

where r_p refers to the protons in the target, and it is assumed (a) that classical trajectories are sufficient for the problem i.e. that

$$\frac{Z Z' e^2}{\hbar v} \gg 1, \tag{22.2}$$

and (b) that the energy loss of the incident proton in passing near the nucleus is small i.e. that $\Delta E \ll E$. These two conditions limit the validity of a simple treatment, particularly for light nuclei. The interaction potential is then expanded into a series of multipole terms and the cross section calculated by means of time dependent perturbation theory. For electric 2^l-pole excitation it is found that

$$\sigma_l = A_l(E, Z)\, B_e(l)\, g_l(\xi). \tag{22.3}$$

$g_l(\xi)$ is an integral over space coordinates for the incident particle, ξ being a parameter representing the ratio of "collision time" to the nuclear period $\hbar/\Delta E$. $B_e(l)$ is the reduced electric 2^l-pole transition probability and is related to the corresponding radiative transition probability $T_E(l)$ (as used, for instance, by WEISSKOPF, Sect. 19) by the formula

$$T_E(l) = \frac{8\pi(l+1)}{l[(2l+1)!!]^2} \frac{1}{\hbar} \left(\frac{\omega}{c}\right)^{2l+1} B_e(l) \frac{2J_0+1}{2J+1} \tag{22.4}$$

where J_0 and J are the spins of the ground state and excited state, and

$$(2l+1)!! = 1 \cdot 3 \cdot 5 \cdots (2l+1). \tag{22.5}$$

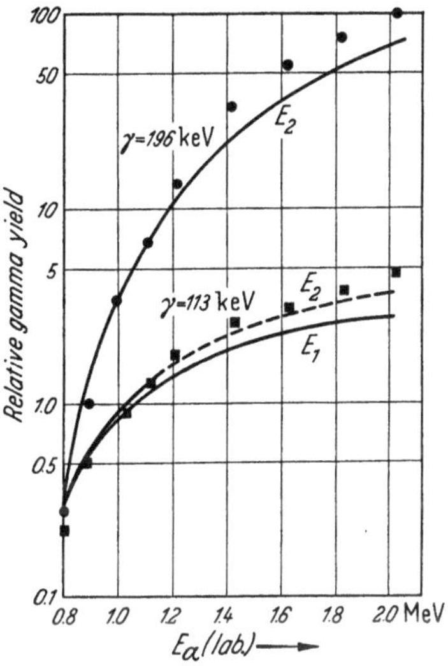

Fig. 19. COULOMB excitation of radiative transitions in F^{19} by alpha particles (TEMMER and HEYDENBURG [43J]).

Determination of the cross section for COULOMB excitation is therefore equivalent to the measurement of the lifetime of the state so excited, and the cross section is larger the shorter the life.

COULOMB excitation of light nuclei by alpha particles has been investigated and discussed by TEMMER and HEYDENBURG[2], who point out several technical advantages of this method over the use of protons. One of their excitation functions is shown in Fig. 19. The use of different types of bombarding particle, e.g. H^1, H^2, He^3 for determining the multipolarity of electrically excited transitions has been pointed out by BJERREGAARD and HUUS[3]; the method is to compare, for the same value of $\xi = \frac{Z Z' e^2 \Delta E}{\hbar m v^3}$ the cross sections obtained with particles of different m/Z.

23. Photodisintegration. The simplest disintegration of a nucleus by gamma radiation is a process inverse to the capture reactions of the type $(p\gamma)$ $(\alpha\gamma)$ $(d\gamma)$

[1] TER-MARTIROSYAN: Zhr. eksper. teor. Fiz. **22**, 284 (1952). — K. ALDER and A. WINTHER: Phys. Rev. **91**, 1578 (1953), **96**, 237, (1954).

[2] G. M. TEMMER and N. P. HEYDENBURG: Phys. Rev. **96**, 426 (1954).

[3] J. H. BJERREGAARD and T. HUUS: Phys. Rev. **94**, 204 (1954).

and $(n\gamma)$. Precise information about resonant states is obtainable from capture reactions because of the high intensity and controllable energy of the heavy particle beams. The capture radiation produced by such reactions may itself be used to induce transmutations and has the advantage of being a line spectrum or at most a group of lines, but the intensity is low. Despite this, much work has been done using the 17.6 and 14.8 MeV gamma radiation produced by the capture of protons in lithium

$$Li^7 + H^1 \rightarrow Be^{8*} \rightarrow Be^8 + \gamma.$$

Other capture and similar reactions can produce useful line sources of gamma radiation, but most of them give energies below the photodisintegration threshold of all nuclei except H^2 and Be^9. The most useful sources, in addition to $Li^7(p\gamma)$, are the $(p\gamma)$ reactions in H^3, Be^9 and B^{11} and the $(p\alpha\gamma)$ reactions in N^{15} and F^{19}.

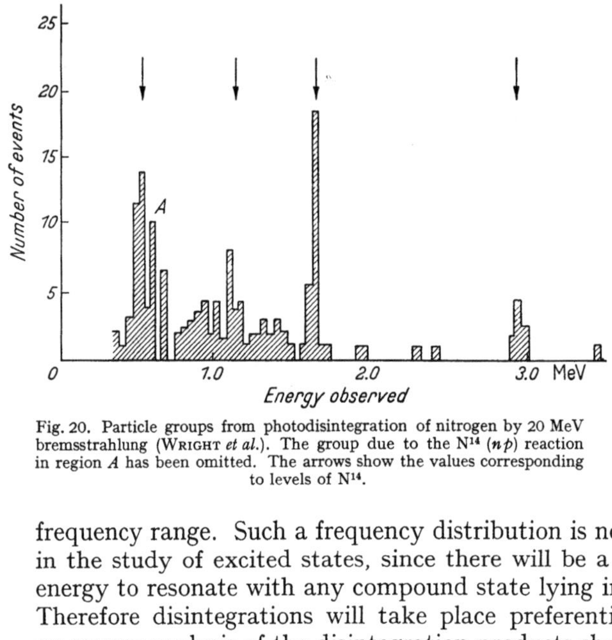

Fig. 20. Particle groups from photodisintegration of nitrogen by 20 MeV bremsstrahlung (WRIGHT *et al.*). The group due to the N^{14} (np) reaction in region *A* has been omitted. The arrows show the values corresponding to levels of N^{14}.

Much stronger sources of radiation are available from electron betatrons and synchrotrons, of which machines with peak energy up to 30 MeV are probably the most useful for nuclear reaction studies. Gamma radiation is produced as a bremsstrahlung spectrum by the impact of electrons on a lead target; this spectrum gives an intensity distribution which is roughly uniform over the frequency range. Such a frequency distribution is not necessarily a disadvantage in the study of excited states, since there will be a component of just the right energy to resonate with any compound state lying in the available energy range. Therefore disintegrations will take place preferentially with these quanta and an energy analysis of the disintegration products should show groups corresponding to excited states of the compound nucleus. This is shown in the work of WRIGHT *et al.*[1] on the photodisintegration of nitrogen by bremsstrahlung of about 20 MeV. The disintegration particles were detected in a nitrogen filled expansion chamber which is a very convenient instrument for neutron and photon induced reactions because it may be made relatively insensitive to the primary beam. Fig. 20 shows the energy distribution of the proton plus recoil tracks and well displays the resonance energies of the compound nucleus. Above the sharp levels the cross section for (γp) and (γn) reactions shows a rise to the "giant resonance" of photonuclear reactions. The bremsstrahlung spectrum can be used to study the giant resonance region if the maximum energy of the electron accelerator is controlled and excitation functions can be followed by the "photon difference" method. In this way it has been possible to exhibit sharp resonances as breaks in an otherwise smooth excitation curve (KATZ *et al.*[2]). Because of the rather low cross sections for photonuclear reactions, detectors

[1] I. F. WRIGHT, D. R. O. MORRISON, J. M. REID and J. R. ATKINSON: Proc. Phys. Soc. Lond. A **69**, 77 (1956).

[2] L. KATZ, R. N. H. HASLAM, R. J. HORSLEY, A. G. W. CAMERON, R. MONTALBETTI: Phys. Rev. **95**, 464 (1954).

with an integrating property, such as the nuclear emulsion are particularly valuable. The disintegration of the light and heavy elements of the emulsion (C, N, Ag, Br) has been studied in great detail and in some cases, e.g. $C^{12}(\gamma\, 3\alpha)$ information on reaction mechanisms has been obtained. For the study of (γp) reactions with the collimated beam from an electron accelerator photographic plates may be arranged round the target and both energy and angular distributions obtained in a single exposure. A review of the use of the nuclear emulsion technique in photodisintegration work has been given by TITTERTON [*42*]. The neutrons from (γn) reactions may be slowed down and detected in a boron trifluoride filled counter. The residual nucleus in either the (γp) or (γn) reaction will be radioactive, and the activation method is a convenient method of following an excitation function.

It has not yet been possible to excite the narrow bound levels of light nuclei by radiative absorption without excess COMPTON scattering from the accompanying bremsstrahlung spectrum. The line spectrum of a nucleus cannot be absorbed in exact resonance by similar nuclei because the loss of energy by recoil considerably exceeds the level width, at least for bound levels. In a number of special cases in heavier nuclei (DAVEY and MOON[1]), it has been possible to restore the recoil energy by the DOPPLER effect. Extension of these techniques to light nuclei seems not impossible, and the detection of nuclear resonant scattering would then give a direct measure of the radiative width of the level concerned (in terms of the thermal energy of the target). Another interesting line of attack is to attempt a direct observation of nuclear resonance absorption by passing a bremsstrahlung beam through a long absorber and measuring the energy distribution before and afterwards. Preliminary experiments have given encouraging results, in particular an absorption line due to the giant resonance has been found (KOCH[2]). This experiment is the direct analogue of the absorption method in optical spectroscopy; in this case the spectrometer was a crystal of sodium iodide, 5″ in diameter and 9″ long. Elastically scattered quanta associated with the nuclear absorption have also been detected. A full account of the nuclear photoeffect is given by BISHOP and WILSON in Vol. XLI of this Encyclopedia. A brief review of some of the experimental results relating to nuclear energy levels is given in Part D of this article.

24. Direct measurement of lifetimes of nuclear states. The lifetime of a nuclear state is related to the observed width of the state by the relation

$$\tau_{(\text{mean life})} = \hbar/\Gamma.$$

Methods for measuring Γ have already been described; they lead to values from about 1 MeV for highly excited, particle-unstable states in the lightest nuclei to 10^{-4} eV for slow neutron widths in heavy nuclei. In the latter case, however, radiation is the main process of de-excitation, and total widths of 0.1 eV or more are observed. The range of lifetimes which is therefore important in practice for nuclear processes (excluding beta-decay and isomeric transitions) is from about 10^{-13} sec to the time of about 10^{-22} sec required for a nucleon to traverse a nuclear diameter. Particular circumstances may however arise where much longer lifetimes are encountered even in processes involving charged particles. Thus if a bound state of very high angular momentum is formed by a preceding transition, a slow decay may be observed; this is the isomeric type of transition

[1] W. G. DAVEY and P. B. MOON: Proc. Phys. Soc. Lond. A **66**, 956 (1953).
[2] H. W. KOCH: Proceedings of the 1954 Glasgow Conference London: Pergamon Press 1955.

familiar in moderately heavy nuclei. Or if a particular transition is forbidden, e.g. radiation in the $J=0$ to $J=0$ transition, the lifetime is lengthened and the range of interest extends perhaps down to 10^{-8} sec.

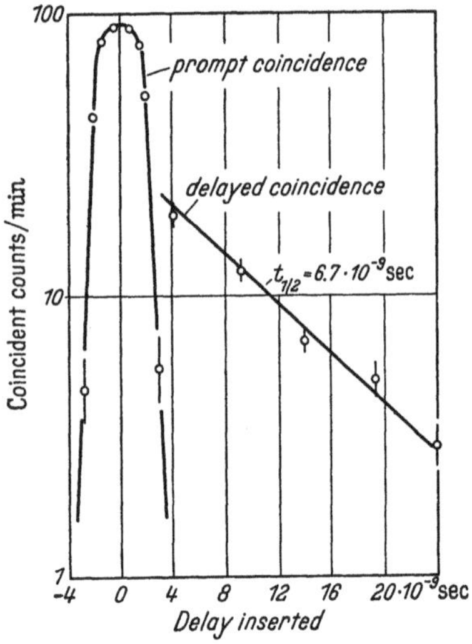

Fig. 21. Delayed coincidence curve for lifetime of K^{41*} excited in decay of A^{41} (Elliott).

Lifetimes of more than 10^{-11} sec may be determined by the delayed coincidence method (Bell *et al.*[1] and [*21*]) if there is a preceding radiation. The method involves fast circuit techniques and high gain photomultipliers; a good example is the measurement of the lifetime of the first excited state of K^{41} by Elliott [*26*]. Fig. 21 shows the delayed coincidence curve obtained, from which a lifetime of $(9.5\pm0.6)\times10^{-9}$ sec was deduced. A lifetime of $(3.6\pm2.5)\times10^{-11}$ sec for an $E2$ transition in Mg^{24} has been measured by this method by Coleman [*26*]. Shorter lifetimes may be measured by the recoil method (Devons[2] *et al.*) in which the distance travelled by an excited nucleus before de-excitation is measured. Since typical velocities of recoil of excited nuclei produced in heavy particle reactions are 10^8 to 10^9 cm/sec the distances involved may be only 10^{-4} cm and extremely accurate measurements are necessary. The apparatus used by Devons *et al.*[2] for measuring the lifetime of the pair-emitting state in O^{16} is shown in Fig. 22. The target is displaced until the electron-positron pair from a recoiling

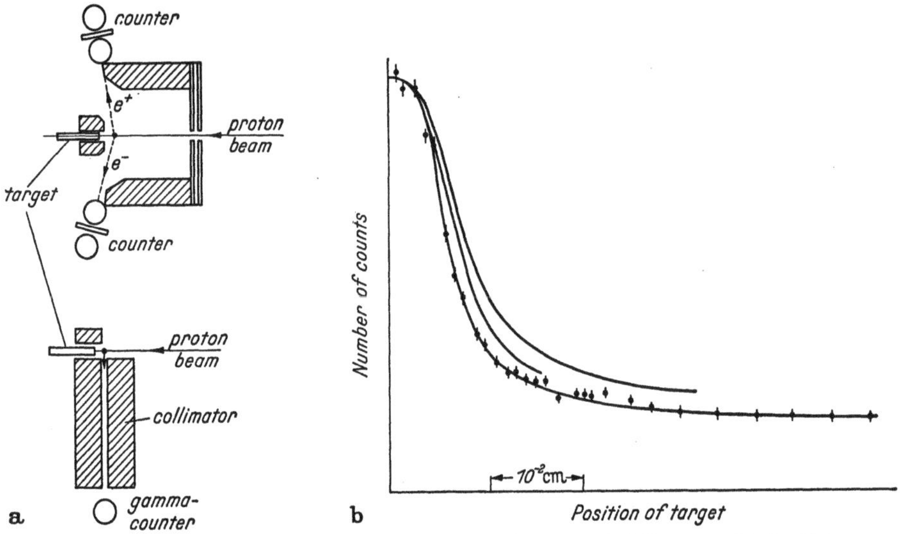

Fig. 22 a and b. Determination of lifetime of states of O^{16} by recoil-distance method (Devons *et al.* [*26*]). (a) Apparatus for pair emitting and gamma emitting states. (b) Results for 6.13 MeV level.

O^{16} nucleus can just reach a pair of counters in coincidence. From the counting rate observed as a function of target displacement a half-lifetime of $(5.0\pm0.5)\times10^{-11}$ sec was deduced. Similar experiments using a sodium iodide crystal behind an accurately machined tungsten collimator have been made by Thi-

[1] R. E. Bell, R. L. Graham and H. E. Petch: Canad. J. Phys. **30**, 35 (1952).
[2] S. Devons, G. Goldring and G. R. Lindsey: Proc. Phys. Soc. Lond. A **67**, 134 (1954).

RION and TELEGDI [26] to measure the lifetimes of gamma emitting nuclei. Lifetimes between 2.5×10^{-10} sec and 8×10^{-8} sec have been so determined.

For shorter lives a modification of the recoil method, used by ELLIOTT and BELL[1] for the reaction $B^{10}(n\alpha)\,Li^7$, has been developed. In this the recoiling nucleus is brought to rest in targets of different stopping power, so that it is moving for different lengths of time after production. The radiation from the moving nucleus experiences a DOPPLER shift in frequency because of the motion, and the amount of shift depends on whether the nucleus has or has not been slowed down in a time comparable with the mean life for radiation[2]. ELLIOTT and BELL found a lifetime of 0.75×10^{-13} sec for the 478 keV state of Li^7. DEVONS *et al.*[3] have improved this method by measuring the mean shift of gamma-ray

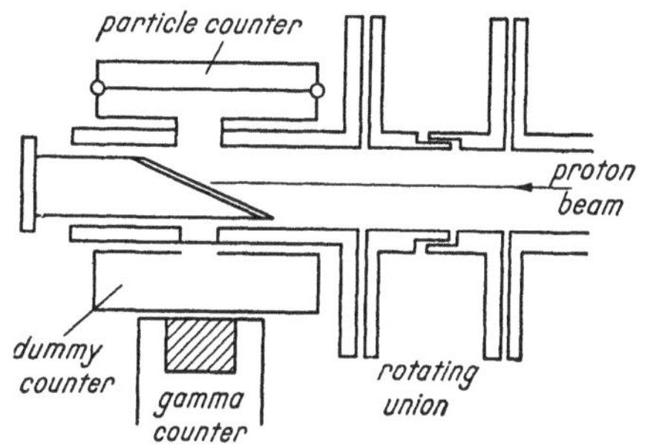

a

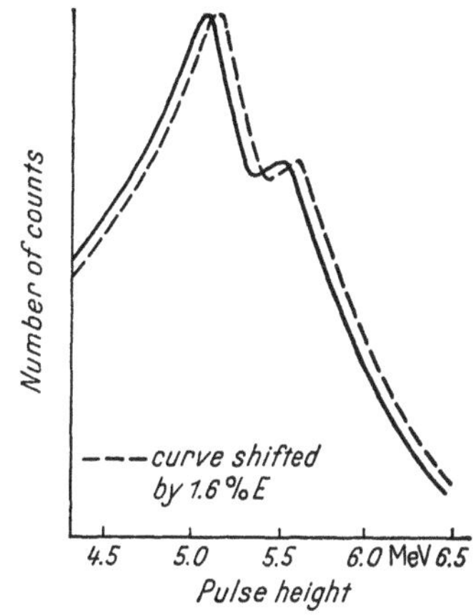

b

Fig. 23 a and b. Determination of lifetime of gamma emitting state of O^{16} by recoil-DOPPLER shift method (DEVONS *et al.* [26]) (a) Coincidence apparatus; (b) observed gamma spectrum, showing maximum possible shift [$F^{19}(p\alpha\gamma)$, 340 keV resonance].

energy for slowing down in a given material for a more definite angle of recoil. This angle was defined by observing only in coincidence with an associated disintegration particle from the primary transmutation in a fixed direction. Fig. 23 shows the apparatus used for measurements on the $F^{19}(p\alpha\gamma)\,O^{16}$ radiations, and the shift obtained. The half-life deduced was $\tau > 5\cdot 10^{-12}$ sec; the accuracy of the determination is limited by uncertainty in the range-velocity relations for heavy ions in the target materials. If improved range velocity data become available this general method should become extremely powerful for the range of lifetimes between 10^{-11} and 5×10^{-15} sec.

Methods are thus available, given suitable primary reactions, for measuring gamma ray lifetimes from the "conventional" isomeric region of 10^{-9} sec down to 10^{-14} to 10^{-15} sec. This corresponds to a total width of 1 eV. For virtual levels widths of this order and greater are easily derived from cross section measurements. For bound levels nuclear resonant scattering and COULOMB excitation together are particularly appropriate for measuring widths in the eV range. Both these methods are effective only for transitions connected with the ground state, although they may induce cascades from the states which they excite.

No example of the radioactive emission of heavy particles, or the determination of the lifetime of a state decaying predominantly by particle emission is yet

[1] L. G. ELLIOTT and R. E. BELL: Phys. Rev. **76**, 168 (1949).

[2] R. G. THOMAS and T. LAURITSEN: Phys. Rev. **88**, 969 (1952).

[3] S. DEVONS, G. MANNING and D. ST. P. BUNBURY: Proc. Phys. Soc. Lond. A **68**, 18 (1955).

known for the light nuclei, owing to the low COULOMB barriers (see however Be^8, Sect. 76).

Existing measurements of the lifetimes of radiative transitions in the light nuclei are shown in Table 3.

Table 3. *Direct measurements of lifetimes of nuclear states.*
The methods are described in Sect. 24. The figures are taken from [*26*].

Nucleus	Level MeV	Method	Mean life τ sec	Multipolarity	$\vert M\vert^2$ (Sect. 19)
Li^7	0.48	a	$7.7 \pm 0.8 \times 10^{-14}$	$M1$	2.0[1]
Be^7	0.43	a	$2.7 \pm 1.0 \times 10^{-13}$	$M1$	2.0[1]
B^{10}	0.72	b	$7 \pm 2 \times 10^{-10}$	$E2$	2.0
C^{12}	4.4	a	$2.6 \pm 0.9 \times 10^{-14}$	$E2$	4.6
N^{14}	2.31	a	$\lesssim 2 \times 10^{-13}$	$M1$	> 0.02
O^{16}	6.06 (π)	b	$7.2 \pm 0.7 \times 10^{-11}$	$E0$	—
	6.13	a, b	$7 \times 10^{-12} \lesssim \tau \lesssim 1.4 \times 10^{-11}$	$E3$	7.3
	6.91	a	$\lesssim 2.5 \times 10^{-14}$	$E2$	> 0.3
	7.12	a	$\lesssim 1.5 \times 10^{-14}$	$E1$	$> 2.3\ 10^{-4}$
O^{17}	0.87	b	$2.5 \pm 1 \times 10^{-10}$	$E2$	1.0
F^{19}	0.11	b	$1.0 \pm 0.25 \times 10^{-9}$	$E1$	$6.4\ 10^{-4}$
F^{19}	0.20	b, c	$1.0 \pm 0.2 \times 10^{-7}$	$E2$	2.4
Ne^{20}	1.63	a	$7.6 \pm 3.3 \times 10^{-13}$	$E2$	11.5
Mg^{24}	1.38	c	$3.6 \pm 2.5 \times 10^{-11}$	$E2$	0.43
Si^{28}	1.78	a	$(\gtrsim 9 \times 10^{-13})$	$E2$	—
K^{41}	1.37	c	$9.5 \pm 0.6 \times 10^{-9}$	$M2$	0.05

(a) Recoil-DOPPLER method.
(b) Recoil-distance method.
(c) Delayed coincidence method.

IV. Angular distribution and correlation experiments.

25. Introduction. In Part B only total scattering and reaction cross sections were discussed. Much useful information can also be obtained from the study of differential cross sections (angular distributions) and of angular correlations of successive radiations. Two kinds of information result from these studies:

(a) total angular momenta and parities may be deduced in many cases from the symmetry properties of the reaction;

(b) particular reaction mechanisms may sometimes be inferred from a detailed analysis of angular distribution patterns into basic distributions determined by special assumptions such as those of jj or LS coupling.

In simple cases the straightforward observation of angular correlation patterns provides unambiguous information of type (a). In general, however, complex parameters including the phase of several different nuclear wave functions have to be introduced and the theoretical patterns are not uniquely determined. As pointed out by CHRISTY [*27*] this situation arises whenever three or more non-zero angular momenta have to be added together to give a fourth and the resultant can be reached in more than one way. Information of type (b) is then obtained by extraction of the arbitrary parameters from the experimental information and comparison with the values predicted by particular interaction mechanisms.

A detailed discussion of angular distribution phenomena is given in the article by DEVONS in Vol. XLI of this Encyclopedia and in the review articles of BLATT and BIEDENHARN and BIEDENHARN and ROSE [*28*], [*29*]. No extensive treatment will be given here but the main results will be outlined for use in

[1] More reliable theoretical widths are available for these nuclei (see Sect. 58).

interpreting experimental results. The differential cross sections for elastic scattering in which special features arise because of interference with potential scattering, have been discussed in Sect. C II.

We distinguish between *single stage reactions* of the type

$$X + a \rightarrow Y + b \tag{25.1}$$

in which no definite compound state is involved, and *two stage reactions* such as

$$X + a \rightarrow C \rightarrow Y + b \tag{25.2}$$

in which the quantum properties of an intermediate state of definite life are important. In many reactions the residual nucleus is formed in an excited state and may emit a radiation q

$$X + a \rightarrow C \rightarrow Y^* + b \rightarrow Y + b + q. \tag{25.3}$$

Such *three stage processes* have also been analysed in detail[1]. In all these calculations extensions may be made to include the polarisations of the reacting components. In the majority of experiments unpolarised nuclei and particle beams are used, or the decay of randomly distributed nuclei is investigated. The polarisation correlation of the incident and emitted radiations may then be predicted. The conditions for non-vanishing polarisation of particles produced in a nuclear reaction have been stated by BLIN-STOYLE[2] and by SIMON and WELTON[3]; spin-orbit coupling is required. Because of the transverse nature of electromagnetic waves, polarisation effects in correlations involving radiation are generally strong; they are discussed by HAMILTON[4] and in [*29*]. It has been possible recently to make experiments with initial systems which are oriented by imposed magnetic fields rather than by the mechanism of a preceding nuclear reaction; the radiation patterns observed from such oriented nuclei are discussed by BLIN-STOYLE and GRACE in their article in Vol. XLI of this Encyclopedia.

In the case of all reactions of the type (25.1) to (25.3) the absorption of particle a may be replaced formally by the emission of a particle or quantum; thus the theory for (25.3) also predicts for instance the triple correlation of gamma rays arising from the chain

$$C \rightarrow Y^{**} + \gamma_1 \rightarrow Y^* + \gamma_1 + \gamma_2 \rightarrow Y + \gamma_1 + \gamma_2 + \gamma_3. \tag{25.4}$$

Single stage processes usually lead to angular distributions with a strong forward maximum. The most important examples of this class are the stripping and pickup reactions discussed in Sects. 51 to 53.

26. Angular correlations in two-stage processes. We consider first the simple case in which the reaction (25.2) proceeds through a single well defined state of angular momentum J and parity. If it happens that the magnetic substates of the compound angular momentum J (labelled by the z component M) are formed with equal probability, then the compound nuclei are effectively isotropically distributed and there is no correlation between the direction of incidence of particle a and of emission of particle b. In general, however, the first reaction results in a non-uniform population of magnetic substates and since for each transition from these states to the states of the final system there may be an

---

[1] Reference [*29*]. — L. C. BIEDENHARN, G. B. ARFKEN and M. E. ROSE: Phys. Rev. **83**, 586 (1951). — J. E. SEED and A. P. FRENCH: Phys. Rev. **88**, 1007 (1952).

[2] R. J. BLIN-STOYLE: Proc. Phys. Soc. Lond. A **64**, 700 (1951).

[3] A. SIMON and T. A. WELTON: Phys. Rev. **90**, 1036 (1953).

[4] D. R. HAMILTON: Phys. Rev. **74**, 782 (1948).

anisotropic pattern, there is usually some deviation from isotropy in the final sum. Let the spins of the initial nuclei X, a be $j_1 j_2$ (z-components $m_1 m_2$) and of the final nuclei $J_1' J_2'$. Then the channel spin[1] j of the initial system is given by

$$|j_1 - j_2| \leq j \leq j_1 + j_2 \tag{26.1}$$

so that there are $(2j_1+1)(2j_2+1)$ channel spins j each of which has $(2j+1)$ magnetic substates m defined with respect to an arbitrary axis, which we take for convenience to be the direction of the incident beam. The unpolarised nature of the initial system corresponds to an incoherent mixture of all channel spins with equal weight.

The probability amplitude for the formation of the compound nucleus C from an initial channel spin state (j, m) is written (Sect. 6)

$$H_{\mathrm{xc}} = (l\, l_z j\, m \mid J\, M)\, A\, e^{i\alpha} \tag{26.2}$$

where $(l l_z j\, m \mid J M)$ is the WIGNER coefficient for the transformation from the (l, j) representation to the J-scheme, and $A\, e^{i\alpha}$ includes barrier penetrability, and reduced width, each expressed as an amplitude. The product of two factors such as (26.2) leads to the expressions for total cross section given in Sect. 9. Since the incident particles are described by a plane wave, the various l-components are coherent, and their amplitudes must be summed for a given spin state (j, m). For a single compound level the phase difference between waves of differing orbital momentum l is due to the COULOMB field and may be calculated exactly (Sect. 16). The contributions to the cross section from different spin states are incoherent and their intensities may be added.

For a given J the l-values permitted are given by

$$|J - j| \leq l \leq J + j \tag{26.3}$$

but because of conservation of parity, and of the rapidly decreasing probability of the emission of waves with high orbital momenta, usually only two values of l, differing by 2 units, are required. We always have $M = m$ and $l_z = 0$ by choice of the z-axis as the direction of incidence of particle a.

The compound state $(J\, J_z)$ has a finite lifetime before the emission of b and it is necessary for the interpretation of an angular correlation experiment that this shall be much shorter than the time required for perturbing fields to destroy the orientation of the nuclei. For charged particle reactions in light nuclei this is usually so, but in $(\gamma - \gamma)$ correlations the precession times due to magnetic coupling with the electronic structure (h.f.s. splitting) may be about 10^{-7} to 10^{-11} sec which covers many useful lifetimes. For the present purpose we neglect these effects. Externally applied magnetic perturbation of an angular correlation pattern has been used by TREACY[2] to determine the magnetic moment of an excited state of F^{19}.

The de-excitation of the compound state may take place by particle or by quantum emission. In both cases the parity rule limits the l-values. The differential cross sections have the form

(i) *Particle emission*

$$\sigma(\vartheta, \varphi) = \frac{\pi \lambda\!\!\!^{-2}}{(E - E_{\mathrm{res}})^2 + \frac{\Gamma^2}{4}} \times Z \tag{26.4}$$

---

[1] In Part. B and C II the symbol S is used for channel spin, following reference [7]. In angular correlation theory the symbol j is more usual.

[2] P. B. TREACY: Nature, Lond. **176**, 923 (1955).

where

(i) for particle emission

$$Z = \sum_{j\,m} \sum_{j'\,m'} \left| \sum_{l} (l\,0\,j\,m \,|\, J\,M)\, A_l\, e^{i\alpha_l} \sum_{l'} (J\,M \,|\, l'\, l'_z\, j'\, m')\, A_{l'}\, e^{i\alpha_{l'}}\, Y_{l'}^{l'_z}(\vartheta\,\varphi) \right|^2 \tag{26.5}$$

and

(ii) for quantum emission

$$Z = \sum_{j\,m} \sum_{J'\,M'} \left| \sum_{l} (l\,0\,j\,m \,|\, J\,M)\, A_l\, e^{i\alpha_l} \sum_{l'} (J\,M \,|\, E_{l'} \,|\, J'\,M', \vartheta\,\varphi)\, A_{l'}\, e^{i\alpha_{l'}} \right|^2 \tag{26.6}$$

where $(J\,M\,|E_{l'}|\,J'\,M', \vartheta\,\varphi)$ is the matrix element for the emission of $2^{l'}$ pole radiation between states of $J\,M$ and $J'\,M'$. These matrix elements, and the transformation coefficients (WIGNER coefficients) are given for some useful cases by CONDON and SHORTLEY[1].

These expressions are complicated but simplify in many practical applications, particularly when spinless particles only are concerned. The interpretation of angular distribution patterns is also simplified by a number of general results for unpolarised beams and non-relativistic reactions. These are now listed and apply equally to particle-particle, particle-quantum or quantum-quantum correlations; angular correlations in beta-decay processes are not considered here.

(i) *Axial symmetry.* The distribution of b is independent of azimuth about the direction of a.

(ii) *Isotropy.* Angular distributions are isotropic when the incident wave, compound state or emitted wave has spherical symmetry so that all sense of direction is lost. This is so if $l=0$ for the incident or emitted particle or if $J=0$ or $\frac{1}{2}$ for the compound state, in which case its magnetic substates are equally populated. Photon emission must remove at least one unit of angular momentum but isotropy may then result from angle dependent terms from different substates cancelling out.

(iii) *Symmetry.* Anisotropic distributions may be either symmetric or asymmetric with respect to the origin of the radiation. If only one compound state is involved in the reaction or if all compound states have the same parity, the intensity distribution of emitted particles or quanta will be symmetrical under reflection in the origin and may be expressed in even powers of $\cos\vartheta$. Odd powers of $\cos\vartheta$ in heavy particle and photon distributions can only result from interference between orbital wave functions of opposite parity.

(iv) *Energy dependence.* Variation of pattern with the energy of a usually indicates the intervention of more than one compound state, since then the resonance denominators must be included in the summation under the modulus in (26.5) and (26.6).

(v) *Limitations on complexity of pattern.* One of the most useful general results is stated by YANG[2] as follows: "If only incoming waves of orbital angular momentum l contribute appreciably to the reaction, the angular distribution of the outgoing particles in the centre of mass system is an even polynomial with an exponent not higher than $2l$." This is true even if the compound state or outgoing waves have greater l-values. If the outgoing wave has an orbital momentum $l' < l$, the complexity of the pattern is determined by l'. Multipole radiation of order l behaves in this respect as a particle with orbital angular

[1] E. U. CONDON and G. H. SHORTLEY: The theory of atomic spectra. Cambridge: The University Press 1935 and 1951. See also D. S. LING and D. L. FALKOFF: Phys. Rev. **76**, 1639 (1949).

[2] C. N. YANG: Phys. Rev. **74**, 746 (1948).

momentum l, so that if in any process dipole radiation is emitted before or after another particle, the correlation function cannot have terms higher than $\cos^2\vartheta$.

The angular correlation and distribution patterns observed with radiation give no information on the parity changes in the transition. This may be found by observing the polarisation direction correlation, since for a given multipole the electric vector for an electric transition is at right angles to that for a magnetic transition. The polarisation expected at a given angle may be directly predicted from the change of m value in the transition; thus a transition $J=1$, $M=0 \to J'=0$, $M'=0$ represents a dipole aligned with the z-axis, and the radiation is plane polarised in planes containing the radius vector and Oz.

The procedure just outlined gives a physical picture of the method of analysis of the simpler angular correlations. The labour of performing summations over the magnetic quantum numbers $m\,m'$ is however considerable and since these quantum numbers are not observed, there is no point in requiring them to appear explicitly in the calculation. BLATT and BIEDENHARN [28] showed how the magnetic quantum numbers could be eliminated from the general expression for differential scattering or reaction cross sections by use of RACAH coefficients. Their method was developed and applied to the angular correlation problem by BIEDENHARN and ROSE [29] who express the gamma-gamma angular correlation in the form

$$W(\vartheta) = A_0 + A_2 P_2(\cos\vartheta) + A_4 P_4(\cos\vartheta) \tag{26.7}$$

where

$$A_\nu = F_\nu(L_1 j_1 j)\, F_\nu(L_2 j_2 j) \tag{26.8}$$

with L_1 and L_2 representing the angular momentum carried by the two quanta and $j_1\, j_2$ and j the angular momenta of the initial, final and intermediate state. The F factors are tabulated. This method may be used for correlations involving particles with spin by interpreting $L_1\, L_2$ as orbital momenta and $j_1\, j_2$ as channel spins of the initial and final state. Known factors multiply the terms of (26.8) and arbitrary constants (real) must be used to weight each channel spin.

The extension of formulae (26.5) and (26.6) to the case of overlapping states of the compound nucleus is straightforward. The arbitrary parameters are complex if the overlapping levels can be formed from the same channel spin, but only real parameters are needed if the different levels correspond to different channel spins.

If many levels are excited in the intermediate state, i.e. if $\Gamma \gg D$, there is a superposition of very many random phases and in the limit these phases may be disregarded again. The properties of angular distributions determined by continuum theory have been discussed by WOLFENSTEIN[1] and by HAUSER and FESHBACH[2] who find that the main features of the distribution of two-stage reactions are (a) symmetry about 90° and (b) peaks in the forward and backward directions. The symmetry is due to the assumption that owing to the large mixture of random phases, interference terms between states of opposite parity should cancel out. If the incident energy is so large that statistical theory applies to the residual as well as the compound nucleus the product particles are isotropically distributed. The predictions of this theory have not been fully tested, but it seems to have some relevance to inelastic scattering distributions observed with 10 to 20 MeV protons.

27. Experimental results and typical analyses.—Two-stage processes. The experimental investigation of two-stage angular distributions requires the intro-

---

[1] L. WOLFENSTEIN: Phys. Rev. **82**, 690 (1951).

[2] W. HAUSER and H. FESHBACH: Phys. Rev. **87**, 366 (1952).

duction of an angle variable into the experimental equipment. A typical apparatus used in angular distribution experiments is that of HEYDENBURG *et al.*[1] for a study of the $Li^7(p\alpha)$ and $Li^6(d\alpha)$ reactions (Fig. 24a). The alpha particles were detected in a proportional counter and the angular distribution was expressed in the form

$$Y(\vartheta) = Y(90°)\,(1 + A\cos^2\vartheta + B\cos^4\vartheta \ldots). \tag{27.1}$$

The coefficients A and B were found as functions of energy over the range 0 to 3.5 MeV and the distributions were analysed in terms of overlapping levels. Fig. 24b shows a plot of the yields obtained at various energies as a function of $\cos^2\vartheta$. In analysing these results in a first approximation, since B is fairly small, the distribution for $Li^7(p\alpha)He^4$ is taken to be (INGLIS[2])

$$Y(\vartheta) = 1 + A(E)\cos^2\vartheta, \tag{27.2}$$

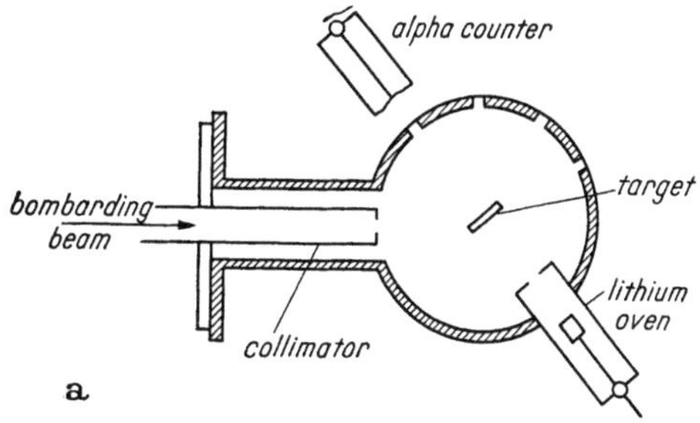

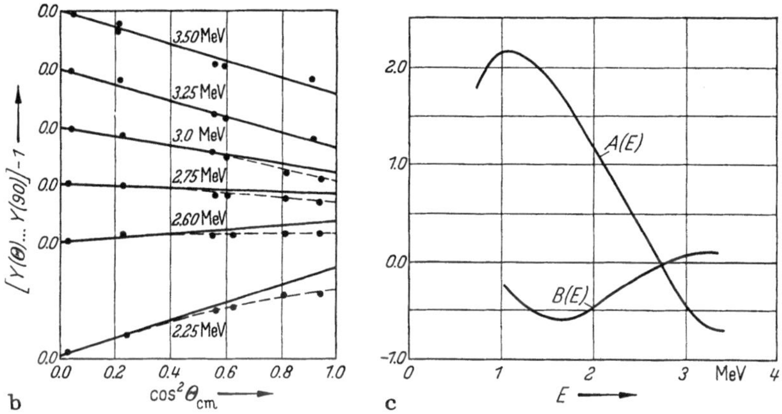

Fig. 24 a—c. Angular distribution for $Li^7(p\alpha)$ reaction. (a) Apparatus; (b) results; (c) $A(E)$, $B(E)$, as functions of energy (HEYDENBURG *et al.*).

and the fact that A is a function of energy immediately indicates that two levels of the compound nucleus Be^8 are involved. For low energies we assume that the incident protons have $l = 1$. Be^8 can only be formed in states of even J and even parity if breakup into two identical particles is to be allowed, and we assume that $J = 0$ or 2, for the two interfering levels. The vector coupling coefficients for the outgoing system are unity (channel spin zero) and for the initial system are of the form

$$(1010\,|\,00) = \frac{1}{\sqrt{2}}; \qquad (101 \pm 1\,|\,2 \pm 1) = \frac{1}{\sqrt{2}} \tag{27.3}$$

to which correspond outgoing waves with angular dependence

$$Y_0^0 = \frac{1}{\sqrt{2}}, \quad Y_2^{\pm 1} = \pm \frac{1}{2}\sqrt{15}\,\sin\vartheta\cos\vartheta\, e^{\pm i\varphi} \tag{27.4}$$

[1] N. P. HEYDENBURG, C. M. HUDSON, D. R. INGLIS and W. D. WHITEHEAD: Phys. Rev. **74**, 405 (1948).

[2] D. R. INGLIS: Phys. Rev. **74**, 21 (1948).

and similar terms. The angular distribution function then becomes, including resonance denominators in the parameters A, B, C,

$$\left.\begin{aligned} Y(\vartheta) = &\left|\frac{1}{\sqrt{2}} A\, e^{i\alpha} \frac{1}{\sqrt{2}} + \sqrt{\frac{2}{3}}\, B\, e^{i\beta} \frac{1}{2} \sqrt{\frac{5}{2}} (3\cos^2\vartheta - 1)\right|^2 + \\ &\qquad\qquad (\text{states } J = 0{,}2 \text{ formed from } l = 1, j = 1, m = 0) \\ &+ 2\left|\frac{1}{\sqrt{2}} B\, e^{i\beta} \frac{1}{2} \sqrt{15} \sin\vartheta \cos\vartheta\right|^2 + \\ &\qquad\qquad (\text{state } J = 2 \text{ formed from } l = 1\ j = 1\ m = \pm 1) \\ &+ 2\left|\frac{1}{\sqrt{6}} C\, e^{i\gamma} \frac{1}{2} \sqrt{15} \sin\vartheta \cos\vartheta\right|^2 + \\ &\qquad\qquad (\text{state } J = 2 \text{ formed from } l = 1\ j = 2\ m = \pm 1) \\ &+ 2\left|\sqrt{\frac{2}{3}}\, C\, e^{i\gamma} \frac{1}{4} \sqrt{15} \sin^2\vartheta\right|^2 \\ &\qquad\qquad (\text{state } J = 2 \text{ formed from } l = 1\ j = 2\ m = \pm 2)\,. \end{aligned}\right\} \quad (27.5)$$

The first terms give the interference behaviour, but in this case, since the parity of the two interfering states is the same only even powers of $\cos\vartheta$ occur. The final distribution can be put in the form (27.2) and a formula for the energy dependent term $A(E)$ extracted which agrees with the experimental observations.

The angular distribution of capture radiation from proton bombardment of light nuclei was studied by DEVONS and HINE [23] with the apparatus shown in Fig. 25a. The gamma radiation was detected with GEIGER counters; in later forms of this apparatus sodium iodide crystals have also been used. In all experiments of this sort, it is important to reduce scattering of the gamma radiation being studied by surrounding material and the target mounting has to be made as light as possible. The angular distribution of the high energy radiation from the $Li^7(p\gamma)$ reaction (resonant at $E_p = 441$ keV) is shown in Fig. 25 which gives also the ratio of intensity at 15° to intensity at 145° as a function of energy. The method of comparing radiation intensities at two angles near the extreme values of $\cos\vartheta$ as a function of energy reveals immediately any interference between states of opposite parity. In the $Li^7(p\gamma)$ reaction the narrow resonant state in Be^8 is formed by p-waves and has even parity (1^+) and there must be a wider or possibly non-resonant state of odd parity at a higher excitation energy whose influence is felt at 441 keV. A complex parameter is required as before to specify the ratio of the probabilities of formation of the two interfering states from the same j value. This particular experiment determines this complex parameter and clearly demonstrates the change in phase as the incident proton energy passes through the resonance energy. In addition the experiment very beautifully tests the form of the resonance denominator $E - E_{\mathrm{res}} + \frac{1}{2} i\Gamma$ at an energy many half-widths from the resonance peak.

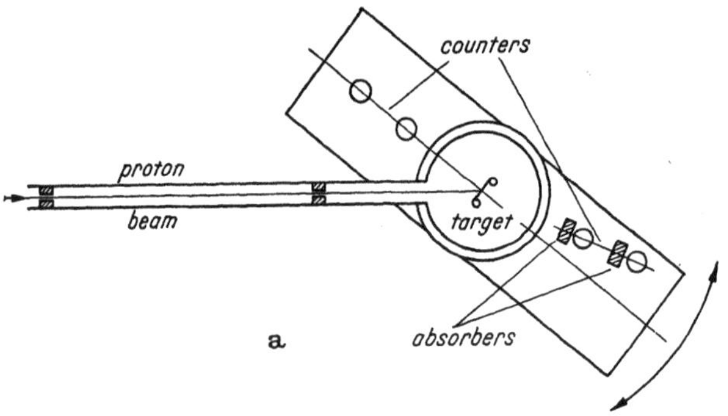

Fig. 25a—c. Study of angular distribution of gamma radiation from proton capture reactions (DEVONS and HINE [23]). (a) Apparatus.

The correlation between successive gamma quanta arising for instance in the de-excitation of an excited nucleus formed by radioactive decay is relatively

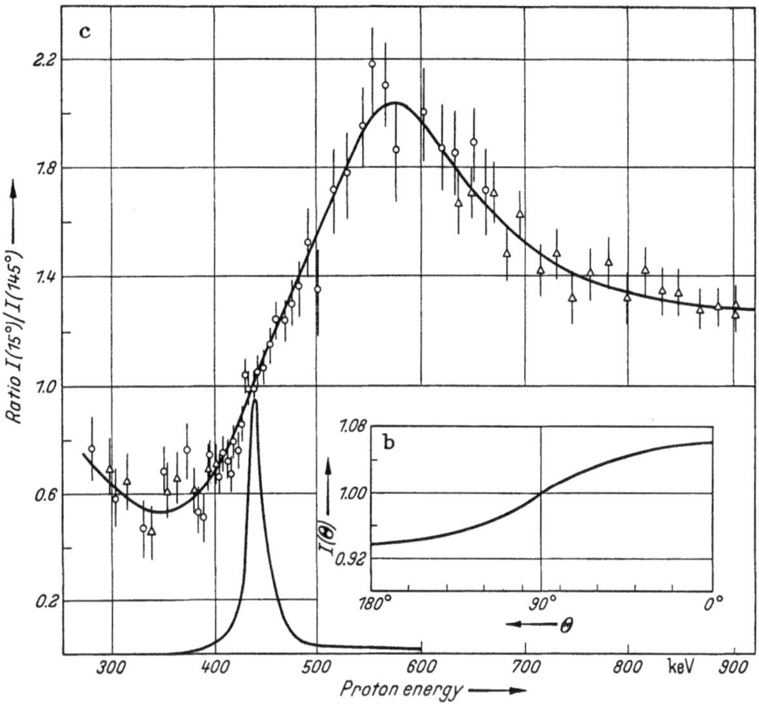

Fig. 25 b and c. (b) Angular distribution for Li^7 $(p\gamma)$ at $E_p = 500$ keV, thick target; (c) ratio of intensity at 15° to intensity at 145° for Li^7 $(p\gamma)$.

simple, both experimentally and theoretically. The techniques are discussed in a review article by FRAUENFELDER [21] and typically call for a pair of scintillation counters of the type shown in Fig. 26a which can move round a lightly mounted active source. The results obtained by KRAUSHAAR *et al.*[1] for the cascade in A^{38} are shown in Fig. 26b; by

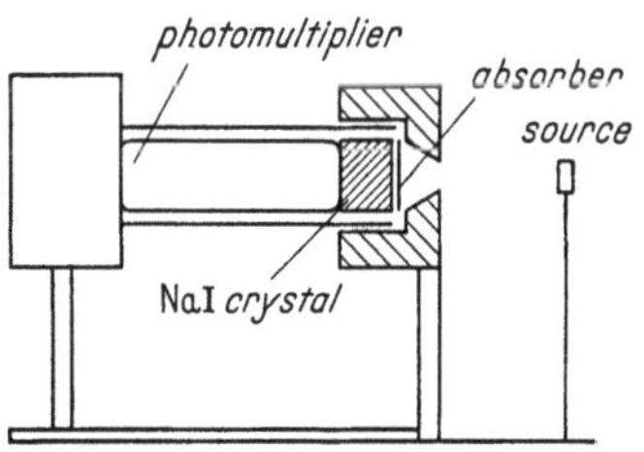

a

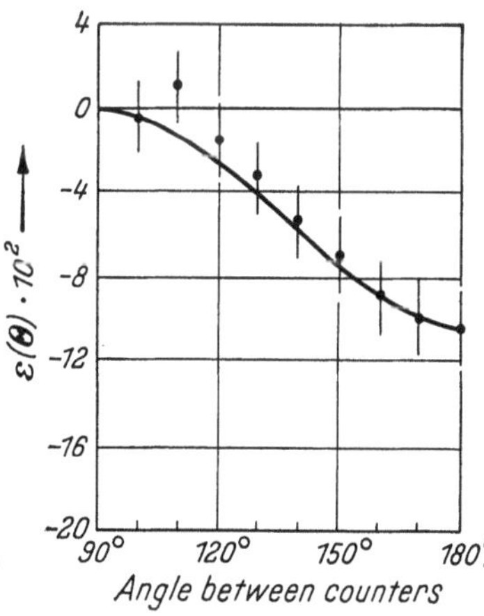

b

Fig. 26 a and b. Angular correlation of successive gamma quanta (KRAUSHAAR *et al.*). (a) Apparatus; (b) results for A^{38}.

comparison with theoretical curves they indicate spin assignments which agree with evidence from β-decay transition probabilities.

[1] J. J. KRAUSHAAR, J. W. MIHELICH and A. W. SUNYAR: Phys. Rev. 95, 456 (1954).

One of the most successful applications of the theory of angular correlations to light nuclei is presented by the work of BARNES, FRENCH and DEVONS[1] and of ARNOLD[2] on the second state of O^{16} at 6.13 MeV. This state is excited in the reaction

$$F^{19} + H^1 \to Ne^{20} \to O^{16*} + He^4$$
$$O^{16*} \to O^{16} + \gamma.$$

The angular distribution of the alpha particles with respect to the proton beam at the resonance energy of 340 keV is isotropic and the intermediate nuclei Ne^{20} can be considered as distributed at random. This converts the problem into one of two-stage correlation rather than three-stage, although at other resonances (SEED and FRENCH[3]) this simplification cannot be made. The directional correlation of the 6 MeV gamma radiation with respect to the direction of the alpha-particles is markedly anisotropic (Fig. 27) and requires terms up to $\cos^6 \vartheta$ for its expression. This shows that the relative orbital momentum of the alpha particle and of the O^{16*} nucleus must be at least 3, and also that the radiation must be at least of octupole order so that the spin of the O^{16} nucleus must be at least 3. If it is assumed that the Ne^{20} state has $J=1$, the simple assumption that l and j are both 3 enables the observed pattern to be predicted. The required transformation coefficients are

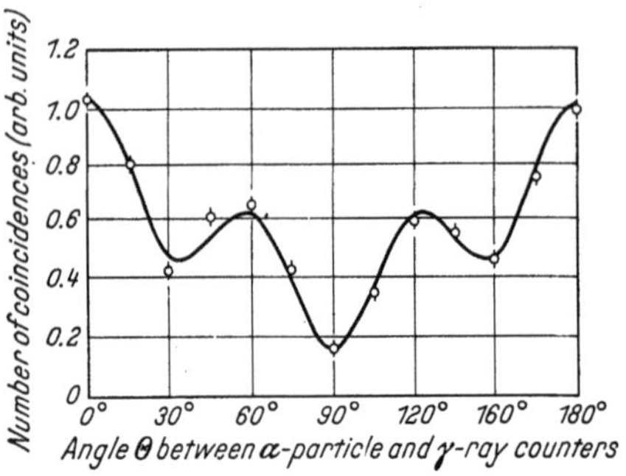

Fig. 27. Correlation between alpha particles and gamma radiation in the $F^{19}(p\,\alpha\,\gamma)\,O^{16}$ reaction (BARNES *et al.*).

$$(1M\,|\,3\,0\,3\,m) \quad \text{with} \quad M = m = 0 \quad \text{or} \quad \pm 1$$

and of these only those with $m = \pm 1$ are finite. The transitions $(3\ \pm 1) \to (0\ 0)$, give the distribution

$$W(\vartheta) = 225 \cos^6 \vartheta - 305 \cos^4 \vartheta + 111 \cos^2 \vartheta + 1.$$

Polarisation experiments (Sect. 36) have shown that the radiation is electric so that the O^{16} state is of type 3^-.

The results of other correlation experiments will be mentioned in Chap. D.

D. Experimental evidence on nuclear reactions.

In this part the atomic masses given by WAPSTRA[4] [*30*] are used. The nuclear reaction energies adopted are either given by WAPSTRA or calculated from his mass values. A list of individual determinations of nuclear disintegration energies is given by VAN PATTER and WHALING [*31*]. Extensive use has also been made of the review articles by AJZENBERG and LAURITSEN [*32*], and by ENDT and KLUYVER [*33*].

I. Reactions of protons with light nuclei.

28. Reactions of protons with H^1, H^2, H^3; Table 4. The elastic scattering of protons by protons is one of the fundamental reactions of nuclear physics and has

[1] C. A. BARNES, A. P. FRENCH and S. DEVONS: Nature, Lond. **166**, 145 (1950).
[2] W. R. ARNOLD: Phys. Rev. **80**, 34 (1950).
[3] J. E. SEED and A. P. FRENCH: Phys. Rev. **88**, 1007 (1952).
[4] Cf. WAPSTRA's article in vol. XXXVIII of this Encyclopedia.

been thoroughly studied at many energies up to 3000 MeV with the object of establishing the law of force between nucleons. Less direct, but still useful, information can be obtained from the proton-deuteron and proton-triton collisions. Nucleon-nucleon collisions are elastic until the threshold for meson production is reached at about 300 MeV. The inelastic reactions possible at low energies between protons and the isotopes H^2 and H^3 are shown in Table 4. The reaction $H^1(p\,\varepsilon^+)\,H^2$ is not forbidden, but the calculated cross section of about 10^{-49} cm² at 1 MeV is too small for investigation in the laboratory although it is large enough to make the reaction important astrophysically (SALPETER[1]). The ^1S state of the deuteron is not bound, and attempts to excite it by inelastic scattering always lead to disintegration of the deuteron, as shown in the reaction $H^2(p\,n)\,2\,H^1$. The theory of this process has been worked out by FRANK and

Table 4. *Reactions of protons with H^1, H^2, H^3.*

Reaction	Energy release Q	Cross section	
	MeV	cm²	Reference
$H^{1,2,3}(p\,p)\,H^{1,2,3}$	—	See articles on two and three body problem in Vol. XXXIX of this Encyclopedia	
$H^1(p\,\varepsilon^+)\,H^2$	1.42	$2.6\,E_{\mathrm{MeV}}^{-1}\exp(-0.99\cdot E_{\mathrm{MeV}}^{-1/2})\times 10^{-49}$	SALPETER
$H^2(p\,n)\,2\,H^1$	-2.227 ± 0.002	0.23×10^{-24} at $E_p=9.66$ MeV	FRANK and GAMMEL
$H^2(p\,\gamma)\,He^3$	5.50 ± 0.03	$0.74\,E_{\mathrm{MeV}}^{0.72}\,10^{-29}$	FOWLER *et al.*
$H^3(p\,\gamma)\,He^4$	19.7 ± 0.2	2×10^{-28} average 0 to 1 MeV	ARGO *et al.*
$H^3(p\,n)\,He^3$	-0.764 ± 0.001	0.56×10^{-24} at $E_p=2.4$ MeV	JARVIS *et al.*

GAMMEL[2] who give the angular distribution of the protons emitted in transitions to the continuum states. Experimental information on the details of the deuteron disintegration is rather scanty but observations of the total cross section and angular distribution have been made[2].

The capture of protons by deuterons was discovered by FOWLER *et al.*[3] who detected a non resonant radiation of approximately 6 MeV energy from thick targets of heavy ice bombarded by protons of energy up to 1.5 MeV. The angular distribution (GRIFFITHS and WARREN[4]) is nearly a pure $\sin^2\vartheta$ pattern relative to the proton beam and thus indicates that a compound nucleus in the true sense, involving a distribution among magnetic substates, is not formed. A pure $\sin^2\vartheta$ distribution is obtained if the reaction is produced by protons with $l=1$ leading to $E\,1$ radiation. Presumably the spin-orbit forces are not sufficient in this reaction to separate out the $p_{3/2}$ and $p_{1/2}$ states of the incoming proton and the reaction proceeds by a "direct" mechanism. The radiation from a dipole is plane polarised and this has been demonstrated for the reaction by WILKINSON[5], using the photo-disintegration of deuterium as an indicator of the direction of the electric vector. The small isotropic component of the angular distribution may be explained by spin-orbit coupling or by magnetic dipole transitions following the capture of s-wave protons (GRIFFITHS and WARREN). The differential cross section for elastic scattering of protons by deuterons has been obtained for proton energies up to 10 MeV. Some of the results, which are summarised by MASSEY [*36*] and by

[1] E. E. SALPETER: Phys. Rev. **88**, 547 (1952).
[2] R. M. FRANK and J. L. GAMMEL: Phys. Rev. **93**, 463 (1954).
[3] W. A. FOWLER, C. C. LAURITSEN and A. V. TOLLESTRUP: Phys. Rev. **76**, 1767 (1949).
[4] G. M. GRIFFITHS and J. B. WARREN: Proc. Phys. Soc. Lond. A **68**, 781 (1955).
[5] D. H. WILKINSON: Phil. Mag. **43**, 659 (1952).

Allred *et al.*[1], were derived by bombarding hydrogen with deuterons. The scattered particles were detected in a nuclear emulsion.

The capture of protons by tritons is probably another example of a direct transition not involving a compound nucleus in the sense of spatial orientation. The excitation curve[2,8] shows no resonances of the compound state for radiation near 90° from 1.0 up to 7.0 MeV bombarding energy. The angular distribution[3] is of the form $A + B\sin^2\vartheta$ with $A \sim 0.1\,B$ which again corresponds to the capture of protons with $l=1$ and the emission of dipole radiation, and ought to show complete polarisation. The capture radiation has a line spectrum of energy determined by the Q-value of the reaction and the incident proton energy[4]. Using the LS coupling picture, it is possible to account fairly well for the yield curve of the H^3 $(p\,\gamma)$ reaction [and of the inverse reaction He^4 (γp)] by the electric dipole transition ${}^1P \rightleftharpoons {}^1S$, which automatically gives the $\sin^2\vartheta$ angular distribution[5]. The small isotropic component in the angular distribution may be due to the quadrupole transition ${}^1D - {}^1S$ or to the intervention of spin-orbit or other non-central effects. It is not possible to obtain a predominantly $\sin^2\vartheta$ distribution with jj coupling [*27*].

The assumption that the $H^1 + H^3$ reactions proceed mainly through P states is confirmed by the observations of Jarvis *et al.*[6] on the reaction $H^3(p\,n)$ He^3. Spherically symmetrical angular distributions are found for very low proton energies but above 1 MeV the angular distribution requires a $\cos^2\vartheta$ term which increases in strength as the proton energy increases to 2.4 MeV ,showing the influence of incident p-waves. There is a broad maximum in the yield curve at a proton energy of 3 MeV[7] but there is no certain evidence of a resonance in the compound nucleus He^4 in the region of 20 MeV from these experiments. Vlasov *et al.*[8], however, postulate a 2^- level in He^4 to account for their results and propose that a 1^- level should be found at a rather higher excitation. Interference between states of opposite parity in the H^3 $(p\,n)$ reaction is indicated by some lack of symmetry about 90° in the angular distribution. Because of its low negative Q-value and fairly high cross section the H^3 $(p\,n)$ reaction, using a gas or zirconium-tritium target, has become very important as a source of neutrons of variable energy. The homogeneity of energy was demonstrated by Vlasov *et al.*

The elastic scattering of protons by tritons[9] shows a minimum at an energy near the H^3 $(p\,n)$ threshold, as predicted by Wigner[10].

29. Reactions of protons with He^3, He^4. The only interaction observed between protons of energy up to 10 MeV and the helium isotopes is elastic scatter-

---

[1] J. C. Allred, A. H. Armstrong, R. O. Bondelid and L. Rosen: Phys. Rev. **88**, 433 (1952). — R. J. S. Brown, G. D. Freier, H. D. Holmgren, W. R. Stratton and J. L. Yarnell: Phys. Rev. **88**, 253 (1952). — Burrows et al.: Reference [*40*].

[2] C. E. Falk and G. C. Phillips: Phys. Rev. **83**, 468 (1951). — H. B. Willard, J. K. Bair and J. D. Kington: Phys. Rev. **90**, 865 (1953).

[3] H. V. Argo, H. T. Gittings, A. Hemmendinger, G. A. Jarvis and R. F. Taschek: Phys. Rev. **78**, 691 (1950).

[4] R. S. Rochlin: Phys. Rev. **84**, 165 (1951).

[5] B. H. Flowers and F. Mandl: Proc. Roy. Soc. Lond., Ser. A **206**, 131 (1951).

[6] G. A. Jarvis, A. Hemmendinger, H. V. Argo and R. F. Taschek: Phys. Rev. **79**, 929 (1950).

[7] H. B. Willard, J. K. Bair and J. D. Kington: Phys. Rev. **90**, 865 (1953).

[8] N. A. Vlasov, S. P. Kalinin, A. A. Ogloblin, L. N. Samoilov, V. A. Sidorov and V. I. Chuev: J. exp. theor. Phys. USSR **28**, 639 (1955). See also A. I. Baz and Ya. A. Smorodinskii: J. exp. theor. Phys. USSR. **27**, 382 (1954).

[9] M. E. Ennis and A. Hemmendinger: Phys. Rev. **95**, 772 (1954).

[10] E. P. Wigner: Phys. Rev. **73**, 1002 (1948).

ing. The p-He^3 differential cross section has been obtained by SWEETMAN[1] for centre-of-mass energies of 3.72 and 6.45 MeV; it is in marked disagreement with the theoretical predictions of SWAN[2], showing too many particles at backward angles. No evidence for a state of Li^4 analogous to the possible 2^- state of He^4 (Sect. 28) has been obtained.

The scattering of protons by He^4 is of particular interest because the proton should enter a p-state if the compound nucleus Li^5 is formed. This nucleus is unstable in its ground state [defined by reactions such as $Li^6(dt)$] against disintegration into $p + He^4$ and scattering anomalies in helium observed for a proton energy of about 2.5 MeV could correspond to the formation of the Li^5 ground

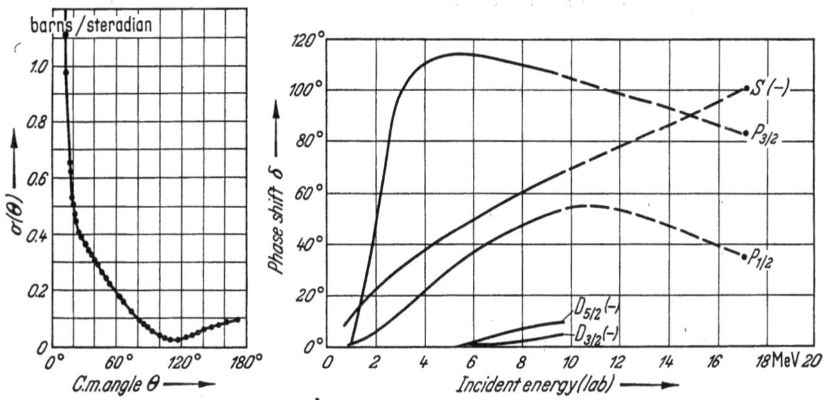

a b

Fig. 28 a and b. (a) Differential cross section in centre of mass system for elastic scattering of 9.48 MeV protons by He^4 (from PUTNAM). (b) Phase shifts for $He^4(pp)$ scattering.

state. The scattering experiments are technically straightforward and precision measurements are available for energies up to 17.5 MeV[3]. Some observations have been made at higher energies [32]. The general form of the differential cross section for a proton energy of about 10 MeV is shown in Fig. 28a and the phase shifts which best represent the experimental information are shown in Fig. 28b. The analysis of the results up to 10 MeV by DODDER and GAMMEL[4] follows the procedure outlined in Sect. 18, and suggests that the broad level at a proton energy of 2.5 MeV is a $^2P_{3/2}$ state since the phase shift δ_1^+ is near 90° at this energy. Another broad level of type $^2P_{1/2}$ is required at a higher excitation. The hypothesis that this is an inverted doublet is supported by measurement of the polarisation of protons scattered by helium[5]; the polarisation after a first scattering is revealed as an azimuthal asymmetry after a second scattering. The spin-orbit doublet in Li^5 and in the mirror nucleus He^5 are technically important for polarisation studies, since a helium scattering experiment may be used as an indicator of polarisation in an incident proton or neutron beam.

---

[1] D. R. SWEETMAN: Phil. Mag. **46**, 358 (1955).

[2] P. SWAN: Proc. Phys. Soc. Lond. A **66**, 740 (1953).

[3] 0.95 to 3.58 MeV: G. FREIER, E. LAMPI, W. SLEATOR and J. H. WILLIAMS: Phys. Rev. **75**, 1345 (1949). — 4.8, 5.1 MeV: C. H. BRADEN: Phys. Rev. **84**, 762 (1951). — 5.78 MeV: W. E. KREGER, W. JENTSCHKE and P. G. KRUGER: Phys. Rev. **93**, 837 (1954). — 9.48 MeV: T. M. PUTNAM: Phys. Rev. **87**, 932 (1952). — 17.5 MeV: K. BROCKMAN: Phys. Rev. **102**, 391 (1956).

[4] D. C. DODDER and J. L. GAMMEL: Phys. Rev. **88**, 520 (1952).

[5] M. HEUSINKVELD and G. FREIER: Phys. Rev. **85**, 80 (1952).

No excited state of He^4 is observed in any of these experiments, although with protons of 32 MeV[1] it was possible to excite the nucleus by as much as 23.3 MeV above the ground state.

30. Reactions of protons with Li^6 and Li^7. The processes observed are listed in Table 5. A complete study of the elastic scattering of protons by Li^6 has not yet been made, but Bashkin and Richards[2] have obtained the differential cross section as a function of energy at an angle of 164°; and have compared it with the yield of He^3 particles from the $(p\alpha)$ reaction for proton energies up to 3.6 MeV. Ordinary lithium metal was used as a target and the groups from Li^6 were separated from those due to Li^7 by magnetic analysis. The $(p\,\alpha)$ curve shows a well defined level of width 0.4 MeV whose excitation in the compound nucleus Be^7 is 7.16 MeV.

Table 5. *Reactions of protons with light nuclei* $(A > 4)$.

Column 2 gives the binding energy of a proton in the compound nucleus. Columns 5, 6 give observed Q-values in MeV, taken mainly from reference [*31*]. Column 7 gives other reactions observed. *a* means that the reaction has been observed by detection of activity, *p* that the reaction particles have been observed.

Target nucleus	Excit^n of C. N. MeV	pp	pp'	$p\alpha$	pn	Other reactions
Li^6	5.6	*p*	*p*	4.023	—	$p\gamma$
Li^7	17.24	*p*	*p*	17.346	−1.645	$p\gamma$, pd
Be^9	6.59	*p*	*p*	2.126	−1.852	$p\gamma$, pd, pt
B^{10}	8.70	*p*	*p*	1.148	−4.35	$p\gamma$, pd, pt, pHe^3
B^{11}	15.95		*p*	8.585	−2.762	$p\gamma$, pd
C^{12}	1.95	*p*	*p*	*p*	*a*	$p\gamma$, pd
C^{13}	7.54	*p*	*p*	—	−3.003	$p\gamma$
C^{14}	10.21			—	*p*	$p\gamma$
N^{14}	7.35	*p*	*p*	*a*	−6.0	$p\gamma$, pd
N^{15}	12.11			4.961	−3.55	$p\gamma$
O^{16}	0.59	*p*	*p*	—	—	$p\gamma$
O^{17}	5.61		(γ)	—	—	
O^{18}	7.96			3.961	−2.453	
F^{19}	12.87	*p*	*p*	8.117	−4.039	$p\gamma$
Ne^{20}	2.45	*p*	*p*	—	−16.1	$p\gamma$
Ne^{21}	6.74			—	*a*	$p\gamma$
Ne^{22}	8.80	*p*	*p*	—	—	$p\gamma$
Na^{23}	11.70		*p*	2.378	4.88	$p\gamma$
Mg^{24}	2.30	*p*	*p*	*a*	−14.8	$p\gamma$

Target nucleus	Excit^n of C. N. MeV	pp	pp'	$p\alpha$	pn	Other reactions
Mg^{25}	6.31		*p*	*a*	*a*	$p\gamma$
Mg^{26}	8.25		*p*	—	*a*	$p\gamma$
Al^{27}	11.59	*p*	*p*	1.595	− 5.61	$p\gamma$
Si^{28}	2.72		*p*	—	−14.9	
Si^{29}	5.51			—	—	$p\gamma$
Si^{30}	7.29			—	*a*	$p\gamma$
P^{31}	8.86			1.909	*a*	$p\gamma$
S^{32}	2.42	*p*	*p*	—	−14.0	
S^{33}	5.09			—	*a*	
S^{34}	6.41			—	*a*	$p\gamma$
S^{36}	8.42			—	—	
Cl^{35}	8.50			—	*a*	$p\gamma$, pd
Cl^{37}	10.15			3.2	−1.598	$p\gamma$
A^{36}	1.92			—	—	
A^{38}	6.42			—	—	
A^{40}	7.86	*p*	*p*	—	*p*	
K^{39}	8.34			—	—	pd
K^{41}	9.26			—	—	
Ca^{40}	1.62		*p*	—	−15.5	

The characteristic form of the excitation function for the differential cross section is due to interference with a state of opposite parity; this is a very broad level at 6.4 MeV in Be^7. The width of the narrow level leads to a reduced width for proton emission which is only comparable with that observed in the mirror reaction $Li^6\,(n\alpha)$ if protons with $l=1$ are responsible for the resonance. The spin of the 7.16 MeV state is then $\leqq \frac{5}{2}$. The great width of the lower resonance suggests that it is formed by *s*-waves. The angular distribution of the alpha particles is isotropic at 0.2 MeV but deviates from isotropy at higher energies.

The virtual levels of Be^7 should be able to de-excite by emission of gamma radiation in competition with alpha particles and elastically scattered protons. Capture radiation was observed, with low yield, by Bashkin and Carlson[3] using

[1] J. Benveniste and B. Cork: Phys. Rev. **89**, 422 (1953).
[2] S. Bashkin and H. T. Richards: Phys. Rev. **84**, 1124 (1951).
[3] S. Bashkin and R. R. Carlson: Phys. Rev. **97**, 1245 (1955).

proton energies at which only the broad s-wave level of Be^7 could have been excited. The gamma radiation proceeds directly to the ground state and also through the 430 keV excited state of Be^7 in the ratio 65/35.

Franzen and Likely[1] found a state of Li^6 at 2.2 ± 0.2 MeV by inelastic scattering of protons of energy 18 MeV; the scattered particles were detected in a sodium iodide crystal. The state was also seen in the $Li^7(pd)$ reaction.

The $Li^7(p\alpha)$ reaction was the first artificial transmutation discovered by Cockcroft and Walton[2]. The high yield of the Li^6 and Li^7 reactions has made them especially amenable to study, and a large amount of precise information has

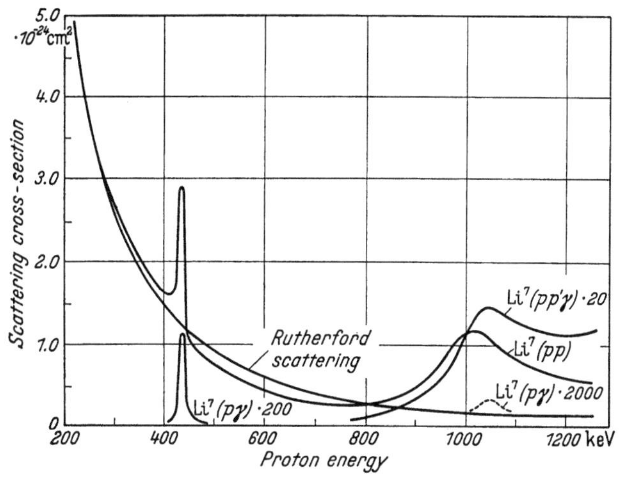

Fig. 29. Scattering of protons by Li^7. The curve gives the cross section for 4π steradians at ϑ (lab) = 137.8° [A. B. Brown, C. W. Snyder, W. A. Fowler and C. C. Lauritsen: Phys. Rev. **82**, 159 (1951)].

been obtained. The $Li^7(p\gamma)$ reaction has a sharp resonance ($\Gamma = 12.2 \pm 0.5$ keV) at a proton energy of 441.5 ± 0.5 keV. The elastic scattering anomaly corresponding to this level and to another less well studied level at 1030 keV ($\Gamma = 168$ keV) was studied in detail by Warters *et al.*[3] who measured the differential cross section and ratio to Rutherford scattering near the resonances. Fig. 29 gives some of the earlier results from their laboratory in which the relative importance of nuclear and resonant scattering are clearly shown. A phase shift analysis of the scattering shows that the 441 keV resonance is of type $J = 1^+$ and is formed by p-wave protons, with channel spins 1 and 2 present in a ratio of 1 to 5. This channel spin ratio is predicted by both LS and jj coupling models [*27*]. Analysis of the elastic scattering curves at the 1030 keV resonance indicates that this is also a 1^+ state. These curves are reported to show a cusp at the threshold of the endothermic $Li^7(pn)$ reaction (Bashkin and Richards).

The 1030 keV resonance in the elastic scattering curve is also apparent in the yield of inelastically scattered protons and of 477 keV gamma radiation from the first excited state of Li^7 produced in the reaction $Li^7(pp')Li^{7*}$. The inelastically scattered protons were found by Mozer *et al.*[4] to be isotropically distributed at the resonance but become anisotropic at higher energies. The behaviour is that of an interference between two waves of opposite parity. In this experiment the inelastically scattered protons were selected magnetically from a thick target of natural

[1] W. Franzen and J. G. Likely: Phys. Rev. **87**, 667 (1952).
[2] J. D. Cockcroft and E. T. S. Walton: Proc. Soc. Lond., Ser. A **137**, 229 (1932).
[3] W. D. Warters, W. A. Fowler and C. C. Lauritsen: Phys. Rev. **91**, 917 (1953).
[4] F. Mozer, W. A. Fowler and C. C. Lauritsen: Phys. Rev. **93**, 829 (1954).

lithium. The gamma radiation is isotropically distributed, as expected for a state of $J=\frac{1}{2}$ (LITTAUER[1]). Elastic and inelastic protons have also been detected in sodium iodide crystals (FRANZEN and LIKELY) and by range methods (GOVE and HARVEY[2]) and identify the states of Li^7.

The resonances in the $Li^7(p\gamma)$ reaction are superimposed on a slowly rising background of non resonant radiation, and interfere with it. The experimental results of DEVONS and HINE which demonstrate this interference for the 441 keV resonance and show that the two states concerned have opposite parity have already been described (Sect. 27). Taken together with the analysis of the elastic scattering, it appears that the resonant state, at an excitation of 17.63 MeV in Be^8 is of type 1^+ and is formed by p-wave protons while the non-resonant radiation is of opposite parity. A state of odd spin or odd parity or both in Be^8 is unable to break up into two alpha particles for reasons of symmetry. WILKINSON[3] ascribes the non-resonant radiation to a direct capture process; the background radiation would on this picture be of odd parity resulting from direct capture of s-wave protons. At energies above about 1 MeV, it is postulated that the radiative process is inverse to the giant resonance phenomenon of photodisintegration. The resonant radiation from the 17.63 MeV state of Be^8 proceeds to the ground state and the 2.9 MeV first excited state of this nucleus and both components have isotropic angular distributions at the resonance[4]. Both these states can break up into two alpha particles; lower energy gamma transitions from the 17.63 MeV state to other levels of Be^8 which can also break up into alpha particles have been suggested by INALL and BOYLE[5] by observation of $(\gamma\alpha)$ coincidences. Gamma-alpha correlations (INALL[6]) suggest spins for the levels of Be^8 concerned; the patterns are either isotropic or of quadrupole type and the levels are 0^+ or 2^+. The corresponding radiation is either $M1$ or a mixture of $M1$ and $E2$. The more intense resonant radiation (both 17.63 and 14.8 MeV components) from the 441 MeV resonance has been shown by DEVONS and GOLDRING [25] to be of order $\Lambda=2$ by observation of the angular correlation of internally produced pairs. This means that the 17.63 MeV radiation is $M1$ and the 14.8 MeV radiation a mixture of $M1$ and $E2$ waves.

Very early observations made it clear that the $Li^7(p\alpha)He^4$ reaction did not show sharp resonances in excitation as a function of proton energy. The yield of the $(p\alpha)$ reaction rises with energy to a broad peak for a proton energy of 3 MeV (HEYDENBURG *et al.*[7]); the angular distribution has already been discussed (Sect. 27). It calls for the assumption of incident p and f waves to produce even final states and for interference between 0^+ and 2^+ states of the compound nucleus. Since these are states of the same parity, no odd powers of $\cos\vartheta$ are found in the angular distribution. The 3 MeV level, with width about 1 MeV, is assigned the spin and parity 2^+ and the 0^+ level is required to be very wide. It was originally supposed that gamma radiation from these levels, in competition with alpha emission might be responsible for the interference pattern in the angular distribution of gamma radiation at the 441 keV resonance, but it is now clear that any radiation from the alpha emitting levels would not only be very weak, but would also have the wrong parity.

[1] R. M. LITTAUER: Proc. Phys. Soc. Lond. A **63**, 294 (1950).
[2] H. E. GOVE and J. A. HARVEY: Phys. Rev. **82**, 658 (1951).
[3] D. H. WILKINSON: Phil. Mag. **45**, 259 (1954).
[4] H. NABHOLZ, P. STOLL and H. WÄFFLER: Helv. phys. Acta **25**, 153 (1952).
[5] E. K. INALL and A. J. F. BOYLE: Phil. Mag. **44**, 1081 (1953).
[6] E. K. INALL: Phil. Mag. **45**, 768 (1954).
[7] N. P. HEYDENBURG, C. M. HUDSON, D. R. INGLIS and W. D. WHITEHEAD: Phys. Rev. **74**, 405 (1948).

The $Li^7(pn)$ reaction was very carefully investigated by TASCHEK and HEMMENDINGER[1]. There is a resonance at a proton energy of 2.25 MeV with a width of 0.2 MeV and the angular distribution near this resonance has a strong $\cos\vartheta$ term suggesting the interference between states of opposite parity which seems to be an important feature of several of the lithium plus proton reactions. The original analysis of the angular distributions by BREIT and BLOCH[2] suggested that the resonant part of the yield was due to capture of s-wave protons and that p-waves contributed the background and the necessary interference term. Since Li^7 and Be^7 have the same type of ground state the orbital momenta of the incident protons and emergent neutrons are the same. In a later examination of the same reaction ADAIR[3] interprets the 2.25 MeV resonance (19.2 MeV state in Be^8) as a 3^+ level formed by p-protons and having isotopic spin 1, in analogy with a known level in the isobaric nucleus Li^8. The assignment of spin 3 is based on the observed cross section at resonance which is 0.25 barn and cannot exceed $\frac{1}{8}(2J+1)\pi\lambda\!\!\!^{-2}$ (Sect. 9) even if Γ_n is equal to Γ_p which is unlikely for energies near the neutron threshold. The background in this interpretation is due to the s-wave interaction. The resonant part of the cross section was shown by ADAIR to be predictable from the reaction constants for the formation of the analogue level in Li^8; this is a good example of the use that can be made of the principle of charge independence in predicting reaction cross sections. The background transitions seem to require two odd parity states but their energy is not precisely determined; one has $T=0$ and cannot be located from the level system of Li^8. ADAIR also points out that a 3^+ level in Be^8 indicates the presence of spin-orbit coupling and that the outgoing neutrons in the p-state interaction should be polarised. This has been demonstrated by ADAIR *et al.*[4].

The $Li^7(pn)Be^7$ reaction has been used extensively as a source of homogeneous neutrons of variable energy and the reaction threshold has been very accurately measured. Effects due to lower energy neutrons produced in transitions to excited states of Be^7 must be corrected for in using this neutron source. The ratio of transitions to the ground state and first excited state of Be^7 is a function of energy which tends to the statistical value $2J+1/2J'+1=\frac{1}{2}$ at high energies but is very much less at lower energies because of barrier transmission factors. A measurement of this ratio for proton energies between 2.5 and 2.9 MeV using a He^3 fast neutron spectrometer is described by BATCHELOR[5]. Other levels of Be^7 have been excited with high energy protons (THOMSON[6]).

The lithium plus proton reactions thus exhibit many characteristic features of nuclear reactions as a whole, namely the existence of sharply defined resonant states from which particle break-up is prohibited, the probable influence of direct transitions not involving a compound state, the existence of very broad states constituting a background with which interference can take place in many reactions, the description of the levels by the BREIT-WIGNER formula, and the implications of the principle of charge independence. The information obtainable on the reduced widths and quantum numbers of the states of Be^8 is discussed in the light of nuclear models in Sect. 76.

31. Reactions of protons with Be^9. The principal reactions of protons with beryllium are shown in Table 5. These reactions pass through intermediate states of the compound nucleus B^{10}; whose levels are shown in Fig. 31 and 85.

[1] R. F. TASCHEK and A. HEMMENDINGER: Phys. Rev. **74**, 373 (1948).
[2] G. BREIT and I. BLOCH: Phys. Rev. **74**, 397 (1948).
[3] R. K. ADAIR: Phys. Rev. **96**, 709 (1954).
[4] R. K. ADAIR, S. E. DARDEN and R. E. FIELDS: Phys. Rev. **96**, 503 (1954).
[5] R. BATCHELOR: Proc. Phys. Soc. Lond. **68**, 452 (1955).
[6] D. M. THOMSON: Phys. Rev. **88**, 954 (1952).

At low proton energies all reactions increase in probability exponentially with energy according to the simple GAMOW relation, except for the elastic scattering which follows the normal COULOMB cross section. The elastic scattering has been observed as a function of energy at 138° (laboratory) for protons of energy up to 1.4 MeV by THOMAS *et al.*[1] using a double focussing magnetic spectrograph to separate scattered protons from reaction particles. There are anomalies in the elastic scattering near 0.33, 0.99 and 1.086 MeV and COHEN[2] associates the upper two resonances with levels of type 2^- and 0^+.

The inelastic scattering of protons by Be^9 shows only a single level of energy 2.428 MeV, for excitations up to 6 MeV[3]. With 32 MeV protons from the Berkeley linear accelerator levels at 6.8 and 11.6 MeV are also excited[4]. The 2.428 MeV state is unstable against disintegration into $Be^8 + n$ but its width is less than 3 keV according to the high resolution electrostatic analyser experiments of BROWNE *et al.*[5]. The variation of the emission of inelastically scattered protons at a laboratory angle of 90° has been determined for the energy range 5.6 to 6.4 MeV by GREENLEES[6] and shows no anomalies. The angular distribution of inelastic protons for an incident proton energy of 7.1 MeV is isotropic over the range 60° to 160° but has a strong forward peak probably to be associated with direct electric excitation (DAVIS[7]). In most of these experiments with high energy protons the $(p\,\alpha)$ and $(p\,d)$ reactions are also observed and excited states in the residual nuclei Li^6 and Be^8 have been identified.

The $Be^9(p\,\alpha)$ and $(p\,d)$ reactions show resonances at 0.33, 0.68 and 0.94 MeV, and an additional level at 0.47 MeV proton energy is found in the deuteron reaction. The cross sections at 138° are given as a function of energy by THOMAS *et al.*[1] and the angular distributions were measured at energies up to 1 MeV by NEUENDORFFER *et al.*[8] using a target chamber arranged to give simultaneous exposures on a set of photographic plates at different angles. The main feature of these results is an angular variation including a $\cos\vartheta$ term indicating strong interference between states of the compound nucleus of opposite parity, but the characteristics of the states are not yet completely defined, although the strength of the 0.33 MeV resonance suggests that it is formed by *s*-wave protons. At 2.56 MeV the $(p\,\alpha)$ reaction shows a strong resonance, which was first detected in the yield of gamma radiation and is ascribed to the process $Be^9(p\,\alpha)Li^6$ with Li^6 formed in a state at 3.58 MeV. The nature of this reaction was established by DAY and WALKER[9] who detected coincidences between alpha particles and gamma radiation at the resonance; their particularly simple apparatus and their excitation curves are shown in Fig. 30. The positron spectrum from the internal pair conversion of the 3.58 MeV gamma radiation was measured by MACKIN[10] whose results fit best with an $M1$ assignment for the transition $Li^{6*} \to Li^6$. Since the spin of Li^6 is 1 this is consistent with a spin of 0^+ for the 3.58 MeV state which in turn explains why this state radiates rather than breaking up into an alpha particle and a deuteron.

[1] R. G. THOMAS, W. A. FOWLER and C. C. LAURITSEN: Phys. Rev. **75**, 1612 (1949).
[2] E. R. COHEN: Phys. Rev. **75**, 1463 (A) (1949).
[3] J. C. ARTHUR, A. J. ALLEN, R. S. BENDER, H. J. HAUSMAN and C. J. MCDOLE: Phys. Rev. **88**, 1291 (1952).
[4] R. BRITTEN: Phys. Rev. **88**, 283 (1952).
[5] C. P. BROWNE, R. N. WILLIAMSON, D. S. CRAIG and D. J. DONAHUE: Phys. Rev. **83**, 179 (1951).
[6] G. W. GREENLEES: Proc. Phys. Soc. Lond. A **68**, 97 (1955).
[7] K. E. DAVIS: Phys. Rev. **88**, 1433 (1952).
[8] J. A. NEUENDORFFER, D. R. INGLIS and S. S. HANNA: Phys. Rev. **82**, 75 (1951).
[9] R. B. DAY and R. L. WALKER: Phys. Rev. **85**, 582 (1952).
[10] R. J. MACKIN: Phys. Rev. **94**, 648 (1954).

The 2.56 MeV resonance is not shown in the alpha particle transitions to other low states of Li^6, including the ground state. This can be explained if the 3.58 MeV state is identified as the first $T=1$ state of Li^6, since then the resonant state of B^{10} at 8.89 MeV would also be predominantly $T=1$, and a break-up to $T=0$ states, such as $Li^6+\alpha$ or Be^8+d would be forbidden, except in so far as mixing of wave functions occurs in the compound state or isotopic spin impurity exists in the wave functions of the $T=0$ nuclei concerned. The conservation of isotopic spin in the $Be^9(p\alpha)$ reaction has been especially discussed by MALM and INGLIS[1]. The magnitude of the resonant cross section is used by MACKIN to predict that $J\geqq 2$ for the 8.89 MeV state of B^{10}, and the observed width of the resonance yields a value of the reduced width for alpha particle emission, which is only reasonable if $J=2^+$.

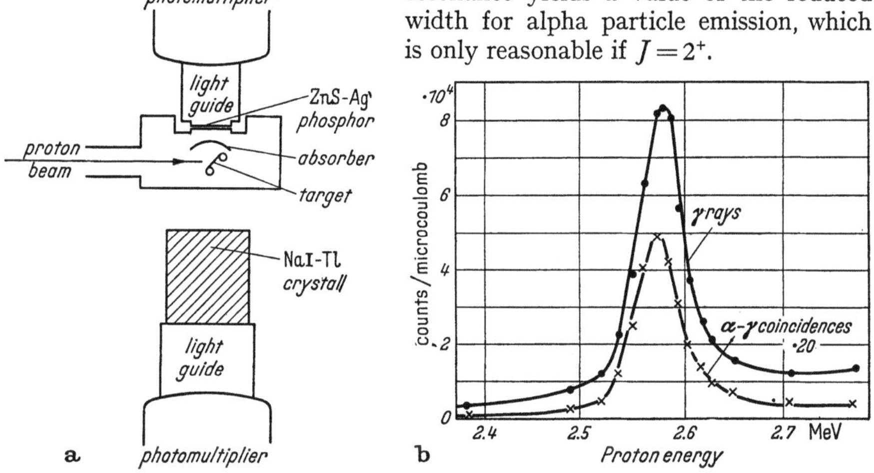

Fig. 30 a and b. ($\alpha-\gamma$) coincidences in Be^9 ($p\alpha\gamma$) Li^6 reaction (DAY and WALKER). (a) Apparatus; (b) results.

The excitation functions for capture gamma radiation from the $Be^9(p\gamma)$ reactions has been extended to 5.3 MeV by the Oak Ridge workers[2]. The main resonances and their characteristics are summarised in Table 6 and some of the cascade transitions are shown in Fig. 31. Below the lowest resonance, there is a yield of radiation which increases slowly with proton energy and can be interpreted as due to the tail of the 0.33 MeV resonance modified by the GAMOW penetrability factor for the incident protons, although the possibility of a lower resonance level has been suggested by TANGEN[3]. There is a strong yield of non-resonant gamma radiation above 2 MeV bombarding energy, perhaps due to direct transitions not involving compound states. The spectrum of gamma radiation at the individual resonances was first accurately studied by FOWLER *et al.*[4] using a lens spectrometer. This established that at the 995 keV resonance most of the radiation was a direct ground state transition with a nearly isotropic distribution, while at the narrower 1.086 MeV level a cascade took place through the first excited state of B^{10}. The spectrum for these two levels was examined again later by HORNYAK and COOR[5] using a sodium iodide spectrometer. The

[1] R. MALM and D. R. INGLIS: Phys. Rev. **95**, 993 (1954).
[2] C.W. SNYDER, H. B. WILLARD, J. K. BAIR, E. D. KLEMA, J. D. KINGTON and F. P. GREEN: Phys. Rev. **85**, 934 (1952).
[3] R. TANGEN: Kgl. norske Vid. Selsk., Skr. **1** (1946).
[4] W. A. FOWLER, C. C. LAURITSEN and T. LAURITSEN: Phys. Rev. **73**, 181 (1948) — T. LAURITSEN, W. A. FOWLER, C. C. LAURITSEN and V. K. RASMUSSEN: Phys. Rev. **73**, 636 (1948).
[5] W. F. HORNYAK and T. COOR: Phys. Rev. **92**, 675 (1953).

largely increased sensitivity of this detector enabled these workers to demonstrate a number of weak cascade transitions.

The internal pair angular correlation for the radiation from the 0.99 MeV level has been shown by DEVONS and GOLDRING [25] to require a transition order (Sect. 19) of 1 and 3 in certain proportions. The spin change (2^- to 3^+) permits these values, but the width that has to be associated with the radiative

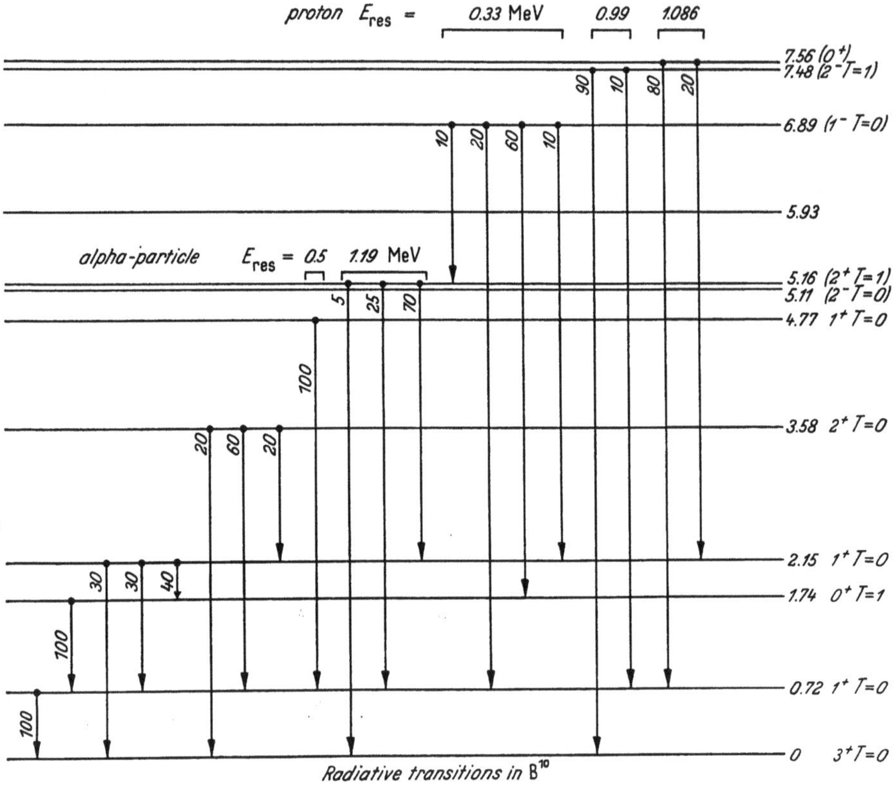

Fig. 31. Radiative transitions in B^{10} [32].

transition for $E3$ or $M2$ ($\Lambda=3$) radiation is unaccountably large, although for $E1$ radiation it is reasonable. If the radiating state has isotopic spin 1, no retardation due to the electric dipole selection rule need be considered.

A similar study of the radiations at the 300 keV resonance was made by CARLSON and NELSON[1] and by WILKINSON and CLEGG[2]. The former workers used a three crystal scintillation spectrometer (Sect. 14). The spectrum of radiation obtained with this instrument from a thick beryllium target bombarded by 315 keV protons shows gamma radiation of energies (6.7 ± 0.15), (6.0 ± 0.1), (5.1 ± 0.1) and (4.7 ± 0.15) MeV, which are associated with cascades from the 6.89 MeV level of B^{10}. The final components of these cascades are located in the low lying levels of B^{10} and were measured with a single crystal scintillation spectrometer; the results are included in Fig. 31 and indicate gamma rays of energies (0.41 ± 0.02), (0.72 ± 0.02), (1.03 ± 0.03) and (1.43 ± 0.03) MeV. Neither the high energy nor low energy spectral components change in relative intensity as the bombarding

[1] R. R. CARLSON and E. B. NELSON: Phys. Rev. **98**, 1310 (1955).
[2] D. H. WILKINSON and A. B. CLEGG: Phil. Mag. **47**, 291 (1956).

energy is moved above and below the 330 keV resonance. Nor is there any evidence for departure from isotropy, suggesting the formation of the level by *s*-wave protons, which is consistent with its large width. The angular correlation of the 1.03 and 0.72 MeV gamma rays was observed and gave

$$\frac{W(180^\circ)}{W(90^\circ)} - 1 = -0.12 \pm 0.04 .$$

The interpretation of these results starts from the assumption that the 6.89 MeV level has spin 1^- or 2^- if *s*-wave proton capture takes place. 2^- is discarded because ground state transitions are relatively weak in contrast with the strong ground state transition seen from the 7.48 MeV level of B^{10}; such high energy transitions as are observed may in fact be ascribed to the tail of this wide level. The resonance cross section for the 0.33 MeV resonance, together with the proton width estimated from the (pd) and $(p\alpha)$ yields indicates a radiation width of 5 eV for the $6.89 \rightarrow 1.74$ MeV transition ($E1$); this is about 10% of the $E1$ width given by the WEISSKOPF formula. It is notable that all the transitions from the 6.89 MeV level to the low states of B^{10} are $E1$ transitions and that two of them may violate the dipole selection rule for isotopic spin. The 6.89 MeV level is likely to have isotopic spin zero since it breaks up into the systems $(Li^6 + \alpha)$ and $(Be^8 + d)$ and the 0.72 and 2.15 MeV levels are similarly $T = 0$ because the $T = 1$ level is definitely identified at 1.74 MeV by comparison with isobaric nuclei and by inelastic scattering experiments. The spin of the 0.72 MeV level is set at 1^+ or 2^+ by the angular correlation experiment, which is consistent with the $M1 - E2$ correlation although the latter may be a mixture of $E2$ and $M1$. The apparent failure of the isotopic spin dipole selection rule may be due to the interference of a 1^- state with $T = 1$ near the 6.89 MeV level of B^{10}.

The nucleus B^9 produced by the endothermic reaction $Be^9(pn)$ is unstable in its ground state, but this state is certainly not more than 2 keV wide, corresponding to a lifetime[1] of greater than 3×10^{-19} sec. The yield of neutrons shows resonances at 2.56 and 4.7 MeV. The neutron spectrum has been studied by AJZENBERG[2] and shows a sharp homogeneous group corresponding to the ground state transition and a group to the first excited state of B^9 at 2.37 ± 0.04 MeV. There is also a continuum of neutrons, perhaps due to the $Be^9(p, pn)Be^8$ reaction which has a Q-value of -1.67 MeV. The 2.37 MeV state has been indicated in neutron threshold measurements; as the proton energy is increased, arrival at a new state of the residual nucleus is accompanied by the emission of a narrow cone of low energy neutrons which may be detected with good efficiency.

32. Reactions of protons with B^{10} and B^{11}. The boron-10 reactions with protons are listed in Table 5; the energy levels of the compound nucleus C^{11} are shown in Fig. 67.

The elastic scattering from B^{10} was studied by BROWN *et al.*[3] at a laboratory angle of 137.8° and for proton energies from 0.9 to 1.6 MeV. The ratio to RUTHERFORD scattering increases over this range and shows irregularities near resonances known at 1.15 and 1.5 MeV in comparison reactions. Between proton energies of 1.5 and 2.6 MeV a steadily increasing yield of 718 keV radiation at $\vartheta = 90^\circ$ was observed by DAY and HUUS[4] using a scintillation spectrometer to select this particular radiation. The radiation is due to inelastic scattering with

[1] P. H. STELSON and W. M. PRESTON: Phys. Rev. **83**, 469 (1951).
[2] F. AJZENBERG and W. W. BUECHNER: Phys. Rev. **91**, 674 (1953).
[3] A. B. BROWN, C. W. SNYDER, W. A. FOWLER and C. C. LAURITSEN: Phys. Rev. **82**, 159 (1951).
[4] R. B. DAY and T. HUUS: Phys. Rev. **95**, 1003 (1954).

excitation of the first state of B^{10} at 0.72 MeV, also observed by CRAIG *et al.*[1] with 2.19 MeV protons. These authors used an accurate electrostatic analyser to measure the Q-value for the reaction; it agrees very well with the measured energy of the gamma radiation. No other levels of B^{10} were observed in this reaction with protons of energies up to 4 MeV. With protons of energy between 6 and 8 MeV, COWIE *et al.* [*34*] found excitation of the 2.14 level of B^{10} and BOCKELMAN *et al.*[2] detected inelastic groups at 90° corresponding to all levels of B^{10} up to that at 3.58 MeV. The comparison of these results with the similar results of the inelastic scattering of deuterons (Fig. 32) reveals the interesting fact that at 90° the deuteron group corresponding to the first $T=1$ state of B^{10} is not found. At first sight this seems to provide excellent confirmation of the conservation of isotopic spin in nuclear reactions since the system $B^{10}+H^2$ is one of zero isotopic spin, but it should be noted (a) that the corresponding proton group is also very weak, and this should not be affected by the isotopic spin rule, (b) that observations have been reported only for one angle, though for several different energies and (c) that the excitation of the compound state of C^{11} is about 15 MeV and there may be considerable mixing of wave-functions belonging to different T-values in these states. This experiment was made possible by the combination of an 8 MeV electrostatic generator with a high resolution magnetic spectrograph employing photographic plate detection. The ratio of proton emission to the 2.15 ($T=0$) and 1.74 ($T=1$) levels should be 3 for high incident energies if spin factors may be neglected; this is because the 1.74 MeV level is a member of a triad of states while the $T=0$ levels have no counterparts in the $T_z=\pm 1$ nuclei B^{10} and C^{10}. This ratio is actually found for 7.6 MeV protons.

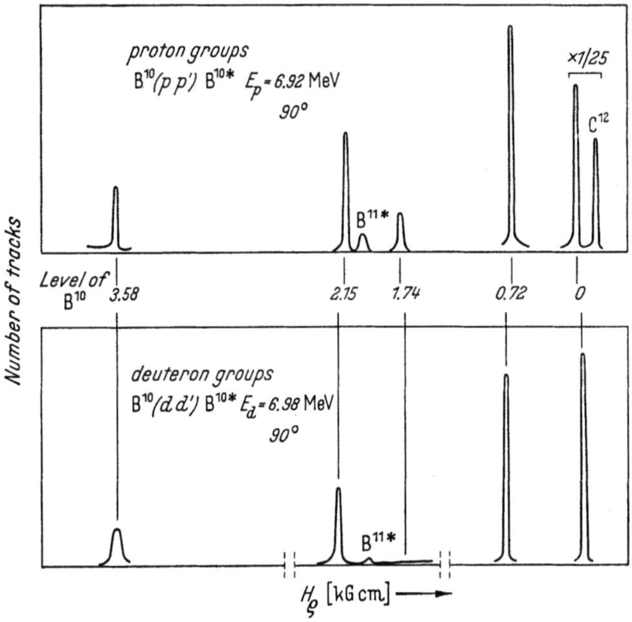

Fig. 32. Inelastic scattering of protons and deuterons by B^{10} (BOCKELMAN *et al.*).

The $B^{10}(p\alpha)\,Be^7$ reaction probability increases exponentially according to the GAMOW penetrability formula (Sect. 7) over the energy range 70 to 200 keV at least (BACH and LIVESEY[3]) and continues to rise smoothly to broad resonances at 1.15 and 1.5 MeV (BROWN *et al.*). Magnetic analysis was essential in these experiments because of the low Q-value of the reaction and consequent short range of the alpha particles in comparison with that of the scattered protons. Absolute cross sections were found by determining the constants of the magnetic

[1] D. S. CRAIG, D. J. DONAHUE and K. W. JONES: Phys. Rev. **88**, 808 (1952).

[2] C. K. BOCKELMAN, C. P. BROWNE, W. W. BUECHNER and A. SPERDUTO: Phys. Rev. **92**, 665 (1953).

[3] G. G. BACH and D. L. LIVESEY: Phil. Mag. **46**, 824 (1955).

analyser. At energies above 1 MeV the yield of lower energy alpha particles to the first excited state of Be^7 is appreciable and both these, and the associated gamma radiation, show a resonance at 1.52 MeV with a width of 0.25 MeV (BROWN *et al.*). The energy of the gamma radiation from Be^{7*} has been accurately measured by THOMAS and LAURITSEN [*39*]; in this reaction the lifetime of the radiating state is about 2.7×10^{-13} sec and as the stopping time of the recoil nucleus is perhaps 5×10^{-13} sec a DOPPLER correction of 1.3 keV to the observed energy (429 ± 1.5) keV is required (Sect. 24). The yield and energy of the gamma radiation has been found by DAY and HUUS[1] using a sodium iodide spectrometer; their excitation function for the gamma radiation of energy (432 ± 3) keV shows a peak at about 1.5 MeV.

The reaction $B^{10}(p\,He^3)\,Be^8$ has been observed with protons of 3.4 MeV (CRAIG *et al.*[2]). Both this reaction and the $B^{10}(p\,\alpha)$ reaction have been studied by REYNOLDS[3] using 18 MeV protons and a differential proportional counter for discriminating against long range scattered protons. The excitation of levels in Be^7 at 0.49, 4.72, 6.27, 7.21 and 14.6 MeV was indicated.

The radiative capture reaction $B^{10}(p\,\gamma)$ was first observed by WALKER[4] using a pair spectrometer, a thick target, and 1.2 MeV protons; the gamma ray energy of (9.47 ± 0.12) MeV is consistent with a ground state transition from the wide C^{11} levels at 9.70 MeV. A similar energy was measured by DAY and HUUS[1] with a scintillation spectrometer; they were able to establish a broad resonance for $B^{10}(p\,\gamma)\,C^{11}$ at an energy of 1.2 MeV, which is near that observed in the $B^{10}(p\,\alpha)$ reaction for ground state alpha particles. The cross section at resonance is low and the presence of residual B^{11} with a high yield of gamma radiation even in the separated isotopic targets makes the experiment difficult. This may account for the apparent disagreement between these results and those of KRONE and SEAGONDOLLAR[5] who report a resonance at 0.78 MeV. The precise allocation of the ground state transition to the levels of C^{11} is not yet certain; no cascade transitions have so far been observed.

Resonances in the $B^{10}(p\,n)\,C^{10}$ reaction have not yet been demonstrated, but AJZENBERG and FRANZEN[6] have observed the neutrons, using protons of energy 17.2 MeV from a small synchrocyclotron and have found an excited state of C^{10} at (3.34 ± 0.2) MeV in addition to wide levels near 5 MeV. The excited state transition seems stronger than that to the ground state, which may be understood since the ground state of C^{10} should have spin zero and the first excited state is probably of spin 2, so that the change of angular momentum from the B^{10} ground state $(J = 3)$ is greater for the C^{10} ground state transition. The $(p\,d)$ reaction has also been observed, and proceeds by a mechanism analogous to deuteron stripping in reverse so that a forward peak is obtained.

The bombardment of B^{11} with protons creates a compound nucleus, C^{12}, with an excitation energy of about 16 MeV, and as in the case of other nuclei of mass $(4n - 1)$, many resonance phenomena are observed. The elastic scattering reaction is under investigation, but full results have not yet been reported; the inelastic scattering from low states of B^{11} was observed by FULBRIGHT and BUSH[7] and by COWIE *et al.* [*34*], but the resonant features were not studied by

[1] R. B. DAY and T. HUUS: Phys. Rev. **95**, 1003 (1954).
[2] D. S. CRAIG, D. J. DONAHUE and K. W. JONES: Phys. Rev. **88**, 808 (1952).
[3] J. B. REYNOLDS: Phys. Rev. **98**, 1289 (1955).
[4] R. L. WALKER: Phys. Rev. **79**, 172 (1950).
[5] R. W. KRONE and L. W. SEAGONDOLLAR: Phys. Rev. **92**, 935 (1953).
[6] F. AJZENBERG and W. FRANZEN: Phys. Rev. **95**, 1531 (1954).
[7] H. W. FULBRIGHT and R. R. BUSH: Phys. Rev. **74**, 1323 (1948).

these authors. HUUS and DAY[1] however find a marked resonance for 2.13 MeV gamma radiation at 2.664 MeV; this corresponds to excitation of the first level of B^{11}. Higher resonances for this radiation and also for the $B^{11}(pn)\,C^{11}$ reaction have been reported, the latter by BLASER *et al.*[2] using cyclotron techniques.

The main resonances in the boron plus proton interaction have been followed in the $(p\alpha)$ and $(p\gamma)$ reactions, both of which excite not only ground states of the residual nucleus Be^8 or C^{12} but also the first excited state with spin 2^+, and both of which share certain resonances. The observed resonance characteristics are shown in Table 6. Full details of the higher resonances reported in [*32*] have not yet been published and we consider here only the well-investigated resonances for proton energies of less than 2 MeV. The lowest resonance level, proton energy 0.163 MeV, breaks up into both ground state and excited state alpha particles and 12 MeV and 16 MeV gamma radiation in competition, with the partial widths shown in Table 6. The angular distribution of the ground state alpha particles near the resonance was observed by THOMSON *et al.*[3] using a proportional counter and pulse analyser to locate the group. These distributions are anisotropic and asymmetric so that the 16.10 MeV level in C^{12} cannot be formed by *s*-wave protons. The ratio of alpha particles emitted at 150° to those seen at 30° confirms interference between two states of opposite parity. Since the yield of alpha particles at the 0.163 MeV resonance is fairly large it is likely that the sharply resonant state is formed by the lowest *l*-waves possible, i.e. $l=1$. With this assumption and also assuming that the interfering state is a broad one formed by protons with $l=0$, THOMSON *et al.*[3] were able, following the procedure outlined in Sect. 27, to predict the variation of $I(150°)/I(30°)$. Since both the sharp and resonant states emit alpha particles to even states of Be^8 it is reasonable to assume spin values 2^+ and 1^- for the sharp (16.10 MeV) and broad (17.22 MeV) states of C^{12}.

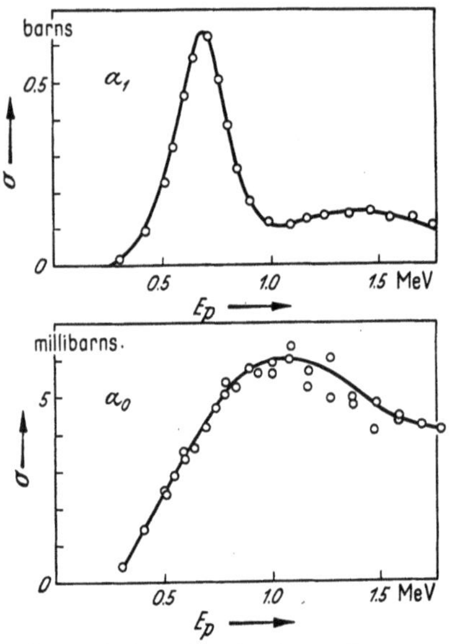

Fig. 33. Excitation functions for α_0 and α_1 groups in $B^{11}(p\alpha)$ reaction (BECKMAN *et al.*).

The nature of the interfering states in C^{12} (Fig. 96) was further investigated by BECKMAN *et al.*[4]. These authors found the excitation curve shown in Fig. 33 for the ground state (α_0) and first excited state (α_1) alpha particles and deduced from these curves the resonance parameters given in Table 6. The yield of excited state alpha particles is much larger than that of all other products at the first 3 resonances, but at the 0.163 MeV resonance the particle widths are so small that some special mechanism for inhibiting the reaction seems necessary; it is likely that the corresponding level in C^{12} has isotopic spin unity, so that some hindrance to break-up into even systems could result. The 0.67 and 1.34 MeV resonances have large proton widths, suggesting *s*-wave formation. The absence

[1] T. HUUS and R. B. DAY: Phys. Rev. **91**, 599 (1953).

[2] J. P. BLASER, F. BOEHM, P. MARMIER and P. SCHERRER: Helv. phys. Acta **24**, 465 (1951).

[3] D. M. THOMSON, A. V. COHEN, A. P. FRENCH and G. W. HUTCHINSON: Proc. Phys. Soc. Lond. A **65**, 745 (1952).

[4] O. BECKMAN, T. HUUS and Č. ZUPANČIČ: Phys. Rev. **91**, 606 (1953).

of α_0 transitions for the 0.67 MeV resonance is ascribed to the spin value 2^-; the presence of these for the higher resonance suggests spin 1^-. Since the radiative width for $E1$ radiation to the first excited state of C^{12} is normal, isotopic spin unity is also likely for these two levels.

The alpha particle spectrum at all resonances is complicated by the fact that the transitions α_1 leading to the broad excited state of Be^8 are accompanied by particles α_{12} due to the break-up in flight of the recoil Be^{8*} nuclei; these form a continuous distribution between certain limits. The analysis of this distribution by DEE and GILBERT[1] to prove the existence of an excited state of Be^8, was the first successful treatment of a three-body reaction. The ground state of Be^8 is also unstable to break-up into alpha particles; angular correlations in the alpha particle reactions have confirmed previous assignments for the Be^8 levels.

With this information on the levels concerned it is possible to give a consistent account of the capture gamma transitions. HUUS and DAY[2] using a scintillation spectrometer, investigated the resonances for the 12 MeV and 16 MeV transitions whose energies had been determined by FOWLER *et al.*[3] using a cloud chamber in a magnetic field, and by WALKER[4] using a pair spectrometer. The 12 MeV transition is always accompanied by the 4.4 MeV radiation from the first excited state of C^{12} involved in the cascade. The 16 MeV radiation is not found at the 0.67 MeV resonance because of the small radiative width for the necessary $M2$ radiation. The angular distributions and correlations of these gamma rays have been extensively studied as a function of energy by scintillation spectrometry. The distributions are given by GRANT *et al.*[5] and by HUBBARD *et al.*[6] for proton energies up to 600 keV and by GIVIN *et al.*[7] and GOVE and PAUL[8] for energies up to 3 MeV. At the 0.163 MeV resonance the angular distribution of the 12 MeV radiation is consistent with $M1$ radiation between two 2^+ levels. Near the 0.67MeV resonance the radiation distribution appears to include a $\cos\vartheta$ term and is certainly not isotropic. This is in apparent disagreement with the alpha particle evidence, but the background effects due to other levels are difficult to disentangle, and on the whole the evidence still favours formation of the 0.67 MeV resonance by s-waves. The course of the cascade of 12 MeV and 4.4 MeV gamma radiation is well established as $2^+(M1)\,2^+(E2)\,0^+$ by the correlation experiments of HUBBARD *et al.*; and of LEWIS[9]. The differential cross section for the 16 MeV radiation is given by GRANT *et al.*[1] in terms of the parameters of two interfering levels of spin 2^+ and 1^- (0.163 MeV and 1.39 MeV resonances). The mixing ratio of channel spins 2 and 1 was found to be 0.45 ± 0.05 which is in good agreement with that required by the alpha particle distributions (THOMSON *et al.*[10]).

The complexity of both alpha particle and gamma ray spectra and the great width of the upper resonances probably account for the fact that neither experimental nor theoretical information on the boron reactions is yet satisfactory. CHRISTY [*27*] concludes that neither the LS nor the jj coupling model is satisfactory for this reaction and implies that intermediate coupling may produce a better result.

[1] P. I. DEE and C. W. GILBERT: Proc. Roy. Soc. Lond., Ser. A **154**, 279 (1936).
[2] T. HUUS and R. B. DAY: Phys. Rev. **91**, 599 (1953).
[3] W. A. FOWLER, E. R. GAERTTNER and C. C. LAURITSEN: Phys. Rev. **53**, 628 (1938).
[4] R. L. WALKER: Phys. Rev. **79**, 172 (1950).
[5] P. J. GRANT, F. C. PLACK, J. G. RUTHERGLEN and W. M. DEUCHARS: Proc. Phys. Soc. Lond. A **67**, 751 (1954).
[6] T. P. HUBBARD, E. B. NELSON and J. A. JACOBS: Phys. Rev. **87**, 378 (1952).
[7] H. H. GIVIN, G. K. FARNEY, T. M. HAHN and B. D. KERN: Phys. Rev. **96**, 1337 (1954).
[8] H. E. GOVE and E. B. PAUL: Phys. Rev. **97**, 104 (1955).
[9] G. M. LEWIS: Phil. Mag. **43**, 690 (1952).
[10] D. M. THOMSON, A. V. COHEN, A. P. FRENCH and G. W. HUTCHINSON: Proc. Phys. Soc. Lond. A **65**, 745 (1952).

Table 6. *Resonant reactions of protons.*

Only reactions in which particle or radiation widths are known are listed and not all known resonances are included. Figures are taken mainly from references [18], [24], [32] and [33].

Target nucleus	Reaction channel s	E_{res} MeV	Excitation of compound nucleus MeV	J	Γ_λ keV	$\Gamma_{\lambda s}$ keV	$\frac{2Ma\gamma_{\lambda s}^2}{3\hbar^2}$ %	E_γ MeV	Γ_γ eV	Multipolarity	$\lvert M\rvert^2$ (Sect. 19)	Other resonant radiations
He^4	$He^4 + p$	2.1	0	$3/2^-$			75					
		5.3	2.5	$1/2^-$			75					
Li^6	$Li^6 + p$	0.8	6.35	$1/2, 3/2^+$			42	6.35	0.7	$E1$	7×10^{-3}	
								5.92	0.4		4.7×10^{-3}	
		1.75	7.16	$\leq 5/2^-$	420	400	12					α, γ
Li^7	$Li^7 + p$	0.441	17.63	1^+	12.2		6.4	17.63	16.7	$M1$	0.14	
								14.73	8.3		0.13	
		1.030	18.14	1^+				18.14	5.3	$M1$	0.04	
		2.22	19.18		200		2.0					
Be^9	$Be^9 + p$	0.33	6.89	1^-	175	23	5.7	6.17	0.7	$E1$	5.9×10^{-3}	α, d
		0.49	7.03									
		0.67	7.19		110							
		0.995	7.48	2^-	88		3.1	7.48	21	$E1$	0.10	α, d
		1.086	7.56	0^+	4		0.11	6.84	4.8	$M1$	0.72	
		2.56	8.89	2^+	39			8.17	8	$M1$	0.75	n, α, γ
B^{10}	$C^{11} + \gamma$	1.15	9.70	$3/2$				9.70	2.3	$M1$	0.12	p, α
B^{11}	$B^{11} + p$	0.163	16.10	2^+	5	0.005	1.9	16.10	3	$E2$	0.84	$\alpha_0\,\alpha_1$
								11.66	70	$M1$	2.1	
		0.675	16.58	2^-	322	150	14	12.13	15	$E1$	0.015	α_1
		1.388	17.23	1^-	1270	1000	25	17.22	40	$E1$	0.014	$\alpha_0\,\alpha_1$
								12.78	20		0.017	
	$Be^8 + \alpha_0$		16.10	2^+	5	0.1	0.0025					
			17.23	1^-	1270	7	0.125					
	$Be^{8*} + \alpha_1$		16.10	2^+	5	5	0.18					
			16.58	2^-	322	150	4.6					
			17.23	1^-	1270	200	5.3					

C^{12}	$C^{12}+p$	0.457	2.37	$^1/_2{}^+$	39.5		44	2.37	0.67	$E1$	0.08	
		1.70	3.51	$^3/_2{}^-$	60		2.5 {	3.51	0.67	$M1$	0.74	
		1.75	3.56	$^5/_2{}^+$	66		17	1.14	0.04	$E1$	0.04	
C^{13}	$C^{13}+p$	0.554	8.06	1^-	32.5		10	8.06	10.2	$E1$	0.03	
		1.15	8.62	0^+	6		1	8.62	0.73	$M1$	0.05	
		1.25	8.70	0^-	500		19	8.70	46	$E1$	0.11	
		1.76	9.18	2^-	2.1		0.8	9.18	10.6	$E1$	0.02	
C^{14}	$C^{14}+p$	1.165	11.29	$^1/_2{}^-$	12	10.4	4	11.29	0.25	$M1$	8.4×10^{-3}	
		1.31	11.43	$^1/_2{}^+$	41	8.2	0.7	11.43	2.3	$E1$	2.3×10^{-3}	
		1.50	11.61	$^1/_2{}^+$	475	470	100	11.61	28	$E1$	0.028	
	$N^{14}+n$		11.29	$^1/_2{}^-$	12	1.6	0.2					
			11.43	$^1/_2{}^+$	41	32.8	1					
			11.61	$^1/_2{}^+$	475	5.0	0.1					
N^{14}	$N^{14}+p$	0.277	7.61	$\leq {}^5/_2$	<2		<15					
		0.70	8.00	$^1/_2, {}^3/_2{}^+$	100		18					
		1.06	8.34	$\leq {}^5/_2$	4.8			8.34	1.3	$M1$	0.11	
		1.55	8.79	$^1/_2{}^+$	50			8.79	0.48	$E1$	10^{-3}	
		1.75	8.98	$^3/_2{}^-$	11			8.98	0.3	$M1$	0.02	
		1.81	9.04		7			9.04	1.04	$M1$	0.07	
		2.60	9.77	$^1/_2, {}^3/_2{}^+$	14		20	9.77	92	$E1$	0.15	
N^{15}	$N^{15}+p$	0.338	12.43	$0^+, 1^-$	94		8.1					α_0
		1.050	13.09	1^-	150		11	13.09	150	$E1$	0.1	$\alpha_0\,\alpha_1$
	$C^{12}+\alpha_0$		12.43	$0^+, 1^-$	94		1.8					
			13.09	1^-	150		1.0					
O^{16}	$O^{16}+p$	3.47	3.86	$^7/_2{}^-$	<3.5		0.31	3.86	0.6	$E1$	0.015	
F^{19}	$F^{19}+p$ $F^{19*}+p'$ $O^{16}+\alpha_0$ $O^{16*}+\alpha_n$ $Ne^{20}+\gamma$	The extensive experimental information is tabulated in reference [*32*], by BARNES [Phys. Rev. **97**, 1226 (1955)] and by BARANGER [Phys. Rev. **99**, 145 (1955)].										
Mg^{24}	$Mg^{24}+p$	0.825	3.08	$^3/_2$	1.5		16					
		1.49	3.72		0.3		19					
		1.62	3.85	$^1/_2$	36		12					
		1.66	3.88		0.1		0.14					
		2.01	4.22		0.15		0.14					
		2.40	4.60		0.3		0.14					
		3.14	5.34	$^3/_2$	200		7.2					

33. Reactions of protons with C^{12}, C^{13} and C^{14}. C^{12} is a very tightly bound nucleus and many of its reactions with protons have high negative Q-values. This means that such reactions have in general been studied only in cyclotron or linear accelerator bombardments, and although Q-values have in some cases been obtained, no resonance features have been established for these reactions for low energy protons. This applies to the processes $C^{12}(p\alpha)$ (REYNOLDS[1]) $C^{12}(pd)$ and $C^{12}(p, pn)$ (AAMODT *et al.*[2]). N^{12} has been produced and identified by ALVAREZ[3] using protons of energy above 20 MeV. The elastic scattering of protons by C^{12} shows resonances at 0.46 MeV and 1.70 MeV; capture radiation competes in both cases. The shape of the 0.46 MeV scattering resonance has been very fully explored by MILNE[4] who plotted excitation curves over the resonances at 13 angles, using an electrostatic generator and a magnetic spectrometer to select the protons scattered by C^{12}. Excitation curves at four angles over a wider range of energy were obtained by JACKSON *et al.*[5] using gaseous targets. Further resonances at higher energies were reported by MARTIN[6] *et al.* The analysis of these results by JACKSON and GALONSKY[7] is simplified by the zero spin of the scattering nucleus, and by the fact that the levels of N^{13} involved (Fig. 68) are near the ground state and are not strongly contaminated by wide distant levels. This latter point is essential for the validity of the one-level dispersion formula calculation of the phase shifts in terms of level parameters. The calculations follow the general method outlined in Sect. 18 and lead to the conclusion that the first resonance is associated with an $S_{1/2}$ state and the second with a close doublet of $P_{3/2}$ and $D_{5/2}$ states. The parameters deduced for these states are shown in Table 6. A large dispersive level shift is obtained for the $S_{1/2}$ level; this is because of its large reduced width, which in common with that of the $D_{5/2}$ level indicates that these levels result from single particle excitation. The $D_{5/2}$ level is presumably one component of a spin-orbit doublet of large spacing.

The capture radiation from the $C^{12}(p\gamma)$ reaction was detected by FOWLER and LAURITSEN[8] and by VAN PATTER[9], and at the 1.7 MeV resonance the angular distribution is $1+A\cos^2\vartheta$ (DAY and PERRY[10]) as would be expected for a $M1$ transition from the 3.51 MeV level of N^{13}. At 0.46 MeV it is isotropic. No evidence of the second ($D_{5/2}$) level proposed from the elastic scattering experiments is seen in this work as a ground state transition, although the width of the gamma resonance at 1.7 MeV is greater (74 keV) than that suggested for either of the levels concerned in elastic scattering, and this difficulty remains to be resolved. At proton energies between 1.2 and 2.5 MeV WOODBURY *et al.*[11] using a scintillation spectrometer observed a weak, soft radiation of energy $0.88\,(E_p-0.45)$ MeV. This energy variation suggests and fits the interpretation that the incident proton induces a direct transition from its continuum state (assumed p-wave) to the S state of N^{13} at 2.37 MeV. This is supported by the fact that the excitation function shows a resonance anomaly near the $P_{3/2}\,D_{5/2}$ levels and that the non-resonant radiation seems to have a small yield at 0° which is not usual with

[1] J. B. REYNOLDS: Phys. Rev. **98**, 1289 (1955).
[2] R. L. AAMODT, V. PETERSON and R. PHILLIPS: Phys. Rev. **88**, 739 (1955).
[3] L. W. ALVAREZ: Phys. Rev. **75**, 1815 (1949).
[4] E. A. MILNE: Phys. Rev. **93**, 762 (1954).
[5] H. L. JACKSON, A. I. GALONSKY, F. J. EPPLING, R. W. HILL, E. GOLDBERG and J. R. CAMERON: Phys. Rev. **89**, 365 (1953).
[6] M. MARTIN, H. SCHNEIDER and M. SEMPERT: Helv. phys. Acta **26**, 595 (1953).
[7] H. L. JACKSON and A. I. GALONSKY: Phys. Rev. **84**, 401 (1951); **89**, 370 (1953).
[8] W. A. FOWLER and C. C. LAURITSEN: Phys. Rev. **76**, 314 (1949).
[9] D. M. VAN PATTER: Phys. Rev. **76**, 1264 (1949).
[10] R. B. DAY and J. E. PERRY: Phys. Rev. **81**, 662A (1951).
[11] H. H. WOODBURY, A. V. TOLLESTRUP and R. B. DAY: Phys. Rev. **93**, 1311 (1954).

compound nucleus formation. The interference between the resonant and non-resonant radiation can be analysed by the methods discussed earlier (Sect. 26) and gives a good fit to the observed differential cross section. The angular distribution of the non-resonant radiation was found by HEIBERG *et al.*[1] to have the dipole pattern $(0.14 \pm 0.03) + \sin^2\vartheta$ as expected. The end state at 2.37 MeV is a virtual level and breaks up with the emission of a 0.46 MeV proton; so that the reaction is effectively of type $C^{12}(p, \gamma p')\,C^{12}$.

The reaction $C^{12}(p\gamma)\,N^{13}$ is of astrophysical interest, and the excitation function has been measured at proton energies near 100 keV by BAILEY and STRATTON[2] and by HALL and FOWLER[3] using high current, low energy accelerators. Cross sections as low as 10^{-34} cm² were measured and were shown to be in satisfactory agreement with those predicted by extrapolation of the single level formula from the 0.46 MeV resonance, using a proton width of the form

$$\Gamma_p = A\,\gamma_p^2 \exp(-b E^{-\frac{1}{2}})$$

to allow for barrier penetrability.

At high proton energies, well out of the resonance region, attention has chiefly concentrated on angular distributions for elastic and inelastic scattering. Typical distributions found with a cyclotron beam of 9.6 MeV, gas target, and photographic plate recording are given by BURCHAM *et al.*[4]. Analysis of the elastic scattering, in which interference between COULOMB and nuclear scattering is seen, gives some information on the constants of the complex nuclear potential including the well shape. Similar information may ultimately be extracted from the inelastic distributions. In the intermediate energy range (5.37 and 5.6 MeV) individual resonances can still be distinguished and are identified by the yield of 4.4 MeV gamma radiation as a function of energy; they coincide with anomalies in elastic scattering (MAEDER *et al.*[5]). The angular distribution of the 4.4 MeV gamma radiation from the first excited state of C^{12} is of the form $1 + 3.6\cos^2\vartheta - 3.6\cos^4\vartheta$ (GOVE and WALL[6]) which is consistent with the transition $2^+(E2)\,0^+$ expected. Inelastic proton groups have been observed with 96 MeV protons, and excitation of 15 MeV gamma radiation by protons of energy up to 340 MeV has been reported [*32*].

The C^{13} resonant reactions have also been fully studied; the cascade transitions are shown in Fig. 34 taken from [*32*]. MILNE[7] measured the differential cross section for elastic scattering of protons by a solid target enriched in C^{13} for energies from 0.45 to 1.60 MeV and for 5 angles. The results show the familiar interference phenomena due to superposition of COULOMB and nuclear scattering with anomalies at 0.556, 1.16, 1.47 and 1.55 MeV. Analysis of these anomalies led to spin and parity assignments, some of which are shown in Table 6; they are in agreement with evidence from the $(p\gamma)$ capture reaction. At higher proton energies work has been confined to observation of inelastic proton groups corresponding to excited states of 3.14 and 4.03 MeV in C^{13} (COWIE *et al.* [*34*]).

The cross section for the radiative capture $C^{13}(p\gamma)$ is also of astrophysical interest and has been measured by WOODBURY and FOWLER[8] for an energy of

[1] S. A. HEIBERG, T. K. ALEXANDER and D. B. JAMES: Canad. J. Phys. **33**, 34 (1955).

[2] C. L. BAILEY and W. R. STRATTON: Phys. Rev. **77**, 194 (1950).

[3] R. N. HALL and W. A. FOWLER: Phys. Rev. **77**, 197 (1950).

[4] W. E. BURCHAM, W. M. GIBSON, A. HOSSAIN and J. ROTBLAT: Phys. Rev. **92**, 1266 (1953).

[5] D. MAEDER, M. MARTIN, R. MULLER and H. SCHNEIDER: Helv. phys. Acta **27**, 166 (1954).

[6] H. E. GOVE and N. S. WALL: Canad. J. Phys. **31**, 189 (1953).

[7] E. A. MILNE: Phys. Rev. **93**, 762 (1954).

[8] E. J. WOODBURY and W. A. FOWLER: Phys. Rev. **85**, 51 (1952).

129 keV; the value of $(5 \pm 1) \times 10^{-33}$ cm² is associated with the tails of higher resonances. The higher resonances were investigated by Seagrave[1] and by Woodbury *et al.*[2] using Geiger counters and scintillation spectrometers respectively. The 0.55 MeV resonance is wide and is presumably formed by *s*-wave protons; the gamma radiation associated with this state is isotropically distributed [*23*], and the spin is established as 1^- from the (pp) reaction. The cas-

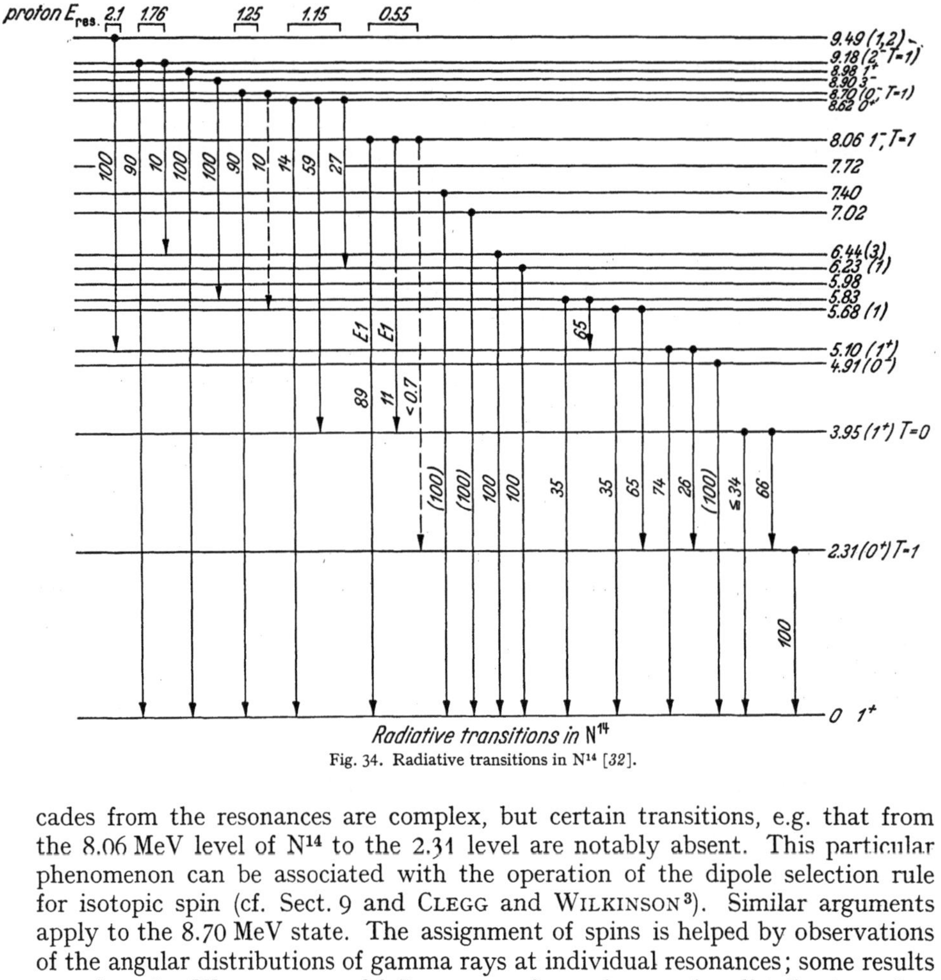

Fig. 34. Radiative transitions in N^{14} [*32*].

cades from the resonances are complex, but certain transitions, e.g. that from the 8.06 MeV level of N^{14} to the 2.31 level are notably absent. This particular phenomenon can be associated with the operation of the dipole selection rule for isotopic spin (cf. Sect. 9 and Clegg and Wilkinson[3]). Similar arguments apply to the 8.70 MeV state. The assignment of spins is helped by observations of the angular distributions of gamma rays at individual resonances; some results are given by Woodbury *et al.*[2] It appears from the angular distributions that all resonances except those at 0.55 and 1.25 MeV are formed by partial waves with $l > 2$.

The 9.18 MeV level in N^{14} decays predominantly to the ground state by radiation with an angular distribution $1 - 0.48 \cos^2\vartheta$ and a width which suggests electric dipole radiation. The spin would then be 2^- and the level would be formed by *d*-wave protons. Christy [*27*] shows that the observed distribution is consistent with the absorption of a $d_{5/2}$ proton (*jj*-coupling) or the creation of a ${}^3F_2$ compound state in LS coupling.

[1] J. D. Seagrave: Phys. Rev. **85**, 197 (1952).
[2] E. J. Woodbury, R. B. Day and A. V. Tollestrup: Phys. Rev. **92**, 1199 (1953).
[3] A. B. Clegg and D. H. Wilkinson: Phil. Mag. **44**, 1269 (1953).

The analysis of the complex cascade transitions in the $C^{13}(p\,\gamma)$ reaction by scintillation spectrometry is simplified by the use of a three-crystal spectrometer (Sect. 14) as in the work of HIRD *et al.*[1]. These authors have also established one particular cascade by coincidence counting. The energy of the main ground state transition has been determined by CARVER and WILKINSON[2] by pulse height analysis of the photoprotons from deuterium in a high pressure ionisation chamber. The $(p\,n)$ reaction with C^{13} has a high threshold and the neutron resonances lie at a much greater excitation in N^{14} than the levels just discussed; they have been observed by BAIR *et al.*[3].

Sufficient radioactive C^{14} can be produced by the $N^{14}(n\,p)$ reaction in nuclear reactors to permit a direct study of its properties as a nuclear reactant. The $(p\,\gamma)$ and $(p\,n)$ reactions have been observed in proton bombardment by BARTHOLOMEW *et al.*[4] and ROSEBOROUGH *et al.*[5]. The extensive series of resonances observed is given in part in Table 6, and neutron and gamma ray yield curves are shown in Fig. 35. An interesting feature is the broad state at about 1.5 MeV proton energy which emits gamma radiation but not neutrons. This state has nearly the single particle width; its decay into $N^{14}+n$ could be forbidden if it were assumed to be a state of isotopic spin $^3/_2$, and if there were little mixing of $T=\frac{1}{2}$ wave functions, since the final state at low bombarding energies is the $T=0$ ground state of N^{14}. A further interesting point is the comparison between the resonance energies and yields of the $C^{14}(p\,n)$ and $N^{14}(n\,p)$ reactions in the centre of mass system: Fig. 35 includes the excitation function for the $(n\,p)$ reaction. The correspondence in resonance energy and relative yield is excellent in view of the possible angular dependence of the neutron yield. BARTHOLOMEW *et al.* further show that the wide state at 11.61 MeV in N^{15} is partly concerned in the slow neutron capture in N^{14}; the gamma spectra for the $N^{14}(n\,\gamma)$ reaction and the $C^{14}(p\,\gamma)$ reaction near the 1.50 MeV resonance are very similar.

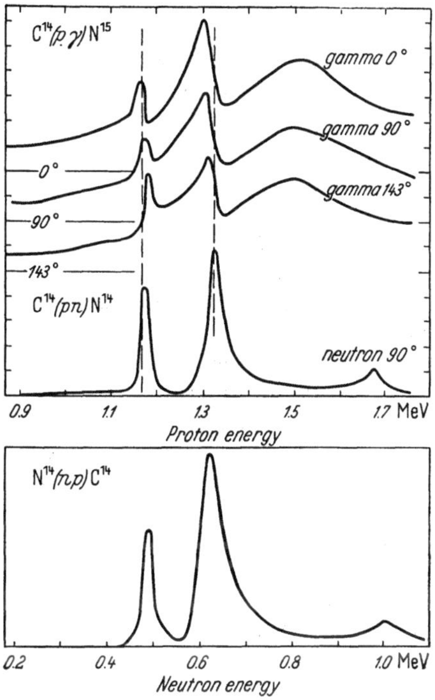

Fig. 35. Neutron and γ-ray yield curves for $C^{14}+p$ reaction. (BARTHOLOMEW *et al.*) with total cross section for N^{14} $(n\,p)$ for comparison.

34. Reactions of protons with N^{14} and N^{15}. The reactions found are shown in Table 5.

The elastic scattering of protons by N^{14} is reported to show anomalies between 1.07 and 1.78 MeV [*32*], and phase shift analyses have been made. With higher proton energies, attention has concentrated on the inelastic excitation of the

[1] B. HIRD, C. WHITEHEAD, J. BUTLER and C. H. COLLIE: Phys. Rev. **96**, 702 (1954).
[2] J. H. CARVER and D. H. WILKINSON: Proc. Phys. Soc. Lond. A **64**, 199 (1951).
[3] J. K. BAIR, J. D. KINGTON and H. B. WILLARD: Phys. Rev. **90**, 575 (1953).
[4] G. A. BARTHOLOMEW, F. BROWN, H. E. GOVE, A. E. LITHERLAND and E. B. PAUL: Canad. J. Phys. **33**, 441 (1955).
[5] W. D. ROSEBOROUGH, J. J. MCCUE, W. M. PRESTON and C. GOODMAN: Phys. Rev. **83**, 1133 (1951).

low states of N^{14} for comparison with the corresponding excitation by deuterons which may be affected by the conservation of isotopic spin (Sect. 9). The presence or absence of certain groups may, however, be due to factors other than isotopic spin; in particular the 2.31 MeV level in N^{14} was only excited with difficulty by FREEMANTLE *et al.*[1] using 9.5 MeV protons although it was seen easily by other groups of workers (BOCKELMAN *et al.*[2]). The differential cross section for elastic scattering is also given by FREEMANTLE *et al.*; it shows the usual interference with COULOMB scattering. The $(p\alpha)$ reaction shows broad resonances at proton energies above the threshold (BLASER *et al.*[3]); observations were made by the stacked foil method using a cyclotron beam, and it is likely that the observed widths are experimental. The (pn) reaction is highly endothermic and resonance features have not been studied, although the Q-value has been measured (AJZENBERG and FRANZEN[4]) and the excited states of O^{14} located.

The chief resonance phenomena in the N^{14} plus proton reactions are found in the proton capture process which has rather a low yield; some of the resonances observed are shown in Table 6 taken from the work of DUNCAN and PERRY[5] who observed the residual O^{15} activity in a gaseous target. The capture gamma radiation was detected and analysed by JOHNSON *et al.*[6] using a scintillation spectrometer. From the observed and reduced resonance widths some assignments of orbital momentum for the incident proton and spin values for the O^{15} states have been made. The reaction cross section can be extrapolated to low energies by superimposing a set of one-level curves and the result obtained is 1.24×10^{-10} barn at 128 keV. Direct measurement of this cross section, which is of interest in the carbon-nitrogen cycle of astrophysics gives a rather larger result. The $N^{14}(pd)$ reaction has been observed with 18.7 MeV protons and appears to be a pick-up process.

There are no reports of the elastic or inelastic scattering of protons in N^{15}. Strong resonances are however found in the $(p\alpha)$ and $(p\gamma)$ reactions. In the former reaction there are, in analogy with the same process in other nuclei of mass $(4n-1)$, strong transitions to both the ground and first excited state (2^+) of the residual even-even nucleus (C^{12}). Resonance information for some of the levels is given in Table 6.

The prominent resonance for capture radiation of 13 MeV at $E_p=1.05$ MeV was discovered in the work of SCHARDT *et al.*[7]. The great width suggests s-wave capture, and the radiative width suggests $E1$ radiation to the ground state of O^{16}. This radiation is not forbidden by the isotopic spin dipole selection rule if the O^{16*} level has $T=1$, but this level actually also breaks up into $C^{12}+\alpha_0$ (ground state) implying $T=0$, with a reasonable partial width Γ_α of 75 keV. The explanation of this discrepancy is not clear, nor is it known why the level shows no cascade branching to the 6.06 MeV 0^+ level of O^{16} whose characteristics resemble those of the ground state. This latter fact was established by GOLDRING[8], using the characteristic angular distribution of the pairs from the 6.06 MeV O^{16} level as an indication of the existence of a preceding transition.

---

[1] R. G. FREEMANTLE, D. J. PROWSE and J. ROTBLAT: Phys. Rev. **96**, 1268 (1954).
[2] C. K. BOCKELMAN, C. P. BROWNE, W. W. BUECHNER and A. SPERDUTO: Phys. Rev. **92**, 665 (1953).
[3] J. P. BLASER, P. MARMIER and M. SEMPERT: Helv. phys. Acta **25**, 442 (1952).
[4] F. AJZENBERG and W. FRANZEN: Phys. Rev. **94**, 409 (1954).
[5] D. B. DUNCAN and J. E. PERRY: Phys. Rev. **82**, 809 (1951).
[6] C. H. JOHNSON, G. P. ROBINSON and C. D. MOAK: Phys. Rev. **85**, 931 (1952).
[7] A. SCHARDT, W. A. FOWLER and C. C. LAURITSEN: Phys. Rev. **86**, 527 (1952).
[8] G. GOLDRING: Proc. Phys. Soc. Lond. A **67**, 930 (1954).

The resonances for the ground state and excited state alpha particle emission were found by SCHARDT *et al.* and by KRAUS[1]. The reduced widths deduced from the results of the former workers are shown in the table. COHEN and FRENCH[2] have measured the angular distribution of ground state alpha particles through resonances and demonstrated interference phenomena from which level assignments can be deduced. KRAUS[3] *et al.* measured the angular distributions of the gamma radiation and alpha particles from the $N^{15}(p\alpha)\,C^{12*}$ reactions for some of the resonances and SEED and FRENCH[4] observed the $(\alpha_1\gamma)$ correlation. The theoretical interpretation of the angular distribution at the 429 keV resonance requires a channel spin ratio of $t \approx 5$; at the 898 keV resonance the ratio is $t=\frac{3}{2}$. CHRISTY [27] shows that the latter ratio is consistent with absorption of a $d_{5/2}$ proton (jj scheme) or the formation of a $^3F_2$ compound state in O^{16} (LS scheme). The 429 keV information indicates only the LS scheme. Spin assignments for higher resonances are discussed by KRAUS.[1]

35. Reactions of protons with O^{16}, O^{17}, O^{18}. As can be seen from Table 5, only a few nuclear reactions of the oxygen isotopes with protons have been studied. Most of the detailed information available on the compound nucleus F^{17} formed from O^{16} with protons comes from the elastic scattering experiment of LAUBENSTEIN *et al.*[5]. The yield of elastic protons was measured as a function of incident energy by using a magnetic analyser to separate protons scattered by oxygen at 165° from protons scattered by other elements in the target. The yield curves show anomalies similar to those seen in other elastic scattering experiments. For the energy range $E_p=0.6$ to 4.5 MeV high resolution was available since an electrostatic generator was used as the proton accelerator. For higher energies, broader resonances, whose width may be experimental, were found by SEMPERT *et al.*[6]. The resonance parameters given in [32] are based on the analyses of LAUBENSTEIN and LAUBENSTEIN[7] and of SEMPERT *et al.*[6]. The energy level diagram of F^{17} (Fig. 71) shows that there is a level just below the dissociation energy into $O^{16}+p$. This bound level has a noticeable effect on the elastic scattering and the analysis shows that it has a reduced width very nearly equal to the single particle limit.

The $(p\gamma)$ reaction was observed by LAUBENSTEIN *et al.* by detection of the F^{17} activity and over the range $E_p=0.8$ to 2.1 MeV by WARREN *et al.*[8] who detected the gamma radiation due to (non-resonant) capture. This shows transitions to both the ground state and the first excited state of F^{17}; the latter has a pure $\sin^2\vartheta$ distribution characteristic of direct transitions and the former is isotropically distributed, as is the radiation which completes the cascade through the $1/2^+$ excited state. The earlier interpretation of the capture radiation as due to the low lying $S_{1/2}$ level seems to be incorrect, except possibly for the high energy (10%) component, since 90% of the radiative transitions pass through the $S_{1/2}$ level.

---

1 A. A. KRAUS: Phys. Rev. **94**, 975 (1954).
2 A. V. COHEN and A. P. FRENCH: Phil. Mag. **44**, 1259 (1953).
3 A. A. KRAUS, A. P. FRENCH, W. A. FOWLER and C. C. LAURITSEN: Phys. Rev. **89**, 299 (1953).
4 J. E. SEED and A. P. FRENCH: Phil. Mag. **43**, 1214 (1952).
5 R. A. LAUBENSTEIN, M. J. W. LAUBENSTEIN, L. J. KOESTER and C. MOBLEY: Phys. Rev. **84**, 12 (1951).
6 M. SEMPERT, H. SCHNEIDER and M. MARTIN: Helv. phys. Acta **27**, 313 (1954).
7 R. A. LAUBENSTEIN and M. J. W. LAUBENSTEIN: Phys. Rev. **84**, 18 (1951).
8 J. B. WARREN, K. A. LAURIE, D. B. JAMES and K. L. ERDMAN: Canad. J. Phys. **32**, 563 (1954).

The elastic and inelastic differential cross sections for the scattering of 9.6 MeV protons by oxygen have been measured (BURCHAM *et al.*[1]) and the former has been analysed[1] according to the optical model (Sect. 5).

The O^{18} (pn) and O^{18} $(p\alpha)$ reactions show resonances [*32*]. A detailed study of angular distributions in the $(p\alpha)$ reaction (COHEN[2]) permits spins to be assigned to two levels of F^{19} and also exhibits the phenomenon of interference between two states of the same spin $(\frac{1}{2})$ but opposite parity.

36. Reactions of protons with F^{19}. The resonant reactions of F^{19} with protons have been studied more, and by a greater variety of methods, than any other nuclear reaction. This is in part due to the high yield of the reactions and in part to the fact that the excitation energy of the compound nucleus is not so high that the levels overlap seriously for low proton energies, nor so low that the number of levels available for study is too limited. The resonances found in the fluorine reactions for proton energies below the (pn) threshold are given in [*32*]; of these resonances some are shared by several competing reactions such as (pp) (pp') $(p\alpha)$ $(p\alpha')$ and $(p\gamma)$, while others are seen only in one or two of these processes. The complexity of the resonance phenomena is shown by the excitation function for gamma rays in the $F^{19}+p$ reaction, observed by WILLARD *et al.*[3] (Fig. 36). There is no complete theory for the way in which the reactions are distributed through the resonances and in particular the usual conservation laws of angular momentum, parity and isotopic spin do not seem sufficient to explain fully all the complex intensity relations which are found. The complete interpretation must await a more detailed theory of the Ne^{20} nucleus; it may then be possible to determine why the transition probability for a particular transition vanishes when it is permitted by all selection rules. The high binding energy of the product system $O^{16}+He^4$ permits several alternative modes of disintegration for the Ne^{20} levels excited in the proton capture process. Four levels in the 6 to 7 MeV region of O^{16} are available in addition to the ground state and of these, one emits electron-positron pairs. The triple correlations in the processes

$$F^{19}+p\rightarrow Ne^{20*}\rightarrow O^{16*}+\alpha\rightarrow O^{16}+\alpha+\gamma \qquad (36.1)$$

have proved a very powerful means of establishing the spins and parities not only of the resonant states of Ne^{20} but also of the O^{16} levels concerned. An internal check of consistency is provided by the $(p\alpha)$ correlations for the ground state transition

$$F^{19}+p\rightarrow Ne^{20*}\rightarrow O^{16}+\alpha \qquad (36.2)$$

which provide, at certain resonances, an alternative route to the same final state.

We consider first the proton elastic and inelastic scattering and the production of capture gamma radiation. This capture radiation is of high energy in contrast with the much more intense and readily observable 6 to 7 MeV radiation resulting from the $(p\alpha\gamma)$ reaction leading to the excited states of O^{16}. The capture radiation was discovered by DEVONS and HEREWARD[4] using the $Cu^{63}(\gamma n)$ reaction as a threshold detector at the 669 keV proton resonance; the most probable transition at this resonance is the $(p\alpha\gamma)$ reaction, but there is a 2%

[1] W. E. BURCHAM, W. M. GIBSON, A. HOSSAIN and J. ROTBLAT: Phys. Rev. **92**, 1266 (1953). — Y. FUJIMOTO and A. HOSSAIN: Phil. Mag. **46**, 542 (1955).

[2] A. V. COHEN: Phil. Mag. **44**, 583 (1953).

[3] H. B. WILLARD, J. K. BAIR, J. D. KINGTON, T. M. HAHN, C. W. SNYDER and F. P. GREEN: Phys. Rev. **85**, 849 (1952).

[4] S. DEVONS and H. G. HEREWARD: Nature, Lond. **162**, 331 (1948).

alternative decay by high energy radiation which has an isotropic angular distribution. Further capture resonances were found by SINCLAIR[1] and by FARNEY *et al.*[2] using sodium iodide scintillation detectors. The energy of the capture radiation was shown to be about 12 MeV by RAE *et al.*[3] using a pair spectrometer and by CARVER and WILKINSON[4] using a deuterium filled ionisation chamber. This

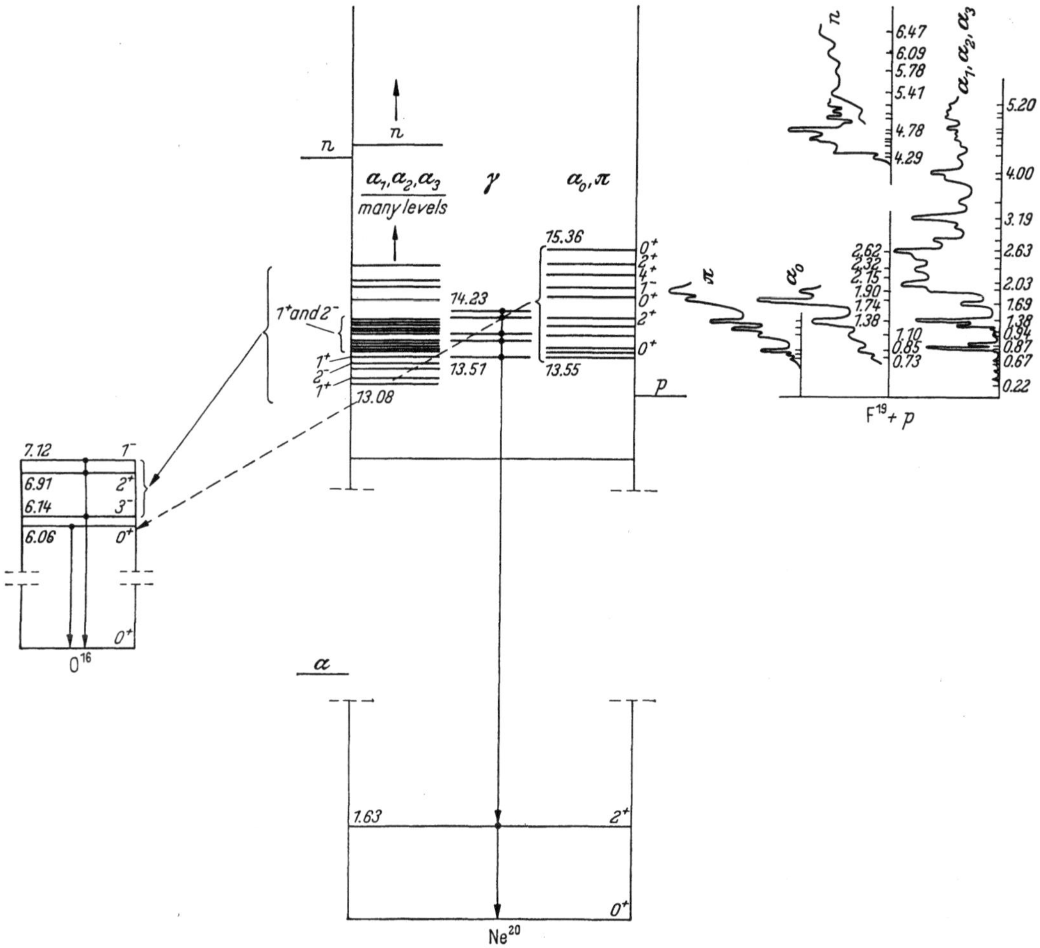

Fig. 36. Resonant reactions of protons with fluorine [32].

indicates a cascade through the 1.63 MeV level of Ne^{20} (Fig. 36); no ground state radiation has been seen. The 1.63 MeV radiation was detected by JONES and WILKINSON[5] and shown to be in coincidence with a high energy radiation. The angular correlation between the components of this cascade was measured by CLEGG *et al.*[6] and found to be

$$W(\vartheta) = 1 - (0.39 \pm 0.08)\cos^2\vartheta.$$

[1] R. M. SINCLAIR: Phys. Rev. **93**, 1082 (1954).
[2] G. K. FARNEY, H. H. GIVIN, B. D. KERN and T. M. HAHN: Phys. Rev. **97**, 720 (1955).
[3] E. R. RAE, J. G. RUTHERGLEN and R. D. SMITH: Proc. Phys. Soc. Lond. A **63**, 775 (1950).
[4] J. H. CARVER and D. H. WILKINSON: Proc. Phys. Soc. Lond. A **64**, 199 (1951).
[5] G. A. JONES and D. H. WILKINSON: Proc. Phys. Soc. Lond. A **65**, 1055 (1952).
[6] A. B. CLEGG, G. A. JONES and D. H. WILKINSON: Proc. Phys. Soc. Lond. A **68**, 538 (1955).

Since the radiating level in Ne^{20} at 13.50 MeV is formed by *s*-wave protons the double correlation analysis can be used and shows that the sequence is $1^+(M1)\ 2^+(E2)\ 0^+$ with a small admixture of $E2$ radiation in the $M1$ transition. There are other levels in Ne^{20} which would be expected, on the basis of spin, parity and estimated proton width to show high energy capture radiation, but radiation from some of these could be discouraged by the electric dipole selection rule for isotopic spin if they were characterised by isotopic spin 0. The high energy radiation at some of these 2^- levels has been sought for unsuccessfully and an upper limit set to the possible mixing of $T=1$ wave functions in the states concerned (WILKINSON and CLEGG[1]).

The elastic scattering of protons by fluorine has been studied by DEARNALEY[2] and by WEBB *et al.*[2] using magnetic analysis. It shows resonance features at energies known to correspond with levels of Ne^{20} from other reactions, and a partial wave analysis[3] permits assignments of spin values to the Ne^{20} states concerned. The experiment also determines the proton width Γ_p at some of the resonances. The resonances for inelastic scattering have also been extensively studied because of their interest for the low states of F^{19}, in particular the levels at 110 keV ($^1/_2{}^-$) and 197 keV ($^5/_2{}^+$). Of these two levels the lower is preferentially excited at the *s*-wave resonances and the latter at the *p*-wave resonances of Ne^{20} (PETERSON *et al.*[4]). The elastic and inelastic proton groups were also observed by GOSSET *et al.*[5]. The excitation function for the gamma radiation from the two excited states of F^{19} was obtained by BARNES[6], who also measured absolute cross sections and angular distributions. Isotropy is found for the 110 keV radiation at all inelastic scattering resonances, suggesting spin $^1/_2$ for the corresponding level, while the distributions for the 197 keV radiation are consistent with a spin $^5/_2{}^+$. The absolute cross sections for gamma ray emission agree with those found by PETERSON *et al.*[4] from measurements of the inelastically scattered protons but in addition there is a finite yield of radiation of both energies between well separated sharp resonances. This is ascribed by BARNES[6] to the COULOMB excitation process and he shows that the $E2$ and $E1$ excitation cross sections ($^1/_2{}^+ \rightarrow {}^5/_2{}^+$) and ($^1/_2{}^+ \rightarrow {}^1/_2{}^-$) are in good agreement with the observed cross section for the $E2$ excitation but much lower than observed for the $E1$ process suggesting a contribution from compound nucleus formation. The angular distributions of the inelastic proton groups to the two fluorine levels were measured by PETERSON *et al.*[4], and the results fit into the general picture of spins and parities shown in Fig. 36. Some of the higher levels of F^{19} have been excited by inelastic scattering of 8 MeV protons, but in these experiments there was insufficient resolution to display the 197 and 110 keV levels [*32*].

In the $F^{19}(p\alpha)$ reaction, a typical resonance level of Ne^{20} breaks up into five channels. The relative intensity of the corresponding groups of alpha particles varies markedly from resonance to resonance. The resonances and widths for some of the low-lying levels have been measured with considerable precision. The excitation functions for gamma radiation of 6 to 7 MeV from the residual state of O^{16} were many years ago followed by STREIB *et al.*[7] and have since been studied by a large number of workers [*32*] including in particular BONNER and

[1] D. H. WILKINSON and A. B. CLEGG: Phil. Mag. **44**, 1322 (1953).

[2] G. DEARNALEY: Phil. Mag. **45**, 1213 (1954). — T. S. WEBB, F. B. HAGEDORN, W. A. FOWLER and C. C. LAURITSEN: Phys. Rev. **99**, 138 (1955).

[3] E. U. BARANGER: Phys. Rev. **99**, 145 (1955).

[4] R. W. PETERSON, C. A. BARNES, W. A. FOWLER and C. C. LAURITSEN: Phys. Rev. **94**, 1075 (1954); **96**, 1250 (1954).

[5] C. R. GOSSET, G. G. PHILLIPS and S. T. EISINGER: Phys. Rev. **98**, 724 (1955).

[6] C. A. BARNES: Phys. Rev. **97**, 1226 (1954).

[7] J. F. STREIB, W. A. FOWLER and C. C. LAURITSEN: Phys. Rev. **59**, 253 (1941).

EVANS[1] and the Oak Ridge group whose excitation function is shown in Fig. 36. Sodium iodide detection has been used in most of the recent work; GEIGER counters or LAURITSEN electroscopes in the earlier studies. The energy of the gamma radiation from the $(p\alpha\gamma)$ reaction was measured by WALKER and McDANIEL[2] using a pair spectrometer and a thick target. They confirmed that the gamma ray energy did not vary with proton energy, as is required by the scheme shown in Fig. 36. Two lines of energy 6.13 and (6.98 ± 0.07) MeV were identified for a proton energy of 1.15 MeV. Gamma ray energies were determined at individual resonances by RASMUSSEN *et al.*[3] using a magnetic lens spectrometer, and the energies at certain resonances agree well with those found by WALKER and McDANIEL. At other resonances, however, the 7 MeV line was found to show an apparent change of energy, which was consistent with the observation of BURCHAM and FREEMAN[4] that the alpha particles emitted from Ne^{20} in transitions to the 7 MeV state of O^{16} actually fell into two groups whose intensity varied from resonance to resonance. The short range alpha particle resonances $(p, \alpha_1\,\alpha_2\,\alpha_3)$ and their association with the corresponding resonances for 6 MeV gamma radiation were demonstrated by BURCHAM and DEVONS[5] and by McLEAN *et al.*[6] using magnetic analysis to separate the alpha particles from the protons scattered from the targets. This work was extended by CHAO *et al.*[7] to several higher resonances; the large variation in relative intensity of all the low energy alpha groups, including that to the pair emitting 6.06 MeV level of O^{16} was clearly shown. The association of the low energy alpha particles with the 6 to 7 MeV gamma radiation was confirmed by the agreement between the sum of the Q-value for the $(p\;\alpha\;\gamma)$ reaction and the energy of the corresponding gamma radiation with the accurately determined energy release in the overall reaction $F^{19}(p\,\alpha)\,O^{16}$ (STRAIT *et al.*[8]).

STREIB *et al.* also found a number of resonances for the emission of the ground state alpha particles (α_0); evidence for this had been seen in the earlier work of BURCHAM and DEVONS. Higher resonances of this type have been reported [*32*]. STREIB *et al.* further demonstrated resonances for the production of a soft radiation which has proved to be electron-positron pairs (and accompanying annihilation radiation) from the first excited state of O^{16} with spin and parity 0^+. The associated short range alpha particles (α_π) are seen in the magnetic spectra of CHAO *et al.* and vary sharply in intensity from one resonance to another. Since the pair emitting state of O^{16} has the same characteristics as the ground state, the ground state alpha particles α_0 and the short range alpha particles α_π would be expected to show the same resonances; this is approximately so, although the 12.36 keV pair-resonance is not found for long range alpha particles. The excitation curve for the $(p\,\alpha\,\pi)$ reaction has recently been studied by PHILLIPS and HEYDENBURG[9] using anthracene scintillation counters in coincidence to detect the electrons; the pair resonances are all wide compared with the $(p\,\alpha\,\gamma)$ resonances at similar energies.

---

[1] T. W. BONNER and J. E. EVANS: Phys. Rev. **73**, 666 (1948).

[2] R. L. WALKER and B. D. McDANIEL: Phys. Rev. **74**, 315 (1948).

[3] V. K. RASMUSSEN, W. F. HORNYAK, C. C. LAURITSEN and T. LAURITSEN: Phys. Rev. **77**, 617 (1950).

[4] W. E. BURCHAM and J. M. FREEMAN: Phys. Rev. **75**, 1756 (1949).

[5] W. E. BURCHAM and S. DEVONS: Proc. Roy. Soc. Lond., Ser. A **173**, 555 (1939).

[6] W. B. McLEAN, R. A. BECKER, W. A. FOWLER and C. C. LAURITSEN: Phys. Rev. **55**, 796 (1939).

[7] C. Y. CHAO, A. V. TOLLESTRUP, W. A. FOWLER and C. C. LAURITSEN: Phys. Rev. **79**, 108 (1950).

[8] E. N. STRAIT, D. M. VAN PATTER, W. W. BUECHNER and A. SPERDUTO: Phys. Rev. **81**, 747 (1951).

[9] G. C. PHILLIPS and N. P. HEYDENBURG: Phys. Rev. **83**, 184 (1951).

The angular distributions in the $(p\,\alpha\,\gamma)$ reaction have been exhaustively studied. For the resonances below 900 keV, DEVONS and HINE [23] found isotropic distributions for all resonances except those at 598 and 874 keV where there is a distribution of the form $1+A\cos^2\vartheta$. Spherical symmetry for the gamma radiation at the 330 keV resonance was reported by JACOBS *et al.*[1] and for the short range alpha particles by VAN ALLEN and SMITH[2]. The angular distributions for the 6 and 7 MeV components at some of the higher resonances where they are produced in comparable amount, were obtained by SANDERS[3] using a deuterium filled ionisation chamber. Angular distributions for the associated short range alpha particles are more difficult to obtain since they must be resolved magnetically from the scattered protons, and then the magnetic analyser itself must be rotated about the direction of the proton beam. This has however been done by PETERSON *et al.*[4] and the correlations are useful in assigning spins to the Ne^{20} states, in particular the 935 keV 1^+ resonance.

The theoretical treatment of the $(p\alpha)$ and $(p\,\gamma)$ angular distributions has been given in full by CHAO[5] at the time of whose work the experimental information on the fluorine reactions was somewhat incomplete, so that formulae for many possible alternative level schemes were given. The assignment of quantum numbers is aided by the consideration that it is necessary to forbid long range alpha particle emission from the $(p\,\alpha\,\gamma)$ resonances in general, which can be achieved if these resonances have odd J with even parity or vice versa, e.g. 0^- 1^+ 2^- etc. CHAO applied his formulae to the angular distributions of the long range α_0 groups studied by the photographic emulsion technique by RUBIN[6]; these distributions contain odd powers of $\cos\vartheta$, and CHAO concluded that the best fit was obtained under the assumption, (consistent with the independent particle model) of even parity for F^{19}. He also suggested the spin value 3^- for the 6.13 MeV level of O^{16}.

The calculations of CHAO were given in different form by SEED and FRENCH[7] who also give explicit formulae for the (3 stage) angular correlation between the short range alpha particles and the subsequent gamma radiation in the reaction $(p\,\alpha\,\gamma)$ for a variety of combinations of spin values for the Ne^{20} and O^{16} states concerned. The striking angular correlation obtained by BARNES, FRENCH and DEVONS for the 330 keV resonance level has already been discussed (see Sect. 27 and Fig. 27). The same type of observation was extended to higher resonances by SEED and FRENCH using a magnetic analyser to select the particular group of alpha particles $(\alpha_3\,\alpha_2\,\alpha_1)$ required. In this very detailed experiment, the assignment of 3^- to the 6.1 MeV level in O^{16} and of 1^+ to the compound state in Ne^{20} at the 330 keV resonance was adopted as a reference point. Spin values for the higher resonances in Ne^{20} were then deduced from the $(\alpha_1\,\gamma)$ correlation patterns. With this knowledge the $(\alpha_2\,\gamma)$ and $(\alpha_3\,\gamma)$ correlations enabled spin values to be assigned to the 6.9 and 7.1 MeV levels in O^{16}. The results of this study, which is one of the best examples of the application of angular correlation theory to nuclear reactions, are shown as spin values in Fig. 36; Typical angular correlation patterns are given in Fig. 37, together with a sketch of the apparatus with which they were observed.

Many auxiliary observations to support deductions based on the angular correlation experiments have been made. The polarisation of the 6.13 MeV gamma

[1] J. A. JACOBS, P. R. MALMBERG and J. S. WAHL: Phys. Rev. **73**, 1130 (1948).
[2] J. A. VAN ALLEN and N. M. SMITH: Phys. Rev. **59**, 501 (1941).
[3] J. H. SANDERS: Phil. Mag. **44**, 1302 (1953).
[4] R. W. PETERSON, W. A. FOWLER and C. C. LAURITSEN: Phys. Rev. **96**, 1250 (1954).
[5] C. Y. CHAO: Phys. Rev. **80**, 1035 (1951).
[6] S. RUBIN: Phys. Rev. **72**, 1176 (1947).
[7] J. E. SEED and A. P. FRENCH: Phys. Rev. **88**, 1007 (1952).

radiation at the 330 keV level was measured by FRENCH and NEWTON[1] and indicated electric octupole radiation. These authors used the sensitivity of the COMPTON scattering process to the electric vector as a means of observing polarisation and studied the direction of polarisation with respect to the alpha particle in the $(p\,\alpha\,\gamma)$ reaction. FAGG and HANNA[2] used the photodisintegration of deuterium as a polarisation sensitive detector and obtained the polarisation of the gamma radiation with respect to the incident proton beam. Their observations agree well with the polarisations predicted by the assumed spins for a 2^- state in Ne^{20} ($E_p = 0.87$ MeV), 3^-, 2^+ and 1^- states in O^{16} as shown in Fig. 36. The order of the 6.13 MeV gamma transition has been established directly by

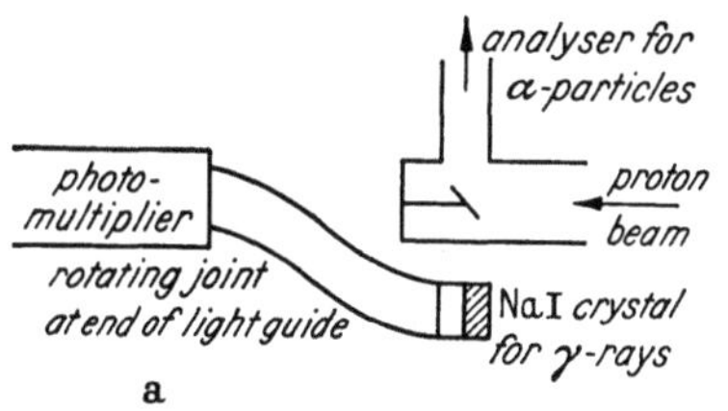

a

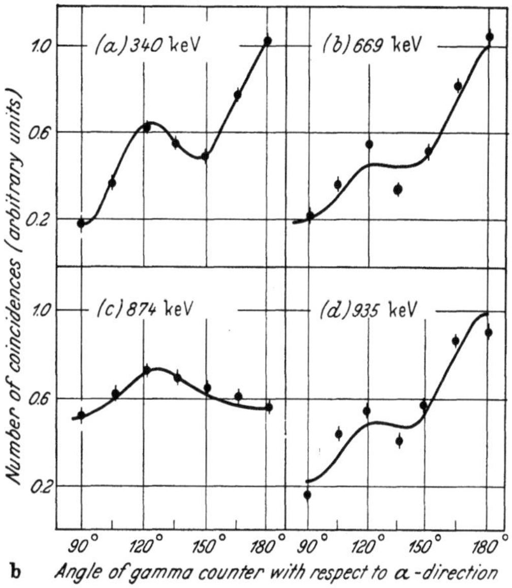

b

Fig. 37 a and b. α—γ angular correlation patterns in the F^{19} $(p\alpha\gamma)$ reaction (SEED and FRENCH). (a) Apparatus; (b) results for several resonance energies.

DEVONS and GOLDRING [25] as $\Lambda = 3$ by measurements of the internal pair angular correlation and the half-life of the level has been set as $5 \times 10^{-12} < \tau < 10^{-11}$ sec (Sect. 24 and Table 3).

The angular correlation between the internal pairs from the $(p\,\alpha\,\pi)$ resonance at 840 keV was observed by DEVONS *et al.* [25], and the half-life of the pair emitting level established as $(5.0 \pm 0.5) \times 10^{-11}$ sec by the recoil distance method (Sect. 24). It is clear that the fluorine reactions have provided an unusually extensive amount of information on the properties of the levels involved, but some difficulties remain. Notable among these is the small proton width and consequent low intensity of capture radiation at the 330 and 935 keV levels of Ne^{20} which seem to have the same properties and to show the same $(\alpha\,\gamma)$ correlations as the 669 keV resonance which gives capture radiation.

Resonances above threshold are observed in the F^{19} $(p\,n)$ reaction; part of the excitation curve for neutrons in this reaction may be seen in Fig. 36.

37. Reactions of protons with Ne^{20}, Ne^{21}, Ne^{22}. The resonant scattering of protons by Ne^{20} has been studied, but full details are not yet available. Resonances are found for both inelastic and elastic processes for protons of energy up to 4 MeV; and the energies, yield and half-widths are listed in [33].

The inelastic scattering chiefly excites the 1.63 MeV level of Ne^{20}, although higher levels have also been seen ([32] and FREEMANTLE *et al.*[3]). These authors

[1] A. P. FRENCH and J. O. NEWTON: Phys. Rev. **85**, 1041 (1952).
[2] L. W. FAGG and S. S. HANNA: Phys. Rev. **92**, 372 (1953).
[3] R. G. FREEMANTLE, D. J. PROWSE, A. HOSSAIN and J. ROTBLAT: Phys. Rev. **96**, 1270 (1954).

used protons of fixed energy from a cyclotron; absolute differential cross sections for elastic scattering and for inelastic scattering to the 1.63 MeV level are given by the latter for 9.5 MeV protons. Some of the results are shown in Fig. 38 in which the characteristic Coulomb—nuclear interference is seen in the elastic distribution, and symmetry about 90° for the inelastic group. The excitation of the Na^{21} nucleus for 9.5 MeV protons is about 12 MeV, and it is probable that the level density is there already so large compared with the experimental resolution of about $\frac{1}{2}$ MeV that specific level structure, which would almost certainly result in an asymmetric distribution, is smoothed out. On the other hand, the inelastic distribution for argon (Freemantle *et al.*) is strongly asymmetric and in this case the nucleus is heavier and the excitation much greater, so that factors other than random phase may determine the question of symmetry. Direct interaction processes of the type invoked by Austern *et al.*[1] in their discussion of the (np) reaction may be important.

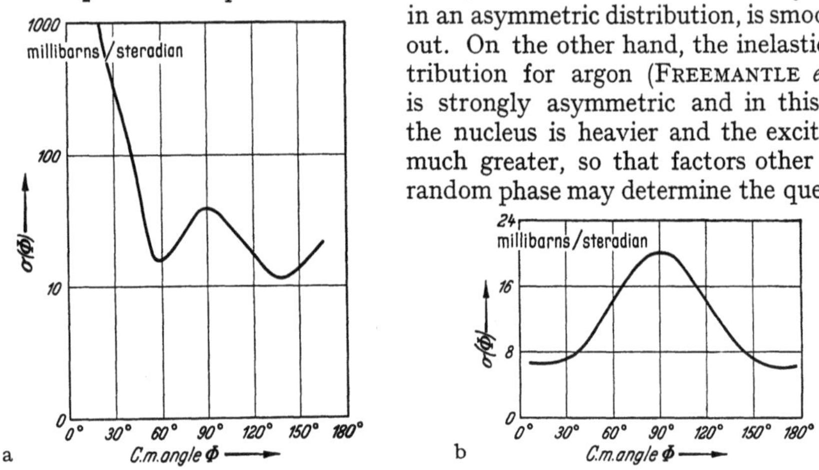

Fig. 38 a and b. Angular distribution of 9.5 MeV protons scattered by neon. (a) Elastic group; (b) inelastic group (1.63 MeV level) (Freemantle *et al.*).

The elastic and inelastic scattering of protons by Ne^{22} has also been studied and information is given in [33]. The level chiefly excited for protons of energy up to 4 MeV is at 1.28 MeV in Ne^{22} and the corresponding gamma radiation has been observed.

The $(p\gamma)$ reaction has also been observed in Ne^{20}, Ne^{21} and Ne^{22} by Broström *et al.*[2] using isotopic targets prepared by bombarding a foil with neon ions in a mass separator. One resonance is found for Ne^{20} and Ne^{21}, and 20 resonances for Ne^{22} for proton energies between 0.5 and 1.5 MeV. The only other reactions of the neon isotopes with protons which have been reported are the (pn) processes leading to Na^{20}, Na^{21} and Na^{22} with thresholds of 16.9, 4.3 and 3.6 MeV respectively [33].

38. Reactions of protons with the nuclei from sodium to calcium, $Z = 11$ to $Z = 20$. The nuclei of the elements from sodium to calcium react with protons in the same general way as do the lighter nuclei, but there are differences due to the higher Coulomb barrier and the greater number of nucleons per nucleus. The barrier effect reduces partial widths for charged particle emission and resonance levels are very narrow, so that direct determination of width below the (pn) threshold becomes difficult. The greater nuclear complexity usually results in a greater level density at comparable excitation than in a light nucleus and the number of resonance levels observed increases with A until the decreasing proton width so reduces their strength that they are unobservable for a given proton energy and beam spread. Elastic and inelastic scattering, and the $(p\alpha)$,

[1] N. Austern, S. T. Butler and H. McManus: Phys. Rev. **92**, 350 (1953).
[2] K. J. Broström, T. Huus and J. Koch: Nature, Lond. **160**, 498 (1947).

$(p\gamma)$ and (pn) reactions have been observed for most of the stable nuclei in the range $Z=11$ to $Z=20$. The experimental results may be classified into (α) those in which emphasis is placed on the excitation of bound states, including the ground state, and on the measurement of Q-values for transitions to these states and absolute differential cross sections for the corresponding particle groups; (β) those in which emphasis is on the formation of virtual levels of a compound state and investigation of the various competing modes of decay for discrete levels. As in the case of the lighter nuclei the techniques used in these two types of experiment may differ considerably.

α) Excitation of bound levels. Table 5 lists the Q-values for ground state transitions in the proton reactions of nuclei between sodium and calcium. Details of individual observations are given in [*33*].

Absolute differential cross sections for elastic scattering of protons of 7 to 9 MeV by Mg^{24} and Al^{27} [*33*] show interference between COULOMB and nuclear scattering, but phase shift analyses have not been worked out. With 22 MeV protons COHEN and NEIDIGH[1] find very definite diffraction maxima and minima for the elastic scattering by many elements, and a definite correlation of the diffraction peaks with the value of $\lambda\!\!\!^{-}/R$. Fig. 39 shows their results for aluminium.

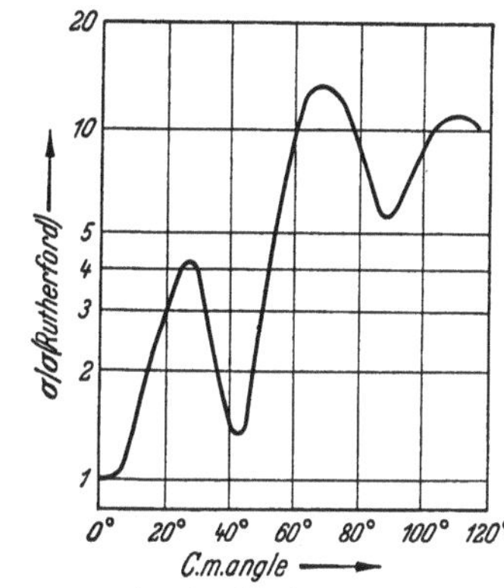

Fig. 39. Angular distribution of 22 MeV protons elastically scattered by aluminium (COHEN and NEIDIGH).

The angular distributions of inelastically scattered protons from Mg^{24} have been reported by several workers, including GOVE and STODDART[2], using protons of energy 7 to 9 MeV. The angular correlation between the inelastically scattered protons to the first excited state of Mg^{24} at 1.38 MeV and the gamma radiation was studied by GOVE and HEDGRAN[3]. Analysis of such distributions, in which it can probably be assumed that the intermediate nucleus Al^{25} $(\rightarrow Mg^{24*}+p\rightarrow Mg^{24}+ +p+\gamma)$ is randomly distributed, owing to the high bombarding energy, should confirm the quadrupole nature of the radiative transition. This has already been established by measurements of the pair internal conversion coefficient (Sect. 2). The variation of the differential cross section for excitation of the 1.38 MeV level of magnesium by protons has been found by GREENLEES[4] for the energy range 4 to 7 MeV and shows broad resonances, which may be a superposition of sharp resonant peaks of the type observed at lower energies. The inelastically scattered protons from aluminium bombarded by high energy (18.3 MeV) protons were shown by GUGELOT[5] to have the distribution predicted by continuum theory. Inelastic scattering was established for 96 MeV protons incident on carbon and sulphur by STRAUCH and TITUS[6].

The angular distribution of neutrons from the (pn) reaction with 22 MeV protons incident on Mg^{24} and Al^{27} is given by COHEN[7]. The neutron spectrum from $Al^{27}(pn)$ with 18 MeV protons was used by GUGELOT[8] to obtain the relative level density of Si^{27}.

[1] B. L. COHEN and R. V. NEIDIGH: Phys. Rev. **93**, 282 (1954).
[2] H. E. GOVE and H. F. STODDART: Phys. Rev. **86**, 572 (1952).
[3] H. E. GOVE and A. HEDGRAN: Phys. Rev. **86**, 574 (1954).
[4] G. W. GREENLEES: Proc. Phys. Soc. Lond. A **68**, 97 (1955).
[5] P. C. GUGELOT: Phys. Rev. **93**, 425 (1954).
[6] K. STRAUCH and W. F. TITUS: Phys. Rev. **95**, 854 (1954).
[7] B. L. COHEN: Phys. Rev. **98**, 49 (1955).
[8] P. C. GUGELOT: Phys. Rev. **81**, 51 (1951).

β) Excitation of virtual states. The reactions by which virtual states have been excited are included in Table 5; for the most part observations have been made with electrostatic generators and have been confined to determination of resonance energies. In many cases, widths have not been tabulated and only in a few instances have reduced widths been evaluated. The resonance energies are given by ENDT and KLUYVER [*33*]; in this section only certain features of particular general interest are referred to.

Because of the high negative Q-values for the (pn) reactions (Table 5), particularly for the more abundant isotopes, resonances in these reactions have been found only under conditions of relatively poor resolution, e.g. with cyclotron beams (BLASER *et al.*[1]). Above the (pn) threshold the total width of resonance levels is in any case expected to increase since neutron emission is not hindered by the COULOMB barrier, and most of the narrow resonances observed are seen below this energy.

Many resonance levels decay by emission of elastic protons, inelastic protons, gamma radiation and alpha particles in competition. A striking feature of the extensive level systems found in reactions such as $Na^{23}(px)$ and $Al^{27}(px)$ is the fact that for some levels the emission of a particular radiation, for instance alpha particles, is apparently rigorously forbidden. This can be explained in those cases in which the final state is an alpha particle together with an even-even nucleus by assuming that the resonance level is a single level with J odd, π even or J even, π odd. Similarly the decay of levels of zero spin by single quantum emission to other states of zero spin is not allowed. The decay of virtual states into even systems such as $He^4 + Ne^{20}$ may also be hindered if the resonant state has isotopic spin 1. This however is unlikely to be a strong selection rule for the highly excited states involved (see Sect. 9).

Certain cases in which emission of a particle appears to be forbidden may be explained by low barrier penetrability for the orbital momentum required. The widths of the levels vary between 0.1 keV and 50 keV and the spacing does not decrease rapidly with increasing excitation energy.

The gamma radiation observed from the resonances may be full energy capture radiation to the ground state of the residual nucleus. More generally it is cascade radiation though a series of bound levels and often there are strong lines connected with bound states of nuclei excited in the (pp') and $(p\alpha')$ reactions. Scintillation and pair spectrometry have been used to obtain energy spectra at particular resonances, and intensities of competing transitions have been estimated. Cascade schemes have been verified by gamma-gamma coincidence methods. Angular distribution and correlation measurements of these cascades and for inelastically scattered particles enable spins and parities to be assigned and throw light on the validity of LS or jj coupling for the reaction.

The most prolific nuclei for these studies are those of mass $(4n-1)$ such as Na^{23}, Al^{27} and P^{31}. Proton capture in this type of nucleus leads to an excitation of about 10 MeV in the compound state at which the level spacing may only be about 30 keV.

Resonant elastic scattering of protons by Mg^{24}, observed by MOORING *et al.*[2], was analysed by KOESTER[3] to indicate levels in the compound nucleus Al^{25}. The reduced widths obtained from the observed level widths fall into two groups,

---

[1] J. P. BLASER, F. BOEHM, P. MARMIER and P. SCHERRER: Helv. phys. Acta **24**, 465 (1951).

[2] F. P. MOORING, L. J. KOESTER, E. GOLDBERG, D. SAXON and S. G. KAUFMANN: Phys. Rev. **84**, 703 (1951).

[3] L. J. KOESTER: Phys. Rev. **85**, 643 (1952).

one of about 20% of the WIGNER limit, and the other much smaller. The level characteristics of the group with large reduced widths agree with the sequence of single particle states of the shell model. The other group, of narrow levels, probably involve a type of core excitation, perhaps as a result of pairing of nucleons in some of the higher states at the expense of the partly filled $d_{5/2}$ shell. The capture gamma radiation in this reaction has been studied for the lower resonances.

The resonant reactions of sodium with protons are typical of a case in which no one reaction is much stronger than all others. Fifty-two resonances for alpha particle emission or gamma ray emission or both in the proton energy range 1.0 to 2.5 MeV are tabulated by STELSON and PRESTON[1], who give the observed widths. As in the case of the F^{19} reactions, the intensity of gamma radiation from the $(p\,\alpha\,\gamma)$ and $(p\,p')$ reactions is much greater than that of capture radiation. The analysis of the widths of resonances in the energy range 0.7→1.5 MeV into partial widths for the $(p\,\gamma)$, $(p\,\alpha\,\gamma)$ and $(p\,p)$ processes is given by NEWTON[2] who assumes that the main contribution to each level width arises from the re-emission of protons. Spin assignments for some of the levels of Mg^{24} are suggested by angular correlations and distributions obtained in this work and also by the alpha particle distributions measured by STELSON[3] for the ground state transition in the $(p\,\alpha)$ reaction. Interference phenomena were observed in this work and unknown parameters enter the theoretical analysis. The gamma ray cascades from resonance levels excited by protons of energy below 700 keV were analysed by FLACK *et al.*[4] using a single crystal scintillation spectrometer and for the 310 keV resonance by CARLSON *et al.*[5] with a three crystal instrument. Ground state transitions were not observed at any resonance but cascades through the well known 1.38 MeV and 4.24 MeV levels of Mg^{24} were always found and confirmed by coincidence counting. The order of the cascade was established because the energy of the highest energy gamma ray changed from resonance to resonance. Angular distributions of, and correlations between, the gamma rays were measured by GRANT *et al.*[6] and lead to the spin and parity values shown in Fig. 99, when taken in conjunction with the knowledge that certain of the resonance levels do not emit alpha particles to the ground state of Ne^{20}. The interpretation suggests values of the ratio of channel spin 2 to channel spin 1 which favour LS coupling for at least three resonances. GRANT *et al.* have also compared their values of $|M|^2$ with those predicted by the WEISSKOPF formula (Sect. 19) for the multipolarity established by the correlation experiments, and they conclude that both $E1$ and $M1$ transitions are much slower, by a factor of 10^3 to 10^4. The capture of protons by Mg^{24} has also been studied by scintillation spectrometry.

Similar reactions are observed between aluminium and protons, and the analysis follows the same lines. The elastically scattered protons, inelastically scattered protons, two groups of alpha particles from the $(p\alpha)$ reaction and gamma radiation all show resonances at $\vartheta = 165°$ (SHOEMAKER *et al.*[7] 1951, see also Fig. 46) for proton energies up to 4 MeV, some of the levels being still sharp

[1] P. H. STELSON and W. M. PRESTON: Phys. Rev. **95**, 974 (1954).
[2] J. O. NEWTON: Phys. Rev. **96**, 241 (1954).
[3] P. H. STELSON: Phys. Rev. **96**, 1584 (1955).
[4] F. C. FLACK, J. G. RUTHERGLEN and P. J. GRANT: Proc. Phys. Soc. Lond. A **67**, 973 (1954).
[5] R. R. CARLSON, E. H. GEER and E. B. NELSON: Phys. Rev. **94**, 1311 (1954).
[6] P. J. GRANT, J. G. RUTHERGLEN, F. C. FLACK and G. W. HUTCHINSON: Proc. Phys. Soc. Lond. A **68**, 369 (1955).
[7] F. C. SHOEMAKER, J. E. FAULKNER, G. M. B. BOURICIUS, S. G. KAUFMANN and F. P. MOORING: Phys. Rev. **83**, 1011 (1951).

even at the highest excitation investigated (15 MeV) in the compound nucleus Si^{28}. The gamma radiation at resonances above 2 MeV is mainly from excited states of the residual nuclei Mg^{24} and Al^{27}. The widths observed in this work are experimental, since the proton width is still only of the same order as the experimental resolution for proton energies of about 2 MeV and alpha particle widths are still less. Real widths in the Al^{27} reaction may have been measured by HUNT and JONES[1] who quote resonance energies near 300 keV with a precision of 0.1% and give widths of (0.7 ± 0.3) keV for two of the levels observed by means of the $(p\gamma)$ reaction. The spectrum of capture radiation was determined by RUTHERGLEN *et al.*[2] at several of the resonances using scintillation spectrometry, and shows again that there are no direct transitions to the Si^{28} ground state, but only cascades through the first excited state (2^+) at 1.8 MeV. The angular distributions of the more prominent components were measured. These can all be described by a formula $W(\vartheta) = 1 + A\cos^2\vartheta$ so that no proton waves with $l > 1$ seem necessary to reach the Si^{28} levels near 12 MeV; this can be understood because of the high spin of the target nucleus Al^{27}. The analysis takes account of the fact that alpha particle emission competes at some but not all of the resonances. The angular distributions can in general be fitted by a mixture of channel spins 3 and 2 in nearly the ratio suggested by the LS coupling scheme (GRANT[3]). The suggested spin and parity assignments for one transition have been confirmed by a polarisation measurement using the photoelectric disintegration of the deuteron as a polarisation sensitive indicator[4].

In the nuclei between $Z = 11$ and $Z = 20$ remaining to be discussed resonances have been observed in the following reactions.

$Mg^{25}(p\gamma)Al^{26}$; $Mg^{26}(p\gamma)Al^{27}$
$Si^{29}(p\gamma)P^{30}$; $Si^{30}(p\gamma)P^{31}$
$P^{31}(p\gamma)S^{32}$
$S^{32}(pp)S^{32}$; $S^{34}(p\gamma)Cl^{35}$
$Cl^{35}(p\gamma)A^{36}$; $Cl^{37}(p\gamma)A^{38}$; $Cl^{37}(p\alpha)S^{34}$; $Cl^{37}(pn)A^{37}$

Although a large number of resonance levels has been found, little has so far been done which compares in usefulness with the studies made on the Na^{23}, Mg^{24}, Mg^{25} and Al^{27} reactions. Many of the lower resonances of the nuclei listed above were discovered by TANGEN[5]; further details of his work and that of later workers are given by ENDT and KLUYVER [*33*].

The $(p\gamma)$ reactions in Mg^{25} and Si^{29} have been used to determine the excitation of the first level with isotopic spin 1 in the nuclei Al^{26} and P^{30}, in which this level lies near the ground state (KAVANAGH *et al.*[6], ENDT *et al.*[7]). In Al^{26} the $T = 0$ ground state has a high spin and a very long life, and is located by observation of capture radiation. The excited state decays by positron emission to Mg^{26} with a period of 6.7 sec, and resonances for production of this activity do not always coincide with those for the ground state transition, thus establishing the ground state to be 220 keV lower than the positron emitting state.

---

[1] S. E. HUNT and W. M. JONES: Phys. Rev. **89**, 1283 (1953).
[2] J. G. RUTHERGLEN, P. J. GRANT, F. C. FLACK and W. M. DEUCHARS: Proc. Phys. Soc. Lond. A **67**, 101 (1954).
[3] P. J. GRANT: Proc. Phys. Soc. Lond. **67**, 392 (1954).
[4] I. S. HUGHES and P. J. GRANT: Proc. Phys. Soc. Lond. **67**, 481 (1954).
[5] R. TANGEN: Kgl. norske Vid. Selsk., Skr. **1946**, No. 1.
[6] R. W. KAVANAGH, W. R. MILLS and R. SHERR: Phys. Rev. **97**, 248 (1955).
[7] P. M. ENDT, J. C. KLUYVER and C. VAN DER LEUN: Phys. Rev. **95**, 580 (1954).

II. Reactions of neutrons with light nuclei.

The present section excludes discussion of the two- and three-body problems, and reduces to a minimum the treatment of resonant processes, since these topics are dealt with in other articles in this Encyclopedia (Vol. XXXIX and XL). General information on neutron induced reactions is contained in references [*6*] and [*7*] and total cross sections are given by ADAIR and by HUGHES and HARVEY [*35*]. The reactions observed are listed in Table 7.

39. Reactions of neutrons with H^1, H^2, H^3, He^3 and He^4. The scattering and radiative capture of neutrons by protons are fundamental reactions and have been very fully studied. The radiative capture of neutrons by deuterons has a cross section of $(5.7 \pm 1.0) \times 10^{-28}$ cm² for thermal velocities[1] in agreement with theory [*36*]. No bound state of H^4 is expected theoretically to exist[2] and there is no evidence for the formation of any such state in any reaction.

Differential cross sections for the scattering of neutrons by deuterons have been measured at several energies up to 14 MeV[2]. Results are summarised by ALLRED *et al.*[3] and by MASSEY [*36*] who compare them with those for the similar (pd) reaction and with theory[4]. There is evidence for the deuteron disintegration $H^2(n, p)\,2n$. Total cross sections are also known as a function of energy.

The $He^3(np)$ reaction has a cross section of about 5000 barns for thermal neutrons and the reaction cross section follows a $1/v$-law up to neutron energies of about 3 MeV[5]. The observed value agrees with that calculated from the cross section of the inverse reaction (Table 4).

The scattering of neutrons by He^4 is the mirror reaction to the $(p\alpha)$ scattering process discussed in Sect. 29. The variation of total cross section with energy shows a resonance maximum at 1.15 MeV (BASHKIN et *al.*[6]). Angular distributions of the scattered neutrons were measured by ADAIR[7] at several energies between 0.4 and 2.73 MeV, by SEAGRAVE[8] at energies between 2.6 and 14.3 MeV and by ALSTON *et al.*[9] at 15.7 MeV. In the observations at the highest energies, there is evidence for a *d*-wave phase shift; at lower energies, the phase shift analysis also follows that discussed for the $(p\alpha)$reaction and can be described by the same level parameters (ADAIR[7]), as would be expected from the principle of charge independence. The *s*-wave phase shift has been calculated by BRANSDEN and McKEE[10].

40. Excitation of bound states by radiative capture of slow neutrons. Radiative capture is mainly important for thermal neutrons for which there is the technical advantage that very large intensities are available from reactors. The energy spectrum of thermal capture radiations has been observed by KINSEY and his collaborators [*37*] for a large number of nuclei. In this work the neutron capturing sample was placed in a high flux near the core of a nuclear reactor, and collimated

---

[1] L. KAPLAN, G. R. RINGO and K. E. WILZBACH: Phys. Rev. **87**, 785 (1952).
[2] P. SWAN: Proc. Phys. Soc. Lond. A **66**, 1066 (1953).
[3] J. D. SEAGRAVE: Phys. Rev. **97**, 757 (1955). — J. C. ALLRED, A. H. ARMSTRONG and L. ROSEN: Phys. Rev. **91**, 90 (1953).
[4] R. S. CHRISTIAN and J. L. GAMMEL: Phys. Rev. **91**, 100 (1953).
[5] J. H. COON: Phys. Rev. **80**, 488 (1950).
[6] S. BASHKIN, F. P. MOORING and B. PETREE: Phys. Rev. **82**, 378 (1951).
[7] R. K. ADAIR: Phys. Rev. **86**, 155 (1952).
[8] J. D. SEAGRAVE: Phys. Rev. **92**, 1222 (1953).
[9] M. H. ALSTON, A. V. CREWE, W. H. EVANS, L. L. GREEN and J. C. WILLMOTT: Proc. Phys. Soc. Lond. A **67**, 657 (1954).
[10] B. H. BRANSDEN and J. S. C. McKEE: Phil. Mag. **45**, 869 (1954).

gamma radiation from the sample was studied with a pair spectrometer with a line width of about 100 keV; Fig. 40a shows the experimental arrangement.

The results for the light elements studied are represented in the energy level diagrams of Fig. 41; Fig. 40b shows a typical pair spectrum observed. From a thorough analysis of the instrumental factors it is possible to deduce the actual intensity of a given line as a percentage of the total number of neutrons captured.

The results are analysed taking into account the fact that absorption of a thermal neutron can cause no change in parity. This suggests for Be^{10} (Fig. 41) that the ground state and excited state capture transitions are both $E1$ with capture taking place from states of spin 1^- or 2^- in the continuum. In C^{13} the observed transitions can be similarly analysed, taking into account other information on the spins of levels. The ground state transition and the other capture gamma ray (energy too low to be observed in the pair spectrometer), should be $E1$ and the 3.68 MeV radiation $M1$ or $E2$. The radiation needed to excite the 3.86 MeV level, and the subsequent radiation would both have to be $M2$. The 3.09 MeV state (Fig. 68) is a single particle s-wave state and the absorption of a thermal neutron can provide no change of orbital momentum to this state so that no capture radiation should be found. In N^{15} nine radiations were identified, including the ground state transition. They can be fitted into the known level spectrum of the N^{15} nucleus, and show that only certain of these levels are excited in observable degree by radiative capture, presumably because of unfavourable spin changes.

Only one capture gamma ray is observed in F^{20}, corresponding to the direct ground state transition, but its intensity suggests that other radiations, of lower energy, are also present. The capture radiations from Na^{24}, Mg^{25} and Mg^{27} are more complex but can be fitted into the level schemes for these nuclei known from (dp) and $(d\alpha)$ reactions. The gamma radiation from Al^{28} shows a very large number of levels again in accord with the high level density of this nucleus known from the $Al^{27}(dp)\,Al^{28}$ reaction. For the silicon isotopes the radiation spectrum is again simpler. Of these elements F^{20} and Al^{28} show strong ground state transitions, with the energy of the radiation equal to the neutron binding energy in the isotope, while Na^{24}, Mg^{25} and Si^{29} have only weak ground state radiations. For sodium, this may perhaps be due to the necessity for quadrupole radiation to the ground state, and for the other elements, shell model arguments may make the absence of the ground state transition plausible.

The same genera pattern of capture gamma rays is found in the heavier nuclei P^{32}, S^{33}, Cl^{36} and K^{40}. The neutron binding energies deduced from the ground state transitions are shown in Table 7. The individual gamma ray lines, whose relative intensities were measured, can in each case be associated with level spectra derived from the (dp) reaction, which leads to the same radiating nucleus. In cases in which the (dp) reaction is not known the $(n\gamma)$ lines lead to a level system. Completion of the picture of capture gamma radiation requires extensive studies with a detector not limited by a low energy cut-off, as is the pair spectrometer. Work by HAMERMESH and HUMMEL[1] and by REARDON *et al.*[2] suggests that scintillation spectrometry may be applied; these authors found gamma rays of energies 0.78, 1.15, 1.59 and 2.0 MeV in Cl^{36} and all of these may be combined with higher energy pair spectrometer lines to give the same neutron binding energy. This is shown in Fig. 41, in which the interlocking of the two sets of observations is clear. An extensive study of neutron capture gamma ray

[1] B. HAMERMESH and V. HUMMEL: Phys. Rev. **88**, 916 (1952).
[2] W. A. REARDON, R. W. KRONE and R. STUMP: Phys. Rev. **91**, 334 (1953).

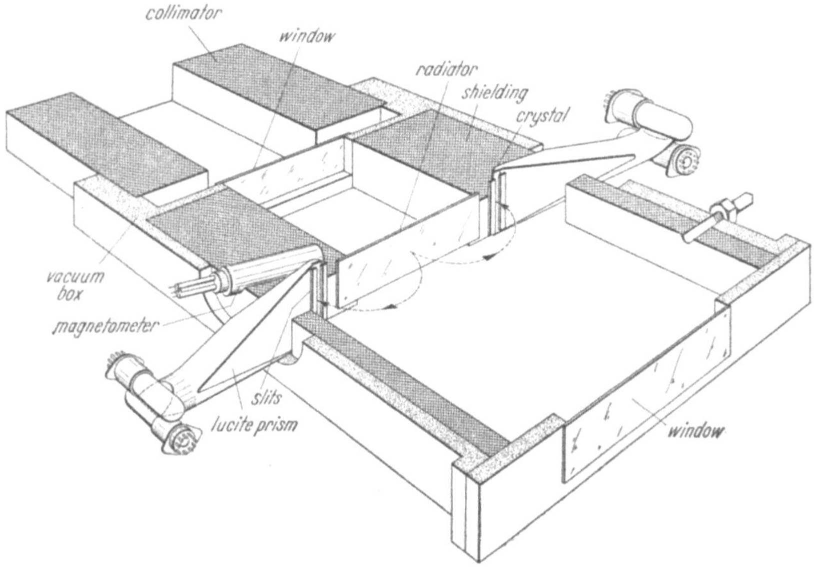

a

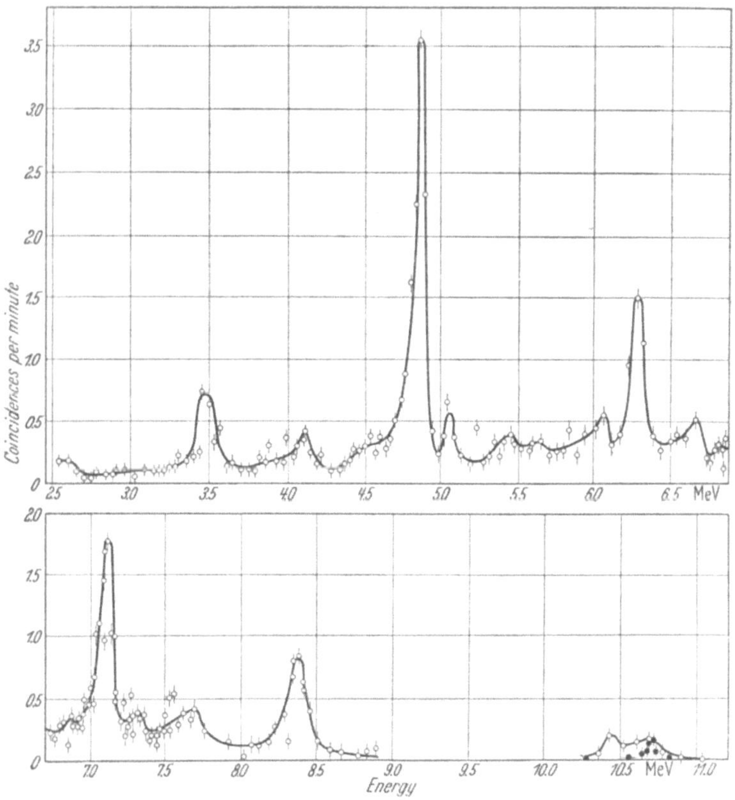

b

Fig. 40 a and b. Pair spectrometer study of $(n\gamma)$ capture radiations (KINSEY [37]). (a) Apparatus; (b) results for silicon.

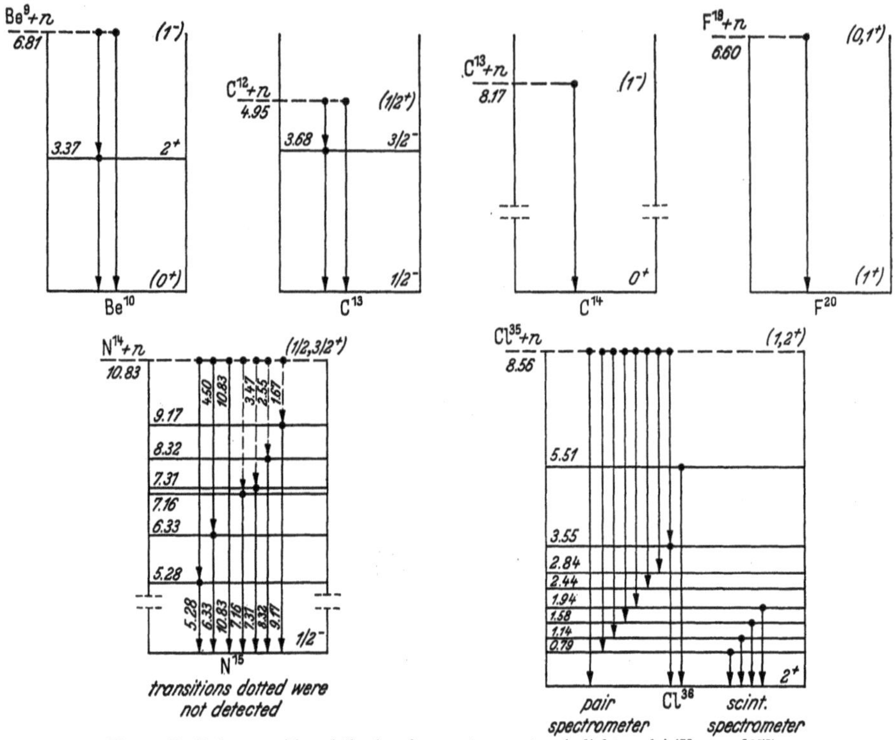

Fig. 41. Radiative transitions following slow neutron capture in light nuclei (KINSEY [37]).

Table 7. *Reactions of neutrons with light nuclei* ($A \geq 3$).

Column 2 gives the binding energy of a neutron in the compound nucleus. Columns 3, 4 give observed Q-values in MeV taken mainly from reference [31]. Column 5 gives other reactions observed; the (nn) process is known, in total cross section, for nearly all nuclei. a: activity observed; p: reaction particles observed.

Target nucleus	Excitation of compound nucleus MeV	np	$n\alpha$	Other reactions	Target nucleus	Excitation of compound nucleus MeV	np	$n\alpha$	Other reactions
He^3		0.764	—		Mg^{24}	7.32	a	—	nn', $n\gamma$
He^4	−1.0	—	—	—	Mg^{25}	11.10	a	—	$n\gamma$
Li^6	7.25	a	4.797	nd, nt	Mg^{26}	6.43	—	—	$n\gamma$
Li^7	2.04	—	—	nd, nn', nt, $n\gamma$	Al^{27}	7.72	a	a	nn', $n\gamma$
Be^7	18.89	p	p	—	Si^{28}	8.47	a	—	$n\gamma$
Be^9	6.81	—	a	$n\gamma$, $n2n$	Si^{29}	10.61	a	—	$n\gamma$
B^{10}	11.46	a	2.786	nd, $n2\alpha$, nn'	Si^{30}	6.59	—	a	
B^{11}	3.36	—	a		P^{31}	7.93	−0.94	a	$n\gamma$, $n2n$
C^{12}	4.95	a	—	nn', $n\gamma$, $n2n$	S^{32}	8.65	−0.93	1.16	$n\gamma$, nn', nt
C^{13}	8.17	—	a	$n\gamma$	S^{33}	11.4	a	—	
C^{14}	2.35	—	—		S^{34}	7.0	a	a	
N^{14}	10.83	0.624	−0.27	$n2n$, nt	S^{36}	5.7	—	—	
				nd, $n\gamma$, nn'	Cl^{35}	8.6	0.52	1.02	$n2n$, $n\gamma$
N^{15}	2.49	—	a	$n\gamma$	Cl^{37}	6.1	a	a	
O^{16}	4.14	a	−2.38	nn', $n2n$, nd	A^{36}	8.9	—	2.0	
O^{17}	8.06	a	a		A^{38}	6.6	—	—	
O^{18}	3.96	—	—		A^{40}	6.1	—	−2.5	
F^{19}	6.60	a	a	nn, $n\gamma$, $n2n$	K^{39}	7.8	—	—	$n\gamma$
Ne^{20}	6.78	—	−0.75		K^{40}	10.0	—	—	$n\gamma$
Ne^{21}	10.36	—	—		K^{41}	7.4	a	a	$n\gamma$
Ne^{22}	5.19	—	—		Ca^{40}	8·3	—	—	$n\gamma$
Na^{23}	6.96	a	a	nn', $n\gamma$					

spectra in the energy range 0.3 to 3 MeV has been made by BRAID[1] using a two-crystal COMPTON spectrometer (Sect. 14). The existence of cascades was proved by RECKSIEDLER and HAMERMESH[2] by a coincidence method.

The absolute emission probabilities for specified transitions observed in $(n\,\gamma)$ capture reactions have been obtained by KINSEY [*37*] in certain favourable cases in which the resonance responsible for thermal neutron capture can be identified. The multipolarity is known from level characteristics in many cases. The data can then be compared with theoretical estimates of radiation widths (see Sect. 19).

41. Resonant reactions of neutrons with light nuclei. Most of the resonances observed in light nuclei are due to scattering, but when reactions are energetically possible, these may also show resonance. Some of the reactions studied are listed in Table 8; total cross sections have also been measured as a function of energy by the transmission method for the normal isotopic mixture of all the elements listed.

Resonant capture of thermal neutrons occurs only rarely in light nuclei and the radiative widths Γ_γ cannot easily be obtained: this is due to the fairly wide

Table 8. *Resonant reactions of neutrons.*

Only reactions in which particle widths are known are listed and not all resonances are included. The radiation widths are for *thermal* neutron capture at an excitation equal to the neutron binding energy. The figures are taken mainly from references [*18*], [*24*] and [*32*].

Target nucleus	Reaction channel s	E_{res} MeV	Excitation of compound nucleus MeV	J	Γ_λ keV	$\Gamma_{\lambda s}$ keV	$\frac{2Ma\gamma_{\lambda s}^2}{3\hbar^2}$ %	E_γ MeV	Γ_γ eV	Multipolarity	$\|M\|^2$ Sect. 19
He^4	He^4+n	1.15	0	$3/2^-$	1000		75				
Li^6	Li^6+n	0.255	7.46	$5/2^-$		114	42				
	$H^3+\alpha$	0.255	7.46	$5/2^-$		60	0.04				
Li^7	Li^7+n	0.256	2.28	3^+	40		15				
Be^9	Be^9+n	0.62	7.37	3^+	30		1.5	6.81	7.0	$E1$	0.04
		0.81	7.54	(2)	≤ 11		$\ll 0.44$	3.44	0.9		0.04
B^{11}	$B^{11}+n$	0.43	3.76	2^+	40		3.6				
		1.28	4.53	3^-	130		28				
		1.78	4.99	1	65		1.2				
		2.45	5.61	2	120		5.3				
		2.58	5.73	3	60		2.6				
C^{12}	$C^{12}+n$	2.08	6.87	$3/2, 5/2^+$	11		0.23	4.95	1.3	$E1$	0.02
		2.95	7.67	$3/2^+$	120		1.4	1.27	0.6		0.48
C^{13}								8.17	0.5	$E1$	$1.5\cdot 10^{-3}$
N^{14}	$N^{14}+n$*	0.499	11.30	$1/2^-$	7.5		0.5	10.83	4.6	$E1$	$5.4\cdot 10^{-3}$
		0.640	11.43	$1/2^+$	43		0.8	4.50	4.0		0.067
								3.47	2.5		2.8
								2.55	0.9	$M1$	2.6
								1.67	0.4		4.0
F^{19}								6.60	0.2	$M1$	0.03

* More extensive data are available.

[1] T. H. BRAID: Phys. Rev. **102**, 1109 (1956).
[2] A. L. RECKSIEDLER and B. HAMERMESH: Phys. Rev. **96**, 109 (1954).

level spacing at the neutron binding energy. The thermal scattering length, however, which may be obtained in magnitude from thermal cross sections and in sign from interference phenomena, gives an indication of the presence of levels near the neutron binding energy. As shown by FERMI and MARSHALL[1] the scattering length is positive or negative according as the level energy lies above or below the neutron energy and the presence of bound levels can thus be detected. Radiative capture of fast neutrons is difficult to detect directly with available neutron intensities because Γ_γ/Γ_n is small and because of background problems. It may be demonstrated by observation of the active nucleus produced (see F^{19}).

If absorption widths such as Γ_α or Γ_p happen to be large in comparison with the term $(E - E_{res})$ in the single level formula (9.16), then the cross section for an $(n\,\alpha)$ or $(n p)$ process becomes

$$\sigma = 4\pi \lambda\!\!\!^{-2} g \frac{\Gamma_n}{\Gamma} \tag{41.1}$$

where $\Gamma \sim \Gamma_{\alpha,p}$ and is not a rapidly varying function of energy. This is usually so for reactions with high positive Q-values, and in such cases, since the neutron width is proportional to velocity, the cross section follows the "$1/v$ law". Familiar examples are the $He^3(np)$, $Li^6(n\alpha)$, $B^{10}(n\alpha)$, and $N^{14}(np)$ reactions at low neutron energies.

The Li^6 resonances were observed by JOHNSON *et al.*[2]; the peaks in both total cross section and reaction cross section are shown in Fig. 15. A similar peak is found in the total cross section of Li^7 (STELSON and PRESTON[3]). The inelastic scattering of neutrons by Li^7 has been studied by FREEMAN *et al.*[4], who detected radiation from the 478 keV level. The shape of the excitation curve just above threshold depends on the orbital momenta of the ingoing and outgoing neutron waves and in this case indicates a strong s-wave contribution to the reaction. This type of reaction would be expected to show resonances of the compound nucleus as for other inelastic processes, and evidence for this is given by KIEHN and GOODMAN[5] for scattering from Al^{27}.

Sufficient Be^7 ($4 \cdot 10^{-9}$ gm) can be produced by cyclotron bombardments to enable reaction cross sections to be measured. HANNA[6] found a cross section of $(5.3 \pm 0.8) \times 10^{-20}$ cm² for the $Be^7(np)$ reaction with thermal neutrons and a cross section of less than 10^{-24} cm² for the $(n\alpha)$ reaction. These results are consistent with odd parity for Be^7. Comparison with the cross section for the inverse reaction $Li^7(pn)$ suggests an odd parity level near the neutron binding energy in Be^8.

The scattering of neutrons by Be^9 is accompanied, at the 2.6 MeV resonance, by a peak in the $(n\alpha)$ reaction. Assignments of spin are based partly on angular distribution measurements.

The $B^{10} + n$ reactions show resonance in both total cross section [*38*] and the $(n\alpha)$ reaction at 1.9 MeV. At thermal energies the reaction proceeds mainly to the first excited state of Li^7 at 478 keV, despite the greater energy release for the ground state transition; a recent accurate measurement[7] gives the branching ratio as (6.52 ± 0.05)%. This ratio varies with incident neutron energy[8]. The

[1] E. FERMI and L. MARSHALL: Phys. Rev. **71**, 666 (1947).
[2] C. H. JOHNSON, H. B. WILLARD and J. K. BAIR: Phys. Rev. **96**, 985 (1954).
[3] P. H. STELSON and W. M. PRESTON: Phys. Rev. **84**, 162 (1951).
[4] J. M. FREEMAN, A. M. LANE and B. ROSE: Phil. Mag. **46**, 17 (1955).
[5] R. M. KIEHN and C. GOODMAN: Phys. Rev. **95**, 989 (1954).
[6] R. C. HANNA: Phil. Mag. **46**, 381 (1955).
[7] J. A. DE JUREN and H. ROSENWASSER: Phys. Rev. **93**, 831 (1954).
[8] H. BICHSEL, W. HÄLG, P. HUBER and A. STEBLER: Helv. phys. Acta **25**, 119 (1952); B. PETREE, C. H. JOHNSON and D. W. MILLER: Phys. Rev. **83**, 1148 (1951).

possibility that it could be explained by assuming a spin of $^5/_2$ for the first state of Li^7 was discussed by DEVONS[1], but the observation of an isotropic angular correlation between the alpha particles to the excited state and the associated gamma radiation[2] eliminates this possibility. The assignment of spin $^1/_2$ to the 478 keV state of Li^7 is in agreement with all other available evidence, including the measurement of the lifetime of this state by recoil methods. The difficulties in the interpretation of this reaction and their possible explanation in terms of accidental cancellation of matrix elements are discussed by INGLIS[3]. The $B^{10}(n\alpha)$ and $Li^6(n\alpha)$ reactions, because of their high cross section for charged particle emission, are especially useful as neutron detectors. B^{11} and C^{12} also show resonances in total cross section [*38*], some of which have been analysed to give spins and reduced widths for levels of B^{12} and C^{13}. There is evidence for an *s*-level in C^{13} just below the neutron binding energy.

The charged particle reactions of neutrons with N^{14} have been extensively studied because of their low *Q*-values, and because of the convenience of nitrogen as a gas for filling detecting instruments. The *Q*-values are found by irradiating a nitrogen-filled ionisation chamber with homogeneous neutrons and analysing the resulting distribution of pulse sizes. The resonance energies are found by irradiating a nitrogen-filled ionisation chamber with neutrons with a continuous energy distribution such as those from a radon-beryllium source or a $H^2 + H^2$ source suitably moderated. STETTER and BOTHE[4] and others thus observed many sharp peaks at energies corresponding to a high probability for the $(n\alpha)$ and (np) disintegrations; the two types of disintegration may be distinguished by varying the chamber pressure. Interpretation of such experiments is sometimes difficult because peaks in the pulse height distribution also arise owing to the excitation of states of residual nuclei. The resonance energies, at least for the range of excitation up to about 2 MeV above the neutron binding energy, are therefore best determined by using homogeneous neutrons of variable energy. In this way JOHNSON and BARSCHALL[5] found sharp resonances in the (np) and $(n\alpha)$ reaction; some of their results have already been discussed in connection with the inverse reaction and are shown in Fig. 35. These resonances may be compared with similar resonances in the total cross section of nitrogen for neutrons, obtained by JOHNSON *et al.*[6]. The reason for the lack of correspondence between certain of the peaks is that for some the neutron width is large compared with the charged particle width, and for others the reverse is true.

A comprehensive study of the resonances in total cross section for nitrogen is presented by HINCHEY *et al.*[7] who apply the general scattering theory (Sect. 17) to assign spins and parities to eleven levels and also derive level spacings for decay by proton, neutron or alpha particle emission. Additional information can be obtained from differential cross section measurements (FOWLER and JOHNSON[8]). Both directly scattered neutrons and nitrogen recoil nuclei were detected, following in the latter case the technique introduced by BALDINGER *et al.*[9]. Higher

---

[1] S. DEVONS: Proc. Phys. Soc. Lond. A **62**, 580 (1949).
[2] B. ROSE and A. R. W. WILSON: Phys. Rev. **78**, 68 (1950).
[3] D. R. INGLIS: Phys. Rev. **81**, 914 (1951).
[4] W. STETTER and W. BOTHE: Z. Naturforsch. **6**a, 61 (1951).
[5] C. H. JOHNSON and H. H. BARSCHALL: Phys. Rev. **80**, 818 (1950).
[6] C. H. JOHNSON, B. PETREE and R. K. ADAIR: Phys. Rev. **84**, 775 (1951).
[7] J. H. HINCHEY, P. H. STELSON and W. M. PRESTON: Phys. Rev. **86**, 483 (1952). See also D. SPEISER: Helv. phys. Acta **27**, 427 (1954).
[8] J. L. FOWLER and C. H. JOHNSON: Phys. Rev. **98**, 728 (1955).
[9] E. BALDINGER, P. HUBER and H. STAUB: Helv. phys. Acta **11**, 245 (1938).

resonance levels of N^{15} have been explored by MEIER *et al.*[1] using the H^2+H^2 reaction as a source of neutrons.

BOCKELMAN *et al.* [38] using *p*-T neutrons, have observed resonances in the total cross section of oxygen for energies up to 3.4 MeV and list the resonance parameters for the narrow O^{17} levels including reduced widths. Their work is extended to higher energies by RICAMO *et al.*[2] using the H^2+H^2 neutron source. ADAIR[3] studied the angular distribution of elastically scattered neutrons near the narrow resonances corresponding to levels at 4.56, 5.08 and 5.39 MeV in O^{17} and

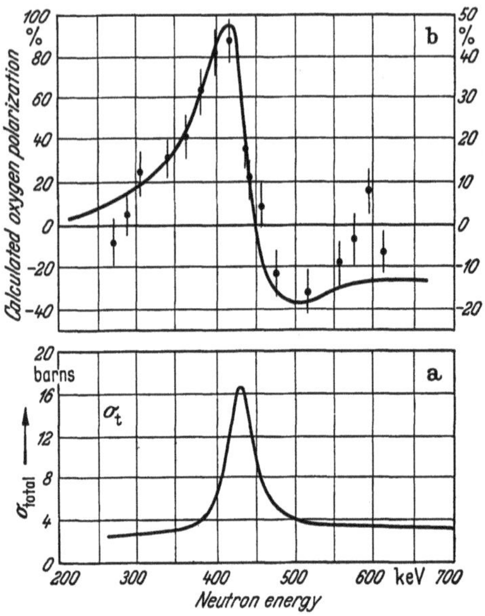

Fig. 42a and b. Polarisation of neutrons from the Li^7 (*pn*) reaction demonstrated by scattering in oxygen (WILLARD *et al.*). (a) Total cross section of oxygen; (b) polarisation.

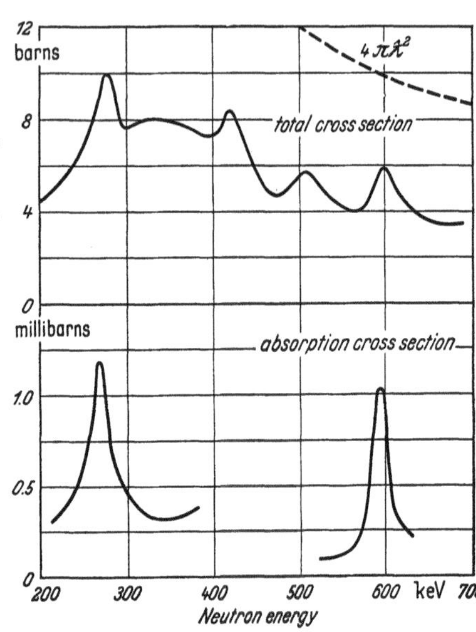

Fig. 43. Comparison of total neutron cross section and neutron capture cross section for fluorine (HENKEL and BARSCHALL).

determined parities for these states. His work identifies the $D_{3/2}$ member of the $D_{5/2,3/2}$ spin orbit doublet, the $5/2$ component being the ground state. The spin-orbit splitting is about 5 MeV. The resonance level at 430 keV is assigned the character $P_{3/2}$, and interference between waves scattered from this resonance with *s*-wave background radiation has been used by WILLARD *et al.*[4] to demonstrate the polarisation of a beam of neutrons from the $Li^7(pn)$ reaction. Fig. 42 shows the way in which the left-right asymmetry indicative of polarisation varies through the *p*-wave resonance of O^{17}. The $(n\alpha)$ resonances corresponding to levels in O^{17} above 6.8 MeV have been observed by GIERKE[5] using an ionisation chamber filled with oxygen and irradiated by a continuous distribution of high energy neutrons.

Resonances in total cross section have been observed for the interaction of neutrons with all the main isotopes of elements from fluorine to calcium excepting neon and argon. Detailed analysis has only been possible for levels which are well separated, but some resonance parameters have been obtained. Observations

---

[1] R. C. MEIER, R. RICAMO, P. SCHERRER and W. ZÜNTI: Helv. phys. Acta **26**, 451 (1953).
[2] R. RICAMO, W. ZÜNTI, E. BALDINGER and P. HUBER: Helv. phys. Acta **23**, 508 (1950).
[3] R. K. ADAIR: Phys. Rev. **92**, 1491 (1953).
[4] H. B. WILLARD, J. K. BAIR and J. D. KINGTON: Phys. Rev. **95**, 1359 (1954).
[5] G. VON GIERKE: Z. Naturforsch. **8a**, 567 (1953).

of the (np) $(n\alpha)$ and $(n\,2n)$ reactions have also been made for many of these elements. SIKKEMA[1] and JOHNSON *et al.*[2] found resonances for the $Ne^{20}(n\alpha)$ reaction.

The $(n\gamma)$ reaction in F^{19} and Al^{27} has been observed by HENKEL and BARSCHALL[3] by detection of the resulting activity of F^{20} and Al^{28}. The correlation of the resonances in total and absorption cross section is shown in Fig. 43. The ratio of the corresponding peak cross sections is approximately equal to Γ_γ/Γ_n.

42. Reactions of high energy neutrons with light nuclei. In many experiments made with the H^2+H^2 or H^3+H^2 neutron source emphasis has been placed on the measurement of Q-values, activation cross sections or angular distributions which do not depend sharply on bombarding energy. The excitation of compound nuclei produced by neutrons of 14 MeV energy is in any case so large that level structure would not be apparent. The elastic scattering of 14 MeV neutrons by nitrogen was shown by SMITH[4] to follow the pattern expected on the basis of the optical model (Sect. 5). Inelastic processes in oxygen and nitrogen were observed by LILLIE[5], using an expansion chamber, and MARTIN[6] followed the variation of the cross section for the $O^{16}(np)$ reaction over the energy range 12 to 18 MeV. The elastic scattering of high energy neutrons in oxygen also exhibits a diffraction pattern type of angular distribution[7]. Inelastic excitations of the 6 and 7 MeV levels of O^{16} were reported by THOMPSON and RISSER[8]. The energy release in several of these reactions has been measured using ionisation chamber techniques as in the work of HURLIMAN and HUBER[9] on the $S^{32}(n\alpha)$ and $S^{32}(np)$ reactions, of FOLKIERSI[10] for $Cl^{35}(n\alpha)$ and of BELLAMY and FLACK[10] for A^{40} $(n\alpha)$. An extensive list of cross sections for the activation of elements by 14 MeV neutrons producing the (np), $(n\alpha)$, and $(n\,2n)$ reactions is given by PAUL and CLARKE[11] who compare their results with the predictions of statistical theory. This is not expected to apply well to nuclei as light as those considered in this article, and some deviations which may be interpreted as resulting from direct interaction processes were found. The theory of such processes in the (np) reaction has been given by AUSTERN *et al.*[12]. Many other inelastic processes such as (nd), (nt) have been reported by workers using 14 MeV neutrons (FRYE[13]). The $Li^6(nd)$ and $B^{10}(nd)$ reactions have been analysed with the assumption of an inverse stripping mechanism (Sect. 53).

The energy distribution of inelastically scattered neutrons from Al^{27} has been examined by STELSON and GOODMAN[14] and by GRAVES and ROSEN[14] and compared with the predictions of statistical theory (Sect. 8). O'NEILL[15] detected low energy neutrons by a time of flight method using 14 MeV neutrons and carbon and aluminium scatterers; he was able to show that the angular distributions were

[1] C. P. SIKKEMA: Nature, Lond. **165**, 1016 (1950).
[2] C. H. JOHNSON, C. K. BOCKELMAN and H. H. BARSCHALL: Phys. Rev. **82**, 117 (1951).
[3] R. L. HENKEL and H. H. BARSCHALL: Phys. Rev. **80**, 145 (1950).
[4] J. R. SMITH: Phys. Rev. **95**, 730 (1954).
[5] A. B. LILLIE: Phys. Rev. **87**, 716 (1952).
[6] H. C. MARTIN: Phys. Rev. **93**, 498 (1954).
[7] J. P. CONNER: Phys. Rev. **89**, 712 (1953).
[8] L. C. THOMPSON and J. R. RISSER: Phys. Rev. **94**, 941 (1954).
[9] T. HURLIMAN and P. HUBER: Helv. phys. Acta **28**, 33 (1955).
[10] A. FOLKIERSKI: Proc. Phys. Soc. Lond. A **65**, 1006 (1952). — E. H. BELLAMY and F. C. FLACK: Phil. Mag. **46**, 341 (1955).
[11] E. B. PAUL and R. L. CLARKE: Canad. J. Phys. **31**, 267 (1953).
[12] N. AUSTERN, S. T. BUTLER and H. MCMANUS: Phys. Rev. **92**, 350 (1953).
[13] G. M. FRYE: Phys. Rev. **93**, 1086 (1954).
[14] E. R. GRAVES and L. ROSEN: Phys. Rev. **89**, 343 (1953). — P. H. STELSON and C. GOODMAN: Phys. Rev. **82**, 69 (1951).
[15] G. K. O'NEILL: Phys. Rev. **95**, 1235 (1954).

isotropic within about 15%. The ring scattering method was used by POOLE[1] who detected neutrons, and by GRIFFITH[2] who detected gamma radiation, for studying inelastic scattering. Inelastic cross sections have been measured by the sphere transmission method for a number of elements at neutron energies of 1.0, 4.0, and 4.5 MeV by BEYSTER *et al.*[3] The nuclear level densities obtained from experiments such as these agree with those indicated for nuclei of similar mass by the neutron spectra from (pn) reactions.

The $(n, 2n)$ reaction is expected to show an increased cross section for high neutron energies at the expense of the (n, n') process. This is probably so for heavier elements, but the small number of nucleons involved in nuclei even as heavy as calcium makes it rather doubtful that thermal equilibrium is actually obtained for light nuclei, particularly with 14 MeV neutrons and particularly when it is known that direct interaction processes can contribute appreciably to the total cross section.

The total cross sections for several light nuclei for 14 to 18 MeV neutrons have been measured by COOK and BONNER[4] in addition to many earlier workers. They find that the cross sections observed are described by the formula

$$\sigma = 2\pi (R + \lambda\!\!\!^{-})^2$$

but that there are large deviations from the simple relation $R \propto A^{\frac{1}{3}}$ in the light elements. This has also been noted in (dp) reactions (see Sect. 52).

III. Reactions of alpha particles with light nuclei.

43. Scattering of alpha particles by helium. The nucleus Be^8 is unstable in its ground state against disintegration into two alpha particles. According to TREACY[5], the disintegration energy is (90 ± 5) keV and the half-lifetime is $< 4 \cdot 10^{-15}$ sec, by the recoil method. This corresponds to a width of about 1 eV. If therefore the energy of an alpha particle beam can be defined with this sort of precision, the scattering of alpha particles of 180 keV by helium should reveal a resonance anomaly as the incident energy is varied through that corresponding to the Be^8 ground state. Positive results have not yet been obtained in this particular experiment, but there are in Be^8 (Fig. 95) several excited states with spin and parity even which can disintegrate into two alpha particles and these have shown resonance effects.

DEVONS[6], using an annular scattering technique, sought for possible resonance anomalies at angles corresponding to zeros of $P_l(\cos\vartheta)$ by varying the energy of the incident alpha particles from Ra C′ and ThC′ sources and found indications of a broad resonance requiring $J = 2$ at an energy of about 6 MeV. This corresponds to the known properties of the 2.9 MeV level of Be^8. WHEELER[7] subsequently analysed the results of this and similar early experiments with low intensity sources and concluded that the level spin was 0. HAEFNER[8] has recently examined this information again with results agreeing with those of WHEELER, but with the suggestion that the data available were not really precise enough for a reliable phase shift analysis.

[1] M. J. POOLE: Phil. Mag. **43**, 1060 (1952); **44**, 1398 (1953).
[2] G. L. GRIFFITH: Phys. Rev. **98**, 579 (1955).
[3] J. R. BEYSTER, R. L. HENKEL, R. A. NOBLES and J. M. KISTER: Phys. Rev. **98**, 1216 (1955).
[4] C. F. COOK and T. W. BONNER: Phys. Rev. **94**, 651 (1954).
[5] P. B. TREACY: Proc. Phys. Soc. Lond. A **68**, 204 (1955).
[6] S. DEVONS: Proc. Roy. Soc. Lond., Ser. A **172**, 559 (1939).
[7] J. A. WHEELER: Phys. Rev. **59**, 16 (1941).
[8] R. R. HAEFNER: Rev. Mod. Phys. **23**, 228 (1951).

No recent experiments with strong sources of artificially accelerated alpha particles seem yet to have been completed which relate to the 2.9 MeV level of Be^8, although the necessary energy is well within the range of electrostatic accelerators. The alpha particle beams from cyclotrons are less easy to control in energy, but STEIGERT and SAMPSON[1] were able to obtain differential cross section measurements for the energy range 12.88 to 21.62 MeV. Some of their results are shown in Fig. 44 together with the resulting phase shifts. These were deduced from the general scattering theory for identical particles making use of geometrical constructions. The s-wave phase shift passes through 90° at an energy which indicates

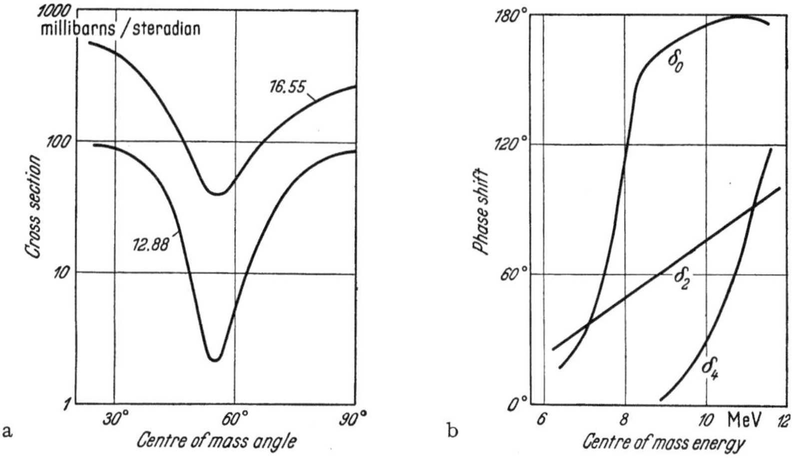

Fig. 44 a and b. Scattering of alpha particles by helium (STEIGERT and SAMPSON). (a) Differential cross section at 12.88 and 16.55 MeV; (b) phase shift.

a Be^8 level at (7.65 ± 0.08) MeV with a width of (1.2 ± 0.4) MeV (and spin zero). The d-wave phase shift does not pass sharply through 90° in the region examined but the g-wave anomaly indicates a spin 4^+ level at (10.8 ± 0.4) MeV with width (1.2 ± 0.4) MeV. Evidence for both these levels is seen in other experiments, such as $Li^7(dn)$, $Li^8(\beta^-)\,Be^8$ and $C^{12}(\gamma\, 3\alpha)$.

The differential cross section for 30 MeV alpha particles scattered by helium (GRAVES[2]) shows minima at about 40° and 70° in the centre of mass system and requires s, d and g waves for a theoretical fit.

44. Excitation of bound states in alpha particle reactions. The main reactions of alpha particles with light nuclei are summarised in Table 9.

α) *Inelastic scattering.* The inelastic scattering of alpha particles, like that of deuterons, is interesting because of the charge symmetry of the particle. According to the simple predictions of the theory of isotopic spin, bombardment by alpha particles cannot lead to states of the intermediate nucleus with isotopic spin different from that of the target nucleus and in the case of inelastic scattering, only states with the same isotopic spin as the ground state can be excited. In practice this absolute rule may be modified because of isotopic spin impurity of the reacting nuclei or because of the COULOMB perturbation of the intermediate state. Experiments have been made on the nuclei Li^6, B^{10} and N^{14} in which a low lying state with $T=1$ is expected, but full details are not yet available. There are in these nuclei more highly excited states to which isotopic spin $T=1$ has been

[1] F. E. STEIGERT and M. B. SAMPSON: Phys. Rev. **92**, 660 (1953).
[2] E. GRAVES: Phys. Rev. **84**, 1250 (1951).

Table 9. *Reactions of alpha-particles with light nuclei* ($A > 4$).

Column 2 gives the binding energy of an alpha-particle in the compound nucleus. Columns 4 and 5 give observed Q-values in MeV taken mainly from reference [*31*]. Column 5 gives other reactions observed. *a*: activity observed; *p*: reaction particles observed.

Target nucleus	Excitation of compound nucleus MeV	$\alpha\alpha$	αp	αn	Other reactions	Target nucleus	Excitation of compound nucleus MeV	$\alpha\alpha$	αp	αn	Other reactions
Li^6	4.45		—	—	$\alpha\gamma$	Mg^{25}	11.14		*p*	—	
Li^7	8.67		—	—	$\alpha\gamma$, $\alpha\alpha'$	Mg^{26}	10.63		*a*	—	
Be^9	10.66		−6.92	5.71	αd, $\alpha\alpha'$	Al^{27}	9.68		2.27	−2.9	
B^{10}	11.61		4.08	*p*	αd	Si^{28}	6.94		*p*	*a*	
B^{11}	10.99		0.75	*p*		Si^{29}	7.12		*a*	—	
C^{12}	7.15	*p*	−4.96	*a*	$\alpha\alpha'$	Si^{30}	7.90		—	—	
C^{13}	6.34		—	2.06		P^{31}	7.01		*p*	*a*	
C^{14}	6.24		*a*	—		S^{32}	6.65		−2.04	*a*	αd
N^{14}	4.41	*p*	−1.16	*p*	$\alpha\gamma$, $\alpha\alpha'$	S^{33}	6.84		—	—	
N^{15}	3.99		—	—	$\alpha\gamma$	S^{34}	7.19		—	*a*	
O^{16}	4.75	*p*	−8.12	—		S^{36}	6.76		*a*	—	
O^{17}	7.36		—	—		Cl^{35}	7.21		0.81	*a*	
O^{18}	9.66		—	—		Cl^{37}	6.20		—	*a*	
F^{19}	10.49	*p*	*p*	*p*	$\alpha\alpha'$	A^{36}	7.04		—	—	
Ne^{20}	9.31	*p*	−2.38	—		A^{38}	6.31		—	—	
Ne^{21}	9.89	*p*	—	—		A^{40}	8.93		*a*	*p*	$\alpha p n$
Ne^{22}	10.64	*p*	*a*	*p*		K^{39}	4.82		*p*	—	
Na^{23}	10.20		1.75	*a*	$\alpha\alpha'$	K^{41}	7.97		*p*	—	
Mg^{24}	10.00	*p*	−1.60	*a*		Ca^{40}	—		*p*	—	

assigned and in the case of N^{14} the absence of groups corresponding to these levels in the spectrum of inelastically scattered alpha particles has been noted. In odd mass nuclei and nuclei of mass number $4n$, (for which $T=1$ states are highly excited) inelastic scattering of alpha particles from low-lying levels follows the normal pattern of charged particle reactions. In Li^7 Gove and Harvey[1], using 31 MeV alpha particles in a collimated beam from a cyclotron, observed a reduced energy group corresponding to the 4.79 MeV level. The resolution in this experiment was insufficient to show the first excited state of Li^7 at 478 keV, but gamma radiation corresponding to this level was detected by Heydenburg and Temmer [*43*] and by Li and Sherr[2] using singly charged beams of alpha particles of energy up to 4 MeV from an electrostatic accelerator. The latter authors, following Jones and Wilkinson[3], demonstrated the Doppler shift of gamma ray energy with angle of observation relative to the alpha particle beam and deduced an upper limit of 3×10^{-13} sec for the lifetime of Li^{7*}.

A cyclotron beam of 22 MeV alpha particles was used by McMinn *et al.*[4] to excite a 2.63 MeV level of Be^9. Lower energy alpha particles have been used to excite the low lying levels of F^{19} and Na^{23} by the Coulomb interaction process[5,6] [*43*]. This process has been studied in considerable detail for F^{19} which has a ground state spin of $^1/_2{}^+$ and low lying levels of spin $^5/_2{}^+$ (197 keV) and $^1/_2{}^-$ (110 keV). These states can be reached by $E2$ and $E1$ excitation respectively, and Coulomb excitation of these levels by protons has already been discussed (Sect. 36). Sherr *et al.*[6] found excitation functions for the production of gamma radiation of these

[1] H. E. Gove and J. A. Harvey: Phys. Rev. **82**, 658 (1951).
[2] C. W. Li and R. Sherr: Phys. Rev. **96**, 389 (1954).
[3] G. A. Jones and D. H. Wilkinson: Phil. Mag. **43**, 958 (1952).
[4] W. O. McMinn, M. B. Sampson and V. K. Rasmussen: Phys. Rev. **84**, 963 (1951).
[5] G. A. Jones and D. H. Wilkinson: Phil. Mag. **45**, 230 (1954).
[6] R. Sherr, C. W. Li and R. F. Christy: Phys. Rev. **96**, 1258 (1954).

energies (cf. Fig. 19) which are notably different from the excitation function for the compound nucleus reaction $F^{19}(\alpha p')$ Ne^{22*}. The resonant yield of 197 and 110 keV gamma radiation decreases rapidly with respect to that of the $(\alpha p')$ reaction because of decreasing barrier penetrability for the low energy alpha particles, and the non-resonant yield is attributed to COULOMB excitation. From the cross section for excitation, the lifetime for the transition was calculated and compared with the value found directly. Angular distributions were also compared with theory; this established the nature of the decay process following COULOMB excitation. For F^{19} the facts are best fitted by assuming $E2$ excitation and decay for the 197 keV level and $E1$ excitation and decay for the lower level.

β) (αp) reactions. There are many measurements of the Q-values for (αp) and (αn) reactions to the ground state and low lying excited states of light nuclei. The existence of excited states and of resonance levels was first established in the experiments of CHADWICK, BLACKETT and others using naturally radioactive sources. Several of these measurements have now been refined, although progress has been relatively slow by comparison with that in proton induced reactions, because of the difficulty of obtaining doubly charged beams of alpha particles from ion sources for use in accelerators. Using 22 MeV alpha particles from a cyclotron MCMINN *et al.*[1] observed four groups of protons ascribed to the reaction $Be^9(\alpha p)$, giving three levels in B^{12}; they also established the reaction $Be^9(\alpha, d)$ B^{11}, in which there should be no change of isotopic spin between the inital and final nucleus. The level of 2.18 MeV in B^{11} excited in this reaction should have the same isotopic spin as the ground state. The $B^{10}(\alpha p)$ reaction has a Q-value of 4.13 MeV and several groups of protons have been observed[2,3] as well as resonances in the excitation function of this and companion reactions (Sect. 45). SHIRE *et al.*[2] detected the gamma radiation associated with the lower energy proton groups. The gamma ray of energy (0.21 ± 0.03) MeV corresponding to transition between the 3.86 and 3.68 MeV levels of C^{13} (Fig. 68) is found in such large intensity in comparison with the direct 3.86 MeV transition that $E1$ character seems necessary. Angular distributions and correlations were made to confirm the spin sequence shown in Fig. 68; this was already known from other work except that alternatives $^3/_2{}^+$ and $^5/_2{}^+$ for the 3.86 MeV level could not be distinguished. The $B^{11}(\alpha p)$ C^{14} ground state reaction was also observed by FRYE and WIEDENBECK[3] and by SHIRE *et al.*[2] and the $B^{10}(\alpha, d)$ reaction, which has a positive Q-value, by the latter authors. The other (αp) reactions observed are noted in Table 9. Among these the bombardment of C^{14} to produce N^{17} is one of the few examples of the use of an artifically prepared nuclide as a target[4]. The historically interesting reaction $N^{14}(\alpha p)$ O^{17} shows resonances (Sect. 45) and an excited state at about 0.86 MeV has been demonstrated by the nuclear emulsion technique in recent work[5]. The angular distribution of protons reaching to this state is isotropic. The $O^{16}(\alpha p)$ F^{19} reaction has been investigated with 21 MeV alpha particles[6] and three excited states of F^{19} were found; this type of experiment, using cyclotron beams, has insufficient resolving power to display the triplet of states near and including the ground state of F^{19}. The $F^{19}(\alpha p)$ Ne^{22*} reaction was detected by HEYDENBURG and TEMMER [43] by observation of the associated gamma radiation from the 1.28 MeV

[1] W. O. MCMINN, M. B. SAMPSON and V. K. RASMUSSEN: Phys. Rev. **84**, 963 (1951).
[2] E. S. SHIRE, J. R. WORMALD, G. LINDSAY-JONES, A. LUNDEN and A. G. STANLEY: Phil. Mag. **44**, 1197 (1953).
[3] G. M. FRYE and M. L. WIEDENBECK: Phys. Rev. **82**, 960 (1951).
[4] K.-H. SUN, B. JENNINGS, W. E. SHOUPP and A. J. ALLEN: Phys. Rev. **82**, 267 (1951).
[5] R. R. ROY: Phys. Rev. **82**, 227 (1951). — E. HJALMAR and H. SLÄTIS: Phys. Rev. **89**, 1151 (1953).
[6] M. L. BULLOCK and M. B. SAMPSON: Phys. Rev. **84**, 967 (1951).

first excited state in Ne^{22} by means of a scintillation spectrometer. Association of proton groups in the (αp) reaction with the subsequent gamma radiation from excited states has been demonstrated for the $Al^{27}(\alpha p)$ reaction by ALLEN *et al.*[1], for $Na^{23}(\alpha p)$ by MAY and FOSTER[2], and for $F^{19}(\alpha p)$ by FOSTER *et al.*[3]. Decay schemes involving cascade transitions are suggested from this work. The gamma radiation resulting from a number of (αp) and (αn) reactions has been studied by BREEN and HERTZ[4] in the course of preparation of neutron sources in which up to 10 C of polonium were used in combination with various light elements.

γ) (αn) reactions. Excitation of bound states by the (αn) reaction has been less fully exploited than excitation by the (αp) process, because of the low intensity of natural alpha particle sources and the difficulty of making precise measurements of neutron energy with high efficiency. The reaction had been observed for nearly all nuclei up to $A = 40$ before beams of alpha particles of accurately defined energy became available, but in general only cross sections for the reaction for alpha particles of given energy, or rather rough reaction thresholds and excitation curves were determined[5]. The most thoroughly examined reaction is that by which the neutron was discovered, $Be^9(\alpha n)\,C^{12}$, which has a high positive Q-value (5.71 MeV), and gives a plentiful yield of neutrons even for alpha particles with energy as low as 1 MeV. The neutron spectrum for homogeneous alpha particles of energy 1.4 MeV was observed in a nuclear emulsion by BRADFORD and BENNETT[6], who counted forward recoil protons and found groups corresponding to the ground state and the first excited state of C^{12} at 4.4 MeV. Other experiments with higher energy alpha particles[7] reveal the 7.5 MeV level of C^{12} as well. The gamma radiation from the 4.4 MeV level excited in this reaction has been observed by many workers, including TERRELL[8] who used a pair spectrometer and found only this line for a Po-Be neutron source. MILLS and MACKIN[9], using a scintillation spectrometer and 3 MeV alpha particles, demonstrated the DOPPLER shift in gamma ray energy and set an upper limit to the lifetime of the C^{12} state as 3×10^{-13} sec. Using a magnetic lens spectrometer they detected the internal conversion pairs and showed that the conversion coefficient was that expected for $E2$ radiation. A low intensity gamma radiation of energy 3.16 MeV observed by BEGHIAN *et al.*[10] may be associated with a transition between the 7.65 and 4.43 MeV levels of C^{12} and gamma-gamma coincidence measurements[11] support this. HARRIES[12] found internal conversion pairs of an energy corresponding to 7.0 MeV and suggested a spin of 0^+ for this level. With 21.7 MeV alpha particles a 15.1 MeV gamma radiation was found by RASMUSSEN *et al.*[13] and is ascribed, on account of the small particle width of the level, to excitation of the first $T = 1$ state of C^{12} by the $Be^9(\alpha n)\,C^{12*}(\gamma)\,C^{12}$ reaction.

[1] R. C. ALLEN, J. E. MAY and W. RALL: Phys. Rev. **84**, 1203 (1951). See also R. R. ROY and C. GODEAU: Phil. Mag. **44**, 1184 (1953).
[2] J. E. MAY and B. P. FOSTER: Phys. Rev. **90**, 243 (1953).
[3] B. P. FOSTER, G. S. STANFORD and L. L. LEE: Phys. Rev. **93**, 1069 (1954).
[4] R. J. BREEN and M. R. HERTZ: Phys. Rev. **98**, 599 (1955).
[5] I. HALPERN: Phys. Rev. **76**, 248 (1949).
[6] C. E. BRADFORD and W. E. BENNETT: Phys. Rev. **78**, 302 (1950).
[7] W. H. GUIER, H. W. BERTINI and J. H. ROBERTS: Phys. Rev. **85**, 426 (1952).
[8] J. TERRELL: Phys. Rev. **80**, 1076 (1950).
[9] W. R. MILLS and R. J. MACKIN: Phys. Rev. **95**, 1206 (1954).
[10] L. E. BEGHIAN, H. H. HALBAN, T. HUSAIN and L. G. SANDERS: Phys. Rev. **90**, 1129 (1953).
[11] R. G. UEBERGANG: Austral. J. Phys. **7**, 279 (1954).
[12] G. HARRIES: Proc. Phys. Soc. Lond. A **67**, 153 (1954). See however R. D. BENT, T. W. BONNER, J. H. MCCRARY and W. A. RAMSAY: Phys. Rev. **100**, 771 (1955) who did not find pairs.
[13] V. K. RASMUSSEN, J. R. REES, M. B. SAMPSON and N. S. WALL: Phys. Rev. **96**, 812 (1954).

Among the other (αn) reactions, that with C^{13} was produced by JONES and WILKINSON[1] using 1 MeV alpha particles, and the neutron energy was found from the spectrum of recoil protons in a high pressure ionisation chamber. The (αn) reaction in F^{19} shows a step in its excitation curve at about 3 MeV corresponding to the threshold for neutron transitions to the first excited state of Na^{22} (HEYDENBURG and TEMMER [*43*]). The yield of the corresponding 592 keV gamma radiation, observed in a scinitillation spectrometer, shows the same threshold. As in other threshold determinations for neutron production using variable energy alpha particles, account must be taken of possible resonance effects in the compound nucleus distorting the yield curve. The (αn) reactions in heavier nuclei have been detected mainly by the radioactivity of the product nucleus [either the (αp) or (αn) reaction must lead to an active product] or by observation of the slowed down neutrons in a boron-trifluoride counter. PECK[2] bombarded Al^{27} with 0.03 μA of 7.6 MeV alpha particles and was able to distinguish neutron groups by recoil proton measurements in a nuclear emulsion. Quantitative information on the energy changes in (αn) reactions is relatively scarce.

45. Excitation of virtual states in alpha particle reactions. The resonance parameters for certain levels excited by alpha particles in light nuclei are given in table 10. The capture of low energy alpha particles by Li^6 excites states of the B^{10} nucleus in a region which has been thoroughly studied by the $(d n)$ reaction. JONES and WILKINSON[3], using alpha particles of energy up to 1.5 MeV found that the 4.77 and 5.16 MeV levels (see Fig. 31) were excited, and suggested that their failure to observe the 5.11 MeV level in this reaction was due to the operation of the isotopic spin selection rule for electric dipole transitions. This implies $T=1$ for the 5.16 MeV level and its formation in the $Li^6+\alpha$ reaction is thought to be due to isotopic spin impurity in the states concerned. Intensity considerations permit the assignment of level spins shown in Fig. 31.

In Li^7 BENNETT *et al.*[4] observed three resonances for the $(\alpha\gamma)$ reaction and showed that the strength of these resonances did not differ greatly from that observed in proton capture reactions. Their work was extended by JONES and WILKINSON[5], who analysed the spectrum of capture radiation by scintillation spectrometry. The large number of transitions found from the three capture levels can be fitted into the level scheme of B^{11} known (mainly) from the $B^{10}(d p)$ reactions. Angular distribution and angular correlation measurements in addition to arguments based on intensities, enabled spin and parity assignments to be made, as shown in Fig. 67, with little ambiguity. The assignments are in excellent agreement with those expected from the shell model, and lead to reasonable values for the splitting of the P, D, and F spin-orbit doublets. No anomalies in the $(\alpha\alpha)$ or $(\alpha\alpha')$ reactions have yet been reported at the capture resonances, but HEYDENBURG and TEMMER [*43*], and LI and SHERR[6], observing 478 keV gamma radiation from the inelastic scattering of alpha particles in Li^7, found resonances at 1.91, 2.46, and 3.06 MeV. The latter authors evaluated reduced widths for the 1.91 MeV resonance and thus inferred certain limits on its spin. Higher resonances in the Li^7 reactions were reported by HAXEL and STUHLINGER[7] who

[1] G. A. JONES and D. H. WILKINSON: Proc. Phys. Soc. Lond. A **66**, 1176 (1953).
[2] R. A. PECK: Phys. Rev. **73**, 947 (1948).
[3] D. H. WILKINSON and G. A. JONES: Phys. Rev. **91**, 1575 (1953). — Phil. Mag. **45**, 703 (1954).
[4] W. E. BENNETT, P. A. ROYS and B. J. TOPPEL: Phys. Rev. **82**, 20 (1951).
[5] G. A. JONES and D. H. WILKINSON: Phys. Rev. **88**, 423 (1952).
[6] C. W. LI and R. SHERR: Phys. Rev. **96**, 389 (1954).
[7] O. HAXEL and R. STUHLINGER: Z. Physik **114**, 178 (1939).

Table 10. *Resonant reactions of alpha-particles.*

Only reactions in which particle or radiation widths are known are listed and not all resonances are included. Figures are taken mainly from references [*18*], [*24*] and [*32*].

Target nucleus	Reaction channel s	E_{res} MeV	Excitation of compound nucleus MeV	J	Γ_λ keV	$\Gamma_{\lambda s}$ keV	$\frac{2Ma\gamma_{\lambda s}^2}{3\hbar^2}$ %	E_γ MeV	Γ_γ eV	Multi-polarity	$\lvert M\rvert^2$ (Sect. 19)
Li^6	$B^{10}+\gamma$	0.5	4.77	1^+				4.05	0.15	$M1$	0.11
								6.16	0.03	$M1$	0.01
		1.18	5.16	2^+				4.44	0.15	$M1$	0.08
								3.01	0.42	$M1$	0.74
Li^7	$B^{11}+\gamma$	0.4	8.92	—				8.92	—		—
								9.19	—		—
		0.82	9.19	$^5/_2{}^-$	4			4.73	0.4	$M1$	0.18
								2.38	0.07		0.25
								9.28	0.8		$1.9\cdot 10^{-3}$
		0.96	9.28	$^5/_2{}^+$	7			4.82	1.9	$E1$	0.031
								2.47	0.6		0.073
B^{10}	$B^{10}+\alpha$	1.51	12.69	3^-	14	1.7	5.2				
		1.64	12.79	4^+	14	1.0	7.5				
	$C^{12}+d$		12.69	3^-	14	0.93	0.21				
			12.79	4^+	14	2.0	3.2				
	$C^{13}+p$		12.69	3^-	14	0.62	0.01 *				
			12.79	4^+	14	0.18	0.01				
	$N^{13}+n$		12.69	3^-	14	4.3	0.16 *				
			12.79	4^+	14	0.59	0.14				
C^{12}	$C^{12}+\alpha$	3.24	9.58	1^-	860		85				
		3.58	9.84	2^+	1		0.15				
		4.28	10.36	4^+	36		25				
		5.47	11.25	0^+	3300		75				
		5.82	11.51	2^+	106		3				
		5.96	11.62	3^-	1600		70				
		7.04	12.43	1^-	230		4				
O^{16}	$O^{16}+\alpha$	2.49	6.74	0^+	24		19				
		3.04	7.18	3^-	10		31				
		3.09	7.22	0^+	5		1.0				
		3.38	7.45	2^+	10		4.1				
		3.89	7.85	2^+	3		0.5				
Ne^{20}	$Ne^{20}+\alpha$	2.48	11.40	1^-	0.5		6				
Ne^{22}	$Ne^{22}+\alpha$	3.25	13.39	3^-	2.5		15				

* Later work gives a ratio of 5.7 for γ_n^2/γ_p^2 at this resonance.

observed thresholds for extra neutron production as the alpha particle energy was varied in the reaction $Li^7(\alpha n)B^{10}$.

The bombardment of beryllium with alpha particles shows resonances in the (αn) and $(\alpha n')$ reactions leading to the ground state and first excited state of C^{12}. The ground state reaction Q-value is 5.6 MeV and the high yield of neutrons makes the observation of capture radiation at the (αn) resonances very difficult (ALLAN and SARMA[1]). The resonances in neutron (or associated gamma ray) yield were obtained by HALPERN[2], using 0.5 C of polonium, and by TAL-

[1] H. R. ALLAN and N. SARMA: Proc. Phys. Soc. Lond. A **68**, 535 (1955).

[2] I. HALPERN: Phys. Rev. **76**, 248 (1949).

BOTT and HEYDENBURG [43], using artificially accelerated particles. In the former experiment the residual range of alpha particles incident on a beryllium target was varied by alteration of the gas pressure in a reaction vessel. The angular distributions expected in the $Be^9(\alpha n)$ reactions have been listed by HAEFNER[1]. Further resonances in this reaction have been reported, but full details are not yet available [32]; work is also proceeding on the angular distributions and correlations in the reaction, but interpretation is likely to be complex owing to the rather large width of some of the C^{13} levels at 11 MeV excitation.

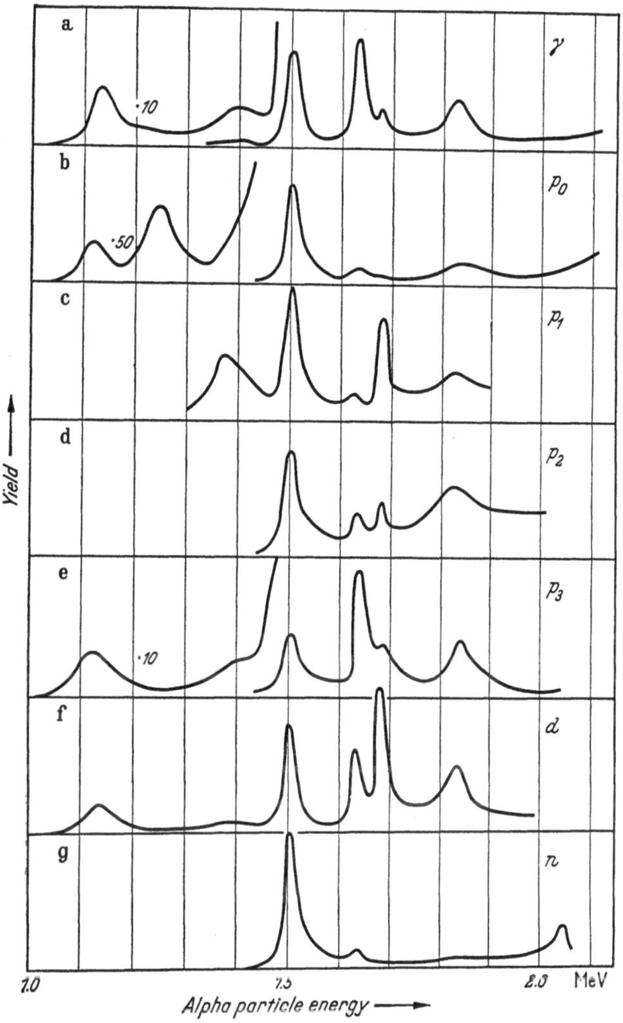

Fig. 45 a—g. Yield curves for $B^{10} + \alpha$ reactions (SHIRE *et al.*).

The resonant reactions of alpha particles with boron have been very fully studied by SHIRE *et al.*[2], who observed protons, deuterons, neutrons and gamma rays from B^{10} and protons from B^{11}. Their yield curves for B^{10} are shown in Fig. 45; that for gamma radiation is extended to higher energies by TALBOTT and HEYDENBURG [43] and that for neutrons by WALKER[3] with low resolution, using a polonium alpha particle source. The levels in N^{14} excited by these reactions are at an excitation of about 12 MeV, and there is a general background between resonances, but total widths of from 5 to 50 keV were determined for the levels excited by alpha particles of energies between 1.0 and 2.0 MeV. The gamma radiation in this work was detected with a sodium iodide crystal, the charged particles by a magnetic analyser, proportional counter or differential ionisation chamber, and the neutrons by a zinc sulphide plastic phosphor. Angular distributions of the more intense groups were measured, and assignments of spins and parities made to the resonance levels in N^{14} and to the levels of C^{13} excited

[1] R. R. HAEFNER: Rev. Mod. Phys. **23**, 228 (1951).
[2] E. S. SHIRE, J. R. WORMALD, G. LINDSAY-JONES, A. LUNDEN and A. G. STANLEY: Phil. Mag. **44**, 1197 (1953).
[3] R. L. WALKER: Phys. Rev. **76**, 244 (1949).

by the emission of the lower energy groups of particles. Some of these assignments were confirmed by angular correlation work[1]. The reaction cross sections observed enabled partial level widths to be deduced for the resonance levels corresponding to alpha particles of energy 1.51 and 1.64 MeV. The alpha particle width was taken to be the smaller of the two alternative values. The ratio of the reduced width for protons to the C^{13} ground state to the single particle width is much smaller than that for the other proton groups. The ratio of the reduced widths for proton and neutron emission to the ground states of the mirror nuclei C^{13} and N^{13} (γ_p^2/γ_n^2) was studied by SHIRE and EDGE[2] who detected the protons in an ionisation chamber and measured the neutrons by means of the SZILARD-CHALMERS reaction in potassium permanganate solution. The ratio γ_n^2/γ_p^2 was found to be (5.7 ± 0.5) MeV at the 1.51 MeV alpha particle resonance; according to the assumption of charge independence this ratio should be unity for mirror reactions. The experimental result has not been explained, except qualitatively by the suggestion of isotopic spin impurity in the compound states. In all other respects (excitation function, angular distribution) the two reactions $B^{10}(\alpha p)\,C^{13}$ and $B^{10}(\alpha n)\,N^{13}$ appear to be very similar, and at the 2.16 MeV resonance energy the ratio γ_n^2/γ_p^2 has fallen to the expected value of about 1. Some of the very detailed information obtained from the $B^{10}+\alpha$ reactions is included in Table 10.

SHIRE and EDGE also investigated the $B^{11}(\alpha p)\,C^{14}$ and $B^{11}(\alpha n)\,N^{14}$ reactions, which are not of mirror type, but which are both resonant. The 2.05 MeV resonance appears in the (αn) reaction but not in (αp), perhaps because of the inversion of the (J, T) values between the two final nuclei. A similar situation for the levels of N^{15} is known from the C^{14} reactions with protons (Sect. 33); the levels in the $B^{11}(\alpha x)$ reactions agree with some found in $C^{14}(p n)$.

The (αp) (αn) and $(\alpha\alpha')$ reactions with C^{12} have high negative Q-values and only the elastic scattering is known to show resonances. It was studied by HILL[3] and later by BITTNER and MOFFAT[4] using a doubly charged beam of alpha particles of energy up to 7.6 MeV from a 3.8 MeV electrostatic accelerator. Their results, which resemble those shown for oxygen in Fig. 16, and are analysed by the same method (Sect. 18), give the resonances listed in Table 10. Some of the levels approximate to the single particle type, and there is good correspondence between the observed level structure and the predictions of the alpha particle model (Sect. 4). Narrow resonances in $C^{13}(\alpha n)$ are reported by JONES and WILKINSON[5].

The $(\alpha\gamma)$ reaction in N^{14} was observed by PRICE[6], and narrow resonances with 1.53 and 1.62 MeV alpha particles were found. The capture gamma radiation was analysed by scintillation spectrometry and shows transitions to an excited state of F^{18} at both resonances and also a ground state transition at the lower resonance. The (αn) reaction was studied by FÜNFER[7] with natural alpha particles and showed resonances, as do the $(\alpha\alpha)$ and (αp) reactions. The latter two reactions have been studied with artificially accelerated alpha particles for energies up to 3.5 MeV (HEYDENBURG and TEMMER [*43*]) and share resonance energies of 2.94, 3.14 and 3.5 MeV. Higher scattering resonances are shown in the early work of DEVONS[8].

---

[1] A. G. STANLEY: Phil. Mag. **45**, 430 (1954).
[2] E. S. SHIRE: and R. D. EDGE: Phil. Mag. **46**, 640 (1955).
[3] R. W. HILL: Phys. Rev. **90**, 845 (1953).
[4] J. W. BITTNER and R. D. MOFFAT: Phys. Rev. **96**, 374 (1954).
[5] G. A. JONES and D. H. WILKINSON: Proc. Phys. Soc. Lond. A **66**, 1176 (1953).
[6] P. C. PRICE: Proc. Phys. Soc. Lond. A **67**, 849 (1954).
[7] E. FÜNFER: Ann. d. Phys. **32**, 313 (1938).
[8] S. DEVONS: Proc. Roy. Soc. Lond., Ser. A **172**, 127 (1939).

The elastic scattering of alpha particles by O^{16} follows the general pattern observed with other spinless nuclei and is discussed in Sect. 18.

Similar high resolution work on the elastic scattering of alpha particles by neon is reported by GOLDBERG *et al.*[1]. In this work an excitation of 12 MeV in

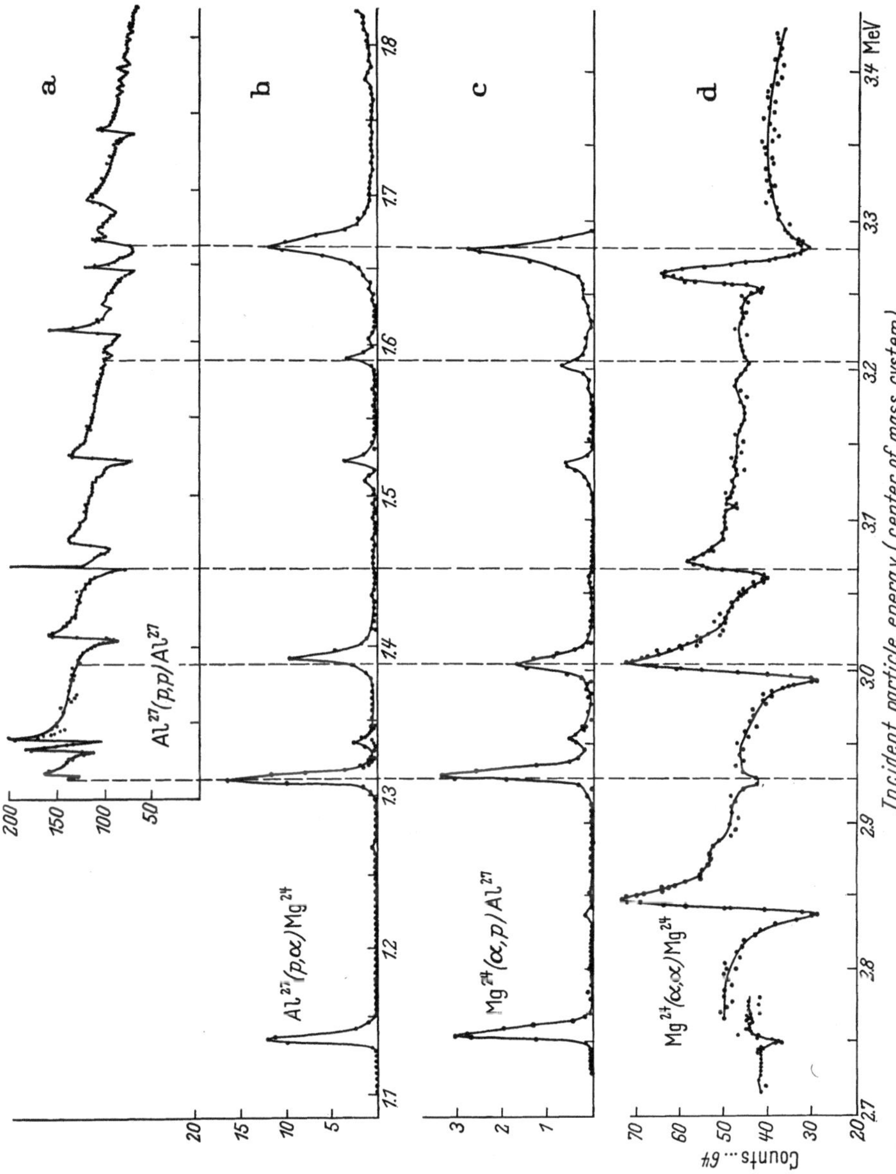

Fig. 46 a—d. Reactions proceeding through compound nucleus Si^{28} (KAUFMANN *et al.*).

Mg^{24} is reached, at which the level spacing shown by the $Na^{23}(p\gamma)$ reaction is about 50 keV. The larger spacing found in the $(\alpha\alpha)$ experiment is due to the fact that this reaction can only excite levels of "natural parity", i.e. 1^- 2^+ 3^- etc., and may only strongly excite levels of $T=0$. Some states of Mg^{26} were also identified in this work.

[1] E. GOLDBERG, W. HAEBERLI, A. I. GALONSKY and R. A. DOUGLAS: Phys. Rev. **93**, 799 (1954).

The $Mg^{24}(\alpha\alpha)$ and $Mg^{24}(\alpha p)$ reactions were also studied by the Wisconsin group[1] using separated isotopic targets. These reactions are interesting because the inverse $(p\alpha)$ reaction has also been studied with high resolution in the same laboratory. The results for the several reactions proceeding through the compound nucleus Si^{28} and observed at an angle of 160° are shown in Fig. 46. It is seen that there is good correspondence of levels (allowing for the interference shift of scattering peaks) and relative intensities in the $(p\alpha)$ and (αp) reactions, and the absolute values are within a factor of 3 of those predicted from the principle of detailed balance. Since the system $Mg^{24}+\alpha$ exists only in natural parity levels, more states are expected to show in the $Al^{27}(pp)$ excitation curve than in the others given in Fig. 46.

Many other (αp) and (αn) reactions which are listed in Table 9 as observed were found in early experiments to show resonance phenomena and will in the future undoubtedly be further studied with high resolution.

IV. Reactions of deuterons with light nuclei.

46. Reactions of deuterons with H^2, H^3 and He^3. The elastic scattering of deuterons by deuterons has been measured for energies of 1 to 3.5 MeV, 8 MeV, and 10.8 MeV[2] in the laboratory system. Because of the symmetry of the initial system all the H^2+H^2 reactions must be symmetrical about 90° in centre of mass coordinates. Some of the distributions obtained are represented in Fig. 47a. RUNGE[3] has given a theoretical treatment of this problem, using the BORN approximation, which agrees with the 10.8 MeV results qualitatively. The minimum at small angles predicted by this theory and due to cancellation between the nuclear and COULOMB scattering has not been observed experimentally. Similar experiments have been conducted with H^3 and He^3 for deuterons of energies between 0.96 and 3.22 MeV from an electrostatic generator, and for 10 MeV deuterons from a cyclotron[4]. The particles scattered from a gas target were counted in the lower energy experiment by a proportional counter and were recorded in the latter in nuclear emulsions. Fig. 47b shows the results for 10 MeV deuterons; at this energy the subsidiary peak in the differential cross section is well developed in comparison with COULOMB scattering which is more evident at lower energies. Fig. 47b shows that the results for He^3 are nearly indistinguishable from those for H^3. This provides some evidence for charge independence of internucleon forces, since the systems $(d\,H^3)$ and $(d\,He^3)$ differ only by a neutron-proton exchange.

The reactions of deuterons with deuterons are $H^2(dn)\,He^3$ and $H^2(dp)\,H^3$. Because of the low barrier due to the large radius of the deuteron these processes have high yields for low energies[5]. Early experiments showed that the angular distribution of protons in the $H^2(dp)$ reaction was anisotropic in the centre-of-

---

[1] S. G. KAUFMANN, E. GOLDBERG, L. J. KOESTER and F. P. MOORING: Phys. Rev. **88**, 673 (1952).

[2] 1 to 3.5 MeV: J. M. BLAIR, G. FREIER, E. LAMPI, W. SLEATOR and J. H. WILLIAMS: Phys. Rev. **74**, 1594 (1948). — 8 MeV: H. B. BURROWS *et al.*: Reference [*40*]. — 10 to 10.8 MeV: J. C. ALLRED, K. W. ERICKSON, J. L. FOWLER and E. J. STOVALL: Phys. Rev. **76**, 1430 (1949). — L. ROSEN and J. C. ALLRED: Phys. Rev. **88**, 431 (1952).

[3] R. J. RUNGE: Phys. Rev. **85**, 1052 (1952).

[4] 0.96 to 3.22 MeV: W. R. STRATTON, G. D. FREIER, G. R. KEEPIN, D. RANKIN and T. F. STRATTON: Phys. Rev. **88**, 257 (1952). — R. J. S. BROWN, K. F. FAMULARO, H. D. HOLMGREN, D. RANKIN and T. F. STRATTON: Phys. Rev. **96**, 80 (1954). — 10 MeV: J. C. ALLRED, A. H. ARMSTRONG, A. M. HUDSON, R. M. POTTER, E. S. ROBINSON, L. ROSEN and E. J. STOVALL: Phys. Rev. **88**, 425 (1952).

[5] 0.63 mb at 25 keV.

mass system, even at low deuteron energies. BRETSCHER *et al.*[1], and later WENZEL and WHALING[2], and PRESTON *et al.*[3] expressed the differential cross section in the form $1 + A\cos^2\vartheta$ for energies below 200 keV and showed that higher even powers of $\cos\vartheta$ were necessary as the bombarding energy increased. It was also found that, although the $(dn)/(dp)$ branching ratio is nearly unity, there is a small but significant difference between the total cross sections for the two

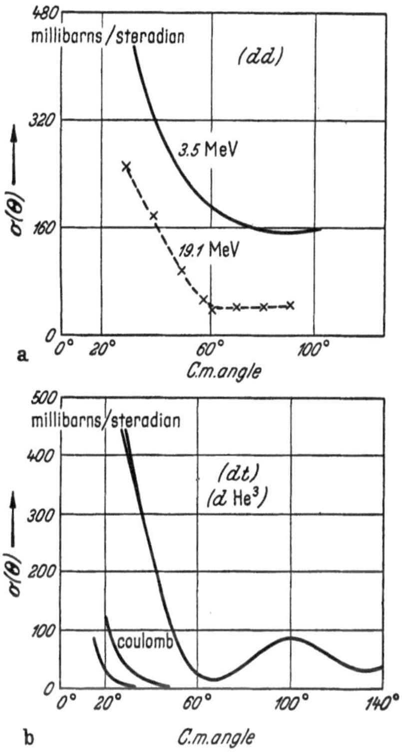

Fig. 47.

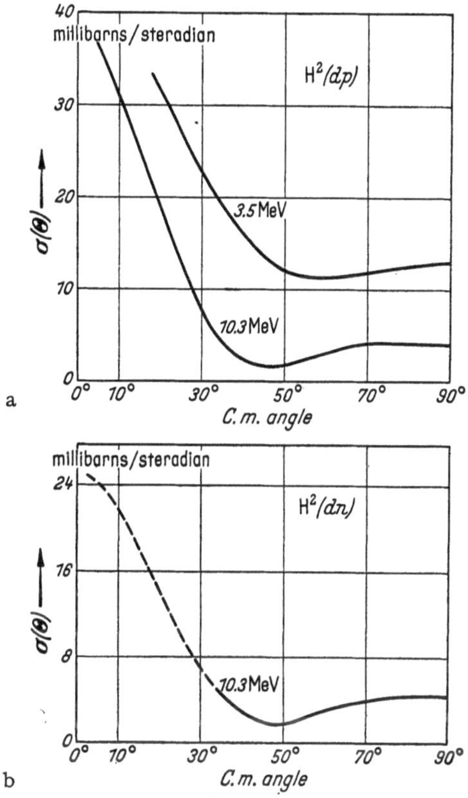

Fig. 48.

Fig. 47 a and b. (a) Angular distribution of $d+d$ elastic scattering at 3.5 and 19.1 MeV [BLAIR *et al.*, R. G. FREEMANTLE (unpublished)]. (b) Angular distribution of (dt) and $(d\,He^3)$ elastic scattering, showing calculated COULOMB scattering (ALLRED *et al.*).

Fig. 48 a and b. (a) Angular distribution of $H^2(dp)$ reaction at 3.5 and 10.3 MeV (ALLRED *et al.*). (b) Angular distribution of $H^2(dn)$ reaction at 10.3 MeV (ERICKSON *et al.*).

reactions in the energy range below 500 keV and that the angular distributions for the two reactions also differ significantly. The excitation function for the (dp) reaction and for the sum of the (dp) and (dn) reaction cross sections follows a GAMOW function below 100 keV bombarding energy. These experiments were conducted by passing a beam of deuterons into a thin gas target and detecting the protons or He^3 particles in a proportional counter, a magnetic analyser with scintillation counter, or a nuclear emulsion.

Differential cross sections for both (dp) and (dn) reactions were obtained by BLAIR *et al.*[4] for deuteron energies from 1 to 3.5 MeV. BURROWS *et al.* [40]

[1] E. BRETSCHER, A. P. FRENCH and F. G. P. SEIDL: Phys. Rev. **73**, 815 (1948).
[2] W. A. WENZEL and W. WHALING: Phys. Rev. **88**, 1149 (1952).
[3] G. D. PRESTON, P. F. D. SHAW and S. A. YOUNG: Proc. Roy. Soc. Lond., Ser. A **226**, 206 (1954).
[4] J. M. BLAIR, G. FREIER, E. LAMPI, W. SLEATOR and J. H. WILLIAMS: Phys. Rev. **74**, 1599 (1948).

using 8 MeV deuterons from a cyclotron, and ALLRED *et al.*[1] using 10.3 MeV deuterons found similar angular distributions; some of the results are shown in Fig. 48. Terms in $\cos^8\vartheta$ are necessary to express the 10 MeV distribution analytically.

The (dn) angular distributions were obtained by HUNTER and RICHARDS[2] over the energy range 0.5 to 3.7 MeV; a thin gas target was used and an energy-insensitive long counter was used for detection. These support the conclusion that at low deuteron energies the angular distribution should change fairly rapidly with deuteron energy because of changing (centrifugal) penetrability, but that at higher deuteron energies the penetrability of the outgoing neutrons should be the determining factor and the shape of the distributions should be relatively energy independent. Fig. 48b is the 10.3 MeV curve given by ERICKSON *et al.*[3] who counted He^3 particles for large angles and used induced radioactivity in copper to survey the neutron emission in the forward direction. As with the (dp) reaction terms in $\cos^8\vartheta$ are required for the analytical expression of the angular distribution.

The results on the H^2+H^2 reactions for the energy range up to 4 MeV may be expressed by the formula $\sigma_\vartheta=\sigma_0(1+A\cos^2\vartheta+B\cos^4\vartheta+\cdots)$ in which A, B and C are functions of energy. BEIDUK *et al.*[4] interpret the variation of A, B, C in terms of penetration of the centrifugal barriers by the different partial waves necessary to connect the initial and final states ${}^1S\to{}^1S$, ${}^3P\to{}^3P$, ${}^1D\to{}^1D$, and they find that transitions of the type ${}^3P\to{}^1P$ and ${}^3P\to{}^3F$ must be included. This can only be so if spin and orbital momenta are not individually conserved, i.e. if spin-orbit coupling forces, as assumed in the nuclear shell model, exist between the nucleons composing the deuterons. The theoretical treatment for deuteron energies higher than 4 MeV might proceed in a similar way, with the introduction of higher partial waves and more spin-orbit coupling, but FAIRBAIRN[5] has suggested that both (dp) and (dn) reactions may proceed by a stripping process. The theoretical expression derived for the H^2+H^2 collision well describes the results obtained with deuterons of energy greater than 8 MeV, but does not well fit results at lower energies, presumably owing to COULOMB effects. The operation of spin-orbit forces in the H^2+H^2 reactions leads to polarisation of the outgoing particles. (See article by BLIN-STOYLE and GRACE in Vol· XLI of this Encyclopedia.)

The reactions of deuterons with H^3 and He^3 are the mirror processes

$$H^2+H^3\to He^5\to He^4+n \tag{46.1}$$

and

$$H^2+He^3\to Li^5\to He^4+p. \tag{46.2}$$

Each of these reactions exhibits an apparent resonance of the compound nucleus at low deuteron bombarding energies. The experimental methods used are the same as those discussed for the elastic scattering studies. For energies below 45 keV JARVIS and ROAF[6] find that the difference between the GAMOW plots for the two excitation functions is due to the double charge of the He^3 nucleus. The yield curve for the $H^2(tn)$ reaction, which has been much used as a source of high energy neutrons (Sect. 12), is given by BRETSCHER and FRENCH[7] to 125 keV

[1] J. C. ALLRED, D. D. PHILLIPS and L. ROSEN: Phys. Rev. **82**, 782 (1951).
[2] G. T. HUNTER and H. T. RICHARDS: Phys. Rev. **76**, 1445 (1949).
[3] K. W. ERICKSON, J. L. FOWLER and E. J. STOVALL: Phys. Rev. **76**, 1141 (1949).
[4] F. M. BEIDUK, J. R. PRUETT and E. J. KONOPINSKI: Phys. Rev. **77**, 622 (1950).
[5] W. M. FAIRBAIRN: Proc. Phys. Soc. Lond. A **67**, 990 (1954).
[6] R. G. JARVIS and D. ROAF: Proc. Phys. Soc. Lond. A **66**, 310 (1953).
[7] E. BRETSCHER and A. P. FRENCH: Phys. Rev. **75**, 1154 (1949).

and by ARGO *et al.*[1] to 1700 keV. If a compound nucleus resonance near 100 keV is assumed, it has to be described as $J = {}^3/_2{}^+$. FLOWERS[2] however is able to describe the course of the cross section curve by considering barrier penetration for different partial waves. The simplest and most plausible transition ${}^2S \rightarrow {}^2S$ is inadequate to provide the experimentally observed cross section, but with spin-orbit coupling the transition ${}^4S \rightarrow {}^2D$ can be invoked to give a satisfactory description. This requires only *s*-waves so that the cross section at low energies should be isotropic, as indeed is found. At higher energies anisotropy is found (STRATTON and FREIER[3]), indicating the intervention of higher partial waves. The angular distribution at 10 MeV, given by BROLLEY *et al.*[4] may be interpreted as a stripping distribution (BUTLER and SYMONDS[5]) as may be the distribution

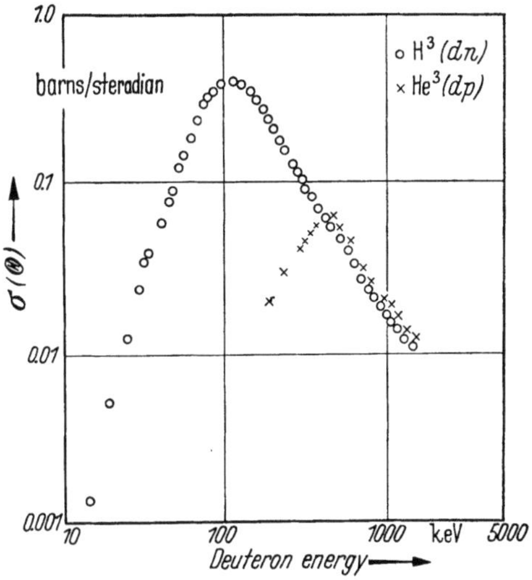

Fig. 49. Cross section for $H^3(dn)$ and $He^3(dp)$ reactions (BONNER *et al.*).

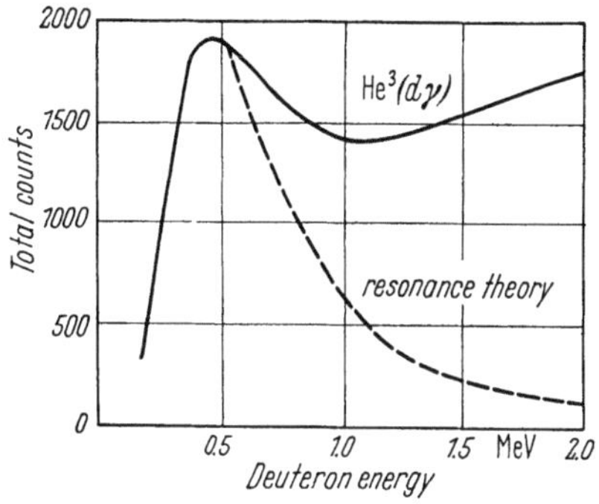

Fig. 50. Excitation functions for radiative capture of deuterons in He^3 (BLAIR *et al.*).

for the mirror (dp) reaction with He^3. This latter reaction was found[6] to show a broad maximum in cross section at about 430 keV with an isotropic angular distribution. If this is a resonance level of Li^5 the cross section at resonance implies a state with $J = {}^3/_2{}^+$ as for the mirror nucleus He^5, but probably an explanation in terms of direct transitions can be given as in the case of $H^3(dn)$. The similarity of the $H^3(dn)$ and $He^3(dp)$ reactions is shown in Fig. 49. Assuming a resonance, the deuteron reduced widths for these reactions are given [*32*] as 42 and 21% of the single particle value respectively while the nucleon widths are 1.2 and 0.4%; the excitation of the compound nucleus is about 16.8 MeV in each case. Radiative capture of deuterons at this level, with an $E1$ transition to the ground state was observed by BLAIR *et al.*[7]. They passed a deuteron beam into a chamber containing He^0 gas and detected radiation in a large sodium iodide crystal. Their excitation curve, for a gamma ray of energy (16.6 ± 0.2) MeV is shown in Fig. 50 together with a theoretical curve calculated using reduced

[1] H. V. ARGO, R. F. TASCHEK, H. M. AGNEW, A. HEMMENDINGER and W. T. LELAND: Phys. Rev. **87**, 612 (1952). See also J. P. CONNER, T. W. BONNER and J. R. SMITH: Phys. Rev. **88**, 468 (1952), who used the $H^3(dn)$ reaction.

[2] B. H. FLOWERS: Proc. Roy. Soc. Lond., Ser. A **204**, 503 (1951).

[3] T. F. STRATTON and G. D. FREIER: Phys. Rev. **88**, 261 (1952), [$H^3(dn)$].

[4] J. E. BROLLEY, J. L. FOWLER, E. J. STOVALL: Phys. Rev. **82**, 502 (1951), [$H^3(dn)$].

[5] S. T. BUTLER and J. L. SYMONDS: Phys. Rev. **83**, 858 (1951).

[6] T. W. BONNER, J. P. CONNER and A. B. LILLIE: Phys. Rev. **88**, 473 (1952). — J. L. YARNELL, R. H. LOVBERG and W. R. STRATTON: Phys. Rev. **90**, 292 (1953). — W. E. KUNZ: Phys. Rev. **97**, 456 (1955); see also D. P. HALE and R. D. PRESENT: Phys. Rev. **99**, 1366 (1955) for a theoretical discussion of the mirror reactions.

[7] J. M. BLAIR, N. M. HINTZ and D. M. VAN PATTER: Phys. Rev. **96**, 1023 (1954).

widths obtained from the (dp) reaction. The radiative width is 11 eV which gives (Sect. 19) $|M|^2 = 7.3 \times 10^{-3}$. A resonance effect is noticed in the elastic scattering of deuterons by He^3 as an interference with the COULOMB scattering (FREIER and HOLMGREN [1]).

The mirror reaction $H^3(d\gamma)\,He^5$ has not been found; SAWYER and BURKHARDT [2] quote a cross section of less than 5×10^{-28} cm² for a deuteron energy of 160 keV, which is not far from resonance in the $H^3(dn)$ reaction. The $H^3(dp)\,H^4$ reaction has not been observed [3].

47. Scattering of deuterons by He⁴. The elastic scattering of deuterons by helium provides a very convenient way of studying the energy levels of the Li^6 nucleus because of the low binding energy of a deuteron in this nucleus. Only $T=0$ levels of Li^6 should be so excited and in addition no level of Li^6 can be formed in this reaction if it has zero spin and even parity because of the conservation laws for these quantities. The known levels of Li^6 are shown in Fig. 84; there is good evidence for the spin assignments shown in the case of the 2.189 and 3.57 MeV levels from the $Be^9(p\alpha)\,Li^6$ reaction, and the 3.57 MeV level lies near the position expected for the first $T=1$ state of Li^6. Scattering experiments were made for a range of energies up to 3.5 MeV by BLAIR *et al.* [4] but these, like early work with natural alpha particle sources, did not reveal any anomalies corresponding to levels of Li^6. LAURITSEN *et al.* [5] however, using deuterons of more precisely defined energy, and a differentially pumped gas target, were able to demonstrate resonant scattering from the 2.189 MeV level; their results are shown in Fig. 51. Later work [6] has shown no resonant scattering at 174° from the next $(T=1)$ level at 3.57 MeV, as expected from the isotopic spin rules, but scattering from levels at 4.5 $(J=2^+)$ and 5.4 MeV $(J=1^+)$. Analysis of the lowest resonance shows that the observed width of 23 keV cannot be produced by particles incident with more than two units of orbital momentum. The possibility of s, p and d-waves is then considered and tried in the phase shift formula for particles of spin 1 to find the best prediction for the differential cross section at 156° and 120°. Only $l=2$, $J=3^+$ gives reasonable agreement with the experimental results. Using these values, a fairly good detailed fit to the experimental curves could be obtained with an interaction radius of 5.0 to 5.4×10^{-13} cm which is rather larger than that predicted by the usual formula. The fit is improved when account is taken of coherent scattering from the ground state level, which may have a large reduced width. The width of the level is in agreement with the upper limit found from the $Be^9(p\alpha)$ reaction; this is one of the few cases in which the same finite width may be observable both in an intermediate state and as an

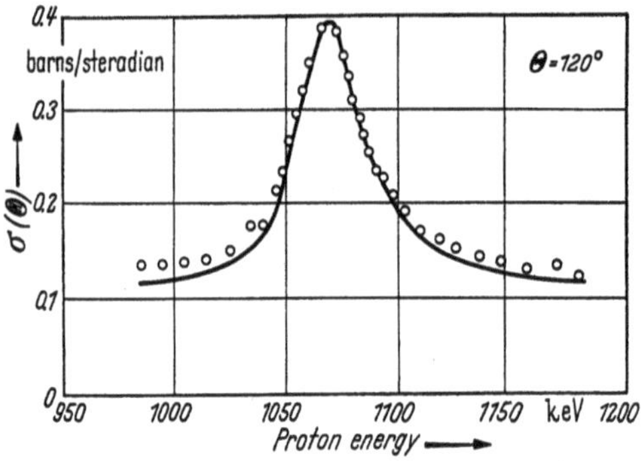

Fig. 51. Differential cross section for d-He^4 scattering at 120° c. m. (LAURITSEN *et al.*).

[1] G. FREIER and H. HOLMGREN: Phys. Rev. **93**, 825 (1954).
[2] G. A. SAWYER and L. C. BURKHARDT: Phys. Rev. **98**, 1305 (1955).
[3] K. G. MCNEILL and W. RALL: Phys. Rev. **83**, 1244 (1951).
[4] J. M. BLAIR, G. FREIER, E. E. LAMPI and W. SLEATOR: Phys. Rev. **75**, 1678 (1949).
[5] T. LAURITSEN, T. HUUS and S. G. NILSSON: Phys. Rev. **92**, 1501 (1953).
[6] A. GALONSKY and M. T. MCELLISTREM: Phys. Rev. **98**, 590 (1955).

end product of a reaction. The reduced widths for the $d + He^4$ system at the observed resonances are about half the single particle value [32].

The $(d - He^4)$ interaction for 10.4 MeV deuterons from a cyclotron has been studied by ALLRED *et al.*[1] and for 13.7 and 19 MeV deuterons by FREEMANTLE *et al.*[2]. Their results demonstrate the increasing importance of higher partial waves as the bombarding energy increases. Evidence was obtained in this work for the $He^4(dp)He^5$ reaction with a Q-value of (-3.10 ± 0.05) MeV.

48. Scattering of deuterons by nuclei with $A > 4$. No sharp resonance phenomena in the elastic or inelastic scattering of deuterons have been reported for these nuclei, but for inelastic scattering by Mg^{24}, GREENLEES[3] found a broad peak in the excitation function for particles observed at 70° at a deuteron energy of about 5.5 MeV. The theoretical discussion of deuteron inelastic scattering given by HUBY and NEWNS[4] on the basis of a stripping mechanism can be adapted to explain these results. This theory also gives a partial description of the angular distribution found by LEVINE *et al.*[5] for the inelastic scattering of 14.5 MeV deuterons by lithium with excitation of the first level of Li^7 at 478 keV.

The inelastic scattering of deuterons by the nuclei Li^6, B^{10}, and N^{14} has been studied with the object of verifying the position of the first state of $T = 1$ of these nuclei. BOCKELMAN *et al.*[6] compared proton and deuteron scattering from B^{10} and N^{14} and demonstrated the absence for 90° inelastic deuteron scattering of groups corresponding to the 1.74 and 2.31 MeV levels respectively of these nuclei. A full examination of the whole angular range remains to be completed, but there is at first sight some evidence in this work for the conservation of total isotopic spin. Some of the results are shown in Fig. 32.

The differential cross section for elastic scattering of 8 MeV deuterons by Be^9 was obtained by EL-BEDEWI [40]; it shows, in addition to the rise at small angles, a subsidiary peak at 64° in the centre-of-mass system.

The interaction of 19 MeV deuterons with C^{12} was studied by FREEMANTLE *et al.*[7] who found differential cross sections for elastic scattering and for inelastic scattering from the 4.43 and 9.61 MeV states of C^{12}. All distributions show a rise at forward angles, that for elastic scattering being due to COULOMB effect and those for the inelastic process presumably being due to the stripping mechanism. The observation of forward peaks in inelastic scattering is difficult because the bombarding beam may enter the counter used for detecting the lower energy inelastic groups. The 7.65 MeV level of C^{12} was not excited in this work; since this cannot be a $T = 1$ state the absence of inelastic deuteron groups must not necessarily always be ascribed to isotopic spin selection rules.

GUGGENHEIMER *et al.*[8] observed the elastic scattering of 6.3 MeV deuterons from O^{16} and N^{14} and MIDDLETON and TAI[9] obtained both elastic and inelastic distributions from Ne^{20} with 8 MeV deuterons. All these distributions, like those from carbon, show a rise at forward angles and subsidiary maxima which

---

[1] J. C. ALLRED, D. K. FROMAN, A. M. HUDSON and L. ROSEN: Phys. Rev. **82**, 786 (1951).

[2] R. G. FREEMANTLE, T. GROTDAL, W. M. GIBSON, R. MCKEAGUE, D. J. PROWSE and J. ROTBLAT: Phil. Mag. **45**, 1090 (1954).

[3] G. W. GREENLEES: Proc. Phys. Soc. Lond. A **68**, 97 (1955).

[4] R. HUBY and H. C. NEWNS: Phil. Mag. **42**, 1442 (1951).

[5] S. H. LEVINE, R. S. BENDER and J. N. MCGRUER: Phys. Rev. **97**, 1249 (1955).

[6] C. K. BOCKELMAN, C. P. BROWNE, W. W. BUECHNER and A. SPERDUTO: Phys. Rev. **92**, 665 (1953).

[7] R. G. FREEMANTLE, W. M. GIBSON and J. ROTBLAT: Phil. Mag. **45**, 1200 (1954).

[8] K. M. GUGGENHEIMER, H. HEITLER and C. F. POWELL: Proc. Roy. Soc. Lond., Ser. A **190**, 196 (1947).

[9] R. MIDDLETON and C. T. TAI: Proc. Phys. Soc. Lond. A **65**, 752 (1952).

shift with energy. Gamma radiation from inelastic scattering of deuterons in F^{19} is reported by THOMPSON [39] and the level system of Na^{23} has also been studied [32]. The well known level at 1.37 MeV excitation in Mg^{24} was found in inelastic deuteron scattering by GREENLEES *et al.*[1], and by HOLT and YOUNG[2]. The latter authors obtained the angular distribution, which is similar to those found for other elements, and by application of their theory HUBY and NEWNS were able to suggest even parity and spin 1, 2, or 3 for the Mg^{24} level ($J=2^+$). The levels of Al^{27} excited by 10 MeV deuterons were studied by KELLER[3] who magnetically analysed the inelastically scattered groups. His results show that not all of the levels of Al^{27} known from inelastic proton scattering are excited by deuterons.

Theoretical interpretation of deuteron scattering phenomena is still incomplete, and experimental evidence is neither as extensive nor as accurate as is desirable.

49. The ($d\,\alpha$) reaction. The ($d\,\alpha$) reaction in the lighter nuclei generally has a high positive Q-value and is therefore useful for a survey of excited states of residual nuclei. The reaction does not change isotopic spin in the absence of perturbations. The reactions observed are listed in Table 11.

A broad resonance at 347 keV was found in the bombardment of Li^6 by deuterons by WHALING and BONNER[4], and there is some evidence for the same resonance from the ($d\,p$) reaction, but the interpretation of such peaks at very

Table 11. *Reactions of deuterons with light nuclei* ($A > 4$).

Column 2 gives the excitation of the compound nucleus. Columns 5, 6, 7 give observed ground state Q-values in MeV taken mainly from [31]. *a*: activity observed; *p*: reaction particles observed.

Target nucleus	Excitation of compound nucleus MeV	dd	dd'	dp	dn	$d\alpha$	Target nucleus	Excitation of compound nucleus MeV	dd	dd'	dp	dn	$d\alpha$
Li^6	22.26	*	*p*	5.027	3.40	22.39	Mg^{25}	17.14			8.880	5.58	7.019
Li^7	16.68	*	*p*	−0.192	15.0	14.2	Mg^{26}	13.75			4.207	5.68	*a*
Be^9	15.82	*p**	*p*	4.588	4.35	7.153	Al^{27}	17.84		*p*	5.494	9.08	6.694
B^{10}	25.18	*p*	*p*	9.235	6.6	17.86	Si^{28}	11.76			6.246	0.29	1.416
B^{11}	18.67			1.137	13.8	8.018	Si^{29}	15.68			8.388	3.27	5.994
C^{12}	10.26	*p*	*p*	2.722	−0.281		Si^{30}	13.00			4.364	4.92	3.120
C^{13}	16.15	*		5.942	5.40	5.164	P^{31}	15.28			5.704	6.81	8.158
C^{14}	10.47			0.15	8.18	0.25	S^{32}	11.51			6.422	0.25	*a*
N^{14}	20.72	*p**	*p*	8.614	5.12	13.40	S^{33}	15.57			8.67	*a*	—
N^{15}	14.03			0.23	10.9	7.681	S^{34}	12.75			*a*	—	*a*
O^{16}	7.53	*p*		1.919	−1.630	3.115	S^{36}	12.31			—	—	—
O^{17}	13.79			5.821	*a*	9.807	Cl^{35}	15.11			6.28	—	9.1
O^{18}	12.33			1.730	5.7	4.237	Cl^{37}	14.56			4.0	—	*a*
F^{19}	17.40	*	*p*	4.373	10.8	10.038	A^{36}	11.70			6.58	—	—
Ne^{20}	11.27		*p*	4.528	−0.17	2.791	A^{38}	11.99			*p*	—	—
Ne^{21}	16.93			8.137	—	6.432	A^{40}	12.99			3.84	5.97	*a*
Ne^{22}	13.53			2.964	—	2.62	K^{39}	14.48			5.576	—	*a*
Na^{23}	16.80		*p*	4.727	9.23	6.902	K^{41}	15.97			5.12	—	—
Mg^{24}	11.42		*p*	5.097	0.07		Ca^{40}	—			6.138	—	4.650

* (dt) reaction also observed.

[1] G. W. GREENLEES, A. E. KEMPTON and E. H. RHODERICK: Nature, Lond. **164**, 663 (1949).
[2] J. R. HOLT and C. T. YOUNG: Nature, Lond. **164**, 1000 (1949).
[3] K. K. KELLER: Phys. Rev. **84**, 884 (1951).
[4] W. WHALING and T. W. BONNER: Phys. Rev. **79**, 258 (1950).

high excitations of the compound nucleus is not certain; it was not confirmed by HIRST *et al.*[1]. No other resonances have been reported in $(d\alpha)$ reactions. The angular distribution of the $Li^6(d\alpha)$ reaction was studied by HEYDENBURG *et al.*[2] and treated theoretically by RESNICK and INGLIS[3]. The distribution is of the form $1+A(E)\cos^2\vartheta+B(E)\cos^4\vartheta$ for energies up to 3.5 MeV as expected from the symmetry of the final system, and indicates the effectiveness of *s*- and *d*-waves. Odd waves are not permitted by symmetry rules. A straightforward application of angular distribution theory, similar to that for the $Li^7(p\alpha)$ reaction (Sect. 27) suggests that two states of the compound nucleus Be^8 are necessary to account for the observed features of the angular distribution, a sharp state with $J=2$ (which may be that observed by WHALING and BONNER) and a broad peak with $J=0$ formed by *d*-waves.

The $Li^7(d\alpha)\,He^5$ reaction is complicated by the instability of He^5. The spectrum of alpha particles from this reaction therefore displays a broad high energy group (α_1) and a continuous distribution of particles (α_2) from the break-up of the recoiling He^5. FRENCH and TREACY[4] observed coincidences between α_1 and α_2 using a differential ionisation chamber as a detecting instrument, and from the range distribution of the α_2 particles were able to infer the $(\alpha_1-\alpha_2)$ correlation function. This was found to be anisotropic, showing that the He^5 ground state spin is not $^1/_2$. Coincidences between neutrons and alpha particles in this reaction have also been observed. The width of the ground state group is about 0.5 MeV.

The $Be^9(d\alpha)\,Li^7$ reaction has been much used for exciting the states of Li^7 and the 0.48 MeV and 4.62 MeV states have been seen with moderate bombarding energies[5] and possibly the 7.65 MeV state with 14 MeV deuterons[6]. The angular distribution for the ground state group is isotropic for deuterons of energy from 0.3 to 0.7 MeV[7], and the correlation between the alpha particles to the first excited and the subsequent radiation is isotropic[8]. This confirms the spin $J=\frac{1}{2}$ for the first excited state of Li^7. The alpha particle spectrum in the bombardment of beryllium by deuterons includes a continuous distribution of particles from the three body process $Be^9(d\,2\alpha)\,H^3$.

The $(d\alpha)$ reaction in B^{10}, with a Q-value of 17.8 MeV, is a convenient way of surveying the levels of Be^8. The alpha particle distribution is complicated by the presence of particles from the break-up of Be^8 in its ground state and excited states, and there is some disagreement on the number of excited states actually seen [*32*], but all observers agree on the appearance in this reaction of the 2.9 MeV level with a width of about 1 MeV. TREACY[9] finds that the alpha distribution corresponding to this level can be fitted by the BREIT-WIGNER formula for a single level with $J=2^+$. No gamma radiation has with certainty been ascribed to this reaction; the excess of excited state transitions over those to the ground state was correctly interpreted by GOLDHABER[10], before the spin of B^{10} was definite, to indicate a high spin for this nucleus. The $(d\alpha)$ reaction in B^{11} excites the ground state and 2.43 MeV state of Be^9. Although the observed width of

[1] F. HIRST, I. JOHNSTONE and M. J. POOLE: Phil. Mag. **45**, 762 (1954).
[2] N. P. HEYDENBURG, C. M. HUDSON, D. R. INGLIS and W. D. WHITEHEAD: Phys. Rev. **74**, 405 (1948).
[3] R. RESNICK and D. R. INGLIS: Phys. Rev. **76**, 1318 (1949).
[4] A. P. FRENCH and P. B. TREACY: Proc. Phys. Soc. Lond. A **64**, 452 (1951).
[5] R. W. GELINAS and S. S. HANNA: Phys. Rev. **89**, 483 (1953).
[6] H. E. GOVE and J. A. HARVEY: Phys. Rev. **82**, 658 (1951).
[7] I. RESNICK and S. S. HANNA: Phys. Rev. **82**, 463 (1951).
[8] R. G. UEBERGANG and N. W. TANNER: Austral. J. Phys. **6**, 53 (1953).
[9] P. B. TREACY: Phil. Mag. **44**, 325 (1953).
[10] M. GOLDHABER: Phys. Rev. **74**, 1194 (1948).

the latter state is less than 7 keV (VAN PATTER *et al.*[1]) it has been shown by time-of-flight coincidence counting to decay mainly by neutron emission (DISSANAIKE and NEWTON[2]). No other states up to 5.1 MeV were observed in Be^9 by this reaction which is expected to exhibit only those levels which have the same isotopic spin, $T=\frac{1}{2}$, as B^{11}.

The $C^{12}(d\alpha)\,B^{10}$ reaction, rather exceptionally, has a negative Q-value. ASHMORE and RAFFLE[3] using a beam of 7 MeV deuterons from a cyclotron, coupled with magnetic analysis of the disintegration particles, found excitation of the ground state and 0.72 MeV state of B^{10}. The $T=1$ state at 1.74 MeV was not observed, but neither was the nearby $T=0$ state at 2.15 MeV. The reaction $C^{13}(d\alpha)\,B^{11}$ apparently does not share the resonance peak at 0.65 MeV shown by the $C^{13}(dp)$ reaction (CURLING and NEWTON[4]).

In the $N^{14}(d\alpha)\,C^{12}$ reaction DUNBAR *et al.*[5] found groups corresponding to C^{12} levels at 4.43 and 7.65 MeV as well as the ground state transition; the 7.65 MeV level, which is weakly excited, is unstable to alpha particle emission. The 9.61 MeV level of C^{12} was observed by MALM and BUECHNER[6], who also studied the $N^{15}(d\alpha)\,C^{13}$ reaction with a magnetic spectrograph. STANLEY[7] confirmed the spin of 2^+ for the 4.43 MeV level of C^{12} by angular correlation measurements.

The $O^{16}(d\alpha)\,N^{14}$ process has been frequently quoted as an example of isotopic spin conservation because of the non-appearance of the 2.1 MeV ($T=1$) level of N^{14} in the experiments of all workers so far, although higher levels have been seen[8]. The Q-value for the $(d\alpha)$ reaction with the rare isotope O^{17} has been measured by a magnetic analysis technique using a target prepared by bombarding a foil with O^{17} ions in an isotope separator (PAULI *et al.*[9]) and similar work is reported for O^{18} [*32*] indicating a group of low lying states in N^{16}. The reaction $F^{19}(d\alpha)\,O^{17}$ gives the low excited states of O^{17} in agreement with results from the (dp) reaction and $Ne^{20}(d\alpha)\,F^{18}$ is another reaction in which a low lying $T=1$ state should in principle not be excited. Unfortunately, the level spectrum of F^{18} is not well known and the $(d\alpha)$ work of MIDDLETON and TAI[10] does not specifically fail to excite any level known to exist from other reaction studies. The $(d\alpha)$ reactions of the neon isotopes were also studied using targets prepared in a mass separator, and MILEIKOWSKY and WHALING[11] thus discovered the low lying excited states of F^{19} at 110 and 197 keV.

Observations of the $(d\alpha)$ reaction in Na^{23}, Mg^{25}, Al^{27}, Si^{29}, P^{31}, Cl^{35} are confined to Q-value measurements for ground and excited states except in the case of Al^{27} for which SCHELBERG *et al.*[12] using a magnetically resolved cyclotron beam of 11 MeV deuterons obtained a number of angular distributions. The even mass nuclei from Ne^{22} to Ca^{40} have rather small Q-values and the COULOMB barrier for the emergent alpha particles makes the yield for the heavier nuclei

---

[1] D. M. VAN PATTER, A. SPERDUTO, K. HUANG, E. N. STRAIT and W. W. BUECHNER: Phys. Rev. **81**, 233 (1951).

[2] G. A. DISSANAIKE and J. O. NEWTON: Proc. Phys. Soc. Lond. **65**, 675 (1952).

[3] A. ASHMORE and J. F. RAFFLE: Proc. Phys. Soc. Lond. A **64**, 754 (1951).

[4] C. D. CURLING and J. O. NEWTON: Nature, Lond. **165**, 609 (1950).

[5] D. N. F. DUNBAR, R. E. PIXLEY, W. A. WENZEL and W. WHALING: Phys. Rev. **92**, 649 (1953); see also D. CARTWRIGHT, L. L. GREEN and J. C. WILLMOTT: Phil. Mag. **45**, 742 (1954)

[6] R. MALM and W. W. BUECHNER: Phys. Rev. **81**, 519 (1951).

[7] A. G. STANLEY: Phil. Mag. **45**, 807 (1954).

[8] D. S. CRAIG, D. J. DONAHUE and K. W. JONES: Phys. Rev. **88**, 808 (1952). This paper gives references to earlier work.

[9] R. PAULI, K. AHNLUND and C. MILEIKOWSKY: Ark. Fysik **8**, 213 (1954).

[10] R. MIDDLETON and C. T. TAI: Proc. Phys. Soc. Lond. **64**, 801 (1951).

[11] C. MILEIKOWSKY and W. WHALING: Phys. Rev. **88**, 1254 (1952).

[12] A. D. SCHELBERG, M. B. SAMPSON and R. G. COCHRAN: Phys. Rev. **80**, 574 (1950).

low even when the energy of the incident deuteron is comparable with the potential barrier. They have been established by detection of the positron activity of the final nucleus in deuteron bombardments with cyclotron beams, and a few excitation functions have been obtained.

The $Si^{28}(d\alpha)Al^{26}$ reaction has been studied with deuterons of energy from 5 to 7 MeV, and with magnetic analysis of the alpha particles by BROWNE[1] in order to identify $T=0$ levels of Al^{26}. The energy release in the reaction $Si^{30}(d\alpha)Al^{28}$ has also been measured by magnetic spectroscopy [*31*].

50. The (dn) and (dp) reactions; determination of energy levels. Because of high positive Q-values (Table 11) the (dp) and (dn) reactions have, since 1935, been an important source of information on excited states of residual nuclei. These are best located, when possible, by high resolution magnetic analysis of proton groups from the (dp) reaction, and by observation of the subsequent cascades of gamma radiation. Neutron spectroscopy is less accurate but also in the (dn) reaction the gamma radiation from residual nuclei may be measured with high precision. The levels excited by the (dp) reaction may also be observed in the cascade transitions following slow neutron capture and there is in general very good agreement between the two sets of observations. The experimental techniques are discussed in Sects. 13 and 14.

Individual energy levels located by the (dp) reaction are too numerous to mention in this article and are given in [*32*] and [*33*]. They are mainly due to the work of BUECHNER and his collaborators[2]. The associated gamma radiation has been measured by THOMAS and LAURITSEN [*39*] using a COMPTON lens spectrometer, by TERRELL and PHILLIPS [*39*], RUTHERGLEN *et al.* [*39*], and BENT *et al.* [*39*] using pair spectrometers, and by THOMPSON [*39*] with a scintillation spectrometer.

Li^6, Li^7: The energy of the first excited state of Li^7 at 478 keV is accurately known from the (dp) reaction, which also excites a level at 4.61 MeV (GELINAS and HANNA[3]). The excitation energy of the first level is confirmed by measurement of the gamma radiation [*39*]. The angular correlation between the group of protons leaving Li^7 in its first excited state and the subsequent radiation was found by NEWTON[4] to be isotropic, in agreement with the assignment of $J=\frac{1}{2}$ to the state. This correlation has been confirmed by CLASS and HANNA[5], who also showed that the angular distribution of the gamma radiation with respect to the deuteron beam was isotropic. The first excited state of Be^7 at 430 keV was found by GIBSON and GREEN[6] in the (dn) reaction after its discovery in the $B^{10}(p\alpha)Be^7$ disintegration, and the energy of the gamma radiation from this state has also been accurately measured [*39*]. Angular correlation measurements between the neutrons and gamma radiation are consistent with $J=\frac{1}{2}$ for the first excited state of Be^7 (THIRION[7]). Both the (dp) and (dn) reactions in Li^6 show broad resonant features as the deuteron energy varies (BAGGETT and BAME[8]). The lifetimes of the first excited state of Li^7 and Be^7 have been measured by using the DOPPLER shift method (Sect. 24 and Table 3).

[1] C. P. BROWNE: Phys. Rev. **95**, 860 (1954).

[2] The (dp) reactions in light elements have also been accurately studied by L. M. KHROMCHENKO using magnetic analysis (Sect. 13): Dokl. Akad. Nauk. SSSR. **93**, 451 (1953); **94**, 1037 (1954); **98**, 761 (1954); and (with V. A. BLINOV): J. exp. theor. Phys. USSR. **28**, 741 (1955).

[3] R. W. GELINAS and S. S. HANNA: Phys. Rev. **86**, 253 (1952).

[4] J. O. NEWTON: Proc. Phys. Soc. Lond. **64**, 938 (1951).

[5] C. M. CLASS and S. S. HANNA: Phys. Rev. **87**, 247 (1952).

[6] W. M. GIBSON and L. L. GREEN: Proc. Phys. Soc. Lond. A **63**, 494 (1950).

[7] J. THIRION: Ann. Phys., Paris **8**, 489 (1953).

[8] L. M. BAGGETT and S. J. BAME: Phys. Rev. **85**, 434 (1952).

The proton groups from the reaction $Li^7(dp)$ define excited states of Li^8 at 0.97 and 2.28 MeV; the reaction shows broad resonances in excitation function[2] which are shared by the (dn) reaction. The neutron spectra have been much studied in order to investigate states of Be^8. Green and Gibson[1] obtained the spectrum shown in Fig. 52a by the nuclear emulsion technique and Trail and Johnson[2], using a counter telescope, found the distribution given in Fig. 52b. Levels in Be^8 have also been found by the slow neutron threshold method (Sect. 13)

*Be*⁹: The reactions $Be^9(dp)$ Be^{10} and $Be^9(dn)$ B^{10} have high cross sections and have been used to identify the $T=1$ and $T=0$ states of the nuclei of mass 10 (Fig. 85). The cross sections for these two reactions to analogous levels of the two isobars should be in the ratio 1:2 when energy dependent factors are removed. The states of Be^{10}

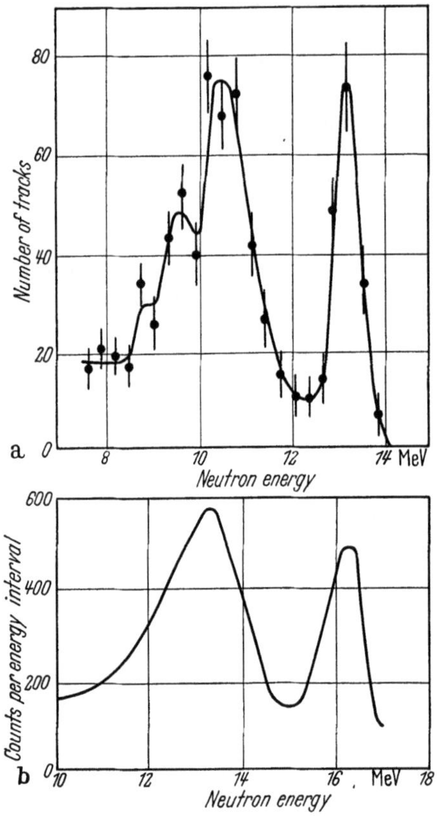

Fig. 52a and b. Neutron spectrum from $Li^7(dn)$ reaction. (a) Nuclear emulsion method (Green and Gibson). (b) Counter telescope method (Trail and Johnson).

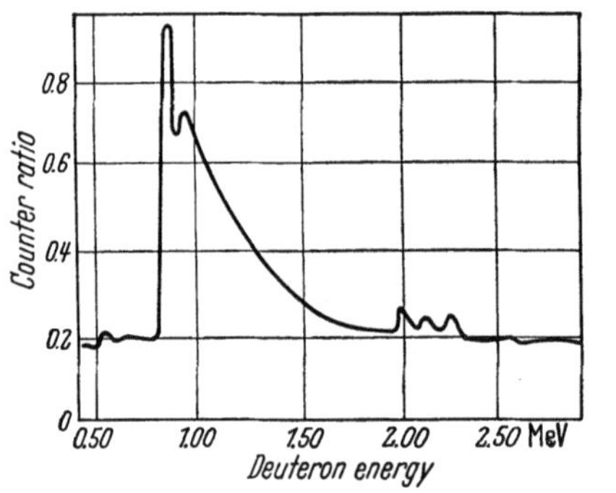

Fig. 53. Levels of B^{10} by slow neutron threshold method (Bonner and Cook).

up to 7.5 MeV were observed by Jung and Bockelman[3] using 7 MeV deuterons and magnetic analysis, and the neutron spectra, defining states of B^{10}, have recently been reported by Ajzenberg[4] who extends the results of earlier workers [*32*]. Higher states of B^{10}, including the close doublet at 5.11 to 5.16 MeV have been resolved by Bonner and Cook, using the slow neutron threshold method of detection (Sect. 13); Fig. 53 gives their results.

The gamma radiation emitted in the bombardment of Be^9 by deuterons was studied by Rasmussen *et al.*[5] and by Chao *et al.*[6] using a lens spectrometer; the lines found may be associated mainly with levels of B^{10}. The radiation was also examined by Shafroth and Hanna[7] using a scintillation spectrometer.

[1] L. L. Green and W. M. Gibson: Proc. Phys. Soc. Lond. A **62**, 407 (1949). See also W. M. Gibson and D. J. Prowse: Phil. Mag. **46**, 807 (1955), who find a 2.1 MeV level in Be^8.

[2] C. C. Trail and C. H. Johnson: Phys. Rev. **95**, 1363 (1954).

[3] J. J. Jung and C. K. Bockelman: Phys. Rev. **96**, 1353 (1954).

[4] F. Ajzenberg: Phys. Rev. **88**, 298 (1952).

[5] V. K. Rasmussen, W. F. Hornyak and T. Lauritsen: Phys. Rev. **76**, 581 (1949).

[6] C. Y. Chao, T. Lauritsen and V. K. Rasmussen: Phys. Rev. **76**, 582 (1949).

[7] S. M. Shafroth and S. S. Hanna: Phys. Rev. **95**, 86 (1954).

They were able to confirm the B^{10} decay scheme by coincidence measurements. The angular correlation between protons in the $Be^9(dp)$ reaction and gamma radiation from the 3.37 MeV excited state of Be^{10} together with stripping information suggests a spin of 2^+ for this state (COHEN *et al.*[1]). This assignment is also in agreement with the measurement of the internal pair conversion coefficient [*32*]. The lifetime of the 0.72 MeV state of B^{10} has been measured (Table 3, Sect. 24).

B^{10} B^{11}: The deuteron disintegration of B^{10} was one of the earliest transmutations by deuterons to be studied and gives a very striking proton spectrum. The most comprehensive results for both reactions are those of ELKIND[2] who used deuterons of energy up to 8.5 MeV from an electrostatic accelerator, a magnetic spectrograph and nuclear emulsions. His results, together with those of VAN PATTER *et al.*[3], cover the B^{11} spectrum from the ground state up to an excitation of 10.3 MeV (12 levels) and the B^{12} spectrum to 3.38 MeV (5 levels). Several of the B^{11} levels are unstable to particle emission and can be excited by the $Li^7(\alpha\gamma)$ reaction. The radiative transitions between the levels of B^{11} were studied by RUTHERGLEN *et al.* [*39*] and by BAME and BAGGETT[4], using pair spectrometry of external and internal pairs respectively. Association of the proton groups with subsequent gamma radiation was established by $(p-\gamma)$ coincidence measurements[5], and angular correlation measurements by THIRION[6] suggest $J=\frac{1}{2}$ for the 2.14 MeV state of B^{11}. These measurements were made with relatively low energy deuterons; the excitation functions for energies up to 2.0 MeV show no very marked irregularities. The radiative transitions in B^{12} have also been investigated [*39*].

Neutron spectra for the $B^{10}(dn)$ and $B^{11}(dn)$ reactions were obtained by GIBSON[7] and by JOHNSON[4] using targets of separated isotopes and the nuclear emulsion technique. This work establishes the level systems of C^{11} and C^{12} up to excitations of about 9 MeV and 16 MeV respectively. Radiative transitions between these levels have also been observed in the bombardment of boron by deuterons.

C^{12}, C^{13}, C^{14}: The (dp) and (dn) reactions with C^{12} show many resonances for deuterons of energy up to 3 MeV [*32*]. A recent survey of the C^{13} states was made by SPERDUTO *et al.*[8]. The (dn) reaction with C^{12} (which also shows resonances[9]) is unusual in having a negative Q-value so that only rather few states in the residual N^{13} nucleus have been excited. Neutron spectra are given by GROSSKREUTZ[10] and by MIDDLETON *et al.* [*41*]. The C^{13} reactions[8] show the excited states of C^{14} below 7 MeV and of N^{14} below 8 MeV (BENENSON[11]).

The gamma radiation from the bombardment of carbon by deuterons has been extensively studied [*39*] and MACKIN *et al.*[12]). The lines observed correlate

---

[1] L. COHEN, S. S. HANNA and C. M. CLASS: Phys. Rev. **94**, 419 (1954). See also A. G. STANLEY: Phil. Mag. **45**, 807 (1954).

[2] M. M. ELKIND: Phys. Rev. **92**, 127 (1953).

[3] D. M. VAN PATTER, W. W. BUECHNER and A. SPERDUTO: Phys. Rev. **82**, 248 (1951).

[4] S. J. BAME and L. M. BAGGETT: Phys. Rev. **84**, 891 (1951).

[5] C. D. CURLING and J. O. NEWTON: Nature, Lond. **166**, 339 (1950).

[6] J. THIRION: Ann. Phys. Paris **8**, 489 (1953).

[7] W. M. GIBSON: Proc. Phys. Soc. Lond. A **62**, 586 (1949). — V. R. JOHNSON: Phys. Rev. **86**, 302 (1952).

[8] A. SPERDUTO, W. W. BUECHNER, C. K. BOCKELMAN and C. P. BROWNE: Phys. Rev. **96**, 1316 (1954).

[9] J. E. RICHARDSON: Phys. Rev. **80**, 850 (1950).

[10] J. C. GROSSKREUTZ: Phys. Rev. **76**, 482 (1949).

[11] R. E. BENENSON: Phys. Rev. **90**, 420 (1953).

[12] R. J. MACKIN, W. B. MIMS and W. R. MILLS: Phys. Rev. **98**, 43 (1955).

well with the proton and neutron groups. In C^{14} no ground state transition is found from the 6.89 MeV level for which a spin 0 is consequently suggested. The 6.09 MeV transition in this nucleus is of type $E1$ according to the observed positron spectrum from internal pairs.

C^{14}, in the form of a few milligrams of enriched barium carbon has been used as a target, and the (dp) and (dn) reactions have been observed[1]. The excitation curve for the former reaction, analysed in terms of penetrabilities, suggests a spin of $^5/_2{}^+$ for the ground state of C^{15}; the neutron groups from the (dn) reaction are those expected from the known level system.

N^{14}, N^{15}: The levels of N^{15} obtained from the (dp) reaction (SPERDUTO *et al.*[2]) agree well with those indicated in the $(n\gamma)$ capture process. The gamma radiation observed in the (dp) reaction is also consistent with the level system [*32*].

Neutron groups from the (dn) reaction were observed by ROSE *et al.*[3]. A more extensive survey was made by EVANS *et al.* [*41*], who obtained 7 groups of neutrons of which the last three correspond to virtual levels of O^{15} known in the $N^{14}(p\gamma)$ reaction.

The $N^{15}(dp)$ and (dn) reactions have been observed[4] but further work is required.

O^{16}, O^{17}, O^{18}: The levels of O^{17} have been explored by many workers. The most extensive results from the (dp) reaction were obtained by BURROWS *et al.*[5] using 8 MeV deuterons and photographic detection; their Q-values are means of observations at five different angles and agree well with those deduced from other reactions. Gamma radiation from the first excited state at 0.873 MeV was observed by THOMAS and LAURITSEN [*39*]; they were successful in measuring the internal conversion line in a lens spectrometer. The conversion coefficient indicates an $E2$ transition and suggests $J = {}^1/_2{}^+$ for the first excited state of O^{17} since the ground state spin is known to be $^5/_2$, and from the shell model (and stripping reactions) the parity is even. The spin for the excited state is confirmed by the isotropic $(p-\gamma)$ angular correlation observed by THIRION[6]. The lifetime of this state has been measured (Table 3, Sect. 24).

Q-values for the (dp) reactions of the rare isotopes O^{17}, O^{18} are reported by AHNLUND *et al.*[7] who used occluded gas targets prepared in an isotope separator.

Neutron spectra for the reaction $O^{16}(dn)$ are given by MIDDLETON *et al.* [*41*] and by AJZENBERG[8]. The $O^{18}(dn)$ reaction has been detected (SEALE[9]).

F^{19}: The excited states of F^{20} from the $F^{19}(dp)$ reaction were observed by WATSON and BUECHNER[10] and the intercombination gamma ray lines have been seen. The most extensive neutron spectrum, giving the levels of Ne^{20}, is still the early work of POWELL[11]. Neutron threshold measurements indicate a doublet in Ne^{20} at 11.69, 11.87 MeV [*32*].

---

[1] E. L. HUDSPETH, C. P. SWANN and N. P. HEYDENBURG: Phys. Rev. **80**, 643 (1950). — J. A. RICKARD, E. L. HUDSPETH and W. W. CLENDENIN: Phys. Rev. **96**, 1272 (1954).

[2] A. SPERDUTO, W. W. BUECHNER, C. K. BOCKELMANN and C. P. BROWNE: Phys. Rev. **96**, 1316 (1954).

[3] W. B. ROSE, E. L. HUDSPETH and N. P. HEYDENBURG: Phys. Rev. **87**, 382 (1952).

[4] L. D. WYLY: Phys. Rev. **76**, 104 (1949) (dp). — D. C. WORTH: Phys. Rev. **78**, 378 (1950) (dn).

[5] H. B. BURROWS, C. F. POWELL and J. ROTBLAT: Proc. Roy. Soc. Lond., Ser. A **209**, 478 (1951).

[6] J. THIRION: Ann. Phys., Paris **8**, 459 (1953).

[7] K. AHNLUND, S. THULIN and R. PAULI: Ark. Fysik **8**, 489 (1954). — K. AHNLUND and C. MILEIKOWOSKY: Ark Fysik **8**, 161 (1954).

[8] F. AJZENBERG: Phys. Rev. **83**, 693 (1951).

[9] R. L. SEALE: Phys. Rev. **92**, 389 (1953).

[10] H. A. WATSON and W. W. BUECHNER: Phys. Rev. **88**, 1324 (1952).

[11] C. F. POWELL: Proc. Roy. Soc. Lond., Ser. A **181**, 344 (1943).

Ne^{20} to Ca^{40}: Observations of (dp) and (dn) reactions have been made for most of the stable nuclei with A between 20 and 40. High resolution magnetic spectra are available for the (dp) reactions with Ne^{20}, Ne^{22}, Na^{23}, Mg^{24-26}, Si^{28-30} and K^{39} [32], [33]. The proton spectrum for $Al^{27}(dp)\,Al^{28}$ is shown in Fig. 8. Lower resolution proton spectra have in many cases been obtained in the study of stripping reactions, and are referred to in Sect. 52. Neutron spectra from the (dn) reaction in these nuclei have also been obtained sometimes in connection with the stripping reaction. Gamma radiation associated with both (dp) and (dn) reactions has been observed ([33] and ALBURGER[1]). SMITH and ANDERSON[2] detected the 31 keV radiation between the ground state doublet levels of Al^{28} and concluded that it was a dipole transition because of the absence of internal conversion electrons. There is a similar ground state doublet in K^{40}; this was predicted from the intensity of the ground state radiation in the $K^{39}(n\gamma)$ reaction [37] in comparison with expectation from probable spin values, and was demonstrated by means of the (dp) reaction[3].

51. Deuteron stripping reactions; theory. The deuteron is a loosely bound structure in which the component neutron and proton spend most of their time at a distance apart of about 2.18×10^{-13} cm which is greater than the range of nuclear forces. If such a structure passes near a nucleus one or other of the component particles may be stripped off and absorbed by the target nucleus leaving the other to continue nearly in the forward direction. The interest of this process for nuclear spectroscopy arises from the fact that when the absorbed particle forms a definite quantum state of the residual nucleus, it must take in a certain definite orbital momentum and it must cause a certain definite change in kinetic energy. The change of energy and of orbital momentum are impressed on the wave function of the scattered nucleon, and the angular distribution of this particle determines in particular the l-value absorbed and hence the parity change between the initial and final nucleus. The probability of the stripping process is determined by the ease with which the stripped nucleon can be assimilated and thus by the reduced width of the nuclear state. The stripping reaction is the main method for determining directly the reduced width of bound levels. The spin change in the stripping reaction is determined within certain limits by the conservation law. In order that the properties of discrete states may be studied the deuteron energy must not be so high that particle groups from the stripping reaction cannot be resolved. On the other hand, it must not be so low that COULOMB effects become very important. These conditions restrict the useful range of deuteron energies to be between about 3 and 30 MeV, depending on the charge of the target nucleus. Since a state of given spin may sometimes be formed by absorption of a nucleon of more than one l-value stripping experiments may indicate a preference for a particular orbital momentum and hence can be used as a test of nuclear models.

The theory of stripping reactions has been reviewed by HUBY[4] who discusses also the original treatment of excitation functions for deuteron induced activities given by OPPENHEIMER and PHILLIPS[5] and high energy stripping reactions (SERBER[6]) in which interest is concentrated more on the break-up of the

[1] D. E. ALBURGER: Phys. Rev. **75**, 51 (1949).
[2] R. D. SMITH and R. A. ANDERSON: Nature, Lond. **168**, 429 (1951).
[3] W. W. BUECHNER, A. SPERDUTO, C. P. BROWNE and C. K. BOCKELMAN: Phys. Rev. **91**, 1502 (1953).
[4] R. HUBY: Progress in Nuclear Physics, Vol. 3. London: Pergamon Press 1953.
[5] J. R. OPPENHEIMER and M. PHILLIPS: Phys. Rev. **48**, 500 (1935).
[6] R. SERBER: Phys. Rev. **72**, 1008 (1947).

deuteron than on properties of residual states. The most useful forms of stripping theory have been given by BUTLER[1], and by BHATIA *et al.*[2]. If the reaction considered is written

$$X + (b + c) \rightarrow Y + b \tag{51.1}$$

where $(b+c)$ represents a deuteron and c is the nucleon absorbed by X to form Y in a certain state, then the differential cross section for the process may be written

$$\sigma(\vartheta) = \gamma_c^2 F_{lc}(\vartheta) \tag{51.2}$$

where γ_c^2 is the nucleon reduced width for the system $X+c$ and $F_{lc}(\vartheta)$ is the angular distribution factor characteristic of the absorption of the nucleon c with angular momentum $l\hbar$.

The calculation of $F(\vartheta)$ given by BUTLER is essentially the solution of a boundary condition problem; it may be represented schematically in a simple one-dimensional case as in Fig. 54. In this figure region I represents the incident deuteron as a whole, and regions II and III represent the capture of one or other of the nucleons. The boundary conditions are that the total wave function and its derivative shall be continuous at $r_n = R$ or $r_p = R$ where R is the nuclear radius, and the wave function which fits these conditions contains the angular distribution of the particle [cf. Eq. (51.1)]. In BUTLER's theory all COULOMB forces are neglected, compound nucleus formation is disregarded and interaction between neutron and proton after stripping and between the particle b and the residual nucleus are disregarded. Despite the far reaching nature of these assumptions the angular distributions so calculated agree well with experiment; in particular the marked forward peaks, which are difficult to explain by the compound nucleus mechanism, arise naturally because the particle b, which does not enter the nucleus, can carry off a large angular momentum.

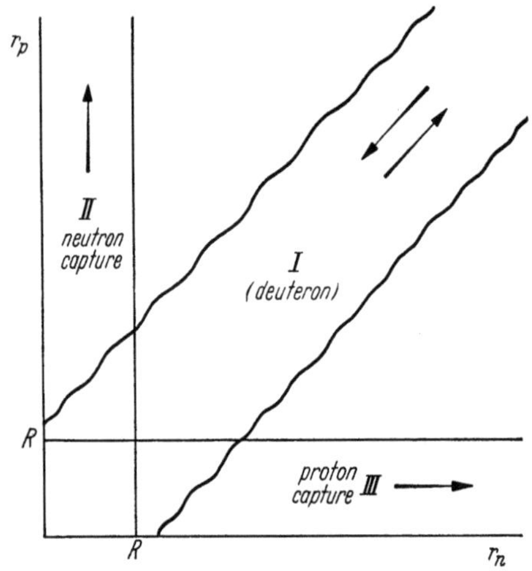

Fig. 54. One dimensional representation of deuteron stripping process.

The theory of BHATIA *et al.* is based more intimately on the BORN approximation than the BUTLER theory. The two theories give similar results, and both have been widely used in interpreting experimental results. The theory of BHATIA *et al.* gives the differential cross section in the form

$$\sigma(\vartheta) = \frac{2j'+1}{2j+1} \Pi(K) \sum_{l_c} \Omega_{l_c} [j_{l_c}(kR)]^2 \tag{51.3}$$

where j, j' are the spins of the initial and final nuclei, $\Pi(K)$ is a function determined by the internal motion of the deuteron, k is the wave number of the captured particle c and $(2j'+1)\Omega_{l_c}$ is a "nucleon capture probability" proportional to the reduced width γ_c^2. The summation over l_c, which usually reduces in practice to one term, is incoherent. Eq. (51.2) and (51.3) are equivalent. The formulae of BHATIA *et al.* require a rather larger radius of interaction for best fit to the experimental results than does the BUTLER theory. They also avoid a singularity in the BUTLER theory which occurs when the particle c is captured with approxi-

[1] S. T. BUTLER: Proc. Roy. Soc. Lond., Ser. A **208**, 559 (1951).

[2] A. B. BHATIA, K. HUANG, R. HUBY and H. C. NEWNS: Phil. Mag. **43**, 485 (1952).

mately zero energy, i.e. for Q-values of about -2.2 MeV, the binding energy of the deuteron. This makes the BUTLER formula difficult to apply in certain (dn) reactions (C^{12}, Sect. 52).

The selection rules determining the spin j' of the nucleus Y in the final state are

$$j + l_c + \tfrac{1}{2} \geq j' \geq \text{ minimum of } |j \pm l_c \pm \tfrac{1}{2}|. \tag{51.4}$$

The reduced width is obtained from the maximum differential cross section using formulae such as (51.3). A list of reduced widths obtained from stripping reactions is given by FUJIMOTO *et al.*[1].

The evidence presented in Sect. 52 shows that with a suitable choice of nuclear radius R stripping theory has proved extremely successful in the determination of spins and parity changes. This is basically because the main feature of the stripping pattern, namely the forward peak, is determined by high angular momentum components of the incident wave, for which particle b is so far from the nucleus that the assumptions made in the theory have little effect. The explanation of other details of the stripping distributions, such as the intensity observed at backward angles, requires an extension of the theory. This is also demanded by the fact that reduced widths calculated from formulae (51.2), (51.3) are generally about 5 times smaller than would be expected.

Modifications have been made to the BUTLER theory to include COULOMB effects and interaction of the particle b with the final nucleus by many authors, notably HOROWITZ and MESSIAH[2], and TOBOCMAN and KALOS[3]. A method of eliminating such effects from the calculation of reduced widths by fixing attention on the high orbital momentum components of the system has been given by BOWCOCK[4]. This method gives excellent agreement between reduced widths of bound states calculated from stripping patterns and of analogous virtual states [see $Mg^{24}(dp)$, Sect. 52]. The effects of compound nucleus formation and resonances in the final state $b+Y$ have also been discussed; they can lead to polarisation of the particle b, which is not expected according to the unmodified BUTLER theory. A recent formulation of stripping theory by GRANT[5] discusses some of these effects.

The reduced widths and parity for formation of ground states confirm the sequence of single particle orbits proposed by the nuclear shell model (Sect. 3, and BETHE and BUTLER[6]). Excited states of a given nucleus which have nucleon capture probabilities comparable with that found for the ground state may be interpreted as single particle states. Comparison of capture probabilities in (dp) and (dn) reactions may indicate the first state of $T=1$ in a nucleus of $T=0$ (see Al^{27}, Sect. 52). The relative capture probabilities or reduced widths for two different nuclear states (preferably with the same l_c) can be used (AUERBACH and FRENCH[7]) to determine the intermediate coupling parameter a/K (Sect. 3 and [*10*]).

Further information about spin values may be obtained by observation of the angular correlation between particle b in the stripping reaction and the radiation from the state excited by the capture of particle c. There is very little experimental

---

[1] Y. FUJIMOTO, K. KIKUCHI and S. YOSHIDA: Progr. Theor. Phys. **11**, 264 (1954).
[2] J. HOROWITZ and A. M. L. MESSIAH: J. Phys. Radium **14**, 695 (1953).
[3] W. TOBOCMAN and M. H. KALOS: Phys. Rev. **97**, 132 (1955).
[4] J. E. BOWCOCK: Proc. Phys. Soc. Lond. A **68**, 512 (1955).
[5] I. P. GRANT: Proc. Phys. Soc. Lond. A **67**, 981 (1954); A **68**, 244 (1955).
[6] H. A. BETHE and S. T. BUTLER: Phys. Rev. **85**, 1045 (1952).
[7] T. AUERBACH and J. B. FRENCH: Phys. Rev. **98**, 1276 (1955).

data of this sort, but the theory has been given by several authors, including SATCHLER and SPIERS[1]. The angular correlation should be symmetrical about the direction of momentum of the captured particle c, as in an ordinary nucleon capture reaction and should depend on the nuclear coupling scheme in the same way as does the angular distribution in a capture reaction.

52. Deuteron stripping reactions; experimental. Although excitation functions for deuteron induced activity were shown by OPPENHEIMER and PHILLIPS to indicate a stripping mechanism, the significance of this process for the determination of nuclear properties was not realised until the publication of the work of HOLT and YOUNG[2] and of BURROWS *et al.*[3] on the angular distributions of typical (dp) reactions. Most of the existing experimental work has been done with cyclotron beams of energies between 4 and 20 MeV although sufficiently high energies are now available from electrostatic accelerators for similar work with higher energy resolution. A large number of stripping reactions have been studied with 8 MeV deuterons from the cyclotron of the University of Liverpool [*40*], [*41*] and the target arrangements for these experiments are shown in Fig. 55. The results of stripping experiments are presented in numerical form in Table 12 and special features are considered below. Absolute cross sections and reduced widths are known only in a small number of cases, but relative values of the nucleon capture probability are generally available.

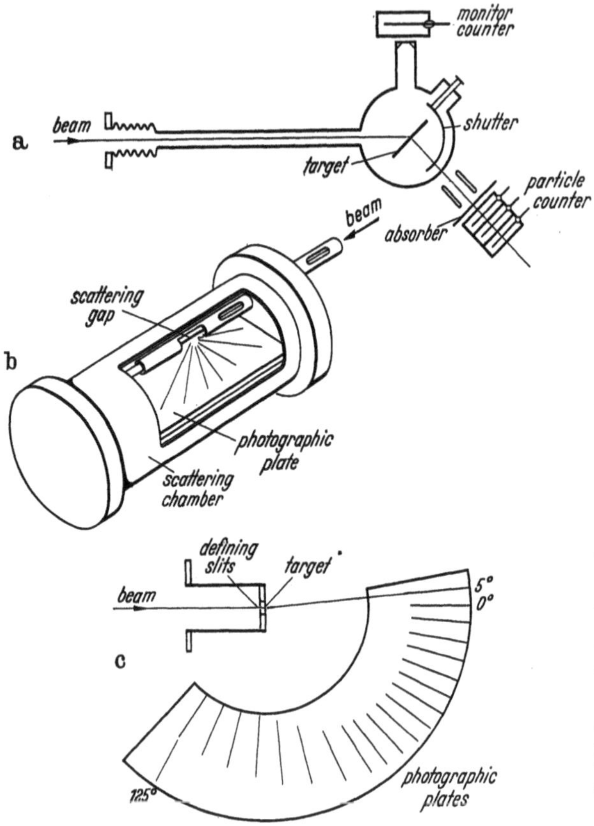

Fig. 55 a—c. Target arrangements for studying the deuteron stripping process [*40*], [*41*]. (a) (dp) reaction, counter method; (b) (dp) reaction, emulsion method; (c) (dn) reaction, emulsion method.

Li⁶ Li⁷: Angular distributions for the $Li^6(dp)$ reaction[4] for energies between 1 and 15 MeV show clearly the gradual growth of the stripping peak. Angular distributions for the $Li^7(dp)$ reaction are also given by LEVINE *et al.*[4]

[1] G. R. SATCHLER and J. A. SPIERS: Proc. Phys. Soc. Lond. A **65**, 980 (1952). See also G. R. SATCHLER: Proc. Phys. Soc. Lond. A **66**, 1081 (1953); **67**, 471 (1954).

[2] J. R. HOLT and C. T. YOUNG: Proc. Phys. Soc. Lond. A **63**, 835 (1950).

[3] H. B. BURROWS, W. M. GIBSON and J. ROTBLAT: Phys. Rev. **80**, 1095 (1950).

[4] W. E. NICKELL: Phys. Rev. **95**, 426 (1954). — HOLT and MARSHAM: Reference [*40*]. — S. H. LEVINE, R. S. BENDER and J. N. MCGRUER: Phys. Rev. **97**, 1249 (1955).

Table 12. *Stripping and pick-up reactions.*

Column 6 gives the nucleon capture probabilities $\left(\frac{\Lambda}{2j_i+1}\right.$ or $\Omega(2j_f+1)$ in the notation of BHATIA *et al.*$\left.\right)$. They are in general in arbitrary units and should not be compared between different nuclei. The dimensionless reduced widths are taken from the paper of FUJIMOTO *et al.*, with a slight correction to adjust the interaction radius to the value given in Sect. 9.

A. The (dp) stripping reaction.

Target nucleus and nucleon absorbed	Final nucleus	Excited state MeV	l_n	Spin and parity (final)	Nucleon capture probability	$\frac{2Ma\gamma^2}{3\hbar^2}$ %	Suggested shell model orbit
Li^6+n	Li^7	0.0	1	$3/2^-$	1		$p_{3/2}$
		0.478	1	$5/2\ 3/2\ 1/2^-$	0.73		$p_{1/2}$
Li^7+n	Li^8	0.0	1	$3-0^+$	7.1		
		0.97	1	$3-0^+$			
Be^9+n	Be^{10}	0.0	1	$3-0^+$		5.3	
		3.37	1	$3-0^+$		0.84	
$B^{10}+n$	B^{11}	0.0	} 1	$9/2-3/2^-$	5.0		
		2.14			0.9		
		4.46			1.6		
		5.03			0.5		
$B^{11}+n$	B^{12}	0.0	1	1^+	10		$p_{1/2}$
		0.95	1	$3-0^+$	10		
		1.67	0	$2,\ 1^-$	17		
		3.38	1	$3-0^+$	11		
		3.76	(1)	2^+	2		
		4.53	2	3^-	27		
$C^{12}+n$	C^{13}	0.0	1	$1/2^-$	19	10.7*	$p_{1/2}$
		3.09	0	$1/2^+$	22	13.8	$s_{1/2}$
		3.68	1	$3/2,\ 1/2^-$	2.7	0.7	$p_{3/2}^{-1},\ s_{1/2}^2$
		3.86	2	$5/2^+$	102	16	$d_{5/2}$
$C^{13}+n$	C^{14}	0.0	1	0^+			
		6.1	0	$1,\ 0^-$			
$N^{14}+n$	N^{15}	0.0	1	$1/2^-$		10	$p_{1/2}^{-1}$
		8.32	0	$3/2\ 1/2^+$		55, 215	$s_{1/2}$
$O^{16}+n$	O^{17}	0.0	2	$5/2^+$	62	5.1	$d_{5/2}$
		0.87	0	$1/2^+$	26	12.5*	$s_{1/2}$
$O^{18}+n$	O^{19}	1.47	0				
$F^{19}+n$	F^{20}	0.0	0, 2	1^+		0.04, 0.63	
		0.65	2	$3-1^+$			
		2.05	2	$3-1^+$			
		3.49	} 0	$1,\ 0^+$			
		3.53					
$Ne^{20}+n$	Ne^{21}	0.0	2	$5/2,\ 3/2^+$			
		0.33	2	$5/2,\ 3/2^+$			
		1.68	0,1	$1/2^+;\ 3/2\ 1/2^-$			
		2.79	0	$1/2^+$			
		3.73	} 1,2	$3/2,\ 1/2^-;\ 5/2,\ 3/2^+$			
		4.71					
		5.44					
		5.74					
		7.30					

Table 12. (continued.)

Target nucleus and nucleon absorbed	Final nucleus	Exited state MeV	l_n	Spin and parity (final)	Nucleon capture probability	$\frac{2May^2}{3\hbar^2}$ %	Suggested shell model orbit
$Na^{23}+n$	Na^{24}	0.0	2	$4-0^+$			
		0.47, 0.56	0, 2	$2, 1^+;\ 4-0^+$			
		1.34	0	$2, 1^+$			
$Mg^{24}+n$	Mg^{25}	0.0	2	$5/2, 3/2^+$	15.4	3.9	$d_{5/2}$
		0.58	0	$1/2^+$	6.2	8	$s_{1/2}$
		0.97	2	$5/2, 3/2^+$	6.2	2.3, 3.4	$d_{3/2}$
		1.61	isotropic	—			
		1.96	2	$5/2, 3/2^+$	5.5		
		2.57, 2.74, 2.80	0	$1/2^+$			
		3 40	1	$3/2, 1/2^-$	8.7	7.6*	$p_{3/2}$
$Mg^{25}+n$	Mg^{26}	0.0	(2)	0	(2.1)	(20)	
		1.83	0, 2	$3, 2^+, 5-0^+$	0.65, 12.3	1.5	
		2.97	0	}	5.7		
		3.97	0	} $3, 2^+$	6.1		
		4.35	0	}	5.6		
		6.15	0	}	3.3		
$Mg^{26}+n$	Mg^{27}	0.0	0	$1/2^+$	2.8	3.9	
		0.89	2	$5/2, 3/2^+$	4.6		
		3.50	0	$1/2^+$			
$Al^{27}+n$	Al^{28}	0.0, 0.031	0	$3, 2^+$	6.3	10, 15	
		0.97, 1.01	0, 2	$3, 2^+;\ 5-0^+$	0.6; 6.7		
$Si^{28}+n$	Si^{29}	0.0	0	$1/2^+$	9.0	2.3	$s_{1/2}$
		1.28	2	$5/2, 3/2^+$	15	2.1	$d_{3/2}$
		2.03	2	$5/2, 3/2^+$	5.7		$d_{5/2}^{-1}\,s_{1/2}^2$
		2.43	isotropic	—			
		3.07	2	$5/2, 3/2^+$	1.9		
		3.62	3	$7/2, 5/2^-$	19		$f_{7/2}$
		4.93	1	} $3/2, 1/2^-$	16		$p_{3/2}$
		6.38	1	}	5.4		$p_{1/2}$
$P^{31}+n$	P^{32}	0.0	2	1^+			$d_{3/2}$
$S^{32}+n$	S^{33}	0.0	2	$5/2, 3/2^+$	8.0		$d_{3/2}$
		0.85	0	$1/2^+$	1.5		$s_{1/2}^{-1}\,d_{3/2}^2$
		2.90	3	$7/2, 5/2^-$	20		$f_{7/2}$
		3.26	}		13		$p_{3/2}$
		4.21	} 1	$3/2, 1/2^-$	1.6		
		4.89	}		0.8		
		5.72	}		9		$p_{1/2}$
		6.48	1, 2	$3/2, 1/2^-;\ 5/2, 3/2^+$	2.4		
$Cl^{35}+n$	Cl^{36}	0.0	2	2			$d_{3/2}$
$A^{40}+n$	A^{41}	0.0	3	$7/2, 5/2$		1.0	$f_{7/2}$
$Ca^{40}+n$	Ca^{41}	0.0	3	$7/2, 5/2^-$	25	1.5	$f_{7/2}$
		1.90	1	} $3/2, 1/2^-$	14	2.1	$p_{3/2}$
		2.42	1	}	5.3	1.7	$p_{1/2}$
		3.96	}		8.8		
		4.76	} 2	$5/2, 3/2^+$	8.4		
		5.72	}		6.9		

* These transitions have been analysed by an improved method by BOWCOCK.

Table 12 (continued). *B. The (dn) stripping reaction.*

Target nucleus and nucleon absorbed	Final nucleus	Excited state MeV	l_p	Spin and parity (final)	Nucleon capture probability	$\frac{2Ma\gamma^2}{3\hbar^2}$ %	Suggested shell model orbit
Be^9+p	B^{10}	0.0 to 3.58	1	3^+			
		5.14	0	$2, 1^-$			
		5.93	1	$3-0^+$			
		6.38	2	$4-0^+$			
$B^{10}+p$	C^{11}	0.0	1	$^9/_2-{}^3/_2^-$			
$B^{11}+p$	C^{12}	0.0	1	$3-0^+$			
		4.4	1	$3-0^+$			
$C^{12}+p$	N^{13}	0.0	1	$^3/_2, {}^1/_2^-$			$p_{1/2}$
		2.37	0	$^1/_2^+$			$s_{1/2}$
		3.51, 3.56	2	$^5/_2, {}^3/_2^+$			$d_{5/2}$
$C^{13}+p$	N^{14}	0.0 to 3.95	1	$2, 1, 0^+$			
		4.91	0	$1, 0^+$			
$N^{14}+p$	O^{15}	0.0	1	$^5/_2-{}^1/_2^-$			
		5.27	2	$^7/_2-{}^1/_2^+$			
		6.14	1	$^5/_2-{}^1/_2^-$			
		6.84	0	$^3/_2, {}^1/_2^+$			
		7.48, 8.42, 9.06	1	$^5/_2-{}^1/_2^-$			
$O^{16}+p$	F^{17}	0.0	2	$^5/_2, {}^3/_2^+$			
		0.51	0	$^1/_2^+$			
$F^{19}+p$	Ne^{20}	0.0	0	$1, 0^+$	7.0*		$d_{5/2}^2\, s_{1/2}^2$
		1.63	2	$3, 2, 1^+$	44		
		7.3	0, 2	—	3.9, 100		
		9.2	1	$2, 1, 0^-$	5.4		
$Na^{23}+p$	Mg^{24}	0.0, 1.37	—	$0^+, 2^+$	*		
		4.12, 4.23	2, 0	$4^+; 1, 2^+$	11; 2.4		
		7.5	0, 2	$2, 1^+; 4, 3, 2^+$	3.6		
		8.4	0	$2, 1^+$	3.5		
$Mg^{24}+p$	Al^{25}	0.0	2	$^5/_2, {}^3/_2^+$		0.26	$d_{5/2}$
		0.45	0	$^1/_2^+$		1.6	$s_{1/2}$
		0.95	2	$^5/_2, {}^3/_2^+$		0.27	$d_{3/2}$
$Al^{27}+p$	Si^{28}	0.0	2	0^+	7.3*		$d_{5/2}$
		1.78	0	$3, 2^+$	0.7		
		6.11, 6.65	0	$3, 2^+$	2.0		
		7.55			2.0		
		8.18	0	$3, 2^+$	5.0		
		9.16			14.0		
		10.2			1.0		
$P^{31}+p$	S^{32}	0	0	0^+	2.3*		$s_{1/2}$
		2.25	2	$3, 2, 1^+$	7.8		
		3.81 to 4.50	0	$1, 0^+$	1.1		
		5.83	2	$3, 2, 1^+$	13		

* Absolute values, units 10^{-50} c.g.s.

Table 12 (continued). *C. Pick-up reactions.*

dt, $d\,He^3$				pd				nd			
Target nucleus and incident particle	Final nucleus	Excited state MeV	l	Target nucleus and incident particle	Final nucleus	Excited state	l_n	Target nucleus and incident particle	Final nucleus	Excited state	l_p
Li^6+d	He^5			He^4+p	He^3			Li^6+n	He^5	0.0	1
	Li^5									—	1
Li^7+d	He^6	0, 1.71	1	Li^7+p	Li^6	0.0	1	Li^7+n	He^6		
	Li^6	0, 2.19	1								
Be^9+d	Be^8	0.0	1	Be^9+p	Be^8	0.0	1	$B^{10}+n$	Be^9	0.0	1
		2.9				2.9				2.43	1
						4.0					
	Li^8					5.1		$N^{14}+n$	C^{13}		
								$O^{16}+n$	N^{15}		
$C^{13}+d$	C^{12}	0.0	1	$B^{10}+p$	B^9	0.0	1				
$N^{14}+d$	N^{13}					2.4	1				
$F^{19}+d$	F^{18}										
				$B^{11}+p$	B^{10}	0.0	1				
				$N^{14}+p$	N^{13}	0.0	1				
				$F^{19}+p$	F^{18}	0.0	0				

Be^9: Angular distributions of some of the proton groups from the $(d\,p)$ reaction were obtained by RESNICK and HANNA[1] (E_d up to 900 keV), CANAVAN[1] (E_d up to 2MeV), FULBRIGHT *et al.*[1] ($E_d=3.6$ MeV), EL BEDEWI[1] ($E_d=8$ MeV), and by EBY[2] ($E_d=11.9$ MeV). The excitation function for the ground state proton group shows irregular features which may represent the last traces of discrete level structure. At the higher energies stripping peaks are found.

Angular distributions for the $(d\,n)$ reaction have also been obtained and analysed by stripping theory by AJZENBERG[2] using nuclear emulsions and by BROMLEY[2], who used a high pressure helium filled ionisation chamber as a neutron detector. Neutron spectra and angular distributions for several of the groups for a deuteron energy of about 1 MeV, are given by PRUITT *et al.*[3], and by GREEN *et al.*[3]. These authors conclude that there is some evidence for the stripping reaction even at these low bombarding energies, but that the compound nucleus mechanism is certainly also important, in particular the presence of $\cos\vartheta$ terms shows interference between levels of opposite parity.

B^{10}, B^{11}: The angular distributions already begin to show stripping features between 500 keV and 2 MeV, but for deuterons of 3 MeV there is still a relatively large contribution from compound nucleus formation[4]. EVANS and PARKINSON[5]

[1] F. S. EBY: Phys. Rev. **96**, 1355 (1954). — F. A. EL BEDEWI: Reference [*40*]. — H. W. FULBRIGHT, J. A. BRUNER, D. A. BROMLEY and L. M. GOLDMAN: Phys. Rev. **88**, 700 (1952). — F. L. CANAVAN: Phys. Rev. **87**, 136 (1952). — I. RESNICK and S. S. HANNA: Phys. Rev. **82**, 463 (1951).

[2] F. AJZENBERG: Phys. Rev. **88**, 298 (1952). — D. A. BROMLEY: Phys. Rev. **88**, 565 (1952).

[3] J. S. PRUITT, C. D. SWARTZ and S. S. HANNA: Phys. Rev. **92**, 1456 (1953). — L. L. GREEN, J. P. SCANLON and J. C. WILLMOTT: Proc. Phys. Soc. Lond. A **68**, 386 (1955).

[4] W. H. BURKE, J. R. RISSER and G. C. PHILLIPS: Phys. Rev. **93**, 188 (1954). — W. W. PRATT: Phys. Rev. **93**, 816 (1954). — C. H. PARIS, F. VALCKX and P. ENDT: Physica, Haag **20**, 573 (1954).

[5] N. T. S. EVANS and W. C. PARKINSON: Proc. Phys. Soc. Lond. **67**, 684 (1954).

made a detailed study of six proton groups from the $B^{10}(dp)$ reaction and found that most of the angular distributions were compatible with stripping theory. Their curves do, however, show a much greater intensity at angles near the stripping minimum than is expected on the simple theory. The variation of pattern with deuteron energy (over about 2 MeV) does not however suggest that resonances of the compound nucleus are important. The experiment was made with 8 MeV deuterons from a cyclotron, and the proton energy spectrum was obtained in a sodium iodide crystal. In the $B^{11}(dp)$ reaction, states of B^{12} are excited whose reduced widths are known from the total neutron cross section. The neutron capture probability in the stripping reaction is roughly proportional to the reduced widths.

The neutron groups observed by JOHNSON[1] in the $B^{10}(dn)$ reaction with 3.4 MeV deuterons have forward angular distributions. PARIS and ENDT[2] obtained similar distributions for 500 keV deuterons and assigned a number of l_p values which have not yet been confirmed by high energy experiments. BURKE *et al.*[3] exhibit the development of a forward peak in the $B^{10}(dn)$ reaction as the deuteron energy increases from 0.7 to 1.4 MeV. There is general agreement on the l_p value of 1 for the ground state group. For $B^{11}(dn)$ GIBSON[4] and GRAUE[4] give angular distributions indicating $l_p=1$ for the neutron group corresponding to the first excited state $J=2^+$ of C^{12}. WARD and GRANT[5] used a plastic scintillator to detect the neutrons from B^{11} bombarded by 600 keV deuterons and were able to resolve the ground state and first excited state groups in an integral bias curve. The angular distribution for the ground state group a the stripping curve fairly well but has a deficiency of forward neutrons. By contrast, the distribution for the 4.4 MeV state of C^{12} is nearly symmetrical about 90° and seems to be characteristic of compound nucleus formation. These conclusions differ from those reached by GIBSON for a deuteron bombarding energy of 8.1 MeV. In the reaction to the excited state, it appears that for low energy deuterons the intermediate nucleus is probably formed in a state of $J={}^7/_2{}^+$. The angular distribution for neutrons decaying to a $J=2^+$ state can then be calculated for the channel spins ${}^5/_2$ and ${}^3/_2$ and the mixing ratio is found to be very near that predicted by the jj coupling interaction with $j_n={}^5/_2$, although a fit can be obtained in the LS scheme if the spin of B^{11} is considered to be due to a configuration rather than to the odd $p_{3/2}$ particle.

C^{12}, C^{13}: The characteristic angular distributions for stripping reactions with different l_n were used by ROTBLAT[6] to establish the existence of the doublet states of C^{13} at 3.68 and 3.86 MeV. The stripping pattern for 3.29 MeV deuterons was carefully studied by HOLMGREN *et al.*[7] who found the general characteristics of an $l_n=1$ pattern but with an increase of intensity at angles greater than 80° of the same sort as was observed in the B^{10} reactions. The patterns at 8 MeV are given in [*40*], and at 19 MeV by FREEMANTLE *et al.*[8]. At the latter energy the stripping features for the ground state group are becoming lost, but the excited state patterns agree with expectation. Stripping patterns for the $C^{13}(dp)$ C^{14}

[1] V. R. JOHNSON: Phys. Rev. **86**, 302 (1952).
[2] C. H.. PARIS and P. M. ENDT: Physica, Haag **20**, 585 (1954).
[3] W. H. BURKE, J. R. RISSER and G. C. PHILLIPS: Phys. Rev. **93**, 188 (1954).
[4] W. M. GIBSON: Phil. Mag. **44**, 297 (1953). — A. GRAUE: Phil. Mag. **45**, 1205 (1954).
[5] A. WARD and P. J. GRANT: Proc. Phys. Soc. Lond. A **68**, 637 (1955).
[6] J. ROTBLAT: Phys. Rev. **83**, 1271 (1951).
[7] H. D. HOLMGREN, J. M. BLAIR, B. E. SIMMONS, T. F. STRATTON and R. V. STUART: Phys. Rev. **95**, 1544 (1954).
[8] R. G. FREEMANTLE, W. M. GIBSON and J. ROTBLAT: Phil. Mag. **45**, 1200 (1954).

reaction are given by BROMLEY and GOLDMAN[1] and by BENENSON[2]. An extensive study of stripping patterns in the C^{12} (dp) and C^{13} (dp) reactions is reported by McGRUER *et al.*[3], who used 14.8 MeV deuterons and magnetic analysis.

BENENSON[2] also gives angular distributions for the $C^{13}(dn)$ N^{14} reaction. A clear pattern for the ground state transition in this reaction was obtained by BROMLEY[1], whose work establishes that the parity of the ground states of C^{14} and N^{14} is the same (even). GREEN *et al.*[4] suggest the superposition of stripping and compound nucleus patterns in the angular distribution obtained for the 5 MeV levels of N^{14} by the nuclear emulsion method with 0.86 MeV deuterons, while patterns for lower levels at this energy seem to be fitted by compound nucleus theory. The stripping results suggest odd parity for the 4.91 MeV level of N^{14} and this state does not radiate to the 2.31 MeV ($J=0^+$ $T=1$) level, but only to the ground state. Assignment of spin 0^- or $T=1$ or both to the 4.91 MeV level would explain these facts, but since it has no analogue in the C^{14} nucleus it almost certainly has $T=0$.

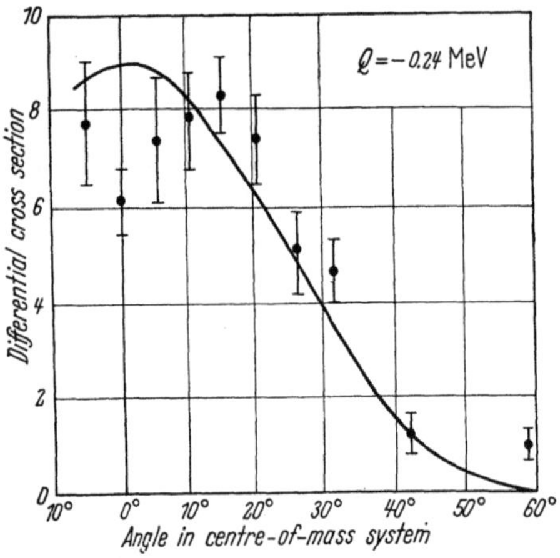

Fig. 56. Angular distribution of neutrons from $C^{12}(dn)$ N^{13} reaction (MIDDLETON *et al.* [41]).

One of the stripping patterns given by MIDDLETON *et al.* [41] for the $C^{12}(dn)$ reaction is shown in Fig. 56. The theoretical curves for this reaction calculated on the BUTLER theory completely fail to represent the experimental results for the 2.37 MeV level of N^{13}. This is likely to be a single particle (s-wave) level, and the analogue level in C^{13} observed by the (dp) reaction gives rise to a normal stripping pattern. This confirms the theoretical expectation that the BUTLER theory does not apply when the reaction Q-value is about that observed for the N^{13} level in question (-2.62 MeV), which is nearly equal to the binding energy of the deuteron. The BORN approximation theory gives a satisfactory pattern under these circumstances, with the use of a larger radius than is required in the corresponding (dp) reaction.

N^{14}, N^{15}: The angular distributions obtained for the $N^{14}(dp)$ reaction by JONGERIUS *et al.*[5] at deuteron energies up to 600 keV and by WYLY[6] at 1.5 to 3.05 MeV suggest a mixture of compound nucleus and stripping effects. GIBSON and THOMAS[6] using a gas target and nuclear emulsion detection were able to assign l_n values for the ground state transition with 8 MeV deuterons, but for other states, except perhaps that at 8.32, the curves are not even approximately fitted by the BUTLER theory. EBY[7] using 11.9 MeV deuterons, a solid target and counter detection found stripping distributions for the ground state transition and for protons corresponding to the unresolved doublet at 5.3 MeV. No definite assign-

---

[1] D. A. BROMLEY and L. M. GOLDMAN: Phys. Rev. **86**, 790 (1952). D. A. BROMLEY: Phys. Rev. **88**, 565 (1952).

[2] R. E. BENENSON: Phys. Rev. **90**, 420 (1953).

[3] J. N. McGRUER, E. K. WARBURTON and R. S. BENDER: Phys. Rev. **100**, 235 (1956).

[4] L. L. GREEN, J. P. SCANLON and J. C. WILLMOTT: Proc. Phys. Soc. Lond. A **68**, 386 (1955).

[5] H. M. JONGERIUS, F. P. G. VALCKX and P. M. ENDT: Physica, Haag **20**, 29 (1954). — L. D. WYLY: Phys. Rev. **76**, 104 (1949).

[6] W. M. GIBSON and E. E. THOMAS: Proc. Roy. Soc. Lond., Ser. A **210**, 543 (1951).

[7] F. S. EBY: Phys. Rev. **96**, 1355 (1954).

ment for the spins of this doublet can be made from the stripping results, but a $(p\gamma)$ correlation experiment by STANLEY[1] gives a dipole pattern and hence suggests $J=1/2$ and $3/2$ for the two components.

The neutron groups from the analogous reaction $N^{14}(dn)\,O^{15}$ (EVANS *et al.*, [*41*]) have angular distributions which tend to fall below the theoretical curves for $l_p=1$.

O^{16}, O^{17}, O^{18}: The striking distributions obtained in the $O^{16}(dp)\,O^{17}$ reactions with 8 MeV deuterons [*40*] stimulated the development of stripping theory.

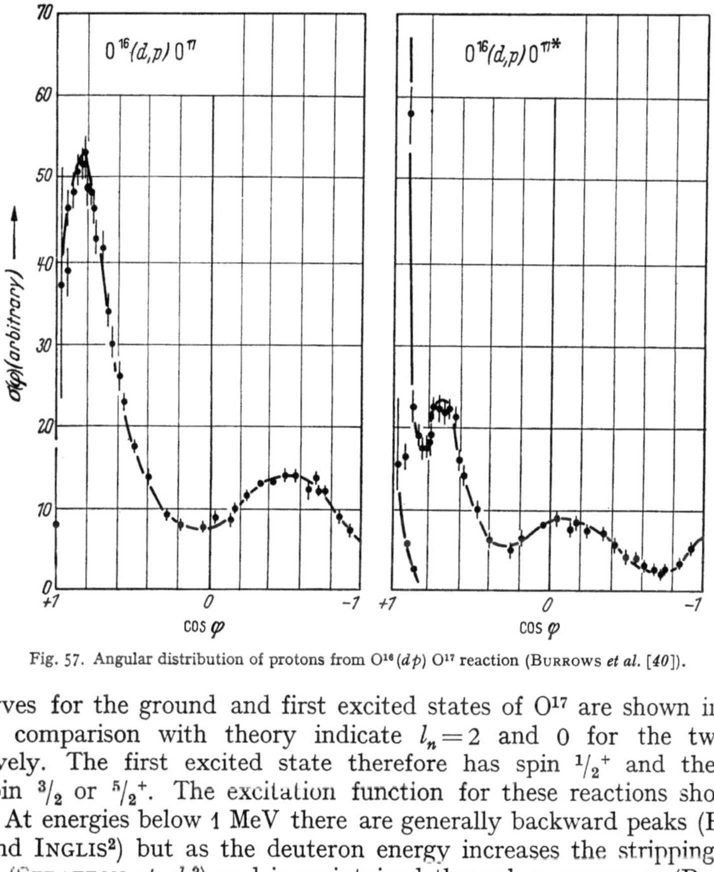

Fig. 57. Angular distribution of protons from $O^{16}(dp)\,O^{17}$ reaction (BURROWS *et al.* [*40*]).

The curves for the ground and first excited states of O^{17} are shown in Fig. 57 and by comparison with theory indicate $l_n=2$ and 0 for the two states respectively. The first excited state therefore has spin $1/2^+$ and the ground state spin $3/2$ or $5/2^+$. The excitation function for these reactions shows resonances. At energies below 1 MeV there are generally backward peaks (HEYDENBURG and INGLIS[2]) but as the deuteron energy increases the stripping pattern emerges (STRATTON *et al.*[3]) and is maintained through a resonance (BERTHELOT *et al.*[4]) without essential modification. The patterns for 19 MeV deuterons were obtained by FREEMANTLE *et al.*[5] using the nuclear emulsion method and are in general, but not detailed, agreement with the BUTLER theory although measurements were not made at the peak of the stripping distributions. Reduced widths calculated from the 8 MeV and 19 MeV data are given by FAIRBAIRN[6]. These are

---

[1] A. G. STANLEY: Phil. Mag. **45**, 807 (1954).

[2] N. R. HEYDENBURG and D. R. INGLIS: Phys. Rev. **73**, 230 (1948).

[3] T. F. STRATTON, J. M. BLAIR, K. F. FAMULARO and R. V. STUART: Phys. Rev. **98**, 629 (1955).

[4] A. BERTHELOT, R. COHEN, E. COTTON, H. FARAGGI, T. GRJEBINE, A. LEVEQUE, V. NAGGIAR, M. ROCLAWSKI-CONJEAUD and D. SZTEINSZNAIDER: J. Phys. Radium **16**, 241 (1955).

[5] R. G. FREEMANTLE, W. M. GIBSON, D. J. PROWSE and J. ROTBLAT: Phys. Rev. **92**, 1268 (1953).

[6] W. M. FAIRBAIRN: Proc. Phys. Soc. Lond. A **67**, 564 (1954).

considerably smaller than the single particle value, which would be expected for the ground and first excited states of O^{17}, and also vary with deuteron energy. This suggests again that the simple theory needs modification to take account of the proton-nucleus interaction. The reduced widths for the $O^{16}(dp)\ O^{17}$ reaction obtained by the method of BOWCOCK[1] from the observed stripping curve, are also somewhat smaller than the single particle value. For the $O^{18}(dp)\ O^{19}$ reaction (1.47 MeV level) STRATTON *et al.* found a very clear stripping pattern.

The analogous reaction $O^{16}(dn)\ F^{17}$ shows stripping patterns for the ground and first excited state of F^{17} which agree with those found for the corresponding O^{17} states. (AJZENBERG[2], MIDDLETON *et al.* [41]). A good fit with the BUTLER theory is obtained with a Q-value of -2 MeV in this case.

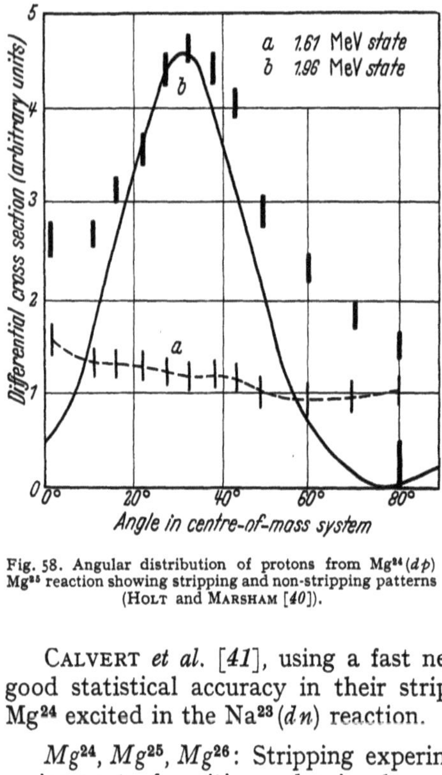

Fig. 58. Angular distribution of protons from $Mg^{24}(dp)$ Mg^{25} reaction showing stripping and non-stripping patterns (HOLT and MARSHAM [40]).

F^{19}: BROMLEY *et al.*[3] studied both the (dp) and (dn) reactions. The ground state stripping pattern might be fitted by $l_n=0$ and $l_n=2$ patterns superimposed. If this is so, and is not a complex angular distribution due to interference, it follows that the ground state spin of F^{20} is 1^+ and is a mixed state.

Ne^{20}, Ne^{22}: Stripping patterns were obtained by MIDDLETON and TAI [40].

Na^{23}: Angular distributions are given for 1 MeV deuterons by TAKEMOTO *et al.*[4], for 3 MeV by SHAPIRO[5], and for 10 MeV by BRETSCHER *et al.*[6]. These experiments give very similar patterns, with all the stripping features shifted to smaller angles in the 10 MeV observations. They agree in ascribing a mixture of $l_n=0$ and $l_n=2$ to the unresolved states at 0.47 and 0.56 MeV.

CALVERT *et al.* [41], using a fast neutron spectrometer, were able to obtain good statistical accuracy in their stripping patterns for the excited states of Mg^{24} excited in the $Na^{23}(dn)$ reaction.

Mg^{24}, Mg^{25}, Mg^{26}: Stripping experiments with 8 MeV deuterons [40] permit assignment of parities and spin alternatives to several residual states although not all can be resolved. A feature of interest is the appearance of patterns requiring $l_n=0$ and $l_n=2$ to be superimposed, indicating that the final state is described by a mixture of two wave functions with the same resultant spin, as in the case of Na^{23}. High resolution energy distributions show no trace of an unresolved doublet at the corresponding state. At least one state in Mg^{25} (Fig. 58) shows no sign of a stripping peak. The stripping curve for the 3.40 MeV excited state of Mg^{25} has

---

1 J. E. BOWCOCK: Proc. Phys. Soc. Lond. A **68**, 512 (1955).

2 F. AJZENBERG: Phys. Rev. **83**, 693 (1951).

3 D. A. BROMLEY, J. A. BRUNER and H. W. FULBRIGHT: Phys. Rev. **89**, 396 (1953). — D. A. BROMLEY: Phys. Rev. **88**, 565 (1952).

4 S. TAKEMOTO, T. DAZAI and R. CHIBA: Phys. Rev. **91**, 1024 (1953).

5 P. SHAPIRO: Phys. Rev. **93**, 290 (1954).

6 M. M. BRETSCHER, J. O. ALDERMAN, A. ELWYN and F. B. SHULL: Phys. Rev. **96**, 103 (1954).

been analysed by BOWCOCK[1] and gives an absolute reduced width of $\gamma^2 = 0.41 \times 10^{-18}$ erg cm. For this level a direct comparison can be made with the analogue level in Al^{25} which is unbound and for which $\gamma^2 = 0.37 \times 10^{-18}$. The excellent agreement confirms both the existence of analogue states and the method of analysis used by BOWCOCK. The $Mg^{26}(dp)\,Mg^{27}$ reaction shows that the 15th neutron to be added in building up a nucleus goes into an $s_{1/2}$ state as expected on the shell model at the completion of the $d_{5/2}$ shell.

Angular distributions of neutrons emitted in the $Mg^{24}(dn)\,Al^{25}$ reaction were observed by GOLDBERG[2] using a thin isotopic target. Assignments of l_p values are not entirely straightforward, because of the negative Q-values involved for the higher levels. The proton capture probability appears to be high for the s-wave level at 0.45 MeV, and there is general correspondence between the levels of Mg^{25} and of Al^{25}. The cross sections for the $Mg^{24}(dn)$ and $Mg^{24}(dp)$ reactions at corresponding stripping peaks are in the approximate ratio 1 to 5; in these cases the cross section at the maximum of the stripping peak is compared[3]. The stripping patterns are less definite than desirable, but better statistics require the use of a fast neutron spectrometer.

Al^{27}: The $Al^{27}(dp)\,Al^{28}$ reaction was one of the first for which accurate stripping data became available. The states of Al^{28} are very numerous, but below 2 MeV there appear to be only 6, and 4 of these form two close doublets. The angular distributions were obtained by HOLT and MARSHAM [*40*] and their results have been supplemented by BUECHNER[4] who was able to obtain distributions for transitions to the individual doublet components by use of magnetic analysis.

Accurate angular distributions for the $Al^{27}(dn)\,Si^{28}$ reaction have been obtained by CALVERT *et al.* [*41*] using a fast neutron spectrometer. 10 neutron groups were resolved and l_p values assigned to many of them. From the intensities at the peak of the stripping distributions, proton capture probabilities were calculated, and evidence for excitation of the first $T = 1$ state in Si^{28} was obtained. This state is analogous to the ground state of Al^{28} and is therefore approximately a single particle state. It has $l_p = 0$, and the proton capture probability is again a factor of 5 smaller than the corresponding $l_n = 0$ neutron capture probability for the ground state (dp) reaction. A factor of 2 arises from the branching ratio for different T_z values; the remaining factor must be ascribed to COULOMB effects.

Si^{28} Si^{29} Si^{30}: The (dp) stripping reaction was studied by HOLT and MARSHAM [*40*] and the results are listed in Table 12. The states with high neutron capture probability may be associated with the single particle levels of the shell model for nucleons added in $s_{1/2}$, $d_{3/2}$, $f_{7/2}$, $p_{3/2}$, and $p_{1/2}$ states. The 2.03 MeV state, with low capture probability, is ascribed to the configuration $(d_{5/2})^{-1}\,(s_{1/2})^2$ which postulates core excitation and pairing of two nucleons in the $s_{1/2}$ level. This is supported by the fact that the first excited state of Si^{28} has an energy near that of the Si^{29} state concerned.

P^{31}: PARKINSON *et al.*[5] using 7.2 MeV deuterons studied the stripping pattern for the (doublet) ground state transition and found clear evidence for $l_n = 2$. This is the value required by the shell model since the d-orbits are filling at the 17th neutron; the alternative $l_n = 0$ transition could lead to the same value for the spin of the final nucleus P^{32} but is unacceptable to the shell model. Since $l_n = 0$

[1] J. E. BOWCOCK: Proc. Phys. Soc. Lond. A **68**, 512 (1955).

[2] E. GOLDBERG: Phys. Rev. **89**, 760 (1953).

[3] See J. M. CALVERT, A. A. JAFFE and E. E. MASLIN: Phys. Rev. **101**, 501 (1956) for a recent comparison of reduced widths obtained from (dp) and (dn) reactions in Be^9, C^{12}, N^{14}, O^{16} and Si^{28}.

[4] W. W. BUECHNER, M. MAZARI and A. SPERDUTO: Phys. Rev. **101**, 188 (1956).

[5] W. C. PARKINSON, E. H. BEACH and J. S. KING: Phys. Rev. **87**, 387 (1952).

transitions are much more probable than $l_n = 2$, this experiment, as pointed out by BETHE and BUTLER[1], is a sensitive test of the contamination of the wave functions of pure shell-model states by other wave functions.

EL BEDEWI and EL WAHAB[2] observed the angular distributions for several groups of neutrons from the $P^{31}(dn)$ S^{32} reaction. A state of type 1^+ in the S^{32} nucleus is suggested by this work. The reduced widths for the S^{32} states are all comparable, in contrast to the situation encountered in (dn) reactions leading to odd mass nuclei. The possible configurations in jj coupling for the S^{32} levels are discussed by these authors, and it is pointed out that the ground state cannot have the configuration $(s_{1/2})^3$ since this would lead to two states of spin zero for $l_p = 0$ whereas there is known to be a radiative transition between the two states with this value of l_p. The reaction has also been studied with a fast neutron spectrometer.

S^{32}: The (dp) reaction was studied by DAVISON[3] using 3.2 MeV deuterons from a cyclotron and range analysis. $(p-\gamma)$ coincidences were found for some of the groups of protons leading to excited states. The stripping patterns are given by HOLT and MARSHAM [*40*] who also measured Q-values and deduced the energy levels shown in Table 12. The neutron capture probabilities, shown in Table 12, demonstrate the existence of single particle levels and states of more complex excitation. The state at 0.85 MeV in S^{33} may be due to pairwise filling of the $d_{3/2}$ shell as discussed for Si^{29} and in fact its energy is close to that of the first excited state of S^{32}, obtained by promoting a $s_{1/2}$ nucleon to the $d_{3/2}$ state.

The neutron groups and stripping distributions are given by MIDDLETON *et al.* [*41*]; some of the latter agree fairly well with the curves predicted both by the BUTLER and BORN approximation theories, but for the states with negative Q-values greater than 2 MeV no agreement with either theory was achieved. In particular the sharp forward minimum is not predicted, although the form of the curve at larger angles is more satisfactory. The states of Cl^{33} observed and tentative spin assignments are in satisfactory agreement with those for the mirror nucleus S^{33} obtained from the (dp) reaction.

Cl^{35} Cl^{37}: The excited states of the nuclei Cl^{36} and Cl^{38} have been investigated with fairly low resolution by SHRADER and POLLARD[4] using deuterons accelerated in a cyclotron and range analysis. KING and PARKINSON[5] showed that the stripping pattern for the ground state (dp) reaction is fitted by $l_n = 2$. Since the spin of Cl^{36} is known to be 2 and that of Cl^{35} is $^3/_2$ l_n values of 0, 2 and 4 are permitted by the selection rule for J. The shell model however requires the 19th neutron to enter a d-orbit, and this is confirmed by the experiment. The admixture of $l_n = 0$ waves appears to be less than 4%.

A^{40}: Range distributions for $A^{40}(dp)$ were obtained by DAVISON *et al.*[6] using 3.4 MeV deuterons and by GIBSON and THOMAS [*40*] with 8 MeV particles. The ground state transition seems to require $l_n = 3$ as expected if the $f_{7/2}$ shell is filling.

Ca^{40}: The level system of Ca^{41} is known from magnetic analysis of protons from the $Ca^{40}(dp)$ reaction [*33*].

Angular distributions for some of the levels were obtained by HOLT and MARSHAM [*40*], but the resolution of their equipment was insufficient in general to separate the many narrow doublets of the spectrum. Despite this, many of the

---

1 H. A. BETHE and S. T. BUTLER: Phys. Rev. **85**, 1045 (1952).
2 F. A. EL BEDEWI and M. A. EL WAHAB: Proc. Phys. Soc. Lond. A **68**, 754 (1955).
3 P. W. DAVISON: Phys. Rev. **75**, 757 (1949).
4 E. F. SHRADER and E. C. POLLARD: Phys. Rev. **59**, 277 (1941).
5 J. S. KING and W. C. PARKINSON: Phys. Rev. **88**, 141 (1952).
6 P. W. DAVISON, J. O. BUCHANAN and E. C. POLLARD: Phys. Rev. **76**, 890 (1947).

levels, or unresolved groups of levels, show fairly clear stripping patterns so that the l_n value appears to be the same for the several components of a "line". The ground state and first two levels have l_n values which suggest the order $f_{7/2}$, $p_{3/2}$, $p_{1/2}$ and neutron capture probilities near to the single particle value.

The conclusions of stripping experiments with 8 MeV deuterons on elements lighter than calcium are discussed by HOLT and MARSHAM [*40*]. In general the main features of the observed angular distributions are predicted by theory but isotropic backgrounds are often present. The nuclear radius used was usually, in the BUTLER theory

$$(1.7 + 1.22\,A^{1/3}) \times 10^{-13}\,\text{cm} \quad \text{for} \quad (d\,p) \text{ reactions}$$

and

$$(2.7 + 1.22\,A^{1/3}) \times 10^{-13}\,\text{cm} \quad \text{for} \quad (d\,n) \text{ reactions}$$

but a better fit could be obtained for some of the lighter elements with an empirically determined radius. A plot of the empirical radius against $A^{1/3}$ shows an anomaly near mass numbers 9 to 12; a similar anomaly is found in total neutron cross sections (Sect. 42).

V. Other reactions.

53. Pick-up reactions (inverse stripping). The process by which a deuteron in passing near a nucleus picks up a nucleon without amalgamation to form a compound state is formally analogous to a stripping process. The differential cross sections for stripping and pick-up processes are related by the principle of detailed balancing and the cross sections for $(d\,t)$ and $(d\,He^3)$ reactions may therefore be obtained from those for the stripping reactions $(t\,d)$ and $(He^3 d)$ which have been calculated by NEWNS[1] and by BUTLER and SALPETER[1]. The cross section for the $(p\,d)$ and $(n\,d)$ reactions may be obtained from the deuteron stripping theories.

Examples of these reactions are listed in Table 12 and are discussed below.

$Li^6(dt)\,Li^5$, $Li^6(dHe^3)\,He^5$: These reactions have been observed [*32*]; the ground state groups are wide and both Li^5 and He^5 break up in the second stage of the reaction.

$Li^7(dt)\,Li^6$; $Li^7(dHe^3)\,He^6$: The $Li^7(d\,t)\,Li^6$ reaction was observed by GOVE and HARVEY[2], who found groups of tritons for the ground and first excited state of Li^6. The ground state angular distributions obtained by LEVINE *et al.*[3] with 14.5 MeV deuterons are shown in Fig. 59. These, and similar distributions for excited states, are well fitted by theoretical curves using a radius considerably larger than that required for the $(d\,p)$ reaction, perhaps owing to the use of GAUSSian wave functions for the He^3 or H^3 particle. The same large radius was required by HOLT and MARSHAM [*40*] to describe the angular distribution of tritons from $Li^7(d\,t)$ for a deuteron energy of 8 MeV.

The excitation function for the production of tritium by the reaction $Li^7(d\,t)$ was followed by MACKLIN and BANTA[4] from 0.4 to 4.0 MeV. The tritium was recovered and estimated by counting beta disintegrations in a proportional counter.

$Be^9(dt)\,Be^8$, $Be^9(dHe^3)\,Li^8$: The $(d\,t)$ reaction to the ground state of Be^8 has long been known; the absolute cross section was measured for a range of

[1] H. C. NEWNS: Proc. Phys. Soc. Lond. A **65**, 916 (1952). — S. T. BUTLER and E. E. SALPETER: Phys. Rev. **88**, 133 (1952).
[2] H. E. GOVE and J. A. HARVEY: Phys. Rev. **82**, 658 (1951).
[3] S. H. LEVINE, R. S. BENDER and J. N. MCGRUER: Phys. Rev. **97**, 1249 (1955).
[4] R. L. MACKLIN and H. E. BANTA: Phys. Rev. **97**, 753 (1955).

deuteron energies up to 8 MeV by WOLFGANG and LIBBY[1] by tritium estimation in a proportional counter. The angular distribution of ground state tritons at low bombarding energies has been measured by many workers [*32*]. For 8 MeV deuterons HOLT and MARSHAM [*40*] obtained a pattern which, while not in precise agreement with pick-up theory, enabled an angular momentum transfer of 1 to be assigned. The 2.9 MeV state of Be^8 is also observed in this reaction; both this state and the ground state break up into alpha particles with a very short lifetime (GELINAS and HANNA[2], CUER and JUNG[2]). — The $(d\,He^3)$ reaction has been observed by detection of the alpha activity of Li^8 (WINN[3]).

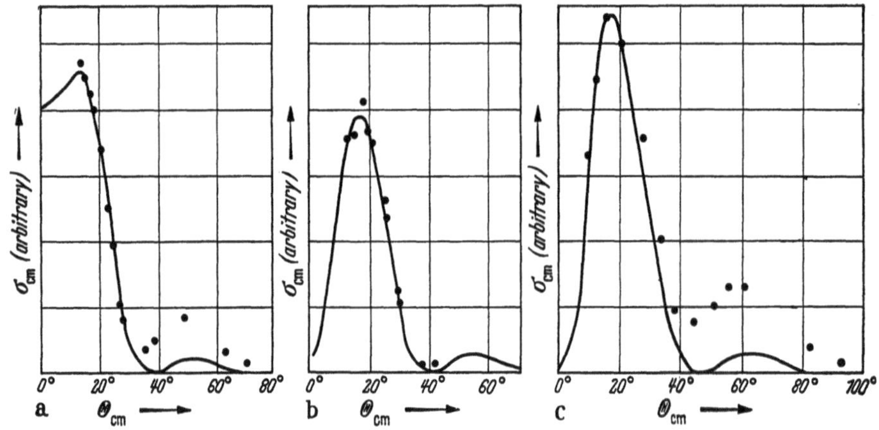

Fig. 59. Angular distributions of pick-up reactions. (a) $Li^7(dt)$; (b) $Li^7(d\,He^3)$; (c) $N^{14}(pd)$.

$C^{13}(dt)\,C^{12}$: The differential cross section for this reaction was measured by HOLMGREN *et al.*[4] for deuterons of energy 3.29 and 2.19 MeV and analysed by pick-up theory. At the lower energy the pick-up peak moves to larger angles and the whole distribution becomes broader. The radius required is comparable with that used in the stripping theory for $C^{12}(dp)\,C^{13}$.

$He^4(pd)\,He^3$; This reaction was observed by BENEVISTE and CORK[5] using 32 MeV protons. The angular distribution agrees with that predicted by theory for $l_n = 0$.

$Li^7(pd)\,Li^6$: The reaction was observed with 18 MeV protons by FRANZEN and LIKELY[6] using a sodium iodide crystal and pulse height analysis. At this energy both $T=0$ and $T=1$ states of Li^6 were excited with good yield, as would be expected because of the high energy of excitation of the intermediate state. The angular distribution suggests $l_n = 1$ [*32*] for pick-up deuterons to the ground state.

$Be^9(pd)\,Be^8$: This reaction has a small positive Q-value and has been studied in detail at low energies (see Sect. 31) for which resonance phenomena are important. The angular distributions show a forward tendency developing as the proton energy increases to 1 MeV. The 2.9, 4.0 and 5.1 MeV energy levels of Be^8

[1] R. L. WOLFGANG and W. F. LIBBY: Phys. Rev. **85**, 437 (1952).
[2] P. CUER and J. J. JUNG: Phys. Rev. **89**, 1151 (1953). — R. W. GELINAS and S. S. HANNA: Phys. Rev. **89**, 483 (1953).
[3] M. WINN: Proc. Phys. Soc. Lond. A **67**, 946 (1954).
[4] H. D. HOLMGREN, J. M. BLAIR, B. E. SIMMONS, T. F. STRATTON and R. V. STUART: Phys. Rev. **95**, 1544 (1954).
[5] J. BENEVISTE and B. CORK: Phys. Rev. **89**, 422 (1953).
[6] W. FRANZEN and J. G. LIKELY: Phys. Rev. **87**, 667 (1952).

were found in this reaction by ARTHUR *et al.*[1] using 8 MeV protons from a cyclotron with magnetic analysis. With 22 MeV protons COHEN *et al.*[2] found an angular distribution which appeared relatively insensitive to proton energy. It appears to be consistent with $l_n = 1$ [*32*].

$C^{12}(pd)\,C^{11}$: This reaction has a high negative Q-value but has been observed by BRITTEN[3] using 31.5 MeV protons from a linear accelerator.

$N^{14}(pd)\,N^{13}$: STANDING[4] observed the deuterons from this reaction using a scintillation detector and pulse height analysis. The angular distribution of ground state deuterons obtained with 18.7 MeV protons is shown in Fig. 59 and may be fitted by a pick-up curve with $l_n = 1$. No trace was found of deuterons corresponding to the first excited state of N^{13}.

$Li^6(nd)\,He^5$: FRYE[5], using 14 MeV neutrons and nuclear emulsion detectors showed that the angular distribution of the ground state deuteron peak agreed with inverse stripping theory for $l_p = 1$. The excited state of He^5 was less well defined in this experiment but again the angular distribution was consistent with $l_p = 1$.

$B^{10}(nd)\,Be^9$: Angular distributions of pick-up deuterons were obtained by RIBE and SEAGRAVE[6] using 14 MeV neutrons and a coincidence counter telescope. The distributions indicate $l_p = 1$ for the ground and first excited states of Be^9.

Evidence for the other pick-up reactions shown in Table 12 is given in [*32*].

54. Reactions induced by H^3, He^3 and heavy ions. The reaction of H^3 and He^3 with H^1, H^2 and He^4 have already been discussed. There remain for consideration the reactions of these nuclei with themselves and with nuclei of $A > 4$. H^3 and He^3

Table 13. *Reactions observed with* H^3 *and* He^3 *in nuclei with* $A > 4$. (References [*32*] and [*33*].)

Target nucleus	Reaction and final nucleus								
	tp	tn	$t\alpha$	td	He^3p	He^3n	$He^3\alpha$	He^3d	He^3t
Li^6	Li^8		He^5	Li^7	Be^8		Li^5		
Li^7		Be^9	He^6		Be^9		Li^6	Be^8	
Be^9					B^{11}	C^{11}	Be^8		
B^{10}	B^{12}		Be^9		C^{12}		B^9	C^{11}	
B^{11}					C^{13}			C^{12}	
C^{12}	C^{14}						C^{11}	N^{13}	
C^{13}									N^{10}

Also $O^{16}(He^3\,\alpha)\,O^{15}$; $O^{16}(t\,p)\,O^{18}$; $Ne^{20}(t, n)\,Na^{22}$; $Mg^{26}(t\,p)\,Mg^{28}$; $Al^{27}(t\,p)\,Al^{29}$; $Al^{27}(t\,He^3)\,Mg^{27}$; $Si^{28}(He^3\,p)\,P^{30}$; $Si^{30}(He^3\,p)\,P^{32}$;

share with the neutron and proton the property of inducing mirror reactions in a given target nucleus, and it is therefore convenient to consider both processes together. The reaction data are summarised in Table 13; most of the accurate work so far published has been done with high tension accelerators, although

[1] J. C. ARTHUR, A. J. ALLEN, R. S. BENDER, H. J. HAUSMAN and C. J. McDOLE: Phys. Rev. **88**, 1291 (1952).
[2] B. L. COHEN, E. NEWMAN, T. H. HANDLEY and A. TIMNICK: Phys. Rev. **90**, 323 (1953).
[3] R. BRITTEN: Phys. Rev. **88**, 283 (1952).
[4] K. G. STANDING: Phys. Rev. **94**, 731 (1954); **101**, 152 (1956). See also J. B. REYNOLDS and K. G. STANDING: Phys. Rev. **101**, 158 (1956) for the angular distribution of deuterons from (pd) reactions in Li, Be, B, F and Al.
[5] G. M. FRYE: Phys. Rev. **93**, 1086 (1954).
[6] F. L. RIBE and J. D. SEAGRAVE: Phys. Rev. **94**, 934 (1954).

H^3 and He^3 accelerated in cyclotrons have also been used to induce activities, notably in Be, C and O (KUNDU *et al.*[1], FREMLIN[1]).

α) *H^3+H^3, H^3+He^3, He^3+He^3.* Each of these interactions can lead to a two stage process of the form

$$H^3 + H^3 \rightarrow He^5 + n \qquad (54.1)$$

$$\rightarrow He^4 + 2n \qquad (54.2)$$

or an instantaneous three body reaction may take place

$$H^3 + H^3 \rightarrow He^4 + 2n. \qquad (54.3)$$

In the former case a discrete, although possibly wide, group of particles is expected in addition to a continuous distribution due to break-up of the He^5 or Li^5 in flight. In the latter case only a continuous distribution of particle energies is expected. If there is such a particle as a di-neutron, reaction (54.2) gives homogeneous groups; the theory of such a reaction is discussed by MIGDAL[2]. The reactions have been studied with H^3 and He^3 beams of energies up to 2 MeV. Targets were gaseous He^3 in the case of electrostatic generator experiments, and for the lower bombarding energies the tritium targets were made by adsorbing tritium on zirconium. The He^3 targets were produced by bombarding a metal surface with a beam of He^3 ions. Charged particles were detected in a sodium iodide crystal arranged for pulse height analysis and neutrons were observed in a nuclear emulsion.

If charge independence holds, the general features of these three interactions should be very similar, except for charge dependent effects. The COULOMB barrier is presumably responsible for the comparatively low total cross section for He^3+He^3 compared with the other reactions. The disintegration products of H^3+H^3 were examined by ALLEN *et al.*[3] and no evidence for di-neutron production, which would have lead to a homogeneous group of alpha particles, was obtained. For higher triton energies LELAND and AGNEW[4] found some evidence for production of He^5 in both ground and excited states by observation of the associated neutron spectrum. The differential cross section for the reaction at 0° given by AGNEW *et al.*[5] for tritons up to 2 MeV shows a steady rise.

The H^3+He^3 interaction gives rise to the three reactions

$$H^3 + He^3 \rightarrow He^4 + H^2, \qquad (54.4)$$

$$\rightarrow He^4 + p + n, \qquad (54.5)$$

$$\rightarrow He^5 + p, \qquad (54.6)$$

and the energy spectrum found by MOAK[6] shows these features, namely discrete groups of deuterons, alpha particles and protons corresponding to reactions (54.4), (54.6) together with a continuous distribution of particles due to (54.5). There is no evidence for an excited state of He^5.

---

[1] D. N. KUNDU, T. W. DONAVEN, M. L. POOL and J. K. LONG: Phys. Rev. **89**, 1200 (1953). — J. H. FREMLIN: Proc. Phys. Soc. Lond. A **65**, 762 (1952).

[2] A. B. MIGDAL: J. exp. theor. Phys. USSR. **28**, 3 (1955), translated in Soviet Physics (J. E. T. P.) **1**, 2 (1955).

[3] K. W. ALLEN, E. ALMQVIST, J. T. DEWAN, T. P. PEPPER and J. H. SANDERS: Phys. Rev. **82**, 262 (1951).

[4] W. T. LELAND and H. M. AGNEW: Phys. Rev. **82**, 559 (1951).

[5] H. M. AGNEW, W. T. LELAND, H. V. ARGO, R. W. CREWS, A. H. HEMMENDINGER, W. E. SCOTT and R. F. TASCHEK: Phys. Rev. **84**, 862 (1951).

[6] C. D. MOAK: Phys. Rev. **92**, 383 (1953).

For $He^3 + He^3$ ALMQVIST *et al.*[1] and GOOD *et al.*[1] studied the proton spectrum arising from the processes

$$He^3 + He^3 \to Li^5 + p, \tag{54.7}$$

$$\to He^4 + 2p. \tag{54.8}$$

The protons were detected and their energy measured, by nuclear emulsions and by scintillation spectrometry. There is evidence for formation of Li^5 in its ground state and for the three body process, but an excited state of Li^5 is not seen.

β) $Li^6(tp)\,Li^8$, $Li^6(He^3p)\,Be^8$ and other Li^6 reactions. The (tp) reaction was observed by MOAK *et al.*[2] and by ALLEN *et al.*[2] and the (He^3p) reaction was used by KUNZ *et al.*[3] to study the levels of Be^8. Evidence for only two levels in Be^8 up to an energy of about 11 MeV was found.

Many companion reactions are possible, e.g.

$$\left.\begin{aligned} Li^6 + H^3 &\to Be^9 + \gamma, \\ &\to Be^8 + n, \\ &\to Li^7 + d, \\ &\to 2\alpha + n, \end{aligned}\right\} \quad (54.9) \qquad \left.\begin{aligned} Li^6 + He^3 &\to B^9 + \gamma, \\ &\to B^8 + n, \\ &\to Be^7 + d, \\ &\to 2\alpha + p. \end{aligned}\right\} \quad (54.10)$$

Many of these have been detected by momentum or energy analysis of the product particles and the Q-values for the reactions have been checked.

γ) $Li^7(t\alpha)\,He^6$, $Li^7(He^3\alpha)\,Li^6$ and other Li^7 reactions. ALLEN *et al.* found clear evidence for the formation of an excited state of He^6 in the $Li^7(t\alpha)$ reaction, and MOAK *et al.*[4] obtained similar but not so clear evidence of excited states of Li^6 in the companion reaction. ALMQVIST *et al.*[5] using 240 keV tritons found that the intense group of alpha particles associated with the excited state transition in the $(t\alpha)$ reaction had an isotropic angular distribution while the ground state particles were distributed as $1 - (0.66 \pm 0.06)\cos^2\vartheta$. This reaction was discussed by CHRISTY [*27*] who assumes that it proceeds through a compound state of $J = 2^+$ and suggests that the observations are consistent with a LS coupling scheme. The reaction probably proceeds through a resonance at 0.84 MeV triton energy (CREWS[6]).

MOAK *et al.*[4] also observed the $Li^7(He^3p)\,Be^9$ reaction, which is of some interest because very few levels of Be^9 have been reported. Pulse spectra obtained in a sodium iodide crystal suggested Be^9 levels at 1.18, 2.43, 3.2 and 4.0 MeV, of which only the 2.43 MeV level can be identified in reactions such as $Be^9(pp')$ or $B^{11}(d\alpha)$.

Fast neutrons from the $Li + t$ reactions have been detected by ALMQVIST[7] but the particular reactions concerned were not identified.

δ) Reactions of H^3 and He^3 with heavier nuclei. Some of the reactions observed are listed in Table 13. Results for these reactions, which are not yet published, are given in [*32*]. The He^3 reactions are particularly valuable as a means of studying the excited states of $T = 0$ nuclei such as F^{18} by a reaction in which

[1] E. ALMQVIST, K. W. ALLEN, J. T. DEWAN and T. P. PEPPER: Phys. Rev. **91**, 1022 (1953). — W. M. GOOD, W. E. KUNZ and C. D. MOAK: Phys. Rev. **94**, 87, (1954).

[2] K. W. ALLEN, E. ALMQVIST, J. T. DEWAN and T. P. PEPPER: Phys. Rev. **96**, 684 (1954). — C. D. MOAK, W. M. GOOD and W. E. KUNZ: Phys. Rev. **85**, 928 (1952).

[3] W. E. KUNZ, C. D. MOAK and W. M. GOOD: Phys. Rev. **91**, 676 (1953).

[4] C. D. MOAK, W. M. GOOD and W. E. KUNZ: Phys. Rev. **96**, 1363 (1954).

[5] E. ALMQVIST, T. P. PEPPER and P. LORRAIN: Canad. J. Phys. **32**, 621 (1954).

[6] R. W. CREWS: Phys. Rev. **82**, 100 (1951).

[7] E. ALMQVIST: Canad. J. Phys. **28**, 433 (1950).

states of more than one T-value can be excited. Comparison of the states observed in the bombardment of nuclei of mass $4n+2$ by H^3 and He^3, with emission of protons, give a direct comparison of the level systems of $T=1$ and $T=0$ nuclei, e.g. $B^{10}(H^3p)\,B^{12}$, $B^{10}(He^3p)\,C^{12}$.

The excitation function for the $O^{16}(tn)\,F^{18}$ reaction is given by JARMIE[1], who detected induced activity, and the triton induced activities in Mg and Al were studied by IWERSEN *et al.*[2]. In these experiments tritons from the (dd) reaction and from the $Li^6(n\alpha)\,H^3$ reaction induced by pile neutrons were used.

ε) Heavy ion reactions. The technique of accelerating heavy ions such as C^{12}, N^{14} and O^{16} in cyclotrons has advanced considerably recently and is summarised in a review article by WALKER[3]. An external beam of 28 MeV nitrogen ions from the Oak Ridge cyclotron was used by WYLY and ZUCKER[4] and by REYNOLDS and ZUCKER[5] to induce activities in light nuclei, mainly, it is assumed, by direct two-body or fission reactions of the type

$$C^{12} + N^{14} \rightarrow F^{18} + Be^8. \tag{54.11}$$

The excitation functions for the reactions $C^{12}(N^{14}2p)\,Na^{24}$ and $Mg^{25}(p,2p)\,Na^{24}$ were compared by COHEN *et al.*[6]. These reactions involve the same compound nucleus Al^{26} and after correction for barrier penetration the results were comparable for the same excitation of the compound nucleus. For $(N^{14}\alpha)$ reactions, the heavy ion reactions are enhanced with respect to similar reactions passing through the same compound nucleus, and some form of direct interaction probably takes place. This is strongly indicated by the work of CHACKETT *et al.*[7] on the reactions of N^{14} with Al^{27} which also gives evidence for the stripping reaction $N^{14}\rightarrow N^{13}+n$.

55. Photoprocesses. The nuclear photoeffect is reviewed in Vol. XLI of this Encyclopedia, and only a brief account of those photoprocesses which give information on the lower states of nuclei will be presented here. No general account is given of the "giant resonance" phenomenon of photodisintegration, and only brief mention is made of the energy and angular distributions of photodisintegration products. Both sorts of observation establish the importance of direct interaction processes in addition to the compound nucleus mechanism.

α) (γn), (γp) and (γt) reactions.

He^3, He^4: There are no reports of the photodisintegration of He^3. He^4 has been irradiated in an expansion chamber by GAERTTNER and YEATER[8] using 100 MeV radiation from a betatron. FULLER[8] used 50 MeV bremsstrahlung to irradiate He^4 in a high pressure chamber equipped with nuclear emulsion detectors. These experiments establish both the (γn) and (γp) reactions and suggest a peak in the (γp) excitation function at about 27 MeV and a $\sin^2\vartheta$ angular distribution with respect to the gamma ray beam, i.e. a maximum in the direction of the electric vector. These features are in agreement with the theoretical treatment of the electric dipole transition ${}^1S\rightarrow{}^1P$ given by GUNN and IRVING,

[1] N. JARMIE: Phys. Rev. **98**, 41 (1955).
[2] E. IWERSEN, W. S. KOSKI and F. RASETTI: Phys. Rev. **91**, 1229 (1953).
[3] D. WALKER: Progress in Nuclear Physics, Vol. 4. London: Pergamon Press 1955.
[4] L. D. WYLY and A. ZUCKER: Phys. Rev. **89**, 524 (1953).
[5] H. L. REYNOLDS and A. ZUCKER: Phys. Rev. **96**, 1615 (1954).
[6] B. L. COHEN, H. L. REYNOLDS and A. ZUCKER: Phys. Rev. **96**, 1617 (1954).
[7] K. F. CHACKETT, J. H. FREMLIN and D. WALKER: Phil. Mag. **45**, 173 (1954). — K. F. CHACKETT and J. H. FREMLIN: Phil. Mag. **45**, 735 (1954). — G. A. CHACKETT, K. F. CHACKETT and J. H. FREMLIN: Phil. Mag. **46**, 1 (1955).
[8] E. R. GAERTTNER and M. L. YEATER: Phys. Rev. **83**, 146 (1951). — E. G. FULLER: Phys. Rev. **96**, 1306 (1954).

and FLOWERS and MANDL[1], whose excitation curve is shown in Fig. 60. This describes both the (γn) and (γp) reactions; the second curve is for the $(\gamma\, 2d)$ electric quadrupole process which may have been observed by FULLER. The $He^4(\gamma p)$ reaction, in agreement with the $H^3(p\gamma)$ results (Sect. 28), gives no support for a 21 MeV resonance level in He^4. The angular distribution shows a small interference term which may be due to neglect of the magnetic dipole transition at energies near threshold. The photoneutron cross section for He^4 is given by FERGUSON *et al.*[2] for energies up to 26 MeV. These results are in general agreement with theory; the neutrons were detected in a paraffin covered boron trifluoride counter.

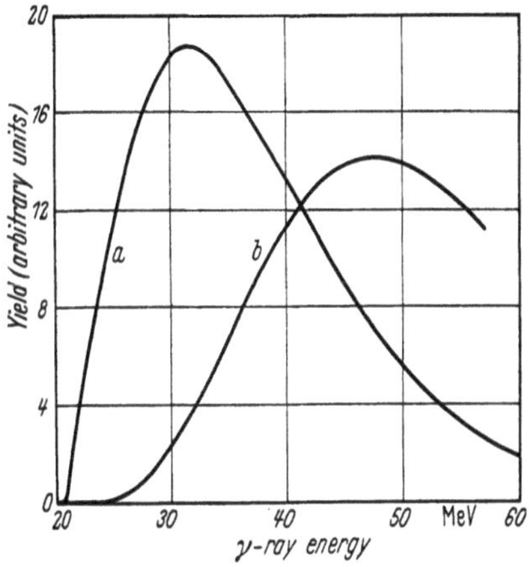

Fig. 60. Photodisintegration of He^4. (a) $\rightarrow He^3 + n$, $H^3 + p$ (dipole); (b) $\rightarrow H^2 + H^2$ (quadrupole) on different scale.

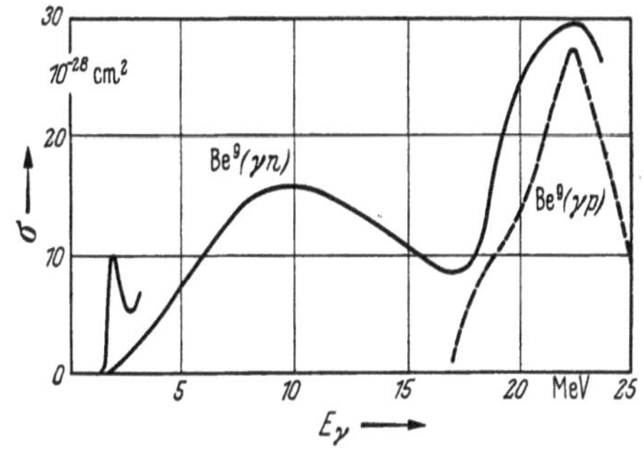

Fig. 61. Photodisintegration of Be^9.

Li^6, *Li*^{7} [42]: Lithium can easily be introduced into a photographic emulsion and the $Li^6(\gamma n)$, (γp) and (γt) reactions leading to the ground state of Li^5 and He^5 have been observed but the excited state transitions are not identified. In Li^7 the same two reactions are observed and the break-up of excited states of Li^6 into $He^4 + H^2$ is noted. The $Li^7(\gamma p)\, He^6$ reaction shows some evidence for the excitation of the 1.71 MeV state of He^6. The excitation function for the (γp) reaction has been determined both by the emulsion method and by observation of the He^6 activity.

The $Li^7(\gamma t)$ and (γn) reactions are suitable for comparing levels of $T = \frac{1}{2}$ and $T = \frac{3}{2}$ in the Li^7 nucleus (Sect. 58). Angular distributions for the tritons from this reaction suggest spin values for some of the states excited (STOLL[3]).

Be^9: The (γp) reaction was studied by detection of the Li^8 β-activity (TUCKER and GREGG[4]). The (γn) reaction has been very fully investigated because of its low threshold and of its importance as a photoneutron source. Accurate measurements of threshold can be made by using bremsstrahlung produced by electrons accelerated in an electrostatic generator (NOYES *et al.*[5]). The result is (1.662 ± 0.003) MeV. The excitation function for the reaction is given near threshold by HAMERMESH and KIMBALL[6] and for higher energies by NATHANS and

[1] B. H. FLOWERS and F. MANDL: Proc. Roy. Soc. Lond., Ser. A **206**, 131 (1951). — J. C. GUNN and J. IRVING: Phil. Mag. **42**, 1353 (1951).

[2] G. A. FERGUSON, J. HALPERN, R. NATHANS and P. F. YERGIN: Phys. Rev. **95**, 776 (1954).

[3] P. STOLL: Helv. phys. Acta **27**, 395 (1954).

[4] B. L. TUCKER and E. C. GREGG: Phys. Rev. **87**, 907 (1952).

[5] J. C. NOYES, J. E. VAN HOOMISSEN, W. C. MILLER and B. WALDMAN: Phys. Rev. **95**, 396 (1954).

[6] B. HAMERMESH and C. KIMBALL: Phys. Rev. **90**, 1063 (1953).

HALPERN[1]. The angular distribution near threshold (2.76 MeV) is $1.67 + \sin^2\vartheta$ (HAMERMESH *et al.*[2]). The excitation functions are shown superimposed in Fig. 61; the lower energy points are obtained with line spectra of gamma radiation and the higher energy results by means of the photon-difference method with a betatron. This has insufficient resolution to exhibit the sharp peak near threshold found in the low energy work.

The photodisintegration of beryllium has been discussed theoretically by GUTH and MULLIN[3] who adopt a simple model in which a neutron is assumed to move in a potential well provided by the rest of the nucleus and to be initially in a $P_{3/2}$ state. Magnetic dipole transitions are assumed to be weak and the cross section near threshold is due to the electric dipole $P_{3/2} \to S_{1/2}$ process, with the $S_{1/2}$ level near threshold and an isotropic angular distribution. For higher energies the electric dipole process leads to D states of the continuum; with an angular distribution of the form $a + b \sin^2\vartheta$. This accounts for the peak in the excitation function at about 10 MeV and for the appearance of a non-isotropic term in the angular distribution even near threshold. The 22 MeV peak in the excitation function is the "giant" resonance which must be attributed to core excitation, since it is shared by the (γp) reaction.

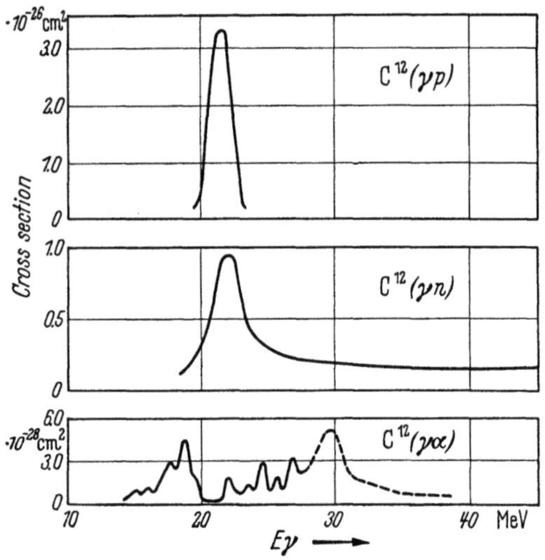

Fig. 62. Cross sections for (γp) (γn) $(\gamma \alpha)$ reactions in C^{12}.

The (γn) transitions were shown by CARVER *et al.*[4] to proceed mainly to the 2.9 MeV excited state of Be^8 for 6.1 MeV photons; in this experiment coincident alpha particles from a beryllium foil irradiated in a proportional counter were detected.

The $Be^9(\gamma t)\,Li^6$ reaction has been reported [*32*] in a beryllium loaded nuclear emulsion.

B^{10}, B^{11}: Boron is another element which can be incorporated easily in a nuclear emulsion. Neutrons from the (γn) reactions have been observed for both isotopes and for B^{11} the (γt) reaction has been studied in the nuclear emulsion [*42*]. This work indicates levels of Be^8 including a level at 2.2 MeV, some of which are not confirmed by other experiments.

C^{12}, C^{13}: The yield of C^{11} and of neutrons from the $C^{12}(\gamma n)\,C^{11}$ reaction[5] is shown in Fig. 62. Fine structure in this yield is associated[6] with photon absorption into discrete levels of C^{12} at an excitation of about 20 MeV and is superimposed on a typical broad resonance peak. The sharp levels are in a region where similar levels are known from the $B^{11}(pn)$ reaction; if they are excited by the $E1$ process, they should be of spin 1^- and isotopic spin 1.

[1] R. NATHANS and J. HALPERN: Phys. Rev. **92**, 940 (1953).
[2] B. HAMERMESH, M. HAMERMESH and A. WATTENBERG: Phys. Rev. **76**, 611, (1949).
[3] E. GUTH and E. J. MULLIN: Phys. Rev. **76**, 234 (1949).
[3] J. H. CARVER, E. KONDAIAH and B. D. MCDANIEL: Phil. Mag. **45**, 948 (1954).
[5] R. MONTALBETTI, L. KATZ and J. GOLDEMBERG: Phys. Rev. **91**, 659 (1953).
[6] L. KATZ, R. N. H. HASLAM, R. J. HORSLEY, A. G. W. CAMERON and R. MONTALBETTI: Phys. Rev. **95**, 464 (1954).

N^{14}, N^{15}: The photo-disintegration of nitrogen has already been discussed (Sect. 23).

Heavier nuclei: The (γn) (γp) and (γt) reactions have in many cases been observed. The available information comprises threshold energy, and giant resonance data (width and integrated cross section). In the cases where the (γn) reaction leads to an unstable product, the intensity of neutron emission can be compared with the induced activity. In at least one case $\left(Al^{27}(\gamma n)\,Al^{26}\right)$ a significant discrepancy is found and is attributed to formation of Al^{26} in its very long lived ground state.

β) (γd) and $(\gamma \alpha)$ processes. These processes have been observed only in elements lighter than neon, and the (γd) process, as distinct from $(\gamma n p)$ is only significant for Li^6, Be^9 and B^{10}.

Li^6, Li^7: The reaction $Li^6(\gamma d)\,He^4$ has a very low cross section for gamma radiation of energies up to at least 17.6 MeV [*42*]. This may be understood in terms of the electric dipole selection rule for isotopic spin in nuclei with $T_z = 0$, since the $E1$ process creates states with $T = 1$ and these cannot break up into the even system $He^4 + H^2$ except as a result of isotopic spin impurity. They can however break up into $p + He^5$ and $n + Li^5$ as observed.

B^{10}, B^{11}: The isotopic spin prohibition of the $B^{10}(\gamma d)\,Be^8$ reaction seems to be less strong than in the case of Li^6, possibly because of greater impurity of the states involved, or of a longer lifetime in these states [*42*].

C^{12} [*42*]: The reaction $C^{12}(\gamma d)\,Be^8$ has been observed by many workers because of the ease with which it can be investigated by the nuclear emulsion technique. The event appears as a three-pronged, coplanar star, and the angular distribution between the alpha particle components has been established.

The excitation function for star production can be obtained by plotting the number of events with a given total energy as a function of energy. This gives the result shown schematically in Fig. 62 which also includes the excitation functions for the (γn) and (γp) reaction over the same energy range. By grouping the tracks in pairs it can be shown that for low gamma energies the reaction takes place in two steps

$$C^{12} + \gamma \rightarrow Be^{8*} + \alpha, \tag{55.1}$$

$$Be^{8*} \rightarrow 2\alpha \tag{55.2}$$

involving the 2.9 MeV level of Be^8. At energies above 26 MeV the transition is mainly to a group of levels near 17 MeV, which is the region in which $T = 1$ levels of C^{12} are expected according to Coulomb energy calculations. The mechanism suggested for the $C^{12}(\gamma\alpha)\,Be^8$ reaction is that up to 20 MeV only absorption of $E2$ and $M1$ photons takes place because no suitable states ($T = 1$, $J = 1^-$) are available for $E1$ absorption. The lowest $T = 1$ state is expected by analogy with isobaric states to have $J = 1^+$. Between 20 and 26 MeV however the giant dipole resonance does take place, because presumably of the availability of suitable levels in C^{12} at the corresponding excitation. These $T = 1$ levels can break up with emission of protons or neutrons and the resonance is observed in the (γp) and (γn) reactions which compete successfully with (γd). Break-up into $Be^8 + He^4$ is prohibited, except by isotopic spin impurity, for $T = 1$ states, and the alpha particle emission remains relatively weak until an energy of about 26 MeV is reached, when Be^8 can itself be left in its first state of $T = 1$.

Alpha particles can then be emitted with greater probability, and the $(\gamma\alpha)$ cross section rises. The ultimate fall-off at higher energies is due to competitive processes.

The angular distribution of the alpha particles with respect to the gamma ray beam is consistent with $J=2^+$ for the first excited state of Be^8.

O^{16} [42]: Somewhat similar results are reported by several workers for the photodisintegration of oxygen with alpha particle emission, although the information is more complicated. The reaction appears to proceed through excited states of C^{12} and then either to the ground state or the 3 MeV state of Be^8. Resonances in the excitation function appear between 20 and 30 MeV, but the giant resonance of the (γn) reaction does not appear in the $(\gamma\alpha)$ process. At about 25 MeV, the excitation of $T=1$ states in C^{12} becomes possible and the $(\gamma\alpha)$ cross section remains fairly large, although the (γn) and (γp) processes should be falling owing to passage through the resonance.

The $O^{16}(\gamma\alpha)C^{12}$ reaction leaving C^{12} in its ground state has also been observed in the photographic emulsion.

E. Energy level schemes.

56. Introduction. In this part the observed energy levels of isobaric nuclei with A between 5 and 40 are compared with theoretical predictions. In the level diagrams the displacement of the ground states of an isobaric group is given by their relative decay energy corrected for COULOMB energy and $(n-H^1)$ mass difference. The degree of correspondence between the level structure of isobaric multiplets in general is a measure of the extent to which nuclear forces are charge independent. The "isobaric corrections" used are given in Table 14; they are based on a nuclear radius of $R=1.45\times10^{-13}A^{\frac{1}{3}}$ cm for the calculation of COULOMB energy. This is rather a crude approximation to the electromagnetic radius, as emphasised by PEASLEE[1] and by CARLSON and TALMI[2], and a more accurate semi-empirical means of making the isobaric correction without assuming a nuclear radius is employed by INGLIS [10] and by WILKINSON[3]. Properties of the ground state such as beta-decay ft value [44], and magnetic moment are discussed in other volumes of this Encyclopedia and are not fully tabulated here[4]. For nuclei of the p-shell, comparison is made in the first place with the level systems given by INGLIS [10] for intermediate coupling; more exact calculations have recently been published by KURATH[5]. The diagrams show the (J, T) values for the lowest states of the configuration p^n, (described by LANE and THOMAS [19] as "normal parity" states), the level order for a chosen intermediate coupling parameter a/K and the spectroscopic symbol for a coupling near the LS extreme.

Evidence for the coupling scheme from static moments, transition probabilities and reduced widths, particularly for analogous levels, is presented in some cases. The discussion of reduced widths is taken mainly from the article of LANE and THOMAS [19].

No excited states are known in the nucleon (if the nucleon-meson isobar state at 200 MeV be excepted) nor in H^3 and He^3. The ^1S state of the deuteron is not bound, H^4 is unlikely to be stable (Sects. 39 and 46) and the existence of a bound excited level of He^4 is not yet certain (Sects. 28 and 55).

---

[1] D. C. PEASLEE: Phys. Rev. **95**, 717 (1954).
[2] B. C. CARLSON and I. TALMI: Phys. Rev. **96**, 436 (1954).
[3] D. H. WILKINSON: Phil. Mag. (to be published).
[4] For beta-deca cf. Vol. XLII and for magnetic moments Vols. XXXVIII and XXXIX.
[5] D. KURATH: Phys. Rev. **101**, 216 (1956).

Table 14. *Lowest analogous states in charge multiplets.*

The isobaric correction is calculated following the procedure given by AJZENBERG and LAURITSEN [*32*]. The COULOMB energy difference is

$$E_c = \frac{z\,(z-1) - z'\,(z'-1)}{A^{1/3}} \times 0.596 \text{ MeV}.$$

The isobaric correction is

$$E_I = E_c - (n - H') = E_c - 0.78 \text{ MeV}.$$

The position of the first analogue state is given by

$$(M_z - M_{z'})\,c^2 - E_I \qquad \text{where } z = z' + 1$$

and the atomic masses are given by WAPSTRA [*30*].

Isobars	E_c MeV	First analogue state MeV	Isobars	E_c MeV	First analogue state MeV
		A. Odd isobars.			
$Li^5 - He^5$	1.39	−0.45	$Si^{27} - Al^{27}$	5.21	0.40
$Be^7 - Li^7$	1.87	−0.22	$Al^{27} - Mg^{27}$	4.83	6.64
$B^9 - Be^9$	2.29	−0.44	$P^{29} - Si^{29}$	5.49	0.25
$Be^9 - Li^9$	1.72	15.0	$Si^{29} - Al^{29}$	5.10	8.29
$C^{11} - B^{11}$	2.68	0.08	$S^{31} - P^{31}$	5.75	0.47
$N^{13} - C^{13}$	3.04	−0.04	$P^{31} - Si^{31}$	5.38	6.17
$O^{15} - N^{15}$	3.38	0.11	$Cl^{33} - S^{33}$	6.01	0.22
$N^{15} - C^{15}$	2.90	10.76	$S^{33} - P^{33}$	5.63	5.10
$F^{17} - O^{17}$	3.71	−0.16	$A^{35} - Cl^{35}$	6.25	−0.06
$O^{17} - N^{17}$	3.24	11.26	$Cl^{35} - S^{35}$	5.90	5.29
$Ne^{19} - F^{19}$	4.02	0.02	$K^{37} - A^{37}$	6.49	0.42
$F^{19} - O^{19}$	3.57	7.57	$A^{37} - Cl^{37}$	6.13	4.53
$Na^{21} - Ne^{21}$	4.35	−0.05	$Cl^{37} - S^{37}$	5.70	9.12
$Ne^{21} - F^{21}$	3.91	9.04	$Ca^{39} - K^{39}$	6.75	0.89
$Mg^{23} - Na^{23}$	4.61	0.27	$K^{39} - A^{39}$	6.38	6.17
$Na^{23} - Ne^{23}$	4.20	7.81	$A^{39} - Cl^{39}$	6.03	8.21
$Al^{25} - Mg^{25}$	4.95	0.07	$Sc^{41} - Ca^{41}$	6.95	−0.21
$Mg^{25} - Na^{25}$	4.52	7.49	$Ca^{41} - K^{41}$	6.60	5.38
			$K^{41} - A^{41}$	6.25	9.05
		B. Even isobars $(4n+2)$.			
$Li^6 - He^6$	1.31	4.08	$Na^{22} - Ne^{22}$	4.28	0.66
$C^{10} - B^{10}$	2.77	1.66	$Al^{26} - Mg^{26}$	4.89	0.08
$B^{10} - Be^{10}$	2.21	1.99	$P^{30} - Si^{30}$	5.42	0.32
$O^{14} - N^{14}$	3.46	2.49	$Cl^{34} - S^{34}$	5.94	−0.36
$N^{14} - C^{14}$	2.97	2.34	$S^{34} - P^{34}$	5.58	9.90
$Ne^{18} - F^{18}$	4.09	0.9	$K^{38} - A^{38}$	6.51	−0.15
$F^{18} - O^{18}$	3.64	1.18	$A^{38} - Cl^{38}$	6.10	10.14
		C. Even isobars $(4n)$.			
$B^8 - Be^8$	2.38	16.2	$Si^{28} - Al^{28}$	5.15	9.02
$Be^8 - Li^8$	1.79	16.99	$Al^{28} - Mg^{28}$	4.75	5.78
$N^{12} - C^{12}$	3.12	15.4	$Cl^{32} - S^{32}$	6.08	7.73
$C^{12} - B^{12}$	2.60	15.19	$S^{32} - P^{32}$	5.68	6.61
$O^{16} - N^{16}$	3.31	12.93	$P^{32} - Si^{32}$	5.30	4.62
$Na^{20} - Ne^{20}$	4.39	11.7	$A^{36} - Cl^{36}$	6.19	6.12
$Ne^{20} - F^{20}$	3.95	10.22	$Cl^{36} - S^{36}$	5.82	3.88
$Al^{24} - Mg^{24}$	4.97	9.82	$Sc^{40} - Ca^{40}$	7.05	7.72
$Mg^{24} - Na^{24}$	4.57	9.31	$Ca^{40} - K^{40}$	6.69	7.24
$P^{28} - Si^{28}$	5.55	9.02	$K^{40} - A^{40}$	6.28	4.01

I. The odd isobars.

57. $A = 5$: Fig. 63. Phase shift analysis of the $(p\alpha)$ and $(n\alpha)$ scattering experiments (Fig. 28) indicates that the ground and first excited states of these nuclei form an inverted doublet $^2P_{3/2}$, $^2P_{1/2}$. These states are formed by addition of a p-nucleon to the alpha particle core and if the p-phase shifts are calculated from resonance theory (ADAIR[1]), the reduced widths obtained are near to the single particle value. There is no distinction in this case between LS and jj coupling. The simplicity of the system suggests also a direct treatment in terms of the scattering of a particle by a potential well. The phase shifts calculated by this method by SACK *et al.*[2] agree best with experiment for a GAUSSIAN radial dependence of the potential with a strong spin-orbit term. The $^2P_{1/2}$ state is too broad to show clearly in reactions such as $Li^6(t\alpha)$, $(He^3\alpha)$.

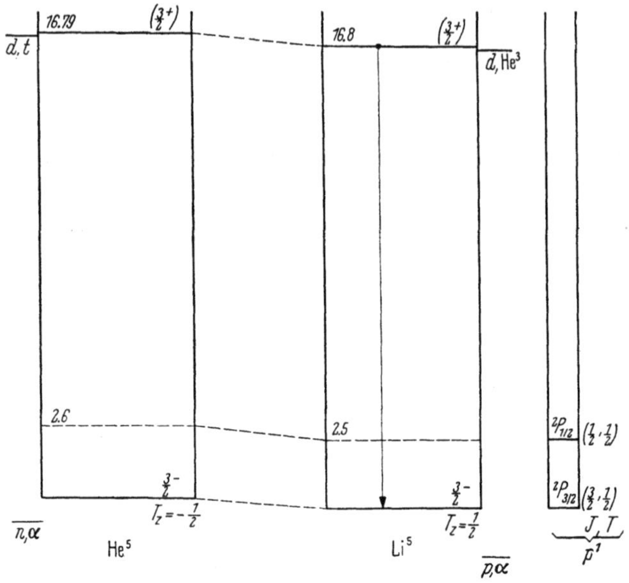

Fig. 63. Levels of nuclei with $A = 5$.

The $1/2^+$ state corresponding to the s-wave scattering phase shift has not been identified and is probably at an excitation of more than 20 MeV [*19*]. The single particle levels $^2D_{5/2}$, $^2D_{3/2}$ corresponding to the d-wave phase shifts have also not been found. Both nuclei have a $3/2^+$ state at about 16.7 MeV, seen as a resonance in the $H^3 + d$ and $He^3 + d$ reactions, but the nucleon reduced width of these states is only about 0.01 so that they presumably arise from a more complex configuration, perhaps $s^3_{1/2}\, p^2_{3/2}$. The reduced width of these states for deuteron emission is large.

58. $A = 7$: Li^7, Be^7, Fig. 64. The levels shown in Li^7 up to 7.46 MeV are known from inelastic scattering of all the usual bombarding particles, and from the (dp) $(d\alpha)$, $(n\alpha)$ and (td) reactions. The decay of Be^7 proceeds to the ground state and 0.47 MeV level. The level at 7.46 MeV is known as a resonance in the Li^6 + neutron cross section. Other levels are suggested by the excitation functions for the (γt) and (γn) reactions; of these the levels shown at 9.6 MeV and above appear in the (γn) reaction, but not all of them are seen in (γt), perhaps because they have isotopic spin $T = \frac{3}{2}$. Extrapolation from heavier isobaric groups and the calculations of [*32*] suggest the occurrence of $T = \frac{3}{2}$ levels at an excitation of about 15 MeV in Li^7 and Be^7. The Be^7 levels are known from the $(p\alpha)$ (pn), $(p\gamma)$ and (dn) reactions and from elastic scattering of protons in Li^6.

The spin and parity assignments shown are based on the analysis of resonance reactions for the 7 MeV levels; these levels have comparable reduced widths. The 6 MeV levels must be formed by s-nucleons if the large thermal neutron scattering length, and the $1/v$ proton absorption in Li^6 are to be simply explained;

[1] R. K. ADAIR: Phys. Rev. **86**, 155 (1952).

[2] S. SACK, L. C. BIEDENHARN and G. BREIT: Phys. Rev. **93**, 321 (1954).

this leads to the assignment shown, with the assumption of even parity for Li^6. The spin of the first excited state of Li^7 is checked by angular correlations in (pp') (dp) $(d\alpha)$ and $(n\alpha)$ reactions and its parity by the stripping reaction. The spin and parity of the ground state of Be^7 follow from the superallowed character of the electron capture decay to the Li^7 states. The spin $\frac{1}{2}$ for the first excited state of Be^7 is likely, apart from the requirements of symmetry between the iso-

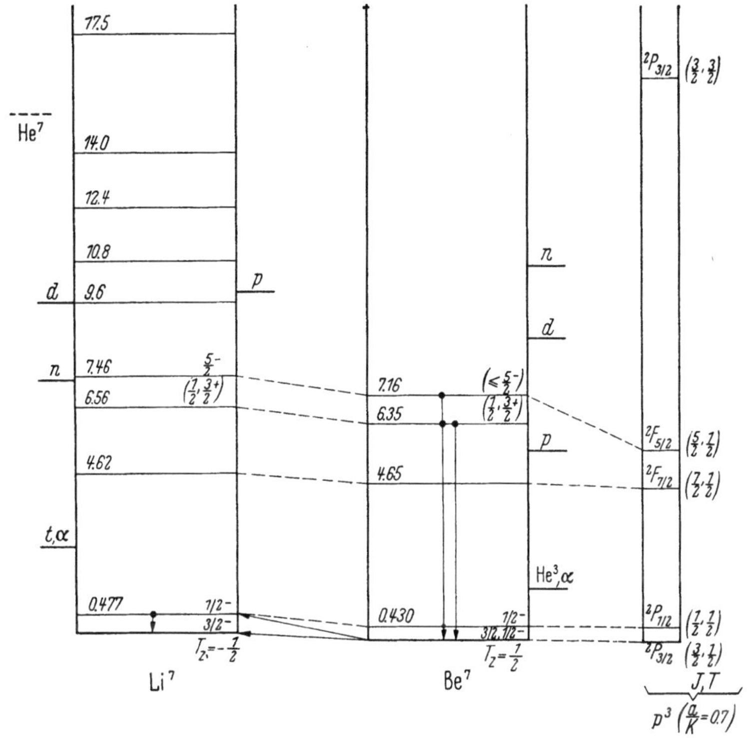

Fig. 64. Levels of nuclei with $A = 7$.

bars, because of the isotropic distribution of 430 keV radiation in the $Li^6(dn)$ reaction. The lifetime of the first excited state of Li^7 and Be^7 is known (Table 3, Sect. 24).

Li^7 and Be^7 have 3 nucleons in the p-shell and belong to the $(\frac{1}{2}, \frac{1}{2}, -\frac{1}{2})$ supermultiplet, having $T_z = \pm\frac{1}{2}$. For $A = 7$ the COULOMB energy difference is insufficient to permit positron emission, and electron capture is observed in Be^7. The supermultiplet $(\frac{1}{2}, \frac{1}{2}, -\frac{1}{2})$ leads to P and F states, and it is reasonable to associate the ground state doublet with the expected ${}^2P_{3/2}$ ${}^2P_{1/2}$ odd parity states due to spin-orbit coupling with a single p-nucleon. This was for a long time chiefly in doubt because of the difficulty of explaining the $B^{10}(n\alpha)$ branching ratio to the ground and first excited states of Li^7 for thermal neutrons, but this is now ascribed to a special and unexpected misbehaviour of a matrix element (INGLIS[1]). The difference between the $P_{3/2}$ and $P_{1/2}$ spin-orbit doublet splitting in the two isobars is 47 keV; this difference is shown by INGLIS[2] and by MOTTELSON[2] to be largely accounted for by electromagnetic effects and increased nuclear radius associated with the larger charge of Be^7. The main spin-orbit difference

[1] D. R. INGLIS: Phys. Rev. **81**, 914 (1951).

[2] D. R. INGLIS: Phys. Rev. **82**, 181 (1951). — B. R. MOTTELSON: Phys. Rev. **82**, 287 (1951).

of 450 keV almost certainly has a mesonic origin (GAUS[1]). The levels expected in intermediate coupling are shown in Fig. 64. The introduction of a small spin orbit coupling is sufficient to split the ^2P and ^2F multiplets giving the lowest states as $^2P_{3/2}$, $^2P_{1/2}$, $^2F_{7/2}$; the 7.46 MeV level is probably the $^2F_{5/2}$ state. The coupling parameter a/K chosen by INGLIS ([*10*]; see also HUMMEL and INGLIS[2]) to give the best fit is 0.7.

The parameter a/K may also be deduced from the intermediate-coupling calculations of reduced width, beta-decay lifetime, and $M1$ transition probability given by LANE[3]. Of these quantities the reduced width is most sensitive to variation of a/K. Fig. 65 shows the reduced width for the Li^7 level at 7.46 MeV formed from Li^6+n as a function of a quantity $y=(a/K)/(a/K+5)$. It is zero for LS coupling because a pure $^2F_{5/2}$ state cannot emit a p-wave nucleon leaving the residual nucleus Li^6 in its ^3S ground state. LANE concludes that the experimental results for Li^7 and Be^7 are best represented by an intermediate coupling parameter a/K of about 3.5. A similar value is deduced from the ratio of nucleon reduced widths for the ^2P doublet in Li^7 obtained from the (dp) stripping reaction. [*19*]. This value is considerably greater than that given by INGLIS, but the ^2F doublet splitting in Li^7 is predicted incorrectly and larger values of a/K are required to improve this, although at the expense then of the ^2P doublet splitting and the P-F separation.

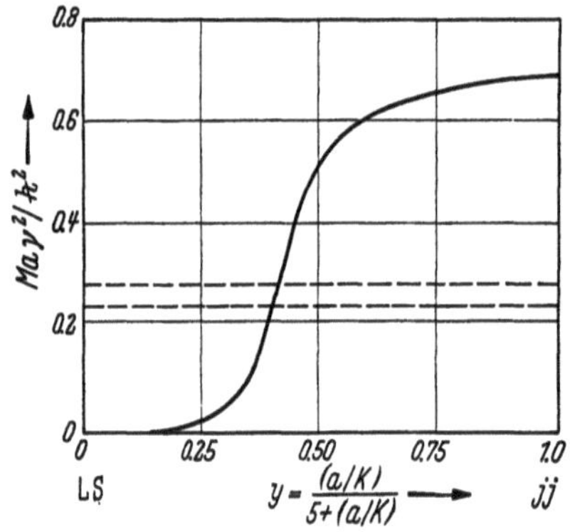

Fig. 65. Reduced width of 7.46 MeV level of Li^7 as function of $y=\frac{a/K}{5+a/K}$ (LANE).

INGLIS [*10*] has also discussed the possible application of a collective model of the alpha particle type to the Li^7 nucleus, particularly with respect to the broad state at 6.56 MeV which finds no explanation in the LS or jj schemes.

The probable occurrence of a $T=\frac{3}{2}$ state in Li^7 at about 15 MeV above the ground state suggests that the nucleus He^7 is unstable with respect to Li^7 by something of this order. This nucleus would be heavy particle unstable (He^6+n, or He^4+3n) and has not been observed.

59. $A=9$: Li^9, Be^9, B^9, Fig. 66. The spin of Be^9 has only recently been measured, but is now known[4] to be $\frac{3}{2}$. The lower bound levels of this nucleus are interesting because they appear rather irregularly in the reactions which are expected to excite them. Thus the 2.428 MeV level is seen in the (pp'), (dd'), $(\alpha\alpha')$, $(d\alpha)$, (nd), (He^3p) and $(t\alpha)$ reactions, but only the last two and perhaps the $(\alpha\alpha')$ reaction[5] appear to excite the 1.8 and 3.1 MeV levels as well. The broad level at 4.8 MeV is reported in the (pp'), (He^3p) and $(t\alpha)$ reactions and higher levels at 6.8 and 11.6 MeV are obtained from the (pp') reaction. All levels are unstable to particle emission and no radiative transitions in Be^9 have been observed. Analysis of the photodisintegration excitation function for Be^9 suggests an $S_{1/2}$ level near the neutron threshold. Spin and parity assignments have only

[1] H. GAUS: Z. Naturforsch. **4**a, 721 (1949).
[2] H. H. HUMMEL and D. R. INGLIS: Phys. Rev. **81**, 910 (1951).
[3] A. M. LANE: Proc. Phys. Soc. Lond. A **68**, 189 (1955).
[4] N. A. SCHUSTER and G. E. PAKE: Phys. Rev. **81**, 886 (1951).
[5] V. K. RASMUSSEN, D. W. MILLER, M. B. SAMPSON and U. C. GUPTA: Phys. Rev. **100**, 851 (1955).

been made for the 2.43 MeV level which has odd parity and $J \geq \frac{3}{2}$ according to the angular distributions observed in the (nd) pickup process. The levels of Be^9 shown between 17.3 and 19.2 MeV are obtained as resonances in the Li^7+d reactions; no information on their symmetry properties is available.

B^9 is unstable to proton emission in its ground state and no beta transition to Be^9 is therefore known. The 2.37 MeV level is presumably analogous to the 2.43 MeV state of Be^9 and is observed as a residual state in the $(p\alpha)$, (pd) and

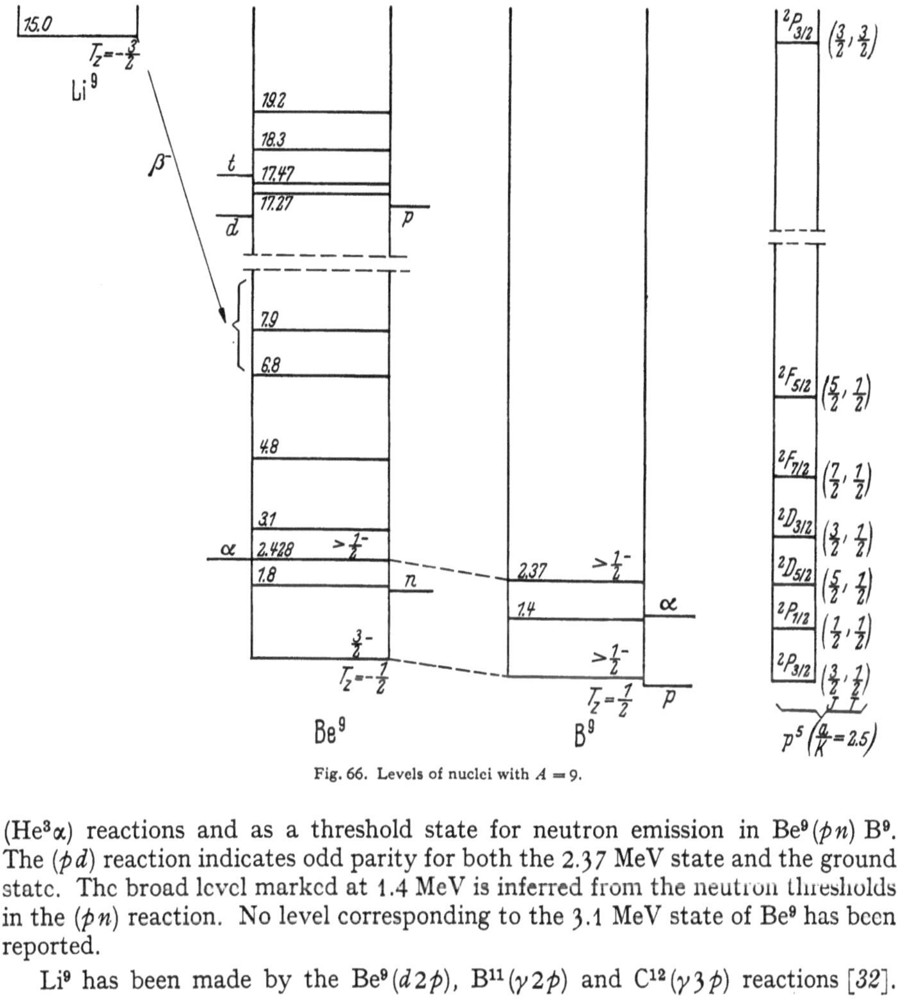

Fig. 66. Levels of nuclei with $A=9$.

$(He^3\alpha)$ reactions and as a threshold state for neutron emission in $Be^9(pn)\,B^9$. The (pd) reaction indicates odd parity for both the 2.37 MeV state and the ground state. The broad level marked at 1.4 MeV is inferred from the neutron thresholds in the (pn) reaction. No level corresponding to the 3.1 MeV state of Be^9 has been reported.

Li^9 has been made by the $Be^9(d\,2p)$, $B^{11}(\gamma\,2p)$ and $C^{12}(\gamma\,3p)$ reactions [32].

The nuclei of mass 9 have 5 p-shell particles and the lowest supermultiplet is $(\frac{1}{2}, \frac{1}{2}, \frac{1}{2})$ to which correspond ^2P, ^2D, ^2F, and ^2G multiplets in this order of energy from the ground state. The lowest states in intermediate coupling are shown in Fig. 66[1]. The ground state of Li^9 belongs to the $(\frac{3}{2}, \frac{1}{2}, -\frac{1}{2})$ supermultiplet in which the probable lowest multiplet is ^2P. In LS coupling the multiplet splitting is expected to be rather small because it reverses at the middle of the p-shell ($A=10$). Comparison of the observed and predicted level schemes (the former almost certainly incomplete) suggests an intermediate coupling para-

[1] D. R. Inglis: Ref. [10] and Rev. Mod. Phys. **27**, 76 (1955); See also J. B. French, E. C. Halbert and S. P. Pandya: Phys. Rev. **99**, 1387 (1955) for a more exact calculation.

meter of $a/K=2.5$ although the $J=\frac{1}{2}$ component of the ground state doublet has not been found and may be unresolved from the $P_{3/2}$ ground state. The magnetic moment of the Be^9 ground state is well predicted by the jj coupling model with account taken of the interaction between the three $p_{3/2}$ holes [*14*].

The B^9 ground state should be approximately a single particle state with a width considerably less than that of the Li^5 ground state because of higher barrier and lower available energy. This is consistent with the observed width of 2 keV from the (pn) reaction.

60. $A=11$: B^{11}, C^{11}, Fig. 67. The levels of B^{11} are obtained from the (dp) (pp') $(n\alpha)$ and $(d\alpha)$ reactions up to about 10 MeV. The 8.92, 9.19 and 9.28 MeV

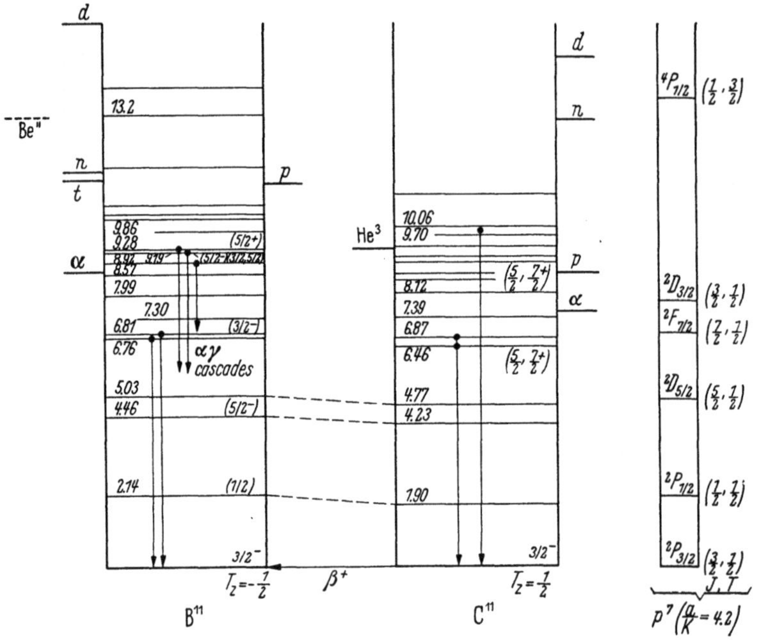

Fig. 67. Levels of nuclei with $A=11$.

states are also observed as resonances in the $Li^7(\alpha\gamma)$ reaction; higher levels appear as resonances in neutron and deuteron induced reactions. The capture radiation from the $Li^7(\alpha\gamma)$ reaction forms a complex spectrum which links the bound levels with the capture resonances. The radiation following the (dp) reaction also verifies the level scheme.

Spin and parity assignments are based on $(\alpha-\gamma)$ and $(\gamma-\gamma)$ correlation measurements in $Li^7(\alpha\gamma)$. The parities of the low states relative to B^{10} are obtained from the stripping reaction, but this is complicated by a large contribution from compound nucleus effects and not all its predictions agree with those of the capture radiation analysis. Reduced widths for the ground state and the 6.81 or 6.76 MeV state are large and suggest that these are single particle levels. Spin $\frac{1}{2}$ for the 2.14 MeV excited state is indicated by the isotropic $(p\gamma)$ angular correlation, but this is inconsistent with $l_n=1$ observed for this state in the (dp) reaction.

The C^{11} levels are mainly deduced from the (dn) reaction up to about 9 MeV, partly from observation of neutron groups and partly as thresholds for slow neutron production as the deuteron bombarding energy is varied. The parities relative to B^{10} are obtained from stripping analysis. Above 9 MeV virtual levels excited in the $B^{10}+p$ reactions are observed.

The $C^{11}\rightarrow B^{11}$ beta transition is superallowed and the positron energy agrees well with that calculated from the Coulomb energy difference. There is an obvious similarity between the level structures of the two nuclei up to 7 MeV.

In the nuclei of $A=11$ the p-shell is more than half full, and the introduction of a small spin orbit coupling suggests a regular doublet and a ${}^2P_{1/2}$ ground state in LS coupling which agrees neither with the observed spin nor with the observed magnetic moment. In jj coupling however, the odd configuration $p^7_{3/2}$ has just one state of $j=\frac{3}{2}$ and excited states are obtained by promoting one nucleon to a $p_{1/2}$ state and coupling it to the states of the $p^6_{3/2}$ configuration. A fair approximation to the observed level sequence can be obtained with an intermediate coupling parameter $a/K=4.2$. The existence of low lying even states in the mass 11 nuclei is not yet certain; they would derive theoretically from configurations such as $p^6_{3/2}\, d_{5/2}$ or $p^6_{3/2}\, s_{1/2}$.

61. $A = 13$: C^{13}, N^{13}, Fig. 68. The bound levels of C^{13} are obtained, with good resolution, from the (dp) (αp) and $(d\alpha)$ reactions and are also seen in the (pp') and $(n\alpha)$ processes which do not however resolve the 3.68 to 3.86 MeV doublet. The group of excited states between 6.87 and 9.0 MeV is found as resonances in the total cross section for neutron scattering by C^{12}, and may also be excited by the (dp) reaction [*32*]. The group of virtual levels above the alpha particle disso-

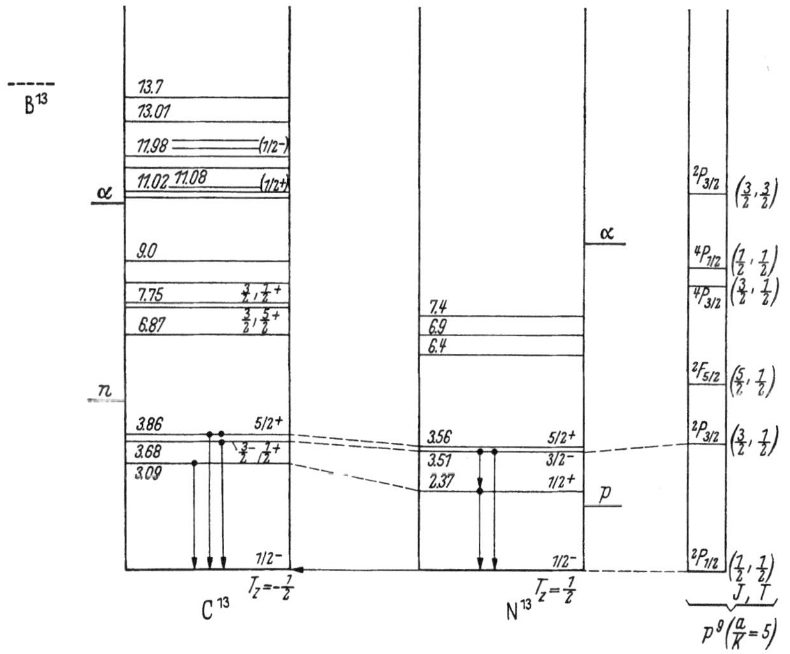

Fig. 68. Levels of nuclei with $A = 13$.

ciation energy represents resonances in the $Be^9\,(\alpha n)$ reaction. In N^{13} the two lower groups of levels also appear; the 2 to 3 MeV levels are obtained from the (dn) reaction, and with better resolution from the $(p\gamma)$, (pp) and (pp') resonant processes. None of these processes directly reveals the 3.56 MeV state whose existence is inferred from analysis of the scattering resonances and by analogy with the C^{13} nucleus. The levels near 7 MeV in N^{13} are found in the (pp) and (pp')

reactions. The 3.51 MeV level exhibits the rather unusual decay scheme of a gamma transition to the 2.37 MeV level, followed by proton emission from this level, which appears as a resonance in the C^{12} $(p\gamma)$ reaction.

The spins and parities and reduced widths of the 3 MeV group of levels are obtained by stripping analysis for C^{13} and by resonance analysis for N^{13}, in which the corresponding levels are virtual. Resonance analysis also gives the J and

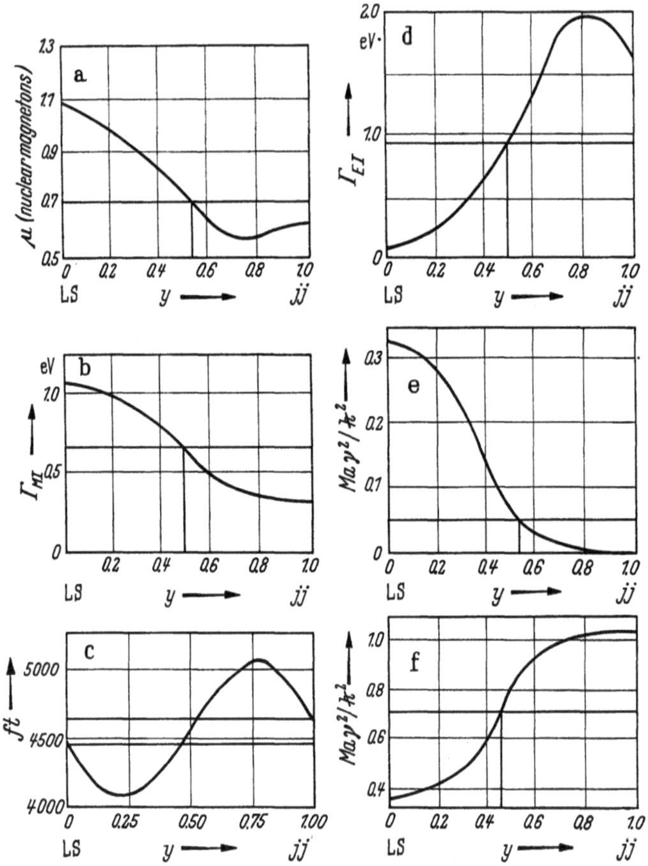

Fig. 69 a—f. Intermediate coupling for $A = 13$. Nuclear properties are plotted against $y = \frac{a/K}{5 + a/K}$ with observed values shown by horizontal lines. (a) Magnetic moment of C^{13}; (b) $M1$ transition in N^{13}; (c) ft value for N^{13} decay; (d) $E1$ transition in N^{13}; (e) reduced width of 3.51 MeV state in N^{13}; (f) reduced width of ground state of N^{13} (ordinate scales in (b) and (f) are not exact). (LANE.)

parity values noted for the 6.87 to 7.75 MeV levels of C^{13} and neutron angular distributions [32] suggest the properties shown for the 11.08 and 11.98 MeV levels.

The beta-decay of the ground state between the isobars is superallowed. Application of the isobaric correction to the level system in general shows that the three lowest N^{13} levels are at a definitely lower excitation than the analogous levels in C^{13}. The N^{13} levels are however unbound, and have larger widths than their counterparts in C^{13}, particularly in the case of the lowest (s-wave) level. The dispersive level shift (Sect. 9) for this state is thus very marked and with the assumption of a suitable radius of interaction the relative displacement of the

3.09 MeV state in C^{13} and the 2.37 MeV state in N^{13} can be accounted for[1]. The level shifts for the $^3/_2{}^-$ and $^5/_2{}^+$ level can also probably be made to account for the difference in excitation of the next two analogous states.

The $A = 13$ nuclei are described in the WIGNER classification by the $(\frac{1}{2}, \frac{1}{2}, \frac{1}{2})$ supermultiplet for which ^2P and ^2F states are expected for 9 p-shell nucleons. These are odd parity states and after splitting by spin-orbit forces the $^2P_{1/2}$ to $^2P_{3/2}$ doublet may be identified with the ground state and 3.68 MeV state of C^{13}. The sequence of odd-parity states obtained with an intermediate coupling parameter $a/K = 5$ is shown in Fig. 68. Intermediate coupling theory for $A = 13$ has been given by LANE[2], who is able to obtain the observed reduced width for the $^3/_2{}^-$ state with about the same intermediate coupling parameter. Extreme jj coupling predicts zero nucleon reduced width for this state $(p^7_{3/2}\, p^2_{1/2} \to p^8_{3/2} + n)$. Fig. 69 shows the dependence of reduced width and other nuclear properties on the coupling parameter for $A = 13$.

The even parity states $^1/_2{}^+$, $^5/_2{}^+$ and $^3/_2{}^+$ may be described as single particle states formed by adding an $s_{1/2}$, $d_{5/2}$ or $d_{3/2}$ nucleon to a C^{12} core. The $^1/_2{}^-$ ground state is also a single particle state formed by adding a $p_{1/2}$ nucleon. In agreement with this interpretation, all these states have large reduced widths. Low energy neutron scattering from C^{12} also confirms the existence of a bound s-wave state [*32*], and the magnetic moment of the C^{13} ground state has the single particle value.

62. $A = 15$: C^{15}, N^{15}, O^{15}, Fig. 70. The lower levels of the N^{15} nucleus are obtained from the (dp) and $(n\gamma)$ reactions with high precision; the observations extend to the virtual levels appearing in the (np), $(n\alpha)$ and (pn), $(p\gamma)$ reactions as well as in (αp), (αn) and total neutron cross section. The dissociation energies for protons, neutrons, and alpha particles all lie within 1 MeV of one another and the N^{15} levels have therefore proved fruitful for testing the principle of detailed balancing for nuclear reactions (Sect. 9). A few levels near 17 MeV are found as resonances in the (dp) and (dn) reactions. The analogous levels in O^{15} are obtained from the (dn), $(p\gamma)$, (pn) and (pp) reactions. Spin and parity assignment for the bound levels are suggested by stripping theory for the (dp) and (dn) reactions and for the virtual levels by analysis of gamma ray angular distributions, neutron angular distributions and by level widths. The values obtained for the virtual levels of N^{15} from the $C^{14}(pn)$ and $C^{14}(p\gamma)$ reactions are listed in [*32*].

The cascade gamma transitions fit well into the level schemes deduced from the heavy particle reactions. $A(p-\gamma)$ angular correlation measurement[3] suggests $J = \frac{1}{2}$ or $\frac{3}{2}$ for the 5.3 MeV doublet in N^{15}. Reduced widths have been evaluated for the virtual levels [*18*] and for some of the bound levels of N^{15}.

The beta decay of the O^{15} ground state is a superallowed transition, as expected. The general pattern of the low lying states of the two isobars is similar, but the level spacing rapidly becomes too small to justify detailed comparison. The $T = \frac{3}{2}$ nucleus C^{15} is known, and its decay energy has been measured both directly and from the Q-value for the $C^{14}(dp)$ reaction. This enables the C^{15} ground state, after isobaric correction, to be plotted on Fig. 70, and shows that a state of $T = \frac{3}{2}$ would be expected in N^{15} and O^{15} at an energy of about 10.9 MeV. Such a state can, on the simplest picture, only decay to $N^{14} + n$ if a state of $T = 1$

[1] J. B. EHRMAN: Phys. Rev. **81**, 412 (1951). — R. G. THOMAS: Phys. Rev. **88**, 1109 (1952).
[2] A. M. LANE: Proc. Phys. Soc. Lond. A **66**, 977 (1953); **68**, 197 (1955); [*19*]. — A. M. LANE and L. A. RADICATI: Proc. Phys. Soc. Lond. A **67**, 167 (1954).
[3] A. G. STANLEY: Phil. Mag. **45**, 807 (1954).

is available in N^{14}, and this is not so until sufficient energy is provided to excite the 2.31 MeV state of this nucleus (see $C^{14}+p$).

The $A=15$ nuclei are described by the $(\frac{1}{2}, \frac{1}{2}, -\frac{1}{2})$ supermultiplet of which the lowest multiplet is ${}^2P$, giving a ground state of ${}^2P_{1/2}$. The 6.33 MeV level of N^{15} has the right spin and parity to be to ${}^2P_{3/2}$ component of the spin-orbit doublet. The odd parity states are those of a single hole and there is no way of obtaining the value of the intermediate coupling parameter. The magnetic

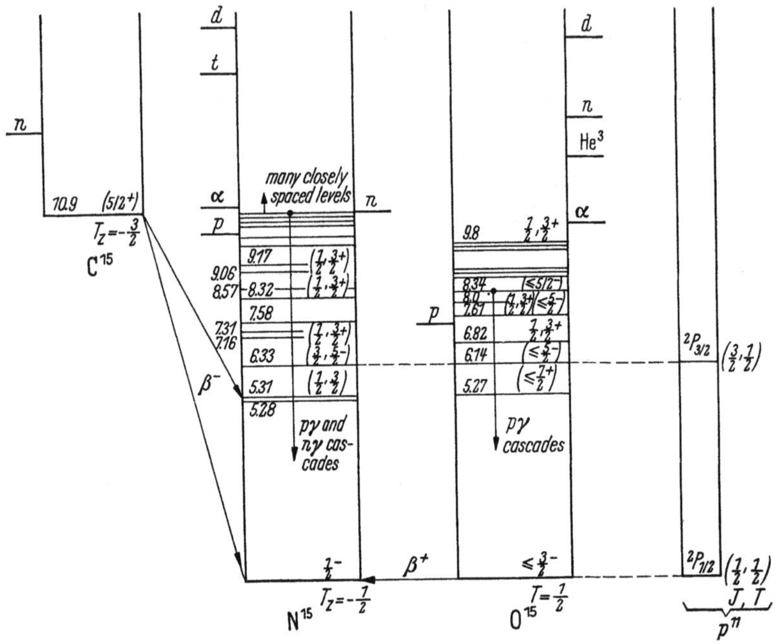

Fig. 70. Levels of nuclei with $A=15$.

moment of N^{15} confirms the single "particle" nature of the nucleus. In jj coupling the energy difference between the ground and 6.33 MeV states is the difference between the configurations $p_{3/2}^8\, p_{1/2}^3$ ($p_{1/2}$ hole) and $p_{3/2}^7\, p_{1/2}^4$ ($p_{3/2}$ hole).

Even parity states may be obtained by elevating one of the $p_{1/2}$ nucleons, giving $p_{1/2}^2\, s_{1/2}$ and $p_{1/2}^2\, d_{5/2}$. The states thus obtained are listed by INGLIS [10] who also suggests that the 5.31 MeV state of N^{15} may be an $s_{1/2}^{-1}$ state (1 s-hole). This would account for its weak excitation in the (dp) reaction. The C^{15} ground state, and its analogous levels, probably belong to the $p_{1/2}^2\, d_{5/2}$ configuration ($T=\frac{3}{2}$) obtained by adding a $d_{5/2}$ nucleon to the $(J, T)=(0, 1)$ state of N^{14}. The other even parity states may be similarly described.

63. $A=17$: N^{17}, O^{17}, F^{17}, Fig. 71. O^{17} and F^{17} are single particle nuclei in the sense that they should show the states of a single particle moving in a central field. Levels due to excitation of the O^{16} core should also appear. The observed spectrum is obtained for O^{17} mainly from the (dp) and $(d\alpha)$ reactions. The virtual levels found in these reactions agree fairly well with those seen in the total neutron cross section of oxygen and as resonances in the (αn) reaction. The F^{17} levels are deduced from the (dn), $(p\gamma)$ and (pp) reactions. Spin and parity assignments are provided by analysis of the stripping angular distributions and from the widths, angular distributions and scattering anomalies observed in the resonance reactions. The reduced widths of the ground state and first excited state are found to be fairly large, as expected for single particle states. The stripping reaction allows spin $^5/_2{}^+$ for the ground state of F^{17} in agreement with the measured value for the isobar.

The spin of the first excited state is confirmed by the isotropic $(p-\gamma)$ angular correlation, and the observed multipolarity ($E2$) of the radiation to the ground state. The lifetime (Table 3) is near that expected for the transition of a single proton. That a neutron transition should give such a value is probably due to collective motion.

The beta transition between the isobars has the superallowed character. The lower levels of the two nuclei show similar structure and spins, although the first

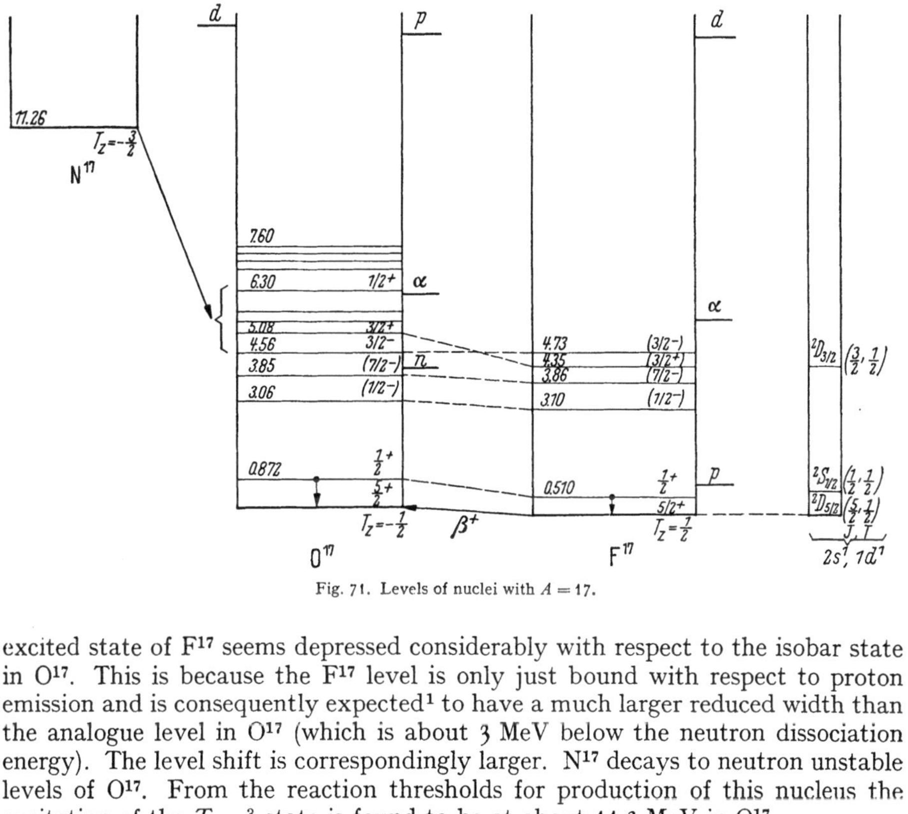

Fig. 71. Levels of nuclei with $A = 17$.

excited state of F^{17} seems depressed considerably with respect to the isobar state in O^{17}. This is because the F^{17} level is only just bound with respect to proton emission and is consequently expected[1] to have a much larger reduced width than the analogue level in O^{17} (which is about 3 MeV below the neutron dissociation energy). The level shift is correspondingly larger. N^{17} decays to neutron unstable levels of O^{17}. From the reaction thresholds for production of this nucleus the excitation of the $T = \frac{3}{2}$ state is found to be at about 11.3 MeV in O^{17}.

O^{17} and F^{17} should be constructed by the addition of an $s_{1/2}$ or $d_{5/2}$ nucleon to the O^{16} core and states corresponding to these nucleons and to the addition of $d_{3/2}$ are expected. All these levels are of even parity, and the lowest even parity states do in fact have spins $\frac{5}{2}$, $\frac{1}{2}$ and $\frac{3}{2}$. The ground state magnetic moment confirms its single particle character. Possibly the level $^7/_2{}^-$ is the single particle state $f_{7/2}$ but LANE and THOMAS [*19*] interpret the odd parity states as due to the coupling of a $p_{1/2}$ hole to the $(J, T) = (1, 0)$ and $(0, 1)$ states of F^{18}. This interpretation is consistent with the fact that all these states have small nucleon reduced widths.

64. $A = 19$: O^{19}, F^{19}, Ne^{19}, Fig. 72. The levels of F^{19} up to 5 MeV have been examined very thoroughly by many reactions, of which the more important are the (pp'), $(\alpha\alpha')$, (nn') and (dd') inelastic scattering reactions, and the $(d\alpha)$, (dn)

[1] R. G. THOMAS: Phys. Rev. **88**, 1109 (1952).

and (dp) processes The decay of O^{19} excites two of the lower levels. Virtual levels have been located in the $(p\alpha)$ and (pn) reactions. In Ne^{19} the only information comes from thresholds for slow neutron production in the $F^{19}(pn)$ reaction; it is difficult to investigate the Ne^{19} states mainly because of the instability of F^{18} which is the first $T=0$ nucleus to be unstable in its ground state after Be^8.

The outstanding feature of the F^{19} level system is the group of three states of small separation of which the ground state is one. The ground state spin is measured to be $\frac{1}{2}$, and even parity is verified by detailed analysis of some of the $F^{19}+p$ resonance reactions. The other states at 0.110 and 0.197 MeV were found

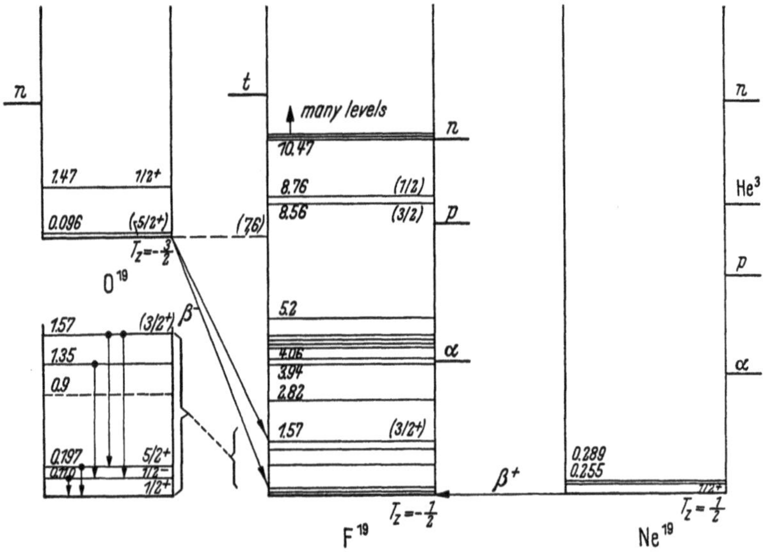

Fig. 72. Levels of nuclei with $A=19$. Recent work by J. M. FREEMAN: Phys. Rev. 99 1446 (1955) and by B. J. TOPPEL, D. H. WILKINSON and D. E. ALBURGER: Phys. Rev. 101, 1485 (1956) establishes a level at 1.45 MeV in F^{19}.

first in the $Ne^{21}(d\alpha)$ reaction by magnetic analysis, and are so low in excitation in comparison with the potential barrier of F^{19} that they can be COULOMB-excited by alpha particles. The shape of the excitation function and the gamma ray angular distributions suggest $E1$ and $E2$ excitation for the 110 and 197 keV states, which are therefore expected to have spin $^1/_2{}^-$ and $^5/_2{}^+$ respectively. This assignment is confirmed by the absence of a $197\to110\to0$ cascade and by the lifetimes (Table 3). The K-shell conversion coefficients for the radiative transitions also agree with the $E1$, $E2$ assignment. Further confirmatory evidence comes from the excitation of the 110 and 197 keV radiations by inelastic scattering of neutrons and from the branching ratio in the O^{19} decay (JONES *et al.*[1]) to excited states of F^{19}. The beta-decay work confirms the allowed nature of the $Ne^{19}\to F^{19}$ ground state transition and together with $\gamma-\gamma$ correlations suggests spins of $^5/_2{}^+$ for O^{19} and $^3/_2{}^+$ for the 1.57 MeV state of F^{19}. The lifetime of the 197 keV state was checked in this work by the delayed coincidence method. The spins of the levels at 8.56 and 8.76 MeV are derived from angular correlations in the $O^{18}(p\alpha)$ reaction. The spin $\frac{1}{2}$ for the 110 keV level is further verified by the fact that the angular distribution of 110 keV gamma rays in the $F^{19}(pp')$ reaction is always isotropic, while the distributions for the 197 keV rays are consistent with $J={}^5/_2{}^+$. Two excited states in O^{19} are known from the (dp) reaction.

[1] G. A. JONES, W. R. PHILLIPS, C. M. P. JOHNSON and D. H. WILKINSON: Phys. Rev. **96**, 547 (1954).

The theoretical picture for the mass 19 nuclei in intermediate coupling has been given by ELLIOTT and FLOWERS [*15*] and by REDLICH[1]. These treatments differ in the type of central potential and in the method of calculation, but lead to comparable results; both allow for the mixing of configurations of the $1d$ and $2s$ shells. This is necessary because it is clear from the O^{17} and F^{17} spectrum that the $d_{5/2}$ and $s_{1/2}$ levels are close together as is indeed expected for the levels of a harmonic oscillator. The $d_{5/2} - d_{3/2}$ splitting is taken from O^{17}. The calculations account

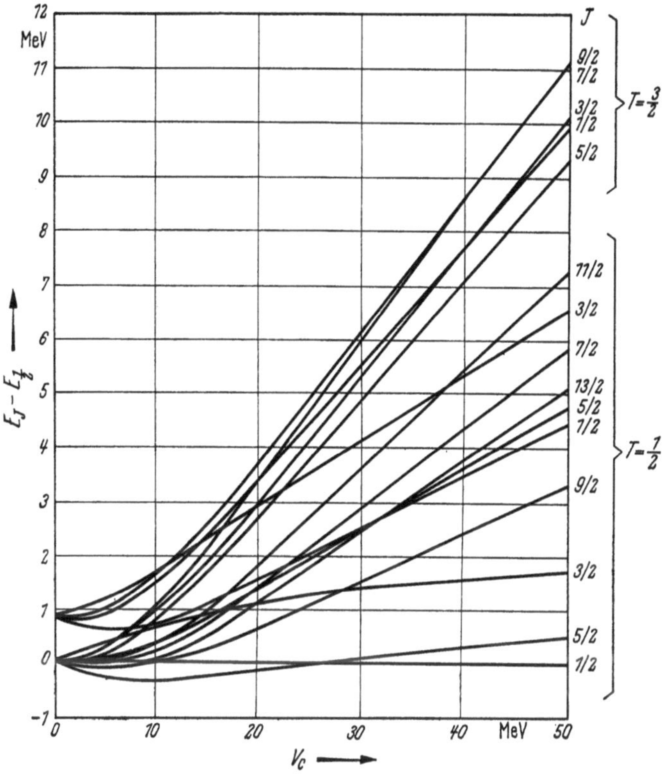

Fig. 73. Levels calculated for $A = 19$ (ELLIOTT and FLOWERS [*45*]).

only for the even parity states since the odd state at 110 keV with $J = 1/2^-$ must arise from excitation of a p-nucleon from the core. The calculated levels for $A = 19$ are shown in Fig. 73 as a function of the strength of the central YUKAWA potential. The level order in F^{19} and Ne^{19} requires the central force strength to be greater than 25 MeV, and good agreement with the experimental position of the $5/2^+$ and $1/2^+$ levels is obtained for $V_c = 74$ MeV. The group of levels shown in Fig. 73 above the $9/2^+$ level may correspond to the group actually seen in F^{19} between 4 and 5 MeV, before the $T = \frac{3}{2}$ levels are reached. It is obvious that there is a remarkable similarity in general structure found by theory and by observation. The increased excitation of the low levels of Ne^{19} compared with those of F^{19} may be accounted for by COULOMB energy which differs between the ground and excited states because of different wave functions.

The analysis gives, for a state of given spin and isotopic spin, the percentage of different configurations which can result in this (J, T) combination. With this

[1] M. G. REDLICH: Phys. Rev. **98**, 199 (1955).

information it is possible to make accurate estimates of nuclear properties dependent on wave function for which only approximate values can be obtained when only the $(J\ T)$ character of the levels concerned is known. Thus the magnetic moment of F^{19} is found to be between 2.80 and 2.87 which is rather higher than the observed value of 2.63 but is a considerable improvement on the SCHMIDT value of 4.79 for $d_{5/2}$ and also on the earlier individual particle value of 3.52 for $d^3_{5/2}$.

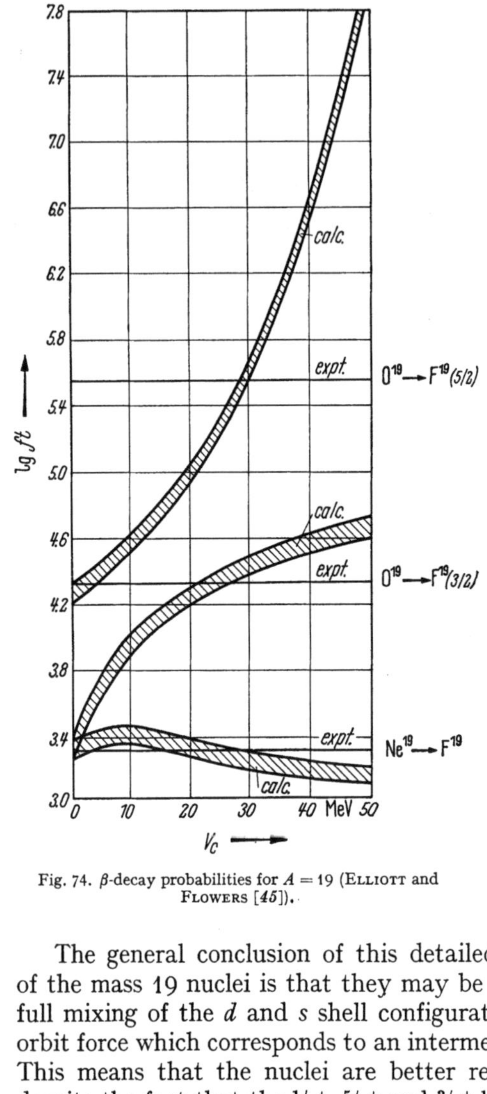

Fig. 74. β-decay probabilities for $A = 19$ (ELLIOTT and FLOWERS [45]).

The $M1$ transition from the 1.57 MeV level of F^{19} to the $^5/_2{}^+$ level should, from J values, be accompanied by a similar $M1$ transition to the ground state, but this is not observed in more than 5% of the intensity of the less energetic radiation. The reason for this is that the wave functions for the $^3/_2{}^+$ and $^5/_2{}^+$ levels are very similar and that both differ considerably from the $^1/_2{}^+$ functions.

The beta decay probabilities calculated using the exact wavefunctions are shown in Fig. 74 as a function of V_c. The superallowed decay of Ne^{19} is not a sharp function of the interaction potential, since there is no change of supermultiplet, but the O^{19} decays give an estimate of V_c which agrees fairly well with that obtained from the level spectrum.

The lifetime for the 197 keV $E2$ transition is near to the single particle value because the configuration change is mainly $d^3 \rightarrow d^2 s$. The calculated value of $5 \cdot 10^{-7}$ sec is 5 times longer than observed and this means that some collective motion of the O^{16} core probably takes part in the transition. Estimates of this effect based on the strength of the $E2$ transition in O^{17} bring the theoretical life time into agreement with observation.

The general conclusion of this detailed and extremely satisfactory analysis of the mass 19 nuclei is that they may be well represented by a model in which full mixing of the d and s shell configurations is assumed, together with a spin orbit force which corresponds to an intermediate coupling parameter of $a/K = 2.1$. This means that the nuclei are better represented in LS coupling than in jj despite the fact that the $^1/_2{}^+$, $^5/_2{}^+$ and $^3/_2{}^+$ levels might well be described as single particle states. The work of ELLIOTT and FLOWERS [45] clearly shows the danger of making such simple interpretations. In particular the complexity of the states results in a small nucleon reduced width, in agreement with observation for the virtual states.

65. $A = 21$: F^{21}, Ne^{21}, Na^{21}, Fig. 75.—$A = 23$: Ne^{23}, Na^{23}, Mg^{23}, Fig. 76. The level systems of the nuclei of mass 21 already show that lack of completeness which characterises many of the isobaric systems of the $2s$ and $1d$ shell and makes detailed comparison with theory unprofitable. The Ne^{21} levels are fairly well known up to 10 MeV from the (dp), $(d\alpha)$ and $(n\alpha)$ reactions and spins and parities can be suggested from stripping data. In Na^{21} however there is no information until the proton binding energy is reached at 2.45 MeV, above which narrow resonance levels for the (pp), (pp') and $(p\gamma)$ reactions are found. The widths of these levels are known, but spin and parity values have not been suggested. No useful comparison with Ne^{21} can be made. The beta decay between the isobars is superallowed, as expected for mirror nuclei.

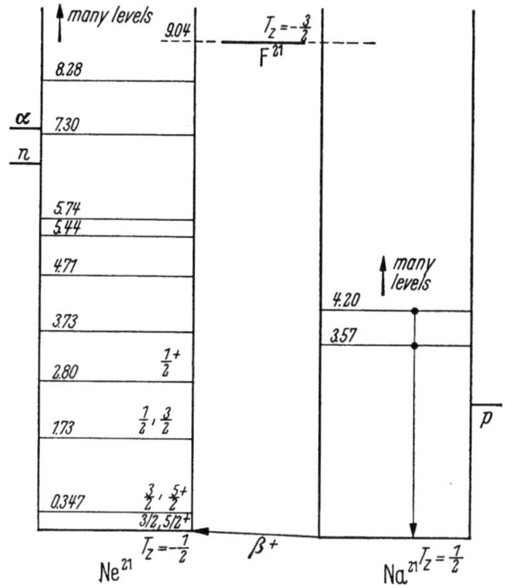

Fig. 75. Levels of nuclei with $A=21$.

There are no published theoretical studies of level structure for mass 21. If the configuration is $s^4_{1/2}\, d_{5/2}$ a similarity to the O^{17} level system might be expected.

Nothing is known of the levels of Mg^{23}. The $T=\frac{3}{2}$ and $T=\frac{1}{2}$ nuclei Ne^{23} and Na^{23} may be compared; the predicted excitation of the first $T=\frac{3}{2}$ level in Na^{23} is in a region where the level spectrum is poorly known. The Ne^{23} levels are found from the (dp) reaction, and those of Na^{23} from inelastic scattering of protons, neutrons, alpha par-

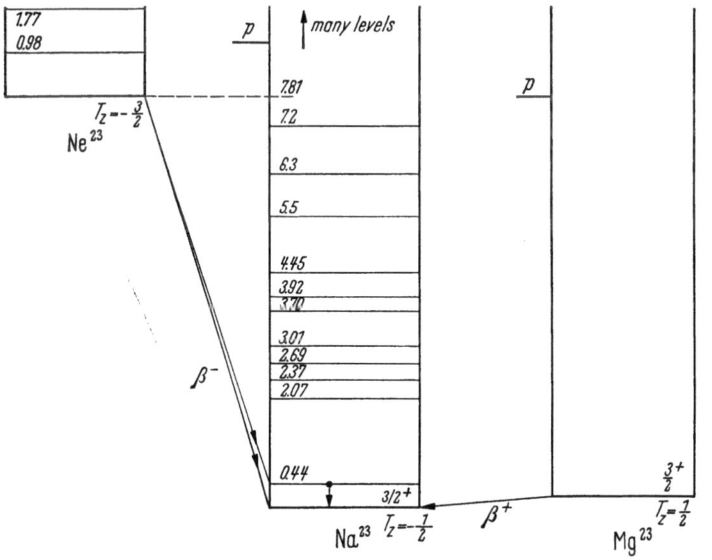

Fig. 76. Levels of nuclei with $A=23$.

ticles and deuterons and from the $(d\alpha)$ reaction. Many virtual levels of Na^{23} appear as resonances in the $Ne^{22}+$ proton and $F^{19}+\alpha$ reactions. The Mg^{23} beta-decay fits into the usual pattern of mirror transitions; but the Ne^{23} decay is probably first forbidden and shows a branching to the 0.44 MeV state of Na^{23} (GERBER *et al.*[1]).

The spin of the ground state of Na^{23} is measured to be $\frac{3}{2}$ instead of the expected $\frac{5}{2}$. This is the first of the "anomalies" in shell structure, and requires that the spin of Na^{23} be determined by the coupling together of the spin vectors of a configuration.

66. $A=25$: Na^{25}, Mg^{25}, Al^{25}, Fig. 77.—$A=27$: Mg^{27}, Al^{27}, Si^{27}, Fig. 78. The level systems of the $T=\frac{1}{2}$ isobars Mg^{25} and Al^{25} correspond well in energy, spin and parity up to about 4 MeV. The Mg^{25} levels are obtained with good resolution from the (dp), (pp') and $(d\alpha)$ reactions, and above the neutron dissociation energy, from the total neutron cross section. The cascade gamma rays observed in the $(n\gamma)$ thermal capture process confirm the level structure. Spins and parities are suggested by intensity relations in the cascades, by analysis of the resonant scattering of neutrons, and by stripping theory for the (dp) reaction. The last two processes also provide reduced widths for many of the levels (Tables 6 and 12), including some which can be identified with analogue levels in Al^{25}. Of these the Mg^{25} level at 3.40 MeV agrees in spin and parity with the 3.09 MeV level of Al^{25} formed in the resonant capture of protons by Mg^{24}. There is excellent agreement between the reduced widths for these levels. The Al^{25} levels are known from the (dn) and the $(p\gamma)$ and (pp) reactions. The cascade gamma transitions following proton capture are potential sources of useful information on transition probabilities, as are the transitions following neutron capture in Mg^{24} for the Mg^{25} levels. The Al^{25} level scheme seems to be well described by the BOHR-MOTTELSON model (Sect. 4).

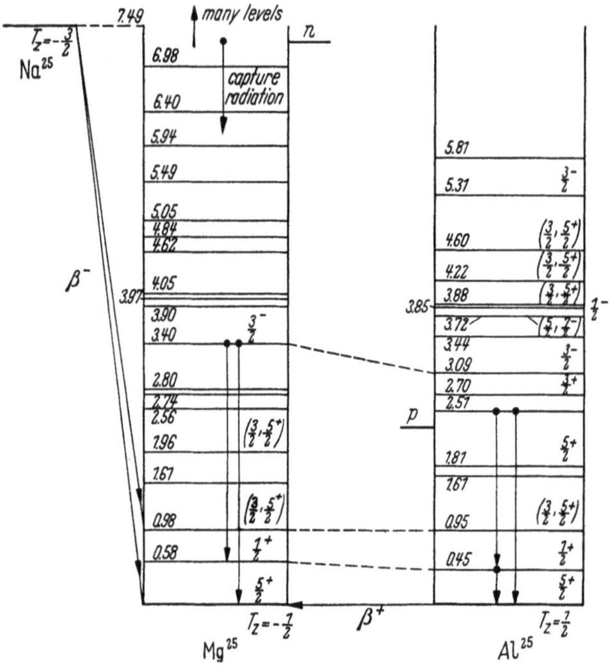

Fig. 77. Levels of nuclei with $A=25$.

The ground state positron decay is superallowed as expected, and the beta-decay of Na^{25} shows branching, presumably to a $\frac{5}{2}$ level in Mg^{25}. The configuration of the nuclei is not known, but the assumption of $d_{5/2}^{-3}$ gives a magnetic moment which is in better agreement with the observed value than the SCHMIDT figure [*14*]. Reduced widths are large for the ground state and levels at 0.58, 0.98 and 3.40 MeV of Mg^{25} and for the ground state, 0.45 MeV state and a number of virtual levels of Al^{25}. These levels may then be formed by addition of a single particle $d_{5/2}$, $s_{1/2}$, $d_{3/2}$, $p_{3/2}$ to existing configurations. The even parity levels presumably arise from such configurations involving only d or s particles.

The levels of Si^{27} are not known; the ground state shows the normal superallowed transition to Al^{27}. The bound levels of this nucleus have been very fully

[1] H. J. GERBER, M. GARCIA MUÑOZ and D. MAEDER: Phys. Rev. **101**, 774 (1956).

and accurately surveyed by high resolution inelastic scattering of deuterons and protons. Confirmatory evidence comes from the $(n n')$, $(d\alpha)$, (dn) and (αp) reactions and from the Mg^{27} decay which is complex. The isobaric correction to the decay energy of Mg^{27} places the ground state of this nucleus at an excitation of 6.64 MeV

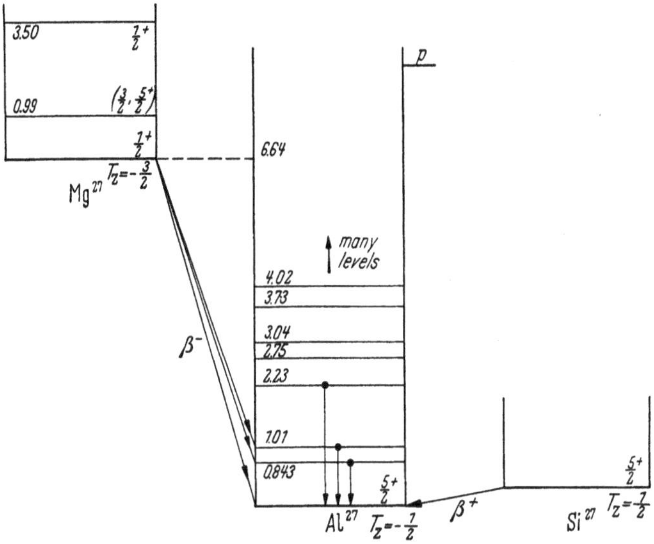

Fig. 78. Levels of nuclei with $A = 27$.

in Al^{27}, in a region in which levels have not been studied. Above the proton dissociation energy virtual levels are obtained from resonances in the $(p\gamma)$ reaction; cascades to the ground state have been observed. The two lowest levels of Al^{27} are excited with comparable probability by the Mg^{27} decay[1]. Levels of Mg^{27} and their spins and parities are deduced from the (dp) stripping reaction.

The magnetic moment of Al^{27} does not agree with the SCHMIDT prediction for the single hole $d_{5/2}^{-1}$ but is well described by the paired configuration $d_{5/2}^{-3}\, s_{1/2}^{2}$. This once more shows the close proximity of levels in the d and s shell. Both configurations correctly give the observed ground state spin.

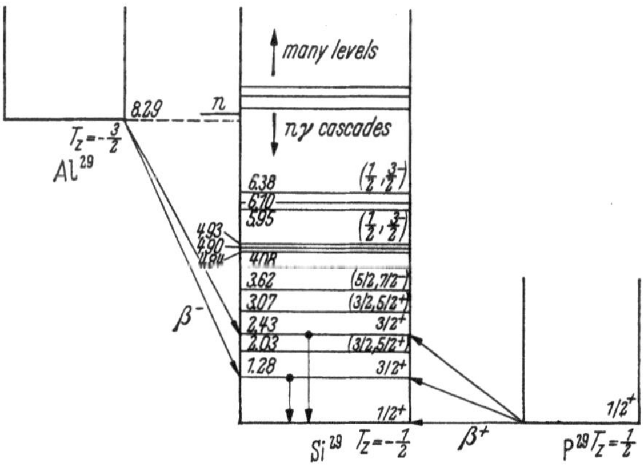

Fig. 79. Levels of nuclei with $A = 29$.

67. $A = 29$: Al^{29}, Si^{29}, P^{29}, Fig. 79.—$A = 31$: Si^{31}, P^{31}, S^{31}, Fig. 80. The levels of Si^{29} are obtained from the (dp) and $(d\alpha)$ reactions, from the complex decay of both other isobars[2] and from the cascade transitions arising from thermal neutron capture. Virtual levels are also known from fast neutron studies. There are practically no levels known in the other $T = \frac{1}{2}$ nucleus

[1] See [33] and also W. S. LYON and N. H. LAZAR: Phys. Rev. **101**, 1524 (1956) who found the 175 keV readiative transition between the two levels.

[2] H. RODERICK, O. LÖNSJÖ and W. E. MEYERHOF: Phys. Rev. **97**, 97 (1955).

to establish mirror properties but the ground state decay is superallowed. The ft values indicate allowed transitions in each case, and from the measured spin of $^1/_2$ for Si^{29} and presumably for P^{29}, spins $^3/_2{}^+$ for the 1.28 and 2.43 MeV states of Si^{29} and for the ground state of Al^{29} follow. Spins, parities and neutron capture probabilities are available for many of the Si^{29} states from the (dp) stripping reaction. The capture probabilities indicate single particle properties for some of the states.

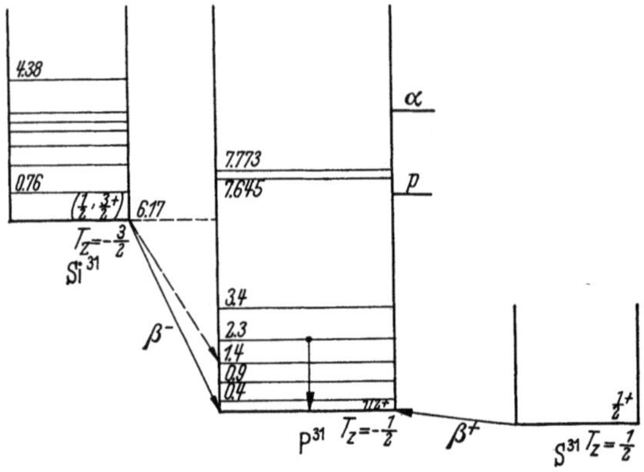

Fig. 80. Levels of nuclei with $A = 31$.

The Al^{29} decay energy and COULOMB correction give the position of the $T=\frac{3}{2}$ level in Si^{29} as 8.29 MeV. This is just below the neutron dissociation energy. An interesting feature of the Si^{29} spectrum is the group of levels at 4.84, 4.90 and 4.93 MeV. This grouping is suggestive of a triplet in LS coupling; in jj coupling INGLIS [10] suggests that it arises from the configuration $d_{5/2}^{-1}\, s_{1/2}^{2}$ with the two s-particles having a spin 1, and coupling with the $\frac{5}{2}$ hole to form states of $\frac{7}{2}$, $\frac{5}{2}$ and $\frac{3}{2}$. The configuration arises from the excitation of one d-nucleon from the possible ground state configuration $d_{5/2}^{12}\, s_{1/2}$ into an s-orbit. The ground state spin of Si^{29} indicates closure of the $d_{5/2}$ sub-shell at $N=14$ and entry of the 15th neutron into an s-orbit.

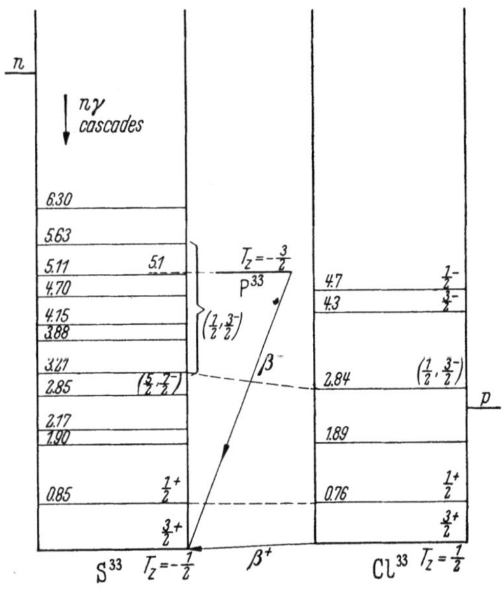

Fig. 81. Levels of nuclei with $A = 33$.

The levels of P^{31} are obtained from the (dn) and (αp) reactions and from resonances in the $(p\gamma)$ process. The spectrum has been but poorly covered; nothing is known of the mirror level system in S^{31}. The Si^{31} nucleus has been more fully studied through the (dp) reaction. The beta-transitions are allowed, and lead to a prediction of the position of $T=\frac{3}{2}$ levels in P^{31}. From the measured spin of P^{31} the spins of S^{31} and Si^{31} are predicted to be $^1/_2{}^+$ and $(^1/_2$ or $^3/_2)^+$.

68. $A = 33$: P^{33}, S^{33}, Cl^{33}, Fig. 81.— $A = 35$: S^{35}, Cl^{35}, A^{35}. The mirror properties of S^{33} and Cl^{33} are well revealed by their level systems which have been obtained mainly by the (dp) and (dn) reactions respectively. The spins and parities of analogous levels, found from the stripping reactions, are also in agreement. Virtual levels are known from the neutron total cross section and from (np), $(n\alpha)$ and $(p\gamma)$ reactions, while cascade transitions from the capture of thermal neutrons have been analysed and accommodated in the known level scheme of S^{33}. The beta decay between the $T=\frac{1}{2}$ isobars is normal; that of P^{33} is of very low energy. The ground state spin of S^{33} is $^3/_2{}^+$ which presumably arises from the configuration

$d_{5/2}^{12}\, s_{1/2}^{4}\, d_{3/2}$; it is curious that Cl^{33} does not show an allowed transition to the 0.85 MeV $1/2^+$ state of S^{33} although there is some evidence for transition to the 2.85 MeV state of this nucleus. The neutron capture probabilities in the (dp) reaction (Table 12) suggest that states at 2.90 and 3.26 and 5.72 MeV have a configuration which is as simple as that of the ground state. A tentative identification of these levels with the single particle states $f_{7/2}\, p_{3/2}\, p_{1/2}$ agrees with the parity change observed. The first excited state, at 0.85 MeV has a small reduced width and would appear to have a more complex configuration; since it is excited with $l_n = 0$, $d_{3/2}^{2}\, s_{1/2}^{-1}$ is a possibility.

The experimental information on the nuclei of mass 35 is fragmentary. The only excited states known are those of Cl^{35} obtained from the S^{32} (αp) reaction. The beta transitions are both simple; that of S^{35} is of very low energy, so that the excitation of the first $T = \frac{3}{2}$ state in Cl^{35} is at an energy approximately equal to the isobaric correction. The spin of the radioactive nucleus S^{35} has been measured and is found to be $\frac{3}{2}$, in agreement with the jj prediction for $d_{3/2}^{3}$. The magnetic moment of Cl^{35} agrees better with the prediction from this same configuration than with the SCHMIDT value for $d_{3/2}$. The $S^{35} \rightarrow Cl^{35}$ decay is allowed, but not superallowed since these nuclei are not a true mirror pair.

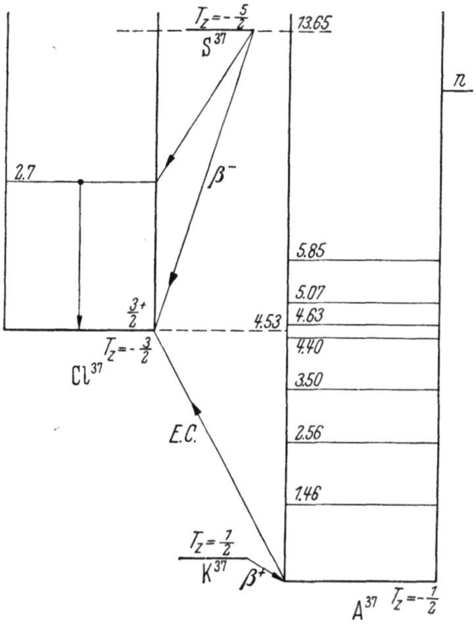

Fig. 82. Levels of nuclei with $A = 37$.

69. $A = 37$: S^{37}, Cl^{37}, A^{37}, K^{37}, Fig. 82. —$A = 39$: Cl^{39}, A^{39}, K^{39}, Ca^{39}.—$A = 41$: A^{41}, K^{41}, Ca^{41}, Sc^{41}, Fig. 83. Levels of A^{37} are obtained, so far without suggestion of spin or parity, mainly from the (dp) reaction, although low levels have also been found from neutron spectra in the (pn) reaction. A level in Cl^{37} is found from the decay scheme of S^{37}. The decay of K^{37} is probably the usual superallowed transition; it is followed by electron capture decay of A^{37}. The A^{37} decay is notable in that L-capture has been established by careful proportional counter experiments (PONTECORVO *et al.*[1]), and that the decay energy has been found by measuring the time of flight spectrum of the Cl^{37} recoil ions (RODEBACK and ALLEN[2]) using electron multiplier detection. The $T = \frac{3}{2}$ level in A^{37} is at a rather low energy, and the $T = \frac{5}{2}$ level, obtained from the complex S^{37} decay, is at about 13.5 MeV excitation. The magnetic moment of Cl^{37} is in disagreement with the assumption of a configuration $d_{3/2}^{-3}$, as was also found for the F^{19} nucleus which occupies a similar position at the other end of the d-shell.

The nuclei of mass 39 are analogous to O^{17} in that single particle (actually hole) levels are expected to occur in Ca^{39}. There is no evidence on this point. The decay to K^{39} appears from the positron spectrum end-point to be governed by the COULOMB energy, and in this case the $T_z = -\frac{1}{2}$ nucleus is again stable. A^{39} also decays, by a forbidden transition, to the $3/2^+$ ground state of K^{39}. Analysis of the shape of the beta spectrum classifies the transition as $\Delta J = 2$, *yes*, so that

[1] B. PONTECORVO, D. H. W. KIRKWOOD and R. C. HANNA: Phys. Rev. **75**, 982 (1949).
[2] G. W. RODEBACK and J. S. ALLEN: Phys. Rev. **86**, 446 (1952).

the A^{39} ground state spin is $^7/_2{}^-$, consistent with the configuration $d_{3/2}^{-2} f_{7/2}$ of the shell model. Levels of A^{39} are found from the Cl^{39} decay, which also undergoes a $\Delta J = 2$, *yes*, transition to the A^{39} ground state. The Cl^{39} spin is therefore $^3/_2{}^+$, consistent with configuration $d_{3/2}^{-3} f_{7/2}^2$.

The nuclei of mass 41 include the mirror pair Sc^{41}, Ca^{41}, which are both unstable and between which the normal superallowed decay takes place. The levels of Ca^{41} have been well explored by the $(d p)$ reaction, and spins and parities are known for levels up to 5.72 MeV. Resonances are observed in the $(n n)$ reaction corresponding to virtual levels of Ca^{41}, and cascade transitions from thermal neutron capture have been fitted into the level scheme obtained from the $(d p)$ reaction. The ground state spin and parity possibilities include the single particle value $^7/_2{}^-$ $(f_{7/2})$, with a large neutron capture probability. Other possible single-particle levels are one component of the 2 MeV doublet $(p_{3/2})$ and the 2.47 MeV state $(p_{1/2})$. Ca^{41} decays by a forbidden electron capture transition to K^{41} which has the measured spin $\frac{3}{2}$ and presumably even parity $(d_{3/2})$. The $T = \frac{3}{2}$ levels of Ca^{41} should be in a region where some levels are known and the spin $^3/_2{}^+$ occurs among the possibilities for these levels. States of K^{41} have been seen in the $(d n)$ reaction and virtual levels are known from the $(p \gamma)$ process. There is some evidence for thermal neutron capture in K^{40} in the investigations of KINSEY *et al.* [*37*] with normal potassium. The 1.37 MeV state of K^{41} is excited preferentially in the A^{41} decay; the ground state transition being first forbidden. The lifetime of the 1.37 MeV state (Table 3) agrees with that expected empirically for an $M2$ transition and indicates $J = {}^7/_2{}^-$ for the 1.37 MeV level. A similar spin for the ground state of A^{41} would account for its forbidden decay to the K^{41} ground state ($\Delta J = 2$, *yes*) and is consistent with stripping data from the A^{40} $(d p)$ reaction. This reaction gives the energy levels of A^{41} up to 4.0 MeV and the $(n n)$ reaction gives resonance levels above the neutron binding energy. The decay energy of A^{41} indicates a $T = \frac{5}{2}$ level at an excitation of about 14 MeV in Ca^{41}.

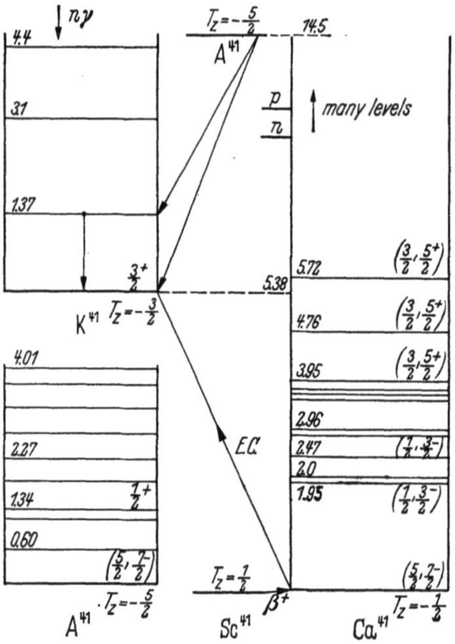

Fig. 83. Levels of nuclei with $A = 41$.

II. The even isobars of mass $4n + 2$.

70. $A = 6$: He^6, Li^6, Fig. 84. There are no bound levels of Li^6, but selection rules for spin, parity and isotopic spin exert an important influence on the modes of decay of the virtual levels. The most extensive information comes from the Li^7 $(He^3\,\alpha)$ reaction [*32*] which gives groups corresponding to all the levels shown in Fig. 84. The ground state and levels at 2.189 and 3.57 MeV are also excited in the $(p \alpha)$, $(p d)$ and $(p p')$ reactions, and the ground state and 2.189 MeV level in the $(d t)$ reaction. These two levels are also seen in the inelastic scattering of deuterons and the 2.189 MeV state appears in the elastic scattering of deuterons by alpha particles, but the 3.57 MeV state has not been found in either the $(d d')$

or ($d\alpha$) reactions. These facts, and the details of the Be^9 ($p\alpha$) reaction, suggest that the 3.57 MeV level is the $J=0$, $T=1$ analogue of the ground state of He^6.

The $T=0$ levels shown at 4.52 and 5.4 MeV are also obtained from ($d\alpha$) scattering experiments. The reduced widths of the levels seen in ($d\alpha$) scattering are all comparable with the single particle value. This suggests that the structure of Li^6 is not greatly dissimilar from that of an alpha particle with an orbital deuteron.

The energy levels of He^6 are found from the ($t\alpha$) and ($d\,He^3$) reactions; the latter indicates a change of parity between Li^7 and the ground state and first

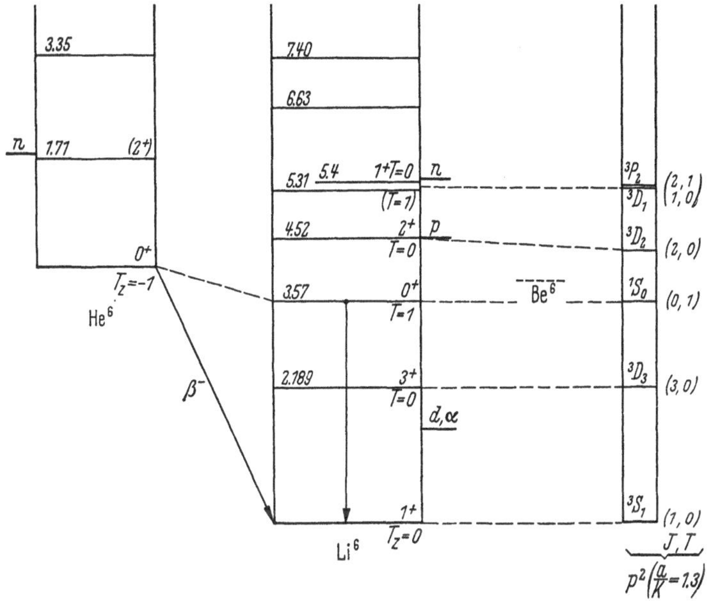

Fig. 84. Levels of nuclei with $A=6$.

excited state (1.71 MeV) of He^6. Virtual states in He^6 and Li^6 are excited by the He^3 (tn) and He^3 (tp) reactions.

The decay of He^6 to the ground state of Li^6 is superallowed and is the most important evidence for the GAMOW-TELLER type of coupling in beta-decay. The FERMI allowed decay which would be expected from He^6 to its isobar state at 3.57 MeV in Li^6 is just energetically impossible. No trace of gamma radiation in the He^6 decay has been found, nor has any transition to the 2.189 MeV level ($J=3^+$) been seen.

Theoretically the Li^6 nucleus, with two p-nucleons outside an alpha particle shell, resembles the deuteron in those features which depend mainly on spin and isotopic spin. Thus the magnetic moments are very similar and low lying states which are respectively symmetrical and antisymmetrical in nucleon spins ($^3S, {}^1S$) are found. The other states permitted for two p-nucleons are listed by INGLIS; for $T=0$ there are $^3S, {}^1P, {}^3D$ and for $T=1, {}^1S, {}^3P, {}^1D$. The order of energy of the $T=1$ levels in LS coupling is shown in Fig. 2, together with the resolution of these states by spin-orbit forces. The states obtained from the configuration $p_{3/2}^2$ in extreme jj coupling as a function of range of central force are also shown in Fig. 2; the state $J=3^+$ always lies lowest. In order to obtain the observed level sequence for Li^6, intermediate coupling is necessary; near the LS extreme the state of spin 1 lies lowest as observed, and the level order for the intermediate coupling parameter chosen by INGLIS [10] is shown in Fig. 84. The states of the excited configuration $p_{3/2}\,p_{1/2}$ overlap those of $p_{3/2}^2$ for small values of a/K. The levels of He^6 are in the order $(J, T) = (0, 1)\ (2, 1)\ (2, 1)$ which is consistent with observation. No odd parity states are known for $A = 6$.

TAUBER and WU[1] show that the magnetic moment of Li^6, calculated with the intermediate coupling parameter which best fits the level spectrum, is in better agreement with observation than that obtained from pure jj coupling. The multiplet splittings in Li^6 can be accounted for[2], at least qualitatively, by the introduction of a tensor force component into the main interaction between nucleons instead of a spin-orbit force.

71. $A = 10$: Be^{10}, B^{10}, C^{10}, Fig. 85. Information for Be^{10} comes from the (dp) reaction and from the total cross section for Be^9 for neutrons; the capture of

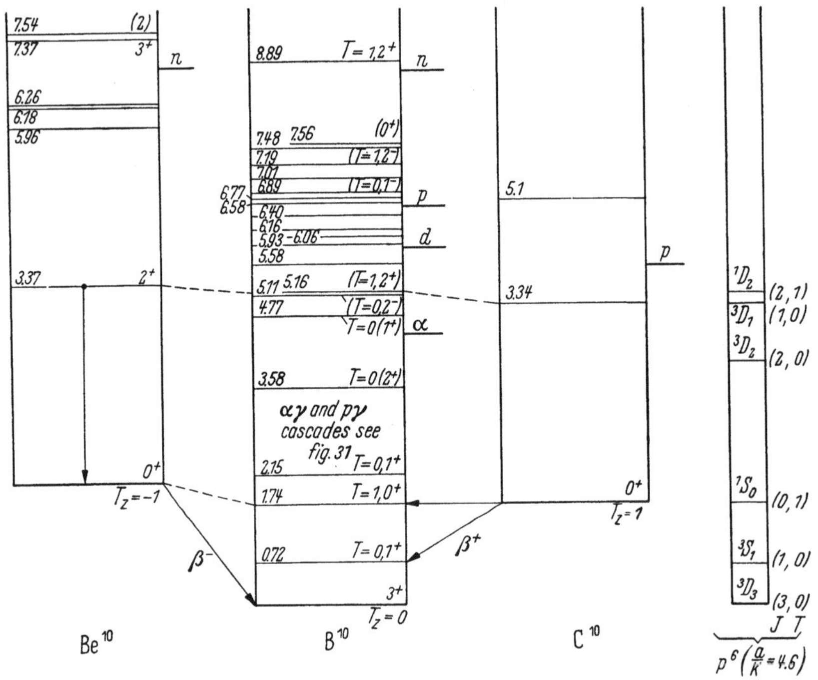

Fig. 85. Levels of nuclei with $A = 10$.

thermal neutrons in Be^9 results in a cascade which verifies the position of the first excited state. The parity of the ground state and states at 3.37 and 7.37 MeV relative to that of Be^9 is established by the stripping reaction. Angular correlation between the protons and 3.37 MeV gamma radiation and the pair internal conversion coefficient of this radiation fix the spin of the 3.37 MeV state as 2^+. Resonances in the Be^9 (nn) reaction give spin values for the 7.37 and 7.54 MeV states and a reduced width for the former. The Li^7 $(t\alpha)$ He^6 reaction yields highly excited states at 17.82 and 18.43 MeV. The angular distributions of the alpha particle groups indicate $J = 2^+$ for the former state [32].

The bound states of B^{10} have been obtained from the (dn), $(d\alpha)$, (pp'), (dd') and (nn') reactions, by the decay of C^{10}, and from cascades observed in the $(p\gamma)$ reaction. Virtual levels near the alpha particle dissociation energy are studied

[1] G. E. TAUBER and TA-YOU WU: Phys. Rev. **93**, 295 (1954).
[2] A. J. M. HITCHCOCK: Proc. Phys. Soc. Lond. A **65**, 661 (1952).

by the Li^6 $(\alpha\gamma)$ reaction and also by the thresholds for slow neutron production in the Be^9 (dn) reaction, in addition to their study by neutron spectroscopy. The proton dissociation energy in B^{10} is at 6.58 MeV and above this energy resonances in the proton induced reactions on Be^9 give further states.

Correction for COULOMB energy differences sets the Be^{10} and C^{10} ground states at an excitation such that either the 1.74 MeV or the 2.15 MeV level of B^{10} could be the analogous state. The positron decay of C^{10} is allowed for transitions to both the 0.72 and 1.74 MeV levels; for Be^{10} decay is possible energetically only to the $J=3^+$ ground state, and is second forbidden. This suggests that the 0.72 and 1.74 MeV states are probably the $(J, T)=(1, 0)$ and $(0, 1)$ components of the lowest multiplet, so that their wavefunction has the same spatial symmetry as that of C^{10} (and of Be^{10}). The assignment of isotopic spin $T=1$ to the 1.74 MeV state is supported by the fact that this state is not observed in the (dd') reaction.

Assignments of spin, isotopic spin and parity to the other levels of B^{10} are based largely on analysis of the radiative transitions between the levels [see Fig. 31 and Be^9 $(p\gamma)$]. The angular distributions in the (dn) reaction, analysed by stripping theory, indicate even parity and $J<3$ for all states up to 4.77 MeV and odd parity, $J=1$ or 2, for one or both of the 5.11 and 5.16 MeV levels. There is no obvious analogue for either the 2.15 or 3.58 MeV states in Be^{10}, so they presumably have $T=0$. Since the 4.77 MeV state is seen in the Li^6 $(\alpha\gamma)$ reaction, it cannot have $J=0$; it radiates only to the 0.72 MeV 1^+ state and therefore is presumably itself of 1^+ type.

The 5.11 and 5.16 MeV levels are resolved in the (dn) (threshold) and (pp') reactions. A level of $J=2^+$, $T=1$ is expected in this region as analogue to the first excited state of Be^{10}, and the 5.16 MeV level, which radiates to 1^+ and 3^+ states, seems to fulfil this requirement. The formation in the $Li^6+\alpha$ reaction is ascribed to isotopic spin impurity. The 5.11 MeV level is not found in the $(\alpha\gamma)$ reaction, perhaps because its most probable mode of decay would be by $E\,1$ radiation, which is discouraged between states of the same isotopic spin in a nucleus with $T=0$. This would be so if its spin were $2^-(T=0)$. The isotopic spin impurity of this state is discussed by RADICATI[1].

Spin and isotopic spin assignments for the B^{10} levels above the proton binding energy are suggested by analysis of the radiative transitions in the $(p\gamma)$ reaction and by comparison with Be^{10}. They are supported by analysis of the $Be^9(pp)$ scattering resonances.

The states of the isobar C^{10} are obtained from the B^{10} $(p\,n)$ reaction. The first state at 3.34 MeV agrees well with the first state of Be^{10} and the second $T=1$ state of B^{10}.

Theoretically the nuclei of mass 10, with 6 p-nucleons, are described by the (1, 0, 0) supermultiplet in which the lowest (S, T) multiplet is (0, 1) or (1, 0). Since the spin of B^{10} is 3, the coupling scheme cannot be pure LS, in which the ^3S state $(J=1)$ is lowest. INGLIS [*10*] chooses an intermediate coupling parameter $a/K=4.6$ to give the best fit with observed levels in Be^{10} and B^{10}, although the $T=0$, 2^+ and 1^+ levels come in the wrong order. The $J=3$ state is also shown to lie lowest in the exact intermediate coupling calculations of ZELDES[2].

Excitation of a $p_{3/2}$ nucleon to the $p_{1/2}$ state gives even states with small nucleon reduced widths [*19*] such as are observed in the nucleon capture processes. The addition of an s-nucleon to the Be^9 ground state would lead to four

[1] L. A. RADICATI: Proc. Phys. Soc. Lond. A **67**, 39 (1954).
[2] N. ZELDES: Phys. Rev. **90**, 416 (1953).

odd states $J=1, 2$ and $T=0, 1$. These are probably to be found near 7 MeV in B^{10}. The large thermal neutron scattering length for Be^9 indicates bound s-wave levels.

72. $A=14$: C^{14}, N^{14}, O^{14}, Fig. 86. In C^{14} the levels shown are found in the $(d p)$ reaction and the spins and parities follow from the stripping patterns and radiative transitions observed in this reaction. Zero spin for the C^{14} nucleus is established by direct observation of the molecular spectrum[1].

In N^{14} the bound levels (up to 7.4 MeV) are obtained from the $(d n)$, $(d \alpha)$, $(p p')$, $(d d')$ and $(n n')$ reactions, and from the gamma radiation observed in these

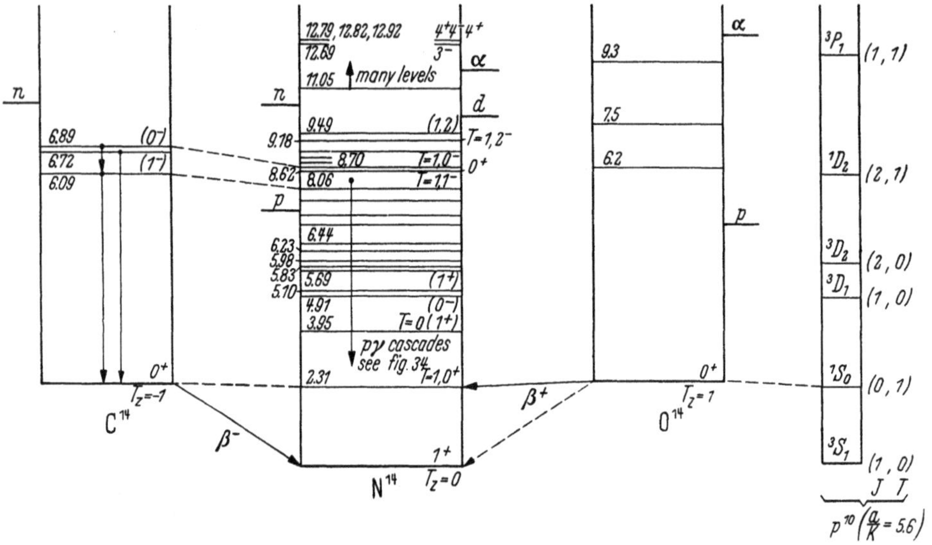

Fig. 86. Levels of nuclei with $A=14$.

reactions and in the cascade transition from higher virtual levels excited in the C^{13} $(p \gamma)$ reaction (Fig. 34). The ground state spin is 1 and the parity is the same as that of C^{12} and C^{14} according to the results of stripping reactions. The 2.31 MeV level is excited in the superallowed decay of the isobar O^{14} and is therefore likely to be the first $T=1$ state in N^{14}, with spin zero; it cannot be reached energetically from C^{14}. The $T=1$ assignment is supported by the difficulty of finding a particle group corresponding to this level in the $(d d')$ or $(d \alpha)$ reactions. The spin is consistent with the condition $J \leq 2^+$ from the $(d n)$ stripping analysis. The 3.95 MeV level has also $J \leq 2^+$ according to the $(d n)$ reaction, and is assigned $J=1^+$ because of the strength of the gamma transition to the 2.31 MeV level. Because of the scarcity of levels of C^{14} at comparable excitation, this and other levels up to 8 MeV should have $T=0$. The other tentative assignments shown in Fig. 86 for levels up to 8 MeV are based on relative intensities of gamma transitions to lower levels of known spin.

The level structure of N^{14} between 8.06 and 10.43 MeV is obtained from the resonances in the C^{13} $(p \gamma)$ and $(p p)$ reactions. Among the levels that at 8.06 MeV has the right J-value (1^-) to agree with the 6.09 MeV level in C^{14} and therefore might have $T=1$. This is supported by the radiative transition probabilities. A similar assignment of $T=1$ is made for the 8.70 and 9.18 MeV levels; the former may then be the analogue of the 6.89 MeV level in C^{14}, and a state of spin 2^- should

[1] F. A. Jenkins: Phys. Rev. **74**, 355 (1948).

in due course be found at an excitation of about 7.4 MeV in this nucleus. The gamma ray angular distribution from the $(p\gamma)$ reaction at the 9.18 and 9.49 MeV levels is consistent with their assigned spins and parities.

Above 10 MeV levels of N^{14} are shown as resonances in the C^{13} (pn) reaction and in C^{12} (dp) and C^{12} (dn). Sharp levels are found above 12.29 MeV from the alpha particle bombardment of boron, and spins and parities are deduced from analysis of angular distributions and correlations in the $(\alpha\gamma)$, (αp), (αd) and (αn) reactions (see Sect. 45). At the 12.69 MeV level the reduced width for protons is only $\frac{1}{5}$ of that for neutrons.

The O^{14} levels are obtained only from neutron groups found in the (pn) reaction and are probably not fully resolved.

The beta transitions between the $A=14$ isobars show some unexplained features. The O^{14} decay to the $T=1$ state of N^{14} at 2.31 MeV is superallowed, as expected since it takes place without change of spatial symmetry. On the other hand the more energetic transition to the ground state ($\Delta J=1$, *No*) is very weak[1], and the decay of C^{14} to this state (which is the only one energetically possible for it) is also strongly forbidden, although the spectrum has allowed shape. The solution to this difficulty may be the introduction of a tensor force which would mix in $^1P_1$ and $^3D_1$ states with $^3S_1$ in N^{14} and $^3P_0$ with $^1S_0$ in C^{14} (JANCOVICI and TALMI[2]), but the magnitude of tensor force required is large and seriously affects the order of energy levels in N^{14}. The difference between the highly forbidden C^{14} decay to the N^{14} ground state and the very weak transition which has been seen to this state in the O^{14} decay[3] is ascribed to extra COULOMB perturbation in the case of O^{14}. The wave function obtained with tensor forces gives a good account of the magnetic moment and electric quadrupole moment of N^{14}. The suggestion of FEINGOLD[4] that the N^{14} ground state is predominantly a P state also leads to difficulties.

The theoretical description of the mass 14 nuclei given by INGLIS [*10*] resembles that for mass 6, since the configuration is $p_{1/2}^{-2}$ instead of $p_{3/2}^{2}$. In LS coupling, the low multiplets are exactly the same, namely ^3S, ^3D, ^1S, ^1D, ^3P, ^1P with isotopic spin 0 for the symmetrical states, and 1 for the antisymmetrical states. In the jj extreme the lowest states of $p_{1/2}^{-2}$ are $(J, T)=(1, 0)$ and $(0, 1)$. This agrees with observation, and the 2.31 MeV separation of the $T=0$ ground state and first $T=1$ level of N^{14} determines the intermediate coupling parameter a/K for a "hole" as 5.6. No accurate comparison for other levels can be made although the 3.95 MeV level has the right spin (1^+) to be the lowest state of the $p_{3/2}^{-1}\,p_{1/2}^{-1}$ configuration. The odd parity levels presumably arise from configurations of the type p^9d and p^9s, formed by coupling a single nucleon to the ground state configuration of C^{13}. The nucleon reduced widths for most of the odd states are large, in agreement with this description. The second $T=1$ state is predicted to be at an excitation of about 8.4 MeV in N^{14} and to be even. In fact, odd levels of $T=1$ are known at about this excitation, corresponding to the lowest excited states of C^{14}.

73. $A=18$: O^{18}, F^{18}, Ne^{18}, Fig. 87. Only one level of O^{18}, near 2 MeV, has been reported from the (dp) reaction, and no levels at all are known in Ne^{18}. At mass 18 the COULOMB energy has become so large that the nucleus of $T=0$ is no longer stable. The positron decays $Ne^{18}\rightarrow F^{18}\rightarrow O^{18}$ are both superallowed and suggest that the ground state spin of F^{18} is 1^+. Decay of Ne^{18} to a $T=1$ state of F^{18}

[1] J. B. GERHART: Phys. Rev. **95**, 288 (1954).
[2] B. JANCOVICI and I. TALMI: Phys. Rev. **95**, 289 (1954).
[3] R. SHERR, J. B. GERHART, H. HORIE and W. F. HORNYAK: Phys. Rev. **100**, 945 (1955).
[4] A. M. FEINGOLD: Phys. Rev. **89**, 318 (1953).

at about 1 MeV would be expected, but has not so far been observed. Ne^{18} is produced by the reaction F^{19} $(p\ 2n)$[1].

The levels of F^{18} are found mainly from the Ne^{20} $(d\alpha)$ reaction which is expected to exhibit only those with $T=0$. Gamma transitions between some of these

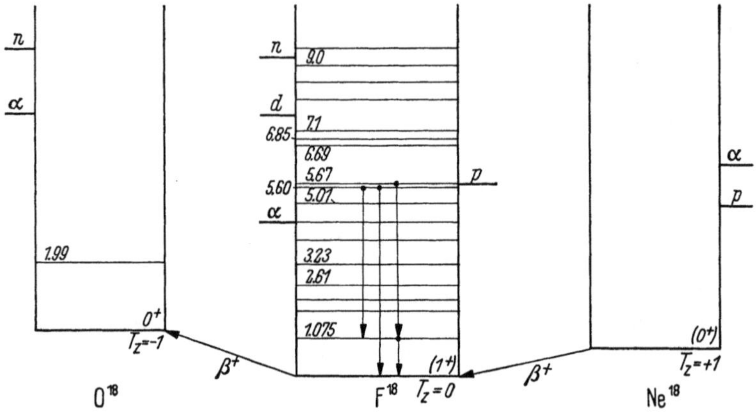

Fig. 87. Levels of nuclei with $A=18$.

levels have also been induced by the N^{14} $(\alpha\gamma)$ reaction. At 6.69 and 6.85 MeV resonances in the $(\alpha\alpha)$ and (αp) reactions give virtual levels of F^{18}, and several more levels in the region above 8 MeV are known from resonances in the O^{16} (dp) cross section. Although all these reactions are nominally concerned only with levels with $T=0$, excitation of $T=1$ states can take place, because of isotopic spin impurity.

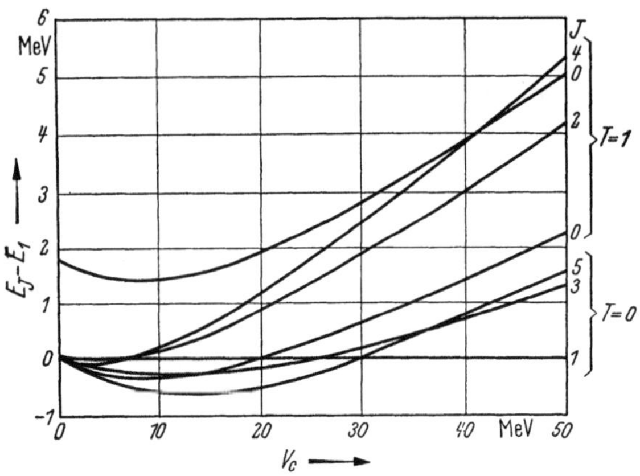

Fig. 88. Levels calculated for $A=18$ (ELLIOTT and FLOWERS [45]).

The theory of the mass 18 nuclei in which d and s states are filled is given by ELLIOTT and FLOWERS [45] using an intermediate coupling approach based on the single particle properties of O^{17}. The calculated levels of even parity are shown in Fig. 88, and agreement with the ground state spin of F^{18} is obtained for a central potential greater than 30 MeV; the first excited state with $T=0$ probably has $J=3$ or 5. The calculations show that the wave functions contain components of several configurations, both in the LS and jj descriptions. The rough COULOMB calculation (Table 14) suggests a $T=1$ state $(J=0)$ at about 1 MeV excitation in F^{18} and Fig. 88 shows that this can be obtained for $V_c \sim 35$ MeV; the $T=0$ states of spin 3 and 5 would then be lower and have not so far been observed. The relative position of the $T=0$ and $T=1$ states depends sensitively on the precise nature of the exchange force and the $T=1$ state can probably be moved to agree with experiment when it is identified.

[1] J. D. Gow and L. W. ALVAREZ: Phys. Rev. **94**, 365 (1954).

The beta transition probabilities for $Ne^{18} \to F^{18} \to O^{18}$ can be calculated from the wave functions. The log ft values are shown in Fig. 89 as a function of the central potential. The V_c value indicated by experiment is rather smaller than that obtained from the level sequence, but the difference between observed and calculated transition probability is not large. The preferred central interaction is very close to that which gives the best account of the properties of the deuteron. The coupling is near the LS extreme, but configurational interaction cannot be neglected. This conclusion has also been reached by REDLICH[1] using a different shape of central potential, which gives wave functions very similar to those obtained by ELLIOTT and FLOWERS, and also a good account of the F^{18} beta-decay.

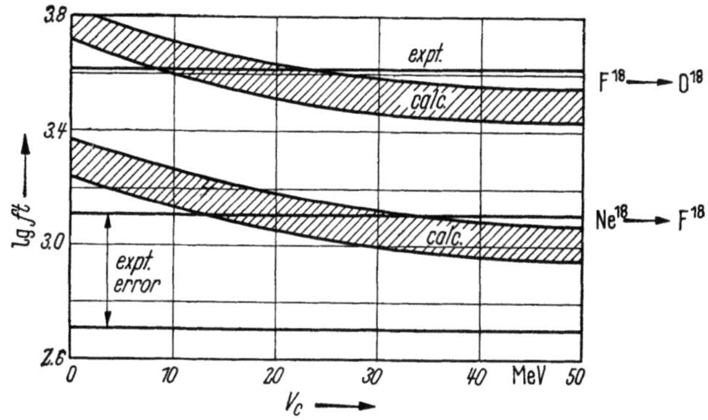

Fig. 89. β-decay probabilities for $A=18$ (ELLIOTT and FLOWERS [45]).

74. $A=22$: Ne^{22}, Na^{22}, Fig. 90.—$A=26$: Mg^{26}, Al^{26}, Fig. 91. The levels of Ne^{22} are obtained from the (pp'), (dp) and (αp) reactions and from the decay of Na^{22}. The gamma radiation following excitation of these states has been observed in coincidence with the protons in the (dp) reaction and confirms the level sequence although it does not give unambiguous spin values. The internal conversion coefficient[2] of the 1.28 MeV gamma radiation agrees with that for an $E\,2$ transition, so that the spin of this state is probably 2^+.

The Na^{22} levels are poorly known. The best established is the 592 keV level from the (αn) reaction which lies approximately in the position expected for the first $T=1$ level. This level could decay to the ground state of Ne^{22} by a $0^+ \to 0^+$ transition if these are analogous levels, but the probability of radiative transition to the Na^{22} ground state is apparently much larger. The Na^{22} ground state has a measured spin of 3 and is expected to have even parity; the $M\,3$ transition would have a lifetime of perhaps 0.1 sec.[3]

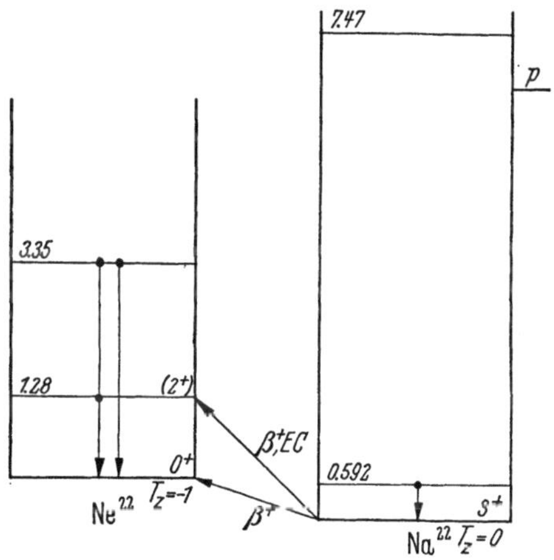

Fig. 90. Levels of nuclei with $A=22$.

Na^{22} decays from its 3^+ ground state to the first excited state of $Ne^{22}(2^+)$ by an allowed transition, in which there is a 10% electron capture branch. There is also a weak second forbidden transition between the Na^{22} and Ne^{22} ground

[1] M. G. REDLICH: Phys. Rev. **95**, 448 (1954).

[2] R. D. LEAMER and G. W. HINMAN: Phys. Rev. **96**, 1607 (1954).

[3] The lifetime is less than 10 μsec [S. H. VEGORS and P. AXEL: Phys. Rev. **101**, 1067 (1956)] and it is therefore unlikely that the 592 keV level is a $T=1$ level of spin 0 if it is also the first excited state of Na^{22}.

states ($\Delta J=3$, *no*). The configurations of Na^{22} are almost certainly a mixture of *d* and *s* states and no detailed calculations have been made. It is known that the 11 protons in Na^{23} couple to give a state of $J={}^3/_2{}^+$ and assuming that the same holds for the 11 neutrons in Na^{22}, the spin of this nucleus is the resultant of two ${}^3/_2{}^+$ configurations, which could be the observed value of 3.

The levels of Mg^{26} are known from the $(p p')$, $(d p)$ and (αp) reactions, from the gamma rays observed in the $(d p)$ reaction and from the radiation following slow neutron capture. Spin and parity values are indicated by the angular distributions in the $(d p)$ reaction combined with branching ratios in the cascade transitions. The Al^{26} levels are derived from the $(d n)$ and $(d \alpha)$ reactions up to 5 MeV; between 7.0 and 7.6 MeV fourteen closely spaced levels are obtained from the $Mg^{25}(p\gamma)$ reaction. Details of the level spectrum are filled in by scintillation analysis of the cascade gamma radiation in the $(p\gamma)$ reaction. In particular, levels at 0.230 and 2.074 MeV are found which are not seen in the $(d\alpha)$ reaction and may therefore have $T=1$.

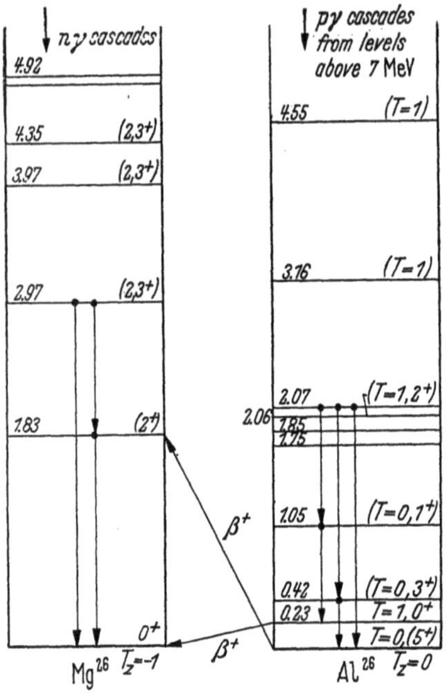

Fig. 91. Levels of nuclei with $A=26$.

The ground states of Mg^{26} and Al^{26} and the first $T=1$ state of Al^{26} are members of equivalent (ST) multiplets according to the WIGNER classification and a superallowed beta decay is expected between these states. It has been established[1] that this takes place from the 0.230 MeV state of Al^{26} to the ground state of Mg^{26}; both these states are taken to have $T=1$, $J=0^+$. The Al^{26} ground state is formed by coupling together two $d_{5/2}$ nucleons and is expected to have spin 5^+ (EDMONDS and FLOWERS [*14*]). It decays to the $J=2^+$ state of Mg^{26} at 1.83 MeV excitation with a half life of 10^6 years. This has been verified by SIMANTON *et al.*[2] using Al^{26} prepared by the $Mg^{25}(d n)$ reaction; they observed electrons of about 1 MeV energy and gamma radiation of energy up to 1.6 MeV.

This interpretation of the Al^{26} decay accounts for the observation that the beta-decay energy does not quite agree with the Mg^{26}-Al^{26} mass difference derived from reaction data. It also explains the excess of neutrons over positrons in the $Al^{27}(\gamma n)\,Al^{26}$ reaction and the appearance of capture gamma radiation effectively without positrons at the 436 and 958 keV resonances in the $Mg^{25}(p\gamma)$ reaction[3]. Measurements of the energy of capture gamma rays at the 730 and 958 keV resonances also show that there is a low lying state in Al^{26}. The 0.416 MeV state in Al^{26} was shown by gamma-gamma coincidence methods to have a short half-life and is therefore not responsible for the 6.7 sec decay now attributed to the 0.230 MeV level. The gamma ray lines obtained by KAVANAGH *et al.* at the 730

---

[1] R. W. KAVANAGH, W. R. MILLS and R. SHERR: Phys. Rev. **97**, 248 (1955).

[2] J. R. SIMANTON, R. A. RIGHTMIRE, A. L. LONG and T. P. KOHMAN: Phys. Rev. **96**, 1711 (1954). — M. J. LAUBITZ: Proc. Phys. Soc. Lond. A **68**, 1033 (1955).

[3] J. S. KLUYVER, C. VAN DER LEUN and P. M. ENDT: Phys. Rev. **94**, 1795 (1954). — S. E. HUNT and D. A. HANCOCK: Phys. Rev. **97**, 567 (1955).

and 958 keV resonances (of which only the first leads to positron emission) are noted in Fig. 91; they suggest a new level in Al^{26} at 2.074 MeV which was not seen in the $(d\alpha)$ reaction. This is possibly the second $T=1$ state of Al^{26} with $J=2^+$. The spins shown in Fig. 91 are consistent with the jj coupling of $d^2_{5/2}$, giving states of $J=0^+, 2^+, 4^+$ $(T=1)$ and $J=1^+, 3^+, 5^+$ $(T=0)$.

75. $A=30$: Si^{30}, P^{30}, Fig. 92.—$A=34$: P^{34}, S^{34}, Cl^{34}, Fig. 93.—$A=38$: Cl^{38}, A^{38}, K^{38}, Fig. 94. The bound levels in Si^{30}, S^{34} and A^{38} are obtained from the (dp) and (αp) reactions and from the decay of the beta-active isobars. Virtual levels in A^{38} are noted as resonances in the Cl^{37} + proton reactions. In P^{30}, the states are deduced from the (dn) reaction and from resonances and spectra in Si^{29} + protons; little is known of the levels of Cl^{34} and K^{38}. The $T=2$ isobars P^{34} and Cl^{38} have also been little investigated although levels in the latter have been found in the (dp) and $(n\gamma)$ reactions.

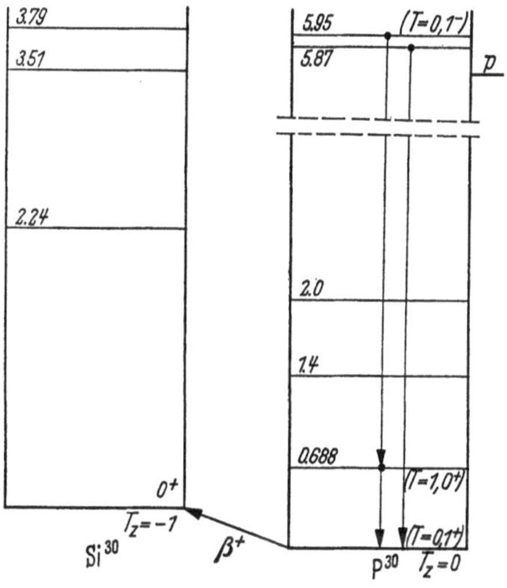

Fig. 92. Levels of nuclei with $A = 30$.

The decay schemes of P^{34} and Cl^{38} have been well studied and show branching to one or more states of the residual nucleus. P^{34} decays to the (0^+) ground state and to the 2.10 MeV level of S^{34} by allowed transitions suggesting a spin of 1^+ for P^{34}. The Cl^{38} decay to the 0^+ ground state of A^{38} is first forbidden and has the beta energy distribution characteristic of the change $\Delta J=2$, *yes*. This fixes the spin of Cl^{38} as 2^-. The transitions to the 2.15 and 3.75 MeV levels of A^{38} are respectively first forbidden and allowed, and this, together with the gamma-gamma correlation[1], gives the sequence 0^+-2^+-3^- for the spins of the ground and first two excited states of A^{38}. The Cl^{38} spin is in agreement with the jj coupling model according to KURATH [*14*] for short range forces, assuming the configuration $d_{3/2}\, f_{7/2}$.

The transitions between the nuclei of $T_z=0$ and those of $T_z=-1$ should all include superallowed decay, although increase of ft values might well be expected as the WIGNER approximation becomes less good. In P^{30} the gamma spectrum from the Si^{29} + proton resonances defines a level at 0.688 MeV which is probably that of $T=1$, $J=0^+$. One of the resonance levels which radiates to this level rather than to the ground state by $E1$ radiation is probably a $T=0$ state. It is not clear why the beta transition to the Si^{30} ground state $(J, T) = (0, 1)$ is somewhat hindered. The 0.688 MeV level has a short life because of $M1$ transitions to the ground state and is therefore unable to make direct 0-0 transitions to Si^{30} as in the case of Al^{26}-Mg^{26}.

In Cl^{34} the superallowed transition of 1.6 sec half-life is the ground state decay. There is an isomeric state at 142 keV which was for some time thought to be the ground state, but which actually feeds it by an $M3$ transition whose internal conversion coefficient has been measured. This means that the ground state of Cl^{34} has spin 0^+ and isotopic spin 1 while the 142 keV level has spin 3^+ and isotopic spin zero. This level also decays by allowed positron emission to

[1] J. J. KRAUSHAAR, J. W. MIHELICH and A. W. SUNYAR: Phys. Rev. **95**, 456 (1954).

the 2.10 and 3.24 MeV levels of S^{34} and this suggests the spin values shown for these levels. KURATH [*14*] considers the configuration $(d_{3/2}\, d_{3/2})$ for Cl^{34} and shows that levels of spin 0 and 3 may be close together for MAJORANA forces.

K^{38} is another case in which it is possible that the $T=1$ state is the ground state. An activity of approximately 1 sec half-life and energy (4.57 ± 0.13) MeV was found in the $K(\gamma n)$ reaction and is now assigned to the ground state, with spin $J=0^+$ and $T=1$. If this is so, the $T=0$ state is that which is known to decay with a 7.7 min period and a maximum positron energy of 2.8 MeV to the

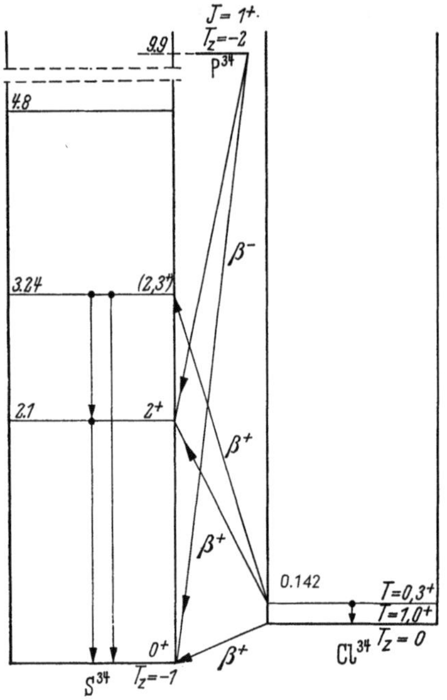

Fig. 93. Levels of nuclei with $A=34$.

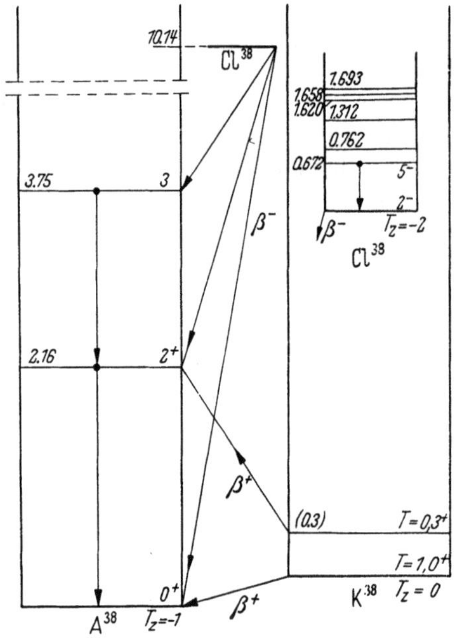

Fig. 94. Levels of nuclei with $A=38$.

2.16 MeV (2^+) level of A^{38}. Since this is an allowed decay the spin of the excited level of K^{38} is 3^+ and its excitation (0.38 ± 0.3) MeV. The expected $M3$ radiative branching to the ground state has not been reported. The configuration of K^{38} in the shell model is $d_{3/2}^{-1}\, d_{3/2}^{-1}$ which has the same states as Cl^{34}, as found (KURATH [*14*)].

The experimental investigation of the $4n+2$ nuclei in which $T=0$ and $T=1$ states lie close together is due to STAHELIN[1]; the transitions are discussed by MOSZKOWSKI and PEASLEE[2].

III. The even isobars of mass 4*n*.

76. $A=8$: Li^8, Be^8, B^8, Fig. 95. The Be^8 nucleus is unstable in its ground state and breaks up into two alpha particles with an energy release of (94.5 ± 1.4) keV. This has been established by the (γn), (pd) and $(p\alpha)$ reactions producing Be^8 as a residual nucleus; references are given by JONES *et al.*[3]. The lifetime is less

[1] P. STAHELIN: Helv. phys. Acta **26**, 691 (1953). See also D. GREEN and J. R. RICHARDSON: Phys. Rev. **101**, 776 (1956) who were unable to find evidence for the expected $M3$ transition in K^{38} and suggest that the $T=0$ level is the ground state in this nucleus, as in P^{30}.

[2] S. MOSZKOWSKI and D. C. PEASLEE: Phys. Rev. **93**, 455 (1954).

[3] K. W. JONES, D. J. DONAHUE, M. T. MCELLISTRTREM, R. A. DOUGLAS and H. T. RICHARDS: Phys. Rev. **91**, 878 (1953).

than $4\cdot 10^{-15}$ sec according to TREACY[1] who could not detect any energy loss before break-up when recoil Be^8 nuclei entered a stopping foil. From the energy spectrum of alpha particles in the break-up of Be^8 produced in the $B^{11}(p\alpha)$ reaction, TREACY[4] was also able to infer an angular correlation and to demonstrate

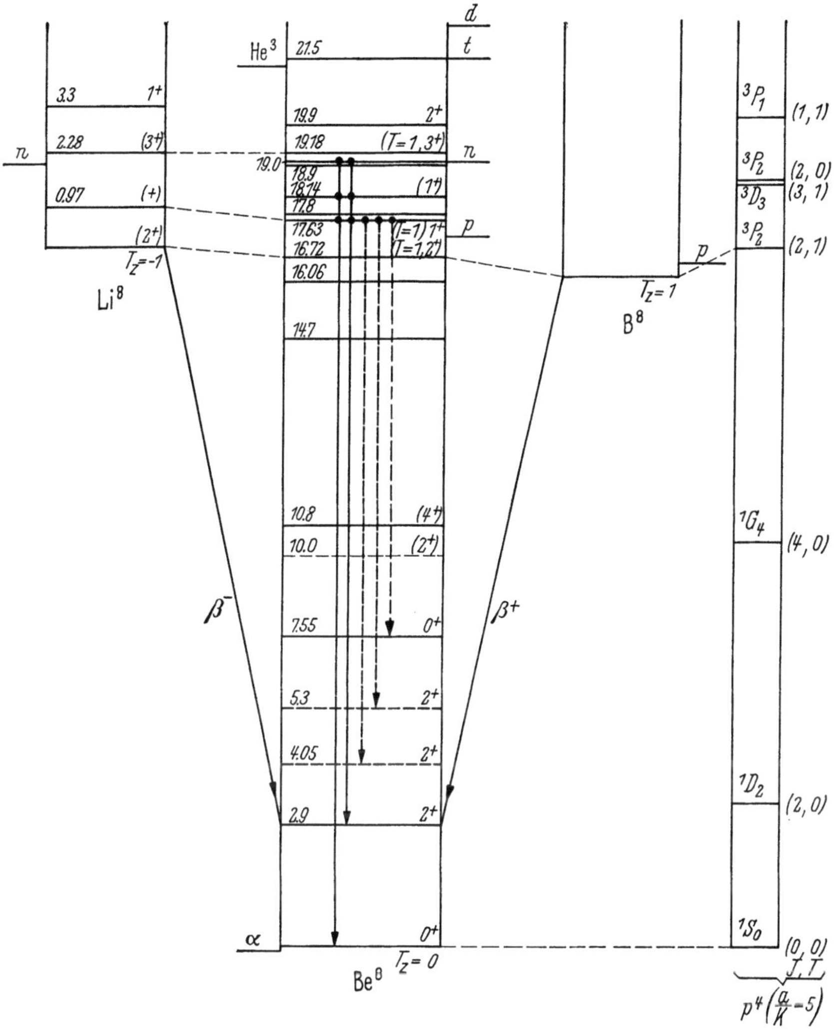

Fig. 95. Levels of nuclei with $A=8$.

that the spin of the ground state is zero, which is consistent with the observed lifetime and energy.

Most of the levels of Be^8 below the proton dissociation energy have even spin and parity and are able to break up into alpha particles. The $4n$ nuclei have $T=1$ states at rather high excitation because the (0, 0, 0) supermultiplet, to which the low states belong, does not contain $T=1$. This means that the change of a neutron into a proton in a $4n$ nucleus necessarily implies the entry of one of the particles into a different spatial state, and hence a decrease in binding energy. The $T=1$ states, if they have the right spin and parity, are not pre-

[1] P. B. TREACY: Proc. Phys. Soc. Lond. A **68**, 204 (1955).

cluded from dissociation into alpha particles, because of isotopic spin impurity, which is rather likely at the excitation of about 15 MeV involved and alpha break-up may be more probable than radiation. This appears to be the case for the first $T=1$ level of Be^8 at 16.72 MeV which is deduced from observations of the $C^{12}(\gamma\alpha)$ Be^8 and $O^{16}(\gamma\alpha)$ reactions and appears as a slow neutron threshold in $Li^7(dn)$. The angular correlation of the alpha particles in the triple stars (Sect. 55) suggests $J=2^+$ for the level.

Of the lower levels of Be^8 the best known is that at 2.9 MeV which has been found in the $(\alpha\alpha)$, $(\gamma\alpha)$, $(p\alpha)$, $(d\alpha)$, $(p\gamma)$, (γn), (pd), (dt) and (dn) reactions as well as in the decay of Li^8 and B^8. The width of this level is about 1 MeV. The spin is 2^+ from the angular correlation of alpha particles in the $(\gamma\alpha)$ stars for photon energies near threshold ($M1$ and $E2$ absorption), from the shape of the alpha particle spectrum to the 2.9 MeV state in the $B^{10}(d\alpha)$ reaction, and from angular correlations in the $B^{11}(p\alpha)$ reaction. The reduced width of the level, obtained from dispersion theory analysis of the $B^{10}(d\alpha)$ peak, gives a value twice the sum rule limit, suggesting that the usual compound nucleus picture does not give a full description of this state. A level at 2.2 MeV is reported from the (dn) and (γt) reactions.

A useful summary of the properties of other levels of Be^8 up to 7.5 MeV is given by TITTERTON[1]. Weak transitions in the $Li^7(p\gamma)$ and $Li^7(dn)$ reactions suggest the levels at 4.05, 5.3 and 7.5 MeV, of which the first two are also seen in the (pd) reaction. They were not observed by HOLLAND *et al.*[2] using the $(p\alpha)$ and $(d\alpha)$ reactions. The 7.5 MeV level is confirmed by $(\alpha\alpha)$ scattering which also demonstrates a level at 10.8 MeV ($J=4^+$) and may be influenced by a wide state near 10 MeV. Levels at 10 MeV and 14.7 MeV appear in the (dn) reaction.

The Be^8 levels above 16.72 MeV are obtained from the Li^7 + proton reactions up to 21.5 MeV and then from broad resonances in the Li^6 + deuteron reactions. The level at 17.63 MeV is the well known resonance for high energy gamma radiation at 441 keV. It is ascribed to $T=1$ tentatively because it apparently takes part in the $E1$ absorption in the $C^{12}(\gamma\alpha)$ process and by analogy with the first excited state of Li^8. The 18.14 MeV state, observed in the $(p\gamma)$, (pp) and (pp') reactions also seems to be formed by p-wave protons and to have $J=1^+$ according to the elastic scattering analysis. Anomalies in elastic scattering of 2.1 MeV protons may imply a 3^+ state at 19.18 MeV and this is possibly a $T=1$ state by analogy with the second excited state of Li^8. The broad state at 19.9 MeV is found in the excitation function of $Li^7(p\alpha)$ and is assigned to $J=2^+$ from the angular distribution analysis for this reaction (Sect. 27).

The beta decay of Li^8 and B^8 involves a change of supermultiplet, although superallowed decay of B^8 to a $T=1$ state might just be possible energetically. The Li^8 decay is mainly to the 2.9 MeV state with an ft value which corresponds to an allowed but unfavoured transition. This, taken together with the evidence from the $Li^7(dp)$ stripping reaction and with theoretical expectation, indicates a spin of 2^+ for the Li^8 ground state. This is also the spin suggested for the possible analogue state in Be^8. It is consistent with the isotropic correlation[3] between the alpha particles from the 2.9 MeV state of Be^8 and the beta rays from Li^8 emitted in the transition to this state. The decay of B^8 also proceeds mainly to

---

[1] E. W. TITTERTON: Phys. Rev. **94**, 206 (1954).

[2] R. E. HOLLAND, D. R. INGLIS, R. E. MALM and F. P. MOORING: Phys. Rev. **99**, 92 (1955). This paper gives references to earlier work.

[3] S. S. HANNA, E. C. LA VIER and C. M. CLASS: Phys. Rev. **95**, 110 (1954).

the 2.9 MeV level of Be^8; the alpha particle spectrum obtained from the products of high energy induced stars in nuclear emulsions resembles that[1] from Li^8.

The Li^8 levels are obtained from the $(d p)$ reaction, from the excitation function for inelastic scattering of neutrons and from the total neutron cross section. There is evidence for s-wave interaction, leading probably to a 1^- level. An odd parity level of this spin also seems to exist in a nearly analogous position in Be^8 according to the results of a study of the $Be^7(n p)$ reaction (Sect. 41).

The lowest states of the (0, 0, 0) supermultiplet in LS coupling are ${}^1S$, ${}^1D$ and ${}^1G$ in this order; all with $T=0$. The lowest state with $T=1$ is ${}^3P$. In jj coupling for forces of short range the lowest states of $p^4_{3/2}$ are $(J, T)=(0, 0)$ $(2, 0)$ $(4, 0)$ followed by (2, 1) which is the same as in the LS extreme. This order is preserved for all degrees of intermediate coupling. INGLIS [10] discusses the Be^8 spectrum in terms of an intermediate coupling parameter $a/K=5$ chosen to fit the ground state and first $T=1$ state of Be^8. The spin 2 and spin 4 states are then in fairly good agreement with the known levels of spin 2 and 4 at 2.9 and 10.8 MeV. The spins of the Li^8 levels are also predicted although the level spacing is wrong. The reduced widths of the 19.18 state of Be^8 and its analogue in Li^8 are in the ratio expected for isobaric states (ADAIR[2]). The reduced width of the 17.63 MeV state supports the assumption of intermediate coupling [19], although there is evidence (Sect. 30) for s-wave proton interaction with Li^7.

The possibility of describing the level system of Be^8 in terms of the rotational states of a simple alpha particle model has been discussed by HAEFNER[3] following earlier discussions by WHEELER and by HAFSTAD and TELLER (Sect. 4). The levels then have energies, assuming a rigidly rotating nucleus in the form of a diatomic molecule, given by

$$E_J = J(J+1)\frac{\hbar^2}{2I}. \tag{76.1}$$

The requirement of BOSE statistics is that the wave function shall not change for a rotation of 180° and this means that $J=0, 2, 4, \ldots$ in agreement with experiment. Probably the lowest states of Be^8 are described by a wave function which incorporates features of both the central model, in which nucleons are freely exchanged, and the alpha-model in which there are clusters with an appreciable lifetime on the nuclear scale. The higher states of Be^8 cannot be described by the alpha model; in some cases spin and parity prevent break-up into the two-alpha system.

77. $A=12$: B^{12}, C^{12}, N^{12}, Fig. 96. The lower levels of C^{12} are found from the $(d n)$, $(d\alpha)$, (αn) and $(p\alpha)$ reactions as well as in the inelastic scattering of protons, deuterons, neutrons, alpha particles and electrons. All levels except that at 4.43 MeV and the ground state are energetically unstable against disintegration into $Be^8+\alpha$.

Above the proton dissociation energy the levels are obtained as resonances in the reactions of boron with protons, notably the $(p\alpha)$, $(p\gamma)$ and $(p n)$ processes. The gamma radiation from the $(p\gamma)$ and (αn) reactions verifies the position of the first excited state. Many levels above 18.3 MeV are also suggested by breaks in the excitation function for the (γn) photoprocess; such levels probably represent $E1$ absorption and may have $J=1^-$ and $T=1$ in accordance with the general behaviour of photodisintegration cross sections. Further levels near 25 MeV appear as resonances in the $B^{10}(d n)$ reaction.

[1] F. C. GILBERT: Phys. Rev. **93**, 499 (1954).
[2] R. K. ADAIR: Phys. Rev. **96**, 709 (1954).
[3] R. R. HAEFNER: Rev. Mod. Phys. **23**, 228 (1951).

The first excited state of C^{12} at 4.43 MeV has spin 2^+. This follows from angular correlation measurements between alpha particles and gamma radiation in the $N^{15}(p\alpha)$ and $N^{14}(d\alpha)$ reactions and from the triple correlations in the $B^{11}(p\gamma_1\gamma_2)$ reaction, leading to a cascade through the level. Supporting evidence comes from the angular distribution of gamma radiation in the (pp') reaction, from the determination of the internal conversion coefficient of the 4.43 MeV transition, and from angular correlations in the $Be^9(\alpha n)$ reaction. Even parity

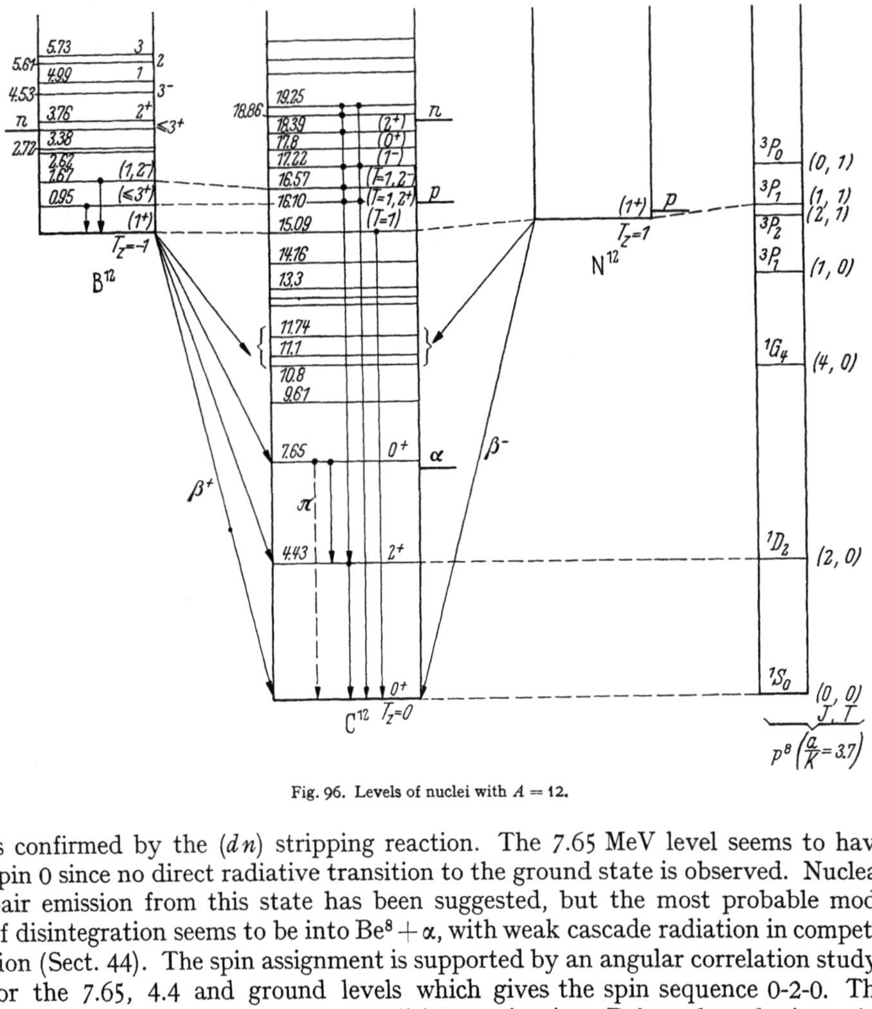

Fig. 96. Levels of nuclei with $A = 12$.

is confirmed by the (dn) stripping reaction. The 7.65 MeV level seems to have spin 0 since no direct radiative transition to the ground state is observed. Nuclear pair emission from this state has been suggested, but the most probable mode of disintegration seems to be into $Be^8 + \alpha$, with weak cascade radiation in competition (Sect. 44). The spin assignment is supported by an angular correlation study[1] for the 7.65, 4.4 and ground levels which gives the spin sequence 0-2-0. The 7.65 MeV level is just unstable to disintegration into $Be^8 + \alpha$ but the intensity of the cascade gamma transition compared with that of the neutron group in the $Be^9(dn)$ reaction leading to this stage suggests that the gamma cascade is the principal mode of decay. The heavy particle break-up must therefore be impeded either by COULOMB barrier or by an unusual admixture of $T=1$ wave functions, or both. No other spin or parity assignments for the levels of C^{12} are available until those excited by proton capture are reached. Near these, $T=1$ levels are expected, and the state analogous to the B^{12} ground state is probably that at 15.09 MeV, below the proton dissociation energy. This level, apart from

[1] J. E. SEED: Phil. Mag. **46**, 100 (1955).

its agreement in energy with the calculated position of the $T=1$ level, is found to radiate 15 MeV gamma rays in the high energy alpha particle bombardment of Be^9 and in the reactions $C^{12}+p$ (up to 340 MeV) and $B^{10}+d$ (up to 50 MeV). Radiation from a level so unstable with respect to alpha particle emission must indicate a strong selection rule which is provided if the state is of isotopic spin unity. This state is probably concerned in the $O^{16}(\gamma\alpha)$ reaction, which also seems to indicate levels at 9.6 and 11 MeV.

The levels at 16.10, 16.57 and 17.22 MeV are ascribed the spin and isotopic spin values shown from analysis of the angular distributions and correlations in the $B^{11}(p\gamma)$ and $(p\alpha)$ reactions (see Sect. 32), using the known properties of the Be^8 states. There is almost certainly a correspondence between these levels and the low lying states of B^{12}. Spin values for higher levels of C^{12} have been suggested, but are only provisional. Reduced widths for nucleon emission have been calculated for several even virtual levels of C^{12} and B^{12} and are in general agreement with the predictions of an intermediate coupling scheme (LANE and THOMAS) [*19*]. The odd-parity levels at 16.57 and 17.22 MeV may be described by the coupling of an $s_{1/2}$ nucleon to the ground state of B^{11}.

No levels are known for N^{12}, but the B^{12} levels are very well established by the (dp) and (αp) reactions and by resonances in total cross section for neutrons. Angular distributions have been measured in both reactions and spin values obtained. The beta transitions from B^{12} and N^{12} are both allowed and proceed partly to excited states; alpha emitting states are observed for N^{12} but not for B^{12} despite intensive search. The spin value of the ground states is thus probably 1^+ which is not in disagreement with the properties of the 15.09 MeV level in C^{12}.

The intermediate coupling picture of the levels of the mass 12 nuclei presented by INGLIS [*10*] reproduces only the general features of the observed levels. There are in LS coupling low states of type ^1S, ^1D and ^1G belonging to the (0, 0, 0) supermultiplet and well separated from each other and from the first levels of $T=1$ which appear in the (1, 1, 0) supermultiplet. In jj coupling the order is $(J, T)=(0, 0)$ $(2, 0)$ $(1, 0)$ and all states except the ground state arise from configurations such as $p^7_{3/2}\,p_{1/2}$ or $p^6_{3/2}\,p^2_{1/2}$. The introduction of intermediate coupling with $a/K=3.7$ does little to shift the levels from their positions in the LS extreme. The density of states after the first $T=1$ state is rather large, in agreement with experiment.

No low level of $J=0^+$ is predicted by these schemes other than the ground state, but it arises naturally as a dilational state in the alpha particle model. This model (Sect. 4) also predicts the 2^+ state but requires low lying states of $J=3^-$ and 1^- which have not been identified. The 0^+ state can also be obtained by double nucleon excitation, e.g. in the configuration $p^6_{3/2}\,p^2_{1/2}$. The transition probability of the 0^+ state for pair emission to the ground state can be estimated from the cross section for excitation of this state with electrons. SCHIFF[1] claims that the value obtained is too small to be explained by the alpha particle model and too large to be accounted for by a jj coupling model with excitation of two p-particles; he suggests that a collective model intermediate in properties between the two extreme models is necessary.

78. $A=16$: N^{16}, O^{16}, F^{16}, Fig. 97. No energy level of O^{16} is found below 6.06 MeV; this reflects the (double) closure of the p-shell at this nucleus. Only four bound levels are known, and although these have been observed only in the (pp'),

[1] L. I. SCHIFF: Phys. Rev. **98**, 1281 (1955). See also P. J. REDMOND: Phys. Rev. **101**, 751 (1956) who suggests simple excitation of a 1s-nucleon.

(nn'), (dn) and $(p\alpha)$ reactions, and in the decay of N^{16} their properties are extremely well known. The 6.06 MeV level is a 0^+ pair emitting state from whose observed lifetime the matrix element for a monopole transition between 0^+ states is found to be 3.8×10^{-26} cm². The spins, parities and lifetimes of the four lowest excited states of O^{16} are known from the reaction of F^{19} with protons (Sects. 24 and 36 and Table 3).

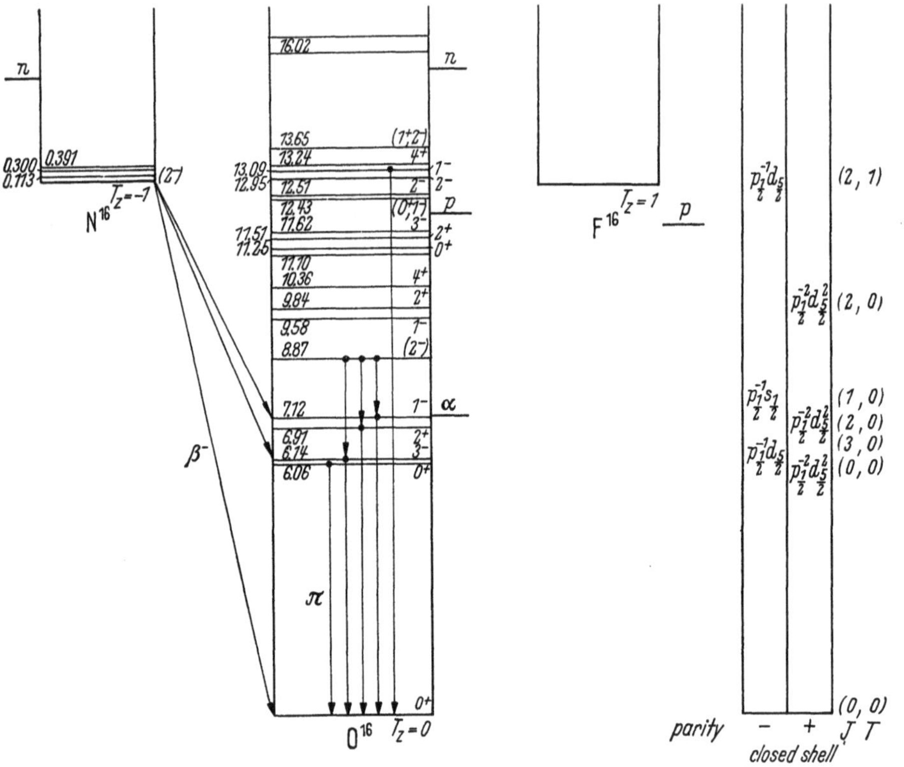

Fig. 97. Levels of nuclei with $A = 16$.

The 8.87 MeV state and several others of higher energy have been observed in the inelastic scattering of 19 MeV protons by O^{16} (HORNYAK and SHERR[1]). The level spectrum above 7.2 MeV and below the proton binding energy, has also been surveyed by the scattering of alpha particles by C^{12}. This can only demonstrate the existence of levels of even spin and even parity or odd spin and odd parity, and a number of such levels, shown in Fig. 97 were found. Reduced widths for alpha emission, many of them large, are known for these levels and suggest that there is some inherent tendency towards an alpha particle structure in O^{16}. Some of these levels are formed as residual states in the $N^{15}(dn)$ reaction and should break up into alpha particles.

From the decay energy of N^{16}, the first state of O^{16} with isotopic spin 1 is expected to lie at 12.9 MeV, just above the proton dissociation energy. The levels below this should be predominantly of $T=0$ character and the 7.12 MeV gamma transition to the ground state is forbidden by the electric dipole selection rule. The level concerned can also radiate by a cascade transition, and if the isotopic spin impurity of the states were nil, this would be the only permitted

[1] W. F. HORNYAK and R. SHERR: Phys. Rev. 100, 1409 (1955).

transition. In the fact the theoretical impurity is sufficient (RADICATI[1]) to make the intensity of the expected cascade transition too small to be detected (WILKINSON and JONES[2]). Other transitions between the first four excited states of O^{16} also show some irregularities of intensity which are in the sense expected from the isotopic spin rules (Sect. 19).

The levels between 12.43 and 13.65 MeV in O^{16} are obtained as resonances in the N^{15} + proton reactions and their angular momenta are derived from angular distributions and correlations in these reactions. The 13.09 MeV level is distinguished by the emission of gamma radiation to the ground state and by the alternative emission of alpha particles to the C^{12} ground state. This makes the assignment of isotopic spin rather uncertain. The other assignments shown in Fig. 97 are obtained from angular distribution and correlation measurements in the $(p\alpha)$ and $(p\alpha\gamma)$ reactions. Higher levels of O^{16} are demonstrated by breaks in the excitation function for the $(\gamma\alpha)$ and (γn) reactions corresponding to resonant absorption. The radiative width of some of these levels has been estimated.

No levels of the proton-unstable nucleus F^{16} are known. N^{16} levels are obtained from the (dp), $(d\alpha)$ and $(n\alpha)$ reactions. The beta decay of N^{16} to the O^{16} ground state is first forbidden, but transitions to the 7.12 and 6.14 MeV negative parity states are allowed, and the subsequent gamma radiation has been detected. The spin of N^{16} is therefore probably 2^-. The isotopic spin allowed transition between the 7.12 and 6.14 MeV levels was not observed[3] and the difference in intensity between the beta-spectrum components and the subsequent radiations is attributed to experimental uncertainties in the measurement of the former.

INGLIS [*10*] attempts to describe the lowest levels of O^{16} in terms of the shell model configurations $p_{1/2}^{-2}\, d_{5/2}^2$, $p_{1/2}^{-1}\, d_{5/2}$ and $p_{1/2}^{-1}\, s_{1/2}$, but the calculations of FLOWERS and EDMONDS [*14*] suggest that for the first of these the states of spin 4, 5, 6 might be lowest. Although a reasonable mixture of excited shell model configurations of even parity gives the observed value of the monopole transition probability for the 0^+ pair-emitting level at 6.06 MeV (ELLIOTT[4]) it is not yet clear why the energy of these configurations is so low. On the other hand the alpha particle model, although it fails to account well for the measured transition probabilities, gives a good description of the excited states. Thus DENNISON[5] is able by means of this model to correlate 16 of the levels in the range 0 to 13.25 MeV with theoretical prediction as to position, angular momentum and parity. In this treatment the O^{16} nucleus is considered as four alpha particles with equilibrium positions at the corners of a regular tetrahedron and the rotation-vibration levels are calculated by the methods of molecular physics. The 0^+ level arises as a dilational vibration and the 3^- state is one of rotation. The 6.91 and 7.12 levels which have spins 2^+ and 1^- originally constituted a difficulty for the alpha particle model since it seemed that the 2^+ level should be accompanied by a 2^- state and that both should arise from a vibrational mode which includes the possibility of inversion of the structure. A careful investigation of the level structure near 7 MeV by PETERSON *et al.*[6], using the $F^{19}(p\alpha\gamma)$ reaction, failed to reveal any 2^- level. In the latest analysis (DENNISON) the difficulty does not arise because of a different interpretation of the levels in terms of vibrational frequencies.

79. $A = 20$: F^{20}, Ne^{20}, Na^{20}, Fig. 98. The states below the proton dissociation energy in Ne^{20}, of which all but the lowest two are unstable against disintegra-

[1] L. A. RADICATI: Proc. Phys. Soc. Lond. A **67**, 39 (1954).
[2] D. H. WILKINSON, and G. A. JONES: Phil. Mag. **44**, 542 (1953).
[3] F. BOEHM, D. C. PEASLEE and V. PEREZ-MENDEZ: Phys. Rev. **90**, 149 (1953).
[4] J. P. ELLIOTT: Phys. Rev. **101**, 1212 (1956).
[5] D. M. DENNISON: Phys. Rev. **57**, 454 (1940); **96**, 378 (1954).
[6] R. W. PETERSON, W. A. FOWLER and C. C. LAURITSEN: Phys. Rev. **96**, 1250 (1954).

tion into $O^{16}+\alpha$, are observed in the (dn), (pp'), (dd') and $(p\alpha)$ reactions, in the resonant scattering of alpha particles by oxygen and in the decay of F^{20}. The stripping pattern for the ground state, obtained from the (dn) reaction, shows that F^{19} and Ne^{20} have the same parity, presumably even.

The lowest level at 1.63 MeV has been shown by angular correlation methods in the $Na^{23}(p\alpha\gamma)$ reaction[1] to have spin 2^+. The next levels, for which no spins are known, are at 4.36 and 5.4 MeV. Above these lie a group of states observed in the $O^{16}(\alpha\alpha)$ reaction for which spins, parities and reduced widths are available

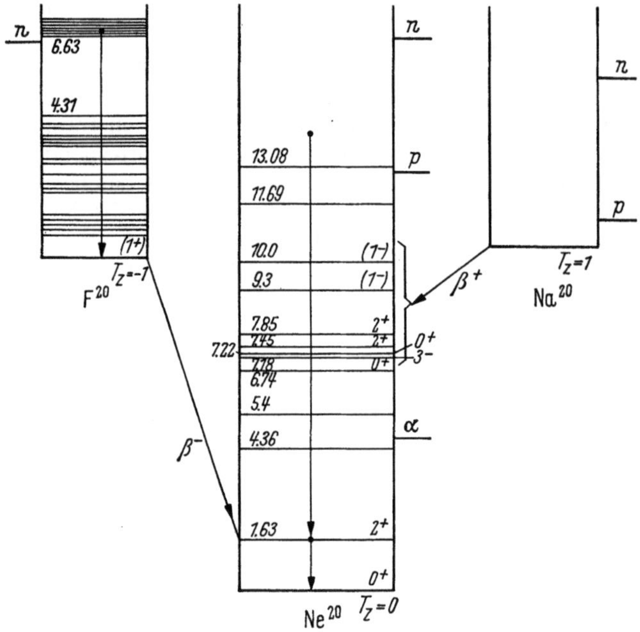

Fig. 98. Levels of nuclei with $A=20$. For levels see reference [32], for Ne^{20} also Fig. 36.

Further levels of which little is known occur in a not very well studied region up to 12.8 MeV, and then begins the extremely complex structure which has been thoroughly examined by means of the $F^{19}+$ proton resonances (Sect. 36 and Fig. 36).

Of these levels that at 13.51 MeV is known from the $F^{19}(p\alpha\gamma)$ reaction to have $J=1^+$ and the reason why the capture radiation at this resonance is to the 1.63 MeV level rather than to the ground state, is not obvious. The correlations observed in the resonance reactions establish the spin values shown in Fig. 36. At energies above 17 MeV, resonances continue and are chiefly observed in the (pn) reaction. The first state of $T=1$ is expected at 10.2 MeV so that the narrow $F^{19}+$ proton resonances must involve both $T=1$ and $T=0$ states. Admixture of $T=1$ wave functions in $T=0$ states may reduce the alpha emission probabilities and account for some of the apparently irregular variations of intensity of a particular radiation from one resonance to another.

No levels of Na^{20} are known, but for F^{20} there is unusually extensive information, from the (dp) and $(d\alpha)$ reactions, from the total neutron cross section of fluorine, from the $(n\gamma)$ reaction, observed by means of the F^{20} yield, and from the inelastic scattering of neutrons by F^{19}, resulting in a yield of 0.1 and 0.2 MeV radiation [32]. These observations confirm the density of levels already established

[1] J. SEED: Phil. Mag. **44**, 921 (1953).

in Ne^{20} and indicate that a high proportion of the Ne^{20} levels above 10 MeV have $T=1$. The density is too great for individual identifications. The (dp) stripping reaction suggests spin 1^+ for the F^{20} ground state; the pattern seems to be a superposition of $l_n=0$ and $l_n=2$ so that the ground state wave function is probably a mixture, as has already been established for other nuclei of similar mass.

The beta decay of F^{20} is a simple allowed (unfavoured) transition to the 2^+ excited state of Ne^{20} at 1.63 MeV; again it is not clear why the equivalent ground state transition does not take place, but the reason probably lies in the particular configurations. Thus if the F^{20} and Ne^{20} states were described by a particularly complex set of configurations while the Ne^{20} ground state was simply s^4, the results might be explained. The Na^{20} decay may contain a weak superallowed component to the first $T=1$ level of Ne^{20}.

The number of states of energy less than 25 MeV in the nuclei of mass 20 has been calculated by CRITCHFIELD and OLEKSA[1] using a simple model which leads to a level density formula for light nuclei.

80. $A=24$: Na^{24}, Mg^{24}, Al^{24}, Fig. 99. The bound levels of Mg^{24} have been very fully studied by inelastic scattering of protons and deuterons and by the (dn) reaction. In addition, the lower levels have been reported in the $(p\alpha)$ and (nn') reactions and are excited in the decay of Na^{24} and Al^{24} and in the cascade transitions from virtual levels.

The 1.368 MeV level is known to have $J=2^+$ from α-γ angular correlation measurement in the $Na^{23}(p\alpha\gamma)$ reaction. This agrees with the 0-2-4 type of gamma-gamma correlation pattern found for the successive transitions resulting from the decay of Na^{24} to the level at 4.122 MeV. Even parity is established for the levels of this cascade by a polarisation-correlation experiment. The radiative transitions are both $E2$ as expected from the spin values; this follows from measurements of the pair internal conversion coefficient (Sect. 21). There is no evidence of a cross-over transition from the 4.122 MeV level to the ground state (BEGHIAN *et al.*[2].)

Magnetic analysis of inelastically scattered protons from Mg^{24} shows that there is another level at 4.24 MeV[3]. A radiation of this energy, corresponding to the ground state transition, appears in the cascade radiation from the $Na^{23}(p\gamma)$ reaction and in the decay of Al^{24} (GLASS and RICHARDSON [*46*]). A gamma ray of energy about 2.88 MeV is also found in the former process and may be associated with the transition from the 4.24 MeV level to that at 1.37 MeV. The $Na^{23}(p\gamma)$ spectrum and the gamma rays from the Al^{24} decay also contain lines which suggest the presence of a level in Mg^{24} at about 5.3 MeV, although the interpretation of the transitions is not yet quite unambiguous. The level is probably also weakly excited in the decay of Na^{24} (BEGHIAN *et al.*) and decays mainly to the 1.37 MeV level. Angular distributions and angular correlations in the $Na^{23}(p\gamma)$ reaction determine the spin of the 4.24 MeV level as 2^+ and of the 5.3 MeV level as 3 with probably odd parity. Isotopic spin selection rules may operate to reduce the relative intensity of the $E1$ transition from the 5.3 MeV to the 1.37 MeV level. The 4.12 MeV level ($J=4^+$) is only weakly excited in the $Na^{23}+$ proton reaction.

The virtual levels of Mg^{24} have been found by elastic scattering of alpha particles in neon and from the $Na^{23}(p\alpha)$ and $(p\gamma)$ reactions. The levels obtained from the $(\alpha\alpha)$ reaction are all of the form $0^+\,1^-\,2^+\ldots$ while those reported from the $(p\gamma)$ reaction are $1^+\,2^-\,3^+$; the latter reaction can of course form states of

[1] C. L. CRITCHFIELD and S. OLEKSA: Phys. Rev. **82**, 243 (1951).
[2] L. E. BEGHIAN, G. R. BISHOP and H. HALBAN: Phys. Rev. **83**, 186 (1951).
[3] H. J. HAUSMAN, A. J. ALLEN, J. S. ARTHUR, R. S. BENDER and C. J. MCDOLE: Phys. Rev. **88**, 1296 (1952).

both $T=0$ and $T=1$. High energy capture radiation is not found at all the resonances in $Na^{23}+p$; for protons of more than 1 MeV there is also radiation from the (pp') and $(p\alpha')$ reactions leading to excited states of Na^{23} and Ne^{20}. It is reported [33] that there are 52 resonant states in the $Na^{23}+p$ reactions between 1.01 and 2.43 MeV proton energy. Above this, energy levels are indicated, though not with high resolution, by the (pn) reaction. The Na^{24} nucleus also shows a high level density. The states are obtained from the $Na^{24}(dp)$

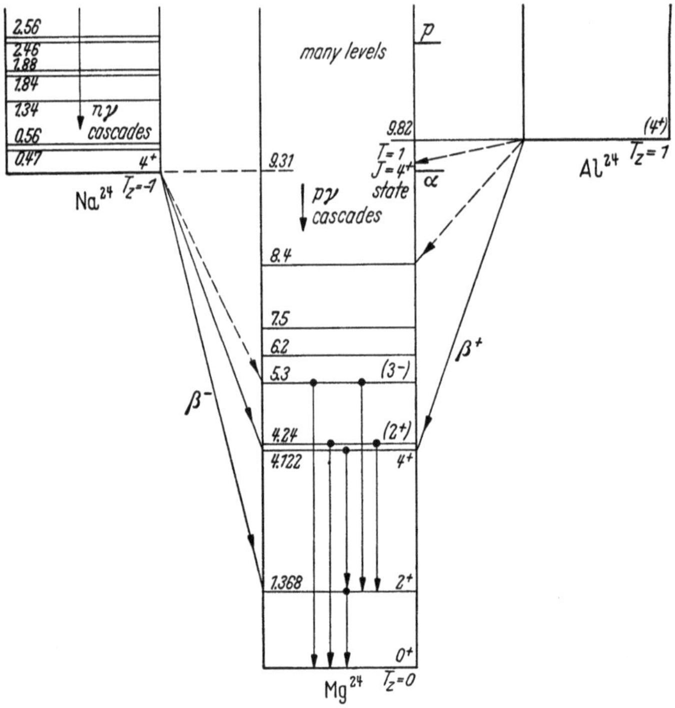

Fig. 99. Levels of nuclei with $A=24$.

reaction and from the $(n\gamma)$ process. With thermal neutrons it has been possible to observe and measure the energy of the gamma ray transitions between levels. The (dp) stripping distribution gives $l_n=2$ which agrees with the value required by the known spin $J=4$ of the Na^{24} ground state and implies even parity. There are 15 resonance levels between 7.0 and 7.9 MeV excitation in Na^{24}; this corresponds (Fig. 99) to an excitation of about 17 MeV in Mg^{24}, a region in which level structure is not well known. The decay of Na^{24} to the 4.12 MeV level of Mg^{24} is an allowed transition ($\Delta J=0$, *no*) although with rather a high ft value; there is a weak transition to the 1.37 MeV level which has $\Delta J=2$, *no* and is presumably second forbidden although the observed ft value does not accord with expectation (TURNER and CAVANAGH[1]). The transition to the 5.3 MeV level of Mg^{24} would be first forbidden ($\Delta J=1$, *yes*) with an energy release of about 200 keV, which would make it very improbable, as is actually found.

The Al^{24} nucleus is made by the (pn) reaction with a threshold of (15.4 ± 0.3) MeV. It decays with a half-life of 2.1 sec by positron emission to alpha unstable levels of Mg^{24} and also to levels below the alpha dissociation energy. The resulting gamma spectrum can be accommodated in the level scheme of Mg^{24} with

[1] J. F. TURNER and P. E. CAVANAGH: Phil. Mag. **42**, 636 (1951).

the suggestion of a level at 9.5 MeV which could be the analogue of the Na^{24} and Al^{24} ground states, and to which it seems likely that there is a superallowed positron transition. The positron transition to the 4.12 MeV level has an ft value consistent with that found for the mirror nucleus Na^{24}. Since the gamma ray spectrum includes a line at 4.22 MeV it seems that the cascades from higher levels lead to this state; since the Al^{24} spin should be 4^+ by analogy with Na^{24}, direct transitions to this level ($J=2^+$) are not favoured and would not have been observed. The possible origin of other gamma rays observed in the Al^{24} decay is indicated in Fig. 99. The position of the $T=1$ level in Mg^{24} calculated from the threshold energy for production of Al^{24} agrees with that deduced from the Na^{24} decay. The Al^{24} positron energy has not been measured but seems to extend to about 8.5 MeV.

The shell model configuration of Na^{24} is almost certainly complex. The possibility of the coupling of an s-neutron spin to that of a partially filled shell to give the doublets which are a notable feature of the Na^{24} level spectrum has been discussed by INGLIS [*10*].

81. $A=28$: Mg^{28}, Al^{28}, Si^{28}, P^{28}, Fig. 100. The bound states of the $T=0$ nucleus Si^{28} have been observed in the (dn) reaction, although the (pp') reaction and the Al^{28} decay confirm the levels at 4.47 and 1.78 MeV respectively. An alternative, and probably more accurate method of locating the lower levels is a study of the energies of the capture gamma rays from the $Al^{27}(p\gamma)$ reaction. The observed transitions can be fitted into the level scheme shown in Fig. 100 which has a general similarity to, but is not identical with, the level system deduced from the (dn) measurements. The $(p\gamma)$ spectra have been observed at discrete resonances and intensities and angular distributions of the individual gamma ray components have been obtained. Certain of these levels emit alpha particles to the ground state of Mg^{24} and must therefore have spins of the form $0^+, 1^-, 2^+ \ldots$. This information, together with the assumption of $J={}^5\!/_2{}^+$ for Al^{27} permits the assignment of spins and parities to the resonance levels and the bound states involved in the cascade transitions. The values suggested are shown in Fig. 100; the parities are confirmed by the angular distributions observed in the (dn) stripping reaction. The 9.16 MeV level seen in the (dn) reaction is identified, because of large proton capture probability, with the first $T=1$ state of Si^{28}. The virtual levels of Si^{28} are very numerous, and only a few have been studied in the detail just discussed. Resonances have been noted in the $(\alpha\alpha)$ and (αp) reactions and probably correspond to some of the 100 resonances observed between 11.807 and 15.555 MeV in the $Al^{27}+$ proton reactions. The density of levels in this region is so great that coincidence of levels of different spins is to be expected by chance. Levels of $T=0$ and $T=1$ are expected both to appear above 9.2 MeV and levels of $T=2$ (which may not radiate to $T=0$ states) at about 15 MeV. The resonance levels above 1300 keV proton energy decay by emission of inelastic protons as well as alpha particles and gamma radiation, and most of the radiation at the higher resonances arises in the (pp') reaction from states of Al^{27}. The complexity of this situation, in which the level spacing approaches the experimental resolution, suggests that interpretation via resonance theory is reaching its limit, and invites comparison of the experimental results for higher excitations and heavier nuclei with continuum theory.

A very similar situation exists in the $T_z=-1$ nucleus Al^{28} where extensive information on the level structure up to about 6 MeV is available from the (dp) reaction. The lower levels are also confirmed and defined by differences between gamma ray lines observed in the $(n\gamma)$ thermal capture reaction. The virtual levels due to fast neutron capture are not resolved, although structure is evident. The level spectrum of Al^{28} is known up to approximately the excitation of the first $T=2$ level. Approximately 50 levels ($T=1$) are seen by the (dp) reaction in this region, whereas certainly more than 100 ($T=0$ and 1) are known in the comparable region in Si^{28}. The results of high resolution angular distribution measurements show that the l_n values for the two components of the ground state doublet of 31 keV spacing are both zero. The neutron capture probability in the stripping

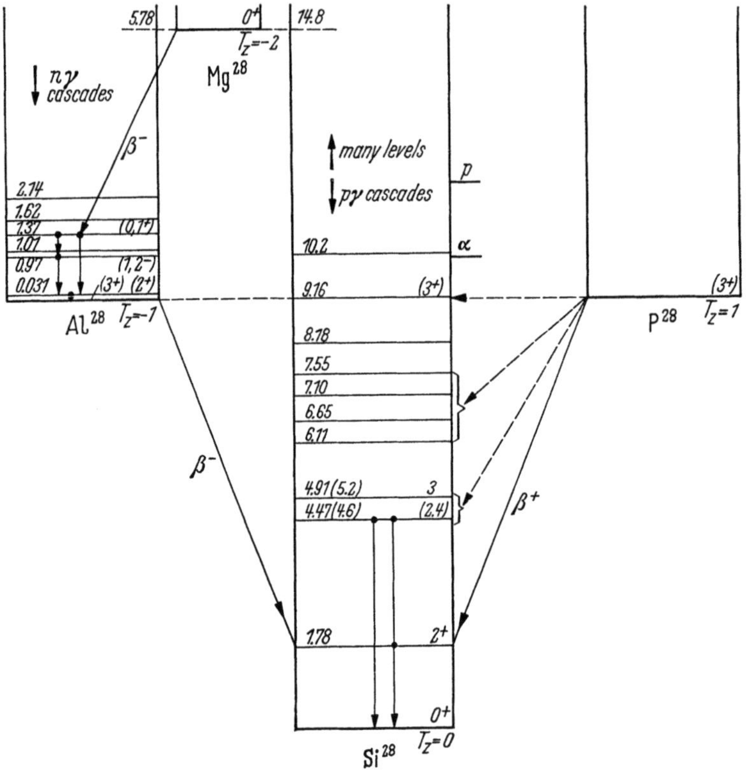

Fig. 100. Levels of nuclei with $A=28$.

reaction for these states is large, and since this may indicate excitation of a single particle state they probably form the jj doublet $d_{5/2}^{-1}\, s_{1/2}$. This postulates the coupling of an s-wave neutron to the hole in the $J=\frac{5}{2}$ proton shell to form states of $J=2^+, 3^+$; from the intensity ratio of the corresponding proton groups it appears that the 3^+ level is the ground state. The occurrence of jj double levels in Al^{28} is discussed by INGLIS [*10*] and by ENGE[1]. INGLIS also discusses an excited double level caused by the raising of the odd $d_{5/2}$ proton to a $d_{3/2}$ level, thus giving $d_{3/2}\, s_{1/2}$ with $J=1^+$ and 2^+. He further suggests triple levels from the configuration $d_{5/2}^{-2}\, s^2$ and it is likely that groups of levels just above 2 MeV may be so described. The same structure should of course be observed in Si^{28} above the 9.16 MeV level, but there is little chance of making reliable comparisons because of the high level density. The 9.16 MeV level of Si^{28} itself does seem to show the expected analogies with the ground state of Al^{28}, namely excitation energy about that predicted, large nucleon capture probability (although smaller than for the $T_z=-1$ isobar) and l-value zero; the doublet nature of the level could not be

[1] H. A. ENGE: Phys. Rev. **94**, 730 (1954).

established with the available resolution. The Al^{28} ground state doublet is also observed in the $Si^{30}(d\alpha)$ reaction.

No levels are known in P^{28} because of lack of suitable reactions; they should be directly comparable with those of Al^{28}. The beta-decay of Al^{28} $(J=3^+)$ takes place by an allowed transition ($\Delta J=1$, *no*) to the 2^+ excited state of Si^{28} at 1.78 MeV; the possible change of configuration is $d_{5/2}^{-1}\, s_{1/2}(J=3)\to d_{5/2}^{-2} s_{1/2}^{2}(J=2)$. The P^{28} nucleus [*33*] probably decays to its isobaric state at 9.16 MeV in Si^{28} by a super-allowed transition and it also decays to many of the lower states, but not to the ground state in observable intensity. The resulting spectrum of gamma radiation can be fitted into the level system of Si^{28}; as in the case of the $Al^{27}(p\gamma)$ cascades, most transitions are to the 1.78 MeV level and thence to the ground state. The ft value for the transition to this level is the same as that from the mirror nucleus Al^{28} to the same level.

The 21 hour β^--emitter Mg^{28} has been produced by the Mg^{26} $(\alpha\, 2p)$, Si^{30} $(\gamma\, 2p)$, Mg^{26} (tp), Si^{30} $(p, 3p)$ and spallation reactions [*33*]. The properties of this nucleus have been carefully studied by SHELINE *et al.*[1], who proved by "milking" experiments that decay into Al^{28} occurs. From these observations the decay scheme shown in Fig. 100 is deduced; the known levels of Al^{28} are adequate to account for the observed transitions, although there are not transitions to all the known levels (e.g. the 1.015 MeV level). By comparison of the intensities of the components of the Al^{28} gamma spectrum excited by Mg^{28} decay, the K-shell conversion coefficient of the 31 keV transition was measured (0.032 ± 0.006) and appears to determine this transition as $M1$; the lifetime of $<2\times 10^{-9}$ sec isin agreement with this conclusion. The allowed beta-transition from Mg^{28} to the 1.375 MeV level of Al^{28} determines the spin of this level as 0^+ or 1^+, and the intensity relations in the cascade suggest $J=1^-, 2^-$ or 2^+ for the 978 keV level. This cannot be checked from the (dp) stripping reaction because the 0.978 MeV component of the 1 MeV doublet does not give a stripping pattern although the other component has $l=2$. The fact that the 1.015 MeV level is not excited in the Mg^{28} decay whereas the lower component is, speaks against the interpretation of these states as the jj doublet $d_{3/2}\, s_{1/2}$ $(J=1^+, 2^+)$. The intensity ratio of the proton groups in the (dp) reaction is further evidence against this hypothesis, since it differs largely from the simple statistical factor expected for levels belonging to the same configuration.

82. $A=32$: Si^{32}, P^{32}, S^{32}, Cl^{32}, Fig. 101. The nuclei of mass 32 resemble those of $A=28$ in general features; all the nucleons concerned are expected to be in the d or s shell in the lowest states.

The S^{32} levels are obtained from the (dn), (pp') and (nn') reactions below the proton binding energy, and above this energy from the resonant reactions of protons with phosphorus. No accurate studies of the cascade radiations in the $(p\gamma)$ reaction have been made, but some spin assignments for resonance levels are available [*33*].

The levels of P^{32} are known more fully from high resolution analysis of the (dp) reaction and from the gamma ray spectrum observed in thermal neutron capture. The virtual levels excited by neutron scattering are not completely resolved. There is a ground state doublet of spacing 77 keV, which could be a jj doublet of the configuration $d_{3/2}\, s_{1/2}$; higher levels also show doublet structure. The allowed shape of the beta spectrum in the decay to S^{32} establishes the spin of the ground state of P^{32} as 1^+, although the ft value is very large, and the 77 keV state is presumably one of 2^+. The (dp) reaction forms P^{32} by adding a 17th neutron to P^{31} $(J=\frac{1}{2})$ to form a spin 1 state. This could be done with orbital momentum zero

[1] R. K. SHELINE, N. R. JOHNSON, P. R. BELL, R. C. DAVIS and F. K. MCGOWAN: Phys. Rev. **94**, 1642 (1954).

but the (dp) stripping reaction to the ground state doublet shows clearly $l_n = 2$ in agreement with the $d_{3/2}$ neutron required by the shell model. The intensity ratio for the ground state doublet components should be proportional then to $2J+1$ and the high resolution (dp) experiment confirms this significant property of a jj doublet. This doublet has also been observed in the (np) reaction. The $T=1$ levels of S^{32} predicted from the P^{32} spectrum are first found above 6.6 MeV, in a region where the S^{32} spectrum has not been well investigated but they extend

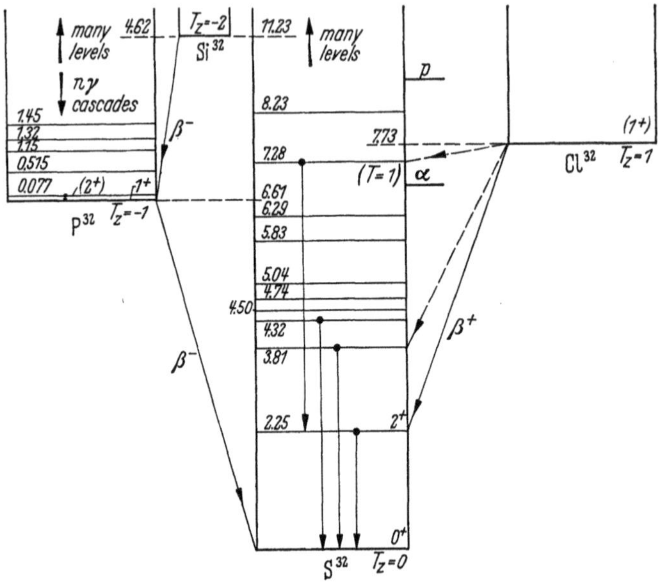

Fig. 101. Levels of nuclei with $A = 32$.

into the distribution of close resonance levels known from the P^{31} + proton reactions. Many of these levels must be presumed to have $T=1$, although isotopic spin is unlikely to be a good quantum number for such high excitations in a nucleus of such high charge.

Cl^{32} has been produced by the (pn) reaction [46] and from the threshold energy the first $T=1$ state in S^{32} should be between 7 and 8 MeV. The gamma radiation observed in the decay of Cl^{32} can be accounted for by transitions between known levels of S^{32} including the $T=1$ level. The ft value for the beta-decay to this level would be that of a super-allowed transition, whereas the ft value for the transition to the 2.25 MeV 2^+ level is that of an ordinary allowed transition. The mirror transition of P^{32} to this level is energetically forbidden and the P^{32} — S^{32} ground state transition seems to be specially hindered because of the configuration change $d_{3/2}\, s_{1/2} \to s_{1/2}^2$. The Cl^{32} spin should be 1^+; weak decay to highly excited alpha emitting states of S^{32} is observed.

The nuclide Si^{32} was found by LINDNER[1] in the spallation products of chlorine bombarded by high energy protons and by TURKEVICH and SAMUELS[2] in neutron irradiated quartz. The beta decay energy is found to be about 100 keV, and the lifetime about 700 years. This locates the lowest $T=2$ level of S^{32} at an excitation of 11.23 MeV. The decay is probably to the ground state of P^{32} by a transition which is hindered in the same way as the P^{32} — S^{32} ground state transition. The configuration change would be $d_{3/2}^2 \to d_{3/2}\, s_{1/2}$.

[1] M. LINDNER: Phys. Rev. **91**, 642 (1953).
[2] A. TURKEVICH and A. SAMUELS: Phys. Rev. **94**, 364 (1954).

83. $A = 36$: S^{36}, Cl^{36}, A^{36}. In the Cl^{35} $(p\gamma)$ reaction many sharp resonances have been observed [33]. Cl^{36} has been studied through the (dp) reaction and through the cascade gamma radiation resulting from the capture of slow neutrons by Cl^{37}. The level systems deduced by these two methods are in general agreement. The fast neutron resonances giving virtual levels of Cl^{36} are incompletely resolved.

The ground state spin of Cl^{36} has been measured as 2, and according to the shell model the parity is even, since the $d_{3/2}$ sub-shell is filling. The formation of Cl^{36} from Cl^{35} $(J = {}^3/_2{}^+)$ therefore only requires the absorption of an s-neutron. The (dp) stripping distribution, however, indicates $l_n = 2$ as required by the shell model.

In the nuclei of mass 32, the $T = 2$ nucleus is just unstable with respect to that of $T = 1$, and for mass 28 this instability is greater. At mass 36 however S^{36} $(T = 2)$ has become stable with respect to Cl^{36} $(T = 1)$, and K-capture disintegration of the latter into the former is energetically possible. The main disintegration of Cl^{36} is to A^{36} by a second forbidden transition ($\Delta J = 2$, *no*) in which two $d_{3/2}$ nucleons with $J = 2$ rearrange to yield $J = 0$. The energy change enables the position of the first state of $T = 1$ in A^{36} to be calculated. The mass of S^{36} is known and this locates the first state of $T = 2$ in Cl^{36} and A^{36}. The magnetic moment of Cl^{36} has been measured directly[1] and supports the configuration $(d_{3/2})_p\,(d_{3/2})_n$, $J = 2^+$ which is needed to give the observed spin.

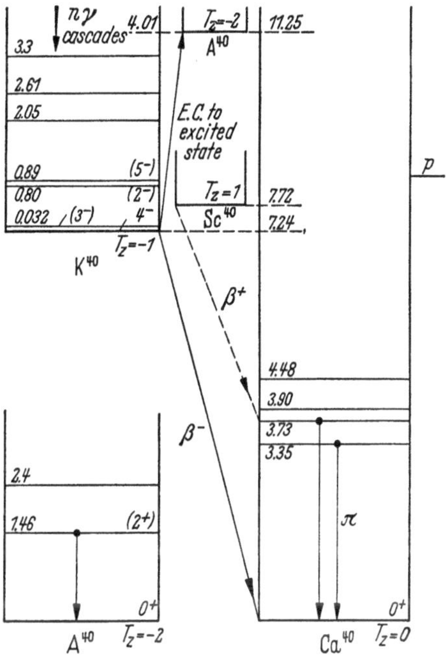

Fig. 102. Levels of nuclei with $A = 40$.

84. $A = 40$: A^{40}, K^{40}, Ca^{40}, Sc^{40}, Fig. 102. In the nuclei of mass 40 the odd parity $f_{7/2}$ shell is filling. The level structure of Ca^{40} is known from the (pp') reaction [33]. One of the levels is thought to be a 0^+ level because of pair emission to the 0^+ ground state (BONNER *et al.*[2]). The K^{40} levels are obtained from the (dp) reaction and from the cascade gamma transitions following capture of thermal neutrons; there are also some resonances in total neutron cross section. The spin of K^{40} has been measured directly and is 4; the configuration $d_{3/2}^{-1}\,f_{7/2}$ suggests odd parity. This configuration would predict spins 2^-, 3^-, 4^-, 5^-, one of which would be the ground state; the others are probably the levels at 32, 800 and 890 keV. The order of the spins shown in Fig. 102 is determined by the intensity of the corresponding proton groups which should be proportional to $2J + 1$ Levels of A^{40} are excited in the (pp') reaction and in the decay of K^{40}.

The nucleus K^{40} is unstable with respect to decay into both its even-even isobars. Because of its high spin the corresponding beta transitions are highly forbidden, and the lifetime is 14.6×10^8 yr. The β^- transition is to the ground state of Ca^{40} and the electron capture transition to the first excited state (probably

[1] P. B. SOGO and C. D. JEFFRIES: Phys. Rev. **98**, 1316 (1955), L. C. AAMODT and P. C. FLETCHER: Phys. Rev. **98**, 1317 (1955).

[2] T. W. BONNER, R. D. BENT and J. H. MCCRARY: Phys. Rev. **98**, 1198A (1955).

$J = 2^+$) of A^{40}; the branching ratio N_K/N_{β^-} is about 0.13[1]. In both cases there is a considerable change of configuration; $d_{3/2}^{-1} f_{7/2} \rightarrow$ closed shell and $d_{3/2}^{-1} f_{7/2} \rightarrow f_{7/2}^2$. The decay energies permit formal identification of the position of $T = 1$ and $T = 2$ levels in Ca^{40}.

The $T_z = +1$ nucleus Sc^{40} has been produced by the (pn) reaction; it decays by high energy positron emission and a gamma ray of 3.75 MeV to the ground state of Ca^{40}. From its threshold for production the position of the isobaric level in Ca^{40} should be at 7.72 MeV. This is in fair agreement with the estimate of this position obtained from the K^{40} decay.

Acknowledgments.

The advice of Drs. M. BOLSTERLI, G. E. BROWN, P. T. MATTHEWS and E. VOGT on the preparation of parts of this article is gratefully acknowledged. The author also wishes to thank Miss M. V. LAWRENCE for help with the calculations, Miss D. LATHAM for preparing sketches and Dr. K. K. DAMODARAN for proof-reading.

General references.

References—Introduction.

[1] BETHE, H. A., and R. F. BACHER: Nuclear Physics A (Stationary states of Nuclei). Rev. Mod. Phys. **8**, 82 (1936). — BETHE, H. A.: Nuclear Physics B (Nuclear Dynamics, Theoretical). Rev. Mod. Phys. **9**, 69, (1937). — LIVINGSTON, M. S., and H. A. BETHE: Nuclear Physics C (Nuclear Dynamics, Experimental). Rev. Mod. Phys. **9**, 245 (1937). — This is the basic review article, which for many years was the only general account of the subject available. Part C contains a very useful discussion of the earlier experimental methods and many tables and graphs of nuclear data.

[2] DEVONS, S.: The Excited States of Nuclei. Cambridge: University Press 1949. — An excellent monograph both on experimental methods and theoretical interpretation.

[3] CAPPELLER, U.: The Energy level schemes of the light nuclei. Ergebn. exakt. Naturw. **25**, 545 (1951); **27**, 125 (1953). — A comprehensive survey of level systems and reaction data for the nuclei up to Cl^{36}. There is an experimental introduction. The information is presented in tables and diagrams.

[4] BURCHAM, W. E.: The low-lying excited states of light nuclei. Progress in Nuclear Physics, Vol. 2. London: Pergamon Press 1952. — A short article discussing experimental methods and results for nuclei up to Mg^{24}.

[5] LAURITSEN, T.: The energy levels of light nuclei. Ann. Rev. Nucl. Sci. **1** (1952). — A survey of energy level schemes for nuclei up to $A = 20$, with an experimental introduction. The diagrams compare the levels of isobaric nuclei.

[6] SEGRÈ, E. (Ed.): Experimental Nuclear Physics, Vol. II. New York: John Wiley 1953.

This volume contains: (a) A Survey of Nuclear Reactions by P. MORRISON. This is a valuable account of nuclear reaction theory presented in an elementary way, with good discussions of typical reactions. Many graphs of penetrability functions are given. — (b) The Neutron by B. T. FELD. A detailed account of neutron physics including detection methods, and nuclear reactions.

[7] BLATT J. B., and V. WEISSKOPF: Theoretical Nuclear Physics, New York: John Wiley 1953. — A comprehensive account of general nuclear theory with a chapter on application to experiments. It is now the standard reference for radiation theory.

General references A.

[8] ROSENFELD, L.: Nuclear Forces. Amsterdam: North-Holland Publishing Company 1948. — This excellent book has become the standard reference in the theory of nuclear interactions.

[9] WIGNER, E. P., and E. FEENBERG: Rep. Prog. Phys. **8**, 274 (1941). This report reviews the fundamental work of the authors on the symmetry properties of nuclear states. Binding energies (for LS coupling) are given in the following papers: FEENBERG, E., and E. P. WIGNER: Phys. Rev. **51**, 95 (1937). — FEENBERG, E., and M. PHILLIPS: Phys. Rev. **51**, 597 (1937).

[10] INGLIS, D. R.: The energy levels and the structure of light nuclei. Rev. Mod. Phys. **25**, 390 (1953). — A clear and readable account of the theoretical interpretation of energy level sequences in p-shell nuclei.

---

[1] G. J. WASSERBURG and R. J. HAYDEN: Phys. Rev. **93**, 645 (1954).

[11] FLOWERS, B. H.: The nuclear shell model. Progress in Nuclear Physics, Vol. 2. London: Pergamon Press 1952. An account of the empirical evidence leading to the shell model, and of the classification of nuclear states in both LS and jj coupling. PRYCE, M. H. L.: Nuclear shell structure. Rep. Progr. Phys. **17**, 1 (1954).

[12] FEENBERG, E.: The shell structure of the nucleus. Princeton: The University Press 1954. — MAYER, M. G., and J. H. JENSEN: Elementary theory of nuclear shell structure. New York: John Wiley 1955.

[13] KLINKENBERG, P. F. A.: Rev. Mod. Phys. **24**, 63 (1952). — A convenient summary of the allocation of nucleons to shell model states.

[14] EDMONDS, A. R., and B. H. FLOWERS: Proc. Roy. Soc. Lond., Ser. A **215**, 120 (1952). — KURATH, D.: Phys. Rev. **88**, 804 (1952); **91**, 1430 (1953). — TALMI, I.: Helv. phys. Acta **25**, 185 (1952). These papers discuss the excited states of nuclei in jj coupling. The papers of KURATH include the calculation of magnetic moments, COULOMB energies and ft values. — FLOWERS, B. H.: Phil. Mag. **43**, 1330 (1952). A calculation of magnetic moments and electric quadrupole moments according to the individual particle model.

General references B.

References [1], [2], [6] and [7] discuss the theory of nuclear reactions.

[15] PEIERLS, R.: The BOHR theory of nuclear reactions. Rep. Progr. Phys. **7**, 87 (1940). — A general review of the theory of the compound nucleus.

[16] FERMI, E.: Nuclear Physics (Lecture Notes). Chicago: The University Press 1950. — Chap. VIII is a very clear analytical account of the theory of nuclear reactions.

[17] FRIEDMAN, F. L., and V. F. WEISSKOPF: The compound nucleus. Essay in: NIELS BOHR and the Development of Physics. London: Pergamon Press 1955. — A brief but excellent account of the general development of nuclear reaction theory in which the "optical" model is presented as a development of the original ideas of BOHR.

[18] VOGT, E.: The widths and spacings of nuclear resonance lines. Report number NDA — 14. of the Nuclear Development Corporation of America. A comprehensive list of reduced widths and level spacings, extending an earlier compilation of TEICHMANN and WIGNER: Phys. Rev. **87**, 123 (1952).

[19] LANE, A. M., and R. G. THOMAS: The Theory of Nuclear Reactions and the Compound Nucleus. Rev. Mod. Phys. to be published in **1957**. — A very full treatment of the dispersion theory of nuclear reactions, particularly valuable for its treatment of reaction constants in various coupling schemes. Comparison with experimental information is made; also LANE, A. M.: The reduced widths of nuclear energy levels, A.E.R.E. Report T/R 1289 (1954). (United Kingdom Atomic Energy Authority.)

General references C.

[20] SEGRÈ, E. (Ed.): Experimental Nuclear Physics, Vol. I. New York. John Wiley 1953. — Includes an authoritative account of a number of experimental methods. Chapters relevant: Detection methods (STAUB), Passage of radiations through matter (BETHE and ASHKIN) Charged Particle Dynamics and Optics (BAINBRIDGE). Many instruments described in this Part are discussed in detail. BAINBRIDGE gives isotopic weights, nuclear reaction Q values.

[21] SIEGBAHN, K. (Ed.): Beta- and Gamma-Ray Spectroscopy. Amsterdam: North Holland Publishing Company 1955. — An excellent collection of authoritative articles dealing with the whole field from both experimental and theoretical standpoints. Chapters on gamma radiation from charged particle reactions and neutron capture gamma rays are included; and full references to earlier review articles are given.

[22] ALLISON, S. K., and S. D. WARSHAW: Rev. Mod. Phys. **25**, 779 (1953). — A useful collection of information on stopping powers, cross sections for charge exchange and general behaviour of ions in their passage through matter.

[23] DEVONS, S., and M. G. N. HINE: Proc. Roy. Soc. Lond., Ser. A **199**, 56, 73 (1949). — These papers were the first to show clearly how the angular distribution of gamma radiation from light nuclei could be interpreted theoretically. Experimental results for proton capture reactions in Li^7, Be^9, C^{12}, C^{13} and F^{19} are given.

[24] WILKINSON, D. H.: Radiative transitions in light elements. Phil. Mag. **1**, 127 (1956). A compilation of observed radiative widths for $Z \leq 10$, with discussion of the shell-model interpretation.

[25] Angular correlation of internally produced electron-positron pairs O^{16} (EO): DEVONS, S., G. GOLDRING and G. R. LINDSEY: Proc. Phys. Soc. Lond. A **67**, 134 (1954). — Be^8, B^{10}, O^{16}: DEVONS, S., and G. GOLDRING: Proc. Phys. Soc. Lond. A **67**, 413 (1954).

[26] The lifetime measurements listed in Table 3 are reported in the following papers: General techiques. Devons, S., G. Manning and D. St. P. Bunbury: Proc. Phys. Soc. Lond. A **68**, 18 (1955). — Li^7, Be^7. Bunbury, D. St. P., S. Devons, G. Manning and J. H. Towle Proc. Phys. Soc. Lond. A **69**, 165 (1956). See also Elliott, L. G. and R. E. Bell: Phys. Rev. **76**, 168 (1949). — B^{10}, O^{17}: Thirion, J. and V. L. Telegdi: Phys. Rev. **92**, 1253 (1953). — C^{12}, O^{16}, Ne^{20}, Si^{28}: Devons, S., G. Manning and J. H. Towle: Proc. Phys. Soc. Lond. A **69**, 173 (1956). — N^{14}: Thirion, J., and R. Barloutaud: C. R. Acad. Sci., Paris **240**, 2136 (1955). — O^{16} (π): Devons, S., G. Goldring and G. R. Lindsey: Proc. Phys. Soc. Lond. A **67**, 134 (1954). — F^{19}: Thirion, J., C. A. Barnes and C. C. Lauritsen: Phys. Rev. **94**, 1076 (1954). — Jones, G. A., W. R. Phillips, C. M. P. Johnson and D. H. Wilkinson: Phys. Rev. **96**, 547 (1954). — Mg^{24}: Coleman, C. F.: Phil. Mag. **46**, 1135 (1955). — K^{41}: Elliott, L. G.: Phys. Rev. **85**, 942 (1952).

[27] Christy, R. F.: Phys. Rev. **89**, 839 (1953). — This paper discusses the dependence of angular distributions on the type of coupling of angular momentum vectors in the nucleus. Several specific examples are given.

[28] Blatt, J. M., and L. C. Biedenharn: Rev. Mod. Phys. **24**, 258 (1952). — The angular distribution of scattering and reaction cross sections.

[29] Biedenharn, L. C., and M. E. Rose: Rev. Mod. Phys. **25**, 729 (1953). Theory of angular correlation of nuclear radiations. See also Sharp, W. T., J. M. Kennedy, B. J. Sears and M. G. Hoyle: Tables of Coefficients for Angular Distribution Analysis. Atomic Energy of Canada Limited Report AECL 97.

General references D.

[30] Wapstra, A. H.: Isotopic Masses. Physica, Haag **21**, 367, 385 (1955). — Gives atomic masses computed from nuclear reaction energies and mass spectroscopic data. These values are used in this article.

[31] van Patter, D. M., and W. Whaling: Nuclear Disintegration Energies. Rev. Mod. Phys. **26**, 402 (1954). — Lists ground state Q-values for heavy particle reactions together with weighted averages. These differ slightly from the values adopted by Wapstra.

[32] Ajzenberg, F., and T. Lauritsen: Energy Levels of Light Nuclei V. Rev. Mod. Phys. **27**, 77 (1955). — An invaluable collection of data on nuclear reactions and energy levels of nuclei up to $Z = 10$. Level diagrams are given for nearly all nuclei. In the Report of the Physics Department, Boston University, dated 30. Sept. 1954, these authors give diagrams of the relative level structure of isobars. These have been made the basis of the diagrams in this article.

[33] Endt, P. M., and J. C. Kluyver: Rev. Mod. Phys. **26**, 95 (1954). — Energy Levels of Light Nuclei. ($Z = 11$ to $Z = 20$). A continuation of the work of Reference [32].

[34] Cowie, D. B., N. P. Heydenburg and G. C. Phillips: Phys. Rev. **87**, 304 (1952). — This paper reports the results of experiments on the inelastic scattering of 7.4 MeV protons by several light nuclei and also gives references to similar work.

[35] Hughes, D. J., and J. A. Harvey: Neutron Cross Sections. Brookhaven National Laboratory Report BNL 325. This is a collection of total and reaction cross sections presented graphically as a function of energy. It extends and brings up to date the earlier report by Adair: Neutron Cross Sections of the Elements. Rev. Mod. Phys. **22**, 249 (1950). This latter report has a very clear introduction in which neutron scattering theory is summarised.

[36] Massey, H. S. W.: The collisions of deuterons with nucleons. Progress in Nuclear Physics, Vol. 3. London: Pergamon Press 1953. — A very clear discussion of both theory and experimental material.

[37] The energy of radiation from the capture of thermal neutrons in light nuclei is reported in the papers of B. B. Kinsey, G. A. Bartholomew and W. H. Walker as follows: Experimental technique. Canad. J. Phys. **31**, 537 (1953). — H^2. Phys. Rev. **80**, 918 (1950). — Be^9, C^{12}. Canad. J. Phys. **31**, 49 (1953). — N^{14}. Phys. Rev. **77**, 723 (1950). — F^{19}, Na^{23}, Mg, Al, Si. Phys. Rev. **83**, 519 (1951) and Canad. J. Phys. **31**, 901 (1953). — P, S, Cl, K, Ca. Phys. Rev. **85**, 1012 (1952). — Radiation widths. Phys. Rev. **93**, 1260 (1954).—The regularities in ($n\gamma$) cross sections are also discussed by L. V. Groshev: Dokl. Akad. Nauk. SSSR. **100**, 651 (1955).

[38] Bockelman, C. K., D. W. Miller, R. K. Adair and H. H. Barschall: Phys. Rev. **84**, 69 (1951). — This paper reports the energy variation of total neutron cross section for Li, Be, B^{10}, B, C and O which show many sharp resonances.

[39] Accurate energy measurements of gamma radiation produced in the nuclear reactions of deuterons and protons with light nuclei are reported in the following papers: (i) Thomas, R. G., and T. Lauritsen: Phys. Rev. **88**, 969 (1952). A lens spectrometer study of gamma radiation from the bombardment of Li^6, C^{12}, C^{13} and O^{16} by deuterons and from

$(p\,\alpha)$ reactions in B^{10} and N^{15}. COMPTON conversion spectra were observed. Results for the reactions $Li^7\,(p\,p')$ and $Be^9\,(d\,n)$ are given by W. F. HORNYAK, T. LAURITSEN and V. K. RASMUSSEN: Phys. Rev. **76**, 731 (1949). — (ii) TERRELL, J., and G. C. PHILLIPS: Phys. Rev. **83**, 703 (1951). A pair spectrometer study of gamma radiation from the bombardment of B^{10} B^{11} N^{14} and F^{19} with deuterons. — (iii) THOMPSON, L. C.: Phys. Rev. **96**, 369 (1954). A scintillation spectrometer study of the radiation from the bombardment of B^{10} B^{11} C^{12} N^{14} F^{19} and Si^{28} with deuterons. — (iv) RUTHERGLEN, J. G., E. R. RAE and R. D. SMITH: Proc. Phys. Soc. Lond. A **64**, 906 (1951). A pair spectrometer study of the radiations from B^{10} and B^{11} bombarded by deuterons and from Al^{27} by protons. — (v) BENT, R. D., T. W. BONNER and R. F. SIPPEL: Phys. Rev. **98**, 1237 (1955). These workers used a magnetic lens pair spectrometer with coincidence counting of pairs, to investigate the reactions $F^{19}+p$, $C^{12}+d$, $C^{13}+d$, $B^{11}+p$, and Li^7+d. In a later paper [Phys. Rev. **99**, 710 (1955)] the same authors, together with J. H. McCRARY and W. A. RANKEN, report on the reactions of deuterons with Be^9, B^{10}, N^{14} and F^{19}.

[*40*] *$(d\,p)$ stripping reactions.* The results of a systematic counter study with 8 MeV deuterons are reported by J. R. HOLT and T. N. MARSHAM in the following papers in the Proc. Phys. Soc. Lond. A: Technique, **66**, 249 (1953); light nuclei and conclusions, **66**, 1032 (1953); Mg, **66**, 258 (1953); Al, **66**, 249 (1953); Si, S, **66**, 467 (1953); Ca, **66** 565 (1953). The photographic emulsion method was used by POWELL, ROTBLAT and others for a study of reactions induced by 8 MeV deuterons. Their angular distribution work is reported in the following papers:
Technique. BURROWS, H. B., C. F. POWELL and J. ROTBLAT: Proc. Roy. Soc. Lond. A **209**, 461 (1951). — H^1, H^2: H. B. BURROWS,, W. M. GIBSON and J. ROTBLAT: Proc. Roy. Soc. Lond., Ser. A **209**, 489 (1951). — He^4, O^{16}: BURGE, E. J., H. B. BURROWS, W. M. GIBSON and J. ROTBLAT: Proc. Roy. Soc. Lond., Ser. A **209**, 534 (1951). — N^{14}, A^{40}: GIBSON, W. M., and E. E. THOMAS: Proc. Roy. Soc. Lond., Ser. A **210**, 453 (1951). — Other papers on $(d\,p)$ stripping reactions with 8 MeV deuterons are Be: EL-BEDEWI, F. A.: Proc. Phys. Soc. Lond. A **65**, 64 (1952). This paper includes an angular distribution for elastically scattered deuterons. — C: ROTBLAT, J.: Phys. Rev. **83**, 1271 (1951). — O^{16}: BURROWS, H. B., W. M. GIBSON and J. ROTBLAT: Phys. Rev. **80**, 1095 (1950). — Ne: MIDDLETON, R., and C. T. TAI: Proc. Phys. Soc. Lond. A **65**, 752 (1952). — P: EL-BEDEWI, F. A., and M. A. EL-WAHAB: Proc. Phys. Soc. Lond. **68**, 754 (1955).

[*41*] *$(d\,n)$ stripping reactions.* C, O, S: MIDDLETON, R., F. A. EL-BEDEWI and C. T. TAI: Proc. Phys. Soc. Lond. A **66**, 95 (1953). — N: EVANS, W. H., T. S. GREEN and R. MIDDLETON: Proc. Phys. Soc. Lond. A **66**, 108 (1953). — F: CALVERT, J. M., A. M. JAFFE, A. E. LITHERLAND and E. E. MASLIN: Proc. Phys. Soc. Lond. A **68**, 1017 (1955). — Na, Al. P: CALVERT, J. M., A. A. JAFFE and E. E. MASLIN: Proc. Phys. Soc. Lond. A **68**, 1008 (1955).

[*42*] TITTERTON, E. W.: The Use of the Photographic Emulsion in Photodisintegration experiments. Progress in Nuclear Physics, Vol. 4. London: Pergamon Press 1955. — This review describes both technique and results.

[*43*] HEYDENBURG, N. P., and G. M. TEMMER: Phys. Rev. **92**, 89 (1953); **94**, 1252 (1954); **96**, 426 (1954). — TALBOTT, F. L., and N. P. HEYDENBURG: Phys. Rev. **90**, 186 (1953).— These papers describe experiments on the interaction of artifically accelerated alpha particles, of precisely defined energy up to about 4 MeV, with Li^7, Be^9, B^{10}, N^{14}, F^{19} and Na^{23}. Evidence for COULOMB excitation of low lying levels of F^{19} and Na^{23} is presented.

General references E.

[*44*] KING, R. W.: Beta disintegration energies. Rev. Mod. Phys. **26**, 327 (1954). A comprehensive table including ft values.

[*45*] ELLIOTT, J. P., and B. H. FLOWERS: The structure of the nuclei of mass 18 and 19. Proc. Roy. Soc. Lond., Ser. A **229**, 536 (1955). — This basic paper is the first to extend intermediate coupling calculations to the $d-s$ shell. It gives a full account of the mathematical technique and of calculations of energy level sequences, static moments and transition probabilities.

[*46*] GLASS, N. W., and J. R. RICHARDSON: Phys. Rev. **98**, 1251 (1955). — BRECKON, S. W., A. HENRIKSON, W. M. MARTIN and J. S. FOSTER: Canad. J. Phys. **32**, 223 (1955). — BOLSTERLI, M., and E. FEENBERG: Phys. Rev. **97**, 736 (1955). Experimental and theoretical study of superallowed beta transitions in nuclei with $A=4n$.

Nuclear Reactions, Levels, and Spectra of Heavy Nuclei.

By

B. B. Kinsey.

With 81 Figures.

A. Introduction.

1. The study of nuclear reactions has allowed us to develop a theoretical framework in terms of which we attempt to understand and collate the experimental data; as yet this framework is far from complete and requires continuous modification and extension as new facts come to light. The volume of data is now so great that it is best digested with reference to this theoretical framework. In the following review, therefore, we adopt the somewhat unorthodox procedure of starting with a brief theoretical description of nuclear reactions and follow with a discussion of the experimental data. This procedure has some advantage in clarity and in ease of exposition; but in using it we neglect the historical development of the subject. However, this has now become so comprehensive that it is impossible in a short space to give an adequate account or even mention of the early and painstaking work which has led to our current ideas of the nucleus. The reader is presented here with a brief summary of the subject as it appears at the present time; we discuss the early work only in so far as it is of immediate importance to the understanding of the argument leaving the reader to obtain this information for himself from general references. To this end we give far more attention to recent experiments than their quality or their ultimate significance may justify.

Theories of nuclear reactions are briefly introduced in Chap. B, starting with the compound nucleus hypothesis. The statistical theory of nuclear reactions is then discussed together with some experiments deliberately designed to test it (Chap. C). We next consider the nature of direct interaction processes and some of the experimental evidence for them (Chap. D). The development of the optical model is then considered, with particular reference to total neutron cross sections and elastic scattering. We then pass on to a more detailed consideration of individual types of nuclear reactions discussing these under the headings of neutron, proton, and deuteron reactions. No separate heading is made for α-particle reactions because the volume of experimental data on this subject does not warrant it; (the few measurements which have been made are covered under other headings). Photonuclear reactions are summarized under a separate heading in Chap. J. We then pass on to a consideration of level densities and experimental data which are dependent on them. Finally we come to the low lying nuclear states. Although the greater part of the experimental data on the low lying energy levels has been inadequately studied, it is nevertheless possible to systematize some of it, particularly among the heaviest elements. This subject is developing now with great rapidity, and it is inevitable that much of what is said here will soon be superseded. We do not try therefore to list all the references to this work but confine our attention to a consideration of broad outlines only. We go from the general to the particular only where necessary to illustrate some theoretical principle.

It is not possible to make a logical distinction between high and low energy nuclear reactions; the following discussion will be confined in the main to reactions involving particles with energies below some 30 Mev. Neutron resonances which are so familiar a characteristic of low energy neutron reactions are fully described elsewhere in this volume and will not be discussed except insofar as they are of immediate importance to the study of level densities. For this reason a discussion of neutron or fission widths are omitted although we describe radiation widths in some detail. Although elastic scattering does not strictly come under our terms of reference, we deal with it here because an understanding of the part which elastic processes play in nuclear reactions is essential to the understanding of reaction cross sections generally. However, we are not concerned with the polarization of the incident particle which sometimes accompanies elastic scattering; nor with the decay of polarized nuclei or the reactions between polarized particles.

Finally in the following we shall often want to refer to the proton and neutron number of a nucleus. In (adjectives) describing the oddness or evenness of these numbers we shall follow the rule that the proton number comes first, the neutron number second; thus an even-odd nucleus is one with an even charge number, and an odd number of neutrons.

B. The compound nucleus.

2. The BOHR hypothesis. During the years immediately following the discovery of the neutron by CHADWICK, experiments with fast neutrons showed that the cross sections for collisions of neutrons with nuclei are comparable to the geometrical areas of nuclei. The discovery by FERMI and his collaborators of selective absorption of neutrons and the demonstration that in the heavy elements this absorption is the result of sharp resonance processes, shows that compound nuclei are formed by neutron capture with lifetimes long compared with the collision time of fast neutron reactions (10^{-21} sec). These facts, together with the surprising preponderance of absorption over scattering in resonance processes, appeared to be inconsistent with the elementary optical model used in the earliest speculations about nuclear resonances[1] and led BOHR[2], in 1936, to propose the compound nucleus theory of nuclear reactions. He suggested that nuclear reactions take place in two stages: first, the formation of a compound nucleus by the adherence of the projectile to the target nucleus; and second, the subsequent decay of this compound nucleus. The cross section for a particular nuclear reaction may then be expressed as the product of two factors; first, the cross section for the formation of the compound nucleus—a quantity, for neutrons, always close to the geometrical area—and second, a quantity which measures the probability that the compound nucleus decays in the mode in question. To calculate this probability it is necessary to specify the compound nucleus more closely. The creation of the compound nucleus implies that the compound nucleus has an independent existence and that it lasts for a time long compared with the duration of the collision. As pointed out by BOHR, in all probability, its lifetime will be long enough for it to lose all memory of its mode of formation when finally it is relieved of its excess energy by the emission of another particle.

[1] For a list of references to this work see reference [3].

[2] N. BOHR: Nature, Lond. **137**, 344 (1936). — N. BOHR and F. KALCKAR: Kgl. danske Vid. Selsk. mat.-fys. Medd. **14**, 10 (1937).

Immediately following the instant of collision this energy is rapidly shared out among the constituents of the nucleus; thus the average energy of any one of them is insufficient to enable that particle to leave the nucleus and the compound nucleus finally emits a particle only when the improbable event occurs that a large fraction of the energy of excitation is again concentrated in one constituent. Clearly in a heavy nucleus, where many particles must participate, the time for this concentration of energy to occur must be long by comparison with the collision time.

The mode of decay then must depend only on the angular momentum and the energy contributed by the incident particle, i.e. the sum of its kinetic energy and its binding energy to the compound nucleus. This we may call the *Bohr hypothesis*. To calculate the probability of a particular reaction it is necessary to make an additional assumption. The simplest is to assume that the emission of the different constituents of the nucleus, neutrons, protons, and more complex bodies like α-particles, may be treated on an equal footing and that the probability of emission of a specified particle may be calculated on a statistical basis. This is the *statistical assumption*. The theory of nuclear reactions founded on this basis is called the *statistical theory*.

Low energy nuclear reactions are essentially resonance processes. The resonances involved in these reactions represent well defined states of the compound nucleus and therefore these reactions fall naturally into the category covered by BOHR's hypothesis. For these low energies, then, a beam of neutrons will independently excite the large number of resonance levels which are comprised within the energy spread of the beam. These levels, because they do not overlap, are separately excited and decay independently of each other and of the way in which they are formed. For higher energies it is less easy to separate the two stages into which BOHR suggested the nuclear process should be divided. As the energy of the incoming particle increases the probability of its re-emission increases, and therefore the level width increases, while the spacing between them decreases. Thus a point is reached where the level widths become comparable with the spacing. For heavy elements this point is reached somewhere near 5 Mev. For higher energies the levels overlap; the decay must then depend on the phase relations between the levels and must depend therefore on the way in which they were formed. That the decay of the compound nucleus formed in such collisions is independent of the mode of formation would be a far reaching assumption, and one that is difficult to justify theoretically. It can be argued that if a sufficiently large number of states are excited in the collision, the phases between them must be distributed at random, so that the decay of the compound nucleus, when it occurs, is not dependent on phase relations. In heavy elements, the width of the levels at the point at which they begin to overlap, is of the order of one electron volt, and the corresponding lifetime, about 10^{-15} sec, is still a million times longer than the collision time. For much higher energies, it is likely that the lifetime of the compound nucleus is still long in the nuclear sense. However, it does not follow that this lifetime is sufficient to ensure that the subsequent decay is independent of the mode of formation. WEISSKOPF[1] has shown that the time corresponding to the average level spacing D, viz., $2\pi\hbar/D$, determines roughly the classical periodic time: if the incident particle enters the nucleus at same specific instant, it will reappear at the same point after a time of the order of this quantity. If the average life of the compound nucleus is much less than this, as must happen when the mean level width Γ is much greater than

[1] V. F. WEISSKOPF: Helv. phys. Acta **23**, 187 (1950).

the spacing D, it is not obvious that the decay of the compound nucleus will not reflect the manner of its creation. We shall not pursue this difficult subject further at this stage; it has been adequately discussed elsewhere [1], [2], [3]. We shall merely note that the theoretical interpretation of nuclear reactions in terms of compound nucleus formation and its subsequent decay seems well justified at low energies (below a few Mev). But for higher energies, say 10 Mev and above, while there is a strong presumption that, in the main, BOHR's hypothesis can be used as a basis of calculation, we note that its use is without adequate theoretical foundation. However it is just in this energy region that some experimental data, accumulated in recent years, is very difficult to reconcile with compound nucleus formation in the simple sense which we have just discussed. Some of these difficulties are undoubtedly caused by the fact that the compound nucleus theory is not concerned with the details of compound nucleus formation; others, however, might arise from just such difficulties of interpretation in the continuum which we have just described.

3. Extension of the compound nucleus hypothesis: direct interactions. We have seen that at low energies the validity of the compound nucleus hypothesis is not in doubt. For higher energies, where the validity of BOHR's hypothesis is most questionable, a natural development of the compound nucleus theory follows from inquiry into the interaction of the incident particle with the constituents of the nucleus. (This development, however, does not throw light on the possible dependence of the decay of the compound nucleus on its formation.) So far we have ignored the mechanics of the formation of the compound nucleus; we have merely assumed that the incident particle, once it hits the nucleus, will stick, and that its energy will be rapidly distributed among the nuclear constituents. This is supported by a naive estimate of the mean free path in nuclear matter. For example, if we take the cross section for internucleon interaction to be 1 b, which is the right order for neutrons and protons with energies of a few Mev, and if we suppose that the density of nuclear matter to be about 10^{38} nucleons per cm^3, the mean free path is about 10^{-14} cm, which is small compared with nuclear dimensions. However, the nuclear constituents obey the FERMI-DIRAC statistics and the collisions which an incident particle can make are severely restricted by the exclusion principle. The actual mean free path, which we shall discuss in Sect. 12, is very much greater than that just mentioned; it is a function of the kinetic energy of the individual particle and it is no longer small compared with nuclear dimensions. (In fact, it is quite possible that an incident particle with an energy of a few Mev will pass through the nucleus without any interaction.) The immediate product of a collision may be a cascade of nucleons. Those nucleons which have energy sufficient to enable them to escape from the nucleus, do so, provided that they suffer no further collisions before they reach the surface. When the energy imparted in collisions is no longer sufficient, the first part of the reaction is at an end and the residual nucleus is left in an excited state. The latter, then, decays when once again its energy is concentrated in a single particle. Thus, the formation of the excited nucleus, in general, should be preceded by a direct reaction. We shall see (Sects. 12 and 13), in the energy range of interest here, that the primary collision rather seldom leads to anything other than the formation of an excited nucleus; that the cascade just described is a rather uncommon event, but one for which the probability increases rapidly with energy. Interest in such direct reactions has been stimulated very recently by the apparent failure of the compound nucleus theory to account for some of the details of reactions in the 10 to 30 Mev range. In this energy range they do not contribute more than a small amount to the reaction cross section; they are significant mainly for reactions, which, for statistical reasons, are unlikely to follow compound nucleus formation. We shall discuss direct reactions in Chap. D.

4. GHOSHAL's experiment. The first attempt to test BOHR's compound nucleus hypothesis in so far as it applies to fast particle collisions, was made by GHOSHAL[1]. This unique experiment was designed to determine whether the various products of the decay of the compound nucleus Zn^{64} were produced in the same relative proportions, when the compound nucleus was formed at the same level of excitation by proton bombardment of Cu^{63} or α-particle bombardment of Ni^{60}. The active products measured by GHOSHAL were Zn^{63}, Cu^{62}, and Zn^{62}; they were formed respectively by the (α, n), (α, pn), and $(\alpha, 2n)$ reactions with Ni^{60}, and by the (p, n), (p, pn) and $(p, 2n)$ reactions with Cu^{63}. According to the BOHR hypothesis, the relative yields of these reactions should be the same regardless of the target provided that the excitation energy of the compound nucleus is the same. For the α-particle reactions, this excitation energy is $B_\alpha + \frac{60}{64}\varepsilon_\alpha$, where B_α is the binding energy of the α-particle to Zn^{64}, and ε_α is the kinetic energy of the α-particle in the laboratory system; or, $B_p + \frac{63}{64}\varepsilon_p$, where B_p and ε_p represent the proton binding energy to Zn^{64} and the proton kinetic energy, respectively. Thus the excitation energies are equal when:

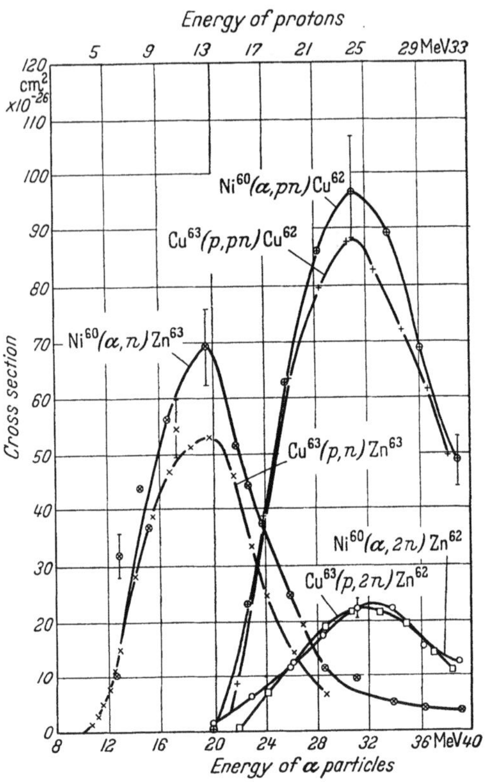

Fig. 1. Yields of the decay products of the compound nucleus, Zn^{64}, according to GHOSHAL. The proton energy scale has been shifted 7 Mev to the right.

$$\varepsilon_\alpha - \frac{21}{20} \cdot \varepsilon_p = \frac{16}{15}(B_p - B_\alpha).$$

The quantity B_p may be deduced from the neutron binding energy of Cu^{64} and the decay energy of that nucleus; it is 7.7 MeV. The binding energy B_α may be deduced from mass measurements; it is about 3.6 Mev. Thus, if the yields of the various reactions are plotted against the α-particle and the proton energy, the two curves correspond to equal excitation energies in Zn^{64} if the proton curve (near 20 Mev) is shifted about 5.4 Mev to the right.

Fig. 1 shows the results obtained by GHOSHAL with the proton curves shifted 7 Mev to the right. It will be seen that the two curves corresponding to the same active product in each case are in the same ratio to each other. Thus the relative yields of disintegration products are the same regardless of the mode of formation, in agreement with BOHR's assumption. It is not clear why the experimental curves require a translation of 7 rather than 5.4 Mev; it remains to be seen whether the discrepancy is significant[2]. It should be noted that energies of excitation

[1] S. N. GHOSHAL: Phys. Rev. **80**, 939 (1950).

[2] Recent mass measurements confirm the value of $B_p - B_\alpha$ given above; see P. C. EASTMAN, N. R. ISENOR, G. R. BAINBRIDGE and H. E. DUCKWORTH: Phys. Rev. **103**, 145 (1956).

covered by this experiment are high enough to ensure that the levels overlap. It is in fact just the region where the validity of the statistical theory is most in doubt. Another interesting feature of this remarkable experiment is the large cross sections of the (α, pn) and (p, pn) reactions relative to the $(\alpha, 2n)$ and $(p, 2n)$ reactions; this subject will be discussed in more detail in Sect. 36.

Very recently the validity of BOHR's hypothesis as applied to reactions in heavy elements has been confirmed by JOHN[1]. He has studied the energy variation of the cross section of (α, xn) reactions with Pb^{206} and compared the results with cross sections of the (p, xn) reactions obtained by KELLY[2] with Bi^{209}. In each case the Po activities measured corresponded to $x=2$, 3 or 4. It was found that the cross sections of the α-particle and proton reactions were in very good agreement provided that the proton energy scale was shifted by (11.3 ± 0.4) Mev, the calculated value being (10.5 ± 0.2) Mev.

5. Definition of cross sections. Before proceeding further in discussing the theory of nuclear reactions, we digress to discuss the definition of nuclear cross sections. For neutrons the most easily measured cross section is the *"total" cross section*, σ_t. This cross section can be obtained from transmission measurements. If a cylindrical neutron detector is set up at some distance from a neutron source with its axis in the direction of the source, and if a cylindrical absorber is placed between the source and detector, then, provided the latter shields the detector from the source at all points, the ratio of the counting rates obtained with the absorber in position, and without it, measures the "transmission" of neutrons through the absorber. The total cross section σ_t is related to the transmission T through the relation $T = \exp(-N\sigma_t)$ where N is the number of atoms per cm^2 in the absorber as seen at the detector. The results obtained depend to some extent on the geometry of the experimental arrangement; if it is desired to measure the sum total of the cross sections of all neutron processes which occur in the absorber, the angle presented by the detector to the absorber must be small so that the neutrons scattered by the absorber have little chance of entering the detector.

The total cross section is significant only in relation to neutrons. For charged particles, it is not possible to make transmission measurements of this kind for the experimental result will always depend on the RUTHERFORD scattering which, in its entirety, cannot be represented by a finite quantity. For high energies ($\sim$100 Mev) it is possible by an adaptation of the transmission method to determine the inelastic part of proton scattering[3]. But for the energies with which we are concerned here, the rate of energy loss of protons and α-particles is too high to make such methods practicable.

The total cross section σ_t may be equated to the sum of the *elastic scattering cross section* (σ_e) and the *inelastic cross section* (σ_n). The former is the sum of the cross sections of all processes which lead to the re-emission of the incident particle without change of kinetic energy in the frame of reference in which the centre of mass is at rest. The latter is the sum of the cross sections of all processes in which the incident particle disappears or is re-emitted with change of energy in the centre of mass system. In this article we shall call this latter cross section the *"nonelastic" cross section*. Thus:

$$\sigma_t = \sigma_e + \sigma_n. \tag{5.1}$$

[1] W. JOHN: Phys. Rev. **103**, 704 (1956).
[2] E. L. KELLY: University of California, 1950 (unpublished).
[3] J. M. CASSELS and J. O. LAWSON: Proc. Phys. Soc. Lond. A **67**, 125 (1954).

The nonelastic cross section is sometimes called the "*reaction*" *cross section*[1]. There still exists some confusion in the exact definition of this expression. However, in this article we shall use the latter expression to denote the cross section for the production of the compound nucleus; this is greater than the nonelastic cross section by the amount required to account for re-emission of the incident particle from the compound nucleus without change of energy, i.e., the "*compound-elastic*" cross section (σ_{ce}). Thus we define the *reaction cross section* by the equation:

$$\sigma_c = \sigma_n + \sigma_{ce} \tag{5.2}$$

As we shall show in Sect. 6, the reaction cross section, defined in this way, is the sum of the cross sections of all processes which involve absorption of the incident particle or its re-emission in a manner incoherent with the incident wave. At low energies, or when only a few states of the residual nucleus can be excited in a reaction, the compound-elastic cross section may be equal to or greater than the nonelastic cross section. At higher energies (a few Mev or more) re-emission of the neutron merely represents one of a large number of possible transitions, and the compound-elastic cross section is negligible compared with the nonelastic cross section; then for all practical purposes the two cross sections are equal.

According to this definition, the elastic cross section as defined by (5.1) is composed of two parts, the compound-elastic cross section, and another part, known as the "*shape-elastic*" *cross section* (σ_{se})

$$\sigma_e = \sigma_{se} + \sigma_{ce} \tag{5.3}$$

As shown in Sect. 6, the two elastic cross sections have a very different origin, for while the amplitude of the shape-elastic part is coherent with the incident wave, the other, the compound-elastic amplitude, is incoherent. No experiments have yet been devised to distinguish between the two cross sections; since the one is incoherent, the total elastic cross section which is the only quantity which can be determined experimentally can be written as the sum of the two (5.3).

As originally pointed out by BOHR, PEIERLS and PLACZEK[2], certain general limitations on the magnitudes of these cross sections must exist. Consider the scattering and absorption of a plane wave of neutrons of unit amplitude, moving in the Z-direction. This plane wave (e^{ikz}) can be expanded in spherical harmonics about some point from which the radius vector is r. For large r,

$$e^{ikz} = \frac{\pi^{\frac{1}{2}}}{kr} \cdot \sum_{l=0}^{\infty} \sqrt{2l+1} \cdot i^{l+1} \left[\exp\left(-i\left(kr - \tfrac{1}{2} l\pi\right)\right) - \exp\left(+i\left(kr - \tfrac{1}{2} l\pi\right)\right)\right] Y_{l0}.$$

If we insert some absorbing and scattering object at this point, the expression above will still represent the wave provided the outgoing part (proportional to e^{ikr}) is changed. Introducing the complex amplitude η_l for the l-th partial wave, the disturbed wave is now represented by:

$$\psi(r) = \frac{\pi^{\frac{1}{2}}}{kr} \sum_{l=0}^{\infty} \sqrt{2l+1} \cdot i^{l+1} \left[\exp\left(-i\left(kr - \tfrac{1}{2} l\pi\right)\right) - \eta_l \exp\left(+i\left(kr - \tfrac{1}{2} l\pi\right)\right)\right] Y_{l0}$$

and the scattered wave is given by the difference between the two expressions. The number of scattered particles is found by integrating (at infinity) the flux of particles determined by the second expression; this gives the elastic scattering

[1] Cf. below ref. 1, p. 210.

[2] Unpublished. See reference [*4*], Chap. VIII, for a full discussion of this subject.

cross section. The nonelastic cross section is found from the number of particles removed from the incident beam. These calculations give:

$$\left.\begin{aligned}\sigma_e^{(l)} &= \pi \lambda\!\!\!^{-2}(2l+1)\,|1-\eta_l|^2\,,\\ \sigma_n^{(l)} &= \pi \lambda\!\!\!^{-2}(2l+1)\,[1-|\eta_l|^2]\,.\end{aligned}\right\} \tag{5.4}$$

We use the nonelastic rather than the reaction cross section in these formulae. This procedure is strictly correct; however, if we take into account the incoherence of the compound-elastic part, it is probably more correct to replace the elastic with the shape-elastic and the nonelastic with the reaction cross section in these equations. For high energy particles the difference is trival. Since the nonelastic cross section cannot be negative $|\eta_l| \ll 1$; its maximum value is obtained when $|\eta_l| = 0$. Then both $\sigma_e^{(l)}$ and $\sigma_n^{(l)}$ are equal to $\pi\lambda\!\!\!^{-2}(2l+1)$. The maximum value of the elastic scattering cross section is obtained when $\eta_l = -1$; it is $2\pi\lambda\!\!\!^{-2}(2l+1)$. Thus the elastic scattering and nonelastic cross sections must lie in the ranges:

$$0 \leq \sigma_e^{(l)} \leq 4\pi\lambda\!\!\!^{-2}(2l+1),$$

$$0 \leq \sigma_n^{(l)} \leq \pi\lambda\!\!\!^{-2}(2l+1).$$

The total cross section is given by:

$$\sigma_t^{(l)} = 2\pi\lambda\!\!\!^{-2}[1-\mathrm{Re}\,\eta_l] \quad (5.4)$$

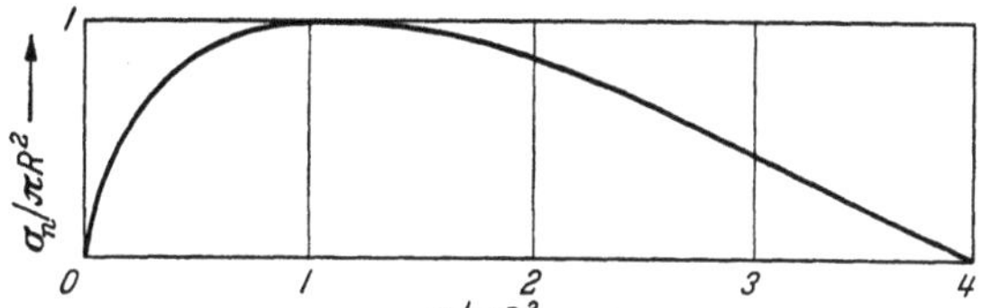

Fig. 2. Elastic and nonelastic cross sections. Possible va ues occur beneath the curve shown.

and can take any value from zero to $4\pi\lambda\!\!\!^{-2}(2l+1)$. In general, (5.4) shows that there cannot be absorption without scattering; and that for any given scattering cross section there is a maximum value for the nonelastic cross section. The possible values of σ_n are those which lie below the curve shown in Fig. 2.

The measured nonelastic and scattering cross sections must correspond to the sum of those partial cross sections which make a significant contribution. If the neutron energy is sufficiently great, the maximum value of the orbital angular momentum (l_m), for which there will be an appreciable contribution, is equivalent to an impact parameter equal to the nuclear radius, R, i.e. $l_m = \frac{R}{\lambda\!\!\!^{-}}$. Thus $\sum_l (2l+1) = \left(1+\frac{R}{\lambda\!\!\!^{-}}\right)^2$. The maximum total and scattering cross sections are therefore $4\pi(R+\lambda\!\!\!^{-})^2$ and the maximum nonelastic cross section is

$$\sigma_n = \pi(R+\lambda\!\!\!^{-})^2. \tag{5.5}$$

For high energies and short wavelengths, the latter becomes equal to the geometrical area πR^2.

In the course of this article many different values of the nuclear radius will appear. We shall assume that the nuclear radius, R, can always be given by the expression $R = A^{\frac{1}{3}} r_0$ although in many instances it is more accurate to add a constant term independent of the atomic weight, A. The value of r_0 is important. Electrical methods of investigation[1] usually lead to $r_0 = 1.2\times10^{-13}$ cm or less for the radius of the charge distribution, while reaction cross sections and elastic scattering, as we shall see, lead to larger values, 1.3 to 1.6×10^{-13} cm. Recent work[2] indicates that the latter measure the radius of the potential well rather

[1] See K. W. Ford and D. L. Hill: Ann. Rev. Nucl. Sci. **5**, 25 (1955).

[2] See ref. 3, p. 232. Also R. A. Berg and L. Wilets: Phys. Rev. **101**, 201 (1956). — L. Wilets: Phys. Rev. **101**, 1805 (1956). — K. A. Brueckner: Phys. Rev. **103**, 1121 (1956). — Nuclear radii have been reviewed recently by Scott: Progr. Nucl. Phys. **5**, 157 (1956).

than the radius of the mass distribution which is now generally believed to be similar to that of the charge distribution. The difference between the two seems to be the result of the tapered edge of the well. Evidence for the latter will be discussed in Sect. 19 to 21.

6. The optical model. It is just in the energy range in which the validity of the compound nucleus hypothesis is not in doubt that the neutron cross sections do not agree in detail with theoretical predictions. This discrepancy appears in the total neutron cross sections and seems to be caused by the elastic scattering. The origin of the discrepancy lies in the fact that in the formulation of BOHR'S hypothesis, on which the statistical theory is based, no account is take of those gross features of the interaction of nucleons with nuclei which are responsible for shell structure. Instead of considering the individual collisions which the incident particle must make within the nucleus, it is possible to consider, alternatively, the interaction of that particle with the nucleus as a whole. This interaction is usually represented by a potential well. Thus the calculation of the scattering and absorption of a nucleon resembles the calculation of the scattering and absorption of a beam of light by a refracting and absorbing sphere. This is the optical model: it can give an account only of the gross features of nuclear reactions; it cannot account for the sharp resonances or for the competition between various nuclear reactions.

The first attempts to account for the behaviour of nuclear reactions were made with the use of an optical model. The failure of this model to account for the sharp resonances was the starting point for BOHR'S compound nucleus hypothesis. In recent years, the inadequacy of the statistical theory which is based on it has renewed interest in the optical model, in particular, by FESHBACH, PORTER and WEISSKOPF[1].

A theoretical difficulty arises when we try to account for the experimental data in terms of the optical model. For low energies the results of cross section measurements must represent an average over many individual resonances; and these resonances do not lie within the scope of the optical model. In principle the cross sections could be obtained theoretically by finding the complex amplitudes given in (5.4). These amplitudes must be given by a wave equation consistent with a "square well" potential of the type

$$\left.\begin{aligned} V &= V_0 + iW = V_0(1+i\xi) & r &< R,\\ V &= 0 & r &> R \end{aligned}\right\} \tag{6.1}$$

and satisfying the usual boundary conditions at the surface of the nucleus. Such a calculation will not give the rapidly varying amplitudes corresponding to the sharp resonances which we know do exist. The experimental cross sections must represent averages over the resonance cross sections. To account for the data it is necessary to find an average cross section and to calculate this from an average value of η. FESHBACH, PORTER, and WEISSKOPF[1], assume that such an average can be obtained by simply averaging over an energy interval large compared with the level spacing, D. In terms if this average the cross sections become:

$$\left.\begin{aligned} \bar{\sigma}_e^{(l)} &= \pi \lambda\!\!\!^{-2}(2l+1)\,\overline{|1-\eta_l|^2}\,,\\ \bar{\sigma}_n^{(l)} &= \pi \lambda\!\!\!^{-2}(2l+1)\,(1-\overline{|\eta_l|})^2 \end{aligned}\right\} \tag{6.2}$$

of which the first equation is equivalent to:

$$\bar{\sigma}_e^{(l)} = \pi \lambda\!\!\!^{-2}(2l+1)\left[|(1-\bar{\eta}_l)|^2 - |\bar{\eta}_l|^2 + \overline{|\eta_l|^2}\right]. \tag{6.3}$$

[1] H. FESHBACH, C. E. PORTER and V. F. WEISSKOPF: Phys. Rev. 96, 448 (1954).

These authors then point out that the difference $\overline{|\eta_l|^2} - |\bar{\eta}_l|^2$, which represents the mean square fluctuation in η_l, is small at high energies where the resonance effects are smoothed out by overlapping of levels; for different reasons, they show that this term is also small at low energies. They identify that portion of (6.3) which contains this difference as the compound elastic scattering, defined in Sect. 5. When this quantity is added to the nonelastic cross section we obtain the reaction cross section, which now becomes

$$\sigma_c^{(l)} = \pi \lambda\!\!\!^{-2} (2l + 1)(1 - |\bar{\eta}_l|^2) \tag{6.4}$$

while the shape-elastic cross section is:

$$\sigma_{se}^{(l)} = \pi \lambda\!\!\!^{-2} (2l + 1)\,|1 - \bar{\eta}_l|^2. \tag{6.5}$$

These two formulae have now the familiar appearance of (6.2). The calculation of η_l will thus determine the average cross sections and it is these which can be compared with the experimental data. It should be noted that (6.4), although written in the form of a nonelastic cross section, now contains the compound-elastic cross section.

An instructive physical picture of the process formulated above has been described by FRIEDMAN and WEISSKOPF [*3*]. The incident wave may be represented by a wave packet whose extent is determined by the precision with which its energy is known. Since we are concerned here with an average effect, the energy spread in the incident beam must be large compared with the level spacing D. Thus the wave passes the nucleus in a time short compared with $2\pi\hbar/D$. As it passes the shape-elastic wave is emitted. The intensity and angular distribution of this scattered wave is as much as the optical model can tell us. However, the compound-elastic wave, which is determined by the fluctuations in amplitude due to the resonance fine structure is emitted afterwards. These authors show that its time dependence, which is given essentially by

$$\int (\eta - \bar{\eta})\, e^{-i\varepsilon t/\hbar} \cdot d\varepsilon$$

integrated over the energy range comprised by the incident wave, is controlled by the decay time of the compound nucleus. The incident and the shape-elastic waves have disappeared from the vicinity of the nucleus long before the compound-elastic scattered wave has come to an end; thus the latter is not coherent with the incident wave, and its cross section is included in the nonelastic term in (6.4). In practice, it cannot be distinguished (at present) from the shape-elastic scattering; since it is incoherent, measurements of elastic scattering determine the sum of the cross-sections, as described in Sect. 5.

Solutions of the SCHRÖDINGER equation with the potential (6.1) in principle, give the elastic and nonelastic cross sections as functions of the parameters V_0, ξ, and the radius R. Such calculations generally require electronic computing machines for their evaluation because many different orbital angular momenta can contribute to the cross sections. We shall show that square well potentials (6.1) give results which are generally in poor agreement with experiment and that when a tapered edge is given to the potential much better agreement is obtained. However, we shall discuss the results of such calculations as we come to them; at this stage we merely note that V_0 is of the order of 40 Mev and that ξ is of the order of a few percent.

C. Statistical theory of nuclear reactions.

7. Theoretical reaction cross sections. Consider the reaction:

$$a + \mathrm{A} \rightarrow \mathrm{C}^*$$

in which an incident particle a collides with a target nucleus A to create an excited compound nucleus C*. (The asterisk indicates the compound nucleus is excited and is not in its ground state.) According to BOHR's hypothesis, the compound nucleus rapidly loses all memory of its formation and in due course decays in any way allowed by the laws of conservation of energy, parity, and angular momentum. Then the probability of any particular mode of decay, say,

$$\mathrm{C}^* \rightarrow \mathrm{B} + b \tag{7.1}$$

will depend only on the excitation of C and can be calculated on a statistical basis.

In light nuclei it is often possible to follow the decay of compound nuclei in resonance reactions for the spacing of levels can be wide compared with the resolving power of existing apparatus. For heavy nuclei, discrete levels in the compound nucleus usually can be detected at the present time only by the interaction of slow neutrons with nuclei. This subject is discussed elsewhere in this series. We are concerned here with compound nuclei in which a great many levels are excited. The theoretical discussion therefore necessitates a statistical approach. Thus if the probability of decay of the compound nucleus C in the direction (7.1) is represented by $F(\mathrm{B}, b)$, and if F_0 is the sum of the probabilities of all possible reactions, then the cross section for the reaction $a\mathrm{A} \rightarrow b\mathrm{B}$ is:

$$\sigma(a, b) = \sigma_c F(\mathrm{B}, b)/F_0. \tag{7.2}$$

In this section we are interested only in the value of the reaction cross section σ_c; the probability ratio in (7.2) is discussed in Sect. 8. The reaction cross sections for neutrons and for charged particles are very different for the latter are affected by the COULOMB barrier. We shall deal with them separately.

α) Neutron reaction cross sections. An approximate value for the reaction cross section was derived in a preceding section (5.5). It is only a rough approximation for it treats all effective partial waves on an equal footing. However, owing to the fact that in its gross behaviour the nucleus behaves towards individual particles like a potential well, particles incident on the nuclear surface may be partly reflected as a result of the change of refractive index at the surface of the nucleus. The transmission coefficient represented by this reflection is a function of the wave numbers of the particle both inside the nucleus (K) and outside (k), and of its angular momentum. Expressions for these transmission coefficients have been obtained by FESHBACH, PEASLEE, and WEISSKOPF[1] on the assumption that the potential barrier can be represented by a square well; they are given in Sect. 7 of the preceding article in this volume. Thus, a more correct expression for the reaction cross section is given by:

$$\sigma_c = \pi \lambda\!\!\!^{-2} \sum_l (2l + 1)\, T_l \tag{7.3}$$

where T_l is the transmission coefficient for angular momentum l. The quantity (7.3) is not very dependent on the internal wave number K and it is reasonable to make the internal wave number $K \approx 10^{13}\ \mathrm{cm}^{-1}$. The expression (7.3) is always

[1] H. FESHBACH, D. C. PEASLEE and V. F. WEISSKOPF: Phys. Rev. **71**, 145 (1947).

less than (5.5). Numerical values of σ_c in units of πR^2 have been plotted by BLATT and WEISSKOPF [4] as a function of kR. For many heavy nuclei, and for neutrons of intermediate energies (say from 2 to 10 Mev), the theoretical cross section (7.3) is less than (5.5) and close to $1.2\pi R^2$.

β) Reaction cross sections for charged particles. For charged particles the calculation of the transmission coefficients is much more complicated. The theory is sketched in the preceding article. If the COULOMB barrier is represented by B, where $B = ZZ'e^2/R$, where Z and Z' are the charges on the incident particle and on the target nucleus respectively, in units of the electronic charge, e, an approximate expression for T_l is:

$$T_l = \sqrt{\frac{B_l - \varepsilon}{\varepsilon}}\,\exp(-2C_l).$$

Here B_l is the barrier height for particles with angular momentum l,

$$B_l = B + \frac{\hbar^2}{2MR^2}\,l(l+1), \qquad (7.4)$$

M is the reduced mass; and C_l is given by the complicated expression

$$\frac{C_l}{\left(\frac{2ZZ'e^2RM}{\hbar^2}\right)^{\frac{1}{2}}} = \frac{1}{2\sqrt{x}} \times$$

$$\times\left[\frac{\pi}{2} + \arcsin\frac{1-2x}{\sqrt{1+4xy}}\right] - \sqrt{1+y-x} +$$

$$+ \sqrt{y}\log\frac{1+2\sqrt{y}\left(\sqrt{y}+\sqrt{1+y-x}\right)}{\sqrt{1+4xy}}$$

in which $y = B_l/B$ and $x = \varepsilon/B_l$. From such formulae the reaction cross section may be calculated by summation of the transmission coefficients, using (7.3). A more accurate computation of the reaction cross section has been made by SHAPIRO[1].

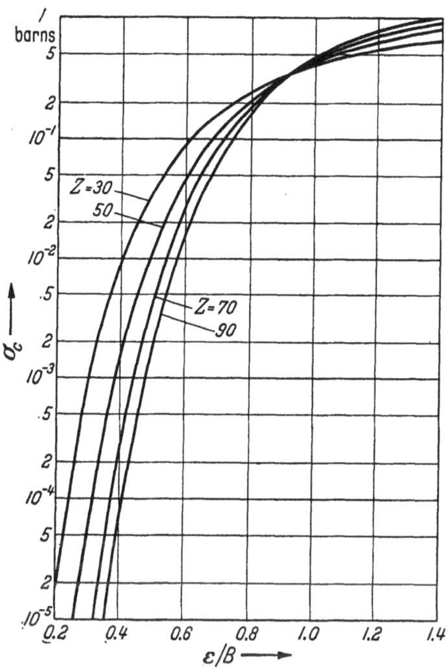

Fig. 3. Reaction cross sections for protons, after SHAPIRO; for $r_0 = 1.5 \times 10^{-13}$ cm.

It should be noted that numerically the barrier height for protons is very simply given [5] by the formula $B = Z/A^{\frac{1}{3}}$ in Mev, if the nuclear radius is chosen so that $R = r_0 A^{\frac{1}{3}}$ with $r_0 = 1.44 \times 10^{-13}$ cm. The contribution of the centrifugal barrier is $10.6\,l(l+1)/A^{\frac{2}{3}}$ in Mev. It is relatively more important in light and intermediate weight nuclei.

For energies below the barrier, the reaction cross section falls very rapidly. For energies equal to the barrier height, the reaction cross section is remarkably independent of atomic weight; in fact, the cross section for most elements lie between 0.3 and 0.5 barn. (see Fig. 3). This cross section is much smaller than the neutron reaction cross section, which for the heavier elements is about 2.5 barn. Above the barrier, the cross section may be calculated with surprising accuracy by elementary classical methods. The result is:

$$\sigma_c = \pi(R + \lambda\!\!\!^{-})^2\left(1 - \frac{B'}{\varepsilon}\right) \qquad (7.5)$$

[1] M. M. SHAPIRO: Phys. Rev. 90, 171 (1953).

with $B' = ZZ'e^2/(R+\lambda)$; (7.5) is in error by only 15% when $B'/\varepsilon = 1.2$, and by less at higher energies.

8. Cross sections for specific reactions. We have seen that the reactions which follow the creation of the compound nucleus should depend only on the excitation of that nucleus and not on its mode of formation. At this stage we shall not inquire further into the validity of that assumption: it is enough to assume that it is valid and to assume that both the number of states excited in the compound nucleus and in the final system into which the compound nucleus decays, is large enough so that the probability of decay into specific disintegration products is calculable on statistical grounds. Furthermore, in the absence of evidence to the contrary, it is necessary to assume that the emission of any kind of particle can be treated on the same footing, the only difference between neutrons and protons, for example, being the effect on the latter of the COULOMB barrier. A statistical theory on these lines was first developed by WEISSKOPF[1], and extended by WEISSKOPF and EWING[2]. In abbreviated form their argument is as follows.

Consider the decay ($C^* \to B + b$) in which the compound nucleus C has initially an excitation energy E_0, and decays to form the nucleus B with energy of excitation E, and the particle b, with kinetic energy ε (see Fig. 4). Then $E = E_0 - Q_{Cb} - \varepsilon$, where Q_{Cb} is the binding energy of the particle b to the compound nucleus, C. The maximum kinetic energy of the particle b corresponds to $E = 0$, i.e. $\varepsilon_m = E_0 - Q_{Cb}$. The probability of decay of the compound nucleus may be obtained by application of the principle of detailed balancing. Suppose the residual nucleus B (at excitation E) and the particle b are contained in a box of volume V. Then the probability per unit time for the reverse reaction, $B + b \to C^*$, is $\sigma_B v/V$ where v is the velocity of b, whose energy lies in the range between ε and $\varepsilon + d\varepsilon$ and σ_B is the cross section of B for this process. The probability of the decay of the compound nucleus ($F_{C\to B}$) is related to the probability of its formation through the equation:

$$F_{C\to B}\,\omega_C = F_{B\to C}\,\omega_B \times (\text{number of states for free particle})$$

where ω_C is the number of states of C per unit energy range and ω_B is the number of states of the residual nucleus into which it can decay. The number of states in a momentum range dp into which the free particle b (proton or neutron) can be emitted is $\frac{8\pi p^2\,dp\,V}{(2\pi\hbar)^3}$ where p is its momentum $(2M\varepsilon)^{\frac{1}{2}}$. Thus the probability for the emission of b, per unit time, is:

$$F_{C\to B+b}(\varepsilon)\,d\varepsilon = \frac{2M}{\pi^2\hbar^3}\cdot\frac{\varepsilon\,\omega_B(E)\,\sigma_B\,d\varepsilon}{\omega_C(E_0)} \tag{8.1}$$

and half this amount for α-particles. This expression takes no account of the spins, the parity, or the statistical weights of the states involved. If these complicating factors are disregarded, the probability for the reaction may be written:

$$F(B,b) \propto \int_0^{\varepsilon_m} \varepsilon\,\sigma_B\,\omega_B(E)\,d\varepsilon$$

with $E = E_0 - Q_{Cb} - \varepsilon$, as defined above. The cross section for the reaction $A + a \to B + b$, is:

$$\frac{\sigma(A\,a \to B\,b)}{\sigma_c} = \frac{\int_0^{\varepsilon_m} \varepsilon\,\sigma_B\,\omega_B(E)\,d\varepsilon}{\sum_{B'}\int_0^{\varepsilon'_m} \varepsilon\,\sigma_{B'}\,\omega_{B'}(E)\,d\varepsilon} \tag{8.2}$$

[1] V. F. WEISSKOPF: Phys. Rev. **52**, 295 (1937).
[2] V. F. WEISSKOPF and D. H. EWING: Phys. Rev. **57**, 472 (1940).

where the summation in the denominator is extended over all possible modes of disintegration. It will be shown in Sect. 10, that this formula accounts satisfactorily for the most distinctive feature of reactions produced by the decay of compound nuclei, viz., the competition between them.

The cross sections given by (8.2) for proton emission are very different to the cross sections for neutron emission. For neutrons, it is sufficient to put $\sigma_B = \pi R^2$, while for protons the cross section is largely determined by the COULOMB barrier. Generally, the cross section for a reaction leading to the emission of a charged particle should always be less then that which leads to the emission of a neutron. In fact, the ratio of the cross sections of the reactions (x, p) to (x, n) where x may be a neutron or a proton or an α-particle, is given by:

$$\frac{\sigma(x, p)}{\sigma(x, n)} = \frac{\int\limits_0^{\varepsilon_{m(p)}} \varepsilon\, \sigma_{B(p)}\, \omega_{B(p)}(E)\, d\varepsilon}{\int\limits_0^{\varepsilon_{m(n)}} \varepsilon\, \sigma_{B(n)}\, \omega_{B(n)}(E)\, d\varepsilon} \tag{8.3}$$

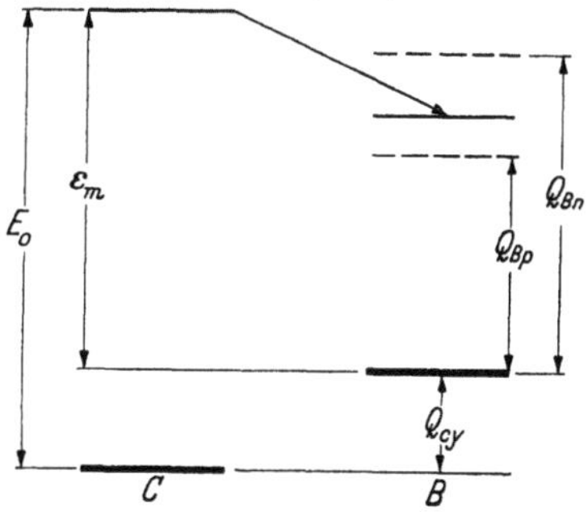

Fig. 4. Decay of compound nucleus in relation to the neutron and proton binding energies of the residual nucleus.

where $\varepsilon_{m(p)}$ and $\varepsilon_{m(n)}$ refer to the maximum kinetic energies of proton and neutron respectively. Usually, these quantities are not equal; but for the sake of argument we shall neglect both their inequality and the fact that the level densities in the two different residual nuclei may not be the same for the same energy. Then the ratio of the cross sections in (8.3) in effect is an average of the barrier penetration, averaged over the proton spectrum. Thus we should expect to find that, for intermediate energies, the yield of the proton reaction to be less than that of the neutron reaction. This prediction, however, is sometimes incorrect; for incident protons of 20 Mev, many instances have been found in which proton emission is much more probable than neutron emission. It is in this energy range that the validity of the compound nucleus theory has been questioned. The origin of this particular discrepancy is not clear; it is discussed in Sect. 36.

Emission of heavy particles, provided their kinetic energy is sufficient, will always compete successfully with the emission of γ-radiation. Therefore the emission of a neutron or a proton which leaves the residual nucleus with excitation energy above its neutron or proton binding energy, will in general lead to the emission of another neutron or proton. Thus a reaction of the type (x, y) may lead to a reaction of the type (x, yn) or (x, yp). The condition for this to occur is that $E_0 - Q_{Cy}$ is greater than Q_{Bn} or Q_{Bp}. Here the Q's represent the binding energies of the particles denoted by the suffixes to the nuclei B and C (see Fig. 4). For heavy elements, even if the energy available to y is much greater than the barrier height, the COULOMB barrier nevertheless will favour the reaction (x, yn) over the reaction (x, yp). However, the proton binding energy of the residual nucleus B is usually less than the neutron binding energy in that nucleus, and in the lightest isotopes of elements of intermediate atomic weight it may be much less; thus there exists a range of excitation of the residual nucleus B when only proton or γ-ray emission will occur, viz., when $Q_{Bp} < E < Q_{Bn}$ (Fig. 4). Then large cross sections of the reaction (x, yp), may be found (see Sects. 29 and 37).

The formulae (8.1) and (8.2) can be simplified by the introduction of the idea of nuclear temperature. In WEISSKOPF'S terminology, let the quantities S_B and

S_C represent the logarithms of the level densities in (8.1). Then that expression becomes:

$$F(\varepsilon)\,d\varepsilon = \frac{2M}{\pi^2\hbar^3}\cdot\sigma\,\varepsilon\exp\left[S_B(E_0 - Q_{Cb} - \varepsilon) - S_C(E_0)\right]d\varepsilon. \tag{8.4}$$

The average energy of the disintegration products of compound nucleus reactions is usually small compared with the excitation energy of the compound nucleus. If therefore we suppose that $E_0 - Q_{Cb}$ is large compared with ε, and if we suppose that at a given excitation energy the level densities in the compound nucleus and the residual nucleus are the same, so that $S_B(E) = S_C(E)$,

$$S_B(E_0 - Q_{Cp} - \varepsilon) = S_C(E_0 - Q_{Cp}) - \varepsilon\,(d\,S/d\,E)_{\varepsilon=\varepsilon_m}$$

approximately. Therefore, writing,

$$\frac{1}{\Theta} = \left(\frac{dS}{dE}\right)_{\varepsilon=E_0-Q_{Cb}} \tag{8.5}$$

the expression (8.4) becomes:

$$F(\varepsilon)\,d\varepsilon = \frac{2M\sigma\varepsilon}{\pi^2\hbar^3}\cdot\exp\left[S_C(E_0 - Q_{Cb}) - S_C(E_0)\right]e^{-\varepsilon/\Theta}\,d\varepsilon. \tag{8.6}$$

If σ does not vary with energy, the energy spectrum of the emitted particles therefore is Maxwellian in shape. For neutrons the peak of the energy spectrum occurs when $\varepsilon = \Theta$ and the average energy is 2Θ. For charged particles, this spectrum will be cut off below the barrier height, because (8.6) contains the cross section σ for the inverse reaction which is very small for energies much below the barrier.

The quantities denoted by S in the expressions above measure the entropies of the systems, and by analogy with thermodynamics, the quantity Θ in (8.5) represents a temperature. This temperature has little physical significance; it is no more than a convenient parameter with which to describe the energy spectrum of the particles emitted. The form of (8.6) would be preserved whatever the law governing the increase of level density with excitation energy should be, provided it increases rapidly enough. It should be noted that Θ is the temperature of the residual nucleus at its maximum excitation. The error committed in neglecting higher terms in the expansion for S is not serious. Although the probability of a particular reaction is determined by the rapidly rising level density, it should be noted also that increasing the number of levels which can be excited in the residual nucleus by raising the excitation energy of the compound nucleus, E_0, does not increase the absolute value of this probability to any great extent for the first exponent in (8.6) to a first approximation is just e^{-Q_{Cb}/Θ_0}. Here Θ_0 is temperature of the compound nucleus, at excitation energy E_0. Thus, to this approximation, the differential cross section for the emission of particles of a particular energy, ε, is independent of the energy of the incident particles.

The expression (8.6) leads to a very simple formula for the ratio of the cross sections of reactions of the type (x, n) and $(x, 2n)$ when the kinetic energy of the incident particle x does not exceed the threshold of the $(x, 3n)$ reaction. The formula, which was obtained by WEISSKOPF and EWING[1], is not strictly valid if the proton binding energy is less than the neutron binding energy. However, if we disregard this point, it is clear that only those states of the residual nucleus B which lie between the ground state and the neutron binding energy Q_{Bn} will result

[1] V. F. WEISSKOPF and D. H. EWING: Phys. Rev. 57, 472 (1940).

in the (x, n) reaction, while those above it will give rise to the $(x, 2n)$ reaction. Thus the ratio of the cross sections is the following:

$$\frac{\sigma(x, 2n)}{\sigma(x, n)} = \frac{\int\limits_{\varepsilon_m - Q_{Bn}}^{0} \varepsilon\, e^{-\varepsilon/\Theta}\, d\varepsilon}{\int\limits_{\varepsilon_m}^{\varepsilon_m - Q_{Bn}} \varepsilon\, e^{-\varepsilon/\Theta}\, d\varepsilon}. \tag{8.7}$$

The numerical value of Θ is usually much smaller than either ε_m or the difference $\varepsilon_m - Q_{Bn}$. Ignoring $\exp(-\varepsilon_m/\Theta)$ relative to $\exp-(\varepsilon_m - Q_{Bn})/\Theta$, this expression becomes:

$$\left.\begin{aligned} \frac{\sigma(n, 2n)}{\sigma(x, n)} &= \frac{e^z}{1+z} - 1 \\ \text{with}\qquad z &= \frac{\varepsilon_m - Q_{Bn}}{\Theta}. \end{aligned}\right\} \tag{8.8}$$

If then the cross sections of other possible reactions are ignored, the sum of the two cross-sections is equal to the reaction cross section, σ_c. Thus:

$$\sigma(n, 2n) = \sigma_c[1 - e^{-z}(1+z)]. \tag{8.9}$$

For energies above the threshold (8.9) shows that the cross section of the $(x, 2n)$ reaction rises rapidly at the expense of the cross section for the (x, n) reaction and soon accounts for the greater part of the reaction cross section. In Sect. 10 we show that, with proper choice of Θ, (8.8) and (8.9) account satisfactorily for the experimental data. However, in a later section it will be shown that the experimental data for protons in the 20 Mev region shows many inconsistencies. We shall postpone comment on these until the data has been examined.

9. Angular distribution of products. The statistical theory assumes that the compound nucleus has lost all memory of its mode of formation by the time the decay product is emitted. It might be expected therefore that the angular distribution of the decay product should be isotropic. However, unless the nucleus is disturbed by an external field, the angular momentum of the compound nucleus and its component in a particular direction must be the same as that of the initial state. It is not obvious therefore that the compound nucleus does not remember the direction of incidence of the ingoing particle. This point has been examined by WOLFENSTEIN[1] who has calculated the angular distribution to be expected assuming that the states in the compound nucleus are excited with random phase relations. For heavy elements, the measurement of the angular distribution of the outgoing particles gives the net effect of large numbers of transitions to the residual nucleus. It appears that, in general, the angular distribution of these products is not isotropic, but symmetrical about a plane at 90° to the direction of the incident particle. Isotropy appears only when the level density of states of spin J in the residual nucleus is proportional to $(2J+1)$ (Sect. 57, 66). The theory has been further developed by HAUSER and FESHBACH[2], who were concerned mainly with the inelastic scattering of neutrons. The experimental data, when the part due to compound nucleus formation can be separated from that due to direct interaction, confirm this theory; for ingoing particles with energies up to 20 Mev, compound nucleus formation does, in general, lead to isotropic angular distributions. For higher energies, the angular distributions appear to

[1] L. WOLFENSTEIN: Phys. Rev. **82**, 690 (1951).
[2] W. HAUSER and H. FESHBACH: Phys. Rev. **87**, 366 (1952).

be concentrated in the forward and backward directions indicating a breakdown of the proportionality of level density with the statistical weight of the spin (see Sect. 66).

10. Experimental confirmation of the statistical theory. We shall now consider some experiments which have been designed to test the statistical theory in so far as it predicts cross sections of various reactions, their inter-relationship, and their energy dependence. In all the experiments discussed below the incident particle was an α-particle. This has the advantage that it can be accelerated in a given cyclotron to a higher energy than can be obtained with protons, therefore making possible the study of the competition of reactions with successively higher thresholds. The cross section for particular reactions were always obtained by measurement of the radioactivity of the residual nucleus. This method has the advantage of experimental simplicity but it is necessarily restricted to those nuclei which give active products with periods in a suitable range for measurement. The experimental procedure adopted consisted in bombarding a stack of foils of the material to be studied. Thus all foils receive the same irradiation; and the energy of the bombarding particles traversing a particular foil can be calculated from the initial energy of the beam and the stopping power of the material.

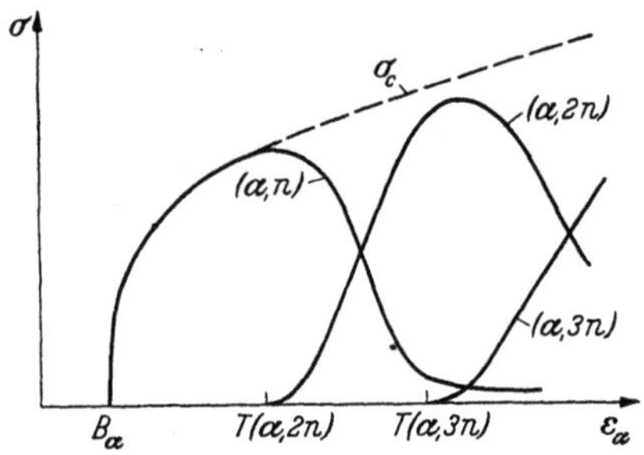

Fig. 5. Behaviour of the cross sections of (α, n), (α, 2n) and (α, 3n) reactions according to the statistical theory; the T's are thresholds, B_α is the barrier height.

We should expect that the reactions (α, n), (α, 2n), etc., should be detected successively as the energy of the α-particles is increased above the thresholds. Thus, if proton emission is disregarded, the cross section for the (α, n) reaction should be equal to the reaction cross section (Fig. 5) for α-particle energies below the threshold of the (α, 2n) reaction. Above this threshold, the (α, 2n) cross section, according to (8.9), should rise rapidly and take over the greater part of the reaction cross section until it, in turn, has to compete with the (α, 3n) reaction. It should be noted that the large contribution to the reaction cross section made by successive reactions, each in its turn, is a result of the rapid rise of the level density of the residual nucleus with excitation energy. The experimental data, as we shall show below, confirm both this rise in level density, and the validity of the statistical theory in so far as it predicts the general shape of the variation of the cross section with energy. We shall see later that anomalies revealed by such experiments are not concerned with the shapes of the cross section curves, but rather with the magnitudes of the cross sections attained by competing reactions.

A search for α-particle reactions suitable for testing the statistical theory was reported by LARK-HOROWITZ and others[1] in 1947. The first quantitative study was made by BRADT and TENDAM[2]. These authors measured the cross sections of the (α, n) and (α, 2n) reactions in Ag and Rh in the range 15 to 20 Mev. The α-particles were obtained from a cyclotron. The results obtained for Rh targets are shown in Fig. 6. They demonstrate the general features discussed above and illustrated in Fig. 5, viz., that the sum of the cross sections for the two reactions is in good agreement with the calculated reaction cross section.

[1] K. LARK-HOROWITZ, J. R. RISSER and R. N. SMITH: Phys. Rev. **72**, 1117 (1947).
[2] H. L. BRADT and D. J. TENDAM: Phys. Rev. **72**, 1117 (1947).

The Ag measurements were repeated in 1953 by BLEULER, STEBBINS, and TENDAM[1]. The method was the same, but in this case attention was paid only to the reaction with Ag^{109}. The (α, n) reaction in this nucleus produces both the 14.5 min activities of the ground state of In^{112}, and the 21 min activity of its isomeric state; both were measured. The $(\alpha, 2n)$ reaction produces the 2.8 day activity of In^{111}. Allowing for the generic relationship between the two activities of In^{112} these authors found that the sum of the cross sections of the two reactions fitted the theoretical reaction cross section calculated for a nuclear radius given by $R = r_0 A^{\frac{1}{3}}$ with $r_0 = 1.6 \times 10^{-13}$ cm. They also found that the ratio of the cross sections was closely given by (8.8) with $\Theta = 1.6$ Mev in fair agreement with the previous result. Very good agreement between the experimental and theoretical values for the ratio of the cross sections was found if the latter were calculated on the basis of

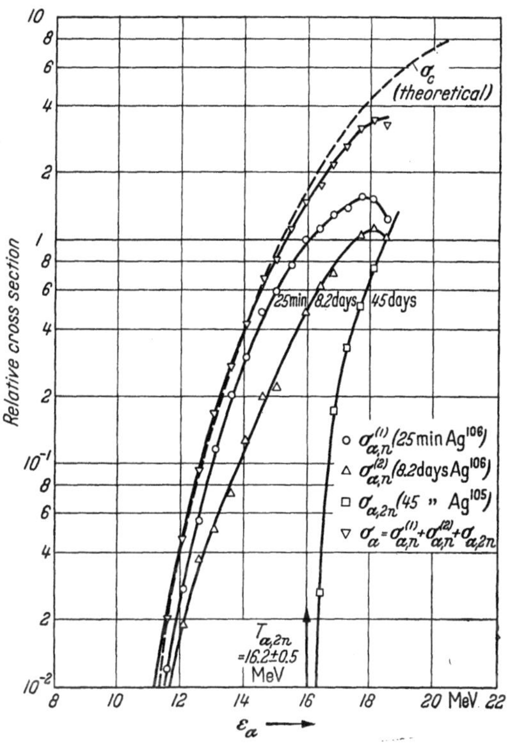

Fig. 6.

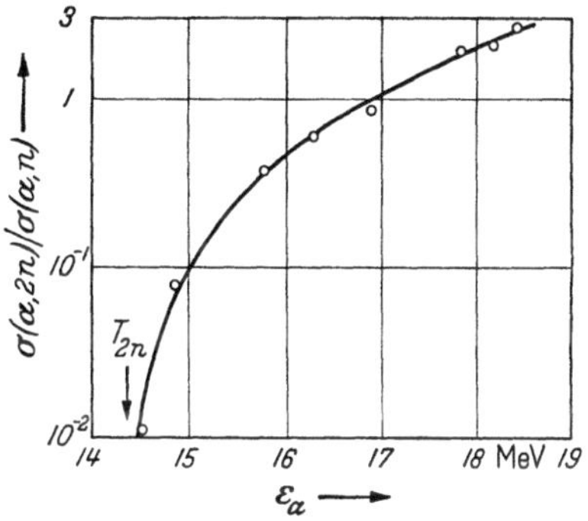

Fig. 7.

Fig. 6. Cross sections for the (α, n) and $(\alpha, 2n)$ reactions with Rh according to BRADT and TENDAM.

Fig. 7. Ratio of the cross sections of the $(\alpha, 2n)$ and (α, n) reactions with Ag^{109} according to BLEULER, STEBBINS, and TENDAM. The full curve is that calculated for $\omega \quad \exp[2(aE)^{\frac{1}{2}}]$, $a = 2.5\ \text{Mev}^{-1}$.

a level density formula of the expected theoretical type, viz., $\exp 2(a\,E)^{\frac{1}{2}}$, the parameter a being adjusted to give the best fit (Fig. 7). In this calculation, the cross sections for reactions involving proton emission were neglected.

Similar measurements with α-particles have been made by KELLY and SEGRÈ[2] and by TEMMER[3]. The former measured the cross sections for the $(\alpha, 2n)$ and $(\alpha, 3n)$ reactions with Bi. The active products were At^{211} and At^{210}, the thresholds for the reactions being close to 20 and 30 Mev respectively. Assuming that other reactions are unimportant, they find good agreement between the sum of these cross sections and the theoretical reaction cross section, calculated on the basis of roughly $r_0 \approx 1.4 \times 10^{-13}$ cm. TEMMER has measured the cross sections of (α, n), $(\alpha, 2n)$ and $(\alpha, 3n)$ reactions in In. The products of these reactions are

[1] E. BLEULER, A. K. STEBBINS and D. J. TENDAM: Phys. Rev. **90**, 460 (1953).
[2] E. L. KELLY and E. SEGRÈ: Phys. Rev. **75**, 999 (1949).
[3] G. M. TEMMER: Phys. Rev. **76**, 424 (1949).

Sb^{118}, Sb^{117}, and Sb^{116}, respectively, and the thresholds lie at 15 and 26 Mev. While the absolute activities of these nuclei were not measured, the rate of increase of the Sb^{118} activity was found to be in agreement with that calculated on the basis of $r_0 \approx 1.5 \times 10^{-13}$ cm.

D. Direct interactions.

11. Introduction. We have pointed out in Sect. 3 that it is natural to assume that the formation of the compound nucleus is preceded by collisions between the incident particle and the nuclear constituents. When the incident particle has sufficient energy, the initial collision may be rapidly followed by another until a nucleon cascade develops. Those components of the cascade which have sufficient energy to leave the nucleus, do so, provided their momenta do not lie in a range forbidden by the exclusion principle; the remainder are retained and their energy must contribute to the excitation energy of the compound nucleus. The cascade continues to develop until no further nucleons are projected with an energy sufficient to escape. The nucleus that remains is the compound nucleus we have already discussed; but it should be noted that this compound nucleus does not necessarily have the charge and neutron number which is to be expected on the old strong coupling hypothesis for one or more nucleons may already have been ejected. Furthermore, the nucleons ejected in this cascade are produced in individual collisions and therefore they are mainly emitted in the forward direction. Thus a forwardly directed component in the angular distribution of the disintegration products, when not accompanied by an equal component in the backward direction, can be attributed to a direct interaction process.

The study of the nucleon cascade which results from a high energy collision was first put on a quantitative basis by GOLDBERGER[1] who, using the Monte Carlo method, calculated the energy spectrum and angular distribution of the products of collisions with 87 Mev neutrons. In this article we are concerned with much lower energies. In this case, the direct process which initiates the compound nucleus is hardly such as to warrant the description of a nucleon cascade. For 20 Mev neutrons and protons, on the average, there probably occurs only one collision for which the participants have sufficient energy to leave the nucleus.

12. Mean free path in nuclear matter. The starting point for the understanding of direct interactions is the mean free path in nuclear matter. The mean free path is the outcome of two conflicting factors: firstly, the internucleon collision cross section, a quantity which falls with increasing kinetic energy; and, secondly, the restrictions imposed by the exclusion principle which, in general, make collisions more probable for increasing energy.

The mean free path in nuclear matter is $\Lambda = 1/\varrho\bar{\sigma}$, where ϱ is the density (about 1.4×10^{38} cm^{-3}) and $\bar{\sigma}$ is an average cross section for internucleon collisions. Generally, $\bar{\sigma}$ is $(N\sigma_{in} + Z\sigma_{ip})/A$ where σ_i stands for a neutron or proton depending on the nature of the incident particle, and N, Z and A have their usual significance.

For energies greater than twice FERMI energy, E_f, ($\sim$28 Mev), GOLDBERGER showed that the average collision cross section is given by the formula

$$\frac{\bar{\sigma}}{\sigma_i} = 1 - \frac{7}{5}\frac{E_f}{E_1} \qquad (12.1)$$

[1] M. L. GOLDBERGER: Phys. Rev. **74**, 1269 (1948).

where σ_i is the free nucleon cross section, and E_1 is the kinetic energy of the incident particle, measured within the nucleus. For energies lower than $2E_f$ this formula does not hold. The mean free path has been computed for various energies by a Monte Carlo study of the collisions of 140 Mev protons with nuclei by MORRISON, MUIRHEAD and ROSSER[1]. The mean free path was found to fall with increasing energy, reaching a minimum near 50 Mev as measured within the nucleus, after which point it increases slowly. At 50 Mev, the mean free path is about 3.5×10^{-13} cm. A numerical calculation has also been made by YAMAGUCHI[2] with rather similar results.

The subject has been studied in more detail recently by HAYAKAWA, KAWAI and KIKUCHI[3] and by CLEMENTEL and VILLI[4]. They showed that the average cross section is given by the formula:

$$\bar{\sigma} = \frac{3}{4\pi P_1 P_f^3}\int \frac{P_1^2+P_2^2-2P_f^2}{|\boldsymbol{P}_1+\boldsymbol{P}_2|}\,\sigma_i(\boldsymbol{P})\,d^3\boldsymbol{P}_2 \quad (12.2)$$

where P_1, P_2, and P_f are the momenta of the incident particle, the target nucleon and the FERMI momentum, respectively. If $P_1^2 \geq 2P_f^2$, the limits of integration are 0 and P_f; but if $P_1^2 \leq 2P_f^2$, the limits are $\sqrt{2P_f^2-P_1^2}$, and P_f. If the internucleon cross section is assumed to be independent of the relative momentum of the colliding particles, (12.2) reduces to a rather simple formula for the average cross section. For $E_1 \geq 2E_f$, we obtain GOLDBERGER's formula (12.1), while if $E_1 \leq 2E_f$,

$$\frac{\bar{\sigma}}{\sigma_i} = 1 - \frac{7}{5}\frac{E_f}{E_1} + \frac{2}{5}\frac{E_f}{E_1}\left(2-\frac{E_1}{E_f}\right)^{\frac{5}{2}}. \quad (12.3)$$

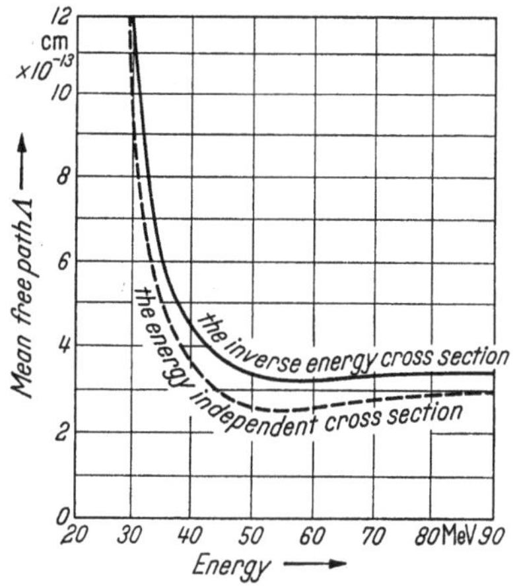

Fig. 8. Calculated mean free path for nucleon collisions within the nucleus, according to HAYAKAWA, KAWAI, and KIKUCHI.

This formula differs from that obtained by YAMAGUCHI, which HAYAKAWA and his collaborators claim to be incorrect. The more realistic assumption that the internucleon cross section varies as $1/P^2$, where P is the relative momentum, gives a more complicated expression. The mean free path, however, is not very different. In this case, if $\sigma_i=\sigma_0/P^2$

$$\bar{\sigma} = 6\,\sigma_0 P_1 I/P_f^3 \quad (12.4)$$

where

$$I = \int_a^b \frac{x(1+x^2-2\alpha^2)}{\sqrt{2(1+x^2)}}\tanh^{-1}\frac{2x\sqrt{2(1+x^2)}}{1+3x^2}\,dx \quad (12.5)$$

with $x=P_2/P_1$, and $\alpha=P_f/P_1$.

Here the limits of integration are 0 and α for $E_1\geq 2E_f$, and $\sqrt{2\alpha-1}$ and α if $E_1\leq 2E_f$. The results are shown in Fig. 8; it is perhaps surprising that the mean free path is so little dependent on the two internucleon cross sections. It should be noted that the minimum mean free path is about 3×10^{-13}, which is roughly equal to the internucleon distance. At higher energies these results are in rough

[1] G. C. MORRISON, H. MUIRHEAD and W. G. ROSSER: Phil. Mag. **44**, 1326 (1953).
[2] Y. YAMAGUCHI: Progr. Theor. Phys., Japan **5**, 332 (1950).
[3] S. HAYAKAWA, M. KAWAI and K. KIKUCHI: Progr. Theor. Phys., Japan **13**, 415 (1955).
[4] E. CLEMENTEL and C. VILLI: Nuovo Cim. **2**, 176 (1955).

agreement with that obtained by TAYLOR[1] (about 4×10^{-13} cm) from an analysis of high energy scattering.

It should be noted that the kinetic energy corresponding to the minimum mean free path is equivalent to about 10 Mev in the laboratory. For lower energies the probability of a direct interaction is small and the nucleus is transparent. This does not mean that the reaction cross section is small because in this energy region the interaction of a nucleon with the nucleus takes on a resonance character.

Since the collision of the ingoing nucleon with one of the constituents of the nucleus is effectively an inelastic process, the probability of these collisions

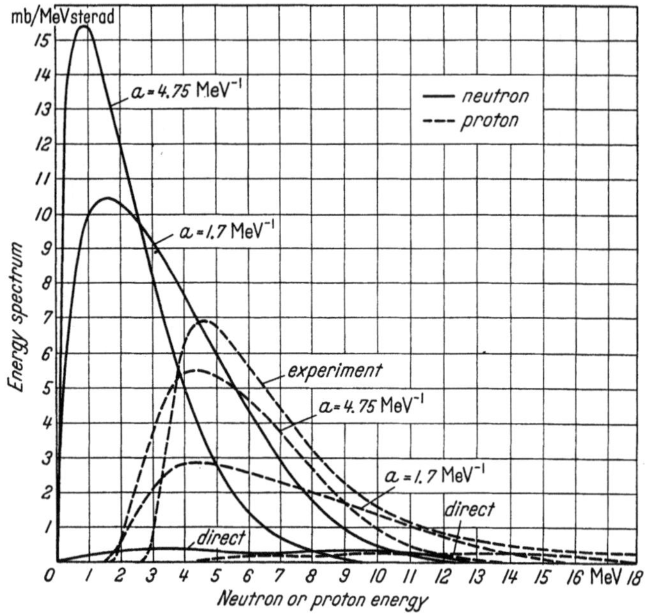

Fig. 9. Cross sections for direct and statistical reactions for 18 Mev protons on Fe^{56}, according to HAYAKAWA, KAWAI and KIKUCHI. The quantity a is that which appears in the level density function, $\exp[2(aE)^{\frac{1}{2}}]$.

can be determined by the imaginary part of the well potential described by the optical model (Sect. 6). Thus the mean free path is related to the imaginary part (W) through the elementary relation[2]: [cf. Eq. (6.1)].

$$W = \frac{\hbar}{\Lambda}\sqrt{\frac{E+V_0}{2m}}\,.$$

The magnitude of this imaginary part has been calculated by CLEMENTEL and VILLI[3], by MORRISON, MUIRHEAD, and MURDOCH[4], and by LANE and WANDEL[5] (see Fig. 21). The experimental data generally is in good agreement with these calculations (see Sect. 21).

13. Cross sections for direct reactions. HAYAKAWA and his collaborators have investigated the energy transfer in direct collisions. It appears that, for protons of energies near 20 Mev, a large fraction of the energy of the incident particle is transferred to the target nucleon. The projected particle therefore

[1] T. B. TAYLOR: Phys. Rev. **92**, 831 (1954).
[2] N. C. FRANCIS and K. M. WATSON: Amer. J. Phys. **21**, 659 (1953).
[3] See ref. 4, p. 221.
[4] G. C. MORRISON, H. MUIRHEAD and P. A. B. MURDOCH: Phil. Mag. **46**, 795 (1955).
[5] A. M. LANE and C. F. WANDEL: Phys. Rev. **98**, 1524 (1955).

has comparatively high energy and the probability of making further collisions, while the incident particle is unlikely to leave the nucleus. These authors have estimated the cross sections of reactions resulting from single collisions with the aid of some simplifying approximations. They take into account the attenuation of the products of collision, the transmission of the incident and the final particles through the nuclear surface, and also the variation of this transmission as a function of angle of incidence.

The results of these calculations are illuminating. The relative yields of the iron reactions for the direct and statistical theories are shown in Fig. 9. It is obvious that quantitatively in this energy range the direct interactions are always relatively unimportant in comparison with the more probable statistical reaction. They are not competitive in the sense of the statistical reactions, for they make only a small contribution to the total reaction cross section; nor do they compete with the compound nucleus formation for their share of the geometrical cross section, for they represent not alternative but preliminary nuclear processes. HAYAKAWA and his collaborators have also estimated the yields of protons of different energies ejected from tin by 31 Mev protons. Their results are compared with the experimental results of EISBERG and IGO[1] in Fig. 10. This figure shows the characteristically flat energy distribution of the products of direct interactions in strong contrast with the low energies of the decay products of the compound nucleus, which, in this instance, might be expected to have an average energy of about 3 Mev. It will be seen that there is good agreement for the forward direction; however, at 90° the direct interaction cross sections are too low[2].

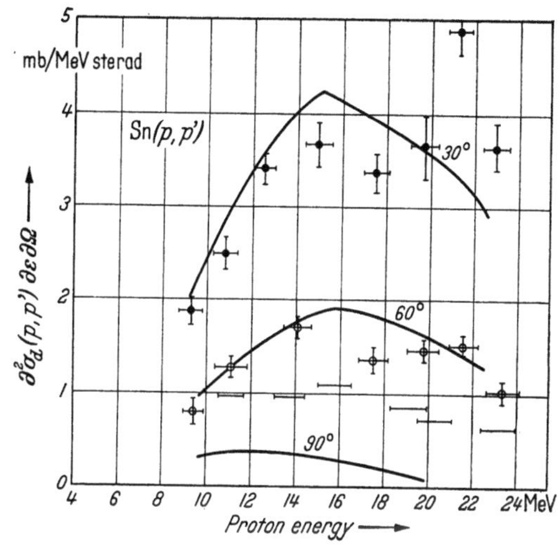

Fig. 10. Differential cross sections for inelastic scattering of 31 Mev protons by tin; the experimental points are those of EISBERG and IGO, the full curves are those calculated by HAYAKAWA, KAWAI, and KIKUCHI.

Direct collisions have been independently investigated by BROWN and MUIRHEAD[3] who have found rather higher cross sections than those obtained by the authors first mentioned. For the (n, p) reactions with 14 Mev neutrons the calculated cross sections which always contain a considerable direct interaction component, follow the general trend found by PAUL and CLARKE (Sect. 28); the direct component varies from a few mb in the heaviest elements to some 40 mb or more in light elements. For proton induced reactions at 18 Mev, their theoretical cross sections do not seem quite adequate to account for the direct component observed.

Although insignificant in their contribution to the reaction cross section the direct processes are readily detected as deviations from the isotropy of the

[1] R. M. EISBERG and G. IGO: Phys. Rev. **93**, 1039 (1954).

[2] According to L. R. B. ELTON and L. C. GOMES (to be published) the results of EISBERG and IGO must be understood as a surface effect because most of the protons produced by collisions in the interior of the nucleus are prevented from leaving it by total internal reflection at the surface.

[3] G. BROWN and H. MUIRHEAD: To be published.

angular distribution. Since the products of direct interaction are expelled in the forward direction, and usually have higher energy than those due to evaporation from the compound nucleus, they may be easily distinguished. BROWN and MUIRHEAD have calculated the angular distribution of 14 Mev neutrons scattered by bismuth and found good agreement between the calculated forward component and the experimental data of ROSEN and STEWART (Sect. 27).

14. Direct interactions: discrete states. The semi-classical theory which we have just outlined gives no account of the yield of direct reactions leaving the residual nucleus in discrete states. MCMANUS and SHARPE[1] have attempted to explain the anomalously high (n, p) cross sections in heavy elements (Sect. 28) by a direct interaction theory in which the incoming neutron is supposed to interact with one of the nuclear protons in the peripheral area of the nucleus, where the proton can be ejected with little probability of collision with nucleons in the main body of the nucleus. According to these authors the direct interaction is, in the main, a surface effect. The subject has been further developed by AUSTERN, BUTLER, and MCMANUS[2]. Disregarding the effect of the COULOMB field, they consider the interaction between an incoming neutron wave and that of a loosely bound proton near the surface of the nucleus. The proton, which is supposed to have an orbital angular momentum l_p, is knocked out of the nucleus, while the neutron is retained in a state with orbital angular momentum l_n. The differential angular distribution which results from this interaction is rather similar to that of the deuteron stripping reactions (Sect. 42). It is given by the formula:

$$\frac{d\sigma}{d\Omega} = \sigma_i \frac{k'}{k} M^2(R) \sum_l \langle l_n l_p 0 0 | l 0 \rangle^2 [j_l(QR)]^2$$

where j_l is a spherical BESSEL function of order l, where l is determined by the selection rule:

$$|l_n + l_p| \geqq l \geqq |l_n - l_p|$$

and must also be consistent with conservation of parity; σ_i is the free particle cross section; k and k' are the wave numbers of incident neutron and outgoing proton respectively; $\langle l_n l_0 0 0 | l 0 \rangle$ is a CLEBSCH-GORDAN coefficient; Q measures the difference in momenta of ingoing and outgoing particles, viz., $Q = |\boldsymbol{k} - \boldsymbol{k}'|$; and $M^2(R)$ is dependent on the nuclear radius and the radial components of the wave functions.

The angular distribution, therefore, is peaked in a way resembling the angular distributions so characteristic of the stripping process. The predicted cross section for a reaction which leads to the residual nucleus in an excited state of low energy is of the order of a few millibarns. As yet there is little experimental evidence to compare with theoretical predictions; however, some measurements show just the angular distribution to be expected. In this theory the type of the incident or ejected particle matters little, and the theory can be applied with equal validity to inelastic scattering. The angular distribution of 17 Mev protons scattered inelastically by Fe^{56} and which leave that nucleus in its first excited state at 840 kev, has been measured by SCHRANK, GUGELOT, and DAYTON[3]. They found that the scattered protons were strongly concentrated in the forward direction (see Fig. 11) and that the observed angular distribution

---

[1] H. MCMANUS and W. T. SHARPE: Phys. Rev. **87**, 188 (1952).
[2] N. AUSTERN, S. BUTLER and H. MCMANUS: Phys. Rev. **92**, 350 (1953).
[3] G. SCHRANK, P. C. GUGELOT and I. E. DAYTON: Phys. Rev. **96**, 1156 (1954).

could be represented by a sum of spherical BESSEL functions of even order. Rather similar results have been obtained by FREMANTLE and collaborators[1] for 9.5 Mev protons scattered in argon. However, some recent measurements by GUGELOT and PHILLIPS[2] on the inelastic scattering of 18 Mev protons by Mg do not show the expected effect so clearly.

HAYAKAWA and YOSHIDA[3] have calculated the cross section for the excitation of surface vibrations by inelastic scattering in even-even nuclei. The angular

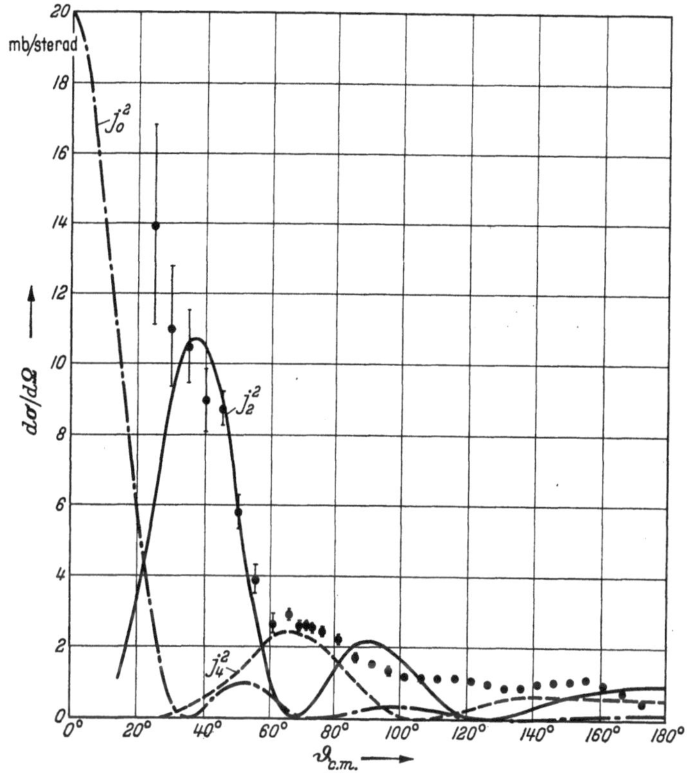

Fig. 11. Angular distribution of 17 Mev protons scattered inelastically by Fe^{56}, leaving that nucleus in its first excited state at 840 kev, after SCHRANK, GUGELOT, and DAYTON. The curves represent the squares of spherical BESSEL functions of even order.

dependence for the excitation of the first excited state (spin 2) is again determined by $[j_2(QR)]^2$ as in the theory of direct interactions outlined above. This process seems to have been detected recently by GUGELOT and RICKEY[4] in the inelastic scattering of 42 Mev α-particles from Mg. These authors point out that direct interaction between an α-particle and a nucleon in the nucleus would probably result in the ejection of the nucleon at these energies. If instead the effect observed is attributed to the excitation of a surface oscillation, this difficulty disappears, and the absolute values of the cross sections can be used to determine the product of the quadrupole moment of the nucleus with the interaction potential of α-particle and nucleus.

[1] R. G. FREMANTLE, D. J. PROWSE, A. HOSSAIN and J. ROTBLAT: Phys. Rev. **96**, 1270 (1954).
[2] P. C. GUGELOT and P. R. PHILLIPS: Phys. Rev. **101**, 1614 (1956).
[3] S. HAYAKAWA and S. YOSHIDA: Progr. Theor. Phys., Japan **14**, 1 (1955).
[4] P. C. GUGELOT and M. RICKEY: Phys. Rev. **101**, 1613 (1956).

E. Total neutron cross sections.

15. Experimental data. We have already pointed out (Sect. 4) that the measurement of total cross sections requires the measurement of transmission of neutrons through absorbing materials in conditions of good geometry. A great deal of experimental data has already been accumulated; but it is only in recent years that systematic studies of total cross sections over a large range of energies and elements have been made. We shall not attempt to list all the references to this data: for references up to 1950, the reader is referred to the admirable review published by ADAIR [6].

Recently, very careful measurements[1] have been made from very low energies (50 kev) to 3 Mev for a large number of elements. In these measurements the

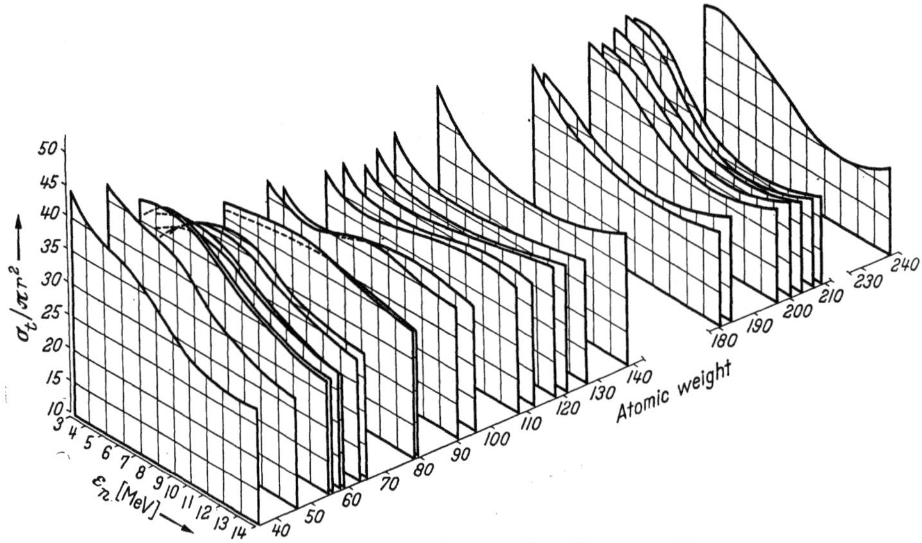

Fig. 12. Total neutron cross sections according to NERESON and DARDEN.

detectors usually consisted of biassed recoil counters, usually helium or hydrogen at high pressures; for neutron sources above 1 Mev, usually the T (p, n), and below, the Li (p, n) reactions have been used.

From 3 to 12 Mev, measurements of the total cross sections in a large number of elements have been made by NERESON and DARDEN[2] using the neutrons emitted by a fast neutron reactor as a source, and a special energy-selective detector. In that instrument, protons projected in the forward direction from a polythene foil were measured by an ionisation chamber connected in coincidence with a proportional counter located between it and the foil, and in anticoincidence with another counter beyond it. The energy of the protons was measured by the size of the pulses they produced in the chamber; the energy resolution was about 10%.

Above 12 Mev, total cross section measurements are less complete. Some measurements have been made by BONNER and his associates[3] in the range

[1] D. W. MILLER, R. K. ADAIR, S. K. BOCKELMAN and S. E. DARDEN: Phys. Rev. **88**, 83 (1952). — M. WALT, R. L. BECKER, A. OKAZAKI and R. E. FIELDS: Phys. Rev. **89**, 1271 (1953). — A. OKAZAKI, S. E. DARDEN and R. B. WALTON: Phys. Rev. **93**, 461 (1954).

[2] N. NERESON and S. DARDEN: Phys. Rev. **89**, 775 (1953); **94**, 1678 (1954).

[3] C. F. COOK and T. W. BONNER: Phys. Rev. **94**, 651 (1954). — T. W. BONNER, F. ALBA, A. FERNANDEZ and M. MAZARI: Phys. Rev. **97**, 985 (1955).

14 to 18 Mev. The neutrons for this energy range were obtained by the T (d, n) reaction. Few accurate measurements have yet been made immediately above 20 Mev; though some measurements have been made at 25 Mev by SHERR[1]. There is a gap in the data extending from this energy to the point where measurements have been made with synchrocyclotrons, viz. 60 Mev and upwards.

The experimental results show some very characteristic features. For the lighter nuclei, and for closed shell nuclei, the low energy cross sections show resonance structure. The cross sections are generally largest at low energies. The average cross section falls with increasing energy, but with characteristic maxima and minima whose position varies continuously with atomic weight. For the lower energies, the results have been presented in three dimensional form by BARSCHALL[2]. Experimental results for neutrons up to 9 Mev are shown in Fig. 40b, p. 99; some higher energy results are shown in Fig. 12.

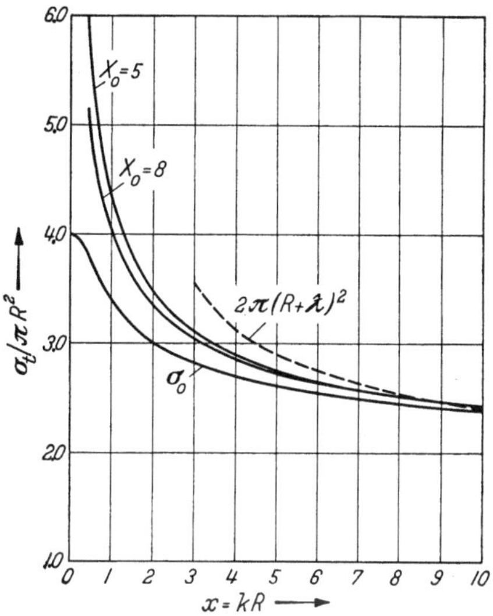

Fig. 13. Theoretical total neutron cross sections according to the strong coupling theory of FESHBACH and WEISSKOPF.

16. Theoretical interpretation: Strong coupling. An attempt to account for the experimental data as it appeared in 1949 was made by FESHBACH and WEISSKOPF[3]. For low energies these authors calculated the total and the reaction cross sections by averaging over the resonance expression obtained by FESHBACH, PEASLEE and WEISSKOPF[4]. For higher energies, when the widths of resonances exceed their spacing, $\Gamma/D \gg 1$, they assumed that the nucleus is strongly absorbing and that the internal wave can be represented by an inwardly converging component only (cf. BETHE[5]). They showed that the low energy calculation becomes equivalent to the high energy method when $\Gamma \gg D$. The results of these calculations are shown in Fig. 13, where the total cross section is plotted against $x = kR$, for different values of $X = KR$ (defined in Sect. 7). It will be seen that the cross section is everywhere a monotonically decreasing function of the energy, through the quantity $x = kR$, and, contrary to what was originally thought, approaches the value of $2\pi R^2$ only at very high energies. In fact, at high energies the best approximation is $2\pi(R + \lambda\!\!\!^-)^2$. The cross section, moreover depends, to some extent (especially at low energies) on the internal wave number. This quantity, K, is given by $K^2 = K_0^2 + k^2$, where K_0 is the value it takes at zero energy. Thus, roughly $K \approx K_0$ which is approximately 1×10^{-13} cm, so $X = KR$ is numerically approximately equal to the nuclear radius. It will be seen that the total cross section is always somewhat higher than that to be expected from the scattering of an impenetrable sphere (the curve labelled σ_0 in Fig. 13, corresponding to $K = \infty$). At low energies the rise in total cross section may be attributed to the barrier transmission of s-waves, for which $T_0 = 4kK/(K + k)^2$. Thus, for $k \to 0$, the cross section becomes $4\pi/kK$, which is inversely proportional

[1] R. SHERR: Phys. Rev. **68**, 240 (1945).
[2] H. H. BARSCHALL: Phys. Rev. **86**, 431 (1952).
[3] H. FESHBACH and V. F. WEISSKOPF: Phys. Rev. **76**, 1550 (1949).
[4] See ref. 1, p. 212.
[5] H. A. BETHE: Phys. Rev. **57**, 1125 (1940).

to the velocity. FESHBACH and WEISSKOPF show that their calculations, in some cases, are in good agreement with the experimental results; but in other cases, there are serious discrepancies. These discrepancies are the more apparent in the recent measurements described in Sect. 15, for there seems no way in which a theory, based solely on the strong coupling hypothesis, can give rise to the maxima and minima which have been observed (Fig. 12).

17. Theoretical interpretation: optical model. The theoretical difficulties just described are explained satisfactorily by the optical model of the nucleus. The previous assumption of strong coupling, requires in effect, that every neutron incident on the nucleus is absorbed; this assumption is replaced by the assumption of weak absorption: a collision, then, does not necessarily involve the absorption of the neutron. The neutron may be imagined to enter the nucleus and make many internal motions before escaping or being absorbed. The large scale resonances in the total neutron cross section, then, receive a natural explanation as size-resonances[1] of the well. The total neutron cross sections calculated by FESHBACH, PORTER and WEISSKOPF[2] on the basis of a square potential well, are shown in Fig. 27, p. 420. It will be seen that in general there is good agreement. However, there remain difficulties in the theoretical interpretation of the widths of the size-resonances which derive from the imaginary part of the potential[3]. The experimental data show the high cross sections at low energies caused by s-wave resonances for atomic weights near 55 and 160; the p-wave resonance for A near 90; and the d-wave resonances at intermediate energies for A near 50 and 140.

18. Scattering lengths. Except for very low energy neutrons, where absorption processes become increasingly important, the total neutron cross section for low energy neutrons (with energies of a few ev) is mainly caused by scattering in s-states. The cross section is given by the usual formula $\sigma = 4\pi \lambda\!\!\!^{-2} \sin^2 \delta_0$. For low energy neutrons the phase difference, δ_0, is always small (provided the neutron energy does not lie too close to a resonance). Thus the cross section is given by the quantity $4\pi a^2$, where $a = -\lambda\!\!\!^{-} \delta_0$ is the displacement of the scattered wave which corresponds to its phase. This quantity, which determines the amplitude of the scattered wave, is called the "*scattering length*". For an impenetrable sphere, the wave must be zero at the surface; the scattering length, then, is equal to the radius (R), and the cross section is $4\pi R^2$. The low energy neutron scattering cross sections provide a rough estimate of the nuclear radius; however it is not an accurate one, because the nucleus, in fact, does not behave like an impenetrable sphere. To derive a better value we require a more realistic model: this is discussed in Sect. 19.

When the wavelength of the neutrons is of the order of the lattice spacing, the cross section determined by the usual transmission methods will not be equal to the free nucleus cross section measured at higher energies because of the existence of interference effects. In the low energy range the total cross section is made up of a coherent and an incoherent part; the former disappears for very low energies, i.e., for neutrons with wavelengths longer than the "cut-off" wavelength (twice the lattice spacing), and for higher energies, for which the coherence is reduced by the thermal motion (0.1 ev and above). In general, the scattering by a nucleus with a finite spin is determined by two scattering lengths, one for

---

[1] We prefer the term "size-resonances" first used by FORD and BOHM, in preference to "giant" resonances which we use in connection with photonuclear processes.

[2] See ref. 1, p. 210.

[3] A. M. LANE, R. G. THOMAS and E. P. WIGNER: Phys. Rev. **98**, 693 (1955). See also E. VOGT: Phys. Rev. **101**, 1792 (1955).

each of the two spin states. These lengths, as we have seen, must have the order of magnitude of the nuclear radius; sometimes, however, they differ very considerably from the nuclear radius, and for a given nucleus, the two scattering lengths may even have opposite sign (e.g., in V^{51}). The measurement of these lengths lies outside the scope of the present article; they are fully described in references [7] and [8]; many scattering lengths are tabulated in reference [7]. Most nuclei have scattering lengths with the same sign; conventionally, this sign is considered to be "positive", i.e., the same sign as that of an impenetrable sphere for which the scattered wave is 180° out of phase with the incoming wave. The few negative scattering lengths, in principle, could be derived from the close proximity of resonances, because, for energies just below the resonance energy, the resonance scattering interferes destructively with the "potential", or hard sphere, scattering and the net scattering length may have opposite sign. However, it seems more likely that most of these negative scattering lengths, in fact, originate in certain resonance properties of the potential scattering which we shall now discuss.

19. Size-resonances. The first reference to size-resonances appears to have been made by FORD and BOHM[1]. These authors pointed out that if the scattering lengths of slow neutrons are plotted against the atomic weight, the distribution obtained shows signs of discontinuities near mass numbers of 55 and 150. In fact, the few negative scattering lengths appear in Li and in the vicinity of mass 55, where, also, both small and large positive scattering lengths are found. In the vicinity of $A=150$, no negative scattering lengths appear, but some low positive scattering lengths occur.

An elementary theoretical formula for scattering lengths based on the optical model may be obtained as follows. In this case, the negative scattering lengths in the neighbourhood of $A=55$, and the erratically distributed positive lengths are supposed to be caused, not by the fortuitous presence of a sharp resonance in the immediate vicinity of the neutron energy, but by the broad size-resonance of the nucleus. If the external wave function is $r\psi=\sin(kr+\delta_0)$ for $r>R$, and the internal wave function is $r\psi=\sin Kr$, for $r<R$, the requirement of continuity of the logarithmic derivative for small k gives[2]:

$$a=R\left[1-\frac{\tan K_0 R}{K_0 R}\right] \tag{19.1}$$

where K_0 is defined in Sect. 16. For $k=0$, $a=R$, and the total cross section becomes $4\pi R^2$. For small values of k, however, a will depend on the potential depth through (19.1); thus, a passes through discontinuities when:

$$K_0 R=(2n+1)\cdot\frac{\pi}{2}. \tag{19.2}$$

The experimental data, which are few, are in fair agreement with the discontinuities given by this formula, when $K_0=\sqrt{2m V_0}/\hbar$, $V_0=45$ Mev, and $r_0=1.4\times10^{-13}$ cm. It is not obvious by inspection that the scanty data constitute evidence for the existence of the size-resonances; however, the statistical validity of this conclusion has been confirmed theoretically by SCOTT[3].

As we have seen above the observed low energy scattering lengths may nevertheless be determined by the proximity of sharp neutron resonances. ADAIR[4]

[1] K. W. FORD and D. BOHM: Phys. Rev. **79**, 745 (1950).
[2] Cf. SACHS: Ref. [9] Chap. 4.
[3] J. M. C. SCOTT: Phil. Mag. **45**, 751 (1954).
[4] R. K. ADAIR: Phys. Rev. **94**, 737 (1954).

has pointed out that a smoother curve is obtained when the square root of the low energy total cross section, smoothed out over several resonances, is plotted against the radius or the mass number. This smoothed out cross section should represent the scattering length with disturbances caused by sharp resonances eliminated from it. The result is shown in Fig. 14. It will be seen that the discontinuities remain; their position, however, is not too good a measurement of the well depth, for the quantity involved is the product $V_0 R^2$, and R is not known with precision. However, it is clear that V_0 must be nearer to 40 Mev than any lower value satisfying (19.2) because it must be sufficient to account for the bound levels of heavy nuclei described by the shell model.

A detailed account of the application of the optical model to the interpretation of the low energy neutron cross sections has been given by FESHBACH, PORTER and WEISSKOPF[1]. There are now two quantities of interest: the scattering length, which will be denoted by R', and the "strength function", $\langle \Gamma_n^{\circ} \rangle / D$, the average neutron width divided by the level spacing. The parameters of the size-resonances at zero energy were found by equating the average amplitude of s-waves with that obtained by averaging over the sharp resonances. The scattering length now becomes:

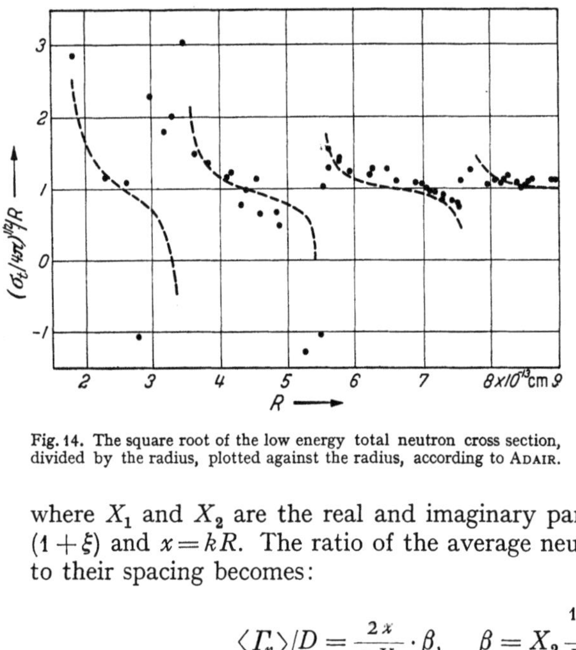

Fig. 14. The square root of the low energy total neutron cross section, divided by the radius, plotted against the radius, according to ADAIR.

$$R' = R\left[1 - \frac{1}{2X_1} \times \frac{\sin 2X_1}{X_2^2 + \cos^2 X_1}\right] \tag{19.3}$$

where X_1 and X_2 are the real and imaginary parts of X where $X^2 = x^2 + K_0^2 R^2 (1+\xi)$ and $x = kR$. The ratio of the average neutron width of sharp resonances to their spacing becomes:

$$\langle \Gamma_n^{\circ} \rangle / D = \frac{2\varkappa}{\pi X_1} \cdot \beta, \qquad \beta = X_2 \frac{1 - \dfrac{\sin 2X_1}{2X_1}}{X_2^2 + \cos^2 X_1}. \tag{19.4}$$

The average neutron widths therefore rise to a maximum value at $K_0 R = (2n+1)\cdot\pi/2$ as before, the value of β being $2/[(2n+1)\cdot\pi/2\cdot\xi]$; and the width at half maximum is $(2n+1)\cdot\pi\xi/2$. The scattering amplitude, R', goes through discontinuities near the resonance points while, between them, it is close to the nuclear radius R.

These predictions have been verified in part by measurements of the quantity $\langle \Gamma_n^{\circ} \rangle / D$. This has been done in two ways: (1) by direct estimates of the average neutron width and the spacing for low energy resonances (for nuclei with $A \sim 150$ by HUGHES and his collaborators[2]); and (2) by measurements of the total cross section for neutrons with energies of the order of a few kilovolts. In the first method the accuracy is necessarily limited by the number of separate

---

[1] See ref. 1, p. 210.

[2] R. S. CARTER, J. A. HARVEY, D. J. HUGHES and V. E. PILCHER: Phys. Rev. **96**, 113 (1954); **99**, 10 (1955).

resonances which it is possible to resolve and for which the parameters can be determined. The second method, which has been developed by HUGHES and PILCHER[1] and by GAYTHER and NICHOLSON[2] is limited by the disadvantage that it is necessary to apply rather difficult corrections to account for the effects of absorption at the resonance peaks. This method has been used by BOLLINGER *et al.*[3] to study nuclei with masses near 50. For thin absorbers, the average value of the total cross section is:

$$\langle\sigma_t\rangle = 2\pi^2 \lambda\!\!\!^{-2} \frac{\langle\Gamma_n\rangle}{D} + 4\pi R'^2. \tag{19.5}$$

In this equation $\lambda\!\!\!^{-2}$ is proportional to ε^{-1} where ε is the neutron energy, and the average neutron width is proportional to $\varepsilon^{\frac{1}{2}}$. Thus, in principle, the value of $\langle\Gamma_n\rangle/D$ may be obtained from the slope of the measured total cross section when the latter is plotted against $\varepsilon^{-\frac{1}{2}}$. The value of the total cross section obtained when the graph is extrapolated to $\varepsilon^{-\frac{1}{2}}\to 0$ gives $4\pi R'^2$, whence the scattering length may be determined.

The scattering length R' may also be determined from direct measurements of the total cross section in thick material[4]. In this case, the neutrons with energies corresponding to the resonance energies are removed by absorption and the measured transmission depends mainly on those neutrons whose energies lies between resonances. These neutrons are attenuated effectively only by potential scattering.

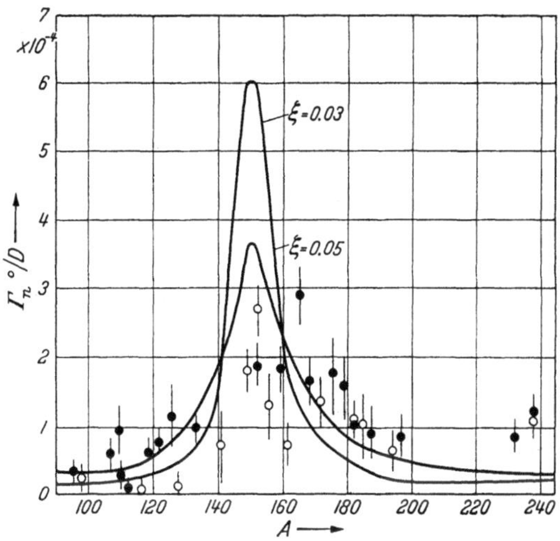

Fig. 15. Experimental data on the strength of neutron resonances. The full curves are theoretical curves calculated on the optical model; the dotted line is that calculated for a black nucleus (see text). Full circles: data of CARTER *et al.*; open circles: data of GAYTHER and NICHOLSON.

The experimental results in the neighbourhood of $A=150$ are shown in Fig. 15 in which the average "reduced" neutron width $\langle\Gamma_n^0\rangle = \langle\Gamma_n\rangle/\varepsilon^{\frac{1}{2}}$ is plotted against the mass number. These results are in fair but not good agreement with the theoretical curve calculated from (19.4) when the parameters required to fit the fast neutron data are used. However, the peak observed is only about one third of the height of that expected theoretically. The results show very clearly the difference between the weak coupling optical model and the former strong coupling model. For the latter the mean neutron width corresponding to neutrons scattered with wave number k is[5]:

$$\frac{\langle\Gamma_n\rangle}{D} = \frac{2k}{\pi K_0}$$

whence the average reduced neutron width is given by

$$\frac{\langle\Gamma_n^0\rangle}{D} = \frac{2}{\pi V_0^{\frac{1}{2}}}.$$

[1] D. J. HUGHES and V. E. PILCHER: Phys. Rev. **100**, 1249 (1955).
[2] D. B. GAYTHER and P. K. NICHOLSON: To be published.
[3] L. M. BOLLINGER *et al.*: To be published.
[4] D. J. HUGHES: To be published,
[5] See ref. 1, p. 212.

For $V_0 = 42$ Mev this quantity is about 1.0×10^{-4}. It will be seen that the experimental results rise above this value at the resonance point and below if off resonance.

The resonance condition (19.2) is that required by a square well and the expressions (19.4) have been calculated on this basis. Better agreement with the experimental data can be obtained by using a well with a sloping edge or taper. Such a well is a more realistic representation of the variation of the potential at the surface of the nucleus and an increasing volume of experimental data, e.g., neutron and proton elastic scattering and nuclear surface energies, demands it. A tapered edge is conveniently given by the following expression:

$$V(r) = \frac{V_0 + i W}{1 + \exp \frac{r - R}{a}} \tag{19.6}$$

which, according to WOODS and SAXON[1], accounts satisfactorily for proton scattering (Sect. 21). The radius R determines the point where the potential is half of its maximum value. The shape of this distribution is shown in Fig. 16 which is plotted for $a = 0.5 \times 10^{-13}$ cm.

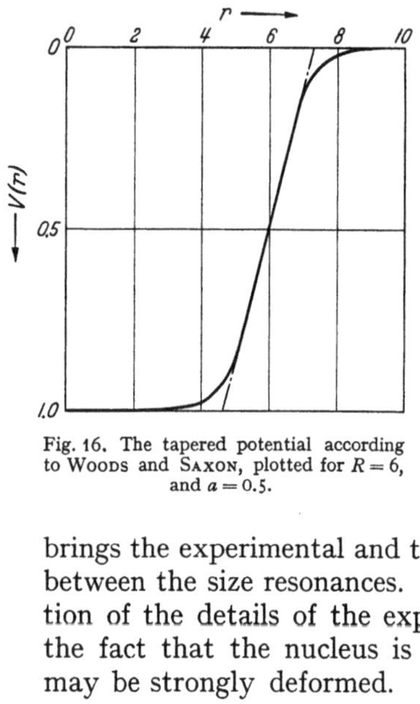

Fig. 16. The tapered potential according to WOODS and SAXON, plotted for $R = 6$, and $a = 0.5$.

With this potential the resonance condition (19.2) does not hold exactly. The position of the resonances in these circumstances has been investigated by LAWSON[2], and by ROSS, MARK and LAWSON[3], and the shape of the resonance curves, by NEMIROVSKI[4]. It appears that the size resonances should be shifted to $A = 56$ and $A = 166$, in fair agreement with experiment (see Fig. 15). In general, the diffuseness of the nuclear edge increases the probability of the neutron of entering or leaving the nucleus because it reduces the reflection coefficient of the surface below that of a square well. Thus the diffuse edge brings the experimental and theoretical results into better agreement for energies between the size resonances. However, difficulties still remain in the interpretation of the details of the experimental data. In part these may be caused by the fact that the nucleus is not spherical; in fact, in the region of $A = 160$, it may be strongly deformed.

F. Elastic scattering.

20. Elastic scattering of neutrons. The total elastic scattering is difficult to measure directly; it can best be obtained by subtracting the nonelastic scattering cross section from the total cross section. For low energies it comprises most of the total cross section, which in the heavy elements may be 10b or more. For higher neutron energies, the nonelastic cross section becomes nearly equal to the theoretical reaction cross section which is approximately $\pi(R + \lambda\!\!\!^{-})^2$. In this energy range, usually above 5 Mev, the total cross section is never very different from $2\pi R^2$; thus the elastic cross section in this energy range is also

[1] R. D. WOODS and D. S. SAXON: Phys. Rev. **95**, 577 (1954).
[2] R. D. LAWSON: Phys. Rev. **101**, 311 (1956).
[3] A. A. ROSS, H. MARK and R. D. LAWSON: Phys. Rev. **102**, 1613 (1956).
[4] R. NEMIROVSKI: To be published.

of the order of πR^2, as we should expect from the inequality theorem (Sect. 5). As we shall see in Sect. 24, the nonelastic cross section seems to be independent of energy above the energy where it reaches its full development. The size-resonance structure of the total cross section which we have discussed in Sect. 17, therefore, appears to be caused by the elastic cross section alone.

Apart from its magnitude, the most interesting characteristic of the elastic cross section is its angular distribution. The measurement of this angular distribution is difficult: in the first place it is not easy to measure the scattered neutrons in the presence of neutrons emitted from the source which can be screened from the detector only with difficulty; secondly, since relatively large quantities of material are required to give a measurable effect, the experimental results require correction for multiple scattering which usually can only be made on a semi-empirical basis.

The arrangement of the experimental apparatus has usually been of two kinds. In one, the scatterer is placed some distance from the source and the detector is screened from the latter by a large block of paraffin; in the other, there is cylindrical symmetry about the axis between the source and detector between which a cylindrical absorber is inserted. The scatterer, in the form of a ring, is then placed between source and detector. It has been known for sometime that the angular distribution is peaked in the forward direction. The earliest accurate measurements of the angular distribution of elastically scattered neutrons were made by AMALDI[1] and his collaborators. These authors used the neutrons produced by the Li (d, n) reaction (about 13 Mev) and a $Cu^{63}(n, 2n)$ detector. They used the ring method, the ring being specially shaped so that scattering at all points occurred effectively at the same angle. Their measurements on Pb showed that the differential cross section passes through a series of maxima and minima as the angle of scattering is increased. In broad outline, such an angular distribution can be explained on the strong coupling model, in which the nucleus behaves like an opaque disc: the observed elastically scattered neutrons, then, represent the wave which must be superposed on the incident wave to give the shadow cast by the disc. According to PLACZEK and BETHE[2], the angular distribution of this "shadow" scattering is given by:

$$\frac{d\sigma_e}{d\Omega} = k^2 (R + \lambda\!\!\!^{-})^4 [J_1(x)/x]^2 \tag{20.1}$$

with

$$x = 2k(R + \lambda\!\!\!^{-}) \sin\frac{\vartheta}{2} \tag{20.2}$$

where J_1 is the BESSEL function of first order, and ϑ is the angle of scattering. The conditions under which this formula are applicable have been further discussed by AKHIESER and POMERANCHUK[3]. The positions of the maxima and minima give values of the nuclear radius in good agreement with currently accepted values.

The angular distribution of elastically scattered neutrons of 3.7 Mev have been measured by WHITEHEAD[4] and SNOWDON, using the ring method. Their results, which were confined to Al, Fe, and Pb, required very considerable correction for multiple scattering. Recently, the angular distribution of 1.0 Mev neutrons has been measured very accurately for a large number of elements by

[1] E. AMALDI, D. BOCCIARELLI, B. N. CACCIAPUOTI and G. C. TRABACCI: Nuovo Cim. 3, 203 (1946).

[2] G. PLACZEK and H. A. BETHE: Phys. Rev. 57, 1075 (1940).

[3] A. AKHIESER and I. POMERANCHUK: J. Phys. USSR. 9, 471 (1945).

[4] W. D. WHITEHEAD and S. C. SNOWDON: Phys. Rev. 92, 114 (1953).

WALT and BARSCHALL[1]. At this energy the scattering in the backward direction, which theoretically is particularly sensitive to the model used, is somewhat easier to measure, because a smaller fraction of the elastic scattering is directed forwards than at higher energies. The results are shown in Fig. 26, p. 419. The distribution is rather flat when A is near 60; it is sharply peaked in the forward direction, and somewhat less so in the backward direction for $A \approx 120$; and for the heaviest elements the backward peak disappears and gives way to a low broad peak in the vicinity of 90°.

The theoretical interpretation of the elastic scattering has been given by FESHBACH, PORTER, and WEISSKOPF[2], on the basis of the optical model. The angular distribution due to the shape-elastic part is given by:

$$\frac{d\sigma_e}{d\Omega} = \frac{\lambda\!\!\!^{-2}}{4} \left| \sum_l (2l+1)(1-\bar{\eta}_l) P_l(\cos\vartheta) \right|^2 \tag{20.3}$$

where $\bar{\eta}_l$ is the average amplitude defined in Sect. 6. The observed cross section, however, includes the compound-elastic cross section, which although smaller than the shape-elastic part is likely to fluctuate from one nucleus to another and is difficult to estimate with precision. The calculated results are shown in Fig. 28a and b (p. 421), for the shape elastic distribution, and for the sum of the shape-elastic and the compound-elastic cross sections respectively. It will be seen that there is fair agreement between theory and experiment.

For higher energies the contribution of the compound-elastic scattering may be neglected and the experimental data can be directly compared with predictions based on the optical model. Recently accurate measurements of the scattering of 14.6 Mev neutrons have been made by CROSS and JARVIS[3]. In their method the usual difficulty of background effects were reduced by measuring delayed coincidences between the α-particles emitted in the backward direction from the source (the $T(d, n)\,He^4$ reaction), and the neutrons (detected after scattering) emitted in the forward direction. The detectors, in each case, consisted of plastic scintillators, the neutron detector being biassed in order to detect only neutrons with energies above 11 Mev. Some of their results are shown in Fig. 17.

A theoretical analysis of the 14 Mev results, based on the optical model has been given recently by CULLER, FERNBACH, and SHERMAN[4]. These authors, whose results were obtained with the aid of an electronic calculating machine, finding the square well inadequate, tried wells with different tapers corresponding to the real and imaginary parts of the potential. They found that practically indistinguishable results were obtained if the tapered wells were replaced, for mathematical convenience, by stepped wells, in which the radial extension of the step was about 0.4 of the linear extension of the taper and in which the depth was a variable parameter. Their final calculations, which gave good agreement with experimental data from Mg to Bi, were made with an imaginary well whose radius differed from that of the real part ($R = 1.22 A^{\frac{1}{3}} \times 10^{-13}$ cm) by 0.74×10^{-13} cm The combination of these two terms is roughly equivalent to the more usual radius $R = 1.45 A^{\frac{1}{3}} \times 10^{-13}$ cm. Recently, BJORKLUND, FERNBACH, and SHERMAN[5] have found that good agreement with the experimental data can be obtained by a real potential of the WOODS-SAXON type (19.6), and an imaginary part of gaussian

[1] M. WALT and H. H. BARSCHALL: Phys. Rev. **93**, 1062 (1954).
[2] See ref. 1, p. 210.
[3] W. G. CROSS and R. G. JARVIS: Phys. Rev. **99**, 621 (1955); full details still to be published. See also J. O. ELLIOT: Phys. Rev. **101**, 684 (1955).
[4] C. CULLER, S. FERNBACH and N. SHERMAN: Phys. Rev. **101**, 1047 (1956).
[5] F. E. BJORKLUND, S. FERNBACH and N. SHERMAN: Phys. Rev. **101**, 1832 (1956).

form concentrated at the surface. Clearly the precise form of the taper on the real part is of no great importance provided its extent is defined. The best form of the imaginary part which will fit equally to both neutron and proton scattering is still to be determined. CULLER, FERNBACH, and SHERMAN have also studied the effect of a spin-orbit interaction confined to the peripheral area defined by the taper described above. It was found to have very little influence on the theoretical angular distribution. The polarization produced by it has not yet been studied.

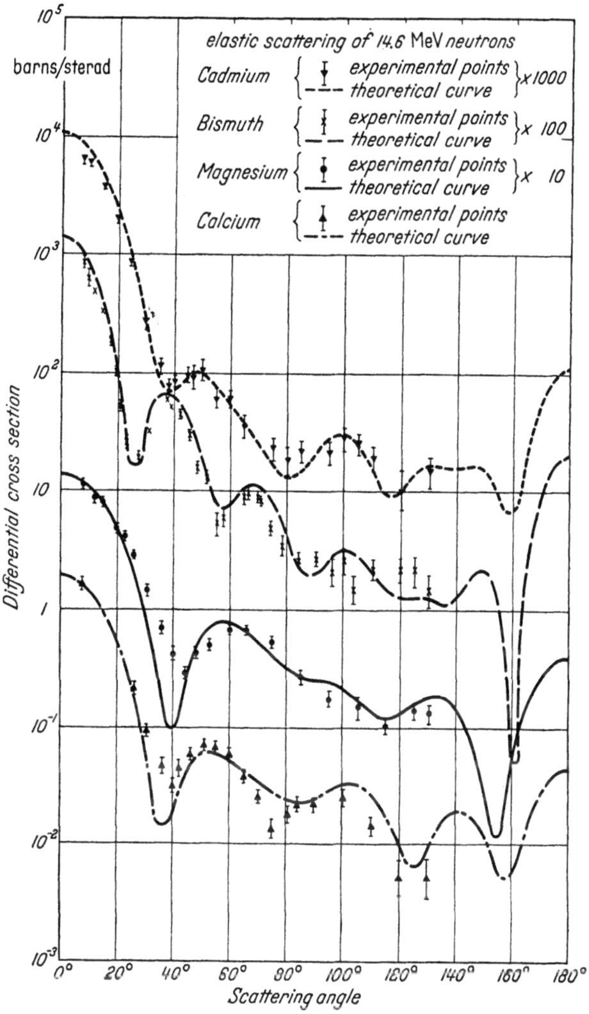

Fig. 17. Elastic scattering of 14.6 Mev neutrons. The experimental points are those of CROSS and JARVIS quoted by BJORKLUND, FERNBACH, and SHERMAN, the full curves are the calculated cross sections.

21. Elastic scattering of protons. The nuclear elastic scattering of protons, although experimentally much easier to study than the scattering of neutrons, requires protons with sufficient energy to surmount the COULOMB barrier. For this reason, the specifically nuclear, as contrasted with the RUTHERFORD scattering, has been investigated only recently. Since the scattering of charged particles always involves the RUTHERFORD scattering, there is no meaning in total scattering cross section because the differential RUTHERFORD scattering cross section goes to infinity at low scattering angles. Thus, only the differential cross section for elastic scattering is of interest.

The nuclear part of the proton elastic scattering was first detected by BURKIG and WRIGHT[1], (with 18.6 Mev protons) and a photographic method. They found that in Al the scattering at high angles was greater than that expected of RUTHERFORD scattering, while for heavier elements, e.g. Pd, and W, the opposite occurs; thus in the latter elements there must be destructive interference between the nuclear and the RUTHERFORD scattering. More recent measurements, using counter methods, by GUGELOT[2] with 18 Mev protons and by COHEN and NEIDIGH[3]

[1] I. W. BURKIG and B. T. WRIGHT: Phys. Rev. **82**, 451 (1951).
[2] P. C. GUGELOT: Phys. Rev. **87**, 525 (1952).
[3] B. L. COHEN and R. V. NEIDIGH: Phys. Rev. **93**, 282 (1954).

with 22 Mev protons, have shown that the differential cross sections contain more structure than was originally obtained by BURKIG and WRIGHT. They results of COHEN and NEIDIGH, in units of the RUTHERFORD cross section, are plotted logarithmically in Fig. 18. They show very clearly the destructive interference of the RUTHERFORD scattering just referred to. The position of the maxima and minima vary continuously from element to element. They represent for protons the diffraction pattern which is so characteristic of neutron scattering. COHEN and NEIDIGH found that the angles at which the maxima and minima occur are proportional to $1/kR$, as one would expect from (20.2).

The first theoretical attempt to interpret the experimental data was made by LELEVIER and SAXON[1] on the basis of the optical model. They found that they could account for the general trend of the results of BURKIG and WRIGHT for Ni and Pt with a square well potential, for which

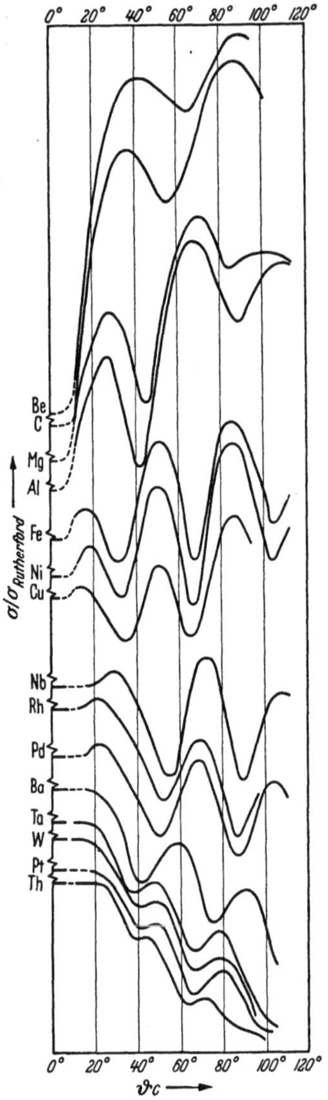

Fig. 18.

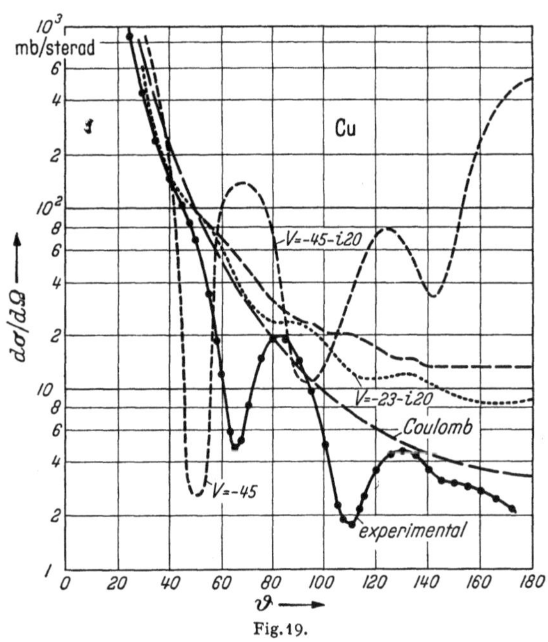

Fig. 19.

Fig. 18. Differential cross sections for the elastic scattering of 22 Mev protons in units of the RUTHERFORD cross section, according to COHEN and NEIDIGH; the vertical scale is logarithmic, but there is no significance to the vertical displacement or position of the curves.

Fig. 19. Differential cross sections for the elastic scattering of 18 Mev protons by Cu, according to DAYTON. The broken curves are theoretical predictions for various complex square potential models.

the real part was 40 Mev, and the imaginary part was 10 Mev. These calculations, however, do not account for the detailed structure subsequently found in the angular distribution. CHASE and ROHRLICH[2] showed that a square well potential was

[1] R. E. LELEVIER and D. S. SAXON: Phys. Rev. 87, 40 (1952).
[2] D. M. CHASE and F. ROHRLICH: Phys. Rev. 94, 81 (1954).

quite inadequate to account for the sharply defined maxima and minima. They found that the general trend of the experimental data could be obtained only by using a large value ($\sim$20 Mev) for the imaginary part of the well potential. However, the use of such a large value obliterates the detailed structure which can be restored only by reducing it, and then the theoretical cross section at high angles is much too great. This dilemma is well illustrated in the results of DAYTON[1] who measured the elastic scattering of 18 Mev protons from Al, Fe, Ni, and Cu (see Fig. 19). CHASE and ROHRLICH suggested that a long tail to the potential distribution might remove the difficulty. This suggestion was followed by WOODS and SAXON[2] who showed that the results for Ni and Pt obtained by COHEN and NEIDIGH could be accounted for satisfactorily by a potential of the type (19.6), with V about 40 Mev; W about 10 Mev; the radius R given, as usual, by $R = r_0 A^{\frac{1}{3}}$ with $r_0 = 1.4 \times 10^{-13}$ cm; and the parameter a, which determines the extent of the taper to the well, about 0.5×10^{-13} cm.

Recently more experimental data has been published[3] for energies extending down to 5.25 Mev, and up to 31.5 Mev. Some results are shown in Fig. 20. This figure shows that at 31 Mev the diffraction effect is well developed only for elements of intermediate atomic weight; for nuclei with high atomic number, and at lower energies, the nuclear diffraction is destroyed by the RUTHERFORD scattering, and for lighter elements, and high energies, by a smearing out of the minima, caused

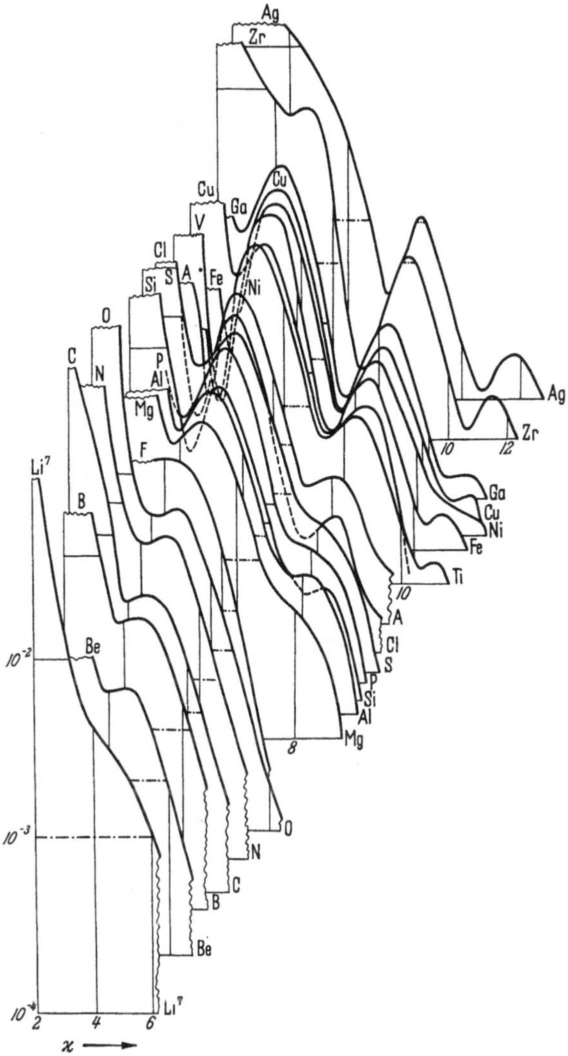

Fig. 20. Elastic scattering of 31 Mev protons according to KINSEY and STONE. The ordinates are in units of the quantity defined by (20.1); the abscissae, in units of (20.2).

[1] I. E. DAYTON: Phys. Rev. **95**, 754 (1954).

[2] See ref. 1, p. 232.

[3] G. R. FISCHER: Phys. Rev. **96**, 704 (1954). — W. BURCHAM, W. M. GIBSON, A. HOSSAIN and J. ROTBLAT: Phys. Rev. **92**, 1266 (1953). — R. G. FREMANTLE, D. J. PROUSE and J. ROTBLAT: Phys. Rev. **96**, 1268 (1954). — Ref. 1, p. 225. — I. E. DAYTON and G. SCHRANK: Phys. Rev. **101**, 1358 (1956). — R. H. CHOW and B. T. WRIGHT: Phys. Rev. **99**, 640 (1955). — B. T. WRIGHT: University of California, UCRL-2422 (1953). — J. LEAHY: To be published. — B. B. KINSEY and T. STONE: Phys. Rev. **103**, 975 (1956). — D. A. BROMLEY and N. S. WALL: Phys. Rev. **102**, 1560 (1956).

probably by the diffuseness of the edge of the nucleus. For 20 Mev protons, the angular distributions are rather similar to those of neutron scattering at 14 Mev, except, of course, that the well developed maxima and minima of neutron scattering in heavy elements is obliterated (for protons) by RUTHERFORD scattering. A comparison of neutron and proton scattering suggests that the angular distributions in light and intermediate elements are indeed very similar when the proton energy exceeds the neutron energy by the COULOMB barrier.

According to MELKANOFF, MOSZKOWSKI, NODVIK, and SAXON[1], who have made machine calculations for a large number of orbital angular momenta, the potential (19.6) together with the COULOMB potential will account for the greater part of this experimental data, with values of r_0 and a which may be fixed independently of atomic weight, and with potentials which vary slowly with the proton energy. They find $r_0 = 1.33 \times 10^{-13}$ cm and $a = 0.5 \times 10^{-13}$ cm, and V and W are given in Table 1.

Table 1. *Parameters of complex potential in* Mev *(after* MELKANOFF *et al.).*

Proton energy	V	W
5.25	52.5	0.9
17	47 ± 1	8.5
31.5	36 ± 1	15.5

The agreement between theory and experiment, however, is confined in the main to low angles of scattering. At higher angles the experimental counting rates are much lower and more difficult to measure, and the differential cross sections therefore are more difficult to estimate. Accurate data at high angles up to 180° has been obtained only by DAYTON and SCHRANK[2] for 17 Mev protons. At high scattering angles, the theory is very sensitive to detail and no doubt requires further development. However, it is remarkable that such good agreement can be obtained from the optical model with such an economy of numerical parameters.

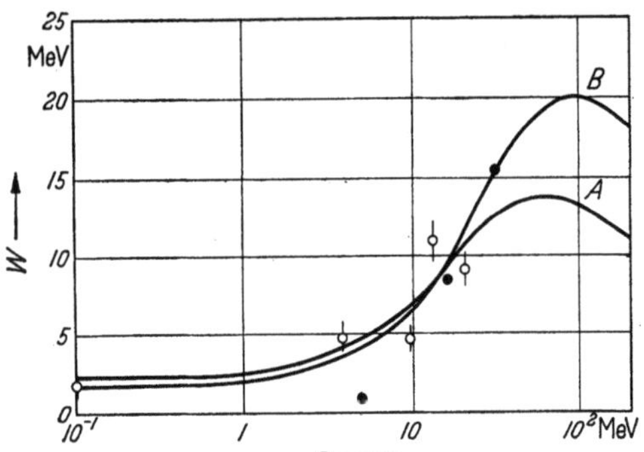

Fig. 21. Imaginary part of the complex potential according to LANE and WANDEL. Curve A, $r_0 = 1.48$; Curve B, $r_0 = 1.20 \times 10^{-13}$ cm. Open circles, data quoted by LANE and WANDEL; full circles, data given by MELKANOFF *et al.*

It will be seen from Table 1 that the value for V for low energy protons is higher than that obtained from neutron size-resonances at low energies (42 Mev) The difference could be caused by the symmetry forces, for an incoming proton sees more neutrons than an incoming neutron sees protons. It could be produced by the energy dependence of the nuclear potential, for it should be noted also that V falls with increasing energy: this is to be expected in view of recent theories of the origin of the nuclear potential[3]. The parameter, a, which determines the diffuseness of the nuclear surface appears to be about the right amount required to account for nuclear surface energies[4].

The magnitude of the imaginary part of the nuclear potential is of interest for as we have seen in Sect. 12, it is directly related to the mean free path of

[1] M. A. MELKANOFF, S. A. MOSZKOWSKI, J. NODVIK and D. S. SAXON: Phys. Rev. **101**, 507 (1956).

[2] See ref. 3, p. 237.

[3] M. H. JOHNSON and E. TELLER: Phys. Rev. **98**, 783 (1955). — K. BRUECKNER: Phys. Rev. **97**, 1353 (1955).

[4] T. SKYRME: To be published.

nucleons in nuclear matter. LANE and WANDEL's[1] calculations are shown in Fig. 21, together with points derived from optical model interpretation of the experimental data. It will be seen that the results of Table 1, except, possibly, at low energies, are in fair agreement with these calculations.

22. Elastic scattering of deuterons. Owing to the difficulty of distinguishing experimentally between protons and deuterons, very little experimental work has been done so far on the elastic scattering of deuterons. The only measurements published are those of GOVE[2] who bombarded Pb and Bi with 15 Mev deuterons. The angular distributions of the scattered deuterons from these two elements are similar in general appearance to those produced by α-particles (see Fig. 22).

23. Elastic scattering of α-particles. Before the artificial acceleration of α-particles became a practical possibility, deviations from the RUTHERFORD law of scattering were found when light elements were bombarded with the natural α-particles emitted by radioactive bodies [*10*]. But owing to the high COULOMB barriers for α-particles in heavy elements, deviations from RUTHERFORD scattering have become apparent only since cyclotrons with α-particle beams of sufficient energy have become available. This subject is of special interest, because, as distinct from the elastic scattering of neutrons and protons, we are concerned here with the scattering of a complex body.

The first measurements with heavy elements to show specifically nuclear rather than RUTHERFORD scattering were those of GOVE already referred to. GOVE studied the angular distribution of 27.5 Mev α-particles scattered elastically from gold. Recently a careful study has been made by FARWELL and WEGNER[3] who bombarded Ag, Ta, Pb, and Th, with α-particles with energies up to 45 Mev, the elastically scattered α-particles being detected by two proportional counters in coincidence. The scattering from these elements was studied at 60°, and at both 90° and 60° for Au. The results of a typical measurement are shown in Fig. 22. For low energies, the cross section falls inversely as the square of the energy, as one would expect from the RUTHERFORD formula; but beyond a certain point, the cross section falls abruptly and exponentially. In some cases the cross section rises slightly and above that to be expected from pure RUTHERFORD scattering before falling exponentially.

At first sight one would expect that the position of the discontinuity should represent the condition that the apsidal distance, i.e., the classical distance of closest approach, should be equal to the sum of the radii of nucleus and α-particle. For a scattering angle ϑ this distance is given by

$$D = \frac{Z Z' e^2}{2\varepsilon}\left(1 + \operatorname{cosec}\frac{\vartheta}{2}\right) \tag{23.1}$$

where ε is the energy of the α-particle in the system in which the centre of gravity is at rest. However, this expression gives distances D which are much greater than the generally accepted value of the sum of the radii. The experimental data has been interpreted by BLAIR[4] as follows. Since the wavelength of 40 Mev α-particles is small compared with the radii of heavy nuclei (about one twentieth) it is legitimate to consider the scattering on a semi-classical basis. In classical

[1] See ref. 5, p. 222.
[2] H. E. GOVE: Phys. Rev. **99**, 1353 (1955).
[3] G. W. FARWELL and H. E. WEGNER: Phys. Rev. **95**, 1212 (1954).
[4] J. S. BLAIR: Phys. Rev. **95**, 1218 (1954). See also C. E. PORTER: Phys. Rev. **99**, 1400 (1955).

mechanics, the angular momentum for scattering at angle ϑ is a precisely defined quantity: it iz $(Z Z' e^2/v) \cot \vartheta/2$; in a semi-classical treatment, only those waves with angular momentum in this immediate vicinity will contribute to scattering at ϑ. However, the critical angular momentum for collision is given by $l' \hbar$, where:

$$\varepsilon = \frac{Z Z' e^2}{R} + \frac{\hbar^2}{2 M R^2} l'(l'+1).$$

For $l < l'$, the α-particle will hit the nucleus, and for these collisions, BLAIR assumed (like FESHBACH and WEISSKOPF) that there is no outgoing wave. For these waves, $\eta = 0$ (see Sect. 5). For $l > l'$, he assumed that the wave is scattered as if by a pure COULOMB field. He then calculated the cross section in units of the RUTHERFORD cross section, as a function of l'/l_c where l_c is the classical angular momentum

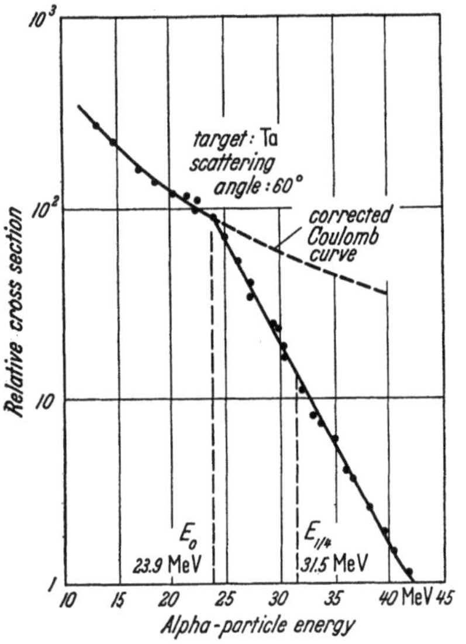

Fig. 22.

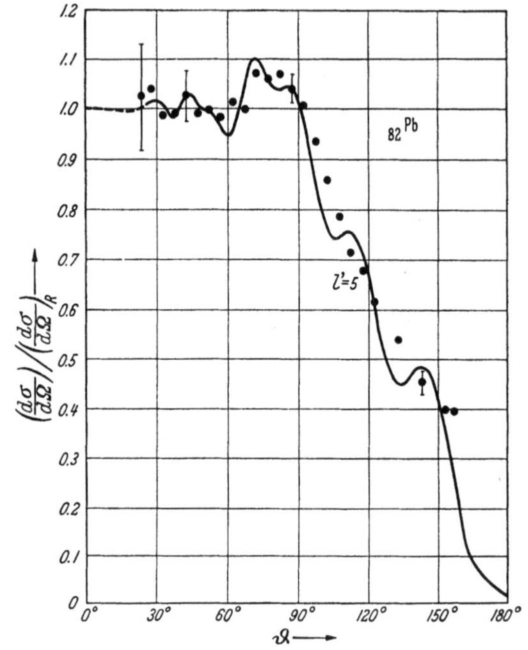

Fig. 23.

Fig. 22. Elastic scattering of α-particles from Ta, according to FARWELL and WEGNER.

Fig. 23. Elastic scattering of 22 Mev α-particles by Pb in units of the RUTHERFORD cross section, according to WALL, LEES, and FORD. The full curve is the theoretical curve with a slowly varying cutoff at $l' = 5$.

corresponding to the angle, ϑ, viz., $l_c = n \cot \vartheta/2$, where $n = Z Z' e/\hbar v$. It then appears that $l' = l_c$, independently of the factor n, at a point where the cross section is a quarter of the RUTHERFORD cross section. Presumably this means that at this point, half of the amplitude of the scattered wave comes from partial waves with angular momenta less than l_c and half with angular momenta greater than l_c. When the former are removed by absorption, the cross section is one quarter of the RUTHERFORD cross section. At this point, then, the quantity D in (23.1) should give the sum of the radii. This indeed appears to be true and may be illustrated by the results obtained for Th; for this nucleus in units of 10^{-13} cm the value D_0 corresponding to the break in the cross section curve is 13.1; $D_{\frac{1}{4}}$, corresponding to the point where the cross section is one quarter of the COULOMB cross section at that energy, is 10.5; the radius, $R = r_0 A^{\frac{1}{3}}$, with $r_0 = 1.5$, is 9.2, and the difference 1.3 is the radius of the α-particle.

This theory also accounts quantitatively for the rise of the cross section above the RUTHERFORD value just before the break; in fact, it tends to predict a bigger effect than is observed. However, it also gives cross sections which are too great at large angles. Both these effects may arise from the sharp cut off in the angular momentum. It does not seem to account for the distinction between the heavy and the intermediate elements which show very marked diffraction effects.

The angular distribution of elastically scattered α-particles for a fixed energy (22 Mev) has been measured by WALL, REES, and FORD[1] for Ag, Au, and Pb. For Pb, their results are shown in Fig. 23. It will be seen that the cross section is equal to the RUTHERFORD cross section for angles up to 60°; it then rises to a maximum and falls rapidly and smoothly thereafter. These authors find that the theoretical method of BLAIR gives, in general, a fair account of the experimental angular distributions, except at large angles, where the theoretical curves show marked maxima and minima which, in fact, do not appear. An improved fit is obtained by assuming that the cut off in the angular momentum is not sharp at $l = l'$. These authors point out that the good agreement found by BLAIR may be partly fortuitous, for the scattering angles of FARWELL and WEGNER's experiments are just those for which the theory predicts minimum oscillations.

Better agreement with experiment over a limited range has been obtained by IZUMO[2] using the optical model. He assumed that the imaginary part of the potential increases linearly with the kinetic energy of the α-particles but it was necessary to assume that the potential depends to some extent on the angle of scattering. No exact calculations on the basis of the optical model have yet been made.

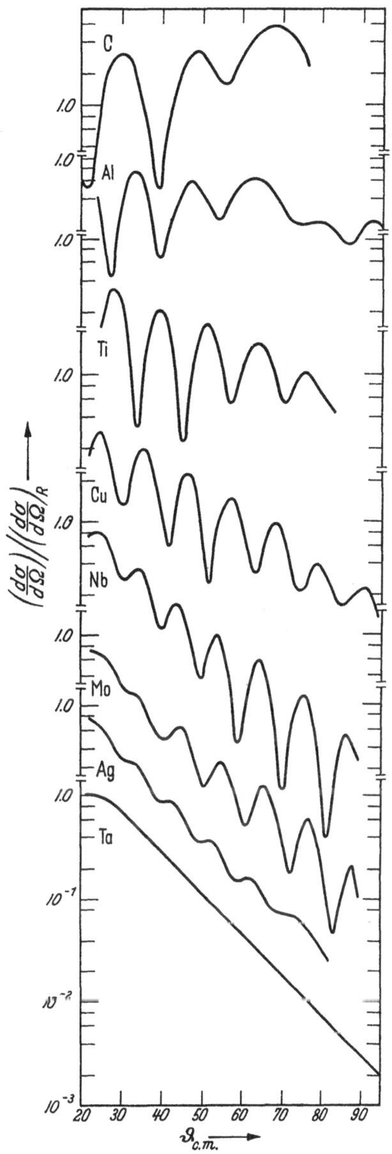

Fig. 24. Elastic scattering of 40 Mev α-particles according to IGO, WEGNER, and EISBERG, in units of the RUTHERFORD cross section.

More recently measurements on Ta, Au, Pb and Th have been made by WEGNER, EISBERG and IGO[3] with 40 Mev α-particles, and by ELLIS and SCHECTER[4] on Au and Pb with 48 Mev α-particles. The latter show that the exponential fall in the cross section in units of the RUTHERFORD cross section extends to four orders of

[1] N. S. WALL, J. R. REES and K. W. FORD: Phys. Rev. **97**, 726 (1954).
[2] K. IZUMO: Progr. Theor. Phys., Japan **12**, 549 (1954).
[3] H. E. WEGNER, R. M. EISBERG and G. IGO: Phys. Rev. **99**, 825 (1955).
[4] R. E. ELLIS and L. SCHECTER: Phys. Rev. **101**, 636 (1955).

magnitude before tapering off to a nearly constant value; at higher angles still, the cross section rises again.

It seems that the smooth fall of the differential cross section with angle is a feature which is characteristic only of the heavy elements. The elastic scattering of α-particles in light and intermediate elements have been studied recently by BLEULER and TENDAM[1] in Al and Cu for 19 Mev α-particles, and for six elements ranging from C to Ag, by IGO, WEGNER, and EISBERG[2] for 40 Mev α-particles. The results show remarkable diffraction effects which are absent in the heavier elements (Fig. 24). The gradual appearance of these diffraction effects with decreasing atomic weight and the disappearance of the exponential slope of the differential scattering cross section are attributed by these authors to a progressive disappearance of the distortion of the α-particle wave by the COULOMB field. If the diffraction maxima and minima can be accounted for by an equation of the type (20.1) the data affords a quite precise determination of the nuclear radii. The latter authors find that the radius is given by $R = r_0 A^{\frac{1}{3}} + b$, where $r_0 = 1.27 \pm 0.07$ and $b = 1.60 \pm 0.23$ in units of 10^{-13} cm.

G. Fast neutron reactions.

24. The nonelastic neutron cross sections. Until recently the production and detection of monoenergetic beams of neutrons was a matter of great difficulty. The (d, n) reaction with deuterium and tritium, and the (p, n) reactions with tritium and lithium, now provide suitable sources of neutrons with energies ranging from a few kev to 20 Mev or more. The technical details of these sources and their use has been discussed by HANSON, TASCHEK and WILLIAMS [*11*]; methods of flux measurement have been discussed by BARSCHALL, ROSEN, TASCHEK and WILLIAMS [*12*], and a very complete account of the techniques involved in the measurement of nonelastic and elastic cross sections has been written very recently by CRANBERG *et al.* [*13*].

We have seen that direct measurements of the transmission of a neutron beam through different materials give, in general, the total cross section, σ_t. The nonelastic cross section, defined in Sect. 5, is equal to the difference between the total cross section and the elastic scattering cross section. It can be obtained by a measurement of this difference, or by modifications of the transmission method.

In the first method, the total cross section is first measured; then the differential elastic scattering cross section is measured, integrated, and subtracted from the total cross section:

$$\sigma_n = \sigma_t - \int \frac{d\sigma_e}{d\Omega} \, d\Omega .$$

The most recent measurements of this kind are those of WALT and BEYSTER[3]. These authors used a collimated beam of 4.1 Mev neutrons produced in the $H^2(d, n)\,He^3$ reaction and measured the elastic scattering of the neutrons produced from targets placed in the neutron beam. Their detector consisted of a number of spherical plastic scintillators separated from each other by quartz plates and placed at the end of a light guide. This method of detection, first described by McCRARY, TAYLOR, and BONNER[4], by limiting the size of each detecting unit,

[1] E. BLEULER and D. J. TENDAM: Phys. Rev. **99**, 1605 (1955).
[2] G. IGO, H. E. WEGNER and R. M. EISBERG: Phys. Rev. **101**, 1508 (1956).
[3] M. WALT and J. R. BEYSTER: Phys. Rev. **98**, 677 (1955).
[4] J. H. McCRARY, H. L. TAYLOR and T. W. BONNER: Phys. Rev. **94**, 808 (1954).

makes it possible to distinguish between pulses produced by neutrons from those produced by secondary electrons. By counting only pulses above a certain size, the γ-ray background is eliminated. By adjustment of the bias on a discriminator, the counts recorded can be made to correspond only to neutrons of undiminished energy.

In the second method the transmission of a neutron beam may be measured in conditions of "poor geometry" [12]. This method also requires a detector sensitive only to neutrons with the maximum energy. The most accurate measurements have been made with a spherical shell. If the shell surrounds the source (Fig. 25 a) and if the radius of the shell is small compared with the distance between the source and detector, then it is easy to show that the transmission of the shell is just $T = \exp(-N\sigma_n)$ where N is the number of atoms of the scattering material per cm², and σ_n, is the nonelastic cross section (see Sect. 5). In using this expression to calculate the cross section, we must assume that the angular distribution of the neutrons emitted by the source is isotropic, and that the energy does not change with angle. For most materials, and for energies in excess of 3 Mev, this means in effect that neither the angular distribution nor the neutron energy must change within the forward hemisphere. The method is suitable for the measurement of nonelastic cross sections for 14 Mev neutrons produced by low energy bombardment of tritium in the reaction $H^3(d, n) He^4$, which has a large cross section for low energies, and which is nearly isotropic and monoenergetic over a wide range of angles. For most other sources, however, this method is quite unsuitable, for both the energy and the yield can vary sharply with the angle of emission. For such sources, BETHE[1] showed that it is legitimate to reverse the positions of source and detector: the shell is now placed round the detector (Fig. 25 b). With this method, a very extensive series of measurements has been made by TAYLOR, LÖNSJÖ, and BONNER[2]. They used 3.5 and 4.2 Mev neutrons from the $H^3(p, n) He^3$ reactions; 4.7 and 7.1 Mev neutrons from the $H^2(d, n) He^3$ reaction, and 12.7 and 14.1 Mev neutrons from the $H^3(d, n) He^4$ reaction. Their detectors consisted of plastic scintillating spheres, of a size graduated to suit the range of the recoil protons to be detected. For higher energy neutrons, they used spheres 1 cm in diameter; for 3.5 and 4.2 Mev neutrons, they used spheres 0.25 cm in diameter. Their results, together with those of WALT and BEYSTER are shown in Fig. 26; they lie between πR^2 and $\pi(R + \lambda\!\!\!^{-})^2$ ($r_0 = 1.4 \times 10^{-13}$ cm) and they are in good agreement with the more accurate calculations of the reaction cross section made by BLATT and WEISSKOPF, based on the summation (7.3). Similar measurements have been

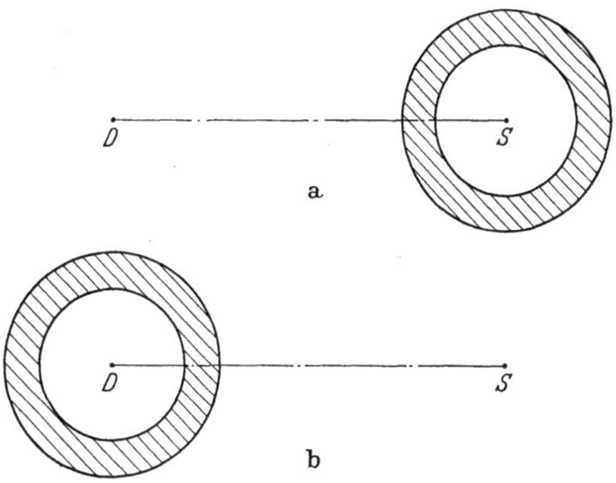

Fig. 25 a and b. (a) Arrangement for measurement of nonelastic cross sections with spherical shell surrounding the source (S). This method requires a neutron source for which both the yield and neutron energy do not vary with the angle of emission. (b) Arrangement for measurement of nonelastic cross sections with spherical shell surrounding the detector (D).

[1] H. BETHE: Unpublished.

[2] H. L. TAYLOR, O. LÖNSJÖ and T. W. BONNER: Phys. Rev. **100**, 174 (1955).

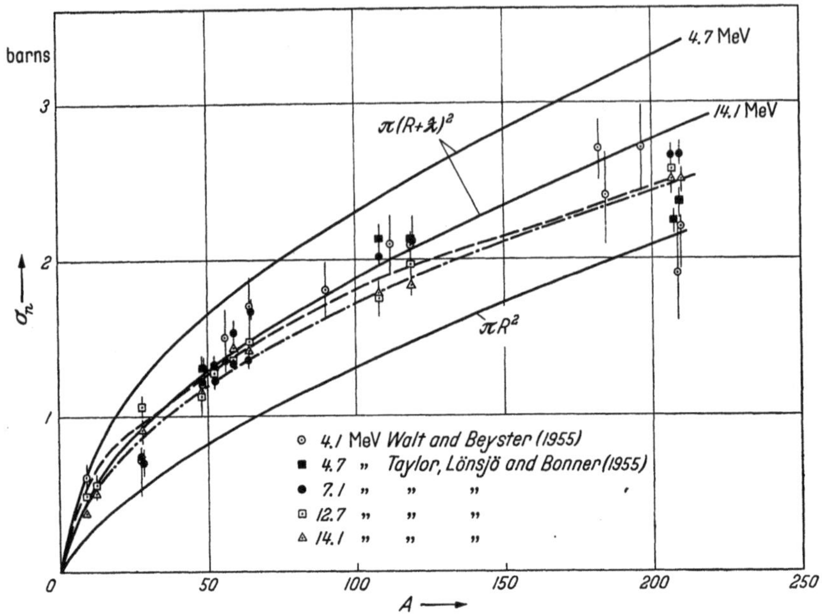

Fig. 26. Nonelastic neutron cross sections plotted against atomic weight. Broken lines: calculated from BLATT and WEISSKOPF's data; upper curve, 4.4 Mev; lower curve, 14 Mev.

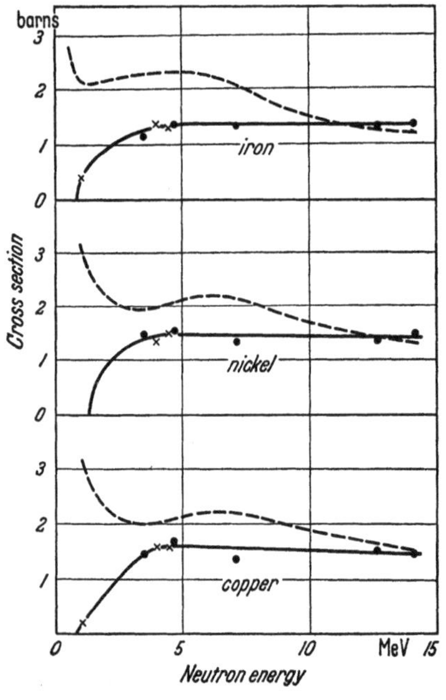

Fig. 27. Nonelastic neutron cross sections plotted against the neutron energy, according to TAYLOR, LÖNSJÖ, and BONNER, full circles; crosses, BEYSTER *et al.* The dotted line is the elastic cross section obtained by subtracting these results from the total neutron cross section.

made by BEYSTER *et al.*[1] for neutrons between 1 and 7 Mev. Fig. 27 shows typical nonelastic cross sections plotted against the neutron energy. It will be seen that the nonelastic cross section attains its full value when the neutron energy is between 3 and 5 Mev. Thereafter the cross section remains nearly constant; it shows none of the slow variations with energy found in total cross section measurements which are attributed to the size-resonances of the optical model. Theoretical nonelastic cross sections calculated on the basis of a square well potential show the expected size-resonances, but (for neutron energies above 5 Mev) seem to be much smaller than the observed cross sections. Somewhat better agreement is obtained by using a tapered potential[2].

At low energies the nonelastic cross section cannot be much greater than the cross section for inelastic scattering, and it will differ from the reaction cross section only by the compound-elastic scattering cross section. At these low energies the number of states which can be excited by

[1] J. R. BEYSTER, R. L. HENKEL and R. A. NOBLES: Phys. Rev. **97**, 563 (1955); **98**, 1216 (1955). — J. R. BEYSTER and M. WALT: Bull. Amer. Phys. Soc. **1**, 176 (1956).

[2] See ref. 1, p. 210, and P. MITTELSTAEDT: Z. Naturforsch. **11**a, 663 (1956).

inelastic scattering will be much smaller than at high energies and will depend on the kind of nucleus and its mass number. Effects of this kind have been found by WALT and BARSCHALL[1] with 1 Mev neutrons.

25. Inelastic scattering of neutrons: (n, n') reactions. We may distinguish between neutron reactions for which the energy of the neutron lies below or above the neutron binding energy of the target nucleus. For the former, inelastic scattering (i.e. the (n, n') reaction) is by far the most probable. Reactions of the (n, p) and (n, α) type occur but do not contribute much to the reaction cross section because in heavy elements the emission of charged particles is inhibited by the COULOMB barrier. However, some important measurements have been made on the energy and angular distribution of neutrons emitted from targets bombarded by 14 or 15 Mev neutrons. For neutrons of this energy, the emission of the first neutron, is often followed by the emission of a second. Thus the $(n, 2n)$ reaction may make a large contribution to the reaction cross section. In general, the energy distribution of the neutrons emitted will be comprised of both the primary and the secondary neutrons. The discussion of these experiments does not properly fall under the present heading; they will be discussed in Sect. 27.

26. (n, n') reactions: low energy inelastic scattering. For many years experimental difficulties associated with the detection and energy measurements of neutrons have prevented the study of low energy inelastic scattering. Thanks to the recent development of experimental methods, there is now a very large body of information on this subject. It has been reviewed recently by FREEMAN [*15*] and the reader is referred to this article for details. We shall sketch here only the bare outline of the subject and we shall make no attempt to describe its development.

The newly developed techniques fall into two categories, (i) methods of measuring both the flux and the energy of the secondary neutrons, and (ii) methods of measurement of the γ-rays emitted by the excited states of the residual nucleus. The latter are useful in the determination of the level scheme for the residual nucleus; usually both methods are needed to elucidate quantitatively the whole sequence of events.

α) Neutron measurements. Some experimenters have made use of the fact that the energy spectrum of the recoil protons produced by collisions with neutrons is of rectangular shape. In principle, groups of neutrons of different energies can be distinguished by steps in the energy spectrum of the recoil protons. Such measurements have been made ELIOT, HICKS, BEGHIAN, and H. HALBAN[2], with 2.5 Mev neutrons, stilbene crystals, and pulse height analysis, and with anthracene crystals by POOLE[3]. These authors deduce the energies of some excited states in the targets they examined, and rough values for the inelastic scattering cross sections. When only one state is excited these cross sections can amount to 1 barn. The remainder of the reaction cross section presumably is accounted for by the compound-elastic scattering. The defect of this method lies in the fact that the spectrum of the neutrons emitted by the target can only be obtained by a differentiation of the pulse height spectrum.

One of the most interesting developments in recent years is the He^3 counter. This has the great advantage that the energy spectrum of the secondary neutrons

---

[1] M. WALT and H. H. BARSCHALL: Phys. Rev. **93**, 1062 (1954).
[2] E. A. ELIOT, D. HICKS, L. E. BEGHIAN and H. HALBAN: Phys. Rev. **94**, 144 (1954).
[3] M. J. POOLE: Phil. Mag. **43**, 1060 (1952); **44**, 1398 (1953).

is obtained directly, and not, as in the recoil methods just described, by differentiation of the experimental data. This instrument, in the original form described by BATCHELOR[1], consists of a proportional counter, 11 in long and 3 in in diameter, filled with 110 cm pressure of Xe (to increase its stopping power) and 5 cm of He, containing 10% of He^3. The neutrons are detected by the $He^3(n,p)H^3$ reaction for which the cross section at 1 Mev is about 1 barn and varies inversely as the neutron velocity. The output of this counter is subjected to pulse height analysis, the energy of a group of neutrons (which appears as a peak in the pulse height spectrum) is determined relative to the smaller peak, produced by thermal, or epithermal neutrons, for which the energy release is about 770 kev. The disadvantage of this instrument is that there exists an upper limit to the neutron energy beyond which the interpretation of the results is ambiguous. This upper limit is set by the energy of elastic recoils of He^3. If ε is the neutron energy, the maximum energy of these recoils is $\frac{3}{4}\varepsilon$. When $\varepsilon > 1$ Mev, the pulse caused by the recoil is greater than 0.77 Mev which is the minimum pulse size produced by the (n, p) reaction; the unwanted pulse therefore, is superposed on the pulse height distribution of the (n, p) reaction and cannot easily be distinguished from it.

With this instrument, interesting measurements have been made recently on the spectrum of neutrons scattered inelastically from U^{238} bombarded by 1 Mev neutrons[2]. A comparison of the peak produced by low energy neutrons in U and in Pb showed that some inelastic scattering occurs which leaves the U^{238} nucleus in a state of low energy (<100 kev). As the neutron energy is raised, no other group of inelastically scattered neutrons appears until the neutron energy reaches 700 kev. At 500 kev the cross section for the production of the low lying state is about 0.8 b, thus accounting for a considerable fraction of the reaction cross section. Above 700 kev, the cross section for the production of this state falls, because it faces increasing competition from the state or states near 700 kev. This is just the kind of phenomenon we should expect, if the low lying states in U^{238} form a rotational system (Sect. 76); for neutron energies up to 1 Mev, only the first state of such a system (with spin 2) can be excited to an appreciable degree, the states with higher spin requiring the absorption and emission of neutrons of much higher angular momentum.

Recently time-of-flight techniques have been developed in order to study inelastic scattering of neutrons. CRANBERG and LEVIN[3] have used pulsed neutron sources to study the inelastic scattering of 2.45 Mev neutrons in a number of elements. From nuclei with atomic weights up to Cu and for nuclei with closed shells they found the energy spectrum of the scattered neutrons possessed the marked structure to be expected of the emission of neutrons which leave the residual nucleus in discrete states. The angular distributions of the discrete neutron groups, in the main, are symmetrical about the 90° degree position, although the anisotropy, generally, does not appear to be as great as the HAUSER and FESHBACH theory demands. The angular distribution of the γ-rays emitted (such as the 840 kev γ-ray from Fe^{56}) were similar. The spectrum of the neutrons emitted by the heavy elements (e.g. Au, U) was continuous and Maxwellian (8.6), the temperature being about (0.3 ± 0.03) Mev. For U, this inelastic spectrum was superposed on a fission neutron spectrum; it had a distinct bump at an energy of about 1.4 Mev; this may reflect the level structure which seems to be responsible for the bumps in the fission cross sections (Sect. 31).

---

[1] R. D. BATCHELOR: Proc. Phys. Soc. Lond. A **65**, 674 (1952).
[2] R. D. BATCHELOR: Proc. Phys. Soc. Lond. A **69**, 214 (1956).
[3] L. CRANBERG and J. S. LEVIN: Phys. Rev. **100**, 434 (1955); **103**, 343 (1956).

β) γ-ray measurements. One of the disadvantages of using stilbene and anthracene counters for the detection of neutrons is that these materials will also detect γ-rays. To detect the latter to the exclusion of neutrons it is better to use NaI crystals, for then low energy γ-radiation may be detected by the presence of a photoelectric peak in the pulse height spectrum. The results obtained, however, are not unambiguous, for peaks are also produced by γ-radiation emitted by I nuclei excited by inelastic neutron scattering, and by other means. However, a careful study of the results, together with coincidence γ-ray counting, is often enough to establish the decay scheme of the nucleus under study.

The energy variation of the cross section for the production of a particular excited state is of great interest, for the experimental results can be compared with the predictions of both the statistical theory and the optical model. In light elements, it is often possible to study the emission of neutrons by a particular well-defined state of the compound nucleus. For intermediate and heavy nuclei the experimental resolution is seldom sufficient to follow a reaction in such detail. For these nuclei, the neutrons which produce the residual nucleus in a particular excited state are derived from a large number of states of the compound nucleus. The angular momentum and parity of these states must be consistent with the angular momentum and parity of the incident neutrons relative to the ground state of the target nucleus, and consistent also with the angular momentum and parity of the residual nucleus. The statistical theory for this process has been given by HAUSER and FESHBACH[1]. They found that the variation of the cross section with energy is sensitive to differences in the spin and parity of the ground state of the target nucleus and the excited state of the residual nucleus. According to these authors the expression (7.3) becomes:

$$\sigma = \frac{\pi \lambda^2}{2(2I+1)} \cdot \sum_l T_l(\varepsilon) \sum_J q_{jl}(2J+1) \frac{\sum\limits_{\varepsilon' l'} q_{j'l'} T_{l'}(\varepsilon')}{\sum\limits_{\varepsilon'' l''} q_{j''l''} T_{l''}(\varepsilon'')} . \tag{26.1}$$

Here the numerator of the term on the far right hand side represents the probability of the re-emission of a neutron with energy ε', angular momentum l' and channel spin j', the denominator being the sum of the probabilities of all possible modes of decay. The middle term represents the sum over all spins (J) of the states of the compound nucleus which lead to the excitation of the residual nucleus in a state with a specified angular momentum. The summation at the left represents the summation over all orbital angular momenta (l) of the incident neutrons with energy ε and channel spin j consistent with J and I, the spin of the target nucleus. The quantities q are 2, 1, and 0, according to whether both values of j (or j'); or one (but not both); or neither, satisfy the relation:

$$|J-l| \leq j \leq |J+l|.$$

The values of the orbital angular momenta l and l' must satisfy the similar relations:

$$|J-j| \leq l \leq |J+j|; \quad |J-j'| \leq l' \leq |J+j'|.$$

For nuclei of intermediate atomic weight, the number of terms in (26.1) which contribute to the cross section for the formation of the first excited state in an even-even nucleus (spin 2) is not large. The theoretical calculation is

[1] See ref. 2, p. 217.

generally found to be in good agreement with experimental results. For example KIEHN and GOODMAN[1] have measured the cross section for inelastic scattering in iron which leads to the formation of the first excited state at 840 kev. The experimental curve rises sharply from the threshold. It is not smooth; there are bumps in it caused by an imperfectly resolved resonance structure. But there is, nevertheless, fair agreement between the smoothed out experimental curve and that obtained by calculation from (26.1).

Many other studies of threshold excitation have been made. The excitation of the 1.63 Mev metastable state of Pb^{207}, measured by STELSON and CAMPBELL[2],

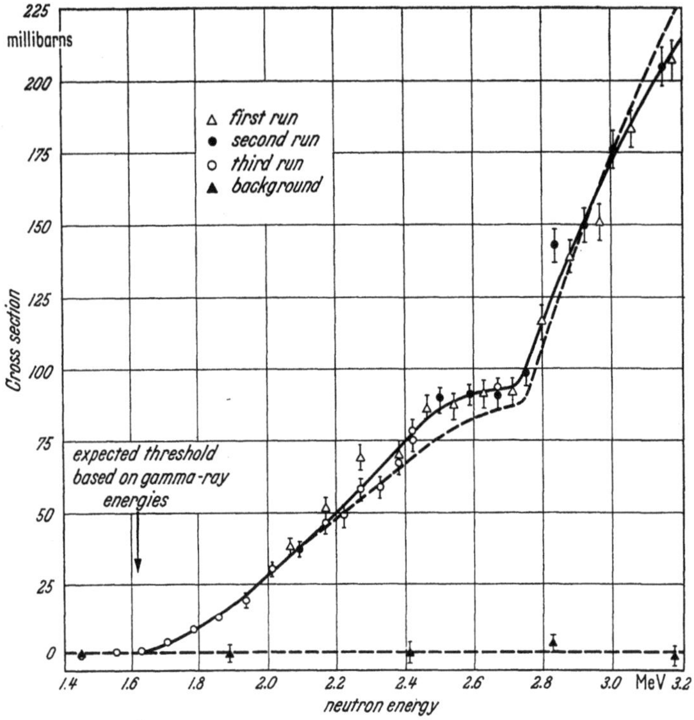

Fig. 28. Excitation of the 1.6 Mev metastable state of Pb^{207} after STELSON and CAMPBELL. The dotted curve is the theoretical curve calculated from the theory of HAUSER and FESHBACH.

is an example. This state has a spin of $\frac{13}{2}$ and it differs in parity from the ground state and in spin by 6 units. The experimental results agree well with theoretical predictions if it is assumed that the 2.34 Mev state does not feed into the metastable state (Fig. 28). The rise in the cross section when the energy exceeds that of the 2.75 Mev state is accounted for on the assumption that the state decays predominantly by γ-ray emission to the metastable state[3]. Similar studies have been made of the excitation of the 396 kev isomeric state of Cd^{111} by FRANCIS, MCCUE, and GOODMAN[4]; of the 335 kev state in In^{115} by MARTIN, DIVEN, and TASCHEK[5]; of the 279 kev state in Au^{197}, by COHEN[6], and by EBEL

[1] R. M. KIEHN and C. GOODMAN: Phys. Rev. **93**, 177 (1954).
[2] P. STELSON and A. CAMPBELL: Phys. Rev. **97**, 1222 (1955).
[3] The levels of this nucleus are shown in Fig. 79.
[4] A. E. FRANCIS, J. J. G. MCCUE and C. GOODMAN: Phys. Rev. **89**, 1232 (1953).
[5] H. C. MARTIN, B. C. DIVEN and R. F. TASCHEK: Phys. **93**, 199 (1954).
[6] S. G. COHEN: Nature, Lond. **161**, 475 (1948).

and GOODMAN[1]; and of Y^{89}, Ba^{137}, by SWANN and METZGER[2]. In and Cd results have been compared with theoretical predictions by MARGOLIS[3].

The validity of the statistical theory for low neutron energies now seems to be sufficiently well established to allow of detailed interpretation of the experimental data when the latter differs from theoretical expectations. Thus GUERNSEY and WATTENBERG[4] conclude from the agreement they obtain between their data and theoretical predictions that the first excited state of Mn^{55} (130 kev) has spin $\frac{9}{2}$, while the 279 kev state of Au^{197} is not the only one to be excited directly.

Agreement between experiment and theory has also been obtained by VAN LOEF and LIND[5] in studies of the excitation of γ-rays by inelastic neutron scattering, for neutrons with energies up to 2 Mev. They calculated the transmission coefficients $T(l)$ [cf. Eq. (7.3)] from the optical model with $V_0 = 42$ Mev, treating ξ as an adjustable parameter [see Eq. (6.1)]. However, the values of ξ required to fit the experimental data were about 0.2, which is much larger than that required to account for total neutron scattering cross sections.

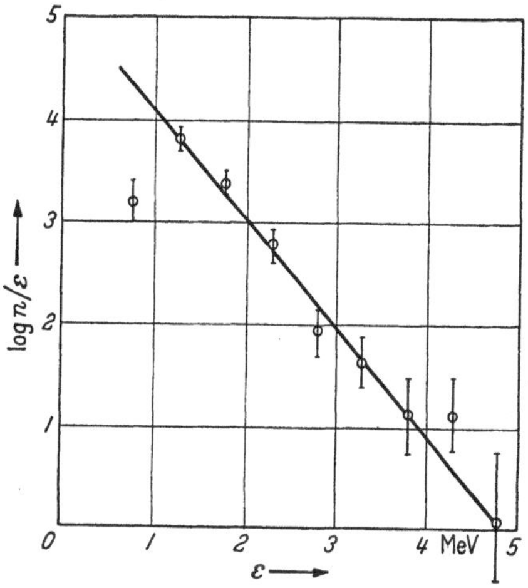

Fig. 29. Energy spectrum of 14 Mev neutrons scattered by Bi, according to WHITMORE and DENNIS.

27. (n, n') reactions: high energy scattering. Under this heading we include scattering by neutrons with energies above the neutron binding energy of the target nucleus. The few experiments which have been published have made use of the neutrons emitted in the $H^3(d, n)He^4$ reaction.

STELSON and GOODMAN[6] have bombarded Al, Fe, and Pb with 15 Mev neutrons; they surrounded the neutron source with a block of the material to be examined; and measured the energy spectrum of the secondary neutrons using a photographic plate as detector. They found that the greater part of the secondary neutrons have a very low energy. For Al, these neutrons had energies below 5 Mev; for Pb, below 3 Mev, and in every case, the energy spectrum was Maxwellian, which, as we have seen, is just the spectrum to be expected on the statistical theory (Sect. 8). The nuclear temperatures were 1.1 Mev for Al, 0.6 Mev for Fe, and 0.8 Mev for Pb.

Similar results were obtained for Pb and Bi by WHITMORE and DENNIS[7], using 14 Mev neutrons (Fig. 29). The temperatures were 0.8 and 0.9 Mev respectively. These authors estimated the nonelastic cross section by finding the reduction in the number of neutrons reaching the photographic plate with the full energy. For Pb and Bi they found 2.6 and 3.3 b respectively, in rough agreement with the cross section measurements discussed in Sect. 24. GRAVES and ROSEN[8], also using 14 Mev neutrons, have measured the energy spectra

[1] A. A. EBEL and C. GOODMAN: Phys. Rev. **93**, 197 (1954).
[2] C. P. SWANN and F. R. METZGER: Phys. Rev. **100**, 1329 (1955).
[3] B. MARGOLIS: Phys. Rev. **93**, 204 (1954).
[4] J. B. GUERNSEY and A. WATTENBERG: Phys. Rev. **101**, 1516 (1955).
[5] J. J. VAN LOEF and D. A. LIND: Phys. Rev. **101**, 103 (1956).
[6] P. H. STELSON and C. GOODMAN: Phys. Rev. **82**, 69 (1951).
[7] B. G. WHITMORE and G. E. DENNIS: Phys. Rev. **84**, 296 (1951).
[8] E. R. GRAVES and L. ROSEN: Phys. Rev. **89**, 343 (1953).

of the secondary neutrons emitted from eleven elements, ranging from C to Bi. The neutron flux was determined by counting the α-particles emitted in the reaction producing the neutrons. In general, their results are in good agreement with those of the authors just mentioned. The temperatures range from 2 or 3 Mev for the light nuclei, to 0.6 Mev for the heavy elements. Again, similar results have been obtained by O'NEILL[1] who measured the neutron flux falling on a scatterer by counting the α-particles emitted by the neutron source, and measured the energy of the secondary neutrons by timing their flight to the detector.

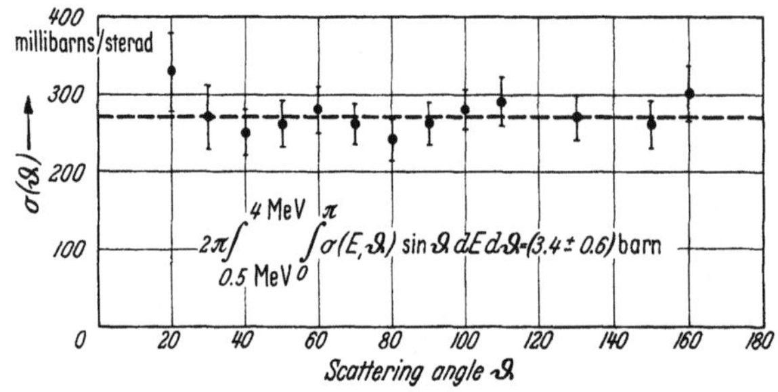

Fig. 30. Angular distribution of low energy neutrons emitted by Bi bombarded by 14 Mev neutrons, after ROSEN and STEWART.

The most interesting feature of the secondary neutrons is their angular distribution. Using 14 Mev neutrons, the angular distribution for Bi was measured by ROSEN and STEWART[2]. They found that the majority of the secondary neutrons, viz., those with energies lying between 0.5 and 4.0 Mev, are emitted isotropically, in agreement with the predictions of the statistical theory (Fig. 30) while the remainder, with energies from 4.5 to 12 Mev, which account for a cross section of about 0.2 b, are emitted mainly in the forward direction (Fig. 31). It is clear that the isotropic component must be caused by the decay of the compound nucleus; the forward component, however, is caused by direct interactions. The observed angular distribution is in agreement with the calculation of BROWN and MUIRHEAD[3].

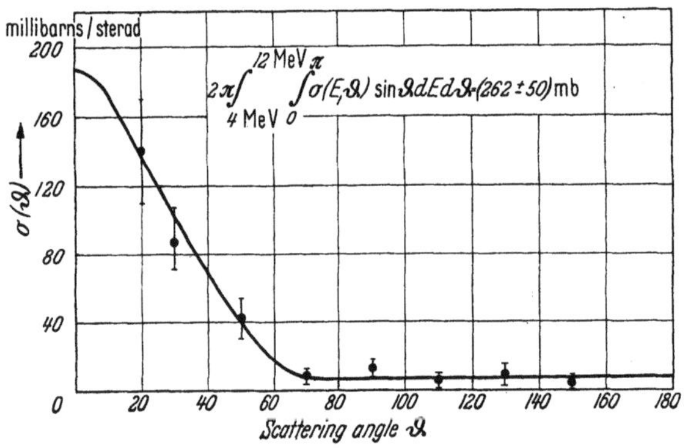

Fig. 31. Angular distribution of high energy neutrons emitted by Bi bombarded by 14 Mev neutrons, after ROSEN and STEWART.

The cross section for the $(n, 2n)$ process has been measured for a number of elements (see Sect. 30) for 14 Mev neutrons, and it is clear that this reaction must make a considerable contribution to the total. In the experiments just described it was not possible to distinguish experimentally between the neutrons emitted in the (n, n') and the $(n, 2n)$ processes. On the average, the second neutron to be emitted must leave the final nucleus with lower excitation energy than the emission of the first. Thus the temperature corresponding to this energy distribution should be different from that of the first neutron, and when the logarithm of the quantity $n(\varepsilon)/\varepsilon$ is plotted against ε it should divide into two linear portions. No such effect has been observed. However, the accuracy of measurement may not be sufficient to demonstrate the expected effect. Moreover, it is possible that the neutron spectrum observed experimentally is degraded to some

[1] G. N. O'NEILL: Phys. Rev. **95**, 1235 (1954).
[2] L. ROSEN and L. STEWART: Phys. Rev. **99**, 1052 (1955).
[3] See ref. 3, p. 223.

extent by multiple inelastic scattering. WHITMORE[1] has looked for such an effect in Pb but has not found it. Possibly this choice of target was unfortunate for the level spacing of low lying states in Pb^{208} is wide.

28. (n, p) and (n, α) reactions. Let Q_{np} be the energy balance in an (n, p) reaction, in which a stable target (ZN) with charge Z and neutron number N is bombarded by neutrons. The product of the (n, p) reaction, $(Z-1, N+1)$, is unstable by β-emission. If the total energy of the β-decay is W, the balance Q_{np} given by:

$$Q_{np} = 0.78\,\text{Mev} - W.$$

The reaction therefore may or may not be exoergic. The (n, α) reactions are usually exoergic, but the Q's of these reactions cannot be expressed by so simple a formula. In general, then, the kinetic energy available to protons and α-particles will not be very different to the energy of the incident neutrons. For neutrons with energies up to the binding energy of the target nucleus, proton emission, and especially α-particle emission, will be inhibited by the COULOMB barrier. In this energy range the cross sections for these reactions would not be expected to contribute much to the reaction cross section. In fact, no comprehensive experimental survey of (n, p) and (n, α) reactions for low neutron energies seems to have been made.

The first detailed investigation of (n, p) and $(n, 2n)$ reactions with neutrons of 9 Mev and above, was made by WÄFFLER[2]. He measured the cross sections for the (n, p) process for a number of elements irradiated by the neutrons emitted in the forward direction from Li and B, bombarded by 500 kev deuterons. In the forward direction the former reaction, in addition to producing many low energy neutrons, produces also neutrons of 10 and 13 Mev; the latter reaction produces neutrons of 9 and 13.5 Mev. The flux of neutrons was determined by a photographic plate method. The cross sections were found by measuring the radioactivities of the product nuclei. Since the neutron sources produced a rather varied energy spectrum, WÄFFLER's results represent average cross sections, and a strict comparison with theoretical values is difficult. His results, however, show that the (n, p) process is consistently more probable than theory will allow.

Very similar results have been obtained by PAUL and CLARKE[3]. These authors measured thirty eight (n, p) and twenty five (n, α) cross sections for 14.5 Mev neutrons by activation methods, the neutron flux being measured by counting the α-particles emitted in the reaction producing the neutrons.

For the light and intermediate weight nuclei, the (n, p) cross sections vary erratically, the highest cross sections being those of Zn^{64} (386 mb), and Pd^{105} (743 mb). For heavier elements, viz., Ba^{138}, La^{139}, Tl^{205}, and Pb^{208}, the cross section is a few millibarns. The ratio of the experimental results to the theoretical values is found to increase roughly exponentially from unity when $A \sim 1$, to 1000 when $A \sim 200$. This behaviour is illustrated in Fig. 32. It should be noted that the ratios for even-even target nuclei lie higher than those for odd-even nuclei. This is to be expected for the (n, p) reaction will produce odd-odd nuclei from even-even targets, and the former, as we shall show in Sect. 58, have the higher level densities. For the lighter elements, $(A < 60)$ where the effect of the COULOMB barrier is less important, the observed cross sections are in rough agreement with those deduced from the statistical theory. For the heavier

[1] B. G. WHITMORE: Phys. Rev. **92**, 654 (1953).
[2] H. WÄFFLER: Helv. phys. Acta **23**, 239 (1950).
[3] E. B. PAUL and R. L. CLARKE: Canad. J. Phys. **31**, 267 (1953).

elements the excessive (n, p) cross sections have been quantitatively accounted for in terms of direct processes by BROWN and MUIRHEAD[1].

The cross sections for the (n, α) reactions are comparable to those of the (n, p) reactions, the highest being those of Ga^{69} (105 mb), Br^{81} (103 mb), and Zr^{90} (194 mb); the average is a few tens of millibarns. The ratio of the experimental to the theoretical values increases at an even higher rate than those for (n, p) reactions: for $A \sim 200$ this ratio is about 10^4 (Fig. 33). For the (n, α)

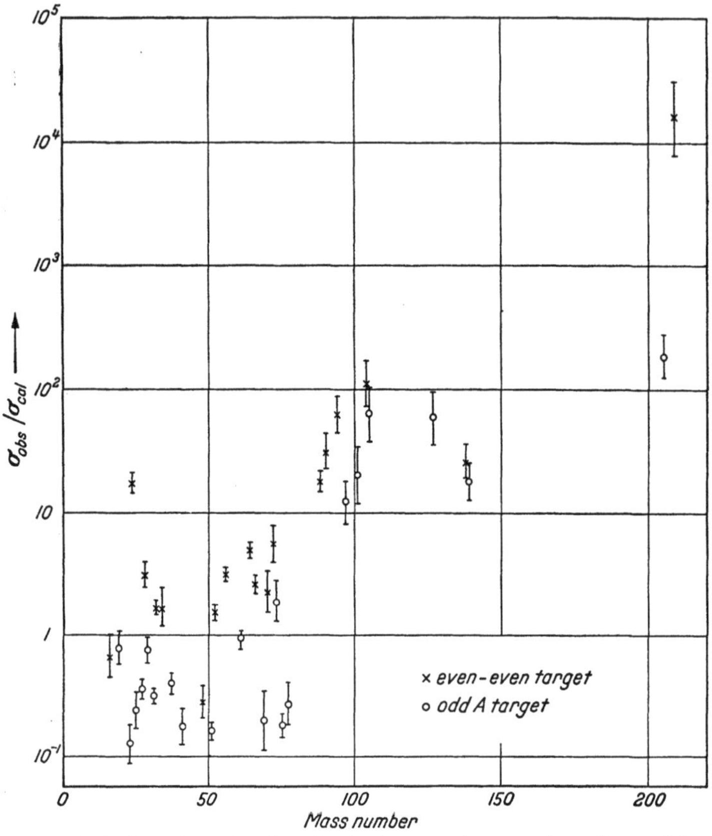

Fig. 32. Ratio of experimental to theoretical (n, p) cross sections, according to PAUL and CLARKE.

reactions there is no dependence on the proton or neutron number of the target nucleus: the emission of an α-particle, like that of the secondary neutron, leaves the parities of these numbers unchanged.

Recent measurements of the (n, α) cross sections of four nuclei (Zn^{68}, Zr^{90}, Zr^{94}, and In^{115}) by BLOSSER *et al.*[2], are consistently lower than the average cross section found by PAUL and CLARKE. Furthermore, the cross sections for the (n, α) process in Zr^{90} and Zr^{94} have been measured by BROLLEY *et al.*[3], for 14 Mev neutrons by measurement of the γ-rays emitted by Sr^{87m} and Sr^{91}; by ARMSTRONG and BROLLEY[4] by measurement of the α-particles by a photographic

[1] See ref. 3, p. 223.

[2] H. G. BLOSSER, C. P. GOODMAN, T. H. HANDLEY and M. L. RANDOLPH: Phys. Rev. **100**, 429 (1955).

[3] J. E. BROLLEY, M. E. BUNKER, D. R. F. COCHRAN, R. L. HENKEL, J. P. MIZE and J. W. STARNER: Phys. Rev. **99**, 330 (1955).

[4] A. H. ARMSTRONG and J. E. BROLLEY: Phys. Rev. **99**, 330 (1955).

plate method; and by RIBE and DAVIS[1], by counter methods. Although γ-ray measurements on the isomer of Sr^{87} do not reveal all transitions which produce this nucleus, the results agree generally in that the (n, α) cross sections of both Zr isotopes are about 4 mb, and not 194 mb (to an accuracy of 40%) as obtained by PAUL and CLARKE. However, the value for the (n, p) cross section in Zr^{90} is in fair agreement with that of PAUL and CLARKE. The energy spectrum of the α-particles appears to be peaked near 17 Mev, which is the barrier height for

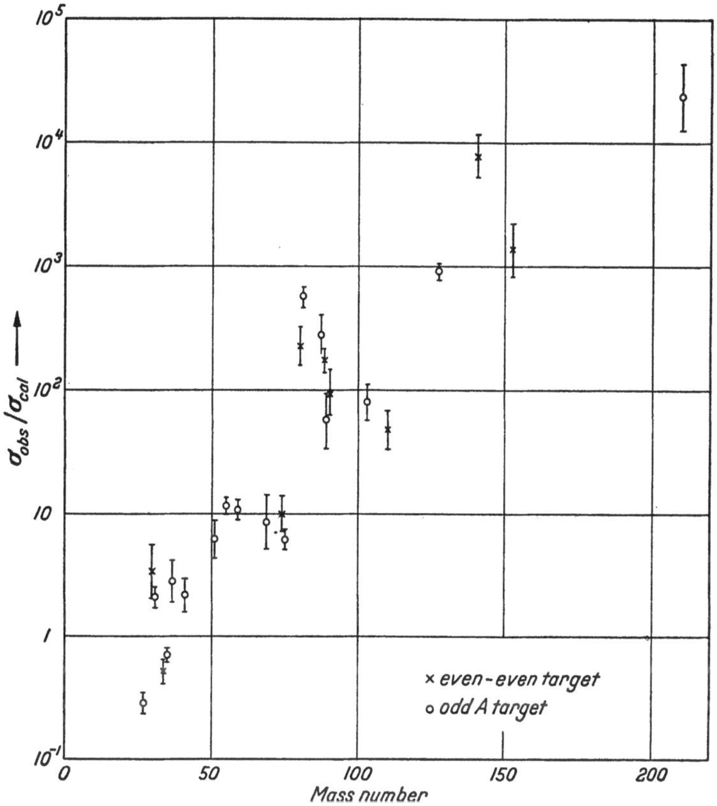

Fig. 33. Ratio of experimental to theoretical (n, α) cross sections according to PAUL and CLARKE.

α-particle emission in Zr. The angular distribution, however, seems to contain an isotropic component, on which another distribution, strongly peaked in the forward direction, is superposed.

The angular distributions of the protons emitted under 14 Mev neutron bombardment have not been measured directly. There have been very few attempts to measure (n, p) cross sections as a function of neutron energy. Some rough measurements of the energy variation of the (n, p) cross sections of Cl^{37} and Sr^{88} have been made by COHEN and WHITE[2]. The former reaction is anomalous but the latter shows the shape to be expected from the statistical theory. Recently (n, p) cross sections for mass numbers near 60 have been measured[3] as a function of neutron energy between 12 and 16 Mev. The results are reported

[1] F. L. RIBE and R. W. DAVIS: Phys. Rev. **99**, 331 (1955).
[2] A. V. COHEN and P. H. WHITE: Nuclear Phys. **1**, 73 (1956).
[3] H. G. BLOSSER and T. H. HANDLEY: Bull. Amer. Phys. Soc. **1**, 263 (1956).

to indicate a strong and continuing competition from the (n, pn) reaction, consistent with the anomalous behaviour of the (p, pn) reaction (Sect. 36).

29. (n, np) reactions. These reactions are difficult to measure by activation methods because the product is usually stable or long-lived. However, upper limits for the cross sections of four nuclei (Se^{80}, Pd^{108}, Ce^{142}, and W^{186}) have been obtained by activation by COHEN, HYDE, and WHITE[1].

The energy spectrum of protons emitted from targets bombarded by 14 Mev neutrons has been measured by ALLAN[2] using photographic plates as detectors and thin foils of separated isotopes as targets, and by COLLI and FACCHINI[3] using scintillation detectors. Some of the protons ALLAN detected

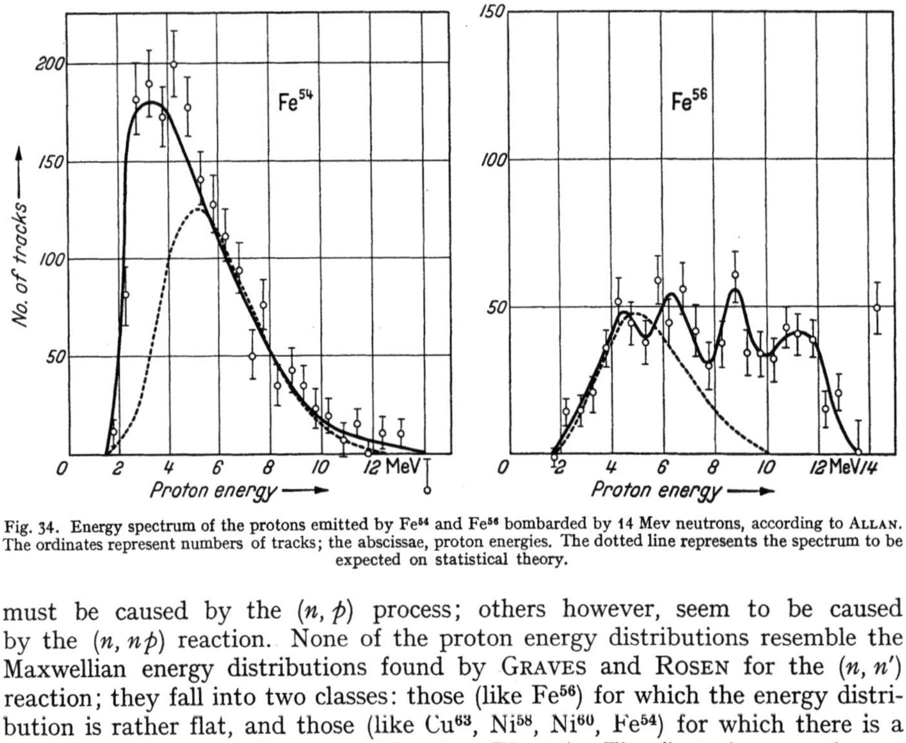

Fig. 34. Energy spectrum of the protons emitted by Fe^{54} and Fe^{56} bombarded by 14 Mev neutrons, according to ALLAN. The ordinates represent numbers of tracks; the abscissae, proton energies. The dotted line represents the spectrum to be expected on statistical theory.

must be caused by the (n, p) process; others however, seem to be caused by the (n, np) reaction. None of the proton energy distributions resemble the Maxwellian energy distributions found by GRAVES and ROSEN for the (n, n') reaction; they fall into two classes: those (like Fe^{56}) for which the energy distribution is rather flat, and those (like Cu^{63}, Ni^{58}, Ni^{60}, Fe^{54}) for which there is a sharp maximum, usually near 3 Mev (see Fig. 34). The first class can be accounted for if, in addition to the statistical energy distribution, there is superposed on it some high energy protons produced by direct interactions[3]. For the second class, if the statistical distribution is calculated assuming a temperature of about 0.8 Mev, as obtained by GRAVES and ROSEN, it is clear that there exists an excess of low energy protons, many of which have energies far below the COULOMB barrier (7 Mev). The low energy part of the proton spectrum is cut off sharply at about 2 Mev. Now the nuclei in the second class are all low mass number isotopes. Their proton binding energies are lower than their neutron binding energies; it is possible therefore to attribute the low energy protons to the (n, np) process, because, when the emission of the first neutron leaves the residual nucleus in states with an excitation energy between the proton and neutron binding energies, only proton or γ-ray emission can occur. The

[1] A. V. COHEN, S. B. HYDE and P. H. WHITE: Nuclear Phys. **1**, 278 (1956).
[2] D. L. ALLAN: Proc. Phys. Soc. Lond. A **68**, 925 (1955), also to be published 1956.
[3] L. COLLI and U. FACCHINI: Nuovo Cim. **14**, 671 (1956).

low energy cut-off (2 Mev) in the proton spectrum corresponds to the energy excess that a proton must have to compete successfully with γ-radiation.

The calculated yields of the (n, p) reactions are in rough agreement with the experimental data when account is taken of the fact that, for even-even targets, the product of the (n, p) reaction is an odd-odd nucleus. This latter effect may be illustrated by a comparison between the data for Fe^{54} and for Cu^{63}. For the former nucleus, for which the yield of the $(n, 2n)$ reaction is negligible, the total proton yield is equivalent to about 680 mb ($\sigma_c \approx 1.2$ b) of which 220 mb is caused by the (n, np) process. For Cu^{63} ($\sigma_c \approx 1.4$ b), the *total* proton yield is 250 mb.

30. $(n, 2n)$ reactions. These reactions occur only when the energy of the incident neutron is greater than the neutron binding energy of the target nucleus. For most nuclei, this energy lies somewhere between 6 and 14 Mev. Owing to the obvious difficulties of distinguishing (n, n') from $(n, 2n)$ reactions there is, as yet, no experimental data on the angular distribution or the energy spectrum of the secondary neutrons. Most experimenters, instead, have been content to determine the cross section for the reaction by measuring the activity of the product nucleus. Such methods, like those used in the measurement of (n, p) and (n, α) reactions, select only such target nuclei which will give radioactive products with periods in a range suitable for measurement. However, recently an attempt has been made to measure the $(n, 2n)$ cross section by a multiplication method in elements not suitable for activation.

WÄFFLER[1] has determined the cross sections for the $(n, 2n)$ reaction for seventeen elements, using the methods described in Sect. 28. With 500 kev deuterons, the lithium reaction produces neutrons of 10 and 13 Mev, and the boron reaction neutrons of 9 and 13.5 Mev. For most targets, the less energetic of the two neutron groups lies below the threshold of the $(n, 2n)$ reaction: the effective neutrons were therefore homogeneous in energy. For Cu^{63}, the cross section (to 10% accuracy) was 314 mb for the lithium reaction (13 Mev neutrons) in accord with the value 360 mb for the boron reaction (13.5 Mev neutrons). Similar measurements, but with a less well defined neutron energy, have been made by COHEN[2]. With 14.5 Mev neutrons, PAUL and CLARKE[3], as described in Sect. 27, have measured the $(n, 2n)$ cross sections for thirty-four nuclei. For Cu^{63} they obtained 482 mb, which in view of the greater neutron energy, is in accord with the cross section obtained by WÄFFLER. The cross sections for other nuclei vary erratically, because the cross section is critically dependent on the neutron binding energy. The results generally are in rough agreement with theoretical expectations (8.9) and show no particular trend with mass number. The $(n, 2n)$ cross sections for the two Cu and Ag isotopes for 14 Mev neutrons have also been measured by FORBES[4]. His results for the Cu isotopes agree with those of PAUL and CLARKE, and those of WÄFFLER, but his results for Ag are three times greater.

The energy variation of the $(n, 2n)$ cross section in Cu has been measured by BROLLEY, FOWLER, and SCHLACKS[5] with neutrons obtained from the $H^3(d, n)$ He^4 reaction. With 10.5 Mev deuterons, they were able to measure the cross

[1] See ref. 2, p. 251.
[2] B. L. COHEN: Phys. Rev. **81**, 184 (1951).
[3] See ref. 3, p. 251.
[4] S. G. FORBES: Phys. Rev. **88**, 1309 (1952).
[5] J. E. BROLLEY, J. L. FOWLER and L. K. SCHLACKS: Phys. Rev. **88**, 618 (1952).

section up to a maximum neutron energy of 27 Mev. Similar but less comprehensive measurements were made by FOWLER and SLYE[1], and by MARTIN and DIVEN[2], for neutrons between 11 and 18 Mev. The results of these authors, together with those mentioned in previous paragraphs for neutrons near 14 Mev, are shown in Fig. 35, where the continuous lines are those calculated from (8.9) with $\Theta = 0.8$ and 2.4 Mev, and BLATT and WEISSKOPF's [4] expression for the reaction cross section (1.4 barn). It will be seen that the experimental results are in agreement with the theoretical curve on the rising part of the curve provided that the temperature is suitably chosen (2.4 Mev). The departure of the experimental data from the theoretical curve above 16 Mev may be caused, in part, by direct collisions[3]. It should be noted that the temperature is not in agreement with that derived from the spectrum of the neutrons emitted (Sect. 27). This subject is discussed in Sect. 59.

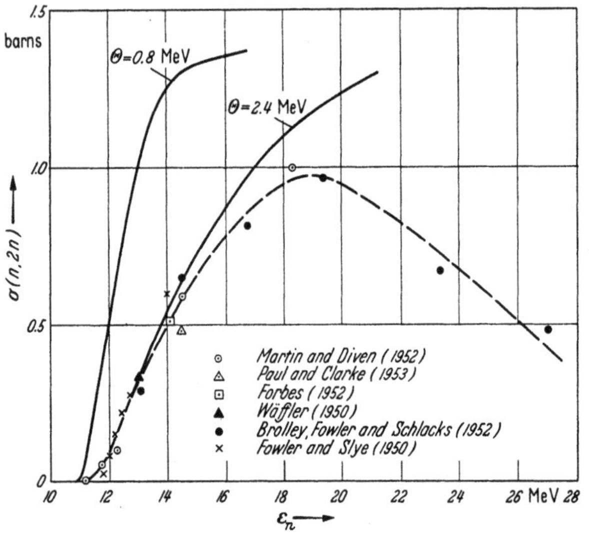

Fig. 35. The $(n, 2n)$ cross section for Cu^{63}. The continuous lines are the cross sections predicted by (8.9) for $\Theta = 0.8$ and 2.4 Mev. The fall in cross sections above 20 Mev is presumably caused by the competition of the $(n, 3n)$ reaction.

Similar curves for the variation of the cross section with energy have been obtained by BROLLEY, FOWLER, and SCHLACKS[4], for Mo^{92} and C^{12}; by MARTIN and DIVEN[2], for Ni^{58}, and Tl^{203}; and by MARTIN and TASCHEK[5], for I^{127}.

By an ingenious combination of transmission and multiplication measurements, GRAVES and DAVIS[6] have succeeded in fixing limits to the cross sections of the competing reactions which occur with 14 Mev neutrons. For the transmission measurements, these authors surrounded the neutron source with a shell of the material to be examined, and used a biassed stilbene crystal as a detector. This measurement gives the nonelastic cross section. To measure the multiplication of neutrons within the shell, they replaced the crystal detector by a "long" counter, with a flat characteristic between 25 kev and 20 Mev. If M is the multiplication, i.e., the ratio of the number of neutrons detected by the "long" counter, with the shell in position, to that measured with the shell removed, and if T is the transmission measured according to the methods described in Sect. 24, they show that, for non-fissionable material:

$$1 + C = \frac{M - T}{1 - T} = \left[1 + \frac{\sigma(n, 2n)}{\sigma_n} + 2\,\frac{\sigma(n, 3n)}{\sigma_n} - \frac{\sigma'}{\sigma_n}\right]$$

[1] J. L. FOWLER and J. M. SLYE: Phys. Rev. **77**, 787 (1950).
[2] H. C. MARTIN and B. C. DIVEN: Phys. Rev. **86**, 565 (1952).
[3] R. NAKASHIMA and K. KIKUCHI: Progr. Theor. Phys., Japan **14**, 126 (1955).
[4] See ref. 5, p. 255.
[5] H. C. MARTIN and R. F. TASCHEK: Phys. Rev. **89**, 1302 (1953).
[6] E. R. GRAVES and R. W. DAVIS: Phys. Rev. **97**, 1205 (1955).

where C is a directly measurable quantity, σ_n is the nonelastic cross section, and σ' is the sum of the cross sections of those reactions which do not reproduce the incident neutron, e.g., radiative capture, (n, p), (n, α) reactions, etc. For most nuclei 14 Mev neutrons have insufficient energy to excite the $(n, 3n)$ reaction. If the term for this reaction is omitted in the above expression,

$$C\,\sigma_n = \sigma(n, 2n) - \sigma'.$$

Now

$$\sigma_n = \sigma(n, n') + \sigma(n, 2n) + \sigma'$$

where $\sigma(n, n')$ refers to the sum of the cross sections of those reactions, which, like the (n, n') reaction, reproduce the incident neutron, e.g. the (n, np) and (n, pn) reactions. From these two expressions it follows that the lower limits of σ' and $\sigma(n, 2n)$ are 0 and $C\sigma_n$, respectively. The upper limits are $\frac{1}{2}\sigma_n(1 - C)$ and $\frac{1}{2}\sigma_n(1 + C)$ respectively. For $\sigma(n, n')$, the limits are 0 and $\sigma_n(1 + C)$. When the $(n, 2n)$ reaction is well developed, the limits on the $(n, 2n)$ cross sections are close together. Thus for Pb, GRAVES and DAVIS found that $\sigma(n, 2n)$ must lie between 1.76 and 2.12 b, and for Bi, between 1.86 and 2.19 b. When $\sigma(n, 2n)$ is known from activation measurements, as in Cu^{63}, it is possible to substitute these values into the expression for the cross sections and for the multiplication, and thus find that σ' is 0.40 b, most of which is presumably due to Cu^{63} (n, p) Ni^{63}, (since the cross section for the corresponding reaction in Cu^{65} is very small) and 0.36 b for the sum of $\sigma(n, n')$ and $\sigma(n, np)$, in rough agreement with ALLAN's results (Sect. 29).

31. Neutron induced fission reactions. In this section we shall discuss fission reactions only to the extent that they compete with other nuclear processes. The experimental data on fission cross sections has been summarized by HUGHES and HARVEY[1].

The variation of the cross sections with neutron energy depends markedly on the proton and neutron number of the target nucleus. Even-odd target nuclei are fissionable at thermal energies; at higher energies the cross section falls slowly and remains nearly constant to the highest energy at which experimental data has been published (usually 3 Mev). In U^{235}, for example, the cross section is 3.7 b at 10 kev, falls to 1.7 b near 100 kev, and remains nearly constant at 1.3 b for energies up to 3 Mev. The cross section for U^{233} is very similar, the cross section of the flat plateau being about 1.8 b.

For even-even target nuclei, fission appears only when the neutron energy is increased beyond a certain threshold, which is the higher for the higher mass isotopes. This phenomenon may be explained on the assumption that the threshold for fission measured in terms of the excitation energy is approximately the same in each isotope, an assumption which is consistent with photofission measurements (thresholds 5 to 6 Mev). The threshold observed in the fission cross section, then, is equivalent to the fission threshold of the compound nucleus. Thus if B_f is the fission threshold of this nucleus, and if B_n is its neutron binding energy, fission will be detected for neutrons with energies above E_f where $E_f = B_f - B_n$. This threshold is not observed when the target nucleus is even-odd; for then the neutron binding energy of the even-even product nucleus is higher than the fission threshold in that nucleus. Now the neutron binding energy of even-odd compound nuclei is greater the lower the mass; therefore the observed threshold for fission in even-even target nuclei decreases with decreasing mass

[1] D. J. HUGHES and J. A. HARVEY: BNL 325, Brookhaven National Laboratory, 1953.

number. Just above the threshold the cross section increases very rapidly; within a few hundred kilovolts above threshold, a plateau is reached. Between the threshold and the attainment of this plateau, the fission cross section fluctuates, in some nuclei (e.g. U^{234} and U^{236}) in a rather regular manner (Fig. 36), in others, as in Th^{232}, more rapidly and less regularly.

Odd charge nuclei have thresholds like even-even target nuclei. For Pa^{231} the threshold is at about 400 kev; in Np^{237} it is at 200 kev. In the former, the cross section shows a short step, about 200 kev, wide, before rising to a plateau of about 1 b. In Np^{237} a smooth plateau at 1.5 b is maintained to an energy of about 5 Mev, beyond which the cross section rises again in another step to attain

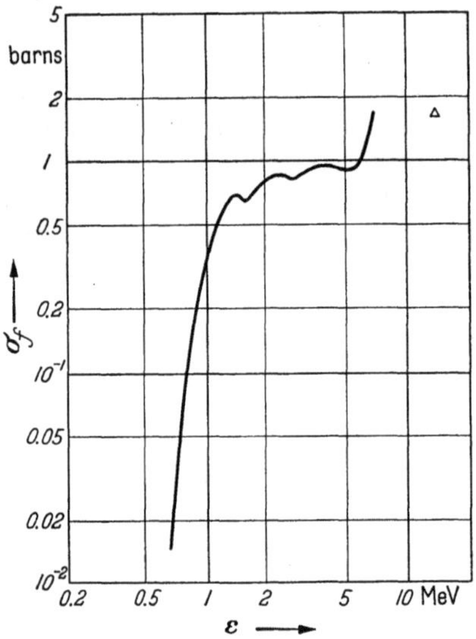

Fig. 36. Fission cross section of U^{236}, after Hughes and Harvey.

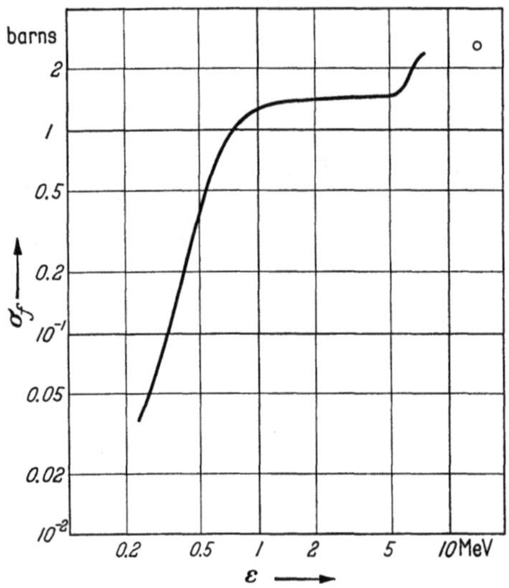

Fig. 37. Fission cross section of Np^{237}, after Hughes and Harvey.

a cross section of about 2.6 b at 14 Mev (Fig. 37). The fission cross section of U^{236} shows a similar step, starting at about 6 Mev (Fig. 36). Above this step the cross section is nearly 1.8 b. Presumably this second step is a common phenomenon and would be observed in the cross sections of other fissionable isotopes if these cross sections could be measured. It is important to note that for the uranium isotopes the fission cross section above the second step comprises the greater part of the reaction cross section.

The most interesting features of the fission cross section and that which most distinguishes it from other types of nuclear reaction, are the appearance of these steps, and the flatness of the cross section curve between them. On the plateau of the first step, i.e., for energies between the threshold and the onset of the second step near 6 Mev, the fission process competes only with inelastic neutron scattering and, to a much less degree, with radiation[1]. At these low energies these processes must all follow compound nucleus formation. Since the fission cross section is usually about 1 b, it is clear that the probability of fission is very similar to that of inelastic neutron emission; in fact, the constancy of the fission cross section shows that if the probability of inelastic neutron emission rises with increasing energy owing to the increase in the number of available states in the target nucleus, then the fission probability must increase at the same rate.

[1] N. Bohr and J. A. Wheeler: Phys. Rev. 56, 426 (1939).

The two processes are radically different in nature; it would indeed be a remarkable coincidence if both these probabilities and their rates of change with energy, were to be so very similar. However, as we have seen in Sect. 8, it is not necessary to assume that the total probability of inelastic neutron emission increases very rapidly with energy. If, as is much more likely, the total neutron width is approximately independent of neutron energy, it follows that the average fission probability is also approximately independent of energy.

The fission cross section in the first step (up to 6 Mev) is clearly caused by the fission of the compound nucleus (mass $A+1$). If Γ_f and Γ_n are the average fission and inelastic neutron widths, respectively, then neglecting the width caused by compound-elastic scattering, the fission cross section in the first step is given by:

$$\sigma_f = \sigma_c \frac{\Gamma_f}{\Gamma_f + \Gamma_n} = \beta_{A+1}\,\sigma_c.$$

It is natural to attribute the step near 6 Mev to fission of the target nucleus. The neutron energy at the step, therefore, should be equal to the fission threshold of the target nucleus; the increase in the fission cross section at this point is due to the production of states in the target nucleus with energies above the fission threshold. Above this threshold (Q_f), by analogy with (8.9), the cross section is:

$$\sigma_f = \sigma_c\left[\beta_{A+1} + \beta_A(1-\beta_{A+1})\left(1-(1+z)\,e^{-z}\right)\right] \tag{31.1}$$

where $z = (\varepsilon_m - Q_f)/\Theta$. The difference between the observed fission cross section and the reaction cross section then is due to inelastic scattering or to secondary processes such as the $(n, 2n)$ reaction. The product of the latter reaction [mass $(A-1)$] will start to contribute to the fission cross section when the neutron energy exceeds the sum of its fission threshold and the neutron binding energy of the target nucleus. Thus we should expect a third step to appear at this point (11 to 12 Mev).

The expression (31.1) is valid only when β_A is small and the fission and $(n, 2n)$ thresholds are identical. However the latter usually exceed the former in even-even nuclei by 1 to 2 Mev. Therefore when the energy of the incident neutrons lies between the fission and the $(n, 2n)$ thresholds, fission in the target nucleus [i.e., the (n, nf) reaction] will increase rapidly with energy because it competes only with radiation following inelastic scattering $(n, n\gamma)$. In this energy range the cross section is still given by (31.1) provided $\beta_A/(1-\beta_A)$ now measures the much larger ratio of the probabilities of fission to radiation. Above the $(n, 2n)$ threshold, the development of the $(n, 2n)$ reaction must offer strong competition, and, according to JACKSON[1], accounts for the reduction in the fission cross section which occurs in Th^{232} just above the 6 Mev step.

It is obvious from an inspection of the fission cross section curves that the temperature is low because the steps attain their full value in a small change of neutron energy. At 14 Mev it is possible to cause fission successively in the nuclei of mass, $A+1$, A, and $A-1$. In this case, putting $z=\infty$, the fission cross section is given by:

$$\frac{\sigma_f}{\sigma_c} = \beta_{A+1} + \beta_A(1-\beta_{A+1}) + \beta_{A-1}(1-\beta_A)(1-\beta_{A-1}). \tag{31.2}$$

If we assume that the factors β are all the same, (31.2) becomes:

$$\frac{\sigma_f}{\sigma_c} = \beta(3 - 3\beta + \beta^2) \tag{31.3}$$

[1] J. D. JACKSON: Unpublished.

or

$$\frac{\sigma_f}{\sigma_c} \approx 3\beta$$

for small β. This equation accounts roughly for the published experimental data. For example, for Th^{232}, the fission cross section for the first step is about 0.15 b; it is 0.35 b at 14 Mev, or a little less than we should expect from (31.3). For Np^{237}, which is the only other nucleus for which published data exists, the fission cross section at 3 Mev is about 1.45 b; at 14 Mev, about 2.7 b. Here the reaction cross section is about $1.2\pi R^2$, or 2.9 b; the cross section at 14 Mev calculated from (31.3) is 2.6 in agreement with experiment.

It follows that the fission cross section approaches the reaction cross section at high energies. If the energy of the incident neutron or proton is just sufficient to eject m neutrons, (31.3) becomes:

$$\frac{\sigma_f}{\sigma_c} = [1 - (1 - \beta)^m]$$

which approaches unity for large m. This has been varified by HARDING[1], by bombarding U^{238} and Th^{232} by 150 Mev protons. For these nuclei β is small; hence we may infer that, in the main, neutrons are evaporated before fission occurs (a phenomenon also demonstrated by HARDING[1]).

H. Proton reactions.

32. Nonelastic cross sections. The measurement of nonelastic cross sections for protons or α-particles presents great difficulties because the usual methods are no longer applicable to particles subject to the RUTHERFORD scattering, for which the differential cross section rises without limit as the scattering angle is reduced. For high energies, as already pointed out, it is possible to measure the number which have lost energy by inelastic processes by "poor geometry" and anticoincidence counting methods, but at low energies, or intermediate energies, the rapid rate at which charged particles lose energy by ionisation makes such methods very difficult, or impractical. Some attempts have been made to determine the nonelastic cross section by measurement of the contributions to this cross section from the various reactions which can occur. For energies below the neutron binding energy, this has been done with some precision by measurement of the (p, n) cross section, which, presumably, greatly exceeds the (p, p') cross section in heavy elements. The results obtained are in good agreement with theory (see Sect. 34). But at higher energies the few attempts that have been made are far from being precise. Recently the reaction cross sections of Pb and Bi have been determined from 10 to 80 Mev by BELL and SKARSGARD[2] by adding the individual cross sections of reactions of the type (p, xn). The results, when plotted as a function of energy, show interesting fluctuations about the smoothly varying cross sections predicted by SHAPIRO (Sect. 7).

33. Proton capture cross sections. At low energies it is not possible to measure the products of proton capture in heavy elements because of the effect of the COULOMB barrier. At higher energies proton capture has, until recently, received very little attention. COHEN's results[3] appear to be very different from those of neutron capture. Unfortunately, the energy ranges over which these cross sections have been measured do not overlap so a strict comparison is not possible.

---

[1] See G. N. HARDING and F. J M. FARLEY, Proc. Phys. Soc. Lond. A **69**, 853 (1956).
[2] R. E. BELL and H. M. SKARSGARD: Canad. J. Phys. **34**, 745 (1956).
[3] B. L. COHEN: Phys. Rev. **100**, 206 (1955).

While the neutron capture cross sections decrease as the energy is increased to 5 Mev (see Sect. 61), at which energy they appear to be approximately independent of atomic weight, proton capture cross sections from 5 Mev upwards are nearly constant. For a given element they appear to be nearly independent of energy, and for a given energy, decrease slowly as a function of atomic weight. According to the statistical theory in its simplest form, the capture cross sections should fall continuously with increasing energy as ever more reactions compete for their share of the reaction cross section. The probable explanation of the lack of energy variation in the proton capture cross sections has been given by MORINAGA[1], who has studied the cross section for the capture of α-particles by Ni^{58}. He found that the yield of Zn^{62} rises from 10 Mev to a flat maximum near 16 Mev which corresponds approximately to the position of the giant resonance of the photonuclear effect. No measurements were made beyond 17 Mev because this corresponds to the threshold of the unwanted Ni^{60} (α, $2n$) Zn^{62} reaction. MORINAGA found that the absolute value of the cross section and the shape of the curve could be accounted for approximately when the radiative probability is calculated from the photonuclear cross sections. He states that a similar calculation for proton capture by Ni^{60} gives results in agreement with COHEN's measurements.

For Bi^{209}, the proton capture cross section has been measured by ANDRE *et al.*[2] over a limited energy range. The cross section rises steadily from 7 to 10.6 Mev; this is caused no doubt, by the rising penetrability.

34. (p, n) reactions. Reactions of the (p, n) type are the most probable when the proton energy exceeds the threshold. Above the neutron binding energy this reaction gives way to others, and at higher energies, as shown in Sect. 36, the probability of secondary proton emission may exceed that of neutron emission.

The product of a (p, n) reaction with a stable target is an unstable body which may decay to the original nucleus by positron emission or electron capture. If Q_{pn} is the energy balance of the (p, n) reaction and if W is the total energy released in positron decay, or electron capture, then:

$$Q_{pn} = -W - 0.78\,\text{Mev}.$$

Thus (p, n) reactions are always endoergic, and by an amount never less than the energy of the neutron decay (0.78 Mev).

The first detailed study of (p, n) reactions was made by BLASER, BOEHM, MARMIER, and PEASLEE[3] with protons with energies up to 7 Mev. These authors found that, as the proton energy is increased to about 1 Mev above the threshold, the cross section for this reaction rapidly approaches the theoretical reaction cross section, when the latter is calculated on the basis of $r_0 = 1.5 \times 10^{-13}$ cm. The agreement between theory and experiment was not always close and the difference often varies, apparently outside the limits of experimental error, from one isotope to another in a single element. Generally the cross sections were the larger for the odd-neutron number target (e.g., Cd^{111}). For low energies one would not expect a dependence on the proton or neutron number unless a competing reaction [e.g. the (p, p') reaction] makes a significant contribution to the total cross section. For odd-charge targets, generally, the reaction cross section of the lightest of two isotopes was the greater (e.g. the cross section of Ag^{107} was

[1] H. MORINAGA: Phys. Rev. **101**, 100 (1956).

[2] C. G. ANDRE, J. R. HUIZENGA, J. F. MELCH, W. J. RAMLER, E. G. RAUH and S. R. ROCKLIN: Phys. Rev. **101**, 645 (1955).

[3] J. P. BLASER, F. BOEHM, P. MARMIER and D. C. PEASLEE: Helv. phys. Acta **24**, 3 (1951).

greater than that of Ag^{109}). Cu^{63} is an exception to this rule: the cross sections for the Cu isotopes are shown in Fig. 38. These authors have compared the experimental with theoretical cross sections for Cu^{63} just above its threshold. In this case, the product nucleus (Zn^{63}) is produced in the ground state or in a few low lying excited states. They show that the theoretical cross section is not very sensitive to the spins of the excited states (which are not known) and that the results generally favour a spin of $\frac{3}{2}$ for Zn^{63}.

An interesting study of the excitation of isomeric states by the (p, n) reaction has been made by Boehm, Marmier, and Preiswerk[1]. For Cr^{52}, the thresholds of the (p, n) reactions leading to the ground state and to the isomeric state of Mn^{52} are high (5.5 and 5.9 Mev respectively). For proton energies just above the higher of the two thresholds, the two states will be excited only by the neutron groups which lead to them directly. The results show that the production of the isomeric state is favoured by a factor of 12; this shows that the isomeric state, and not the ground state, has the lower spin. A similar experiment with Sc^{44} shows just the opposite effect, viz., that the isomeric state in this nucleus has the higher spin.

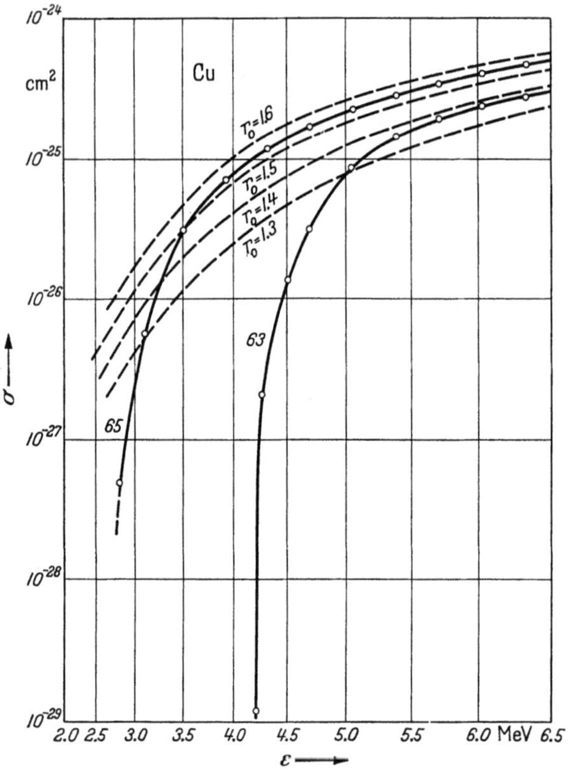

Fig. 38. Cross section for the (p, n) reaction in Cu, after Blaser, Boehm, Marmier, and Peaslee. The dotted lines represent reaction cross sections calculated for different nuclear radii.

A comprehensive study of the cross sections of some 60 nuclei has been made by Blaser, Boehm, Marmier, and Scherrer[2], with a proton beam of 6.8 Mev. The energy variation of the (p, n) cross section was determined by measuring the activities produced in a stack of foils exposed to the proton beam. The results in general are in good agreement with a reaction cross section with $r_0 = 1.5 \times 10^{-13}$ cm. There is little difference in the behaviour of odd-even and even-even targets; this is to be expected because, as pointed out above, the (p, n) reaction is responsible for the greater part of the reaction cross section and is not in effective competition with any other process. There seems to be no general rule for the reaction cross sections of even-even isotopes of a given element: sometimes the reaction cross sections decrease (e.g., Cd), sometimes they increase with mass number (e.g., Sn, Zr). A survey of the (p, n) cross sections has been made recently by Blosser and Handley[3], for 12 Mev protons. At this energy the $(p, 2n)$ reaction can

[1] F. Boehm, P. Marmier and P. Preiswerk: Helv. phys. Acta **25**, 599 (1952).
[2] J. P. Blaser, F. Boehm, P. Marmier and P. Scherrer: Helv. phys. Acta **24**, 441 (1951).
[3] H. G. Blosser and T. H. Handley: Phys. Rev. **100**, 1340 (1955).

make a considerable contribution to the reaction cross section. These authors have estimated this contribution, where it is appropriate, and conclude that the sums of the cross sections are consistent with $r_0 = 1.6 \times 10^{13}$ cm.

The energy distribution of the neutrons emitted from a number of elements bombarded by 16 Mev protons has been measured by GUGELOT[1], using a photographic plate method. For protons of this energy, only a part of the nonelastic cross section is caused by the (p, n) reaction, and most of the neutrons emitted are derived from the (p, pn), (p, np) and $(p, 2n)$ reactions. For intermediate weight nuclei $(A < 100)$ the (p, pn) reactions seem to be much more probable than the other two (Sect. 36). For these nuclei, then, the neutrons observed are, in the main, the second and not the first to be emitted. For heavier nuclei, the (p, np) and $(p, 2n)$ would be the most important. In spite of the complex origin of the neutrons, GUGELOT found that the neutron spectrum was Maxwellian; the temperatures were about 0.8 Mev for the heavy elements (Rh, Au, and Tl) and somewhat higher for the lighter elements (Be, Al, and Fe). These results were obtained after making a final calculated correction for the spectrum of the secondary neutrons from the $(p, 2n)$ reaction. A typical energy spectrum is shown in Fig. 39. These results are in general agreement with those obtained by GUGELOT for the (p, p') reaction (Sect. 35), and by GRAVES and ROSEN (Sect. 27) for 14 Mev neutrons. The cross sections for the neutron production, however, were not measured.

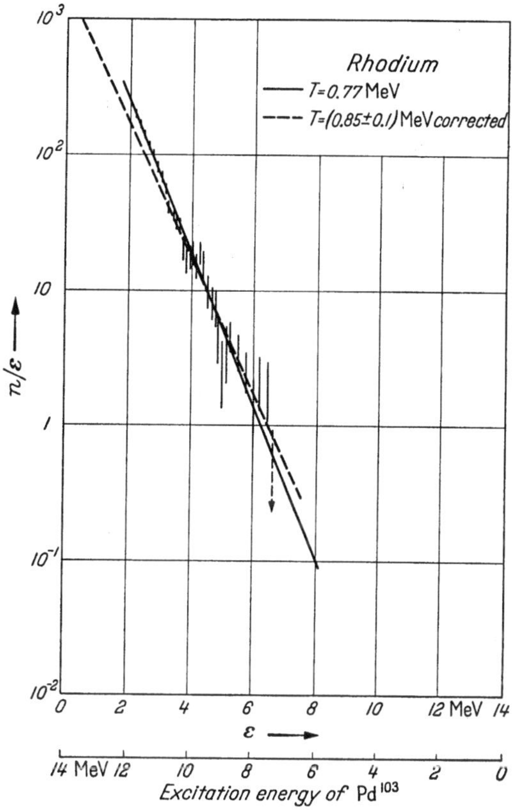

Fig. 39. Energy spectrum $n(\varepsilon)$ of the neutrons emitted by Rh bombarded by 16 Mev protons, after GUGELOT. The dotted line represents the spectrum after correction for the $(p, 2n)$ reaction.

The angular distribution of the neutrons obtained by bombardment with 23 Mev protons has been studied by COHEN[2] using threshold detectors. The activities detected were Ag^{106} caused by the Ag^{107} $(n, 2n)$ reaction (threshold 9.6 Mev) and Mg^{27} from the (n, p) reaction in Al^{27} (threshold 2.0 Mev). If we assume that the statistical theory is valid, we should expect that the neutron angular distribution to be isotropic, or at least symmetrical about the 90° position. These measurements, show, on the contrary, that the angular distribution, like that found by ROSEN and STEWART for Bi (Sect. 27) is far from isotropic, the forward direction being favoured. Thus direct interactions must contribute to the cross

[1] P. C. GUGELOT: Phys. Rev. **81**, 51 (1951). See also D. M. THOMSON: Proc. Phys. Soc. Lond. A **69**, 447 (1956).

[2] B. L. COHEN: Phys. Rev. **98**, 49 (1955).

section. For Al and Mg, moreover, the angular distributions show indications of maxima and minima suggestive of the angular distributions of direct transitions predicted by AUSTERN, MCMANUS, and SHARPE.

35. (p, p') reactions. The cross sections for low energy proton inelastic scattering have not been measured. Presumably these are very small in comparision with the (p, n) cross section, which, as shown in Sect. 34, certainly contributes the major part of the reaction cross section for energies below 7 Mev.

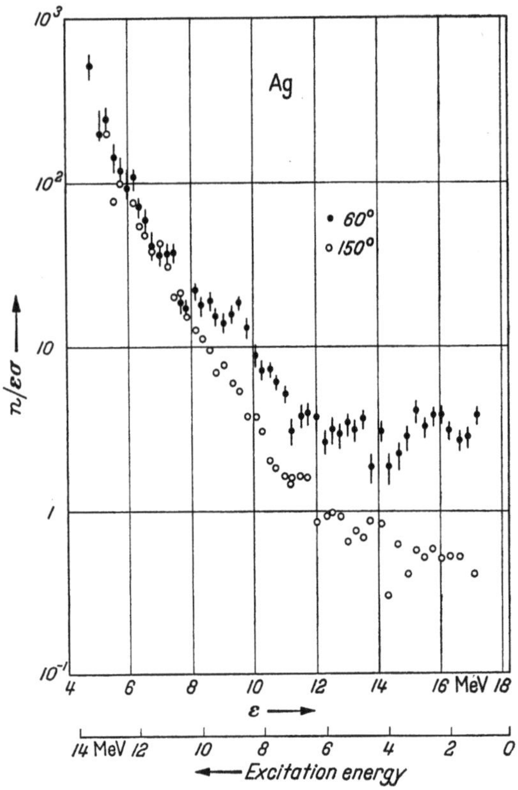

Fig. 40. Energy spectrum of the protons emitted by Ag bombarded by 18 Mev protons, after GUGELOT.

Measurements of the energies of the protons ejected from nuclei by bombardment by 18 Mev protons, have been made by GUGELOT[1]; for 31 Mev protons by EISBERG and IGO[2]; and for Al, and C, by LEVINTHAL, MARTINELLI, and SILVERMAN[3]. While some of the protons detected by these authors were derived from (p, p') reactions, some of them must be of secondary origin. We shall see in the next section that the emission of a proton is indeed a very probable process, when it is not inhibited by the COULOMB barrier; in fact, it appears that the majority of the protons which are detected in these experiments are probably the first to be emitted after the collision.

GUGELOT'S results are similar to those obtained by him for the (p, n) reactions. He has studied the protons emitted by Al, Fe, Ni, Cu, Ag, Sn, Pt, Au, and Pb, at angles of 60° and 150°. The energy spectrum of the emitted protons was roughly Maxwellian; the temperatures, however, depend on the energy of the protons and also on the angle of emission. There was evidence that high energy protons were peaked in the forward direction, like the neutrons in COHEN'S experiments and those of ROSEN and STEWART. Protons emitted with energies below about 6 Mev are emitted isotropically. The temperatures for such protons are about 0.8 Mev for the heavier elements, in agreement with neutron measurements at similar energies. The differential cross section, $\frac{1}{\varepsilon\sigma}\frac{\partial^2\sigma}{\partial\varepsilon\,\partial\Omega}$, is roughly the same for all elements. This is to be expected from statistical theory, as we have seen in Sect. 8. A typical set of data is shown in Fig. 40; it is similar to the neutron spectrum of the (p, n) reaction (Fig. 39).

It is obvious that the statistical theory alone cannot account for the shape of these energy spectra (cf. Sect. 59); in broad outline the slope of the curves is

[1] P. C. GUGELOT: Phys. Rev. **93**, 425 (1954).
[2] R. M. EISBERG and G. IGO: Phys. Rev. **93**, 1039 (1954).
[3] C. LEVINTHAL, E. A. MARTINELLI and A. SILVERMAN: Phys. Rev. **78**, 199 (1950).

inadequate to account for the known differences between the level density near the ground state, and that revealed by slow neutron resonances near the neutron binding energy; moreover the curves themselves have the wrong curvature. However, the slope of the energy spectra, for high excitation energies, i.e., for low proton energies, is sufficient to account for the data on level densities just mentioned. The preponderance of high energy protons, of which there are more at 60° than at 150°, is doubtless due to direct interactions.

For 31 Mev protons, EISBERG and IGO found that the energy distribution of inelastically scattered protons from Pb, Au, Ta, and Sn, was quite flat, and that the majority of them were emitted in the forward direction. Roughly ten times as many protons were emitted at 30° to the direction of the incident protons as at 135°. The total (p, p') cross section was about 15% of the geometrical cross section, πR^2, and rose slowly with atomic weight. These authors point out that direct collisions could quantitatively account for both the energy and the angular distribution if it is assumed that these collisions occur in the periphery of the nucleus. In fact, these results afford a good example of the direct interaction process, and are in agreement with the quantitative predictions of this theory (see Sect. 13).

36. (p, pn) and (p, 2n) reactions. The excitation functions of (p, pn), (p, α), and $(p, \alpha n)$ reactions have been measured recently for some nuclei of intermediate weight, by COHEN, NEWMAN, CHARPIE, and HANDLEY[1]. They found that the cross sections increase with the energy at roughly the rate expected theoretically, assuming that the residual nucleus is left at an average temperature of the order of 2 Mev. However, the absolute value of the cross section for the (p, pn) reaction is about an order of magnitude higher than that calculated on the statistical theory. Thus these (p, pn) reactions are very similar to the (α, pn) and (p, pn) reactions studied by GHOSHAL (Sect. 4, see Fig. 1). More recently, COHEN and NEWMAN[2] have measured the cross sections of some (p, pn) and $(p, 2n)$ reactions for elements of intermediate weight ranging from Ti^{48} to As^{75} for protons of 21.5 Mev. In four cases (Cr^{52}, Fe^{56}, Cu^{63}, and Ga^{69}) it was possible to measure the cross sections of both reactions with the same target; in these four cases the ratio of the (p, pn) cross section to that of the $(p, 2n)$ reaction lay between 1 and 7. The cross sections for the (p, pn) reactions average about 0.5 b (see Table 2). In one case, Zn^{68}, the $(p, 2n)$ cross section was 780 mb, which is exceptionally large. Although the latter figure is not exceptional when considered from the standpoint of the statistical theory, it is large compared with the cross sections for this reaction in nuclei of intermediate atomic weight where strong competition is offered by the (p, pn) reaction. Furthermore, this nucleus (Zn^{68}) seems anomalous in that (p, n) measurements at lower proton energies indicate an exceptionally large reaction cross section[3].

Provided that the proton and neutron binding energies are not too different, the ratio of the cross sections of the (p, pn) and $(p, 2n)$ reactions should be given by (8.3). For mass numbers near 60 this expression (according to COHEN and NEWMAN) should lead to a ratio of 0.07, a figure derived on the assumption that the measured activity is produced entirely from the (p, pn) and not from the (p, np) reaction. The high values obtained experimentally cannot be attributed to the relatively high level densities of odd-odd nuclei, for the effect does not appear to depend on the charge number of the target. Moreover, MEADOWS[4]

[1] B. L. COHEN, E. NEWMAN, R. A. CHARPIE and T. H. HANDLEY: Phys. Rev. **94**, 620 (1954).
[2] B. L. COHEN and E. NEWMAN: Phys. Rev. **99**, 718 (1955).
[3] See ref. 3, p. 261.
[4] J. W. MEADOWS: Phys. Rev. **91**, 885 (1953).

has shown that the effect in Cu persists at high energies. At low energies (10 to 40 Mev) he has confirmed that the cross section for the (p, pn) reaction is greater than that of the $(p, 2n)$ reaction in Cu^{63} and that the cross section for the (p, pn) reaction in Cu^{65} is very similar; but he has also shown that the $(p, p\,3n)$ and $(p, p\,4n)$ cross sections in Cu^{65} are greater than that of the $(p, 4n)$ reaction.

Table 2. *(p, pn) and $(p, 2n)$ cross sections according to* COHEN *and* NEWMAN.

Target	Thresholds (Mev)			Cross sections (mb)	
	(p, n)	(p, pn)	$(p, 2n)$	(p, pn)	$(p, 2n)$
Ti^{48}	4.7	11.7	15.3	—	120
Cr^{52}	5.4	12.0	16.0	425	155
Mn^{55}	0.9	10.3	10.4	620	—
Fe^{56}	5.3	11.5	15.7	760	105
Co^{59}	1.0	10.9	10.8	540	—
Ni^{62}	5	10.6	13.5	—	385
Cu^{63}	4.2	11.0	13.4	590	100
Cu^{65}	2.1	10.4	10.5	500	—
Zn^{68}	3.6	8.9	10.7	—	780
Ga^{69}	4.1	10.4	11.0	360	360
Ga^{71}	1.0	9.0	8.0	260	—
As^{75}	0.6	10.3	10.9	350	—

In Sects. 8 and 29, it was shown that if the neutron binding energy of the intermediate nucleus exceeds the proton binding energy by a sufficient margin, and if the emission of a neutron should leave the nucleus with an excitation energy between these limits, the emission of a proton will follow. These conditions exist for most of the nuclei in Table 2, but it does not seem possible to account for the very large cross sections obtained experimentally by supposing that they are (p, np) reactions, for the high maximum energy of the protons (21.5 Mev) would strongly favour the $(p, 2n)$ process, even in the most favourable case. There remains the possibility that the origin of this anomaly is related to the favoured elastic re-emission of the incident particle which was suggested by BOHR, PEIERLS, and PLACZEK[1] as applicable in the region of overlapping levels. This is unlikely, for the same relative yields of the secondary particles found by GHOSHAL for Zn^{64} are reproduced in the photoreactions (Sect. 53). Of course, the (p, pn) and the (p, d) reactions cannot be distinguished by activation measurements made above the threshold of the latter reaction. Estimates of the (p, d) reaction made by COHEN *et al.* for Cu, using the (d, p) reaction to detect the deuterons, indicate that this reaction does not contribute appreciably to the activities ascribed to the (p, pn) process. If the (p, d) process was, in fact, the origin of the cross sections which have been observed, we should still be left with the difficulty of understanding why such deuteron emission should have a high cross section and show an excitation curve with all the characteristics expected of compound nucleus formation.

Finally it should be pointed out that the anomaly may be a feature peculiar to nuclei of this particular atomic weight. For example, the (p, pn) reactions with Nb^{93} seems to be normal[2] and the yield of the (α, pn) reaction with Ag shows[3] no marked abnormality.

---

[1] N. BOHR, R. PEIERLS and G. PLACZEK: Nature, Lond. **144**, 200 (1939).
[2] R. A. JAMES: Phys. Rev. **93**, 288 (1954).
[3] K. G. PORGES: Phys. Rev. **101**, 225 (1956).

37. (p, 2p) reactions. COHEN, NEWMAN, and HANDLEY[1] have attempted to measure, by activation methods, the sum of the (p, pn) and $(p, 2n)$ cross sections for 21.5 Mev protons for a number of nuclei from F to Ga. When the period of the product of the $(p, 2n)$ process was too short to be measured, it was assumed that the nucleus produced by such a reaction decays by positron emission to the product of the (p, pn) reaction and that the measured value of the latter cross section measures the sum of the two. The sum of these cross sections for nuclei between Ti^{46} and Ga^{69} lies (with two exceptions) between 0.5 b and 0.9 b and must comprise, therefore, the greater part of the reaction cross section (about 1 b).

The first exception (Fe^{54}) is interesting: the cross section was 165 mb. This is exceptionally low; the difference between it and the reaction cross section must be accounted for. (It should be compared with the cross section of 870 mb found for Fe^{56}.) Only (p, pn), $(p, 2p)$, and reactions involving α-particles, are possible in Fe^{54} at this energy. The difference in the cross sections therefore might be accounted for by the $(p, 2p)$ reaction, because the threshold of the $(p, 2p)$ reaction is much lower than that of the (p, pn) reaction in Fe^{54} (9.7 and 14.1 Mev respectively, according to these authors). For Fe^{56} the corresponding quantities are 10.6 and 11.5 Mev respectively. The yield of the latter reaction for Fe^{54} cannot be measured by activation methods, because the product, Mn^{53}, has a very long life. Where the cross sections of the $(p, 2p)$ reactions are known they are usually low (a few millibarns) as is to be expected, for in each case the thresholds of the (p, pn) and the $(p, 2p)$ processes are nearly equal.

The other exception is Ni^{58} for which the thresholds are similar to those of Fe^{54}. For this nucleus the sum of the cross sections of the (p, pn) and $(p, 2n)$ reactions is 240 mb and the difference between this value and the reaction cross section is accounted for by the cross section of the $(p, 2p)$ process (680 mb). The difference between the thresholds of the $(p, 2p)$ and the (p, pn) reactions is 4 Mev. However, the high value of the $(p, 2p)$ cross section is difficult to understand, for the (p, pn) threshold (11.9 Mev) is low enough in relation to the proton energy to reduce the $(p, 2p)$ cross section to a negligible amount. For Cu^{63}, in which the difference between the thresholds (5.7 Mev) is even greater than in Ni^{58}, we have just the opposite situation: the sum of the cross sections of the (p, pn) and $(p, 2n)$ reactions is 700 mb.

I. Deuteron reactions.

38. Energy balance in deuteron reactions. In most nuclear reactions in which the deuteron is the incident particle, the energy balance is favourable. In the (d, p) reaction the deuteron acts as a vehicle for the conveyance of a neutron; the Q-value therefore is equal to the neutron binding energy of the product nucleus less the binding energy of the deuteron:

$$Q_{dp} = Q_n - q_d \tag{38.1}$$

where q_d and Q_n are the neutron binding energies of the deuteron and of the product of the reaction respectively. Since the neutron binding energies are usually approximately 7 Mev, the Q's of the (d, p) reactions are approximately 5 Mev.

In the (d, n) reaction, the deuteron likewise serves as the means of conveyance of a proton into the product nucleus, and therefore the Q value is given by:

$$Q_{dn} = Q_p - q_d$$

[1] B. L. COHEN, E. NEWMAN and T. H. HANDLEY: Phys. Rev. 99, 723 (1955).

Table 3. *Neutron binding energies, from* (d, p), (d, t), (γ, n) *and* (n, γ) *reactions in* Mev. Figures in parantheses show errors beyond the decimal point; letters, to references at the end of the table.

Nucleus	Z	A	(γ, n)	(d, p) and (d, t)	(n, γ)
A	19	40	9.85 (15) (s)		
Ca	20	40	15.9 (4) (q)		
		41		8.37 (01) (c)	
		43		7.94 (01) (c)	
		44		11.14 (02) (c)	
		45		7.42 (01) (c)	
		49		6.15 (01) (c)	
Sc	21	46		9.0 (1) (d)	8.85 (08) (e)
Ti	22	46	13.3 (2) (m)		
		47		8.74 (10) (a)	
				8.68 (05) (g)	
		48		11.1 (4) (a)	
				10.37 (05) (g)	
		49		8.15 (05) (a)	
				8.04 (04) (g)	
		50		10.85 (05) (g)	
		51		6.34 (07) (g)	
V	23	51	11.15 (20) (h)		
		52		7.25 (05) (a)	7.305 (007) (e)
Cr	24	50	13.4 (2)		
		51			9.07 (09) (f)
		52	11.80 (25) (h)		
		53	7.75 (20) (h)		
		54			9.716 (007) (f)
Mn	25	55	10.0 (2) (h)		
			10.15 (20) (r)		
		56		7.2 (i)	7.261 (006) (e)
Fe	26	54	13.8 (2) (q)		
		55		9.34 (05) (a)	
		56	11.15 (20) (h)		
		57	7.75 (20) (a) (h)	7.65 (10) (a)	7.639 (004) (f)
		58			10.16 (04) (f)
Co	27	59	10.25 (20) (h)		
		60		7.66 (20) (a)	
Ni	28	58	11.7 (2) (o)		
		59		9.0 (1) (a)	8.997 (005) (f)
		61			8.532 (008) (f)
		65		< 6.4 (a)	
Cu	29	63	10.8 (2) (h) (n)		
			10.61 (05) (t)		
		64		7.78 (20) (a)	7.914 (006) (e)
		65	9.75 (20) (h)		
		66			7.634 (006) (e)
Zn	30	64	11.65 (20) (h)		
			11.8 (2) (r)		
		65		7.92 (05) (a)	
		66	11.15 (20) (h)		
		67	7.00 (20) (h)		
		68	10.15 (20) (h)		
		70	9.2 (2) (r)		
Ga	31	69	10.1 (2) (h)		
		71	9.05 (20) (h)		

Table 3. (Continued.)

Nucleus	*Z*	*A*	(γ, *n*)	(*d*, *p*) and (*d*, *t*)	(*n*, γ)
As	33	75	10.1 (2) (h) 10.3 (2) (o)		
		76			7.30 (04) (k)
Se	34	78			10.48 (01) (j)
Br	35	79	10.7 (2) (q) 10.6 (2) (h)		
		81	10.2 (2) (q) 9.95 (20) (h)		
Rb	37	86		8.4 (3) (b)	
		88		6.0 (2) (b)	
Sr	38	85		7.48 (30) (b)	
		86	9.5 (2) (h)		
		87	8.4 (2) (h)	8.5 (2) (a) (b)	
		88	11.15 (20) (h)		
		89		6.55 (20) (a) (b)	
Y	39	90		6.64 (05) (b)	
Zr	40	90	12.48 (15) (r) 12.0 (2) (o)		
		91	7.2 (4) (r)	7.16 (05) (a)	
		92		8.73 (10) (a)	8.66 (04) (j)
		93		6.69 (05) (b)	
Nb	41	93	8.7 (2) (h)		
Mo	42	92	13.28 (15) (r)		
		93		8.3 (2) (a) 7.86 (05) (b)	
		96			9.15 (05) (j)
		97	7.1 (3) (r)	6.7 (3) (b)	
		98		8.3 (1) (b)	
Rh	45	104		6.8 (2) (a)	6.792 (014) (k)
Ag	47	108		7.0 (2) (a)	7.27 (02) (k)
		109	9.07 (07) (t) 9.05 (2) (h)		
Cd	48	113	6.5 (2) (h) (r)	6.33 (09) (b)	
		114			9.046 (008) (j)
		115		5.75 (10) (b)	
In	49	115	9.05 (2) (h)		
		116		6.6 (2) (a)	
Sn	50	118	8.95 (25) (h)	9.37 (20) (a)	9.35 (08) (j)
		119	6.51 (15) (r)		
		121		6.2 (3) (a) 6.15 (07) (b)	
		124	8.5 (15) (h) (r)		
		125		5.75 (07) (b)	
Sb	51	121	9.25 (20) (h) (q) 9.45 (20) (o)		
		122		6.64 (20) (a) 6.48 (07) (b)	6.80 (04) (k)
Te	52	125			
I	53	127	9.1 (2) (h)		
		128		6.58 (05) (b)	
Cs	55	133	9.05 (20) (h)		
		134		6.73 (10) (b)	
Ba	56	138			9.23 (07) (j)
		139		5.2 (3) (a)	

Table 3. (Continued.)

Nucleus	Z	A	(γ, n)	(d, p) and (d, t)	(n, γ)
La	57	139	8.8 (2) (h)		
		140		5.1 (1) (b)	
Ce	58	140	9.05 (20) (h)		
		141		5.4 (1) (b)	
		142	7.15 (20) (h)		
		143		5.09 (07) (b)	
Pr	59	141	9.8 (2) (o) (r)		
		142		5.65 (30) (b)	5.83 (03) (k)
Nd	60	143		5.02 (08) (b)	
		150	7.4 (2) (r)		
Sm	62	155		5.6 (3) (b)	
Ta	73	181	7.55 (20) (h)		
		182		6.03 (15) (a)	6.07 (03) (k)
W	74	184			7.42 (02) (j)
Re	75	187	7.3 (2) (h)		
Ir	77	193	7.8 (2) (h)		
Pt	78	194	9.5 (2) (h)		
		195	6.1 (2) (h)	6.14 (20) (a)	
		196	8.2 (2) (h)	8.0 (2) (a)	7.92 (01) (j)
Au	79	197	7.9 (2) (h) (r)		
		198		6.35 (15) (a)	6.494 (008) (k)
Hg	80	201	6.3 (2) (r)		
Tl	81	203	8.8 (2) (h)		
		204		6.52 (15) (a)	6.54 (03) (k)
		205	7.5 (2) (h) (r)		
		206		7.7 (2) (a)	
Pb	82	206	8.25 (10) (p)		
		207	6.85 (20) (h) (r)	6.73 (04) (a)	6.734 (008) (l)
		208	7.44 (10) (p)	7.37 (03) (a)	7.380 (008) (l)
		209		3.87 (05) (a)	
Bi	83	209	7.4 (2) (b) (q)		
		210		4.14 (03) (a)	4.170 (015) (l)
				4.17 (03) (b)	
Th	90	232	6.35 (04) (v)		
		233		4.9 (2) (a)	
U	92	238	5.97 (10) (u)		
		239		4.63 (15) (a)	

(a) J. A. HARVEY: Phys. Rev. **81**, 353 (1951).
(b) N. S. WALL: Phys. Rev. **96**, 664 (1954).
(c) C. M. BRAAMS: Thesis, Utrecht 1956.
(d) W. L. DAVIDSON: Phys. Rev. **56**, 1061 (1939).
(e) G. A. BARTHOLOMEW and B. B. KINSEY: Phys. Rev. **89**, 386 (1953).
(f) B. B. KINSEY and G. A. BARTHOLOMEW: Phys. Rev. **89**, 375 (1953).
(g) G. F. PIEPER: Phys. Rev. **88**, 1299 (1952).
(h) R. SHER, J. HALPERN and A. K. MANN: Phys. Rev. **84**, 387 (1951).
(i) W. D. WHITEHEAD and N. P. HEYDENBURG: Phys. Rev. **79**, 99 (1950).
(j) B. B. KINSEY and G. A. BARTHOLOMEW: Canad. J. Phys. **31**, 1051 (1953).
(k) G. A. BARTHOLOMEW and B. B. KINSEY: Canad. J. Phys. **31**, 1025 (1953).
(l) B. B. KINSEY, G. A. BARTHOLOMEW and W. H. WALKER: Phys. Rev. **82**, 380 (1951).
(m) W. E. OGLE and R. E. ENGLAND: Phys. Rev. **78**, 63 (1950).
(n) J. MCELHINNEY and W. E. OGLE: Phys. Rev. **78**, 63 (1950).
(o) W. E. OGLE, L. J. BROWN and A. N. CARSON: Phys. Rev. **78**, 63 (1950).

(p) H. PALEVSKY and A. O. HANSON: Phys. Rev. **79**, 242 (1950).
(q) J. MCELHINNEY, A. O. HANSON, R. A. BECKER, R. B. DUFFIELD and B. C. DIVEN: Phys. Rev. **75**, 542 (1949).
(r) A. O. HANSON, R. B. DUFFIELD, J. B. KNIGHT, B. C. DIVEN and H. PALEVSKY: Phys. Rev. **76**, 578 (1949).
(s) J. HALPERN, R. NATHANS and P. F. YERGIN: Phys. Rev. **95**, 1529 (1954).
(t) M. BIRNBAUM: Phys. Rev. **93**, 147 (1954).
(u) J. R. HUIZENGA, L. B. MAGNUSSON, P. R. FIELDS, M. H. STUDIER and R. B. DUFFIELD: Phys. Rev. **82**, 561 (1951).
(v) L. B. MAGNUSSON, J. R. HUIZENGA, P. R. FIELDS, M. H. STUDIER and R. B. DUFFIELD: Phys. Rev. **84**, 166 (1951).

where Q_p is the proton binding energy of the product nucleus. Since the proton binding energies are always greater than the binding energy of the deuteron (2.23 Mev), Q_{dn} is also always positive, although it is usually less than Q_{dp}.

The energy balance in (d, α) reactions is also always positive. Thus:

$$Q_{d\alpha} = q_\alpha - q_d - Q_n - Q_p$$

where q_α is the total binding energy of the α-particle and Q_p and Q_n are the energies required to extract successively a proton and a neutron from the target nucleus, in any order. Since $q_\alpha \approx 28$ Mev, it is obvious that the (d, α) reaction is also strongly exoergic. However very little is known about (d, α) reactions in heavy elements.

The reverse reactions, i.e., the pickup processes, (p, d), and (n, d), on the other hand, are endoergic, and for this reason they are more difficult to examine. The (d, t) pickup process, however, is much less endoergic, and for this reaction,

$$Q_{dt} = q_t - Q_n$$

where q_t is the neutron binding energy of the triton, 6.2 Mev, and Q_n is the neutron binding energy of the target nucleus.

The (d, p) and (d, t) reactions have been used by HARVEY[1] and by WALL[2] to obtain the neutron binding energies of a large number of nuclei. Their results, together with those found from the thresholds of the (γ, n) reactions and from (n, γ) reactions, are listed in Table 3. When plotted against the neutron number they show very clearly the discontinuities which result from closed neutron shells. Although there are no exclusive selection rules which govern the intensities of (d, p) reactions it does not always follow that the highest energy proton group produced in the (d, p) reaction is that corresponding to the production of the product nucleus in its ground state, for, as we shall show below, the cross section for the reaction falls rather rapidly with the energy released and can depend rather critically on the spin of the product. For some targets, e.g. Sr^{87}, and Bi^{209}, it is probable that the ground states of the product nuclei were not detected (cf. Sect. 89).

39. The stripping process: energy dependence. The peculiar excitation functions of deuteron reactions were discovered soon after the first use of deuterons as projectiles in nuclear reactions. LAWRENCE, MCMILLAN, and THORNTON[3], found, by activation methods, that the rate at which the cross section of the (d, p) reaction increased with energy was lower than that expected from the usual formula for penetration of charged particles. For Cu^{63} they found measurable activities for energies as low as 2 Mev. These results suggested that the mechanism of the

[1] J. A. HARVEY: Phys. Rev. **81**, 353 (1951).
[2] N. S. WALL: Phys. Rev. **96**, 664 (1954).
[3] E. O. LAWRENCE, E. MCMILLAN and R. L. THORNTON: Phys. Rev. **48**, 493 (1935).

reaction was different from that which was supposed to hold for proton reactions. OPPENHEIMER[1] calculated the probability that the deuteron was dissociated electrically in close collisions in which the neutron is captured and the proton, repelled by the COULOMB field, is set free. In this process the deuteron is subjected to the rapidly varying electric field presented by the nucleus, and is dissociated by those FOURIER components of the field which have quantum energies above the dissociation energy of the deuteron. The cross section of this process, however, is small, and for low energies at least, far too small to account for the effects observed. OPPENHEIMER and PHILLIPS[2] then suggested what has now become known as the stripping process. They took account of the relatively large size of the deuteron (the mean separation is about 2.2×10^{-13} cm) and pointed out that collisions will occur in which the neutron is captured while the proton still lies outside the range of nuclear forces. The effective barrier height for a (d, p) reaction therefore can be considerably lower than that which is to be normally expected. OPPENHEIMER and PHILLIPS found that the cross section for the reaction increases with energy with the exponential:

$$\exp\left[-\frac{2Ze^2}{\hbar}\left(\frac{2M}{q_d}\right)^{\frac{1}{2}} F\left(\frac{E}{q_d}\right)\right]$$

where q_d is the binding energy of the deuteron (2.23 Mev), and F depends on the barrier height and is a decreasing function of its argument. BETHE[3] has shown that the original results of LAWRENCE *et al.*, which seemed at first to demonstrate the validity of this theory, in fact, corroborate it only for the heavier elements. Some refinements have been added to the theory[4]. However, owing to the mathematical difficulties of describing deuterons in the presence of a COULOMB field, there is as yet no very satisfactory theory of stripping reactions at low energies. A qualitative account has been given by PEASLEE[5], which, although lacking in rigour, does give a fair account of both (d, p) and (d, n) reactions. But first it will be better to consider the easier problem of high energy stripping even though this really lies outside the scope of the present article, for in this case the effect of the COULOMB field can be neglected, and some simplifying approximations made.

For high energy deuterons, the neutrons produced in the (d, n) reaction are found to be projected in the forward direction. The angular distribution of the neutrons produced by 190 Mev deuterons was studied by HELMHOLTZ, MCMILLAN, and SEWELL[6], using the activities induced in carbon samples as detectors. The neutrons were mostly contained in an angle of about $\frac{1}{8}$ of a radian from the direction of the incident beam and this angle did not vary much from one element to another. According to these authors the angle contained between positions where the neutron intensity is one half of the maximum does not vary much with the atomic number of the target; it is given by the formula $\vartheta = (0.155 + 0.00060Z)$ radians.

A theory of high energy stripping has been developed by SERBER[7]. He neglected the effect of the COULOMB field, and supposed that the collision occurs instantaneously, the transit time of the deuteron in its motion through the edge

[1] J. R. OPPENHEIMER: Phys. Rev. **47**, 845 (1935).
[2] J. R. OPPENHEIMER and M. PHILLIPS: Phys. Rev. **48**, 500 (1935).
[3] H. A. BETHE: Phys. Rev. **53**, 39 (1938).
[4] E. LIFSHITZ: Phys. Z. Sowjet. **13**, 324 (1938). — P. L. KAPUR: Indian J. Phys. **13**, 87 (1939). — G. M. VOLKOFF: Phys. Rev. **57**, 866 (1940).
[5] D. C. PEASLEE: Phys. Rev. **74**, 1001 (1948).
[6] A. C. HELMHOLTZ, E. MCMILLAN and D. C. SEWELL: Phys. Rev. **72**, 1003 (1947).
[7] R. SERBER: Phys. Rev. **72**, 1008 (1947).

of the nucleus being small compared to the time of its internal motion. The (d, n) reaction occurs whenever the trajectory of the incident deuteron is such that the proton will hit the nucleus and the neutron will pass it by. The (d, p) reaction likewise occurs when the positions of neutron and proton in the deuteron are reversed. The direction and energy of the neutron emitted can then be obtained by compounding the momentum of the neutron in its collective motion in the deuteron, with that derived from its internal motion in that particle. The time for a deuteron directed at the edge of the nucleus to reach the median plane of the nucleus, after striking the surface, is L/v were $L=\sqrt{2RR_d}$; the time for the neutron to move a distance equal to the radius of the deuteron, when the latter is at rest, is of the order of $R_d/\sqrt{q_d/m}$. Hence the condition for the disintegration of the deuteron to be instantaneous is that the first time should be small compared with the second; this leads to the condition:

$$E_d \gg q_d\left(\frac{2R}{R_d}\right)$$

where E_d is the kinetic energy of the deuteron. For heavy elements, $R\approx 8\times 10^{-13}$ cm; thus E_d must be greater than 15 Mev. The angle of emission is given roughly by the ratio of the internal momentum (p_1) of the neutron in the deuteron, when the latter is at rest, to its momentum (p_0) due to the motion of the latter. Here $p_0\approx(ME_d)^{\frac{1}{2}}$, $p_1\approx(q_d M)^{\frac{1}{2}}$, and $\vartheta\approx(q_d/E_d)^{\frac{1}{2}}$, which, for 190 Mev deuterons, is 6°, in rough agreement with experiment. More precisely, the angular distribution is obtained from the internal momentum distribution in the deuteron which in turn is obtained from a FOURIER transform of the wave function (Sect. 42). The result is:

$$\left.\begin{aligned} P(\vartheta)\,d\Omega &= \frac{1}{2\pi}\,\frac{\vartheta_0^3}{(\vartheta_0^2+\vartheta^2)^{\frac{3}{2}}}\,d\Omega \\ \text{where}\qquad \vartheta_0 &= \sqrt{\frac{q_d}{E_d}}\left(1-\frac{E_d}{8Mc^2}\right) \end{aligned}\right\} \tag{39.1}$$

and the energy distribution by

$$P(E)\,dE = \sqrt{q_d E_d}\left\{\pi\left[\left(E-\frac{1}{2}E_d\right)^2+q_d E_d\right]\right\}^{-1}\cdot dE \tag{39.2}$$

which is centred around $\frac{1}{2}E_d$. Both expressions are in fairly good agreement with experiment.

The cross section for the (d, n) process is the average value of that fringing area of the nucleus which can be struck by the proton in the deuteron while the neutron passes by untouched. If the mean distance which separates proton and neutron in the deuteron, $R_d=2.2\times 10^{-13}$ cm, is small compared with the nuclear radius the average cross section is equal to $2\pi R$ times the mean value $(\frac{1}{4}R_d)$ of the projection of this distance (R_d) on a plane perpendicular to its direction of motion. Thus the cross section is $\frac{1}{2}\pi R R_d$; for heavy elements it is about $\frac{1}{8}$ of the geometrical area of the nucleus. Quantitatively, this seems to be about the right cross section; but it seems that no accurate measurements have been made to confirm it.

DANCOFF[1] has calculated the cross section for electric dissociation for high energy deuterons. This cross section, which is proportional to Z^2, should amount to about $\frac{1}{4}$ of the stripping cross section in heavy elements. However, existing experimental data does not make it clear how important this effect really is.

[1] S. M. DANCOFF: Phys. Rev. **72**, 1017 (1947).

For lower energies this cross section should rise, reaching a maximum near 30 Mev and falling rapidly for lower energies.

Low energies, 2 to 15 Mev, lie below SERBER'S criterion for the instantaneous disintegration of the deuteron. Nevertheless, disregarding this point, PEASLEE[1] has attempted to account, at least qualitatively, for the main features of (d, p) and (d, n) reactions. He assumed that the cross section is given by the formula:

$$\sigma = \pi R^2 \xi \varrho$$

where ξ is a "sticking" probability, in fact, a constant to be determined, and ϱ represents the density at the surface [of neutrons in the incoming wave for

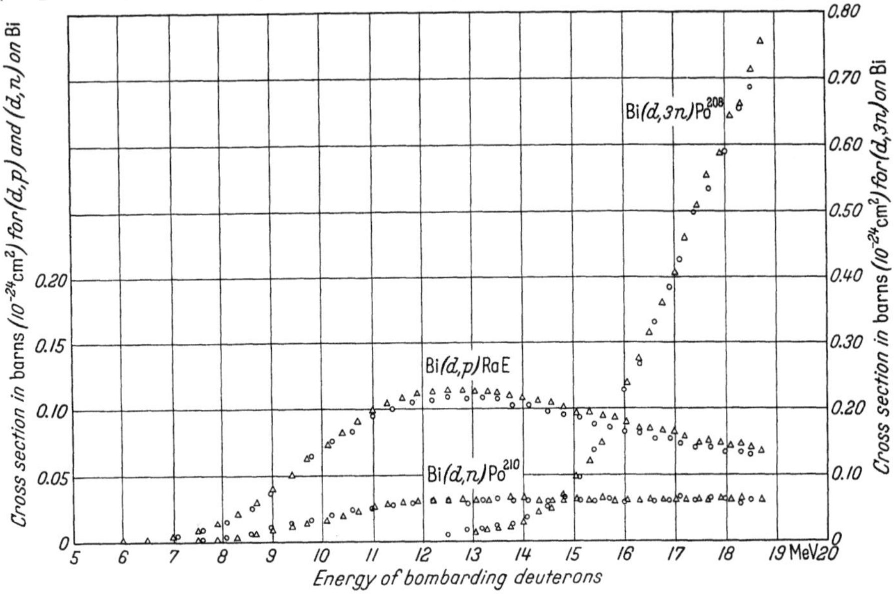

Fig. 41. Cross sections for the (d, p) and (d, n) reactions in Bi after KELLY and SEGRÈ.

the (d, p) reaction, or of protons for the (d, n) reaction]. For deuteron energies below the barrier height, the cross sections of both (d, p) and (d, n) reactions increase with energy, the cross section for the former always being greater than that of the latter because the density of protons at the surface in the latter reaction is always reduced by the effect of the barrier. Above the barrier, both reactions are subject to competition by other reactions and for this reason the cross section falls. The theoretical results agree well with experimental data[2] obtained by measuring the radioactivity of the product nucleus. An interesting case is that of Bi^{209}, which is one of the few examples where the cross section of both the (d, p) and the (d, n) reactions can be determined by measurement of the radioactivities of the product nuclei[3]. For the (d, p) reaction the product nucleus is the β-emitting Bi^{210} (RaE) and for the (d, n) reaction it is the α-emitter, Po^{210}. The results of KELLY and SEGRÈ[4] are shown in Fig. 41. It will be seen that the cross sections both rise as the deuteron energy is increased to the barrier

[1] See ref. 5, p. 272.

[2] E. J. CLARKE and J. W. IRVINE: Phys. Rev. **66**, 231 (1944). — R. S. KRISHNAN and E. A. NAHUM: Proc. Roy. Soc. Lond., Ser. A **180**, 321, 332 (1942).

[3] D. HURST, R. LATHAM and W. B. LEWIS: Proc. Roy. Soc. Lond., Ser. A **174**, 126 (1940). — J. M. CORK, J. HALPERN and H. TATEL: Phys. Rev. **57**, 371 (1940). — J. M. CORK: Phys. Rev. **70**, 563 (1946).

[4] E. L. KELLY and E. SEGRÈ: Phys. Rev. **75**, 999 (1949).

height (about 14 Mev). Beyond this point the cross section for the (d, p) reaction falls while that for the (d, n) reaction remains nearly constant. In general, these cross sections are in agreement with PEASLEE's theory, provided $\xi \sim 0.3$. For other elements the values which have to be chosen for ξ are similar. The cross section for an energy equal to the barrier height is about 100 mb for the (d, p) process; for lighter nuclei it is less.

40. Angular distribution of (d, n) reactions. We have pointed out that the large cross sections of the (d, p) process derive from the fact that delivery of the neutron to the nucleus is effected when the distance between the nucleus and the proton is greater than that of the nuclear radius, thus reducing the effect of the COULOMB barrier. For the (d, n) reaction, however, the distinction between stripping and compound nucleus formation is not so clear, because for stripping in the (d, n) reaction the proton is delivered at the nuclear surface; for deuteron energies below the barrier height, the latter presumably is as effective in impeding the stripping reaction as it is in impeding compound nucleus formation. The experimental data which is described below shows that (d, n) stripping does occur but it is not predominant as in the (d, p) reaction. The relative proportion of stripping to compound nucleus formation is not yet established.

The total yield and the angular distribution of the neutrons emitted by thick targets bombarded by 15 Mev deuterons has been measured by ALLEN, NECHAJ, SUN, and JENNINGS[1]. To detect the neutrons, they measured the P^{32} activity produced by the S^{32} (n, p) reaction; the actual threshold is about 1 Mev, but the effective threshold, when allowance is made for the COULOMB penetration of the protons emitted in this reaction, is probably about 3 Mev. These measurements, therefore, do not comprise all the neutrons emitted but only a fraction of them. These authors found that forward emission was strongly favoured for the lighter elements and less so for the heavy elements for which the neutron emission is nearly isotropic. COHEN[2] has attempted to correct these results theoretically and concluded that the number of neutrons emitted in the backward direction for the heavy elements, is about half of that emitted in the forward direction. FALK[3] has measured the higher energy component of the neutrons emitted when targets are bombarded by 15 Mev deuterons. He used the Ag^{107} $(n, 2n)$ Ag^{106} reaction as a detector (the threshold is about 11 Mev). He found that the angular distributions, even in heavy elements, were peaked either in the forward direction (e.g. Mg, Al, Mo, Ag) or peaked at a small angle to it, usually about 10° (Ti, Co, Cu, Au). Rather similar results have been obtained by SCHECTER[4], with 20 Mev deuterons. COHEN and FALK[5] have measured the energy spectra of the neutrons produced by 20 Mev deuterons using photographic plates as detectors. They found noticeably more low energy neutrons for angles of 90° to the beam than at 0°.

41. Angular distribution of (d, p) reactions. While the angular distribution of the (d, p) reaction in many light elements has been determined, there have been fewer measurements of the angular distribution in heavy elements; like the (d, n) reaction discussed in the preceding section, such measurements require high deuteron energies. The first comprehensive measurements were made by GOVE[6] with 14 Mev deuterons. He detected the protons emitted from C, Al, Ni,

[1] A. J. ALLEN, J. F. NECHAJ, K. H. SUN and B. JENNINGS: Phys. Rev. **81**, 536 (1950).
[2] B. L. COHEN: Phys. Rev. **81**, 632 (1950).
[3] C. E. FALK: Phys. Rev. **83**, 499 (1951).
[4] I. SCHECTER: Phys. Rev. **83**, 695 (1951).
[5] B. L. COHEN and C. E. FALK: Phys. Rev. **84**, 173 (1951).
[6] H. GOVE: Phys. Rev. **81**, 364 (1951).

Ag, Ta, Au, and Bi for angles between 15° and 135°; he used a triple coincidence proportional counter as detector and scanned the energy spectrum by placing absorbers between the counters and the target. Qualitatively, his results are similar to those of FALK but with the difference that the angles for the peak emission were larger. FALK's results were obtained with detectors that comprised the whole range of neutron energies from 11 to near 20 Mev; GOVE's measurements, however, although made with insufficient resolution to reveal the angular distributions corresponding to specific states in the residual nuclei of the heavy elements, were confined to fairly narrow energy ranges. For the lighter elements, GOVE found that the angular distributions were very strongly peaked in the forward direction; for the heavier elements, e.g. Ta, Au, and Bi, the forward concentration was less obvious, and the peak in the angular distribution was near 60°.

42. Angular distribution of (d, p) and (d, n) reactions leading to discrete states. GOVE[1] has pointed out that the order of the LEGENDRE coefficients which are required to describe the angular distributions found by him are higher than the maximum permissible $(2kR)$ for compound nucleus formation. This anomaly receives a natural explanation if stripping is assumed to occur, for then the effective nuclear radius is increased. The peculiar features of the angular distributions of stripping reactions leading to discrete states were first interpreted by BUTLER[2] who showed that it is possible to deduce from them the change in parity in the reaction, and to some extent the change in spin. The experimental data have been obtained mainly in light nuclei, in the first instance, by HOLT[3] and by ROTBLAT and their associates[3]. Most of this work has already been discussed in the preceding article. For heavy elements high deuteron energies are necessary. This introduces experimental difficulties in connection with the energy resolution of the incident deuteron beam. Here we shall briefly sketch the theory of the angular distribution. A more detailed review has been given by HUBY [*17*].

In BUTLER's theory, and in that of BHATIA and his associates[4], the effect of the COULOMB field is disregarded. Although the following is equally applicable to a (d, n) reaction, we assume for the sake of brevity that we are dealing with a (d, p) reaction. The angle between the direction of emission of the proton and that of the incident deuteron beam (ϑ) is determined, in the main, by considerations of conservation of momentum and energy. If $\hbar \boldsymbol{K}$ is the momentum of the proton in the deuteron at the moment of impact, then its momentum after emission will be $\hbar \boldsymbol{k}_p$ where:

$$\boldsymbol{K} = \boldsymbol{k}_p - \tfrac{1}{2}\boldsymbol{k}_d$$

and $\hbar \boldsymbol{k}_d$ is the momentum of the deuteron (Fig. 42). Hence the internal momentum of the proton in the deuteron (in the frame of reference in which the latter is at rest) is given by:

$$K = \left[\left(k_p - \frac{1}{2}k_d\right)^2 + 2k_p k_d \sin^2\frac{\vartheta}{2}\right]^{\frac{1}{2}}. \tag{42.1}$$

---

[1] H. GOVE: Phys. Rev. **81**, 364 (1951).

[2] S. T. BUTLER: Phys. Rev. **80**, 1095 (1950). — Proc. Roy. Soc. Lond., Ser. A **208**, 559 (1951).

[3] H. D. BURROWS, W. M. GIBSON and J. ROTBLAT: Phys. Rev. **80**, 1095 (1950). — J. M. HOLT and C. J. YOUNG: Proc. Phys. Soc. Lond. A **63**, 833 (1950). For later references consult HUBY [*17*].

[4] A. S. BHATIA, K. HUANG, R. HUBY and H. C. NEWNS: Phil. Mag. **43**, 485 (1952).

The probability of finding a proton with momentum $\hbar K$ in the deuteron is given by the formula:

$$P(\boldsymbol{K}) = \frac{1}{(2\pi)^3}\left[\int \mathrm{e}^{-i\boldsymbol{K}\cdot\boldsymbol{r}}\chi(\boldsymbol{r})\, d^3\boldsymbol{r}\right]^2$$

where $\chi(\boldsymbol{r})$ is the wave function of the deuteron, and $\boldsymbol{r}$ is the relative co-ordinate of the neutron and proton. For a deuteron wave function of the simplified type:

$$\chi(\boldsymbol{r}) = \left(\frac{\alpha}{2\pi}\right)^{\frac{1}{2}}\cdot\frac{\mathrm{e}^{-\alpha r}}{r}; \qquad \alpha = \frac{\sqrt{M q_d}}{\hbar} = 0.23\times 10^{13}\ \mathrm{cm}^{-1}$$

the probability $P(K)$ of finding the proton with a momentum K in the range dK is

$$P(K) = \frac{\alpha}{\pi^2\sqrt{\alpha^2+K^2}}. \tag{42.2}$$

The momentum of the neutron captured by the nucleus is $\hbar\boldsymbol{k}$ (Fig. 42). This is also the momentum of the nuclear recoil; it is given by:

$$k = \left[(k_p - k_d)^2 + 4 k_p k_d \sin^2\frac{\vartheta}{2}\right]^{\frac{1}{2}}. \tag{42.3}$$

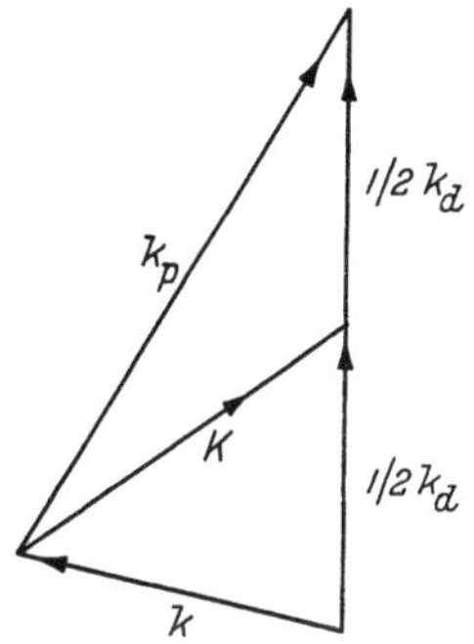

Fig. 42. Momentum relations for the (d, p) reaction according to the stripping theory.

In SERBER's theory, any value of the angle ϑ is possible consistent with conservation of energy and with the above equations for the conservation of momentum. Thus the probability P falls with increasing K, and the probability of emission with a given angle ϑ, therefore, falls with increasing angle. Integration of (42.2) then leads to SERBER's formula. If however the residual nucleus is left in a discrete state, these conditions no longer obtain. The angle of emission ϑ is determined now by the Eqs. (42.1) and (42.3), and the additional requirement that the neutron is captured with a definite angular momentum l. The latter must satisfy the conditions for conservation of parity and of angular momentum:

$$|J_{\mathrm{A}} \pm J_{\mathrm{B}} \pm \tfrac{1}{2}|_{\mathrm{minimum}} \leqq l \leq J_{\mathrm{A}} + J_{\mathrm{B}} + \tfrac{1}{2} \tag{42.4}$$

where J_{A} and J_{B} are the spins of the target and residual nuclei and l is odd or even according to whether there is or there is not a difference in parity between the target and residual nucleus. BUTLER calculated the cross section with the wave function of the ingoing particle and its derivative continuous over the nuclear surface. BHATIA *et al.* have derived a somewhat simpler expression for the cross section by BORN approximation. In either case the cross section depends on three factors: the first (G^2) is proportional to the probability of finding the proton in the deuteron with the momentum K [cf. Eq. (42.2)]; the second is proportional to the probability that the ingoing particle, if it arrives at the nuclear surface, is captured; and the third is proportional to the probability that the ingoing particle is delivered at the surface with the proper linear and angular momentum. For the stripping of tritons, the theory should be very similar; instead of the first factor, we have the probability that the triton contains a deuteron moving with the correct momentum. However, as (t, d) reactions have not been studied in heavy elements we are not concerned with them here. For the (d, p) reaction the formula of BHATIA *et al.* is:

$$\sigma = G^2\frac{M_p^* M_d^*}{2\pi\hbar^4}\cdot\frac{2J_{\mathrm{B}}+1}{2J_{\mathrm{A}}+1}\cdot\frac{k_p}{k_d}\sum_l \Lambda_l [j_l(kR)]^2, \tag{42.5}$$

where $j_l(x)$ is the spherical BESSEL function $\left(\frac{\pi}{2x}\right)^{\frac{1}{2}} J_{l+\frac{1}{2}}(x)$; M_p^* and M_d^* refer to reduced masses; and

$$G = 2(2\pi\alpha)^{\frac{1}{2}}/[\alpha^2 + (k_p - \tfrac{1}{2}k_d)^2 + 2k_p k_d \sin^2\vartheta/2]$$

and Λ_l is the sum of the squares of the relevant matrix elements of the interaction potential between the nucleus and the ingoing particle. Here k is given by (42.3). Effectively, this cross section differs from BUTLER's only in the third term: BUTLER's expression contains, in addition, the spherical BESSEL function $(\pi kR/2)^{\frac{1}{2}} J_{l-\frac{1}{2}}(kR)$ and its derivative.

The theoretical angular distribution, in the main, is determined by the last factor. Thus the characteristic maximum in the experimental angular distribution corresponds to the maximum of $[j_l(kR)]^2$. When the theoretical curve is fitted to the experimental data it is found that the simpler expression given by BHATIA *et al.*, requires a larger nuclear radius (by 1×10^{-13} cm) than the more realistic values demanded by BUTLER's formula. The latter, however, fails when the Q of the reaction is negative (for highly excited states) and near the binding energy of the deuteron.

The theoretical cross section falls with increasing proton energy. It also falls with increasing angular momentum, for the square of the spherical BESSEL function at its first maximum falls with increasing order l. For $l=0$, the maximum in the angular distribution is usually at $\vartheta=0°$. It should be noted that this does not correspond to the maximum of $[j_0(kR)]^2$, for k depends on ϑ through (42.3). In general, the maximum values of $[j_l(kR)]^2$ are 1, 0.19, 0.09, 0.06, and 0.04 for $l=0$, 1, 2, 3, and 4, respectively. This together with the high energy released accounts for the relatively few protons in the group leading to the ground state in a reaction like Mg^{25} (d, p) Mg^{26} where the Q is high and $l=2$; in fact, this group was not discovered until fairly recently. It should be noted that the statistical factor in (42.5) can cause large variations in cross sections. This has been pointed out by WALL[1] and by ENDT and BRAAMS[1]; it can account for the large difference (a factor of 50) between the cross sections for the ground state (d, p) reactions in Ca^{43} and Ca^{40} found by SCHIFFER[2]. In the lighter elements good agreement has been obtained between theory and experiment[3]. However, not all proton groups give angular distributions in accord with that to be expected of the stripping process. Some are nearly isotropic. These may correspond to anomalously low stripping, the background representing compound nucleus formation. Possibly for these states the angular momentum taken into the nucleus by the neutron is high, for then the stripping process will be relatively infrequent. In many cases, an isotropic background appears superposed on the characteristic stripping distribution.

The validity of BUTLER's theory has been discussed by a number of authors. DAITCH and FRENCH[4], and GERJUOY[4] have shown that, in effect, BUTLER's theory is equivalent to the BORN approximation. The validity of the latter is not obvious; it has been discussed by AUSTERN[5]. As pointed out above the BUTLER theory has ignored the effect of the COULOMB field and the interaction

---

[1] N. S. WALL: Phys. Rev. **96**, 670 (1954). — P. M. ENDT and C. M. BRAAMS: Physica **21**, 839 (1955).

[2] J. P. SCHIFFER: Phys. Rev. **97**, 428 (1955).

[3] J. R. HOLT and T. N. MARSHAM: Proc. Phys. Soc. Lond. A **66**, 249, 258, 467, 565 (1953).

[4] P. B. DAITCH and J. B. FRENCH: Phys. Rev. **87**, 900 (1952). — E. GERJUOY: Phys. Rev. **91**, 645 (1953).

[5] N. AUSTERN: Phys. Rev. **89**, 318 (1953).

of the outgoing proton with the residual nucleus. TOBOCMAN[1], and TOBOCMAN and KALOS[1] have calculated the effect of both. It seems that the effect of the COULOMB field is to shift the peak of the angular distribution in the direction of larger angles, while filling in the minimum. The effect of the proton interaction has been calculated on the assumption that the nucleus behaves like a black absorbing sphere, and like a hard scattering sphere. Both models tend to restore the position of the peak, and to fill in the minima. These effects may account for the anomalies found by HOLT and MARSHAM for the heaviest nuclei they measured, viz. Ca and Sr (Figs. 43 and 44). In these nuclei the angular distribution is quite different to those obtained in light nuclei; the minimum in the angular distribution in the forward direction is partly filled in for Ca,

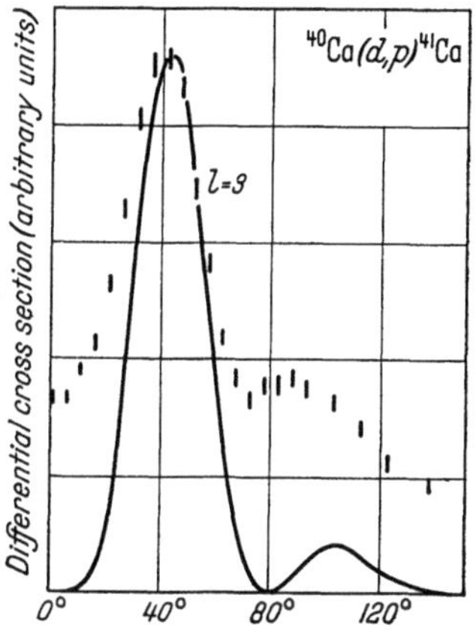

Fig. 43. Angular distribution of the protons producing the ground state in the reaction Ca^{40} (*d, p*) Ca^{41}, according to HOLT and MARSHAM.

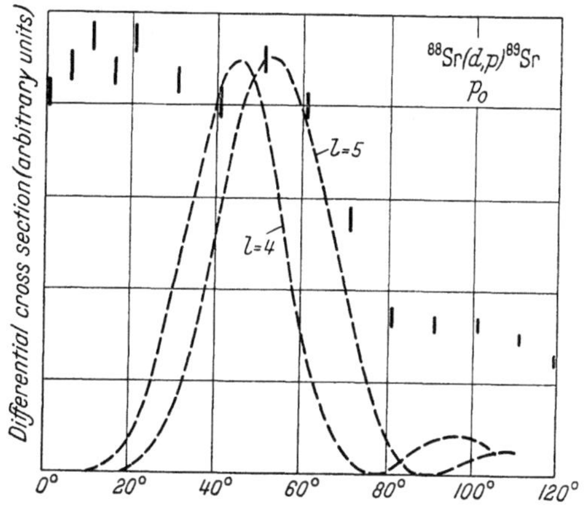

Fig. 44. Angular distribution of the protons producing the ground state in the reaction Sr^{88} (*d, p*) Sr^{89}, according to HOLT and MARSHAM.

and entirely filled in for Sr. For Ca it is clear from shell model considerations that the spin of the product Ca^{41} produced in the Ca^{40} (*d, p*) Ca^{41} reaction, is $\frac{7}{2}$. We should then expect to find a curve corresponding to $l=3$. The peak, in fact, occurs in the right position but the cross section for the forward direction ($\vartheta=0°$) is not negligible. This cannot be accounted for by mixtures of waves with $l=1$ and $l=3$; however, the deuteron energy (8 Mev) is only slightly greater than the barrier height (6 Mev) in this element and the stripping theory, therefore, is not strictly applicable in this instance. The angular distribution is similar for Sr^{88}; for this nucleus the differential cross section is nearly flat up to an angle corresponding roughly to the peak of the theoretical angular distribution with $l=4$, after which point it falls rapidly. In this case, the barrier height is 8.5 Mev which is again close to the deuteron energy. For higher deuteron energies (15 Mev) WALL[2] has found (for Pb^{207} and Y^{89}) better agreement with theory.

The characteristic angular distributions of the stripping process, in some light elements, at least, persist as the energy of the deuteron is reduced far below the COULOMB barrier[3]. In intermediate weight elements, an unsymmetrical peaked

[1] W. TOBOCMAN: Phys. Rev. **94**, 1655 (1954). — W. TOBOCMAN and M. KALOS: Phys. Rev. **97**, 132 (1955).
[2] N. S. WALL: Ref. 1, p. 278.
[3] Cf. M. K. JURIC: Phys. Rev. **98**, 85 (1955), on (*d, p*) and (*d, t*) reactions on Be down to 600 kev.

angular distribution was found by PRATT[1] for $Ti^{48}(d, p)Ti^{49}$ with 2.6 Mev deuterons. However, for targets of Ni^{58} and Ni^{60}, and for proton groups leaving the product nuclei in their ground state, PRATT found no such effect: the angular distribution was symmetrical about the 90° position, which is what one would expect for compound nucleus formation. For the Ni isotopes, the ground state transitions should require only $l=1$; it is difficult to see why stripping of this type was not detected while stripping with $l=3$ was found for Ti^{48}.

43. Purity of states. It has been pointed out that the orbital angular momentum of the neutron captured in the (d, p) process must be consistent with the parity change, if any, between the initial state of the system and its final state. The experimental curve defines the parity change in the transition unambiguously; however, the restriction (42.4) on the values which the orbital angular momentum can take, does not determine precisely the angular momentum change except in special circumstances, e.g., when the spin of the target nucleus is zero. BETHE and BUTLER[2] have suggested that in certain cases where more than one value of l will satisfy (42.4) the experimental data can be used as a test of the shell model. For example consider the reaction $Cl^{35}(d, p)Cl^{36}$, for which the spin of Cl^{35} is $\frac{3}{2}$ and that of Cl^{36} is 2 units. Conservation of angular momentum requires $l=0$, 2 or 4. According to the shell model, the 19th neutron in the latter nucleus is in a $d_{\frac{3}{2}}$ state and its capture in the (d, p) process with Cl^{35} requires $l=2$. This reaction has been investigated by KING and PARKINSON[3] who did, in fact, find that $l=2$ gives a good description of their experimental data, with only a small admixture of $l=0$. Theoretically, as we have seen, the yield of a stripping reaction is very much greater for $l=0$ than for $l=2$; the experimental data therefore is a sensitive measurement of the purity of the state in the shell model sense.

FRENCH and RAZ[4] have suggested that measurements of deuteron stripping and pickup reactions with even-even nuclei (spin 0) can be used to obtain information on departures from jj coupling. Usually the ground state of the product nucleus will have spin $j=l+\frac{1}{2}$ (although there exist some notable exceptions, e.g., Na^{23} and Mn^{55}) and the first excited state will correspond to $j=l-\frac{1}{2}$ (as happens in Ca^{43}). A direct comparison of the cross sections of the stripping and of the pickup process for the two states at the same energy largely eliminates the factors depending on the orbital angular momentum and affords a measure of the mixture of the two states.

44. $(d, p\gamma)$ measurements. The ambiguity in the orbital angular momentum in principle can be removed by study of the angular distribution of the γ-rays which follow the (d, p) reaction. However, the experimental technique required for such measurements is more difficult than that involved in the measurement of the proton angular distribution and no experimental data has yet been published.

The angular distribution can be written[5] in the form:

$$f(\vartheta, \varphi) \propto \sum_l \frac{S_l(\varphi)}{2l+1} |g_l(\vartheta)|^2$$

---

[1] W. W. PRATT: Phys. Rev. **97**, 131 (1955); **95**, 1517 (1954).

[2] H. A. BETHE and S. T. BUTLER: Phys. Rev. **85**, 1045 (1952). — S. T. BUTLER: Phys. Rev. **88**, 685 (1952).

[3] J. S. KING and W. G. PARKINSON: Phys. Rev. **88**, 141 (1942).

[4] J. B. FRENCH and B. J. RAZ: Phys. Rev. **98**, 1523 (1955). — Bull. Amer. Phys. Soc. **1**, 223 (1956).

[5] L. C. BIEDENHARN, K. BOYER and R. A. CHARPIE: Phys. Rev. **88**, 517 (1952). — G. R. SATCHELOR and J. A. SPIERS: Proc. Phys. Soc. Lond. A **65**, 980 (1952). — L. J. GALLAHER and W. B. CHESTON: Phys. Rev. **88**, 684 (1952).

where $S_l(\varphi)$ is the angular distribution of the γ-ray emitted relative to the direction of the ingoing neutron and $|g_l(\vartheta)|^2$ is the angular distribution of the protons with respect to the deuteron beam for a given value of the orbital angular momentum (l). When only one value of the orbital angular momentum is effective (as in the case discussed in the preceding section), the angular distribution is greatly simplified and may be used to determine the spins of the excited state and the ground state of the product nucleus.

45. Pick-up reactions. In principle the differential cross sections for the inverse process, i.e., the (p, d) or the (n, d) reaction, should be given by the formula [*17*]:

$$\frac{\sigma(\vartheta)_{\text{inverse}}}{\sigma(\vartheta)_{\text{direct}}} = \frac{2J_{\text{A}}+1}{2J_{\text{B}}+1} \cdot \frac{2J_a+1}{2J_b+1} \cdot \frac{k_a^2}{k_b^2}$$

where A and B refer to the initial and final nucleus, as before, and a and b to the ingoing and outgoing particles respectively. The shape of the angular distribution of the pickup process therefore is very similar to that of the direct process. This has been verified for (p, d) reactions in light nuclei by STANDING[1] and by REYNOLDS and STANDING[1] with 18 Mev protons. (d, t) reactions have been studied only in light elements. In heavier nuclei experimental studies of (n, d) reactions for 90 Mev neutrons have been made by HADLEY and YORK[2] and for (p, d) reactions for 95 Mev protons by SELOVE[3]. These are essentially high energy reactions and do not come within the scope of this article. It should be noted that there are experimental difficulties involved in the detection and in the measurement of the energy of deuterons in the presence of protons of comparable range and that these difficulties become more serious at low energies. They derive from the fact that the rate of loss of energy of particles of a given charge depends only on the velocity. Protons and deuterons, therefore, may be distinguished only by a measurement of energy losses in conditions which distinguish between their ranges.

Attempts to detect (p, t) reactions have been made by COHEN and HANDLEY[4]. The cross sections and the angular distribution for these reactions on the whole suggest compound nucleus formation rather than pickup.

J. Photonuclear reactions.

46. Experimental methods. The earliest measurements of photoreactions were made by BOTHE and GENTER[5]. Since then accurate measurements of photonuclear cross sections by activation methods have been made by HIRZEL and WÄFFLER[6] using the γ-rays excited by proton bombardment of lithium (about 17.6 Mev together with a lower intensity group with energy about 14 Mev). Since homogeneous γ-radiation with variable energy is obtainable in sufficient intensity only over a very small range of energies, and in an energy region quite unsuitable for studies of photonuclear reactions, measurements in recent years have been made with the continuous X-rays emitted when the electron beam from a betatron or synchrotron is allowed to fall on a target. When the resulting radioactivity is measured as a function of the maximum energy of the betatron,

[1] K. G. STANDING: Phys. Rev. **101**, 152 (1956). — J. R. REYNOLDS and K. G. STANDING: Phys. Rev. **101**, 158 (1956).
[2] J. HADLEY and H. YORK: Phys. Rev. **80**, 345 (1950).
[3] W. SELOVE: Phys. Rev. **101**, 231 (1956).
[4] B. L. COHEN and T. H. HANDLEY: Phys. Rev. **93**, 514 (1954).
[5] W. BOTHE and W. GENTNER: Z. Physik **106**, 236 (1937); **112**, 45 (1939).
[6] O. HIRZEL and H. WÄFFLER: Helv. phys. Acta **19**, 373 (1947).

a curve is obtained from which the differential cross section corresponding to a specific γ-ray energy can be determined. The activity is proportional to:

$$\int_0^{E_0} N(E, E_0)\, \sigma(E)\, dE$$

where $N(E, E_0)$ is the number of photons of energy (E) falling on the target per unit energy when the maximum energy of the spectrum is E_0. The photonuclear cross section $\sigma(E)$ may be obtained by differentiation of this expression. Convenient methods of doing this have been worked out by JOHNS *et al.*[1]. Recently a method of obtaining the differential cross sections using the activity induced in a monitoring target has been described by SAGANE[2], and a rather similar method based on induced activity and mechanical oscillation of target and monitor has been described by NEWKIRK[3]. Although the necessity to differentiate the activation curves obtained involves some loss of accuracy, there seem to be no serious discrepancies, for the cross section at 17.6 Mev derived by differentiation can usually be checked against the results obtained by WÄFFLER and others for irradiation by γ-rays of this energy.

The most noticeable feature of photonuclear reactions is that the total cross section is greatest at relatively low energies. For example, the cross section of the (γ, n) reaction, whose energy variation was first studied for C and Cu by BALDWIN and KLAIBER[4], rises rapidly to a maximum near 20 Mev and then falls off for higher energies. Thus, most of the photonuclear reaction is confined to energies below 25 Mev or so, and only those reactions are of importance for which the thresholds lie below this rather low figure. In heavy elements the most important reactions are the primary reactions (γ, n) and (γ, p), and the secondary reactions (γ, np), $(\gamma, 2n)$, etc. Sometimes (γ, d) reactions are also important; however (γ, α) reactions are much less probable. In many instances the cross sections for these reactions may be studied by measurement of the activities of the radioactive products but this method, as pointed out in another context, is limited to those nuclei for which the product, if radioactive, has a period in a suitable range for measurement. Direct measurement of the neutron yield has been employed by several experimenters; some have used BF_3 counters, others Rh detectors embedded in paraffin. These methods, however, have the disadvantage that the results do not discriminate between primary and secondary neutrons. A combination of such measurements with activation measurements, however, is very useful in determining the contribution of (γ, pn) and $(\gamma, 2n)$ reactions. (γ, p) reactions have usually been studied with photographic plates which readily yield both the energy spectrum and the angular distribution. Owing to the fact that the intensity of x-radiation is usually amply sufficient, the collimation of the beam presents no difficulty, and detecting instruments are relatively easily shielded. Recently, (γ, p) reactions have been studied using photomultipliers and ZnS screens.

47. The giant resonance. It was pointed out in the previous section that the total photonuclear reaction cross section rises to a maximum near 20 Mev. This effect is to be seen in the cross sections for both the (γ, n) and (γ, p) reactions; these cross sections rise to a peak near 20 Mev then fall off rather rapidly. This peak is called the "giant" resonance. In general appearance the variation of

[1] H. E. JOHNS, L. KATZ, R. A. DOUGLAS and R. N. H. HASLAM: Phys. Rev. **80**, 1062 (1950). — L. KATZ and A. G. W. CAMERON: Canad. J. Phys. **29**, 518 (1951).

[2] R. SAGANE: Phys. Rev. **84**, 586 (1951); **85**, 1065 (1952).

[3] L. L. NEWKIRK: Phys. Rev. **86**, 249 (1952).

[4] G. C. BALDWIN and G. J. KLAIBER: Phys. Rev. **73**, 1156 (1948).

cross section as a function of energy resembles cross sections for heavy particle reactions, where, as we have seen, the peaked shape is caused by the competition of reactions which are successively excited as the energy is raised. For photonuclear reactions, however, the peaks in the (γ, n) and (γ, p) reactions cannot be caused by competition from the emission of further neutrons or protons, because the total cross section obtained by adding together the cross sections of all possible reactions is found to be peaked near the peaks of the (γ, n) and (γ, p) cross section curves. The accuracy with which such cross sections can be measured, owing to the necessity of differentiation of the measurements, is not as great as that which obtains for heavy particle reactions. Moreover, usually not all the relevant cross sections can be measured. The results, however, do consistently show that the total cross section is itself peaked. The measurements of WHALIN and HANSON[1] (which are typical) on the (γ, n) and $(\gamma, 2n)$ reactions in Ta, are shown in Fig. 45. A very similar curve has been obtained for Cu^{63} by BERMAN and BROWN (see Sect. 53). For Ta the yield of the proton reaction is much smaller and can be disregarded. It will be seen that the sum of the two cross sections falls when the energy increases above 14 Mev; this fall cannot be due to competition by the $(\gamma, 3n)$ reaction, for the threshold of this reaction lies at 21 Mev. Beyond the peak the cross section falls to some low value. The yield of neutrons at higher energies has been measured by JONES and TERWILLIGER[2] from about 13 Mev to 320 Mev. The slow rise in the neutron yield found by these authors can be attributed, in part, to an increase in the neutron multiplicity as successively more reactions become energetically possible.

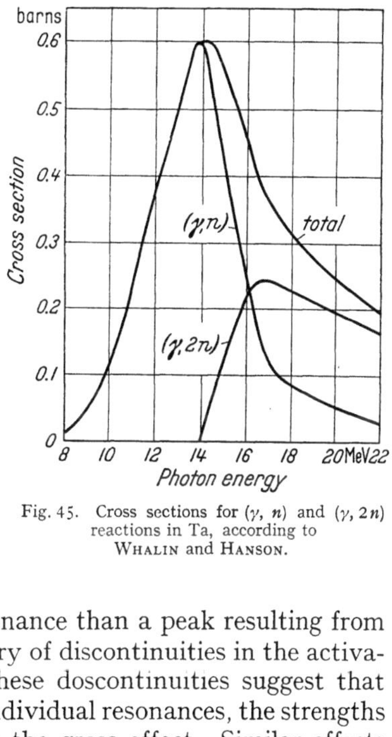

Fig. 45. Cross sections for (γ, n) and $(\gamma, 2n)$ reactions in Ta, according to WHALIN and HANSON.

That the peak has more the nature of a resonance than a peak resulting from competing reactions is confirmed by the discovery of discontinuities in the activation of O^{16} by KATZ and his collaborators[3]; these doscontinuities suggest that the giant resonance, is, in fact, the sum of many individual resonances, the strengths of which change with energy proportionally to the gross effect. Similar effects have been found in Li^7, C^{12} and F^{19}, and recently[4], in P^{31}. It is generally supposed that the discontinuities arise from sharp resonances, but this does not[5] yet seem to be established. Moreover, the increase of the photonuclear cross section above the threshold is not always the smooth curve that one might expect from a simple resonance process. In light elements especially, e.g., in O^{16}, N^{14}, and in F^{19}, the cross section rises abruptly from the threshold and then flattens off to form a step[6].

[1] E. A. WHALIN and A. O. HANSON: Phys. Rev. **89**, 324 (1953).

[2] L. W. JONES and K. M. TERWILLIGER: Phys. Rev. **91**, 699 (1953).

[3] L. KATZ, R. N. H. HASLAM, R. J. HORSLEY, A. G. W. CAMERON and R. MONTALBETTI: Phys. Rev. **95**, 464 (1954).

[4] R. BASILE and C. SCHUHL: C. R. Acad. Sci. Paris **240**, 2399 (1955).

[5] C. TZARA: C. R. Acad. Sci. Paris **242**, 2340 (1956).

[6] R. J. HORSLEY, R. N. H. HASLAM and H. E. JOHNS: Phys. Rev. **87**, 756 (1950). — H. E. JOHNS, R. J. HORSLEY, R. N. H. HASLAM and A. QUINTON: Phys. Rev. **84**, 856 (1951).

The peaks of the cross sections for the (γ, n) and (γ, p) reactions do not coincide. The peak for the (γ, p) reaction appears to lie higher than that of the (γ, n) reaction. The peak for the (γ, α) reaction lies much higher still. This effect is caused, no doubt, by the COULOMB barrier. While the position of the peaks for these reactions changes very little from one element to another, the positions of the peaks of the yields due to inelastic scattering of γ-rays [the (γ, γ') reaction] seem much more erratic; in general these peaks lie much lower.

The systematics of the giant resonances for (γ, n) reactions have been described by MONTALBETTI, KATZ, and GOLDEMBERG[1], and by NATHANS and HALPERN[2]. The data for (γ, p) reactions is much less complete. The position of the peak for the (γ, n) reaction is always very close to the threshold of the $(\gamma, 2n)$ reaction[3]; it increases slowly from about 12 Mev in U^{238} to 19 Mev in Cu and to higher values for the lighter elements (see Fig. 46). Both groups of experimenters agree that the variation of the peak energy with atomic weight can be expressed as $A^{-0.186}$. However, measurements[4] in light nuclei indicate that the peak energy levels off to 22.5 Mev for low mass numbers (He^4, being an exception, with a peak near 24 Mev). At first sight, at least in heavy elements, it might be supposed that the position of the peak follows the neutron binding energy. This is not the case, however, for recent measurements by NATHANS and YERGIN[5] with Mg and Zr isotopes show that the positions of the peaks for Mg^{24} (neutron binding energy, 16.6 Mev) and Mg^{25} (7.3 Mev) are sensibly the same, viz., 19.5 and 20.3 Mev; and for Zr^{90} (12.2 Mev) and Zr^{91} (7.2 Mev), the peaks lie at 16.4 and 16.3 Mev respectively.

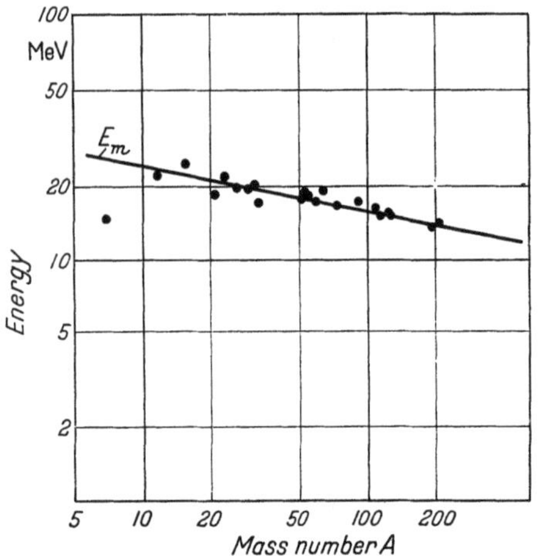

Fig. 46. Peak energies in the (γ, n) reaction, after MONTALBETTI, KATZ, and GOLDENBERG.

The width of the peaks decrease slightly with increasing atomic weight. The width at half maximum, on average, is near 6 Mev. For very light nuclei it appears to be somewhat less, and there seems to be definite evidence[2,5] that it is sharper for closed shell nuclei.

The peak cross section is less easy to measure accurately. According to MONTALBETTI *et al.*, it increases smoothly with atomic weight as $A^{\frac{5}{3}}$, and according to NATHANS and HALPERN, as $A^{\frac{3}{2}}$. For Cu^{63}, the peak (γ, n) cross section is about 100 mb. According to NATHANS and YERGIN[5], there is no interruption of the smooth change of peak cross section with atomic weight near the closed shell nuclei: the passage through the closed shell nucleus (at Zr^{90}) is accompanied by a drop in the width and a corresponding drop in the total integrated cross section. GOLDEMBERG and LOPES[6] have shown that a relation exists between the closed shell numbers and a parameter compounded of the neutron binding energy, the peak energy, and its width.

[1] R. MONTALBETTI, L. KATZ and J. GOLDEMBERG: Phys. Rev. **91**, 659 (1951).
[2] R. NATHANS and J. HALPERN: Phys. Rev. **93**, 437 (1954).
[3] L. EYGES: Phys. Rev. **86**, 325 (1952).
[4] G. E. FERGUSON, J. HALPERN, R. NATHANS and P. F. YERGIN: Phys. Rev. **95**, 776 (1954).
[5] R. NATHANS and P. F. YERGIN: Phys. Rev. **98**, 1296 (1955). See also L. KATZ, R. G. BAKER and R. MONTALBETTI: Canad. J. Phys. **31**, 250 (1953).
[6] J. GOLDEMBERG and J. LEITE LOPES: Phys. Rev. **99**, 1053 (1953).

48. Theories of the giant resonance. Many theories of the origin of the giant resonance have been proposed. We shall not discuss them here in detail for most of them have been reviewed recently by LEVINGER [*18*]. The first was that of GOLDHABER and TELLER[1] who suggested that the origin of the resonance was the classical harmonic oscillation of the protons in the nucleus with respect to the neutrons. Numerical estimates give approximately the correct position of the peak of the giant resonance. Similar but more elaborate calculations have been made STEINWEDEL and JENSEN[2] and by DANOS[2].

WU[3] has shown that the position of the peak is related to the energy separation of the levels of an harmonic oscillator. BURKHARDT[4] and REIFMANN[5] have explored the possibilities of direct absorption by single particle excitation. The giant resonance is then to be identified with the aggregate absorption of many single particle transitions. The difficulty with this theory is that the energy of the peak of the absorption curve is only half what it should be and that the number of nucleons directly ejected into the continuum is far too small. WILKINSON[6] has suggested that the absorption process consists of the elevation of individual nucleons to energetically unstable intermediate states of a single particle type which subsequently decay by nucleon emission or by sharing their energy with the rest of the nucleus. We shall discuss this theory again in Sect. 52. In this model, as in BURKHARDT'S, there remains the difficulty of explaining the position of the peak of the giant resonance; however, the discrepancy may be less serious than appears for there are several features which would tend to increase the position of the peak beyond that given by elementary calculation.

49. The integrated photonuclear cross section. The total integrated photonuclear cross section is an important quantity for it demonstrates that the giant resonance is caused, in the main, by absorption of electric dipole radiation. The total integrated cross section has been calculated by LEVINGER and BETHE[7]. For electric dipole radiation, the cross section for absorption for a particular transition is:

$$\sigma_a = (2\pi^2 e^2 \hbar/M c)\, f_n,$$

where f_n, the "oscillator strength", is a measure of the rate of emission of radiation. The total cross section, integrated over the resonance, is then proportional to the sum of the oscillator strengths, a quantity which, for atoms, is just equal to the number of electrons. For the nucleus, however, an electric dipole moment is developed only as a result of the displacement of the centroid of electric charge from that of mass. Consequently, each proton behaves as if it had a charge of $e(N/A)$ and each neutron as if it had a charge of $-e(Z/A)$. The sum of the oscillator strengths, then, is NZ/A, and the integrated cross section becomes:

$$\int \sigma\, dE = (2\pi^2 e^2 \hbar/M c)\,(NZ/A) = 0.060\, NZ/A \text{ Mev-barn}. \tag{49.1}$$

This is a minimum value for the integrated cross section. It is a very general result and it is independent of any particular nuclear model.

It might be expected that exchange forces between neutrons and protons, if present, would affect the rate of emission or absorption of radiation. The effect

[1] M. GOLDHABER and E. TELLER: Phys. Rev. **74**, 1046 (1948).
[2] H. STEINWEDEL and J. H. D. JENSEN: Z. Naturforsch. **5**a, 413 (1950). — M. DANOS: Ann. Physik **10**, 265 (1952).
[3] C. S. WU: Unpublished, quoted by LEVINGER, reference [*18*].
[4] J. L. BURKHARDT: Phys. Rev. **91**, 420 (1953).
[5] A. REIFMAN: Z. Naturforsch. **8**a, 505 (1953).
[6] D. H. WILKINSON: Proc. Glasgow Conference on nuclear and meson physics, 1954. London: Pergamon Press 1955; Physica, **22**, 1039 (1956).
[7] J. S. LEVINGER and H. A. BETHE: Phys. Rev. **78**, 115 (1950).

of exchange forces on electric dipole emission was first investigated by SIEGERT[1], and later generalised by SACHS and AUSTERN[2]. According to LEVINGER and BETHE, the amount to be added to the integrated cross section (49.1) depends to some extent on the nuclear radius and on the nature of the interaction between neutron and proton and the rest of the nucleus. For a square well they estimated that the total integrated cross section should be:

$$\int \sigma \, dE = 0.060 \cdot NZ/A \, (1 + 0.8\,x), \tag{49.2}$$

where x is the proportion of exchange force in the neutron-proton interaction. A different though comparable result has been obtained by GELL-MANN, GOLDBERGER, and THIRRING[3], by very different methods. This result, which presumably takes account of exchange forces, is $(2\pi^2 e^2 \hbar/M c)\, NZ/A\, (1 + 0.1\, A^2/NZ)$.

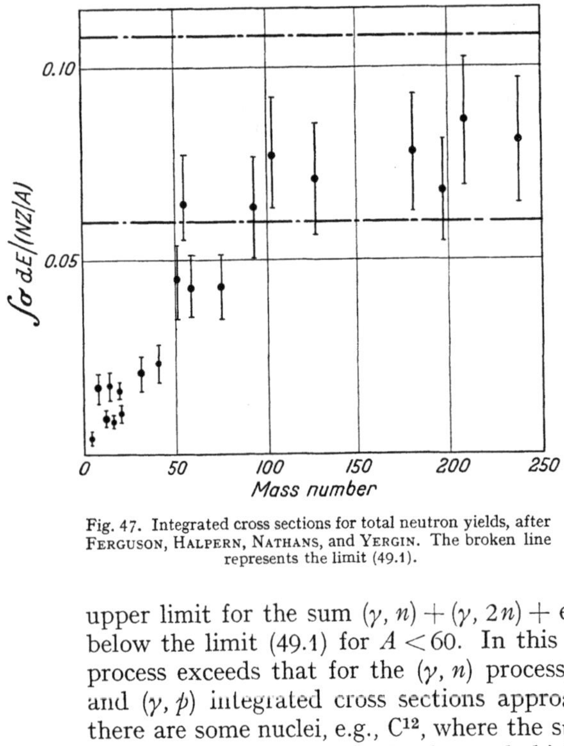

Fig. 47. Integrated cross sections for total neutron yields, after FERGUSON, HALPERN, NATHANS, and YERGIN. The broken line represents the limit (49.1).

Since the cross section falls off very rapidly above the peak of the giant resonance, the integrated cross section may be obtained with sufficient accuracy by integrating the experimental data up to an energy of 25 Mev or so. The integrated cross sections for (γ, n) reactions are usually equal to or greater than (49.1) for atomic weights above $A = 60$. The integrated cross sections for the neutron reactions are plotted[4] in Fig. 47. They refer to the cross sections for neutron production regardless of the neutron origin and therefore represent an upper limit for the sum $(\gamma, n) + (\gamma, 2n) +$ etc. It will be seen that this sum lies below the limit (49.1) for $A < 60$. In this region the cross section for the (γ, p) process exceeds that for the (γ, n) process, and probably the sum of the (γ, n) and (γ, p) integrated cross sections approaches the sum rule limits[5]; however, there are some nuclei, e.g., C^{12}, where the sum of these cross sections seems quite inadequate[5]. Above $A = 60$ the probability of proton emission becomes very much less than neutron emission[6]: for example, in Ag (according to DIVEN and ALMY) the ratio of the two is only 0.023. The integrated neutron yields for the heavy elements shown in Fig. 47, nevertheless, represent upper limits for they include the multiplicity of neutrons emitted in $(\gamma, 2n)$ reactions and the like. For U^{238}, the sum of the integrated (γ, n) and photo-fission cross sections[7] is 3.8 Mev-barn: this is lower than the corresponding point for U^{238} in Fig. 47. For $x = 0$, (49.2) gives 3.4 Mev-barn. Thus for the heavier elements the integrated

[1] A. J. F. SIEGERT: Phys. Rev. **52**, 787 (1937).
[2] R. G. SACHS and N. AUSTERN: Phys. Rev. **81**, 705 (1951). See also reference [*18*].
[3] M. GELL-MANN, M. L. GOLDBERGER and W. E. THIRRING: Phys. Rev. **95**, 1612 (1954).
[4] See ref. 4, p. 284.
[5] H. MORINAGA: Phys. Rev. **97**, 1185 (1955). — S. A. E. JOHANSSON: Phys. Rev. **97**, 1186 (1955).
[6] E. V. WEINSTOCK and J. HALPERN: Phys. Rev. **94**, 1651 (1954).
[7] B. DUFFIELD and J. R. HUIZENGA: Phys. Rev. **89**, 1042 (1953).

cross sections seem to exceed the sum rule limit (49.1) but by an amount which is difficult to estimate. Possibly x lies between 0 and 0.5.

The magnitude of the observed cross sections, in themselves, is good evidence that the photonuclear effect is mainly an electric dipole effect. LEVINGER and BETHE have calculated the electric quadrupole effect and conclude that it can only amount to a few percent of the electric dipole effect. If the giant resonance itself is identified with the electric dipole absorption, some of the photonuclear cross section which corresponds to the tails of this resonance can be ascribed to electric quadrupole absorption. In O^{16} the absorption for x-rays between 13 and 19 Mev is accounted for[1] satisfactorily by electric quadrupole absorption.

50. Energy and angular distribution of (γ, n) reactions. Measurements of the energy distributions of the photoneutrons have received rather less attention than those of the photoprotons. For Cu bombarded by x-rays with a maximum energy of 24 Mev, the neutron spectrum has been measured by a photographic plate method by BYERLY and STEPHENS[2]. Their results, which are shown in Fig. 48, in the main, are in good agreement with the energy spectrum expected on the statistical theory. However, there are more neutrons with energies above 5 Mev than can be accounted for by the statistical theory. Similar results have been obtained by PRICE[3] in Bi. This high energy component is probably caused by a direct photonuclear effect, and is associated with a strong anisotropy in the angular distribution. PRICE's results for Bi have been confirmed by FERRERO *et al.*[4]. The high energy component was separately studied using Al^{27} and Si^{28} detectors which eliminate entirely the statistically distributed neutrons. The high energy component amounted to about 10% of the total neutron yield. With Al^{27} and Si^{28} threshold detectors, PRICE found that the lighter nuclei, e.g. V^{51}, gave an approximately isotropic angular distribution, while the heavier nuclei, e.g. Pr, Au, Pb, and Bi, gave a markedly anisotropic distribution peaked near 90° to the direction of the incident beam. Rather similar results were obtained by POSS[5] for W, Pb and Bi, using threshold detectors, and by PRICE and KERST[6] using Rh detectors.

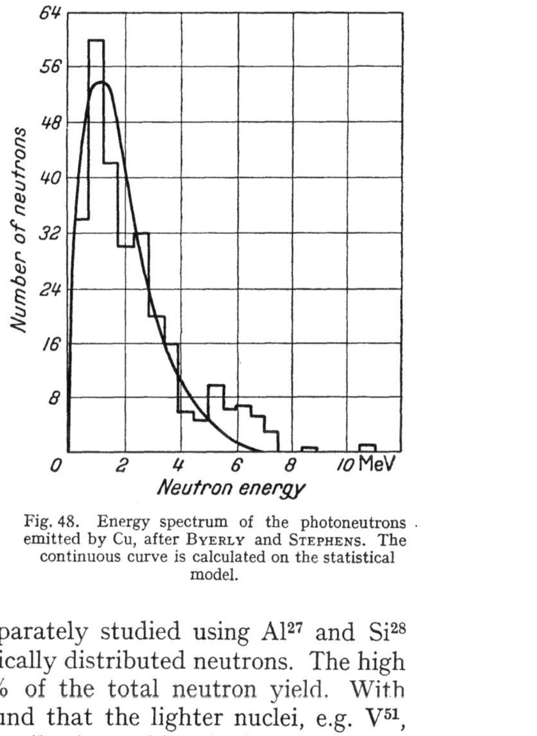

Fig. 48. Energy spectrum of the photoneutrons emitted by Cu, after BYERLY and STEPHENS. The continuous curve is calculated on the statistical model.

51. Energy and angular distributions of (γ, p) reactions. The angular distribution of the photoprotons emitted by C, Al, Ni, and Mo, have been measured with ZnS scintillators by JOHANSSON[7], who found that the curve is peaked near 60°;

---

[1] R. M. SPICER: Phys. Rev. **99**, 33 (1955). — A. S. PENFOLD and B. M. SPICER: Phys. Rev. **100**, 1377 (1955).
[2] P. R. BYERLY and W. E. STEPHENS: Phys. Rev. **83**, 54 (1951).
[3] G. A. PRICE: Phys. Rev. **93**, 1279 (1954).
[4] F. FERRERO, A. O. HANSON, R. MALVANO and C. TRIBUNO: Nuovo Cim. **4**, 418 (1956).
[5] H. L. POSS: Phys. Rev. **79**, 539 (1950).
[6] G. A. PRICE and D. W. KERST: Phys. Rev. **77**, 806 (1950).
[7] S. A. E. JOHANNSON: Phys. Rev. **97**, 434 (1955).

and for Al, by HOFFMAN and CAMERON[1] with photographic plates, who found that the proton spectrum which was peaked near 4 Mev, extended to near 16 Mev, and that the angular distribution, like that found by the previous authors, was unsymmetrical. Angular distributions of the photoprotons emitted by Cu, Ni, and Co have been measured with ZnS scintillators by MANN, HALPERN, and ROTHMAN[2]. The angular distributions from these nuclei were found to differ markedly from one nucleus to another: while the proton spectrum of Co was found to be peaked near 70°, that from Ni fell off continuously with increasing angle. For Co, however, TOMS and STEPHENS[3] found a rather less strongly anisotropic distribution. The energy spectrum of the protons was peaked at 5 Mev and was in good agreement with that calculated statistically after correction for the thickness of the target, except that the usual excess of high energy protons was present (possibly 5%). The charged particles emitted by Cu have been studied by BYERLY and STEPHENS[4], using photographic plates. They found that the proton spectrum was peaked at about 5 Mev, but that, in addition, there was a group of deuterons, (about 30% of the protons) which was peaked sharply at 4 Mev. Again the lower energy protons were isotropically distributed while the few high energy protons were peaked near 90°. The deuteron emission from Cu has been confirmed by SMITH[5] and by SMITH and LASLETT[5] by cloud chamber studies of the emission of charged particles from a copper foil placed in a magnetic field and irradiated by 65 Mev X-rays. It is interesting to note that no deuterons were found by TOMS and STEPHENS[3] in their study of Co; the yield must be less than 2% of the protons in the (γ, p) reaction.

The proton yield for the lighter elements, as we have already pointed out, is usually greater than the neutron yield. This is difficult to account for statistically unless the number of states of the residual nucleus available for the (γ, p) reaction is somehow increased relatively to those left in the (γ, n) reaction. MORINAGA[6] has pointed out that it is improbable that the observed excess of (γ, p) reactions in light elements can be accounted for in terms of differences between neutron and proton binding energies and has suggested that isotopic spin selection rules might be important.

For elements between $A=30$ and $A=60$, MANN, HALPERN, and ROTHMAN, have shown that the proton yield falls off relative to the neutron yield at a rate that is in satisfactory agreement with the statistical theory. But above $A=60$, the proton yields are generally much higher than one would expect from that theory. These high yields are associated with an energy spectrum increasingly dominated by high energy protons.

An interesting study of the (γ, p) reactions with separated Mo^{92} and Mo^{100} has been made by BUTLER and ALMY[7]. In the former nucleus (which contains a closed neutron shell) the neutron binding energy is 13 Mev and the proton binding energy only about 8 Mev. In this case proton emission is greatly favoured, and is, in fact, found to be exceptionally large—about one half of that of the (γ, n) reaction. The energy distribution and the yield are in agreement with statistical calculation and compound nucleus formation. For Mo^{100}, however, the

[1] M. M. HOFFMAN and A. W. G. CAMERON: Phys. Rev. **92**, 1184 (1953).
[2] A. K. MANN, J. HALPERN and M. ROTHMAN: Phys. Rev. **87**, 146 (1952).
[3] M. E. TOMS and W. E. STEPHENS: Phys. Rev. **95**, 1209 (1954).
[4] See ref. 2, p. 287.
[5] W. H. SMITH: Phys. Rev. **85**, 728 (1952). — W. H. SMITH and L. J. LASLETT: Phys. Rev. **86**, 523 (1952).
[6] H. MORINAGA: Phys. Rev. **97**, 444 (1955).
[7] W. A. BUTLER and G. M. ALMY: Phys. Rev. **91**, 58 (1953).

proton binding energy exceeds the neutron binding energy (about 7 Mev) by 2.5 Mev. In this case, the energy distribution is again in fair agreement with statistical calculation (except for an excess of high energy protons) but the yield, which is 20 times smaller than that of Mo^{92}, is about 100 times greater[1] than that calculated.

The energy distributions of the photoprotons emitted by heavier nuclei do not agree with statistical calculation. The photoprotons emitted by In, Ce, Ta, Pb, and Bi, have been studied by TOMS and STEPHENS[2]. Their results for Pb are shown in Fig. 49. For In, the spectrum is peaked at 7 Mev and in Bi and Pb near 11 Mev (Fig. 49). As before the angular distributions show marked variations from one proton energy to another and from one element to another. For protons with energies greater than 11 Mev, the angular distribution is peaked at low angles in In and Ni, near 90° for Ce, and in the forward direction in Pb. For Ag, which lies within this range, DIVEN and ALMY[3] found that the angular distribution was nearly isotropic for protons with energies below 10 Mev and peaked near 90°, for protons with energies between 10 and 14 Mev. The peak in the energy spectrum was near 7 Mev. The energy spectrum of the protons emitted by Rh has been studied by CURTIS *et al.*[4]: it is similar.

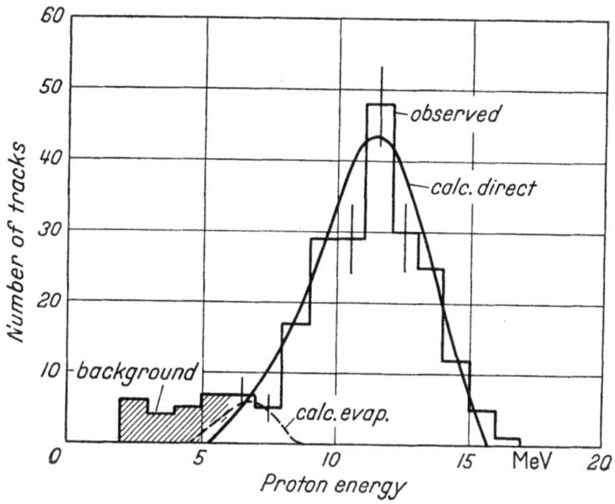

Fig. 49. Energy spectrum of the photoprotons emitted by Pb, according to TOMS and STEPHENS. The full curve is that calculated from direct interaction (Sect. 52), and normalised to the experimental data, the broken curve that calculated from statistical theory.

52. Direct interactions. In general, the energy and angular distributions and the yields of the reactions just described are inconsistent with compound nucleus formation. This has led to the suggestion that photonuclear reactions are the result of direct interactions between the incident radiation and individual nucleons within the nucleus. In brief the evidence for such direct interactions lies in the ejection of high energy nucleons in numbers above that demanded by statistical considerations and in their angular distribution. For (γ, n) reactions this high energy component is not very numerous; for (γ, p) reactions, especially in the heaviest elements, it is predominant. It should be noted that the existence of a statistically distributed energy spectrum in the products of photonuclear reactions is not, in itself, evidence of compound nucleus formation, for there is no particular reason why direct ejection of a nucleon should not leave the residual nucleus in any state consistent with conservation of energy and angular momentum. The evidence for compound nucleus formation comes from the study of the relative yields of photoreactions in which two or more particles (or α-particles) are emitted (e.g. SAGANE's experiment, Sect. 53). But such experiments, as we shall see below, also give evidence of the existence of direct interactions.

[1] R. B. DUFFIELD, L. HSIAO and E. N. SLOTH: Phys. Rev. **79**, 1011 (1950).

[2] M. E. TOMS and W. E. STEPHENS: Phys. Rev. **92**, 362 (1953); **98**, 626 (1955).

[3] B. C. DIVEN and G. M. ALMY: Phys. Rev. **80**, 407 (1950).

[4] N. W. CURTIS, J. HORNBOSTEL, D. W. LEE and E. O. SALANT: Phys. Rev. **77**, 290 (1950).

Thus it is clear that some form of direct interaction must occur in the photonuclear process and it is very probable, especially in heavy elements, that compound nuclei are also formed. What is not so clear is the mechanism of the direct interaction, the extent to which direct interaction and compound nucleus formation occur, and the relationship between them. It is quite reasonable to suppose that direct interaction is an essential preliminary to all photonuclear reactions, in the same sense as direct interaction for heavy particles is the preliminary to compound nucleus formation, because there is a high probability that the nucleon acted upon is absorbed in the process of ejection. In the heavy elements, where proton emission is hindered by the COULOMB barrier, it is clear that direct interaction on the proton must nearly always lead to compound nucleus formation, for the retention of the proton must produce a general excitation of the nucleus.

The first attempts to account for the apparently excessive proton emission by heavy elements, originally discovered by HIRZEL and WÄFFLER[1] with 17 Mev γ-rays, have been made by JENSEN[2] and by COURANT[3]. These authors assume that the primary interaction between a photon and a nucleus is not the absorption of the photon with consequent formation of a compound nucleus, but the absorption of the photon by direct interaction with an individual nucleon within the nucleus. To account qualitatively for the effects observed in the heavy elements the probability of such a direct process need not amount to more than a few percent of compound nucleus formation. In principle, the ejection of the nucleon may leave the residual nucleus in a large range of excited states from which its subsequent decay can be calculated by statistical methods. However, in COURANT'S calculations such refinements are disregarded; the nucleon is supposed to be located in a square well of conventional type; and the probability that it absorbs a photon and is ejected, leaving the remaining nucleons undisturbed, is calculated. The angular distribution which results from this process (which is assumed to be electric dipole absorption) is of the form: $(a+b\sin^2\vartheta)$. For transitions in which the orbital angular momentum (l) of the nucleon is increased by ejection, i.e. $l\to l+1$, $b/a=(l+2)/2l$, while for transitions of the type $l\to l-1$, $b/a=(l-1)/2(l+1)$. Thus the angular distribution should not be very markedly anisotropic; if b/a is greater than 1.5, some nucleons must have been ejected from s-states.

Although it is clear that some kind of direct interaction is necessary to account for the experimental results, COURANT'S theory in its original form seems to be inadequate to account for them for two reasons.

First, the cross sections predicted are too small. According to TOMS and STEPHENS[4], the experimental cross sections for (γ, p) reactions are greater than COURANT'S predictions by factors which very from 4 to 13. COURANT'S model, however, is very crude and, as he has pointed out, there are several factors neglected in the calculations, which, if taken into account would raise the predicted cross sections. However, TOMS and STEPHENS found that the proton energy spectra in the heavy elements is well described by COURANT'S formula, provided it is scaled to fit.

Second, the angular distributions observed for the higher energy neutrons and protons are sometimes sharply peaked about the 90° position giving a ratio

---

[1] See ref. 6, p. 281.
[2] P. JENSEN: Naturwiss. **35**, 190 (1948).
[3] E. D. COURANT: Phys. Rev. **74**, 1226 (1948); **82**, 703 (1951).
[4] See ref. 2, p. 289.

b/a much higher than can be accounted for. More often, as we have seen in Sect. 51, the angular distribution is quite unsymmetrical about the 90° position.

According to WILKINSON[1], all photonuclear reactions are initiated by direct interactions in a process in which a nucleon is promoted into an unbound single particle state which may subsequently emit the nucleon or may, by interaction with the other constituents of the nucleus, generate a compound state by sharing its energy with them. Only those transitions will contribute to the dipole sum in appreciable amount for which the initial state is a closed shell and for which both initial and final states possess nodeless radial wavefunctions. In fact, according to WILKINSON, one such transition can account for the greater part of the dipole sum. This means in effect that only nucleons in the lowest configuration corresponding to a particular orbital angular momentum (l) will contribute and that these nucleons will be promoted always to states for which $l \rightarrow l+1$. The angular distribution of the ejected nucleon is given by $b/a = (l+2)/2l$; this theory, therefore, has the advantage in that it predicts a somewhat more anisotropic angular distribution than that suggested by COURANT, for which l can change in both directions.

The decay of the intermediate state is determined by the relative probabilities of emission and interaction with the other constituents. The former probability will be high, because the reduced width will be of the order of that of single particle states, viz., $\hbar^2/MR$; the latter probability, by analogy with the optical model of the nucleus (Sect. 6) is given by twice the imaginary part of the nuclear potential, or what is equivalent to it, $\hbar v/\Lambda$, where v is the velocity of the nucleon and Λ is the mean free path in nuclear matter (Sect. 12). The ratio of the two will usually be small because in most cases the nucleon is emitted with relatively high angular momentum and this factor greatly reduces the penetrability (T_l) of the nuclear surface. Very roughly the fraction of the number of intermediate states which decay by particle emission will be $T_l \Lambda/R$ and the number of intermediate states formed is calculable from the population of those which are eligible and the absorption cross section for the transition in question.

FERRERO *et al.*[2] have estimated the number of neutrons emitted in the more energetic group from Bi on the basis of WILKINSON's theory and have found good agreement with experiment (roughly 10%). The theory also gives roughly the right angular distribution (for Bi, $b/a \approx 1$). The remainder of the direct interactions, of which about half will concern protons and half neutrons, give rise to compound states which decay with the emission of the usual Maxwellian neutron energy spectrum.

53. $(\gamma, 2n)$ and (γ, pn) reactions. Few of these reactions have been examined in any detail. For Cu^{63}, the $(\gamma, 2n)$ reaction has been measured by a number of experimenters, but notably by BERMAN and BROWN[3] who have measured it up to 30 Mev. The total integrated cross section is about one seventh of that of the (γ, n) reaction. The results obtained for Ta, another element for which the $(\gamma, 2n)$ reaction has been frequently measured, are similar. Measurements of (γ, np) reactions are scarce[4]. For S^{32}, both the (γ, pn) and the (γ, d) reactions have been separately measured by KATZ and PENFOLD[5]. For both of these reactions the peak cross section, which is roughly 1 mb is small compared with the peak

[1] See ref. 6, p. 285.

[2] See ref. 4, p. 287.

[3] A. I. BERMAN and K. L. BROWN: Phys. Rev. **96**, 83 (1954).

[4] For P^{31}: J. HALPERN, R. NATHANS and A. K. MANN: Phys. Rev. **88**, 958 (1952).

[5] L. KATZ and A. S. PENFOLD: Phys. Rev. **81**, 815 (1951).

cross section expected for the (γ, n) reaction (for the neighbour P^{31}, the latter is about 17 mb).

Photoreactions in which two or more particles are emitted, like those produced by heavy particles should be the products of the decay of compound nuclei and their relative cross sections should be calculable by statistical methods. Moreover, the relative yields of the $(\gamma, 2n)$ to the (γ, n) reactions should throw light on the extent to which the (γ, n) reaction is produced by direct interaction. For such studies, however, it is best to take a heavy element for which the (γ, p) or (γ, α) reactions are known to be infrequent. WILKINSON[1] and his collaborators have estimated the relative cross sections of the (γ, n) and $(\gamma, 2n)$ processes in Ta^{181} at 17.6 and 14 Mev and, from their results, have concluded that the direct process must be predominant. This conclusion, however, rests on the assumption that the ratio of the cross sections of these reactions can be calculated on the basis of the usual level density formula, which, as we have seen in Sect. 30, is inconsistent with the cross sections of $(n, 2n)$ reactions and probably, therefore, suspect as a measure of the competition between reactions. WHALIN and HANSON[2] have performed essentially the same experiment from the threshold of the $(\gamma, 2n)$ reaction to 22 Mev; their results (Fig. 45) are not inconsistent with compound nucleus formation. The accuracy of such measurements is not very high for the $(\gamma, 2n)$ reaction in Ta^{181} produces the long-lived Ta^{179} for which the activity is very difficult to measure. The cross section for the $(\gamma, 2n)$ reaction therefore has been obtained by subtracting the yield of the (γ, n) reaction (obtained by activation of Ta^{180}) from that of the total neutron production. Moreover, the choice of Ta is unfortunate because the discovery[3] of stable Ta^{180} shows that the activation of that isotope cannot account for the whole of the (γ, n) reaction.

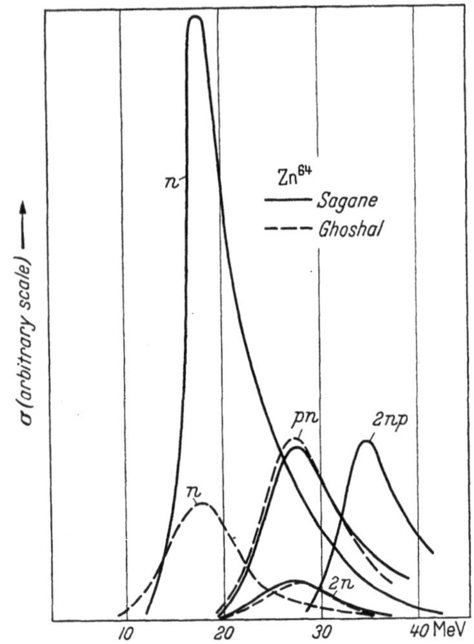

Fig. 50. Relative probabilities of photoreactions from Zn^{64} after SAGANE, and those found by GHOSHAL (Fig. 1).

A more direct and far more revealing experiment has been made by SAGANE[4], who has measured the yields of the active products of the photoreactions in Zn^{64} and Fe^{54}. The cross sections of these reactions in Zn^{64} are shown in Fig. 50, where they are compared with the activities obtained by GHOSHAL[5] (see Fig. 1). It will be seen that for the same energy of excitation the relative cross sections for the (x, pn) and $(x, 2n)$ reactions are the same; only the cross sections of the (x, n) reaction are relatively very different. This is to be expected if the (γ, n)

[1] J. H. CARVER, R. D. EDGE and D. H. WILKINSON: Phil. Mag. **44**, 404 (1953). — J. H. CARVER and H. L. HAY: Phil. Mag. **44**, 1191 (1953).

[2] See ref. 1, p. 283.

[3] F. A. WHITE, T. L. COLLINS and F. M. ROURKE: Phys. Rev. **101**, 1786 (1956).

[4] R. SAGANE: Phys. Rev. **85**, 926 (1952).

[5] See ref. 1, p. 206.

reaction, in the main, is the result of a direct process and the secondary reactions are the products of compound nucleus decay. In fact, Fig. 50 suggests that compound nucleus formation follows direct interaction in only 20% of the photoreactions. The results for Fe^{54} are similar.

54. (γ, α) reactions. Reactions of the (γ, α) type appear to be much less frequent than (γ, n) and (γ, p) reactions. The cross sections, as mentioned in Sect. 47, rise to a peak resembling the giant resonance. The energy of the peak lies higher than the energy of the peaks of the (γ, n) or (γ, p) reactions; this is probably caused by the COULOMB barrier. For Cu^{65} the peak is at 22 Mev (instead of 19 Mev), and for Rb^{87} the peak, according to HASLAM, SMITH, and TAYLOR[1], is at 24 Mev. According to HASLAM and SKARSGARD[2], the width of the peak in the latter reaction is 6.6 Mev, and the integrated yield of the reaction, which starts at 16 Mev, is 4×10^{-4} Mev-barn, which is lower than the yield of the (γ, n) reaction by a factor of the order of 10^4. The general trends of the yields of (γ, α) reactions have been explored by GREENBERG, TAYLOR and HASLAM[3]. They found that the cross section rises from N, and O, to Cu, then falls for heavier elements. For Cu^{63}, BYERLY and STEPHENS[4] found that the α-particle yield was about 1% of that of the (γ, n) reaction and that the energy spectrum was sharply peaked at 8 Mev.

DE LABOULAYE and BEYDON[5] have measured the yield of the Ag^{109} (γ, α) Rh^{105} reaction and found that the yield greatly exceeds that to be expected on statistical grounds. However, it is possible that this result is incorrect for very recently the yields of the (γ, α) reactions have been carefully measured by HEINRICH, WÄFFLER, and WALTER[6], for nine nuclei ranging from V^{51} to Tl^{203} and compared, where possible, with the yields of the (γ, n) reactions, and where this was not feasible (e.g., in V^{51}), against the Cu^{63} (γ, n) Cu^{62} reaction. The results in general are in agreement with that to be expected from statistical calculation although for the heaviest elements they are a little high. For V^{51} the yield of the (γ, α) reaction was comparable with that of the $(\gamma, 2p)$ reaction which suggests that the Rh^{105} activity observed by DE LABOULAYE and BEYDON might be due to the $(\gamma, 2p)$ reaction with Ag^{107}. For the heavy elements the ratio of the yields of the (γ, α) and (γ, n) reactions was about 10^{-4}.

The energy distribution for Br has been measured in a photographic plate by NABHOLTZ, STOLL, and WÄFFLER[7]. Their energy spectrum, which was peaked at 10 Mev, is said to be in agreement with the expectations of theory.

The mechanism of the ejection of α-particles in photoprocesses is not clear. The results of HEINRICH *et al.* indicate that the yield is of the order to be expected from competition within the compound nucleus provided it is assumed that α-particles exist within the nucleus ready to be ejected. But as yet, except possibly in the heavy elements, there is no good evidence that the other particles with which they might be expected to compete, e.g. neutrons and protons, are actually derived from such a compound nucleus. The ejection of tritons in (γ, t) reactions seems to be similar[8].

---

[1] R. N. H. HASLAM, L. A. SMITH and J. G. TAYLOR: Phys. Rev. **84**, 840 (1951).
[2] R. N. H. HASLAM and H. M. SKARSGARD: Phys. Rev. **81**, 479 (1951).
[3] L. H. GREENBERG, J. G. TAYLOR and R. N. H. HASLAM: Phys. Rev. **95**, 1540 (1954).
[4] See ref. 2, p. 287.
[5] H. DE LABOULAYE and J. BEYDON: C. R. Acad. Sci. Paris **239**, 44 (1954).
[6] F. HEINRICH, H. WÄFFLER and M. WALTER: Helv. phys. Acta **29**, 3 (1956).
[7] H. N. NABHOLTZ, P. STOLL and H. WÄFFLER: Phys. Rev. **86**, 1043 (1952).
[8] H. WÄFFLER and F. HEINRICH: Physica, **22**, 1146 (1956).

55. (γ, γ') reactions. Elastic scattering of γ-rays from C and Cu irradiated by bremsstrahlung has been looked for by GAERTTNER and YEATER[1] using a cloud chamber method. The effect, which was not found, must be less than a few percent of the total cross section for resonance absorption. A small residual effect was found for a Pb scatterer by DRESSEL, GOLDHABER, and HANSON[2], who used an activation method for detecting the scattered γ-rays. The effect, if real, corresponds to scattering cross sections of the order of 1 mb per steradian. STEARNS[3] has looked for elastically scattered γ-rays from various materials irradiated with 17 Mev γ-radiation. The differential cross sections, which for Pb and Bi are said to be about 1 mb per steradian, had a strong dependence on the charge number (roughly as $Z^{2.5}$); the inelastic scattering is said to vary even more rapidly with Z. Very recently FULLER and HAYWARD[4] have studied the elastic scattering of a number of nuclei from Na to U when irradiated with bremsstrahlung from 4 to 40 Mev. They used a large NaI crystal and electronic circuits biassed to detect only the radiation from the top of the bremsstrahlung spectrum. When plotted against the energy, the curve of the differential cross section for elastic scattering in heavy elements has two peaks, one at a low energy (below 10 Mev), and another which resembles the giant resonance both in shape and in position. For the lighter elements (up to Ca) the latter is not well developed. The low energy peak is probably caused by competition of γ-rays with neutron emission. At 17 Mev the results are in agreement with those of STEARNS. The cross section corresponding to the giant resonance varies from about 1 mb for Cu to 10 mb in U. The position of the peak varies as $E_m = 82 A^{-\frac{1}{3}}$ (cf. Sect. 47).

Inelastic scattering has been observed by several experimenters by measurement of the activation of a metastable state in the target nucleus. Such measurements, of course, do not measure all the inelastic scattering, for that which produces the ground state without intermediate creation of the metastable state is not usually detected. CAMERON and KATZ[5], have measured the yield of the metastable state in Au^{197}, SIERRA and TELEGDI[6] that in Rh^{103}, and GOLDEMBERG and Katz[7], that in In^{115}. The latter authors estimate that the integrated cross section for the (γ, γ') process in In^{115} is about 0.03 Mev-barn, or about 1% of that of the sum rule limit. The ratio of the yield of the (γ, γ') process to that of the (γ, n) reaction remains very constant above the threshold.

The most interesting feature about these reactions is the energy variation of the cross section. At first sight the latter seem to be peaked in the manner common to photonuclear reactions; the peak energy however is much lower than that of the giant resonance, and seems to be the result of competition with the (γ, n) reaction. For Au^{197}, Rh^{103} and In^{115}, the authors just quoted give peak values at 16, 13, and 9 Mev. More recent work[8,9], indicates that the peak is near the neutron binding energy; for Au^{197} there is another peak corresponding to the giant resonance[8].

56. The COULOMB barrier. The energy spectra of the protons and α-particles emitted in photonuclear reactions are usually found to be peaked at energies

[1] E. R. GAERTTNER and M. L. YEATER: Phys. Rev. **76**, 363 (1949).
[2] R. DRESSEL, M. GOLDHABER and A. O. HANSON: Phys. Rev. **77**, 754 (1950).
[3] M. B. STEARNS: Phys. Rev. **87**, 706 (1952).
[4] E. G. FULLER and E. HAYWARD: Phys. Rev. **101**, 692 (1956).
[5] A. G. W. CAMERON and L. KATZ: Phys. Rev. **84**, 608 (1951).
[6] C. S. DEL RIO Y. SIERRA and V. L. TELEGDI: Phys. Rev. **90**, 439 (1953).
[7] J. GOLDEMBERG and L. KATZ: Phys. Rev. **90**, 308 (1953).
[8] L. MEYER-SCHÜTZMEISTER and V. L. TELEGDI: Phys. Rev. **100**, 961 (1955).
[9] J. L. BURKHARDT, E. J. WINHOLD and T. H. DUPREE: Phys. Rev. **100**, 199 (1955).

appreciably below the usually accepted value for the COULOMB barrier. For example the proton spectrum obtained by TOMS and STEPHENS[1] for Co was peaked at about 5 Mev. If the nuclear radius is given by $R = r_0 A^{\frac{1}{3}}$, with $r_0 = 1.44 \times 10^{-13}$ cm, the barrier height is numerically given by $Z/A^{\frac{1}{3}}$ (see Sect. 7). For Co^{59} this gives $B = 6.9$ Mev. According to TOMS and STEPHENS, their results for Co are in agreement with the calculated energy distribution derived from the usual theoretical parameters for the level densities and corrected for the thickness of the target. There are, as we shall see, some serious discrepancies involved in the level density formulae and the choice of these parameters is to some extent at the discretion of the authors. The available experimental evidence does not point to any very serious discrepancy between the energy spectra obtained experimentally and those calculated theoretically on the basis of the usual barrier heights. However, recently SPICER[2] has measured the energy distributions of the protons emitted by a gaseous argon target. His spectrum, which was peaked at 2.6 Mev, is very difficult to explain on statistical grounds for the calculated barrier height in this nucleus is 5.2 Mev and thickness corrections are not serious.

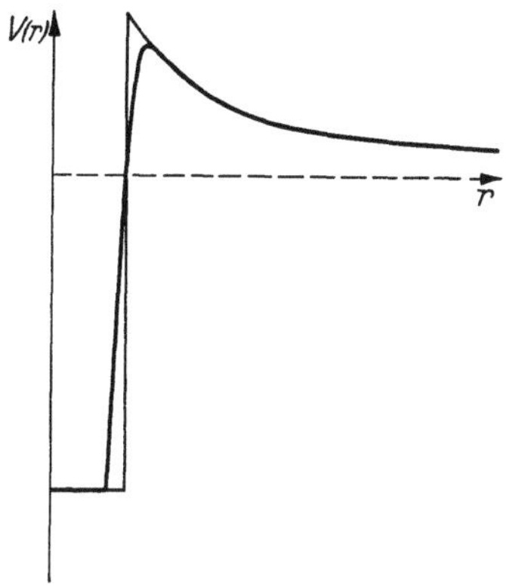

Fig. 51. Reduction of COULOMB barrier for tapered potential well: thin line, square well; heavy line, tapered well.

The apparently low energies of photoprotons have prompted calculations of the barrier height to be expected when the diffuseness of the nuclear surface is taken into account. If the nuclear potential instead of being cutoff abruptly at the surface, is tapered off as described in Sect. 19, the barrier height will be reduced. This is shown in Fig. 51. The effect on the barrier has been examined by SCOTT[3], who found that for tapers of 2×10^{-13} cm considerable reductions in barrier height and corresponding increases in penetrability are to be expected. Except for the single instance just mentioned there seems little evidence for a large reduction in the barrier. The work by BLASER *et al.* (see Sect. 34) shows that the proton reaction cross sections at 7 Mev are correctly given by theoretical formulae provided that r_0 is chosen appropriately. This value of r_0 ($\approx 1.5 \times 10^{-13}$ cm) is somewhat higher than that required to account for the elastic scattering of neutrons; the difference, no doubt, is a measure of the adjustment of the barrier (or rather the equivalent penetrability) required by the tapering of the nuclear potential. Similar results have been obtained by ANDRE *et al.*[4], in measurements on Bi^{209}, up to 10.6 Mev. They found $r_0 = 1.5 \times 10^{-13}$ cm.

For higher energy protons the discovery that (p, pn) reactions have a higher cross section than that expected on the statistical theory has also led to some speculation on the validity of the usual expressions for the barrier height. For higher energies the reaction cross section is much more difficult to measure for more than one reaction may contribute to it. However, in U^{238}, the greater part of the reaction cross section is caused by the fission process, and little error is committed if the reaction and fission cross sections are equated. McCORMICK and COHEN[5] have measured the fission cross sections of U^{238} and Th^{232} for energies

[1] See ref. 3, p. 288.
[2] H. M. SPICER: Phys. Rev. **100**, 791 (1955).
[3] J. M. C. SCOTT: Phil. Mag. **45**, 441 (1954).
[4] See ref. 2, p. 261.
[5] G. H. McCORMICK and B. L. COHEN: Phys. Rev. **96**, 722 (1954).

between 8 and 21.5 Mev. For the former nucleus the cross section is in fair agreement with the reaction cross section calculated on the assumption that $r_0 = (1.55 \pm 0.1) \times 10^{-13}$ cm. There is no evidence here for a reduction of barrier height other than that implied in the nuclear radius. Similar measurements have been made by TEWES[1] by bombardment of Th^{232} with protons with energies from 10 to 31.5 Mev. These results are again in agreement with the theoretical cross sections for $r_0 = 1.5 \times 10^{-13}$ cm. Thus it would appear that the usual cross section formulae are as valid from 10 to 30 Mev as they are at 7 Mev. However, as we have seen in Sect. 8, the cross section at the barrier height and above it is very insensitive to the magnitude of the barrier.

K. Level densities.

57. Theoretical level densities. It is now well understood that an enormous number of levels take part in nuclear reactions; however, as we have pointed out in Sects. 8 and 10, the level density must increase very rapidly with excitation energy. The first theoretical estimate of the level density and its rate of increase with excitation energy was made by BETHE[2] and by OPPENHEIMER and SERBER[3] on the assumption that the constituents of the nucleus behave like free particles and that the nucleus therefore behaves like a gas. The opposite extreme, that the nucleons interact strongly with one another, and that the nucleus as a whole behaves like a liquid drop, was proposed by BOHR and KALCKAR[4]. Quite generally, the level density $\omega(E)$ at energy E, is related to the temperature (Θ) and to the entropy (S) through the relation[5]:

$$\omega(E) = \Theta^{-1}(2\pi\, dE/d\Theta)^{-\frac{1}{2}}\, e^{S}, \tag{57.1}$$

where $dS/dE = 1/\Theta$. If E is supposed to depend on the temperature through the power relation $E = \text{const}\ \Theta^n$, $n = 2$ for the FERMI gas model and $n = \frac{7}{3}$ for the liquid drop model. Substitution of these relations into the expression above shows that the level density must rise exponentially with increasing excitation energy with an argument which differs very little from the one model to the other. If the spin of the levels is disregarded, BETHE's formula for the level density at excitation energy E, is (for the FERMI gas model):

$$\omega(E) = \frac{\sqrt{2}}{12}\,\frac{E_f^{\frac{1}{4}}}{E^{\frac{5}{4}}A^{\frac{1}{4}}}\exp\left[\pi(AE/E_f)^{\frac{1}{2}}\right],$$

where E_f is an average value of the FERMI energy[6]. The level density, however, will depend on the spin. BETHE showed that the density of levels of spin J is given by

$$\omega(E, J) = \omega(E)\left(\frac{dP}{dM}\right)_{J+\frac{1}{2}}$$

where P is the probability of finding a level at excitation energy E with a magnetic quantum number M. $P(M)$ is assumed to be Gaussian in form and depends on the product of the number of nucleons excited and the square of the average

[1] H. A. TEWES and R. A. JAMES: Phys. Rev. **88**, 860 (1952); **98**, 25 (1955).

[2] H. A. BETHE: Phys. Rev. **50**, 332 (1936).

[3] J. R. OPPENHEIMER and R. SERBER: Phys. Rev. **50**, 391 (1936).

[4] See ref. 2, p. 203.

[5] H. A. BETHE: Rev. Mod. Phys. **9**, 69 (1937).

[6] This is defined by $2E_f^{\frac{3}{2}} = E_{fp}^{\frac{3}{2}} + E_{fn}^{\frac{3}{2}}$, where E_{fp} and E_{fn} are the FERMI energies for protons and neutrons respectively. Thus $E_f = (9\pi)^{\frac{2}{3}}\hbar^2/(8Mr_0^2)$ and is inversely proportional to the square of the radius constant.

angular momentum of each particle. In the range of excitation energy in which we are interested only a few particles (n) are excited; the average angular momentum (j) therefore must be derived from the average angular momentum of these particles and that of the vacancies left by them within the FERMI distribution. The probability P is given by the formula:

$$P(M) = [\tfrac{2}{3} \cdot \pi n j (j+1)]^{-\frac{1}{2}} \exp[-3M^2/2nj(j+1)].$$

Calculation shows that the average number of nucleons excited for 8 Mev excitation is of the order of 10. Calculation of the product $nj(j+1)$ shows that the argument of the exponential term is small for small M; equating this term to unity gives the following formula for the density of levels of spin J:

$$\omega(E, J) = \frac{\pi^4 2^{\frac{1}{2}}}{432} \left(\frac{5}{\log 2}\right)^{\frac{3}{2}} (2J+1)\, E_f^{-1}\, x^{-4}\, e^x \tag{57.2}$$

where $x = \pi(AE/E_f)^{\frac{1}{2}}$. Note that the level density for states with a particular spin, in addition to the exponential term, is now proportional to E^{-2}, which has the effect of reducing the level density near the ground state relative to that at higher energies. The main feature of the level density formula, nevertheless, is the exponential term; thus, to a first approximation

$$\omega(E) = \text{const}\; e^{2(aE)^{\frac{1}{2}}} \tag{57.3}$$

with

$$a = A/8.8\ \text{Mev}^{-1} \tag{57.4}$$

if $E_f = 22.0$ Mev, a value which corresponds to $r_0 = 1.51 \times 10^{-13}$ cm. The spacing ($D_J = 1/\omega_J$) of levels of spin J is related to the spacing of levels with zero angular momentum through the formula:

$$D_J = D_0/(2J+1) \tag{57.5}$$

with

$$D_0 = 4.1 \times 10^6\, x^4\, e^{-x}\ \text{ev}. \tag{57.6}$$

As we shall see below there is little experimental data on the level densities of nuclei except in the immediate vicinity of the neutron binding energy. For this energy the expression (57.6) is in good agreement with the data.

A very similar formula was found by VAN LIER and UHLENBECK[1] who took into account the separate identities of neutrons and protons. They calculated the level density in terms of the spacing d of individual particle levels at the top of the FERMI distribution. For one kind of particle the level density is:

$$\omega(E) = \frac{1}{\sqrt{48}\, E} \cdot \exp[\pi(2E/3d)^{\frac{1}{2}}]. \tag{57.7}$$

When both kinds of particles are taken into account and when d is evaluated, their formula reduced to BETHE's (57.2). A level density formula of the same form as BETHE's was obtained by SNEDDON and TOUSCHEK[2] using a quite different method of calculation. The numerical coefficients, however, are not the same and this formula leads to rather lower spacings. None of these calculations take into account factors other than energy and spin. Recently, however, a more elaborate calculation has been made by BLOCH[3].

---

[1] C. VAN LIER and G. E. UHLENBECK: Physica **4**, 531 (1937).
[2] I. N. SNEDDON and B. F. TOUSCHEK: Proc. Cambridge Phil. Soc. **44**, 391 (1947).
[3] C. BLOCH: Phys. Rev. **93**, 1094 (1954).

The density of states for the liquid drop model has been calculated by BETHE[1]. A phenomenological treatment has been given by LANG and LECOUTEUR[2]; the presence of surface oscillations, generally, is to reduce somewhat the dependence of the density on the atomic weight (see Sect. 59).

Experimental data on level densities is very scanty. At low excitation energies (1 Mev) the levels excited by radioactivity are usually determined by strong selection rules; consequently deductions based on such evidence may have little relation to reality. It is difficult to make any estimate of the density of levels of a specific spin in this region. Just above the neutron binding energy we have a clear but again incomplete picture of the level density from the resonances excited by slow neutrons. For excitations between the ground state and the neutron binding energy there is no direct evidence for level densities at all; for higher energies there is some evidence to be obtained from the spectrum of the particles emitted in nuclear reactions. It will be seen that in general this meagre data, which we shall now discuss, is in fair agreement with theoretical predictions.

58. Level densities revealed by slow neutrons. The numbers of neutron resonances which can be identified with time of flight methods is a measure of the level density near the neutron binding energy[3]. In some nuclei as many as a hundred levels have been identified within a 100 eV interval. In such cases therefore, the level density is known to an accuracy of about 10%; in most cases, however, the statistical accuracy of such estimates is much lower. Since s-waves are responsible for the excitation of these resonances, such estimates refer only to levels with the same parity as the ground state of the target nucleus. If I is the spin of the target nucleus in its ground state the spins of the levels which we can detect in this way are $I \pm \frac{1}{2}$, and the observed level spacing (D') is compounded of the separate levels spacings of these two states. If the spin dependence of the level spacing is correctly given by (57.5), the spacing of states with zero angular momentum (D_0) is given by:

$$1/D' = 1/D_{I+\frac{1}{2}} + 1/D_{I-\frac{1}{2}}$$

and

$$D_0 = 2(2I+1)\,D'. \tag{58.1}$$

Experimental values for this quantity[4] are plotted in Fig. 52 where the solid line represents BETHE's formula (57.6) for an excitation energy of 7 Mev (with $E_f = 22$ Mev), and the dashed line LANG and LECOUTEUR's semi-emipirical formula (59.2). It will be seen that both the general trend of the level density and the absolute values are well represented by either formula. It should be noted that the closed shell in the Pb region disturbs the trend of the level densities for atomic weights near 200. For the heaviest nuclei, the observed spacings are greater than theoretical predictions. This discrepancy, however, has little significance for the theoretical curve depends rather critically on the FERMI energy, which in turn depends on the nuclear radius constant, r_0.

The general agreement between theory and experiment is satisfactory. However, no more than this should be expected for the theory as outlined above can predict only general trends; it treats all nucleons on an equal footing and makes no allowance for the profound influence of shell structure. The effect of one

---

[1] See ref. 5, p. 296.

[2] J. M. B. LANG and K. J. LECOUTEUR: Proc. Phys. Soc. Lond. A **67**, 586 (1954).

[3] J. A. HARVEY, D. J. HUGHES, R. S. CARTER and V. E. PILCHER: Phys. Rev. **99**, 10 (1955).

[4] For even mass number target nuclei, D_0 measures the spacings of states of spin $\frac{1}{2}$, and may be calculated from (58.1) except when $I=0$, when $D_0 = D'$.

closed shell (the double closed shell at Pb^{208}) is clearly shown in Fig. 52, but the data from which this Figure is plotted is insufficient to show up the others (for neutron and proton numbers equal to 20, 50, and 82). The effects of these closed shells make their appearance in fast neutron capture cross sections which are a rough measure of the level density (see Sect. 61). It is perhaps surprising that the influence of shell structure is so strong at excitation energies of the order of 7 Mev; but it must be remembered that even at this excitation energy relatively few nucleons are disturbed and their arrangement must depend markedly on the details of their configurations.

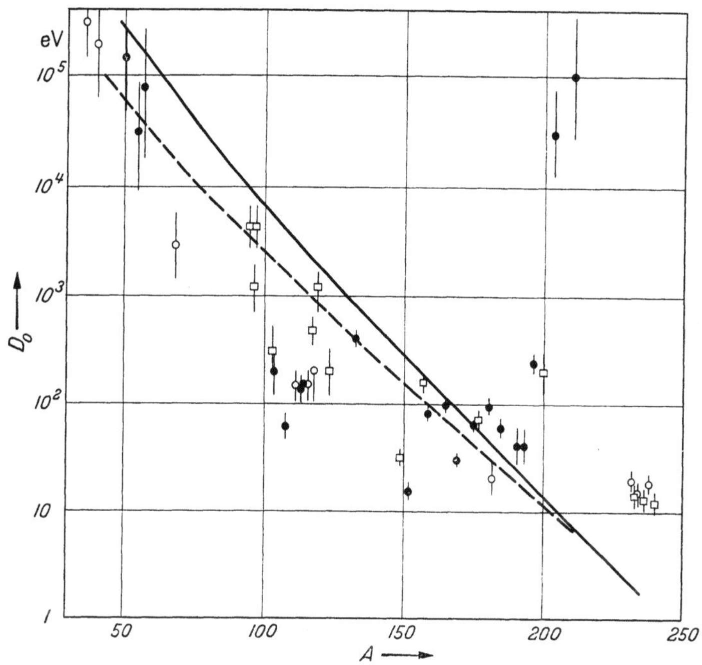

Fig. 52. Level spacings (D_0) as a function of atomic weight, after LEVIN and HUGHES. The full curve corresponds to BETHE's formula (for 7 Mev) the dashed curve is that of LANG and LE COUTEUR. Full circles: odd-odd nuclei; open circles, even-odd nuclei; open squares, even-even nuclei.

In addition to its dependence on shell structure the level density depends also on the oddness and the evenness of the neutron and proton numbers. Fig. 52 shows a tendency for the spacing in odd-odd nuclei to be lower than that of even charge nuclei; the effect is more obvious in the thermal neutron capture cross sections (which, when averaged are roughly proportional to the level density) and in resonance absorption integrals[1], which also depend on this quantity. It is not definitely established that the level densities in odd-odd nuclei are greater than those in even-charge nuclei at all excitation energies but there is good reason to believe that this is indeed the case, for as we have seen (Sect. 28 and 29) this difference appears to be demonstrated in the competition between nuclear reactions and it is apparent also in the neutron capture γ-ray spectra from intermediate elements. In fact, BROWN and MUIRHEAD[2] have concluded, from a comprehensive analysis of the data, that the level densities in odd-odd nuclei,

[1] S. P. HARRIS, C. O. MUEHLHAUSE and G. E. THOMAS: Phys. Rev. **79**, 11 (1950). — H. HURWITZ and H. A. BETHE: Phys. Rev. **81**, 898 (1951).

[2] See ref. 3, p. 223.

on average, are about 12 times greater than those of even-even nuclei, and twice those of even-odd nuclei, for the same excitation energy.

The even-even nuclei have higher neutron binding energies than their neighbours, while the spacing at the neutron binding energy is comparable to that of odd-odd nuclei; this has led to the suggestion[1] that the rate of increase of level density in neighbouring nuclei is similar but with the difference that the level system in even-even nuclei starts at an excitation energy roughly equal to the pairing energy (≈ 1.5 Mev). If this is so the neutron capture γ-ray spectra should differ very little except at their highest energies. We shall see that the shapes of these spectra are not sensitive to the level density; but they do contain definite indications that the normal development of the level system in even-even nuclei is delayed until an energy of the order of the pairing energy is reached. There is confirmation of this in the behaviour of the radiation widths of neutron resonances (Sect. 65). According to NEWTON[2], in a very recent analysis of the experimental data, the dependence of the level spacing on the neutron and proton number, in nuclei far removed from closed shells, can be very largely accounted for when the pairing energy is taken into account. He assumes that the term representing the excitation energy (E) in the theoretical formula should be replaced by $E - c\delta$, where δ is the pairing energy, i.e. the energy by which the ground state is depressed by pairing of the last nucleons of either type; $c = 1$, for even-even nuclei; $c = \frac{1}{2}$, for even-odd and odd-even nuclei; and $c = 0$ for odd-odd nuclei. The pairing energy is difficult to determine from existing experimental data, but roughly seems to be given by the formula $\delta = 22\,A^{-\frac{1}{2}}$ Mev.

NEWTON has also found it possible to account for the effects of shell structure. It will have been noted that the argument of the exponential term in the level density formula is proportional to the square root of the density of single particle states [cf. (57.7)]. NEWTON assumes that the density of these states at the binding energy is proportional to the sum of the statistical weights of the total angular momenta (j_n or j_p) of the neutron or proton in the top occupied levels of the currently accepted shell schemes (see Sect. 85). With some smoothing of these quantities, he finds good agreement between the experimental level spacings and those calculated on the basis of a level density in which the exponential term is now proportional to $(j_n + j_p + 1)^{\frac{1}{2}} A^{\frac{1}{3}} E^{\frac{1}{2}}$.

59. Level densities from temperature measurements. We have already remarked (Sect. 8) that the temperature of the residual nucleus may be obtained from the slope of the energy spectrum of the particles emitted. The temperature thus obtained should be a measure of the level density; thus:

$$\frac{1}{\Theta} = \frac{d}{dE} \log \omega(E) = \frac{d}{dE} \log \frac{n(\varepsilon)}{\varepsilon\sigma}$$

in which σ is the cross section for the reverse process [cf. Eq. (8.6)]. Temperature measurements of this kind have been made by GRAVES and ROSEN for the (n, n') reaction at 14 Mev and by GUGELOT for the (p, n) and (p, p') reactions with 16 and 18 Mev protons (Sects. 34 and 35). However, the measured neutron spectra must contain contributions from the $(n, 2n)$ and $(p, 2n)$ reactions. If calculated corrections[3] are made for these contributions, the temperature (Θ) may be used to determine the constant a in the level density formula $\exp 2(aE)^{\frac{1}{2}}$ (57.3) for the temperature Θ is equal to $(E/a)^{\frac{1}{2}}$. Such corrections however, must be treated

[1] See ref. 1, p. 299.
[2] T. D. NEWTON: Canad. J. Phys. **34**, 804 (1956).
[3] A. TOMASINI: Nuovo Cim. **12**, 134 (1954).

with reserve for they rest on a purely theoretical basis. The results for a, although somewhat higher, are in fair agreement with BETHE's formula, for which $a = A/8.8\ \text{Mev}^{-1}$ (57.4).

The experimental data has been analysed by LANG and LECOUTEUR[1]. Taking into account the data just referred to and that obtained from the evaporation of neutrons from targets bombarded with high energy particles, they conclude that to a first approximation the temperature is indeed proportional to the square root of excitation energy and that it is determined by the formula $E = A\Theta^2/k$, where $k = 10.5$ Mev, and A is the atomic weight[2]. A more accurate definition of the temperature (t) is given by (57.1) which is equivalent to:

$$\frac{d}{dE}\log\omega = \frac{1}{t} - \frac{1}{2}\,\frac{d}{dE}\log\left(t^2\,\frac{dE}{dt}\right)$$

and since $E \propto \Theta^2$ the relation between t and Θ is

$$t = \Theta - \frac{3\Theta^2}{4E}.$$

These authors show that the proper equation of state for a FERMI gas is $E = (A t^2/f) - t$, where f is a constant determined from the data ($f = 8$ Mev). This leads to the following expression for the level spacing:

$$D_0(A, E) = 0.054\, b^3 f^{\frac{1}{2}} A^2 (E + t)^2 \exp\left[-2\left(\frac{AE}{f}\right)^{\frac{1}{2}}\right] \tag{59.1}$$

where b^2 is a measure of the reduction of the moment of inertia below that of a rigid sphere, caused by the interactions between the nucleons. The quantity b was determined from the capture cross section data for neutrons of 1 Mev (Sect. 61) and found to be $b \approx 0.55$. This is in rough agreement with that obtained from the ratio of the deformations calculated from the energies of rotational states and from $E2$ transition probabilities (Sect. 78).

If the interactions between the nucleons are taken into account, surface oscillations may be present. In this case, the energy of the nucleus should be increased by the term $\frac{1}{8} A^{\frac{2}{3}} t^{\frac{7}{3}}$, and the form of the level spacing expression is somewhat different. LANG and LECOUTEUR conclude that the data generally is in better agreement with the assumption that surface oscillations exist and deduce the following formula for the spacing of states with zero angular momentum:

$$D_0(A, E) = 0.054\, b^3 \frac{f^{\frac{3}{2}}}{f'} A^2 (E + t)^2 \exp\left[-2\left(\frac{AE}{f'}\right)^{\frac{1}{2}} - \frac{3}{32}(E f')^{\frac{2}{3}}\right] \tag{59.2}$$

where $f' = 11$ Mev. This expression, which is plotted as the dotted line in Fig. 52, gives quite good agreement with the level spacings at the neutron binding energy.

For low energy particles (up to 20 Mev) it has been claimed[3] that the temperatures derived from the energy spectra do not increase with excitation energy, in contradiction to the findings of LANG and LECOUTEUR. Moreover, the temperatures deduced from α-particle reactions seem higher than those produced by protons

---

[1] See ref. 2, p. 298.

[2] BETHE's formula gives $k = 8.8$ (57.4) if r_0 is assumed to be 1.51×10^{-13}. The value $r_0 = 1.33 \times 10^{-13}$ cm, which corresponds to the radius of the tapered potential well (Sect. 19) gives $E_f = 28$ Mev and $k = 11.3$ in better agreement with LANG and LECOUTEUR's findings. This value fits the experimental data for the level spacings rather better than the results for $E_f = 22$ Mev which are plotted in Fig. 52. The value $r_0 = 1.20 \times 10^{-13}$ cm, which corresponds better to the actual distribution of charge and mass, gives level spacings from 10 to 100 times greater than is observed.

[3] B. L. COHEN: Phys. Rev. 92, 1245 (1953).

and neutrons of similar energies. This has led to the suggestion by BEARD[1] that there are abrupt changes in the numbers of nucleons which participate in nuclear reactions; others[2] have suggested that the nucleus undergoes a "phase" change at low energies which is responsible for temperatures independent of excitation energy. Experimental data, at the present time, however, seems hardly sufficient to support such an hypothesis. Although no measurements have yet been undertaken to test directly the dependence of temperature on excitation energy (as determined from the energy spectrum of the products of reaction), the most recent experimental results on the energy spectrum of inelastically scattered neutrons, suggest that the temperature is indeed proportional to the square root of the energy[3].

The level density formulae which have just been discussed seem to give approximately the correct energy spectra for the particles emitted. However, they do not give an adequate description of competing reactions. In the 10 to 20 Mev range we have seen (Sects. 10 and 30) that a simple formula (8.8) for the relative cross sections of reactions of the type (x, n) and $(x, 2n)$ gives an adequate account of the experimental data provided the temperatures lie between 2 and 3 Mev. This greatly exceeds the temperatures obtained from the energy distributions of the reaction products; the latter temperatures require that the shift in the content of the reaction cross section from the (x, n) to the $(x, 2n)$ process should occur much more rapidly than appears. To some extent the sharpness of the rise of the $(x, 2n)$ cross section will be blunted by proton competition but it does not seem possible that this can account for more than a small amount of the effect observed. Thus it seems that the competition between nuclear reactions is determined at least in part by factors other than those outlined in Sect. 8. If the temperatures obtained from competition are used to derive the constant a in the level density formula, we obtain results in serious disagreement with theoretical predictions[4].

Although the level density formulae give qualitatively the correct energy spectrum for reaction products, there are many discrepancies as we have seen in Sect. 36 and 37. Recent measurements[5] of the energy spectra of the protons emitted in (α, p) reaction with 40 Mev α-particles indicate that the statistical theory is in better agreement with the data when α-particles rather than protons are used as projectiles. This is perhaps to be expected if the α-particle can be regarded as merely a device to convey four nucleons into the target nucleus, each with one quarter of the kinetic energy of the α-particle.

L. Slow neutron reactions.

60. Slow neutron reactions: general. An elementary extrapolation of the parameters of slow neutron resonances in heavy elements hows that for energies up to 2 Mev and more the levels of the compound nucleus are discrete and do not overlap to any great extent. At low energies, then, there can be little doubt that the interaction of a neutron with a nucleus results in the formation of a compound nucleus and the decay of this nucleus does not reflect its mode of

[1] D. B. BEARD: Phys. Rev. **94**, 738 (1954).

[2] D. L. LIVESEY: Canad. J. Phys. **33**, 391 (1955). — L. E. H. TRAINOR and W. R. DIXON: Canad. J. Phys. **34**, 229 (1956). — V. F. WEISSKOPF: Proc. Amer. Acad. Arts. Sci. **82**, No. 7, 360 (1956).

[3] See ref. 3, p. 246; the measured temperatures are all consistent with $\Theta^2 = kE/A$ with $k \approx 9$, as predicted by BETHE's formula (57.4), provided only heavy nuclei are considered.

[4] G. IGO and H. E. WEGNER: Phys. Rev. **102**, 1364 (1956).

[5] R. M. EISBERG, G. IGO and H. E. WEGNER: Phys. Rev. **100**, 1309 (1955).

formation. In this energy range (up to about 2 Mev) the emission of radiation and inelastic scattering are the only processes which occur to any appreciable extent. Reactions of the (n, p) and (n, α) type although exoergic are very greatly retarded by the COULOMB barrier. Some of these reactions have been detected by photographic plates[1].

61. Capture cross sections. α) *Theoretical.* Except at very low energies the probability of elastic or inelastic re-emission of a neutron is always greater than the probability for the emission of radiation. Thus radiative capture cross sections are appreciable only at low energies. For the heavy elements, excepting the closed shell nuclei such as Pb and Bi, the capture cross sections are usually near 0.5 barn at 100 kev; they fall to near 100 mb at 1 Mev, and at an increasing rate for neutrons with higher energies. At 3 Mev such capture cross sections as have been measured are near 25 mb; very few have been measured near 5 Mev. These cross sections are of great interest for they are related to the radiation width of neutron resonances and to the level densities.

A simple expression may be readily obtained for the capture cross section. for s-wave neutrons. To a first approximation the capture cross section is given by the one-level BREIT-WIGNER formula:

$$\sigma(\varepsilon) = \pi \lambda\!\!\!^{-2} g \frac{\Gamma_n \Gamma_\gamma}{(\varepsilon - \varepsilon_r)^2 + \frac{\Gamma^2}{4}} \tag{61.1}$$

where Γ is the total width of the resonance; it is the sum of the radiation width (Γ_γ) and the neutron width (Γ_n). Here ε_r is the resonance energy and g is the statistical weight factor for s-wave neutrons, viz:

$$g = \frac{1}{2}\left(1 \pm \frac{1}{2I+1}\right)$$

where I is the spin of the target nucleus. For many elements, $g \approx \frac{1}{2}$. Integrating (61.1) over the resonance, gives:

$$\int \sigma(\varepsilon)\, d\varepsilon = 2\pi^2 \lambda\!\!\!^{-2} g \frac{\Gamma_n \Gamma_\gamma}{\Gamma}. \tag{61.2}$$

Consider now the effective cross section for a neutron beam with an energy spread $\Delta\varepsilon$. If D is the mean level spacing, the number of levels in this range of energy is $\Delta\varepsilon/D$. The average capture cross section $\langle\sigma_a\rangle$, then, is $\Delta\varepsilon/D$ times the average of the integrated cross section (61.2) divided by the energy interval $\Delta\varepsilon$. Thus:

$$\langle\sigma_a\rangle = 2\pi^2 \lambda\!\!\!^{-2} \langle g\Gamma_n\Gamma_\gamma/\Gamma D\rangle. \tag{61.3}$$

The weighted mean value[2] of g is near $\frac{1}{2}$, and for most purposes it is sufficient to put $g = \frac{1}{2}$. The energy dependence of the quantity in parenthesis in (61.3) is determined in part by the relative values of Γ_n and Γ_γ for $\Gamma_n + \Gamma_\gamma = \Gamma$. At low energies, in heavy elements, $\Gamma_n \ll \Gamma_\gamma$; then:

$$\langle\sigma_a\rangle = 2\pi^2 \lambda\!\!\!^{-2} \left\langle\frac{g\Gamma_n}{D}\right\rangle. \tag{61.4}$$

Since $\lambda\!\!\!^{-2} \propto 1/\varepsilon$ and the neutron width is proportional to the neutron velocity, $\langle\sigma_a\rangle$ must be proportional to $\varepsilon^{-\frac{1}{2}}$, i.e. $\langle\sigma_a\rangle$ is proportional to the reciprocal of

[1] Reactions of the (n, p) and (n, α) type for thermal neutrons have been studied in nuclei of intermediate atomic weight by H. HÄNNI and J. ROSSEL [Helv. phys. Acta **25**, 521 (1952)] and by H. FARAGGI and R. BERNAS [C. R. Acad. Sci. Paris **234**, 1684 (1952); **235**, 425 (1952); J. de Phys. **14**, 160 (1955)].

[2] The weighted mean value of g is $\frac{1}{2}[1 + 1/(2I+1)^2]$.

the velocity. At higher energies, Γ_n will be greater than Γ_γ and then (61.4) reduces to:

$$\langle\sigma_a\rangle = 2\pi^2 \lambda\!\!\!^{-2} \left\langle \frac{g\Gamma_\gamma}{D} \right\rangle \tag{61.5}$$

for which the energy dependence is $1/\varepsilon$. Thus $\langle\sigma_a\rangle$ increases with increasing level density $(1/D)$. However this cross section is not accurately proportional to $1/D$ for (as we shall show in Sect. 65) Γ_γ increases slowly with D. Nor is the cross section really proportional to $1/\varepsilon$, because the capture of neutrons from waves with angular momenta higher than that of s-waves can have a pronounced effect (see δ below).

β) *Experimental data.* All measurements of fast neutron capture cross sections, so far, have made use of the radioactivity of the product nucleus and only such targets have been studied which lead to activities with convenient periods. The earliest measurements are those of HALBAN and KOWARSKI[1] (1938). These authors used a photoneutron source consisting of an active deposit of ThC″ contained in heavy water (200 kev neutrons). GRIFFITHS[2], and DEMENTI and TIMOSHUK[3], used a photoneutron source consisting of radium enclosed in beryllium. Such a source produces neutrons of many different energies, but, in the main, neutrons with an energy of about 90 kev. In both cases, the contribution of thermal neutrons to the activities observed was minimised by enclosing in cadmium foil the material for which the cross sections were to be measured. Measurements have been made for 2.7 Mev neutrons by MESCHERYAKOV[4], and more recently, for fission neutrons (roughly 0.5 Mev) by HUGHES and his collaborators[5].

ALLEN *et al.*[6] have attempted to measure the rate of change of the neutron capture cross section with energy by comparing the relative activities produced in the same targets by the 200 kev neutrons from a ThC″—D_2O source with the 950 kev neutrons from a ThC″—Be source. The results of this measurement should be independent of errors inherent in absolute β-counting. They studied Mn^{55}, Cu^{63}, Cu^{65}, As^{75}, Br^{79} (4 hr and 18 min), Ag^{107}, Ag^{109}, Rh^{103}, In^{115} (54 min) Sb^{121}, I^{127}, W^{186}, and Au^{197}. In all these targets, except In, the activity produced at the higher neutron energy was the lower. In In, the 54 min activity is now known to be caused by the isomeric state; their measurements, therefore, do not include the activation of the ground state and the difference between the activities found at the two neutron energies may depend on changing relative contributions of s and p-waves.

A very comprehensive investigation has been made recently by HUGHES and his collaborators using fission neutrons obtained by absorbing thermal neutrons from a reactor in an uranium plate. These neutrons have an energy of about 1 Mev. In their first measurements, HUGHES, SPATZ, and GOLDSTEIN[7], measured the activities produced in two discs of the material to be examined, one bare, and the other covered tightly with a thick layer of cadmium, located on either side of the uranium plate. The capture cross section was deduced from the ratio of the two activities; from the capture cross section of the material to thermal neutrons; from the calculated ratio of flux of fast neutrons produced by the uranium plate to the flux of thermal neutrons falling on it; and finally, from the flux of thermal neutrons falling on it. These authors have collected the earlier

[1] H. HALBAN and L. KOWARSKI: Nature, Lond. **142**, 392 (1938).
[2] J. H. E. GRIFFITHS: Proc. Roy. Soc. Lond., Ser. A **170**, 513 (1939).
[3] V. S. DEMENTI and D. V. TIMOSHUK: C. R. Acad. Sci. URSS. **27**, 929 (1940).
[4] M. G. MESCHERYAKOV: C. R. Acad. Sci. URSS. **48**, 555 (1943).
[5] See ref. 7 on this page; and ref. 1 and 2, p. 305.
[6] R. ALLEN, G. R. BISHOP, P. DEMERS and H. HALBAN: Nature, **161**, 727 (1948).
[7] D. J. HUGHES, W. D. SPATZ and N. GOLDSTEIN: Phys. Rev. **75**, 1781 (1948).

data, and compared them one against another. These results are quite remarkably inconsistent: for example, it appears that the results of HALBAN and KOWARSKI for Au require to be multiplied by a factor of 17 to make them fit those of HUGHES, SPATZ and GOLDSTEIN, while those of MESCHERYAKOV and of GRIFFITHS, require to be multiplied by factors of 0.13 and 4.9 respectively. The inconsistencies throw doubt on the usefulness of the experimental methods which have been used in this work, and, in particular, on the neutron energies which are supposed to be responsible for the activations observed. Very roughly, the experimental data confirms the predicted $1/\varepsilon$ energy dependence.

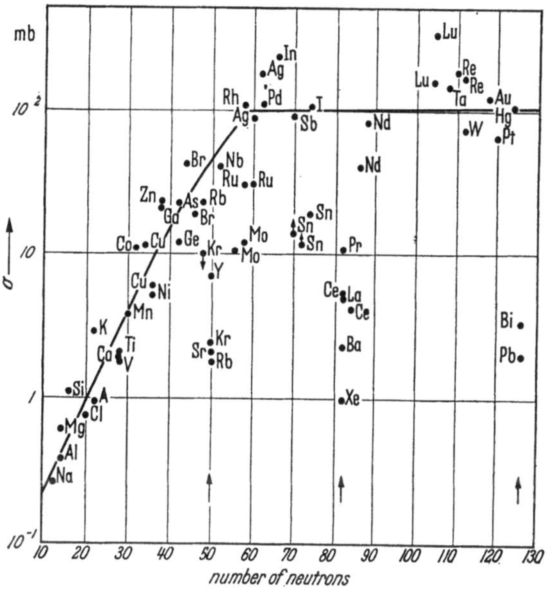

Fig. 53. Capture cross sections for 1 Mev neutrons plotted against neutron number according to HUGHES, GARTH and LEVIN.

γ) Level densities and closed shells. The measurements with fission neutrons have been continued by HUGHES and SHERMAN[1], and by HUGHES, GARTH and LEVIN[2]. These cross sections, plotted as a function of neutron number are shown in Fig. 53. They rise from 0.36 mb for Al, to near 100 mb at Ag. For heavier nuclei, the cross section, in the main, remains constant at about this level. There are certain exceptions where the cross section is much lower than that expected from the general trend. These exceptions occur at, or near, the closed shells and reflect a reduction in level density in these regions. (Similar results have been obtained by NEWSON and ROHRER[3] by a boron absorption method.) When plotted against the neutron number, HUGHES' cross sections are lowest for neutron numbers near 50, 82 and 126. Pb and Bi, for example, have the very low cross sections 2.0 and 3.4 mb respectively. The cross sections, when plotted against the proton number, are also lowest at the closed proton shells, e.g. the Sn isotopes have low cross sections, the proton shell of 50 protons being closed in this element. Assuming the validity of (61.5), HUGHES, GARTH and LEVIN have attempted to calculate the general trend of the level spacing. The result, which is shown in Fig. 54 agrees well with that obtained from the usual level density formula, viz., $\exp 2(aE)^{\frac{1}{2}}$, the value of E being taken to be the mean binding energy, 7 Mev, plus 1 Mev for the kinetic energy of the incident neutrons.

Similar results to those obtained by HUGHES *et al.* have been obtained by ROSE[4] by measurement of the change in the reactivity of a fast reactor when samples of the material to be studied are introduced into it. The change in reactivity, when compared with that produced by a standard sample, very roughly

[1] D. J. HUGHES and D. SHERMAN: Phys. Rev. **78**, 632 (1950).
[2] D. J. HUGHES, R. C. GARTH and J. LEVIN: Phys. Rev. **91**, 1423 (1953).
[3] H. W. NEWSON and R. H. ROHRER: Phys. Rev. **94**, 654 (1954).
[4] H. ROSE: To be published.

measures the capture cross section. The results obtained vary rather more smoothly with atomic weight than those of HUGHES *et al.* and correspond to a somewhat lower neutron energy. The rapid fall in the capture cross sections for the closed shell nuclei was very obvious; it is the more distinct when the cross sections of separated or enriched isotopes were used and the more marked for closed neutron than for closed proton shells. The distinction between odd-odd and even-even or even-odd nuclei, however, was not very distinct, although it was clear that in general the capture cross sections in odd Z nuclei are the higher. An interesting feature of these results is the marked influence of the double closed shell at Pb^{208} which is felt in nuclei so far removed as W and Au. This effect is also apparent in the capture γ-ray spectra (Sect. 62).

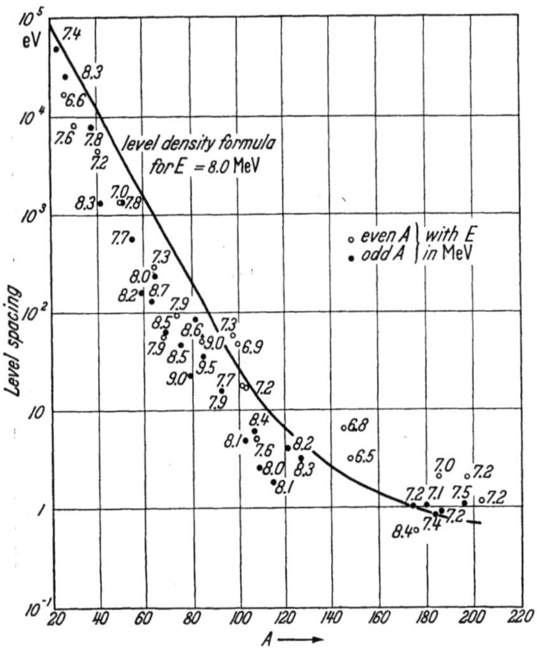

Fig. 54. Level densities derived from neutron capture cross sections at 1 Mev, according to HUGHES, GARTH, and LEVIN. The full line is the theoretical level spacing for 8 Mev excitation. The numbers refer to the excitation energy for specific nuclei.

The results of ROSE and of HUGHES *et al.* show that the level density, for an excitation energy equal to the neutron binding energy, decreases as we approach a closed shell from the direction of smaller neutron numbers, and increases again beyond it. HARVEY[1] has shown that the neutron binding energies for mass numbers just below and including the closed shell nucleus are higher than average, and those for mass numbers just above the closed shell nucleus are lower than average. The capture cross section measurements show that there is indeed a genuine reduction of level densities just short of the closed shell, but it is not obvious from these measurements that the reduction above the closed shell is not simply the result of reduced neutron binding energy. However it is probable that the effect observed above the closed shell is really of the same kind as that observed below, for the reduction in level densities of low lying states, as shown by the position of the first excited state in even-even nuclei, seems to be symmetrical about the closed shell nucleus.

Capture cross section measurements interpreted by (61.5) can give only a very rough measure of the level density. Many nuclei bombarded by 1 Mev neutrons must be excited by inelastic collisions and the effect of this process is not taken into account in (61.5). Moreover neutrons with higher angular momenta must make a contribution to the capture cross sections for energies above 100 kev. An attempt at a more detailed analysis of the energy variation of the capture cross section has been made by MARGOLIS[2] on the basis of the statistical theory and unpublished measurements of the capture cross sections of neutrons of various

[1] See ref. 1, p. 271.
[2] B. MARGOLIS: Phys. Rev. **88**, 327 (1952).

energies in Ag, In, and Au, reported by HUGHES and HARVEY[1]. His results, in which account is taken of the position, spins and parities of the energy levels in these nuclei, are in good agreement with experiment.

δ) Capture cross sections in U^{238} *and* Th^{232}. Recently the capture cross sections of U^{238} and Th^{232} have been measured as a function of neutron energy by ROSE[2] and by HANNA[2] respectively. These measurements are of particular interest because only one low lying excited state can be excited by neutrons with energies up to 700 kev (see Sect. 25) and the theoretical interpretation is therefore greatly simplified. ROSE's results for U^{238} are plotted in Fig. 55.

It will be noted that not only does the capture cross section not fall off with increasing energy according to $1/\varepsilon$ as (61.5) predicts, but it actually rises just before the onset of inelastic scattering near 700 kev. At this point, the total

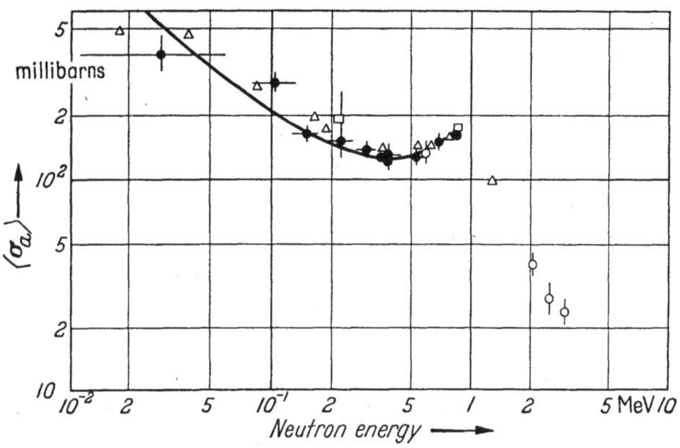

Fig. 55. Capture cross section of U^{238}. Full circles: ROSE's results; for references to other points, see ref. 2, this page. Solid line: cross section calculated by LANE and LYNN.

neutron width must increase rapidly causing the capture cross section to decrease. Both the magnitude of the cross section below this point and the shape of the curve can be understood when due provision is made for capture of neutrons by waves of higher orbital angular momentum (l). For neutron energies above the energy of the low lying excited state, the ratio Γ_n/Γ is no longer unity because that state is excited by inelastic scattering. According to LANE and LYNN[3] the cross section is given by

$$\langle\sigma_a\rangle = \pi^2 \lambda\!\!\!^{-2} \sum_J (2J+1) \sum_{l=|J-\frac{1}{2}|}^{l=J+\frac{1}{2}} \frac{1}{\langle D_J\rangle} \times \left\langle \frac{\Gamma_n^{(lJ)}\Gamma_\gamma}{\Gamma_n^{(lJ)} + \Gamma_{n'}^{(l'J')} + \cdots \Gamma_\gamma} \right\rangle. \qquad (61.6)$$

In this decrease in total width of a state of spin J is given by the denominator of the term on the right and depends on the orbital angular momentum of the ingoing neutron (l) and also on the angular momentum (l') of the inelastically scattered neutron. The term on the right is summed over all possible values of l and l' consistent with the spin J of the compound state, and finally over all possible values of J. LANE and LYNN have shown that this formula (see Fig. 55) gives cross sections in good agreement with experiment for both U^{238} and Th^{232} provided that the level density ($1/D$) is assumed to rise with the excitation energy

[1] See ref. 1, p. 257.
[2] B. ROSE: To be published. — R. HANNA: To be published.
[3] A. M. LANE and E. LYNN: To be published.

according to LANG and LECOUTEUR's formula (59.2) and provided that the radiation width and level density for states with spin $\frac{1}{2}$ are those obtained by HUGHES and his collaborators from the study of slow neutron resonances.

It should be noted that the contribution of waves of high angular momentum, in effect, are weighted proportionally to $(2l+1)^2$; they make therefore a very significant contribution. The good agreement between theory and experiment is evidence that the level density does indeed depend on the spin through the factor $(2J+1)$ (see Sect. 57).

The decrease in capture cross sections with energies above 1 Mev is less easy to account for quantitatively. In addition to the factors already considered, the radiation width must decrease as the neutron energy is increased because the part of the radiation width with which we are concerned here is only that part which leads to bound states. For high energies, say above 5 Mev, the fall in the radiation width due to this cause (it is roughly exponential) may be offset by the increase in matrix elements due to the giant resonance (see Sect. 33).

62. Neutron capture γ-rays. For practical reasons the neutron capture γ-rays which have been studied in most detail are those which are emitted when thermal neutrons are captured. For neutrons of higher energies, the cross sections rapidly become much lower and only very recently have attempts been made to study the γ-rays emitted when neutrons are captured at specific resonances.

The most successful experiments on neutron capture γ-rays have made use of a nuclear reactor as a source of neutrons. Generally two methods of studying the γ-rays have been employed: either a neutron beam has been allowed to pass out of the radiation shield of the reactor and impinge on the target to be studied outside this shield, or the target itself has been placed in the region of high thermal neutron flux and the γ-ray beam has been permitted to emerge. The earliest measurements of the high energy γ-rays have made use of this latter arrangement and the γ-rays were examined with a magnetic pair spectrometer. The use of this instrument[1] and a summary of the results obtained with it is described elsewhere [*19*]. This instrument has the disadvantage that the sensitivity increases very rapidly as a function of γ-ray energy. Recently, measurements have been made with the same general arrangement but with a magnetic spectrometer in which COMPTON electrons projected forward are measured with coincidence counters[2]. This method, which is a development of one originally used by LATYSHEV and his associates[3] to study γ-rays of lower energy, has the very great advantage of a much more constant sensitivity and a resolving power which appears to be as good as that of the pair spectrometer. The sensitivity of both instruments, however, is very low; the successful elucidation of the capture γ-rays derives from the fact that very strong neutron sources were available.

By comparing the intensities of the capture γ-rays with γ-rays of known intensity, e.g. those produced by radioactive decay induced by neutron capture, the absolute intensity of the capture γ-rays were determined. The intensities of the γ-rays emitted by the initial state expressed as the number of photons (I_r) emitted per capture, can be translated immediately into transition probabilities, because if the total radiation width is Γ_γ (which is never very different from 0.1 ev) the partial width of a γ-ray, Γ_r, is given by $\Gamma_r = I_r \Gamma_\gamma$.

For the heavy elements generally the spectrum of the capture γ-rays is very complicated. With instrumental resolutions of the order of a few percent, the

[1] B. B. KINSEY and G. A. BARTHOLOMEW: Canad. J. Phys. **31**, 537 (1953).
[2] L. V. GROSHEV, B. P. ADYASEVICH and A. M. DEMIDOV: Geneva Conf. **2**, 39 (1955).
[3] G. LATYSHEV: Rev. Mod. Phys. **19**, 132 (1947).

spectrum usually can only be resolved at the top and at the bottom end. At the top end the unresolved γ-ray spectrum gives way to the discrete γ-rays which lead directly from the capturing state to the ground state or to low lying excited states when these are sufficiently separated from the ground state. At the bottom end of the spectrum the low energy γ-rays can be distinguished when these are emitted by the transitions between low lying states. Since many γ-rays are emitted in succession during the course of the cascade, these low lying states have a high probability of being excited and the γ-ray transitions between them therefore may be very probable. Fig. 56 shows the spectrum of a typical heavy element obtained with a pair spectrometer. For obvious reasons low energy γ-rays are not detected by this instrument.

For nuclei with intermediate atomic weight the γ-ray spectrum, at least for even charge nuclei, is less complex and many of the γ-rays emitted by the capturing state are separately detected[1]. In this part of the periodic table $(40 < A < 65)$ the difference between the γ-ray spectra emitted by even or by odd charge targets is very marked. The spectra from the latter, which are derived from odd-odd nuclei, are very noticeably more complicated than those derived from even-odd or even-even nuclei. For $65 < A < 180$ only the highest energy γ-rays can be resolved. For still heavier nuclei, the spectra become progressively simpler, and relatively more γ-rays are emitted in the high energy part of the spectrum. For Hg (where most of the capture originates in Hg^{199}) and for Tl, the shape of the spectrum is more like those of light or intermediate weight elements. For the two Pb isotopes (i.e. the targets Pb^{206} and Pb^{207}) by far the greater part of the available energy is emitted in the form of the γ-ray leading directly to the ground state (viz., the "ground state γ-ray") for no other γ-ray has been detected. This remarkable phenomenon is caused in part by the apparent predominance of electric dipole ($E1$) radiation[2]; partly by the lower density of states at and near the closed neutron shell at $N=126$; and partly by the absence of low lying states with suitable spins.

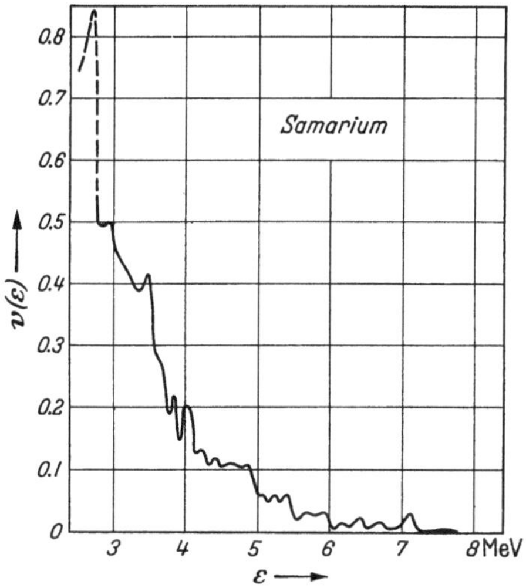

Fig. 56. Neutron capture γ-rays from Sm, after KINSEY and BARTHOLOMEW. The ordinate is the number of γ-rays deduced per unit energy range [see Sect. 64 β].

63. Transition probabilities for neutron capture γ-rays. The multipole orders of the γ-rays produced in direct transitions between the capturing state and the ground state can be inferred from the spins and parities of these states when these are known. As pointed out above, the transition probabilities of these γ-rays can be deduced when the absolute intensities are known. There are few instances where sufficient data is available; some examples are listed in Table 4. It will be seen that the partial radiation widths corresponding to the emission of these γ-rays are roughly proportional to the level spacing in the vicinity of the initial state, a phenomenon predicted by BLATT and WEISSKOPF [*4*]. In general, the

[1] Detailed references are listed in general reference [*19*].

[2] B. B. KINSEY and G. A. BARTHOLOMEW: Physica **18**, 1112 (1952). — Phys. Rev. **93**, 1260 (1954).

Table 4. *Intensities and partial widths of some E1 neutron capture γ-rays.*
The widths and level spacings are taken from LEVIN and HUGHES; γ-ray data from reference [19].

Nucleus	Γ_γ miliivolts	Energy (ε) Mev	Intensity %	Partial width millivolts	Level spacing ev	C mv/Mev⁴
V^{52}	1000	7.5	11	110	10000	21
Mn^{56}	500	7.5	12	60	500	84
Co^{60}	500	7.3	3	15	500	56
Rh^{104}	155	6.8	0.2	0.3	100	10
Ag^{108}	135	7.3	0.4	0.5	33	43
In^{116}	72	5.9	0.7	0.5	16	93
Ta^{182}	49	6.1	0.7	0.35	12	75
W^{183}	70	6.2	3.8	2.6	70	160
Au^{198}	124	6.5	1.5	1.9	60	110
Hg^{200}	250	6.5	5	13	100	95
Tl^{204}	800	6.3	6	48	15000	13
					Mean	65

average partial width can be written

$$\Gamma_r(\varepsilon) = C\,\varepsilon^3 D \tag{63.1}$$

where the constant C is 65 millivolts if Γ_r is also measured in these units and the if the γ-ray energy ε and the spacing D are measured in Mev. The data are plotted in Fig. 57. It will be seen that the few $M1$ γ-rays shown in this Figure have lower partial widths than the average for the $E1$ γ-rays.

It is instructive to compare these partial widths with the transition probabilities estimated by WEISSKOPF[1] for the single particle model of the nucleus. For $E1$ radiation the partial width is estimated to be:

$$\Gamma_{th}(\varepsilon) = 0.047\,R^2\,\varepsilon^3 \quad \text{electron volts}$$

if the nuclear radius is measured in units of 10^{-13} cm and ε is in Mev. This estimate represents an upper limit for the transition probability because in deriving it complete overlapping of the radial wave functions has been assumed. In fact, to be more realistic this upper limit should be reduced by a factor of four to take account of the reduction of the nuclear electric moment (by a factor of 2) caused by the mixture of neutrons and protons (cf. Sect. 49). For $A=125$, to take an average atomic weight, $\Gamma_{th}=0.6\,\varepsilon^3$ (ev) for the single particle model, and for the real nucleus, with its many interacting constituents and far more complex level system:

$$\Gamma_{th} = 0.6\,\varepsilon^3\,\frac{D}{D_s}$$

where D is the spacing of states near the initial state with the same spin and parity and D_s represents the spacing of the levels in the single particle model (∼10 Mev).

[1] V. F. WEISSKOPF: Phys. Rev. **83**, 1075 (1951); for electric 2^l-pole radiation the transition probability is:

$$T_E(l) = \frac{4.4\,(l+1)}{l\,[(2l+1)!!]^2}\left(\frac{3}{l+3}\right)^2\left(\frac{\varepsilon}{197}\right)^{2l+1} R^{2l}\times 10^{21}\ \text{sec}^{-1}$$

and for 2^l-pole magnetic radiation

$$T_M(l) = \frac{1.9\,(l+1)}{l\,[(2l+1)!!]^2}\left(\frac{3}{l+3}\right)^2\left(\frac{\varepsilon}{197}\right)^{2l+1} R^{2l-2}\times 10^{21}\ \text{sec}^{-1}$$

where ε is the energy of the transition in Mev and R is the nuclear radius in units of 10^{-13} cm and $(2l+1)!!=1\cdot 3\cdot 5\ldots(2l+1)$.

Thus the constant C in the empirical expression (63.1), 65 millivolts, should be compared with $0.6\,\text{ev}/D_s$ i.e. 60 millivolts. This good agreement between theory and experiment, although significant only to an order of magnitude, indicates that the emission of these high energy $E1$ γ-rays is not retarded to any appreciable extent. We shall see in a later section (Sect. 84) that low energy $E1$ γ-rays on the contrary are always very highly retarded, and that a theory of radiation widths based on empirical capture γ-ray data gives incorrect results. It is probable that, for reasons not yet clearly understood, the transition probabilities vary much more rapidly with energy than the expression (63.1) indicates.

Theoretically the transition probabilities of $M1$ radiation should be lower than those of $E1$ radiation (for the same transition energy) by a factor of the order of $10(\hbar/McR)^2$. For $A = 100$, this theoretical ratio is about 0.01. The few $M1$ γ-rays represented in Fig. 57 have lower transition probabilities than the average for $E1$ radiation, but the data is hardly sufficient to determine whether the ratio is significantly different from that expected theoretically. Clearly these $M1$ γ-rays are also not appreciably retarded. In this respect, the high energy $M1$ γ-rays are not different to the low energy $M1$ γ-rays (Sect. 71).

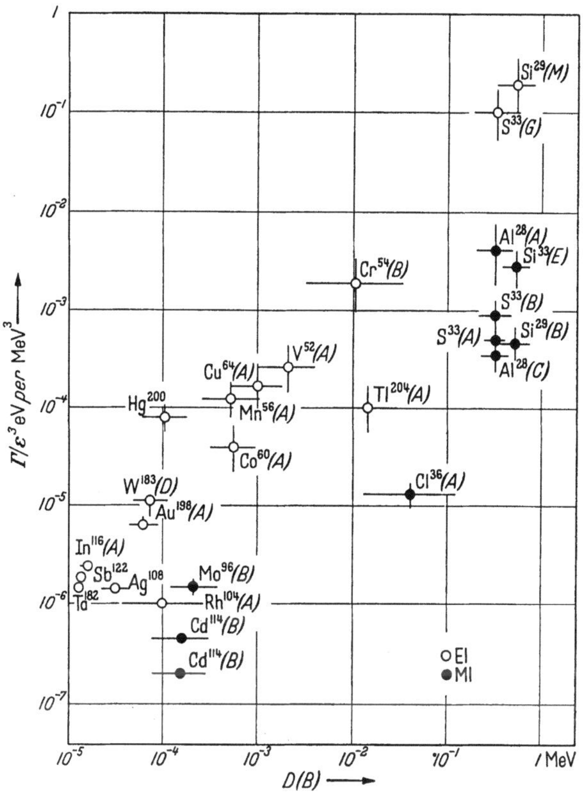

Fig. 57. Partial radiation widths of dipole radiation, normalised to unit energy, plotted against the level spacing. Open circles, $E1$; full circles, $M1$ γ-rays. The letters refer to labels given to the γ-rays in the literature; references to this data are given in [19].

Three instances of capture γ-rays have been found in which it has been possible to compare[1] the transition probabilities of $E1$ and $M1$ radiation in competition with each other. In each case, the transition probability for the $E1$ γ-ray was about 100 times greater than that calculated for the $M1$ γ-ray at the same energy. It is not clear that this ratio is typical of neutron capture radiation generally, because it is possible in each of these three cases, that the predominance of the $E1$ γ-ray is caused by suppression of the $M1$ γ-ray by single particle selection rules (i.e. the $M1$ γ-ray is l-forbidden). However, this possibility is not very likely, because, as will be shown below (Sect. 72) single particle selection rules, in fact, have little influence on $M1$ transition probabilities at low energies.

64. The γ-ray cascade. In a heavy nucleus a very large number of states must lie between the ground state and the neutron binding energy. Consequently

[1] See ref. 2, p. 309.

the compound nucleus formed by neutron capture can be relieved of its energy in a large number of alternative radiative transitions and the gross features of the γ-ray cascade which is produced by neutron capture are largely independent of the characteristics of individual transitions. Thus the radiation width, the number of γ-rays emitted per capture, and their spectrum, are quantities which do not vary much from one nucleus to another. The radiation widths will be discussed in Sect. 65. Here we shall consider only the details of the cascade, the number of γ-rays, the spectrum, and certain features such as the intensities of low energy γ-rays and isomeric transitions which are dependent on the spin of the capturing state.

α) The multiplicity. The numbers of γ-rays which are emitted when a neutron is captured have been measured by MUEHLHAUSE[1] by a comparison of coincidence counting rates to single counting rates in anthracene counters set up near a target in a thermal neutron beam. For nuclei with atomic weights above about 70, usually 3 to 4 γ-rays are emitted per capture. For the heavy elements especially this means only the number of γ-rays which can be readily detected, not necessarily the average number of successive transitions because in many heavy nuclei it is probable that the final transitions have low energy and may be highly internally converted. Thus the number of γ-rays observed, or the *multiplicity*, is only a rough measure of the number of transitions. For odd-charge nuclei, the average multiplicity is less by about one quantum per capture than the average multiplicity for even-charge nuclei. This phenomenon, which at first sight is the opposite of what one might expect, is probably to be explained by the fact that in even charge nuclei, there are more regularities in the spin sequence of the low lying states [*19*].

β) The γ-ray spectrum. For purposes of analysis the γ-ray spectrum may be divided into partial spectra. The "primary" spectrum may be defined as the spectrum of the γ-rays emitted by the capturing state; the "secondary" spectrum that of the *second* γ-ray to be emitted, i.e., the one that follows the emission of the primary γ-ray; the "tertiary" spectrum, that of the third γ-ray and so on. Then if $\nu_n(\varepsilon)$ represents the number of γ-rays of energy ε emitted per unit energy range per capture in the n-th spectrum, in the absence of appreciable internal conversion,

$$\int_0^B \nu_n(\varepsilon)\, d\varepsilon = 1$$

where B is the neutron binding energy. The total spectrum will be given by

$$\nu(\varepsilon) = \sum_n \nu_n(\varepsilon)\,.$$

The number of partial spectra which are effective is approximately equal to the multiplicity. The partial spectra calculated for a simple level density function rising with excitation energy as $\exp 2(aE)^{\frac{1}{2}}$ and a γ-ray transition probability varying as the cube of the energy are shown in Fig. 58 for level spacings of 10 and 100 ev at the binding energy (7 Mev). It will be seen that neither the partial spectra nor their sum depends very much on the level density in the nucleus, and there is in fact little difference between the spectrum of a nucleus with many excited states and one with few. Thus most heavy nuclei emit approximately the same γ-ray spectrum and this spectrum differs from the normal,

[1] C. O. MUEHLHAUSE: Phys. Rev. **79**, 277 (1950).

only when quite radical departures from the usual level densities occur. As mentioned in Sects. 58 and 61 such departures occur for nuclei with atomic weight above 180.

γ) Activation of isomeric states by slow neutrons. After the emission of the second or third successive γ-ray the mean excitation energy is low (probably less than 1 Mev in heavy nuclei). There are now very few possible transitions; low lying states, therefore, have a high probability of being excited, and the γ-rays emitted by transitions between them, or to the ground state, may be quite strong. (For example the γ-ray emitted by the first excited state in Cd^{114} excited by thermal neutron capture, has an intensity of nearly 0.9 quantum per capture.) The intensity of a γ-ray producing a state at low excitation energy will depend on the way the initial state is populated by transitions from above. This effect is clearly shown in cross sections for the production of isomeric states by thermal neutrons. In Table 5 we list, under **A**, examples of nuclei in which the spin of the ground state is closer to the spin of the capturing state than to the spin of the isomeric state, and under **B**, examples where the opposite is true. It will be seen that for category **A** the fraction of the total capture cross section that is measured by the activation of the isomeric state is small, while for category **B** it is large. Numerical calculation of this effect gives results in good agreement with the experimental results.

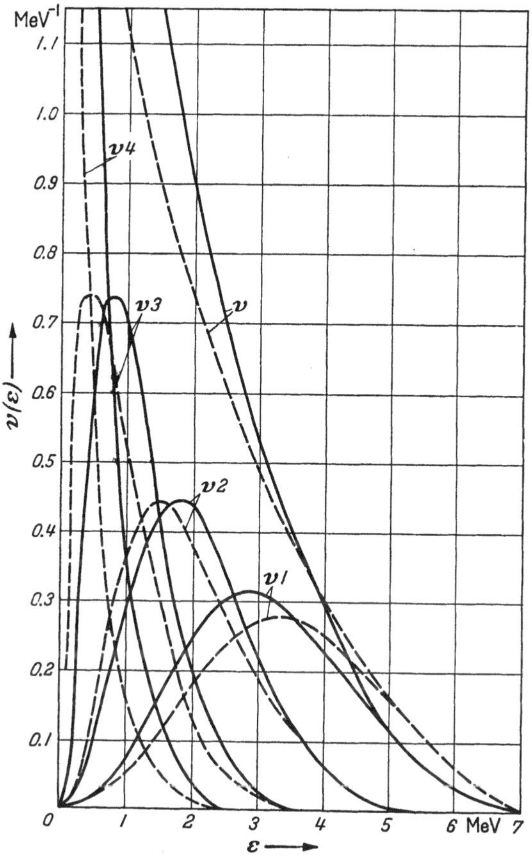

Fig. 58. Calculated capture γ-ray spectrum, for a spacing of 10 ev at the neutron binding energy (full line) and 100 ev (broken line).

If the spin of the initial state is very different from that of the capturing state, we should expect to find that the intensity of a low energy γ-ray depends markedly on which of the two spin states is responsible for capture. For prompt γ-rays this phenomenon has not yet been studied, but the corresponding effect on the yield of isomeric states has been noted by several authors. It seems to have been first detected by CAPRON and VERHOEVE-STOKKINK[1]. In more detail it has been studied by SAILOR[2], who has examined the activation of the isomeric state of In^{116} produced by neutron capture in the first two resonances in targets of In^{115}, and by WOOD[3] in studies of the activation of the isomeric state of Eu^{152}

[1] P. C. CAPRON and A. J. VERHOEVE-STOKKINK: Phys. Rev. **81**, 336 (1950).
[2] Reported by H. H. LANDON and V. L. SAILOR: Phys. Rev. **98**, 1267 (1955).
[3] R. E. WOOD: Phys. Rev. **95**, 453 (1954).

in the first six resonances in Eu^{151}. The difference in the isomeric yields is associated with differences in radiation widths (Sect. 65).

δ) Activation of isomeric states by fast neutrons. For faster neutrons (1 Mev), capture of neutrons of higher angular momentum should increase the yield of isomeric states where the spins of these states differ most from that of the capturing state for thermal neutrons (class **A**); and less sensitively, should decrease it for class **B** (Table 5). The latter would be difficult to detect for the capture by

Table 5. *Yields of ground states and isomeric states produced by thermal neutron capture.* From L. SEREN, H. N. FRIEDLANDER and S. H. TURKEL, Phys. Rev. **72**, 888 (1947); and National Bureau of Standards (USA) Circular Number 499.

Nucleus	Spins			Yield of isomeric state in %
	capturing state	ground state	isomeric state	
		A		
Zn^{69}	$\frac{1}{2}$	$\frac{1}{2}$	$\frac{9}{2}$	22
Br^{80}	1 or 2	1	5	25
Se^{81}	$\frac{1}{2}$	$\frac{1}{2}$	$\frac{7}{2}$	7
Rh^{104}	1	1	4	8
Ag^{110}	1	1	5	2.7
Sn^{124}	3 or 4	3	1	1.2
Te^{127}	$\frac{1}{2}$	$\frac{3}{2}$	$\frac{11}{2}$	8
Te^{129}	$\frac{1}{2}$	$\frac{3}{2}$	$\frac{11}{2}$	10
Te^{131}	$\frac{1}{2}$	$\frac{3}{2}$	$\frac{11}{2}$	<3.5
		B		
Ge^{77}	$\frac{1}{2}$	$\frac{7}{2}$	$\frac{1}{2}$	52
Kr^{85}	$\frac{1}{2}$	$\frac{9}{2}$	$\frac{1}{2}$	61
In^{114}	4 or 5	1	5	35
Sn^{125}	$\frac{1}{2}$	$\frac{11}{2}$	$\frac{3}{2}$	80
Dy^{165}	$\frac{1}{2}$	$\frac{7}{2}$	$\frac{1}{2}$ or $\frac{13}{2}$	55

thermal neutrons already produces a large activation in the isomeric state. An increased yield of isomeric states in class **A** does occur as may be seen from a comparison of the activation cross sections obtained by HUGHES, GARTH and LEVIN[1] with those quoted in Table 5 above.

65. Radiation widths of neutron resonances. *α) Experimental.* The radiation widths of neutron resonances are of interest for these quantities must be compounded of the γ-ray transition probabilities and the level density. Moreover, in principle, they offer a method of testing directly the spin dependence of the level density (57.5).

In the heavy elements, such radiation widths as have been measured vary little from one nucleus to another; they lie in a narrow range, from roughly 25 to 200 millivolts. They are difficult to measure; at present very little accurate data is available. Such small widths can be measured directly only in heavy elements where the radiation width constitutes the greater part of the total width and then only for resonances of very low energies and with neutron spectrometers of high resolution (e.g. crystal spectrometers). Even then the energy variation of the total cross section through the resonance will depart from the

[1] See ref. 2, p. 305.

simple dispersion form because of the distortion caused by thermal motion of the atoms in the target (the DOPPLER effect). In the absence of the latter the total cross section for neutrons of energy E is given approximately by the BREIT-WIGNER formula (61.1); when the DOPPLER effect is taken into account, the cross sections are given by much more complicated formulae. Direct measurements of total widths have been made by LANDON and SAILOR[1] by fitting theoretical resonance cross sections to the experimental data (allowing for the DOPPLER effect). For low energies this method is quite accurate but it becomes rapidly more difficult as the resonance energy is increased; with present equipment measurements of radiation widths for resonances above a few electron volts are not feasible. For higher energies it is necessary to use methods in which the resonance parameters may be determined without having to match the instrumental resolution to the natural width of the resonance. Such methods, which have been described by LYNN and RAE[2], involve, for each resonance, the determination of the following quantities: the spin (which influences the cross section through the statistical factor g); the neutron width Γ_n; and the radiation width Γ_γ. To determine any one of these quantities, all three have to be determined; and this, in general, can be done by measuring three independent quantities, e. g. the transmission of a thick sample as a function of the energy, the rate of neutron capture, and the rate of neutron scattering.

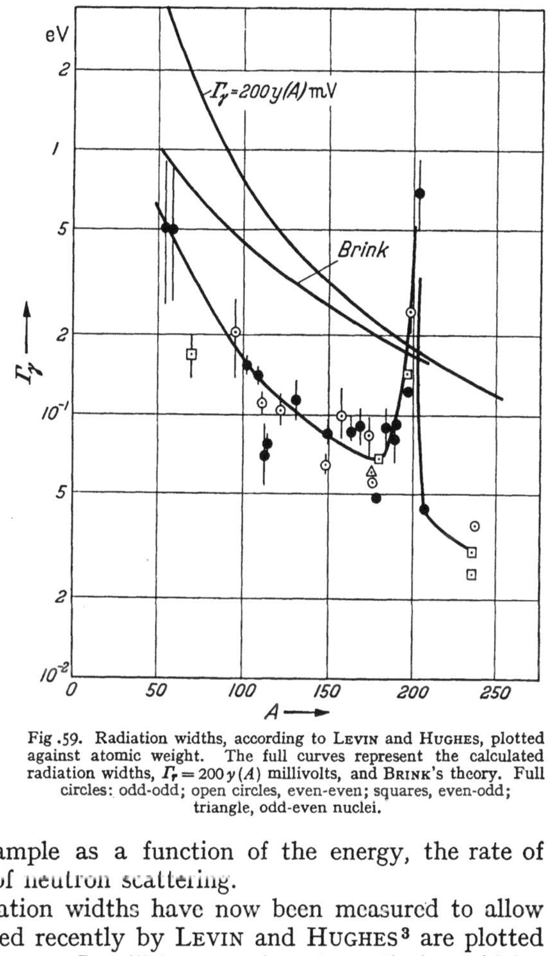

Fig. 59. Radiation widths, according to LEVIN and HUGHES, plotted against atomic weight. The full curves represent the calculated radiation widths, $\Gamma_\gamma = 200\, y(A)$ millivolts, and BRINK's theory. Full circles: odd-odd; open circles, even-even; squares, even-odd; triangle, odd-even nuclei.

A sufficient number of radiation widths have now been measured to allow of systematic study. Those listed recently by LEVIN and HUGHES[3] are plotted against the atomic weight in Fig. 59. It will be seen that the radiation widths, in general, are remarkably constant, although they tend to fall with increasing atomic weight. However, there is a marked tendency for the radiation widths to rise near the atomic weight of Pb; it will be shown below that this is related to the increased level spacing near the neutron binding energy (Fig. 52).

β) Constancy of the radiation widths. Qualitatively the general features of the experimental data are easily accounted for; the details, however, are more difficult. A formal expression for the radiation widths may be obtained as follows.

[1] See ref. 2, p. 313.
[2] E. LYNN and E. RAE: To be published.
[3] J. S. LEVIN and D. J. HUGHES: Phys. Rev. **101**, 1328 (1956). See also H. H. LANDON: Phys. Rev. **100**, 1414 (1955).

The total radiation width is made up of the partial radiation widths of the many possible alternative radiative transitions from the capturing state. We assume first that the capture radiation is all electric dipole radiation. Then the average partial radiation width for a γ-ray of energy ε depends on three factors: (1) it is proportional to the cube of the energy; (2) it is proportional to the spacing of the levels in the immediate vicinity of the initial state with the same spin and parity as the initial state (see Sect. 63); and (3) it should take account of the giant resonance. The third factor, in general, will depend on both the excitation energy (E) of the initial state and on the transition energy (ε); we will denote it by $f(\varepsilon, E)$. It follows that the partial width for emission of a γ-ray of energy ε is:

$$C' f(\varepsilon, E)\, \varepsilon^3 D(B) \tag{65.1}$$

where B is the neutron binding energy and $D(B)$ is the level spacing just referred to, and C' is a constant. This expression is just an elaboration of (63.1) ($C' = C$, if $f = 1$). If $\omega_t(E)$ is the total density of levels at excitation energy E which are accessible to the initial state (spin J) by emission of $E\,1$ radiation, the total radiation width is:

$$\Gamma_\gamma = C' \int_0^B f(\varepsilon, E)\, \varepsilon^3 \omega_t(E)\, D_J(B) \cdot dE$$

where $E = B - \varepsilon$. Now the final states which are obtained by emission of dipole radiation have spins $J+1$, J and $J-1$. Assuming that the level density ω_J of states of spin J is equal to $(2J+1)\omega_0$ where ω_0 is the density of states with zero angular momentum,

$$\omega_t = \omega_{J+1} + \omega_J + \omega_{J-1} = 3(2J+1)\,\omega_0$$

and $D_J(B) = 1/[(2J+1)\,\omega_0(B)]$. Hence

$$\Gamma_\gamma = 3C' y \tag{65.2}$$

where

$$y = \int_0^B f(B-E, E)\,(B-E)^3 \frac{\omega_0(E)}{\omega_0(B)}\, dE. \tag{65.3}$$

Thus the radiation widths generally do not depend on the spin of the target nucleus. This was first noted by HUGHES and HARVEY[1] in a discussion of the experimental data. Moreover, as predicted by BRINK[2], the radiation widths in a particular nucleus should be the same regardless of the spin of the capturing state, and, furthermore, the radiation widths in the nucleus should be constant to within the limits of the statistical fluctuations in the probabilities of individual transitions. If the latter are assume to fluctuate by a factor of ten about the mean (see Fig. 57) then the fluctuations in the total radiation width might amount to 10%.

To calculate y we assume that $f(B-E) = 1$ in (65.3) and equate $\omega(E)$ to $E^{-2}\exp 2(aE)^{\frac{1}{2}}$, with $a = A/8.8$ [cf. Eqs. (57.2) and (57.4)]. The quantity y is then found to change only slowly with the level spacing at the binding energy, increasing slowly as the spacing increases[3]. With increasing atomic weight the radiation width (65.2) falls at approximately the correct rate (see Fig. 59);

---

[1] D. J. HUGHES and J. A. HARVEY: Nature, Lond. **173**, 942 (1954).

[2] D. M. BRINK: Thesis, Oxford 1955.

[3] It should be noted that, in making this calculation, it is assumed that the level density is independent of parity; if this were not the case it would not be possible to use (65.3) without separately specifying the energy dependence of states of opposite parity.

in fact, the radiation width calculated from (65.2) differs from the experimental points only by a constant factor. Thus BETHE's formula is adequate to account for the general trend of the radiation widths, if not their absolute magnitude.

The upward displacement of the experimental points near $A = 200$ (in Fig. 59) can be accounted for by the change in y which results from the change in level spacing in the vicinity of the closed shell at $N = 126$: the higher values of the widths in this region can be attributed entirely to increased level spacing[1].

The lack of distinction between the radiation widths of odd-odd and even-even nuclei (Fig. 59) is interesting because the neutron binding energies of these nuclei usually differ by at least 1 Mev in heavy elements, and y is quite sensitive to changes in binding energy. We have seen that the level densities of these nuclei are comparable near the binding energy and that these level densities are considerably higher than those of even-odd nuclei for the same excitation energy (see Sect. 58). No difference in the radiation widths would be expected if the two types of nuclei had the same binding energy and if the difference in level densities could be represented by a simple multiplicative factor. However, the difference in the binding energies would be expected to increase the radiation widths of the even-even nuclei over those of the odd-odd nuclei by some 30%[2]. The fact that there is no obvious difference between the radiation widths is in accord with the suggestion that the level densities in even-even nuclei start increasing with excitation energy only after an excitation is reached equal to the pairing energy (Sect. 58). The position of the even-odd nuclei in Fig. 59 calls for no special comment. These nuclei, on average, have somewhat lower neutron binding energies than odd-odd nuclei, and this probably accounts for the very low values for the radiation widths of U^{239} and Th^{233}; if corrected to 7 Mev to make their radiation widths comparable to the rest, the observed values would be increased by about 50%.

γ) Numerical estimates. LEVIN and HUGHES[3] have attempted with some success to calculate the absolute values of the radiation widths on the basis of WEISSKOPF's transition probabilities[4].

Since these are known to give rather high values for the widths of individual γ-rays, it is perhaps better to use the empirical data. Thus we require an empirical estimate of the quantity C' in (65.2). This quantity may be obtained from the intensities and widths of neutron capture γ-rays with energies equal to or near the neutron binding energy as described in Sect. 63. Putting $f = 1$, in (65.3), and $C' = C = 65$ millivolts, we obtain $\Gamma_\gamma = 200\, y$ millivolts; this quantity is plotted in Fig. 59. It will be seen that the calculated value of Γ_γ is everywhere too high; this suggests that either the level density rises much more rapidly with excitation energy than is to be expected from BETHE's formula or that the transition probabilities of individual γ-rays increase more rapidly than the cube of the energy. The same conclusions are reached from a study of the ground-state γ-rays listed in Table 4. From (65.2) it follows that the average intensity of a ground-state γ-ray emitted by the capturing state (with energy B) is:

$$I_r = \Gamma(B)/\Gamma_\gamma = B^3 D_J(B)/3\, y.$$

[1] Cf. LEVIN and HUGHES: Ref. 3, p. 315. According to A. G. W. CAMERON (to be published) calculations of the radiation widths using NEWTON's formula (which makes allowance for shell structure), are, within a constant factor, in good agreement with the experimental data.

[2] Near $B = 7$ Mev, y behaves roughly as $\exp(0.23\, B)$ where B is in Mev, regardless of the value of A.

[3] See ref. 3, p. 315.

[4] See ref. 1, p. 310.

This expression leads to values which are consistently lower than those determined experimentally.

BRINK[1] has derived a formula for the radiation width of a γ-ray of energy ε on the basis of the integrated photonuclear cross section (the dipole sum) and the shape predicted for this cross section by STEINWEDEL and JENSEN[2]. This cross section refers to absorption by the ground state of the nucleus; for the radiation width, we are concerned with the emission of radiation by a state already excited to some 7 Mev. BRINK assumed nevertheless that the factor $f(\varepsilon, E)$ in (65.1) can be taken directly from the photonuclear cross section regardless of the fact that the initial state of the nucleus is not the ground state. If E_m is the energy of the peak of the giant resonance, and Γ_g is a constant which determines its width, BRINK found that the radiation width $\Gamma(\varepsilon)$ for a γ-ray of energy ε is given by:

$$\Gamma(\varepsilon)\,\omega(B) = \frac{4}{3\pi}\,\frac{NZ}{A}\,\frac{e^2}{\hbar c}\,\frac{1+0.8x}{Mc^2}\,\frac{\Gamma_g\,\varepsilon^4}{(\varepsilon^2-E_m^2)^2+(\Gamma_g\,\varepsilon)^2}$$

where $\omega(B)$ is the density of states at the excitation energy of the initial state (B) with the same spin and parity as the initial state, and x is the fraction of exchange force (49.2). For neutron capture γ-rays $\varepsilon \ll E_m$ and the last term on the right becomes $\Gamma_g(\varepsilon/E_m)^4$. Thus for low energies this formula makes the partial width proportional to roughly A^{-2} (see Sect. 47) and to ε^4. For Cu, for which $B=8$ Mev, $E_m=19$ Mev, $\Gamma_g=6$ Mev, and $x\approx 0.5$, BRINK's formula gives a partial width of about two thirds of that deduced from the γ-ray data (Table 4). At this energy, then, BRINK's formula is in remarkably good agreement with experiment; the data, however, is hardly good enough to show the expected variation with A. The total radiation width, as before, is given by:

$$\Gamma_\gamma = \int_0^B \Gamma(B-E)\,\omega(E)\,dE\,.$$

The radiation widths obtained by this formula are still too high and they seem to show an insufficient dependence on A (see Fig. 59). The discrepancy may be related to the anomalously by low transition probabilities of low energy $E\,1$ transitions (Sect. 84).

66. Spin dependence of level density. In the previous section we have shown that the radiation widths in a given nucleus should be independent of the spin of the levels. In making this deduction it was assumed that the density of levels with spin J was $(2J+1)$ times the density of levels with zero angular momentum. However, this is not the only condition for the spin independence of the radiation widths; the radiation widths will also be independent of spin if the densities of all levels at a given energy are the same regardless of spin and provided that $J \geq 1$. A crucial test is provided by a study of the radiation widths of the levels produced by slow neutron capture by nuclei with spin $\frac{1}{2}$. In this case the levels have spins 0 or 1 and a constant radiation width independent of spin can be produced only by the condition $\omega_J=(2J+1)\,\omega_0$. This seems to be confirmed by measurements of the radiation widths of the Ag resonances which indicate that the radiation widths are the same to an accuracy of 10% whether the spins are 0 or 1. To some extent this is corroborated by the fact that the majority of the resonances for which the spins have been determined (in either isotope) have spin 1.

[1] See ref. 2, p. 316.
[2] See ref. 2, p. 285.

In some nuclei with large spins there is a clear indication that the radiation widths do depend on the spin. For example the radiation widths of the two lowest levels in In^{116} measured by LANDON and SAILOR[1] are different (72 ± 2 and 81 ± 4 mev for the 1.46 and 3.85 ev levels respectively). Measurements of the three lowest levels in Eu^{152} suggest that the radiation widths fall into two groups with widths of 70 and 93 mev. As remarked in Sect. 64, this grouping is associated with corresponding differences in the activation of the isomeric state. However, not enough is known about the decay of Eu^{152} to determine the spins concerned. These measurements suggest that the density of states with high spin are not in fact proportional to $(2J+1)$. As pointed out in Sect. 57, the spin-dependent part of the level density formula is proportional, in addition, to an exponential term which is a function of both the spin and the excitation energy. According to LANG and LE COUTEUR[2] this term is $\exp[-(J+\frac{1}{2})^2/2c\Theta]$ where $c\hbar^2$ is the moment of inertia of the nucleus [$0.4\, b^2 AMR^2$ (cf. Sect. 59)] and Θ is the temperature of the nucleus at the neutron binding energy (about 0.55 Mev). The In and Eu results are consistent with $2c\Theta = 11$ and 6.5 which gives $b^2 = 0.3$ and 0.1 respectively, values which agree to order of magnitude with those obtained from deformed nuclei (Sect. 78).

We have seen (Sect. 9) that the isotropy of the angular distribution of the particles emitted in nuclear reactions requires the level densities in the residual nucleus to be proportional to $(2J+1)$. We have also seen that this isotropy is a general feature of the lower energy components of the products of nuclear reactions when the incident particles have energies extending up to 20 Mev. For high energy neutrons the isotropy of the low energy products seems to disappear, for the angular distribution of the low energy proton products seem to be peaked in the forward and backward direction[2]. This may also be the result of a failure of the density law (57.5) for as higher energy incident particles are used the higher are the spins of the states which can be excited.

M. Systematics of low energy states.

67. General features. The information on low lying levels which can be most easily obtained experimentally are the energies of the levels, their spins and parities, their spacings and the probabilities of transitions between them. Other characteristics, e.g. the magnetic or electric quadrupole moments, with the exception of the ground state, are more difficult to obtain, and in fact have been obtained only in a few instances. The levels which one can detect depend significantly on the means used to produce them and for this reason some caution is needed in making generalizations. However, a great mass of data is now available and this shows that certain marked regularities exist.

α) Spins and parities. There are marked regularities in the spins and parities of the lowest excited states. In even-even nuclei with very few exceptions, the spins of the first excited states are 2 units (Sect. 68); they have even parity and their energies are related to the position of their neutron and proton numbers relative to closed shells. For the heavy nuclei ($N \geq 90$), except those in the immediate neighbourhood of the closed shell nuclei, the spins of the second excited states are 4 units and their parity is even (Sect. 69). This results from the now well established fact that for both odd and even mass number nuclei the lowest levels form part of rotational systems which originate in the deformation of the

[1] See ref. 2, p. 313.

[2] R. W. WANIEK, L. GARRIDO, R. GOLOSKIE and E. G. SILVER: Phys. Rev. **95**, 586 (1954); **99**, 622 (1955).

nucleus by the nucleons remaining in an incompleted shell (Sects. 75 to 77). For nuclei with $N < 90$ general trends can be discerned but they are less well marked (Sect. 69). The closed shell nuclei show certain peculiarities which are discussed in Sect. 85; the level structure in neighbouring nuclei, i.e., nuclei with one or two nucleons in excess or in defect of the number required to close a shell, are of equal interest and have proved rather more amenable to theoretical interpretation.

Most of the low lying states in even-even nuclei have even parity. Occasionally states of odd parity have been observed. Their spins have been remarked by GLAUBMAN[1]. They are usually odd; in fact, they are most often 3 units. This phenomenon seems to be just one aspect of the more general rule that the lowest states in even-even nuclei have even spins with even parity and odd spins with odd parity[2]. Its origin has been discussed by TALMI[3]: it appears to be a natural consequence of short range forces and the coupling between like particles. The energies of the lowest odd states fall with increasing mass number, as pointed out by MORINAGA[4]; they are most often found near closed shells or sub-shells where a β-transition may shift the configuration most easily from one of even to one of odd parity or vice-versa. In addition to odd states of this type, low lying odd states with spin 1 in the heavy even-even nuclei form a special group: these states seems to have a collective origin and are discussed in Sect. 83.

Although the excited states in odd mass number nuclei for $N < 90$ show localized trends, in general, much less is known about the level structure in such nuclei. Still less is known about the level structure in odd-odd nuclei. In the deformed nuclei with odd mass number ($N \geqq 90$) the level structure is mainly of a rotational type based on levels which it now seems possible to identify from the theoretical order of the levels in a spheroidal potential well (Sect. 81). In these nuclei odd and even rotational systems often follow one another in rapid succession.

β) Transition probabilities. The transition probabilities, generally, show striking regularities. Low energy $E\,1$ transitions, except in the deformed nuclei, are rare and when they occur they are found to have very low transition probabilities, i.e., low in comparison with those to be expected from the single particle model of the nucleus[5]. It should be noted that, in this respect, low energy $E\,1$ transitions are very different to high energy $E\,1$ transitions, which like the $M\,1$ transitions, generally, have normal transition probabilities (Sect. 63). The low energy $M\,1$ transitions are usually slower than the single particle estimates by some 10 to 100 times. The $E\,2$ transitions, on the other hand, are almost everywhere much faster than single particle estimates will allow: a characteristic which seems general over the whole range of heavy nuclei. This phenomenon shows that a considerable fraction of the nuclear charge takes part in $E\,2$ transitions: in fact the lower excited states must have collective properties. Except in closed shell nuclei or in nuclei in that neighbourhood, their energies, in many respects, are closely related to those which we should expect of surface oscillations (Sect. 69).

68. The first excited states in even-even nuclei. The even-even nuclei are remarkable in that the first excited states, with very few exceptions, have a spin

[1] M. L. GLAUBMAN: Phys. Rev. **90**, 1000 (1953).
[2] For example the second excited state in Sr^{88} is 3^- (see ref. 1, p. 356); and the second and third states of Ba^{134} seem to be 1^- and 3^- [E. D. KLEMA: Phys. Rev. **100**, 66 (1955)].
[3] I. TALMI: Phys. Rev. **90**, 1001 (1953).
[4] H. MORINAGA: Phys. Rev. **103**, 503 (1956).
[5] See ref. 1, p. 310.

of 2 units and even parity and energies which vary from one nucleus to another in a very definite and continuous manner. The few exceptions to this rule occur in nuclei for which both neutron and proton shells are closed. For example, the spin and parity of the first excited state in O^{16} and in Ca^{40} (Sect. 86) is 0^+ and that of Pb^{208} is 3^- (Sect. 88).

The spins of the first excited states of even-even nuclei have been discussed by several authors[1]. The spins have been derived from internal conversion measurements, the ratios of K- to L-conversion, measurements of the angular and polarization correlations of γ-rays, angular distributions of γ- and α-rays, β-decay, and the like. In making these deductions it is usually assumed that the spin of the ground state of the even-even nucleus is zero; while there exist only a few examples in which this has been ascertained by direct methods no examples are known contrary to this rule and much indirect evidence has been accumulated which supports it.

Most theories of the nature of the first excited state lead to the conclusion that the spin in 2 units and that the parity is even. Thus if it is assumed that the lowest excited states belong to the configuration of the ground state, the spins of the states belonging to this configuration are quite generally[2] $0, 2, 4 \ldots (2j-1)$, their energies follow in that order, and their parities are even. Here j is the spin corresponding to the configuration of the last pair of neutrons or protons. If configuration mixing occurs, this order may be disturbed. However, it is not entirely clear that this is the origin of the spin of the first excited state, because except in the heaviest nuclei, the spins of few excited states in fact follow this sequence; and in the heaviest nuclei where this sequence is observed it is the consequence of the properties of rotational states.

In nuclei in which both proton and neutron shells are closed, the ground state is unique. The first excited states in these nuclei therefore are not derived from the ground state configuration; thus their spins, as pointed out above, are exceptions to the rule that the spins of even-even nuclei are two units.

The energies of the first excited states in even-even nuclei have been discussed by PREISWERK and STÄHELIN[3] and by SCHARFF-GOLDHABER[3]. The energies are plotted against the neutron number in Fig. 60. It will be seen that there is a marked increase in the energies near the closed shells, especially the double closed shells, e.g. in Ca^{40}, in Zr^{90}, and Pb^{208}. However, the dependence of the energies of the first excited states on neutron and proton number is not symmetrical. For elements with closed proton shells, the energies of the first excited states are higher than those in which neither neutron nor proton shell is closed but not so high as when the neutron shell alone is closed. Between closed shells the energies fall to low values.

The regular behaviour shown in Fig. 60 is in marked contrast to the energies of the excited states in even-odd nuclei which can take much lower values. In even-even nuclei, as we have seen, low values obtain in the heaviest elements but they vary continuously and systematically from one nucleus to another. PREISWERK and STÄHELIN[3] noting this difference have pointed out that the levels of even-even nuclei do not show the continuous displacement with mass number

[1] H. HORIE, M. UMEZAWA, Y. YAMAGUCHI and S. YOSHIDA: Progr. Theor. Phys. (Japan) **6**, 254 (1951). — M. GOLDHABER and A. SUNYAR: Phys. Rev. **83**, 906 (1951). — P. STÄHELIN and P. PREISWERK: Helv. phys. Acta **24**, 623 (1951).

[2] B. FLOWERS: Phys. Rev. **86**, 254 (1952).

[3] P. PREISWERK and P. STÄHELIN: Physica **18**, 1118 (1952). — Nuovo Cim. **10**, 1219, (1953). — G. SCHARFF-GOLDHABER: Phys. Rev. **90**, 587 (1953). (This paper contains a very complete list of references to measurements.)

of the energies of states representing identifiable configurations found in some even-odd nuclei. In fact one might expect that states of spin zero representing different paired configurations in even-even nuclei would approach each other

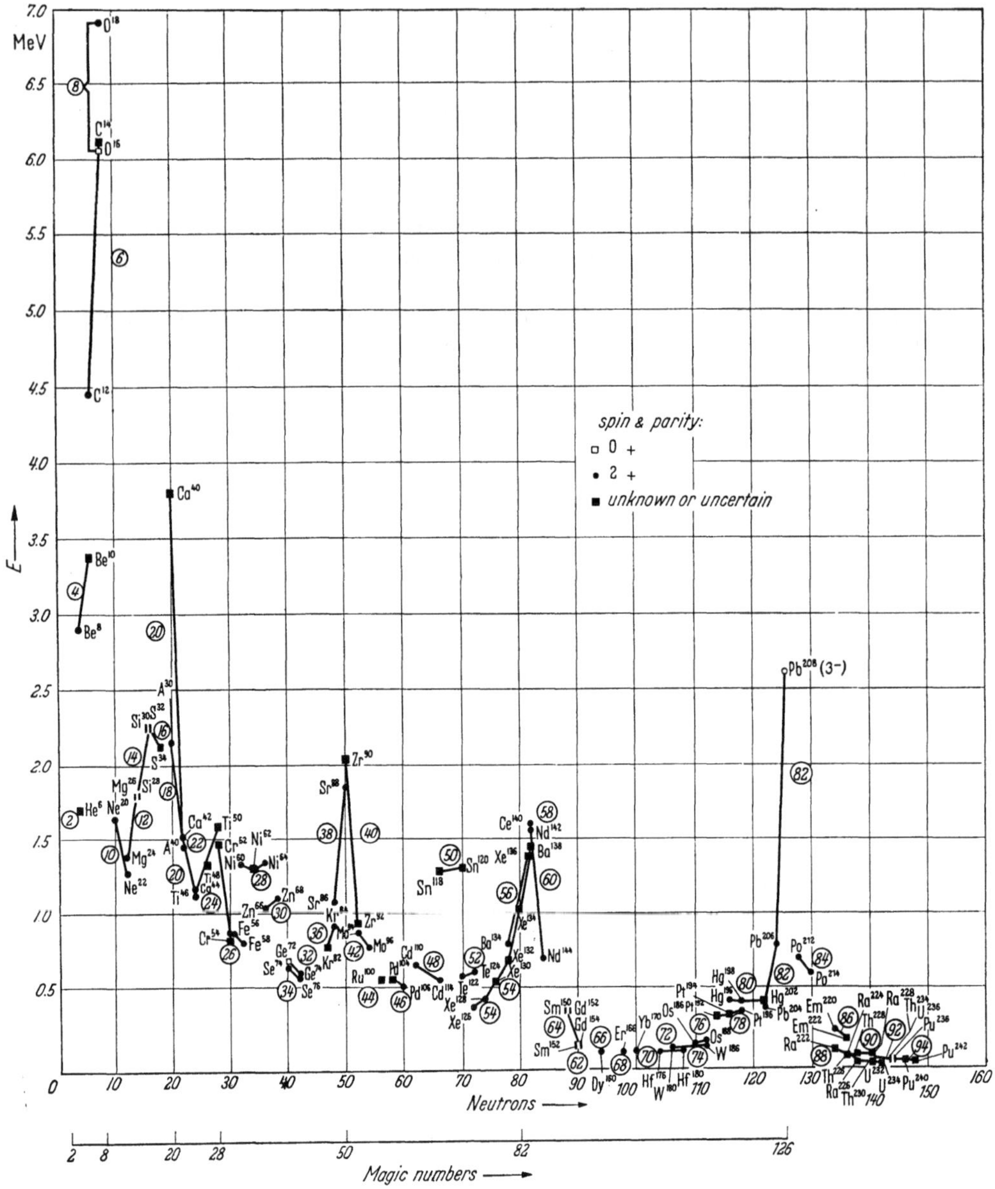

Fig. 60. Energies of the first excited states in even-even nuclei according to SCHARFF-GOLDHABER.

at certain mass numbers. The fact that this does not occur can be explained when the interaction between neutrons and protons are taken into account; they show that the energy difference between such states must be greater than twice this interaction energy. Similar effects are presumably at work in the first excited states and may possibly account for the great difference between the energies of these states for nuclei between, and those with closed shells[1]. However, the extent to which it is legitimate to apply such arguments to the lowest states is doubtful for recent work indicates that many of the lower states have, at least in part, the character of collective behaviour (Sect. 72); in fact, the smooth variation of energy with mass number is in itself an indication of this behaviour.

69. The second excited states in even-even nuclei. The systematics of the second excited states have been discussed recently by NAGASAKI and TAMURA[2] and by SCHARFF-GOLDHABER and WENESER[2]. They find a very noticeable difference between the spins and energies of these states in nuclei with neutron numbers (N) above or below 90. This is the neutron number which corresponds to a rapid increase in the nuclear deformation revealed by the atomic isotope

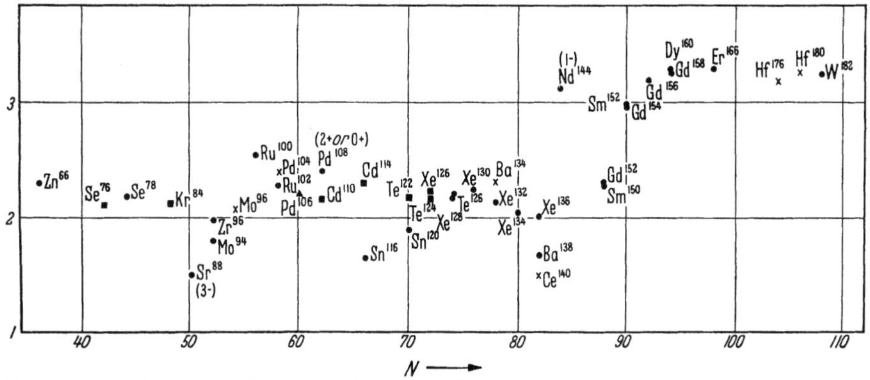

Fig. 61. Ratio of the energy of the second excited states to that of the first, according to SCHARFF-GOLDHABER and WENESER.

shift[3,4] and particularly in the electric quadrupole moments[3] of Eu^{151} and Eu^{153}. For $N \geq 90$ the second excited states have spin 4 and even parity (except in Pb^{208} and other nuclei in this region); their energies (E_2) relative to those of the first excited states (E_1) are just those to be expected from rotational states (Sect. 76), viz., $E_2/E_1 = \frac{10}{3}$, and when they deviate from this ratio they deviate in the manner to be expected from the distortion of the simple rotational energy spectrum by nuclear vibrations. For $N < 90$ the energies of the second excited states are less regular. The ratio E_2/E_1 is usually about 2 and tends to fall as E_1 increases (see Fig. 61); the spins are more often 2 than 4. The very sharp distinction between the two kinds of states is clearly demonstrated at $N = 90$ by the dual decay of Eu^{152}. In this nucleus both the ground state and the isomeric state develop rotational levels in Sm^{152} ($N = 90$) by positron emission and E_2 and E_1 show[5] the regular rotational relationship, $E_1 = 123$, $E_2 = 367$ kev, $E_2/E_1 = 3.0$; while for Gd^{152} ($N = 88$), which is obtained by negatron emission, $E_1 = 344$, $E_2 = 754$ kev, and $E_2/E_1 = 2.2$. Corresponding relationships are shown[6] by low lying levels excited by COULOMB excitation in Eu^{153} ($N = 90$) and in Eu^{151} ($N = 88$).

[1] A. DE-SHALIT and M. GOLDHABER: Phys. Rev. **92**, 1211 (1953).
[2] M. NAGASAKI and T. TAMURA: Progr. Theor. Phys., Japan **12**, 248 (1954). — G. SCHARFF-GOLDHABER and J. WENESER: Phys. Rev. **98**, 212 (1955).
[3] P. BRIX and H. KOPFERMANN: Z. Physik **126**, 344 (1944).
[4] P. BRIX: Z. Physik **132**, 579 (1952).
[5] E. CHURCH and M. GOLDHABER: Phys. Rev. **95**, 626 (1954). — R. E. SLATTERY, D. C. LU and M. L. WIEDENBECK: Phys. Rev. **96**, 465 (1954).
[6] N. P. HEYDENBURG and G. M. TEMMER: Ref. 2, p. 329; also Bull. Amer. Phys. Soc. **1**, 165 (1956).

The specifically rotational character of the lowest states disappears in Ir and Pt as the neutron number is increased. It reappears after the closed shells of Pb^{208} is passed. According to SCHARFF-GOLDHABER[1], the reappearance of rotational features in the heaviest elements is as sudden as its appearance in the rare earth nuclei; but it is connected now with the charge and not the neutron number. The lightest element for which E_2/E_1 is near 3 is Ra ($Z=88$). Thus in this region it seems that the critical number is 88 rather than 90. The explanation of this remarkable phenomenon must lie in some instability in the deformation; the probable cause has been noted by MOTTELSON and NILSSON[2] in terms of the ordering of the levels in a spheroidal potential well (see Sect. 81).

For $N<90$ the second excited states, generally, do not originate in nuclear rotation; however, the ratios of the energies of the first and second excited states suggest that they are produced by nuclear vibrations rather than particle excitations. This is confirmed by the transition probabilities of the $E2$ radiation emitted which are always much higher than one would expect if these states were derived from single particle excitations. KRAUSHAAR and GOLDHABER[3] were the first to note that, when the spins of both first and second excited states are 2 units, the electric quadrupole transition probability between the two states is usually much greater than that of the cross-over transition. In fact, the matrix elements of the cross-over transitions are, with few exceptions, only a few percent of those of the former. GOLDHABER and DE SHALIT[4] have attempted to explain this phenomenon for nuclei near closed shells in terms of selection rules for seniority number, but at the present time it seems more probable that the correct explanation is to be found on the basis of the collective model; in fact, WILETS and JEAN[5] have found that it is quite possible to account for the effect in terms of nuclear vibrations (see Sect. 95).

70. Measurement of transition probabilities. The first estimates of the order of magnitude of the lifetimes of excited states of nuclei were obtained[6] from studies of the competition between γ-ray and α-particle emission in certain α-active bodies (the "long-range" α-emitters, e.g., ThC′, RaC′). The results obtained ($\sim 10^{-12}$ sec) showed that nuclear lifetimes are indeed very short. Such methods, however, are of very limited application: in recent years the efforts of experimenters have been aimed at measurements of lifetimes by the method of delayed coincidences, by resonance fluorescence, and by COULOMB excitation. The latter are of particular value for the measurement of the lifetimes of electric transitions too fast to be determined by delayed coincidences, i.e., 10^{-11} sec or less. Another method, which depends on the DOPPLER shift which appears in the radiation emitted by a recoiling nucleus[7], is also of rather limited application and so far has been applied only to light nuclei. Recently, a broadening caused by recoil has been detected in the L_{I} conversion line (the "A" line) of the 40 kev transition in ThC″ (Tl^{208}) by BURDE and COHEN[8], who showed that the lifetime must be longer than 2×10^{-13} sec.

---

[1] G. SCHARFF-GOLDHABER: Phys. Rev. **103**, 837 (1956).
[2] See ref. 7, p. 347.
[3] J. J. KRAUSHAAR and M. GOLDHABER: Phys. Rev. **89**, 1081 (1953).
[4] See ref. 1, p. 323.
[5] L. WILETS and M. JEAN: Phys. Rev. **102**, 788 (1956); these authors give a rather complete list of the energies and spins of the second excited states in even-even nuclei.
[6] M. DELBRÜCK and G. GAMOW: Z. Physik **72**, 492 (1931). See also E. RUTHERFORD W. B. LEWIS and B. V. BOWDEN: Proc. Roy. Soc. Lond., Ser. A **142**, 347 (1933).
[7] S. DEVONS, G. MANNING and D. ST. P. BUNBURY: Proc. Phys. Soc. Lond. A **68**, 18 (1955).
[8] J. BURDE and S. G. COHEN: Phys. Rev. **101**, 495 (1956).

In the method of delayed coincidences, the coincidence counting rate between detectors responding selectively to genetically related particles, e.g., a β- or α-particle and a γ-ray or an electron or between γ-rays, is plotted as a function of a time delay inserted in the electronic circuits used to detect one of them. The development of this method and of improved electronic techniques has led to the use of resolution times of between 10^{-9} and 10^{-10} sec. We shall not discuss these methods here, for they have been very adequately reported by BELL [19], [20]. Further analysis of the experimental data[1] has made it possible to measure lifetimes of the order of 10^{-11} sec. The limit to the method seems to be determined by the scintillating properties of the phosphor and its physical size.

The resonance fluorescence method has not yet been used as extensively as the method of delayed coincidences. It has been developed by MOON and his co-workers[2]; by METZGER; and by MALMFORS. In this method a beam of γ-radiation emitted by nuclei in the excited states to be studied is allowed to fall on material consisting, at least in part, of the atoms of the element from which it is emitted. With suitable energy-selective counters, a small amount of elastic scattering can be detected, especially in the forward direction. In general this consists of RAYLEIGH, THOMSON, or DELBRÜCK scattering, which, in heavy elements, may amount (for 0.5 Mev radiation) to 10 mb per steradian at angles near 90°. The required resonance radiation must be observed superposed on a background caused by elastic scattering of this type; for this reason, and for practical reasons concerned with screening the detector from the source, the resonance scattering is usually observed in the backward direction, between 90 and 150°.

If the scattering atoms are at rest, and if the source is initially at rest, the necessary resonance condition for the production of resonance scattering will not exist because the energy of the γ-ray emitted will be reduced below that required to excite the nucleus of the scatterer by an amount depending on the nuclear recoil: this energy is usually far greater than the natural width of the state. The resonance condition can be realised by moving the source mechanically with the proper velocity, thereby giving the γ-radiation a DOPPLER shift; by heating the source, thereby giving the γ-ray a DOPPLER broadening; or making use of the DOPPLER shift which results from the recoil of the nucleus caused by the preceding disintegration.

The first method, which was that used by MOON, who first detected the effect, made use of a Au^{198} source electroplated on to the periphery of a mechanical rotor. The details of both first and second methods have been summarized by MALMFORS [19]. The lifetime of the state can be calculated from the measured cross section for the process, the γ-ray energy, and from the DOPPLER effect. It is instructive to compare the lifetimes obtained by different methods. The lifetime of the 411 kev state of Hg^{198} found by GRAHAM and BELL[3] by the method of delayed coincidences was $(1.0 \pm 0.7) \times 10^{-11}$ sec; that found by MOON and DAVEY[4] by mechanical motion was $(2.2 \pm 0.5) \times 10^{-11}$ sec; by MALMFORS[5] (by heating), 3.5×10^{-11} sec; and by METZGER and TODD[6], $(2.2 \pm 0.2) \times 10^{-11}$ sec. In the same way, the lifetime of the 209 kev state of Hg^{199} found by METZGER and TODD[7],

[1] Z. BAY: Phys. Rev. 77, 419 (1950). — T. D. NEWTON: Phys. Rev. 78, 490 (1950). See also Z. BAY, V. P. HENRI and F. MCLERNON: Phys. Rev. 97, 561 (1955).

[2] P. B. MOON: Proc. Phys. Soc. Lond. A 64, 76 (1951). — P. B. MOON and A. STORRUSTE: Proc. Phys. Soc. Lond. A 66, 585 (1953).

[3] R. L. GRAHAM and R. E. BELL: Phys. Rev. 84, 380 (1951).

[4] P. B. MOON and W. G. DAVEY: Proc. Phys. Soc. Lond. A 66, 956 (1953).

[5] K. G. MALMFORS: Ark. Fysik 6, 49 (1952).

[6] F. R. METZGER and W. B. TODD: Phys. Rev. 95, 853 (1954).

[7] F. R. METZGER and W. B. TODD: Phys. Rev. 94, 794 (1954).

with a heating method, was $(3.1 \pm 0.9) \times 10^{-10}$ sec, while that found by BELL, GRAHAM and PETCH[1] by delayed coincidences was 3.6×10^{-10} sec.

The third method mentioned above seems very promising and of more general application than the other two. In this method use is made of the DOPPLER shift which results from the nuclear recoil produced in the preceding transition. For this method to be of use it is obviously necessary that the source should emit the γ-ray before the recoiling nucleus suffers a collision in the source material. Experiments using solid sources have generally been unsuccessful; however, MOON and ILAKOVAC[2] found a strong effect with a liquid source of Zn^{63}. The possibilities of using gaseous sources have been explored by METZGER[3]; he found a very strong effect when the 596 and 835 kev γ-rays emitted respectively by As^{74} and As^{72} are scattered by metallic germanium. The lifetimes for γ-ray emission were found to be $(1.9 \pm 0.3) \times 10^{-11}$ sec and $(4.6 \pm 1.2) \times 10^{-13}$ sec respectively.

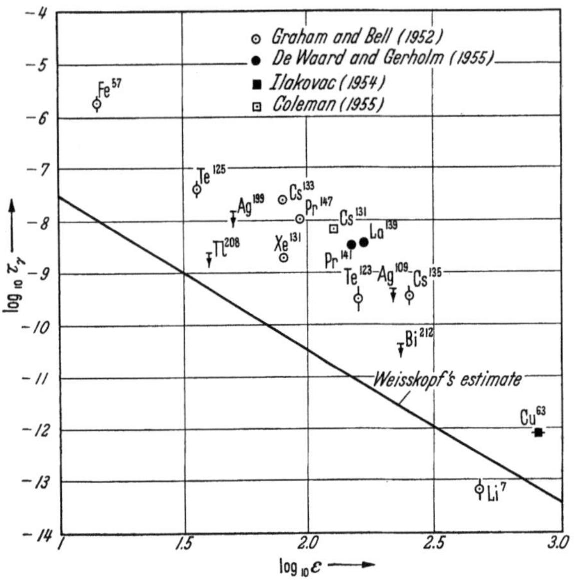

Fig. 62. Lifetimes of states decaying by $M1$ γ-rays. The γ-ray energy (ε) is in kev.

It should be pointed out that measurements on the angular distribution of the resonance radiation may determine the difference in the spin of the target nucleus in its ground state and in its excited state, for the angular correlation with respect to the direction of incidence of the γ-ray beam is just that to be expected from the usual angular correlation of successive γ-rays for which the spin succession is $J_0 \to J \to J_0$. Although it is practical to make measurements only in a limited range of backward directions, the angular distribution is usually sufficiently pronounced to determine the spin of the excited state with certainty at least in the case of even-even nuclei, for which the spin of the ground state is zero[4].

71. Transition probabilities: $M1$ transitions. The lifetimes of a number of predominantly $M1$ transitions have been measured by the method of delayed coincidences by GRAHAM and BELL[5] and by DEWAARD and GERHOLM[6]; and by the resonance fluorescence method (for V^{51}) by SCHOPPER[7] and by others. The experimental data gives the lifetimes of the excited state against both γ-ray emission and against internal conversion. The lifetime against γ-emission alone

[1] R. E. BELL, R. L. GRAHAM and W. PETCH: Canad. J. Phys. **30**, 35 (1952).
[2] P. B. MOON and K. ILAKOVAC: Phys. Rev. **93**, 254 (1954). — Proc. Phys. Soc. Lond. A **67**, 601 (1954).
[3] F. R. METZGER: Phys. Rev. **101**, 286 (1956).
[4] This has been done for the first excited states (spin 2) in Hg^{198} and Hg^{202}. [F. R. METZGER: Phys. Rev. **97**, 1258 (1955); **98**, 200 (1955)].
[5] R. L. GRAHAM and R. E. BELL: Canad. J. Phys. **31**, 377 (1953).
[6] H. DEWAARD and T. R. GERHOLM: Physica **21**, 599, 699 (1955).
[7] H. SCHOPPER: Z. Physik **144**, 476 (1956).

is obtained by multiplying the experimental lifetimes by $(1+\alpha)$ where α is the total internal conversion coefficient. Some results for $M1$ transitions are plotted against the energy in Fig. 62. It will be seen that most transitions are slower than that predicted by WEISSKOPF'S single particle estimate[1]. This is not surprising for WEISSKOPF'S estimates represent upper limits for the transition probability. Generally odd proton nuclei radiate more slowly than odd neutron nuclei. This is the reverse of what we should expect if the magnetic component is due to the reorientation of a spin of a neutron or proton, for the magnetic moment of the proton is numerically larger than that of the neutron[2]. Moreover $M1$ transitions in odd-odd nuclei appear to be the faster.

Recently, the theoretical aspects of magnetic dipole radiation have been very exhaustively discussed by DEWAARD and GERHOLM[3], who have, in addition, remeasured the lifetimes of some of the l-forbidden $M1$ transitions included in Fig. 62, (e.g. Pr^{147}, Cs^{133}, Cs^{135}), and measured some others, e.g.[4] Tl^{203}. These lifetimes, when corrected by the cube of the energy were found to differ very little from one another.

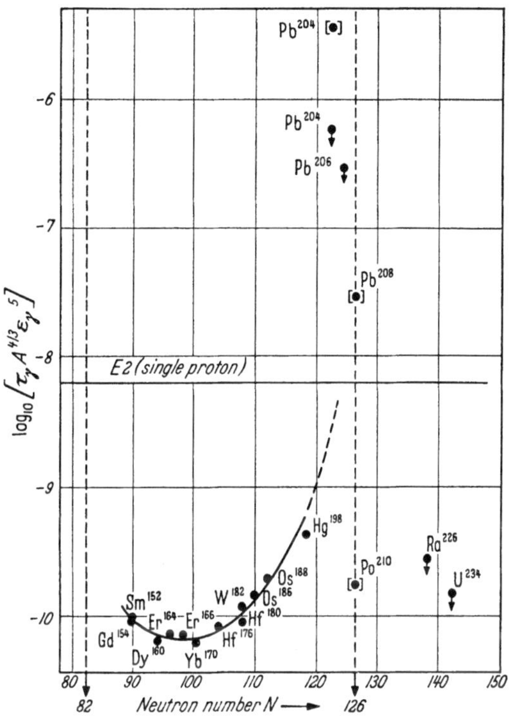

Fig. 63. Lifetimes of states decaying by $E2$ emission, according to SUNYAR.

72. Transition probabilities: *E*2 transitions. We shall discuss here only the data for $E2$ transitions obtained by the methods described in Sect. 70; COULOMB excitation will be discussed in the next section. The lifetimes of $E2$ transitions are difficult to measure for not merely are the γ-ray lifetimes very much shorter than that predicted by the single particle formulae, but for low energy transitions, where the γ-ray lifetimes would otherwise be the longer, the decay is accelerated by internal conversion. Naturally the greater part of the data concerns only the lowest excited states.

The lifetimes of $E2$ transitions in nuclei of intermediate atomic weight have usually proved too short to measure by the method of delayed coincidences. Of nine nuclei between Mg^{24} and Te^{124}, COLEMAN[5] has succeeded in measuring the lifetime of only the first excited state of Te^{122}. The upper limits of the lifetimes of the rest are of the order of 10^{-10} sec. In heavier nuclei, the energies of the first

[1] See ref, 1, p. 310.

[2] According to MOSZKOWSKI [*19*], the ratio of the lifetimes of $M1$ radiations from odd-neutron to odd-proton nuclei should be $[(\mu_p - \frac{1}{2})/\mu_n]^2$ for single particle transitions; cf. J. W. MIHELICH and A. DE SHALIT, ref. 4, p. 371.

[3] H. DEWAARD and T. R. GERHOLM: Nuclear Phys. **1**, 281 (1956).

[4] Cf. R. E. AZUMA and G. M. LEWIS: Phil. Mag. **46**, 1034 (1953).

[5] C. F. COLEMAN: Phil. Mag. **46**, 1135 (1955).

excited states are lower and their lifetimes come within the range of direct electronic measurement. Recently a number of these lifetimes have been measured by SUNYAR[1], who found that they are shorter than those calculated from WEISSKOPF's formula, (except some transitions in the Pb nuclei) by a factor of about one hundred. This phenomenon was first pointed out by GOLDHABER and SUNYAR[2]; it persists even into the closed shell nuclei (Sects. 92 and 96). SUNYAR's results are plotted in Fig. 63. We shall show in Sect. 74 that transitions in nuclei of lower atomic weight are also fast compared with single particle estimates. It is probable that the speed of these transitions is caused by collective motions in which the whole or the greater part of the nuclear charge is effective.

73. COULOMB excitation. The transition probabilities of a large number of $E2$ transitions have been measured in the last few years by COULOMB excitation. The method is suitable mainly for transitions which possess an electric component, and is the more sensitive the faster the transition; it is therefore the more convenient in the region where direct electronic methods of measuring the lifetime of a state are not feasible. Apart from the inevitable complications entailed by the use of a high tension generator, experimentally the method has the advantage of simplicity. A target containing the nuclei to be examined is bombarded by protons or α-particles with energies of a few Mev; these nuclei interact with the rapidly changing electric field of the bombarding particles as they move by and are excited by the appropriate electric multipole moment. The γ-ray or the corresponding internal conversion electrons produced by the decay of the excited state is detected by a NaI crystal, with pulse height analysis. The cross section for the production of this γ-ray is directly related to the transition probability; it is determined from the beam current, the thickness of the target and the efficiency of the detector which may be found by calibration with suitable γ-rays of known absolute intensity.

The possibility of detecting COULOMB excitation was discussed theoretically[3] before the γ-rays produced by this process were identified. It was first found in Ta^{181} independently by HUUS and ZUPANČIČ[4] and by MCCELLAND and GOODMAN[4].

Since the lowest levels of even-even nuclei almost always have the same parity as the ground state and spin 2, it is generally assumed that COULOMB excitation in heavy elements is mainly an electric quadrupole process and that the cross section observed experimentally measures the $E2$ transition probability. Similar assumptions are made for odd-A nuclei. In the majority of nuclei this assumption is justified. However, as BJERREGAARD and HUUS[5] have shown, the multipole order can be determined directly by measurements with bombarding particles with different specific charge, e.g., protons and deuterons (or α-particles). They have verified that the multipole order is indeed two for the excitation of the lowest excited states in the even-even wolfram nuclei.

Proton bombardment in heavy elements produces both bremsstrahlung and a strong emission of K-radiation which can interfere seriously with the measurement of the γ-rays. The yield of this radiation is in fair agreement with theoretical predictions[6]. In part it may be removed by filtration. Recently the use of helium

[1] A. SUNYAR: Phys. Rev. **98**, 653 (1955).
[2] See ref. 1, p. 321.
[3] R. HUBY and H. C. NEWNS: Proc. Phys. Soc. Lond. A **64**, 619 (1951). — C. T. MULLIN and E. GUTH: Phys. Rev. **82**, 141 (1951).
[4] T. HUUS and C. ZUPANČIČ: Dan. Mat. fys. Medd. **28**, No. 1 (1953). — C. L. MCCELLAND and C. GOODMAN: Phys. Rev. **91**, 760 (1953).
[5] J. H. BJERREGAARD and T. HUUS: Phys. Rev. **94**, 204 (1954).
[6] W. HENNEBERG: Z. Physik **86**, 592 (1933). See also C. ZUPANČIČ and T. HUUS: Phys. Rev. **94**, 205 (1954).

ions, either singly charged, or as α-particles, has been found to reduce the emission of X-rays. For example TEMMER and HEYDENBURG[1] using 3 Mev α-particles found that the K-emission from Ta was seven times lower than when that material was irradiated with a similar proton current. Such K-radiation as remains seems to be no more than that expected to follow internal conversion[2]. K-radiation and bremsstrahlung nevertheless are troublesome in the heaviest elements even when α-particles are used.

The calculation of the COULOMB excitation can be treated by a semi-classical approximation. If v is the velocity, and Z_2 and Z_1 are the charges on the target nucleus and the bombarding particle respectively, the quantity $\eta = 2Z_1 Z_2 e^2/\hbar v$ is greater than unity, and the cross sections can be calculated from the classical trajectories of the bombarding particles. In general the total cross section for COULOMB excitation for multipole order l is[3]:

$$\left.\begin{aligned} \sigma_l &= \left(\frac{Z_1 e^2}{\hbar v}\right)^2 \frac{B(l)\, g_l(\xi)}{e^2 a^{2l-2}} \left(\frac{v_f}{v_i}\right)^{2l-2}, \\ a &= \frac{Z_1 Z_2 e^2}{M v_i^2} \end{aligned}\right\} \tag{73.1}$$

where v_i and v_f are the initial and final velocities of the bombarding particle, respectively; v, the velocity in the first term in the right-hand side is equal to v_i for electric radiation, and equal to the velocity of light for magnetic radiation; $B(l)$ is the "reduced" transition probability[4] for radiation of order l, i.e., the square of the multipole moment; and $g_l(\xi)$ is a function of the trajectories of the particles[5], ξ being effectively the ratio of the collision time to the nuclear time, viz.:

$$\xi = \frac{Z_1 Z_2 e^2}{\hbar}\left(\frac{1}{v_f} - \frac{1}{v_i}\right).$$

For $E2$ excitation, (73.1) reduces to:

$$\sigma = \left(\frac{M v}{Z_2 e \hbar}\right)^2 B(E2)\, g_2(\xi) \tag{73.2}$$

with transition probability:

$$T(E2) = \frac{4\pi}{75\hbar}\left(\frac{E}{\hbar c}\right)^5 B(E2). \tag{73.3}$$

The quantity $g_2(\xi)$ is near unity for small ξ and falls rapidly with increasing ξ, in fact, exponentially for $\xi > 0.5$ (see Fig. 64). Thus the cross section for a given multipole order is determined mainly by the value of ξ and for higher energy states this cross section will be measurable only when bombarding particles of higher energy or specific charge are used. Measured cross sections are usually of the order of a millibarn.

---

[1] G. M. TEMMER and N. P. HEYDENBURG: Phys. Rev. **93**, 351 (1953).

[2] N. P. HEYDENBURG and G. M. TEMMER: Phys. Rev. **100**, 150 (1955).

[3] K. A. TER-MARTIROSYAN: J. exp. theor. Phys. USSR. **22**, 284 (1952). — K. ALDER and A. WINTHER: Phys. Rev. **91**, 1578 (1953). Also CERN reports, T/KA-AW-1 (1954) and T/KA-AW-4, 1955.

[4] The reduced transition probability is related to the transition probability T_l per unit time by the formula

$$T_l = \frac{8\pi(l+1)}{l[(2l+1)!!]^2} \cdot \frac{1}{\hbar}\left(\frac{\varepsilon}{\hbar c}\right)^{2l+1} B(l)$$

where ε is the transition energy and $(2l+1)!! = 1 \cdot 3 \cdot 5 \ldots (2l+1)$ (see ref. [4], Chap. XII).

[5] The quantity g is identical with that denoted by f by ALDER and WINTHER.

The cross section for higher multipole orders falls by factors of the order of 100 to 1000 from one multipole order to another. The cross sections for $M1$ excitation are generally somewhat less than those of $E3$ excitation. Recently $E3$ excitation seems to have been observed by JONES and PHILLIPS[1].

For 2 Mev protons and for $Z_2 = 80$, $\eta = 9$. Thus η is never very large compared with unity and the classical approximation may be in error. The exact solution of this problem has occupied the attention of several authors[2]. According to ALDER and WINTHER[3], the value of $g_2(\xi)$ deduced from a quantum mechanical theory does not differ from the classical approximation by more than 1% provided that $\eta \gtrsim 10$. For lower values these authors give suitable corrections; their values for $g_2(\xi)$ are plotted in Fig. 64.

To obtain the reduced transition probability $B(E2)$ for decay from (73.2) it must be remembered that the transition probability for absorption differs from that of decay by the ratio of the statistical weights of the ground and of the excited states:

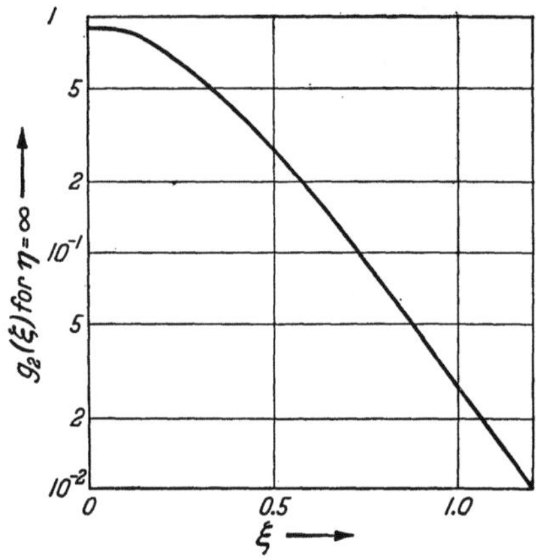

Fig. 64. $g_2(\xi)$ according to ALDER and WINTHER.

$$B(E2)_{\text{decay}}/(2I+1) = B(E2)_{\text{abs}}/(2J+1)$$

where I and J are the spins of the ground state and excited state respectively. For the first excited states in even-even nuclei, these transition probabilities differ by a factor of 5. Owing to internal conversion, only a fraction of the number of states created by COULOMB excitation will decay by γ-emission. For the heavy even-even nuclei, in general, only one state (the first excited state) can be excited; as this state will decay to the ground state with the emission of pure $E2$ radiation, there is uncertainty in the interpretation of the cross sections only to the extent that the conversion coefficients for low energy γ-rays are not always well known. For many even-odd nuclei, however, rotational states (Sect. 76) occur, and the spins of these states differ by one unit. Therefore two states in one rotational family can be directly excited by $E2$ absorption, and two of the three γ-rays, in general, will contain $M1$ components. When the ratio of $M1$ to $E2$ components can be obtained (from angular correlation and polarization measurements or from measurements of conversion) the $M1$ transition probability can be found (Sect. 77).

Detailed calculations on the angular distribution of the γ-rays excited by COULOMB excitation have been given recently by ALDER and WINTHER[4] and by BIEDENHARN, MCHALE, and THALER[5]. The angular distribution is rather similar to that to be expected from the correlation of successive γ-rays: in this instance,

---

[1] G. A. JONES and W. R. PHILLIPS: To be published.

[2] F. D. BENEDICT, P. B. DAITCH and G. BREIT: Phys. Rev. **101**, 171 (1956). — L. C. BIEDENHARN, J. L. MCHALE and R. M. THALER: Phys. Rev. **100**, 376 (1956). — L. C. BIEDENHARN, M. GOLDSTEIN, J. L. MCHALE and R. M. THALER: Phys. Rev. **101**, 662 (1956). — R. L. GLUCKSTERN, J. P. LAZARUS and G. BREIT: Phys. Rev. **101**, 175 (1956).

[3] K. ALDER and A. WINTHER: Dan. Mat. fys. Medd. **29**, No. 18 (1955).

[4] See ref. 3, p. 329.

[5] See ref. 2, above; see also G. BREIT, M. E. EBEL and J. E. RUSSELL: Phys. Rev. **101**, 1504 (1955).

an $E2$ γ-ray is absorbed and an $M1$ or $E2$ γ-ray or a mixture of both is emitted. The direction which determines the angular distribution is the direction of the beam of particles. If ϑ is the angle of emission with respect to this direction, the angular distribution should contain only second and fourth order spherical harmonics. According to ALDER and WINTHER[1] the angular distribution can be represented by:

$$W(\vartheta) = 1 + B_2 a_2(\xi) P_2(\cos\vartheta) + B_4 a_4(\xi) P_4(\cos\vartheta)$$

which differs from the angular correlations of ordinary γ-ray cascades by the a coefficients which depend on the energy through the parameter ξ. The B coefficients have been given by BIEDENHARN and ROSE[2]. According to ALDER and WINTHER, a_4 is very small except when ξ is very small, and a_2 is nearly constant at 0.7.

For Ta^{181} the spin of the ground state is $\frac{7}{2}$, that of the excited state (137 kev) is $\frac{9}{2}$, and the expected angular distribution is very nearly isotropic. Experimentally, the angular distribution of this γ-ray is indeed found to be isotropic[3]. The angular distribution of the 303 kev γ-ray[3], however, although found[3] to be in agreement with ALDER and WINTHER'S calculations for 3 Mev protons, differs at higher energies[4]. For even-even nuclei the anisotropy for the γ-ray produced by the first excited state should be about 30% and nearly independent of the energy. Recent work by MCGOWAN and STELSON[4] on the γ-rays excited in a number of heavy nuclei shows that the a coefficients derived from the experimental data depend quite strongly on the energy in evident disagreement with the curves published by ALDER and WINTHER. According to BENEDICT[5] the latter authors have made an error in the numerical calculation of the a's and that when these quantities are correctly evaluated the theoretical result is in good accord with the experimental data of MCGOWAN and STELSON for the Pt isotopes. The calculations of BIEDENHARN, MCHALE, and THALER also predict a small value for a_4 but a greater variation of a_2 in a useful experimental range, and are in accord with experimental results.

74. $E2$ transition probabilities from COULOMB excitation. Some reduced transition probabilities obtained by COULOMB excitation are listed in Table 6. These quantities are expressed in units of $e^2 \cdot 10^{-48}$ cm^4. The reduced transition probability corresponding to WEISSKOPF'S estimate of the transition probability of $E2$ radiation, in the same units, is given by $3\times 10^{-6} R^4$, where R is the nuclear radius in units of 10^{-13} cm. For $A=180$, $R=8$, the single particle estimate is 0.012; for $A=50$, it is only 0.002. Inspection of Table 6 will show that the reduced transition probabilities for $E2$ transitions are consistently high compared with this estimate. They are high whether the neutron number is above 90 or whether it is below it[4], where, as we have seen, rotational states are not certainly known to exist. This suggests, as pointed out in Sect. 72, that cooperative effects extend over a very wide range of mass numbers.

It is interesting to compare the reduced transition probabilities obtained by COULOMB excitation with those obtained by the delayed coincidence methods.

---

[1] See ref. 3, p. 329.
[2] L. C. BIEDENHARN and M. E. ROSE: Rev. Mod. Phys. **25**, 729 (1953).
[3] J. T. EISINGER, C. F. COOK and C. M. CLASS: Phys. Rev. **94**, 735 (1954).
[4] F. K. MCGOWAN and P. H. STELSON: Phys. Rev. **99**, 127 (1955).
[5] F. D. BENEDICT: Phys. Rev. **101**, 178 (1956). — F. D. BENEDICT and G. TICE: Phys. Rev. **100**, 1545 (1955).
[6] G. M. TEMMER and N. P. HEYDENBURG: Phys. Rev. **96**, 426 (1954).

This is done in Table 7 which is taken from a paper by HUUS, BJERREGAARD and ELBEK[1]; the agreement is good.

Table 6a. *Even-even nuclei.*

Values of B (E2) (for excitation) obtained by COULOMB excitation in units of $e^2 \times 10^{-48}$ cm^4. S. and M.: STELSON and McGOWAN, see ref. 3, p. 339; M. and P.: H. MARK and G. T. PAULISSEN, Phys. Rev. **100**, 813 (1955); H. and T.: HEYDENBURG and TEMMER, see ref. 2, p. 329; H. B. and E.: HUUS, BJERREGAARD, ELBEK, see ref. 1, below; N: J. O. NEWTON, to be published. Note that HEYDENBURG and TEMMER's results are uncorrected for internal conversion, and (for odd mass numbers) for cross-over γ-rays; they are therefore a lower limit.

Nucleus	S. and M. E (kev)	S. and M. B (E2)	H. and T. E (kev)	H. and T. B (E2)	M. and P. E (kev)	M. and P. B (E2)	H., B. and E. E (kev)	H., B. and E. B (E2)
Mo^{98}	770	0.41						
Mo^{100}	540	0.61						
Pd^{104}	575	0.54						
Pd^{106}	520	0.72						
Pd^{110}	380	0.96						
Nd^{146}			455	0.84				
Nd^{148}			300	0.50				
Nd^{150}			128	1.24			132	1.6
Sm^{148}			562	2.06	550	0.74		
Sm^{150}			337	2.32	335	0.52		
Sm^{152}			122	1.36	125	1.1	123	2.3
Sm^{154}			82	0.48	84	2.0	83	6.2
Gd^{154}			123	2.1	123	2.6	124	≈ 2.8
Gd^{156}			89	1.24	89	5.0	90	6.0
Gd^{158}			79	1.02	80	6.0	80	6.5
Gd^{160}			76	≈ 1	76	9.5	76	6.4
Dy^{162} }			76	0.23			82	6.1
Dy^{164} }							75	8.5
Er^{162} }								
Er^{164} }								
Er^{166} }			79	0.48			81	6.8
Er^{168} }								
Er^{170} }								
Yb^{170} }								
Yb^{172} }			78	0.28			77	6.8
Yb^{174} }								
Yb^{176} }								
Hf^{176} }			87	0.56				
Hf^{178} }	90	6.2	90	0.85			95	≈ 6.9
Hf^{180} }			93	0.78				
W^{182} }							100	5.5
W^{184} }	112	≈ 5					112	4.0
W^{186} }							123	3.8
Pt^{194}	330	1.8						
Pt^{196}	358	1.3						
Pt^{198}	403	1.4						
Pb^{206}	810	0.13						
Th^{232}	760	0.13						
	N.							
U^{238}	44	10.7						

[1] T. HUUS, J. H. BJERREGAARD and B. ELBEK: Dan. Mat. fys. Medd. **30**, No. 17 (1956).

Table 6b. *Odd-A nuclei.*

S. and M., M. and P., H., B. and E. and N., as in heading to Table 6a; H. and T.: N. P. HEYDENBURG and G. M. TEMMER, Fe to Ge, Phys. Rev. **96**, 426 (1954); Rh and Ag, Phys. Rev. **95**, 861 (1954); Nd to Ta, see ref. 2, p. 329. F. *et al.*: L. W. FAGG, E. A. WOLICKI, R. O. BONDELID, K. L. DUNNING and S. SNYDER, Phys. Rev. **100**, 1299 (1955).

Nucleus	S. and M.		H. and T.		M. and P.		H., B. and E.		F. *et al.*	
	E (kev)	$B(E2)$	E (kev)	$B(E2)$	E (kev)	$B(E2)$	E (kev)	$B(E2)$	E (kev)	$B(E2)$
Fe^{57}			137	0.05						
Zn^{67}			182	0.04						
Ge^{73}			68	0.04						
Mo^{95}	212	0.04								
Rh^{103}	305	0.2	295	0.18						
	365	0.36	357	0.32						
Ag^{107}	325	0.23	318	0.12			323	≈ 0.15	319	0.16
	427	0.36	413	0.21					419	0.23
Ag^{109}			305	0.13			308	≈ 0.15	306	0.18
			412	0.23					412	0.31
Te^{123}									159	0.02
Te^{125}									466	0.44
									668	0.26
Nd^{145}			70	0.03						
Eu^{151}			310	0.64	300	—				
Eu^{153}			82	0.36	85	1.0	84	1.6		
							195	$\gtrsim 0.6$		
Gd^{155}			145	0.11	140	—				
Gd^{157}			131	0.08	127	—				
Tb^{159}			79	0.19	77	—	58	~ 3.5		
			136	0.04						
Ho^{165}			94	0.054			96	~ 2.5		
			206	0.036			212	~ 0.76		
Tm^{169}			109	1.12			119	3.7		
Yb^{171}			110	0.20						
Yb^{173}			180	0.086						
Lu^{175}			114	0.72			114	3.2		
			250	0.20						
Hf^{177}	}		112	0.77						
	}		250	0.55						
Hf^{179}	} 112	2.4	119	0.67			114	~ 1.5		
	}		260	0.056						
Ta^{181}	137	2.4	136	0.70			136	2.3		
	303	0.6	303	0.15			301	0.64		
W^{183}	295	0.27					46	~ 2.8		
Re^{185}							125	~ 2.7		
Re^{187}							135	~ 1.6		
Ir^{191}							129	0.6		
Ir^{193}							139	0.4		
Pt^{195}	210	~ 0.5					210	0.5		
Au^{197}	191	0.08								
	277	0.31					281	0.28		
Tl^{203}	279	0.13								
	410	0.15								
Tl^{205}	205	0.15								
	410	0.12								
Pb^{207}	570	0.028								
	N.									
U^{233}	40	13								
	92	2.6								
U^{235}	46	7.7								
	103	1.2								
Np^{237}	34	4.9								
	42	2.6								
Pu^{239}	58	5.7								

Table 7. *Comparison of reduced transition probabilities in even-even nuclei obtained by* COULOMB *excitation* (HUUS, BJERREGAARD *and* ELBEK) *and by the delayed coincidence method* (SUNYAR).

Nucleus	Energy kev	$\tau_{\frac{1}{2}}$ 10^{-9} sec	$B(E2)$ delayed coinidences $e^2 \cdot 10^{-48}$ cm^4	$B(E2)$ COULOMB excitation $e^2 \cdot 10^{-48}$ cm^4
Sm^{152}	122	1.4	3.3	2.3
Gd^{154}	123	1.2	3.4	≈ 2.8
Er^{166}	81	1.7	6.0	6.8
Yb^{170}	84	1.57	5.2	6.8
Hf^{180}	93	1.4	4.8	~ 6.9
W^{182}	100	1.3	4.1	5.5

N. Low energy states in deformed nuclei with $N > 90$.

75. General features. We have shown in Sect. 67 that the neutron number 90 is a convenient point of demarcation between the heavy nuclei which have well developed rotational states and the lighter nuclei which do not. In this chapter we shall consider only the excited states in the former. As pointed out in Sect. 69, the nuclei with rotational states divide into two groups; a lighter group with $N \geq 90$, and a heavier group, with $Z \geq 88$. The theoretical interpretation of the highly deformed nuclei is simplified by the assumption of strong coupling between the nucleons which lie outside the closed shells and the nuclear surface (the "unified" model). An understanding of the level structure in nuclei adjacent to the double closed shell nucleus Pb^{208}, e.g., isotopes of Hg, Au, and Pt, is much more difficult because in them the deformation is much smaller and the excited states, although possessed of collective properties, do not show obvious rotational characteristics. We shall not describe the level structure of these latter nuclei in detail.

Most of our present knowledge of the level structure in the deformed nuclei comes from α- and β-decay. The extent to which the product nucleus can be excited in α-decay is rather limited; in many nuclei a much greater range of levels can be excited by β-decay or by electron capture. So far the most detailed information on level structure has come from those β-active sources which can be prepared in high intensity, e.g., by activation in a nuclear reactor. The peculiar characteristics of α-decay, however, are of great interest and can be very informative; in the even-even nuclei for example the spins and parities of the states excited in the product nucleus are restricted by conservation of parity, and in odd-mass number nuclei, the extent to which such states are populated depends markedly on certain characteristics of the level structure. Precise measurements of the energies and intensities of the low energy neutron capture γ-rays and their rates of internal conversion provides a fertile, but, so far, untouched field of research. Studies with heavy particle bombardment hitherto have been of little avail, because (for charged particles) a high energy is required to surmount the COULOMB barrier and because a high degree of resolution is also required to distinguish the many groups of particles ejected from heavy nuclei.

The level structure of odd mass number nuclei is often very complicated and defied analysis until the recent advent of precise methods of energy measurement. The theory of the rotational states is of great assistance in the ordering of such level systems but a reliable interpretation still requires accurate measurements of internal conversion coefficients. In even-even nuclei the level system disclosed by β-decay is usually simpler in appearance: a low-lying band of related rotational levels based on the ground state, of which often only the first one or two are excited and another similar band or bands at a higher level of excitation (~ 1 Mev). The latter decay by transitions between themselves or by transitions to the ground state band, transitions which are sometimes highly converted. The intensities and the multipole orders of transitions between bands are determined, in addition to the usual selection rules for different multipole orders, by selection rules which have their origin in the peculiarities of the coupling between the incomplete shell and the surface. We shall discuss these selection rules in Sect. 80. The apparent absence of states at intermediate energies does not mean that no states exist in this region. No doubt such states exist: some must represent the higher states of the ground state system and because of their high spins are excited only in exceptional circumstances. Occasionally, as in the decay of the isomeric state of Hf^{180} (spin 9 units) as many as four states of the ground state band can be excited.

76. Rotational levels. Although the existence of surface oscillations and collective modes of motion has long been suspected[1], a reliable identification of nuclear energy levels with rotational states was made only quite recently[2]. The importance of this discovery can hardly be exaggerated; it has come about with the realisation that nuclei in general, not excluding even-even nuclei, do not have a spherical shape and that they are deformed by the motion of the nucleons which lie outside the closed shells[3]. The theory of rotational states developed by Bohr and Mottelson[2] not only provides formulae for the energies of the rotational states to which the experimental data can be accurately fitted, but it also accounts quantitatively for the transition probabilities between them. We shall give here only the briefest sketch of the theory for strong coupling between the nucleons and the surface; for details, the reader may consult references [*19*] and [*21*] and the more detailed references quoted below[2].

According to the model originally proposed by Bohr and Mottelson the deformed nucleus (which was supposed to have axial symmetry) was assumed to behave like an irrotational fluid. The rotation of such a nucleus can be regarded as the motion of a wave travelling round the surface of the nucleus. Such a wave carries angular momentum with it. The collective velocity at any point is proportional to the amplitude of the wave, and the moment of inertia, which is less than that of the rigid body, is therefore proportional to the square of this amplitude and proportional to the square of the deformation. Thus, if $\mathfrak{J}_0$ is the moment of inertia of the nucleus calculated on the basis of a rigid nucleus, the actual moment of inertia ($\mathfrak{J}$) is less than $\mathfrak{J}_0$; in fact, for small deformaties $\mathfrak{J} = \mathfrak{J}_0 (\Delta R/R)^2$ where ΔR is the difference between the major and minor axes of the spheroid, and $\mathfrak{J}_0 = \frac{2}{5} M A R^2$. In this we assume that the deformation which has cylindrical symmetry has the lowest energy and that the rotation in which we are interested is about an axis perpendicular to it.

Such a nucleus can also vibrate; the energy required to excite the nucleus into its first vibrational state seems to be of the order of 1 Mev in this mass region. The picture of the rotational process described above, therefore, is valid only if the amplitude of the oscillations is large compared with the zero-point amplitude of the vibration. Moreover the frequency of rotation must be small compared with that of vibration; when the energy of the first rotational state becomes comparable with that of the first vibrational state (1 Mev) the rotational features

[1] N. Bohr: See ref. 2, p. 203. See also S. Flügge: Ann. d. Phys. **39**, 373 (1941). — M. Fierz: Helv. phys. Acta **16**, 365 (1943).

[2] A. Bohr and B. Mottelson: Phys. Rev. **89**, 316 (1953); **90**, 717 (1953). — Dan. Mat. fys. Medd. **27**, No. 16 (1953).

[3] J. Rainwater: Phys. Rev. **79**, 432 (1950).

should disappear. Briefly, there are three kinds of excitation which can contribute: the particle motion, the collective vibrations, and the collective motion, with preservation of shape, which constitutes rotation. The appearance of purely rotational states depends on the wave function being separable into factors which contain exclusively only one of the parameters which describe these motions.

The electric quadrupole moment is defined by the volume integral $eQ_0 = \int \varrho(3z^2 - r^2)\, d\tau$ where z is chosen in the direction of the axis of symmetry and ϱ is the density of electric charge. If ϱ is uniform, $Q_0 = \frac{3}{\sqrt{5\pi}} Z R^2 \beta$, where β is a "deformation parameter". This "intrinsic" quadrupole moment is not identical with the spectroscopically measured electric quadrupole moment because of the nucleon-surface coupling, which causes a precession of the core angular momentum about the direction of the total spin. In the limit of strong coupling, which is all that we are concerned with here, the observed quadrupole moment is equal to Q_0 times the factor $I(2I-1)/(I+1)(2I+3)$. It should be noted that the moment of inertia $\mathfrak{J}$ and the electric quadrupole moment Q_0 are directly related: $\mathfrak{J} = \frac{M^2}{4\mathfrak{J}_0} \cdot \left(\frac{A}{Z}\right)^2 Q_0^2$. This relationship depends on the assumption that the moment of inertia is that of an irrotational fluid, an assumption which now appears to be incorrect (Sect. 78).

In the strong coupling model the angular momentum of the unpaired nucleon precesses about the axis of nuclear deformation with a constant projection $\Omega\hbar$. The total angular momentum $(J\hbar)$ is compounded of the nucleon angular momentum $(j\hbar)$ and the rotational angular momentum $(J_r\hbar)$, and its projection on the axis of symmetry is given by the quantum number K. In the absence of the excitation of vibrational states, rotation is developed only in directions perpendicular to the axis of symmetry and then $K = \Omega$. Associated rotational states correspond to a particular value of K (or Ω) and different rotational systems are developed on the basis of states with different values of K.

To a first approximation the rotational energy is given by $E_r = \hbar^2/2\mathfrak{J} \cdot J_r(J_r+1)$ where $J_r\hbar$ is the rotational angular momentum. In an *even-even* nucleus in which the particle configuration is not disturbed the total angular momentum $(J\hbar)$ is equal to the rotational angular momentum $(J_r\hbar)$ and the energies of the rotational states are given by:

$$E = \hbar^2/2\mathfrak{J} \cdot J(J+1) \tag{76.1}$$

where, for symmetry reasons, J takes only the even values 0, 2, 4, etc. From this equation it follows that the ratio of the energies of the excited states bear a simple relation to each other; the energy of the second to that of the first is $\frac{10}{3}$, the third to the first is 7. The highest rotational state yet detected in an even-even nucleus has $J = 8$ units (Hf^{180}).

Experimental values[1] for these ratios are plotted in Fig. 65. It will be seen that there is good agreement except near the closed shell nuclei where the energies of the first excited states rise rapidly as the shell is closed and the conditions for the existence of rotational states (viz., the frequencies for the rotational motions should be small in comparison with those of vibrations) no longer holds. In general the positions of the rotational states are more accurately determined by[2]:

$$E = \frac{\hbar^2}{2\mathfrak{J}} \cdot J(J+1) - \text{const}\, J^2(J+1)^2 \tag{76.2}$$

[1] F. ASARO and I. PERLMAN: Phys. Rev. **91**, 763 (1953).
[2] A. BOHR and B. MOTTELSON: See ref. 2, p. 335.

where the constant depends on the position of the first vibrational states. By suitable choice of this constant the position of rotational states can often be predicted with great precision.

In an *odd mass number* nucleus, the unpaired nucleons contribute a component of angular momentum equal to Ω along the nuclear axis and the total angular momentum J is given by $J^2 = \Omega^2 + J_r^2$. For $J_r = 0$, $J = \Omega = I$, the spin of the ground state. Thus to a first approximation the excitation energy is given by:

$$E = \frac{\hbar^2}{2\mathfrak{J}} \cdot [J(J+1) - I(I+1)] \tag{76.3}$$

where J can take any value above the minimum, I. However, if $\Omega = \frac{1}{2}$, the orbital angular momentum of the nucleons may have no component along the nuclear

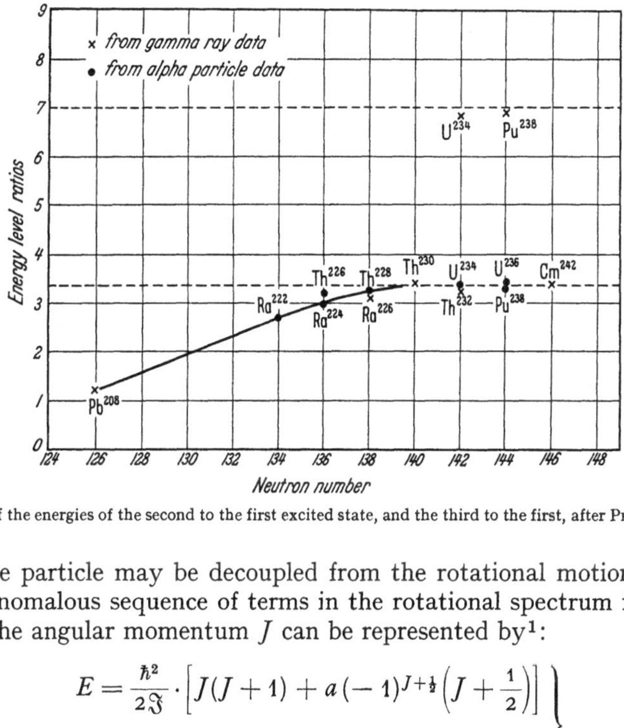

Fig. 65. Ratios of the energies of the second to the first excited state, and the third to the first, after PERLMAN and ASARO.

axis and the particle may be decoupled from the rotational motion. This gives rise to an anomalous sequence of terms in the rotational spectrum for which the term with the angular momentum J can be represented by[1]:

$$\left.\begin{aligned} E &= \frac{\hbar^2}{2\mathfrak{J}} \cdot \left[J(J+1) + a(-1)^{J+\frac{1}{2}}\left(J+\frac{1}{2}\right)\right] \\ \text{with} \quad a &= \sum_j (-1)^{j-\frac{1}{2}} (j+\tfrac{1}{2}) \, |c_j|^2 \end{aligned}\right\} \tag{76.4}$$

where $|c_j|^2$ represents the probability that the unpaired nucleon has an angular momentum j.

There are instances where the vibration term in (76.2) is inadequate to account for the discrepancies between the observed and calculated level structure. It seems possible to account for such differences in terms of an interaction between the particle configuration and the rotational states. A theory of this interaction has been given by KERMAN[2] and applied to the level system of W^{183}.

It will be noted that the energy of the first excited state in the even-even nuclei, $E_1 = 3\hbar^2/\mathfrak{J}$, measures the moment of inertia of the nucleus; the smooth trend of these energies, then, reflects the smooth variation of the moment of

[1] See ref. 2, p. 335.

[2] A. K. KERMAN: Dan. Mat. fys. Medd. 30, No. 15 (1956).

inertia from one element to another. The corresponding quantities for odd mass number nuclei are similar and show the same trend, but they are always a little lower[1].

77. Transition probabilities for rotational levels. The excitation of a high member of the rotational spectrum in an even-even nucleus is followed by a cascade of $E2$ radiation. In an even-odd nucleus the cascade is less simple because $E2$ cross-over transitions are possible and the transition between one level and the next can consist of a mixture of $E2$ and $M1$ radiation. If the transition probabilities were comparable with those to be expected of single particle transitions, $E2$ radiation at these energies (100 kev or less) would be far weaker than $M1$ radiation. The existence of strong $E2$ components and the successful competition of the $E2$ components in the cross-over transitions is a clear indication of the collective nature of these processes where presumably the whole nuclear charge contributes to the emission of quadrupole radiation.

For a pure rotational system, the reduced $E2$ transition probabilities within a specified rotational band are given by[2]:

$$B(E2) = \frac{5}{16\pi} e^2 Q_0^2 \langle J_i 2 K 0 \mid J_i 2 J_f K \rangle^2 \tag{77.1}$$

in which the last term on the right-hand side is a vector addition or CLEBSCH-GORDAN coefficient[3], and J_i and J_f are the spins of the initial and final states respectively. Here $B(E2)$ is defined in (73.3) and Q_0 is the intrinsic electric quadrupole moment. For $K=0$, and for the transition $J+2 \to J$, this formula reduces to:

$$B(E2) = \frac{15}{32\pi} \cdot e^2 Q_0^2 \frac{(J+1)(J+2)}{(2J+5)(2J+3)} \tag{77.2}$$

Thus the intrinsic quadrupole moment may be deduced independently from the transition probabilities or from the energies of the levels. For reasons which will be explained in Sect. 78 the two values of Q_0 derived in this way do not agree to better than a factor of two and only that deduced from transition probabilities is in agreement with those obtained by spectroscopic methods.

For an odd mass number nucleus, the expression (77.1) gives the reduced $(E2)$ transition probability between neighbouring levels in the form:

$$B(E2) = \frac{15}{16\pi} \cdot e^2 Q_0^2 \cdot \frac{K^2 (J+1-K)(J+1+K)}{J(J+1)(2J+3)(J+2)}, \qquad J+1 \to J \tag{77.3}$$

and for the cross-over transition:

$$B(E2) = \frac{15}{32\pi} \cdot e^2 Q_0^2 \cdot \frac{(J+1-K)(J+1+K)(J+2-K)(J+2+K)}{(J+1)(2J+3)(J+2)(2J+5)}, \qquad J+2 \to J \tag{77.4}$$

which reduces to (77.2) when $K=0$. When the reduced transition probabilities for two such transition can be measured, (77.3) and (77.4) give two independent values for the intrinsic electric quadrupole moment. Provided K or $\Omega \neq \frac{1}{2}$ the reduced transition probability for $M1$ radiation is given by:

$$B(M1) = \frac{3}{4\pi} \left(\frac{e\hbar}{2Mc}\right)^2 (g_\Omega - g_r)^2 \Omega^2 \langle J_i 1 K 0 \mid J_i 1 J_f K \rangle^2 \tag{77.5}$$

[1] A. BOHR and B. MOTTELSON: Dan. Mat. fys. Medd. **32**, No. 1 (1955).

[2] See ref. 2, p. 335.

[3] These coefficients have been tabulated by A. SIMON, Oak Ridge National Laboratory ORNL 1718, June 1954 (unpublished).

where the CLEBSCH-GORDAN coefficient in this instance is given by

$$\frac{(J+1-K)\,(J+1+K)}{(J+1)\,(2J+3)}\,.$$

The quantities g_Ω and g_r in (77.5) are the gyromagnetic ratios corresponding to the intrinsic angular momentum of the odd particle and to the rotational motion respectively. The former is a characteristic of the state of the odd particle and the latter is theoretically equal[1] to Z/A. Thus the reduced transition probabilities for magnetic transitions are of the order of the square of the nuclear magneton number, while those of $E2$ transitions are of the order of $e^2 Q_0^2$ or $e^2 \times 10^{-48}\,\text{cm}^4$.

α) $E2$ transitions. The reduced transition probabilities for the low lying states of a large number of nuclei have been obtained by COULOMB excitation (see Table 6). For even-even nuclei, only the first excited state is detected in this way; the cross section for the γ-ray produced in the de-excitation of this state, when multiplied by $(1+\alpha)$, where α is the total internal conversion coefficient, is a measure of the reduced transition probability (73.2). Such a measurement then determines the electric quadrupole moment, Q_0, or the equivalent deformation. The results, generally, are in agreement with those obtained from spectroscopic measurements[2].

For odd mass number nuclei both the first and second states are excited by the $E2$ component in the COULOMB field of the incident particles; the separate measurement of the reduced transition probabilities for the excitation of these two states gives two quantities which are related to the quadrupole moment through the vector addition coefficients (77.1). Such measurements, then, serve as a critical test of the theory of rotational states. The experimental data, at the present time, is scanty and not easy to obtain. For many odd mass number nuclei the second excited state is difficult to detect. When it is observed it is possible to obtain the reduced transition probabilities but only after making corrections to take account of the fact that a cascade γ-ray is produced which feeds the first excited state, and also to allow for the magnetic component in the γ-ray produced by the de-excitation of the first state. If n_{10}, n_{20} refer to the rate of production of the γ-rays between the first and second excited states and the ground state, respectively, and n_{21} refer to the cascade γ-ray, then the rate of production of the first excited state is proportional to $n_{10}\,(1+\alpha_{10}) - n_{21}\,(1+\alpha_{21})$ while that of the second excited state is $n_{20}\,(1+\alpha_{20}) + n_{21}\,(1+\alpha_{21})$. A precise measurement of the transition probabilities, or rather their ratio, therefore, demands a knowledge of the internal conversion coefficients. While the conversion in the $E2$ transition $I+2\to I$ can be obtained from tables with sufficient accuracy, the transitions $I+1\to I$ and $I+2\to I+1$ contain $M1$ in addition to $E2$ components; their conversion coefficients must be either measured directly or inferred from measurements of the $M1$ content obtained from angular distribution measurements. To illustrate this point we may take the case of Ta^{181}. According to STELSON and MCGOWAN[3], for Ta^{181}, n_{21}/n_{20} is 1.74 ± 0.10; from this result and from the angular distribution[4] of the cascade γ-ray, and angular correlation of the 137 kev γ-ray[5] the ratio of the reduced transition probabilities (for $E2$ excitation) for the first (137 kev) and the second (303 kev) states was obtained; it was 4.0 ± 0.2. The theoretical ratio is easily deduced

[1] K. WAY: Phys. Rev. **55**, 963 (1939).
[2] See reference [*24*] for a detailed account of this comparison.
[3] P. H. STELSON and F. K. MCGOWAN: Phys. Rev. **99**, 112 (1955).
[4] See ref. 4, p. 331.
[5] F. K. MCGOWAN: Phys. Rev. **93**, 471 (1954).

from the ratio of (77.3) to (77.4), equating K and I (the spin of the ground state) and multiplying each expression by the appropriate ratio of the statistical weights of initial and final states. It is $I(2I+3)/2(I+1)$. For Ta^{181}, for which $I=\frac{7}{2}$, this ratio is 3.89, in good agreement with experiment. Similar results have been obtained by HUUS, BJERREGAARD and ELBEK[1] from measurements on the conversion electrons. For Ta^{181} they obtained experimentally the ratio 3.7 in agreement with those of STELSON and MCGOWAN; and for Ho^{165} (for which I is also $\frac{7}{2}$) they obtained $\sim$3.3. For Eu^{153}, ($I=\frac{5}{2}$) they obtained 2.85, in close agreement with the theoretical value 2.86.

The significance of these results lies not only in the few instances where the ratios of the reduced $E2$ transition probabilities can be measured and found to be in agreement with theory, but also the fact that the absolute values are in accord with other data on the electric quadrupole moments. There is now an increasing volume of experimental data relating to the relative values of reduced $E2$ transition probabilities for competitive γ-rays emitted by rotational states. We shall not now go into the details of this subject although we shall discuss some instances in Sect. 82.

Most of the reduced $E2$ transition probabilities listed in Table 6 refer to rotational states. As already pointed out these transitions are much faster than would be expected from single particle transitions. When used to derive the intrinsic electric quadrupole moments the results obtained are in good agreement with those deduced from spectroscopic measurements (see reference [*24*]).

The systematics of $E2$ transitions have been discussed by SUNYAR[2], by TEMMER and HEYDENBURG[3], and by FORD[4]. The last two authors plot the square of the nuclear deformation (which is calculated from the transition probability and is proportional to it) against the neutron number (Fig. 66). The deformation rises very sharply to a flat maximum for N near 90, and then falls slowly as the closed shell at Pb is approached. The rapid rise near $N=90$, it should be noted, corresponds to the rapid rate of increase in the nuclear deformation measured by the isotope shift. No corresponding analogue of the sharp rise in the isotope shift just above $N=82$, discussed by BRIX and KOPFERMANN[5], has yet been detected.

β) M1 transitions. The reduced transition probability of the $M1$ component, as we have shown above, can be obtained from that of the $E2$ component in COULOMB excitation, when the ratio of the components is found from angular distributions, angular correlations of successive γ-rays, or from internal conversion measurements. Since this extra information is required there is at present less data available on $M1$ than on $E2$ transitions. For Ta^{181} the reduced $M1$ transition probabilities obtained by STELSON and MCGOWAN[6] for the 137 kev transition and for the 166 kev cascade transition were 0.105 and 0.226 in units of $(e\hbar/2Mc)^2$. From these quantities the square of the difference between the gyromagnetic ratios $|g_\Omega - g_r|^2$ [see Eq. (77.5)] was found to be 0.20 and 0.28, respectively; the difference between the two results, these authors state, is probably not outside experimental errors, and illustrates the experimental difficulties involved in this kind of work.

---

[1] T. HUUS, J. H. BJERREGAARD and B. ELBEK: Dan. Mat. fys. Medd. **30**, No. 17 (1956).
[2] See ref. 1, p. 328.
[3] G. M. TEMMER and N. P. HEYDENBURG: Phys. Rev. **99**, 1609 (1955).
[4] K. W. FORD: Phys. Rev. **90**, 29 (1953).
[5] P. BRIX and H. KOPFERMANN: Phys. Rev. **85**, 1050 (1952).
[6] See ref. 3, p. 339.

The magnetic components can also be obtained by measurements of internal conversion electrons. Experimental methods of measurement of internal conversion electrons produced by COULOMB excitation have been developed by HUUS and BJERREGAARD[1] and used by them to study the rare earth nuclei in some detail[2]. This method has the advantage that the composition of a mixed γ-ray can be determined from the magnitude of the conversion coefficient or from the K/L ratios, although it is still necessary to know the relative intensities of the transitions between the various states.

The results of such measurements, like those of STELSON and McGOWAN, give values in rough agreement with the predictions of (77.5). For Ta^{181}, the value of $|g_\Omega - g_r|^2$ obtained from the 137 kev transition, was 0.30, in agreement with the results of STELSON and McGOWAN. The main interest in this quantity lies in the fact that the separate determination of the quantities g_Ω and g_r is possible when the magnetic moment of the nucleus in its ground state is known. According to the strong coupling model[3], the magnetic moment in units of the nuclear magneton, is related to the gyromagnetic ratios through the expression

$$M = I(I\,g_\Omega + g_r)/(I+1)$$

if $\Omega \neq \frac{1}{2}$, together with an additional term if $\Omega = \frac{1}{2}$. The simultaneous solution of this equation with (77.5) gives two values for the gyromagnetic ratios; the ambiguity arises from ignorance of the sign of the difference between them. According to HUUS, BJERREGAARD and ELBEK[4], existing data is barely sufficient to allow of reliable interpretation of such data, except possibly for Eu^{153} and Ta^{181} where the $M1$ components are respectively weak and strong. In these cases, the results for g_r are low compared with the estimate[5] Z/A.

78. The moment of inertia. We have pointed out (Sect. 76) that the $E2$ transition probabilities and the energies of the rotational states provide independent methods of calculating either the nuclear deformation or the electric quadrupole moments. Several authors have commented on the fact that the values obtained by the two methods do not agree[6]. The results are shown in Fig. 66. It will be seen that the square of the deformations, which differ by about a factor of five, are everywhere roughly in the same ratio. Thus, generally, the electric quadrupole moments obtained from the energies of the levels are about twice as great as those obtained from transitions probabilities. It will be remembered that the former are calculated on the assumption that the moments of inertia are those of an irrotational fluid. FORD[7] has suggested that this assumption may not be justified. If it is assumed that this is indeed the origin of the discrepancy, we must suppose that the moments of inertia lie somewhere between those to be expected of an irrotational fluid and those of a rigid body. BOHR and MOTTELSON[8], in discussing this subject, have pointed out that the moment

[1] T. HUUS and J. H. BJERREGAARD: Phys. Rev. **92**, 1579 (1953).

[2] See ref. 1, p. 340. See also J. H. BJERREGAARD and U. MEYER-BERKHOUT: Z. Naturforsch. **11**a, 273 (1956).

[3] See ref. 2, p. 335.

[4] See ref. 1, p. 340.

[5] See BLIN-STOYLE, reference [*24*].

[6] G. M. TEMMER and N. P. HEYDENBURG: See ref. 3, p. 340. — A. W. SUNYAR: See ref, 1, p. 328. — K. W. FORD: See ref. 4, p. 340. — C. M. McCLELLAND, H. MARK and C. GOODMAN: Phys. Rev. **97**, 1191 (1955).

[7] K. W. FORD: Phys. Rev. **95**, 1250 (1954).

[8] A. BOHR and N. MOTTELSON: Dan. Mat. fys. Medd. **30**, No. 1 (1955). — See also D. R. INGLIS: Phys. Rev. **96**, 1059 (1955). — S. A. MOSZKOWSKI: Phys. Rev. **103**, 1328 (1956). — H. J. LIPKIN, A. DE-SHALIT and I. TALMI: Phys. Rev. **103**, 1773 (1950). — D. R. INGLIS: Phys. Rev. **103**, 1786 (1956). — R. J. BLIN-STOYLE: Nuclear Phys. **2**, 169 (1956).

of inertia of a nucleus consisting of nucleons with negligible interaction must be that of a rigid body, and that the observed moments of inertia represent a position between two limiting conditions.

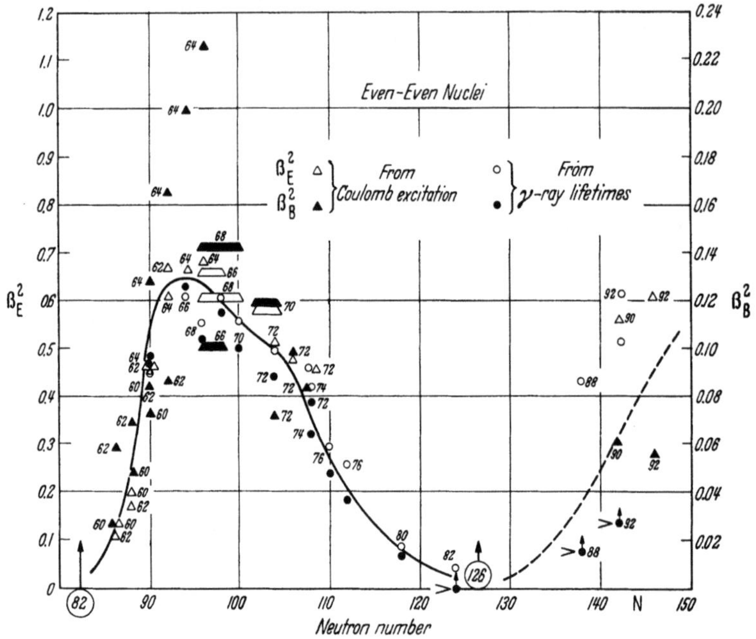

Fig. 66. Deformations in heavy nuclei according to TEMMER and HEYDENBURG, from COULOMB excitation, and from lifetime measurements. Left hand scale: deformations derived from the energies of excited states; right hand scale: from reduced transition probabilities.

79. α-radioactivity. The phenomenon of α-decay is the most characteristic form of radioactivity among the heaviest elements. Many attempts have been made to systematize the experimental data already accumulated[1]. The subject has been recently reviewed by PERLMAN and ASARO [23]; here we shall briefly summarize it and refer the reader to these authors for reference to early measurements and for more details. The aspects of α-decay of most interest in the present connection fall under three headings: the energies of α-decay, the lifetimes, and the excitation of low lying states.

α) Decay energies. Most of the energy available in a transition involving the emission of an α-particle appears in the kinetic energy of the α-particle itself. To obtain the total energy we must add to this that of the recoil of the residual nucleus (usually about 100 kev for a 6 Mev α-particle emitted by a heavy nucleus) and a smaller correction due to the rearrangement of the atomic electrons. When the decay energy between ground states is plotted against the atomic weight, the points corresponding to isotopes of a given element follow a smooth curve (Fig. 67). The decay energy decreases with increasing mass number at a rate in agreement with calculations based on WEISZÄCKER'S mass formula [22]. However, for some elements there is an abrupt discontinuity near the double closed shell nucleus Pb^{208}; in fact, the highest decay energy which appears in

[1] A. BERTHELOT: J. Phys. Radium **3**, 52 (1942). — S. BISWAS and A. P. PATRO: Indian J. Phys. **22**, 539 (1948). — S. BISWAS: Phys. Rev. **75**, 530 (1949). — Indian J. Phys. **23**, 51 (1949). — I. PERLMAN, A. GHIORSO and G. T. SEABORG: Phys. Rev. **77**, 26 (1950). — I. PERLMAN and T. J. YPSILANTIS: Phys. Rev. **79**, 30 (1950). — I. KAPLAN: Phys. Rev. **81**, 962 (1951).

a continuous sequence of isotopes (e.g. in Bi, Po, At or Em) occurs for the nucleus with $N=128$ neutrons. No α-activity has been observed in the Pb isotopes. For light isotopes of elements with higher charge numbers (e.g. Bi, Po, At, Em), α-activity reappears on the left hand side of Fig. 67. α-activity has also been found in some Au and Hg isotopes, and in the rare earth region where the general trends of Fig. 67 are reproduced[1]. Here again, the highest α-decay energy appears in isotopes with $N=84$ and are presumably followed (although not detected) on the low mass number side by an abrupt discontinuity for $N=82$. It should be noted that the only naturally occurring α-emitter in this group (Sm^{147}) has $N=85$; the period of the isotope with $N=84$ (Sm^{146}) is too short compared with the age of the earth[2].

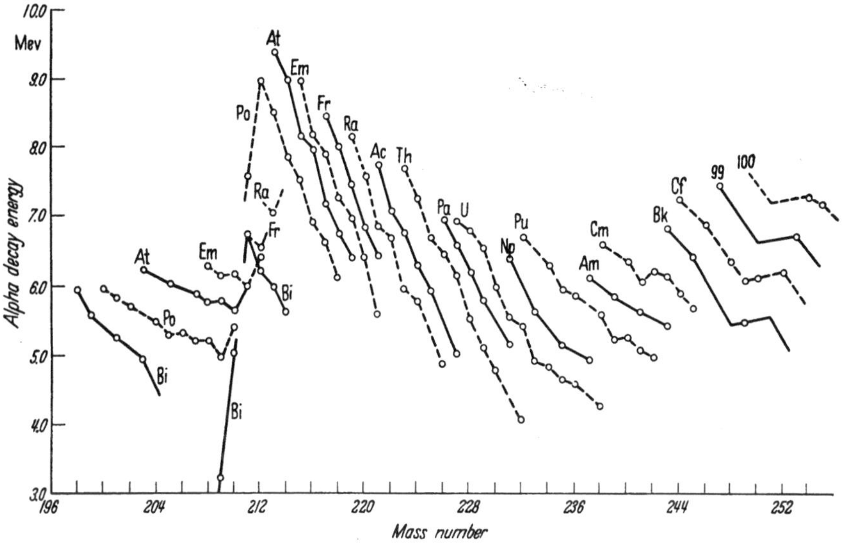

Fig. 67. Energies of α-decay according to PERLMAN and ASARO.

The discontinuity in the energy diagram (Fig. 67) has its origin in the higher binding energies of the closed proton and neutron shells at $Z=82$ and $N=126$. Quantitatively the discontinuities of Fig. 67 have been very simply explained by PRYCE[3] in terms of the interactions between the nucleons in excess or in defect of the number required to close the proton and neutron shells at Pb^{208}. The right hand side of Fig. 67 shows signs of another (but less pronounced) discontinuity which appears to be associated with a closed shell[4] at $N=152$. It first appears in the decay of isotopes of Bk.

β) Lifetimes. A very definite relationship between the lifetime against α-particle emission and the velocity of the α-particle has been known for many years in the form of the lifetime-range relationship: the GEIGER-NUTTALL law. This is, in fact, no more than an expression of the velocity dependence of the barrier penetration for a given value of the orbital angular momentum. The original

[1] J. O. RASMUSSEN, S. G. THOMPSON and A. GHIORSO: Phys. Rev. **89**, 33 (1953).

[2] J. O. RASMUSSEN, F. L. REYNOLDS, S. G. THOMPSON and A. GHIORSO: Phys. Rev. **80**, 475 (1950). — D. C. DUNLAVEY and G. T. SEABORG: Phys. Rev. **92**, 206 (1953).

[3] M. H. L. PRYCE: Proc. Phys. Soc. Lond. A **63**, 692 (1950).

[4] A. GHIORSO, S. G. THOMPSON, G. H. HIGGINS, B. G. HARVEY and G. T. SEABORG: Phys. Rev. **95**, 293 (1954). — A. M. FRIEDMAN, J. E. GRINDLEY, R. F. BARNES, R. SJOBLUM and P. R. FIELD: Phys. Rev. **102**, 585 (1956).

theories of this phenomenon were given by GAMOW[1] and CONDON and GURNEY[1]. More refined theories have been developed by BETHE[2], and by PRESTON[3]. These theories give the dependence of the transition probability on the velocity and on the radius of the nucleus with some precision; but they do not give the absolute transition probability at any energy with equal accuracy for this involves the much more difficult problem of the emission probability in the absence of the COULOMB barrier[4]. For the heaviest elements the decay energies of interest lie between 4 and 10 Mev, all low compared with the barrier heights (about 30 Mev in U^{238}). For these energies the effect of the angular momentum carried away by the α-particle is important though less so than that of the energy, to which the lifetime is extremely sensitive. According to PRESTON, a change in the angular momentum produced by α-decay, if small (say 2 units), may actually increase the probability of decay. To a first approximation all the theoretical formulae can be reduced to a formula of the type

$$\log P = A + \frac{B}{v}$$

where P is the transition probability, A and B are constants which depend on the charge and radius and vary only slowly with the energy, and v is the velocity of the α-particle. For even-even nuclei, for which the lifetime-velocity relation is most regular, the experimental slope B is found to be about 0.9 of that calculated from PRESTON'S formula.

The theoretical formula, when applied to transitions between ground states of even-even nuclei, can be used to determine the nuclear radii. Using BETHE'S formula, PERLMAN and YPSILANTIS[5] found $r_0 = 1.48 \times 10^{-13}$ cm; KAPLAN[5], using PRESTON'S formula, found $r_0 = 1.57 \times 10^{-13}$ cm. The experimental data generally is consistent with nuclear radii which differ by only one or two percent from these figures. It should be noted that these radii are in agreement with those calculated from α-particle reactions (Sect. 10). If the calculated radii are plotted against the atomic weight, it is found that the radius r_0 tends to fall with increasing mass number, a tendency which probably reflects merely an inadequacy of the theory [23]. However, the curve shows a marked discontinuity—of the order of 10%—at $N = 126$, which may represent a real shrinkage associated with the closure of the shell. Recently, TOLHOEK and BRUSSARD[6] have proposed a theory of α-decay based on a high internal potential with a tapered edge. For Po^{214}, the value of r_0 calculated from this theory was 1.13×10^{-13} cm in better agreement with electric measurements of nuclear radii. It should be remembered, however, that the nucleus, in general, is not spherical and in the absence of a thorough theoretical investigation of the emission of α-particles from a deformed nucleus, these radii cannot have any very precise meaning. Moreover, recent measurements of the emission of α-particles from aligned nuclei show that the α-particles are emitted preferentially perpendicular to the axis of deformation[7].

The α-transition probabilities between ground states of even-even nuclei are the greatest which occur. When the logarithm of the lifetime is plotted against

[1] G. GAMOW: Z. Physik **51**, 204 (1928). — E. U. CONDON and R. W. GURNEY: Nature, Lond. **122**, 439 (1928). — Phys. Rev. **33**, 127 (1929).

[2] See ref. 5, p. 296.

[3] M. A. PRESTON: Phys. Rev. **71**, 865 (1947).

[4] See also G. H. WINSLOW: Phys. Rev. **96**, 1032 (1954).

[5] See ref. 1, p. 342.

[6] H. A. TOLHOEK and P. J. BRUSSARD: Physica **21**, 449 (1955).

[7] J. W. T. DABBS, L. D. ROBERTS and G. W. PARKER: Bull. Amer. Phys. Soc. **1**, 207 (1956).

the decay energy (Fig. 68) the points corresponding to a single element fall on a line. The only irregularity which occurs is the position of the lower mass number nuclei of the lighter elements, e.g. Po, Em, for which the α-emission is greatly retarded. This irregularity occurs for $N<128$ and is directly related to the discontinuity in energy in Fig. 67.

γ) Production of excited states. Transitions to excited states occur with less probability than those between ground states. The transitions to first excited states in even-even nuclei (usually about 50 kev in the heavy elements) occur at the rate of about 20% per disintegration and less for lighter nuclei, while those to the second excited states are at the rate of a few transitions per thousand.

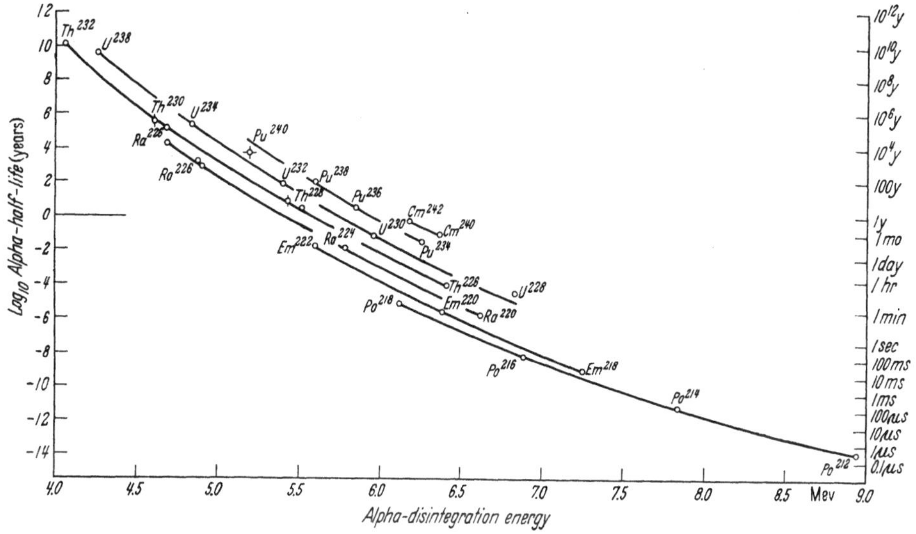

Fig. 68. Relation between the lifetime for α-disintegration and the decay energy, according to PERLMAN, GHIORSO, and SEABORG.

As already pointed out (Sect. 76) the states excited in this process are successively 2^+, 4^+, in the deformed nuclei and the regular changes in transition probability alluded to above demonstrate smooth changes in nuclear properties. However, this smooth pattern of behaviour is broken for masses near 228 by the excitation of states which clearly do not belong to the ground state rotational band (these states are discussed in Sect. 83). Not all of the reduction in transition probability of α-decay leading to excited states can be accounted for by the difference in energy and angular momentum of the α-particle emitted compared with that emitted to the ground state. The ratio of the calculated to the observed transition probability for the production of the excited state has been called by ASARO and PERLMAN[1] a "departure" or "hindrance" factor. For the 2^+ states this is about unity for the lower mass numbers (i.e., there is no departure from the calculated behaviour) but rises to about 2 for the heaviest even-even nuclei. For the 4^+ states, the departure factor is much more marked; it is about 10 for $A \sim 226$ and rises to 500 for Cm^{242}. It seems that this effect is due to the interaction between the α-particle and the quadrupole field and has been accounted for quantitatively by RASMUSSEN[2].

[1] F. ASARO and I. PERLMAN: Phys. Rev. **91**, 763 (1953).

[2] J. O. RASMUSSEN: Quoted by PERLMAN and ASARO, reference [*23*]. See also M. A. PRESTON: Phys. Rev. **75**, 90 (1949).

For odd-A nuclei the transition probabilities are less regularly distributed. The ground state is not always reached by an unhindered transition; and the least hindered α-group often corresponds to a transition to an excited state. On quite general grounds it is clear that the least hindrance must correspond, in general, to a transition without change in angular momentum or parity; presumably inhibited α-particle transitions to ground states entail a change in configuration. Where the least hindered transition can be clearly picked out, the states above it usually form a rotational sequence. The α-particle group producing the first of these states usually has an intensity comparable to that of the corresponding group in an even-even nucleus; the second group is usually much stronger than those which appear in even-even nuclei. The first group has an intensity of the order of 20%, and the second group, three or four percent of that of the uninhibited group. BOHR, FRÖMAN and MOTTELSON[1] have accounted very satisfactorily for the relative intensities of these groups in odd-mass number disintegrations in terms of the angular momentum carried off by the α-particle, the hindrance factors which obtain for the even-even disintegrations and the usual CLEBSCH-GORDAN coefficients. They assume, in effect, that only even angular momenta (l) are involved in these transitions, in that only even angular momenta will leave the product nucleus with the same parity as that of the initial nucleus. In a disintegration in which an initial nucleus with spin j disintegrates to form another with spin j', all even angular momenta will contribute consistent with the usual conservation rules $j+j'>l>|j-j'|$. The transition probability, in general, is given by:

$$P = P(E) \sum_l c_l \langle J_i l K_i K_f - K_i | J_i l J_f K_f \rangle^2$$

where the c's are the reciprocals of the hindrance factors found in even-even nuclei[2]. The c's, as we have seen, are generally functions of the mass number, but roughly c_2 may be taken as 0.7, while c_4 is about 0.01. With these figures, the formula accounts satisfactorily for the observed intensities of the appropriate α-particle groups in Cm^{243}, Am^{243}, Am^{241}, Pu^{239}, and U^{233}.

80. Selection rules. While the energies of the first states excited by COULOMB excitation in the heavy elements follow the trend to be expected of rotational states, excitation by β-decay and the decay of isomeric states often reveal levels which do not so easily fit into a simple rotational system. For the less deformed nuclei, e.g., the isotopes of Pt, Au, and Hg, the excited states do not, as yet, have any obvious connection with collective motions. In many strongly deformed nuclei, however, there appear systems of states rather clearly associated in well defined rotational families. The interpretation of the γ-rays emitted in transitions between levels belonging to different rotational families requires a selection rule in addition to that usually demanded by change of spin and parity. We have seen that the projection of the total angular momentum of the particle motion (Ω) is quantized in the direction of the symmetry axis. In the presence of such an axis of symmetry, this projection must be equal to the projection of the total angular momentum (K) in this direction if only because there can be no rotational angular momentum in that direction. In general different particle excitations are associated with different values of K and a transition from one family to another will involve a rearrangement of the various quantized angular momenta which make up the total. Thus the intensities of the radiations emitted

[1] A. BOHR, P. O. FRÖMAN and B. MOTTELSON: Dan. Mat. fys. Medd. **29**, No. 10 (1955).

[2] A somewhat different formula has been suggested by LANDAU and by TER-MARTIROSIAN, see L. L. GOLDIN, G. I. NOVIKOVA and E. F. TRETYAKOV: Phys. Rev. **103**, 1004 (1956).

obey rules similar to those which obtain in the multiplets of atomic spectra. In fact the reduced transition probability like that for transitions between states of a given rotational band (77.1), is determined by the square of a CLEBSCH-GORDAN coefficient; however, this is zero except when the change in K is equal to or less than the multipole order of the γ-ray emitted[1]. Thus, in addition to the selection rules which determine the multipole order (L) of a γ-ray, L is limited by the change in K; a transition is allowed if $L \geqq \Delta K$. Transitions are allowed (from the point of view of change of K) when they occur within a single family of rotational levels ($\Delta K = 0$); but there will be a minimum allowed multipole order in transitions between families of different K. For example if $\Delta K = 2$, and there is a change in parity, M 2 is allowed but E 1 is forbidden.

Such selection rules are very useful in the interpretation of complex decay schemes. They apply only in so far as the strong coupling model is a good approximation; in effect, their application is limited to the extent to which the nuclear wavefunction can be separated into a part which depends exclusively on the particle configuration and another which depends on the rotation. When this approximation is no longer sufficient, the transition probabilities are more complex. We shall discuss some applications of this selection rule in Sect. 82.

The selection rules which determine the probabilities of β-transitions are likewise more complicated in deformed than in spherical nuclei. For large deformations, the excited states can be classified according to ALAGA[2], in terms of the principal quantum number of the equivalent oscillator, the quantum number of the motion in the direction of the axis of symmetry and the quantum numbers representing the projection of the total (Ω) and the orbital angular momentum on the axis of symmetry. Allowed and first-forbidden β-transitions are divisible into "hindered" and "unhindered" transitions according to whether the β-transition does, or does not violate a selection rule dependent on these four quantum numbers mentioned above; the ft values differ by factors lying between 10 and 100. ALAGA[1] has shown from a study of the decay of rare earth nuclei that these quantities fall into two groups as expected, the identification of the states involved being made on the basis of NILLSON's theory (see below).

81. Levels in a spheroidal well. The order of the levels in a spherical potential well is discussed elsewhere in this series under the heading of shell structure and concerns us here only in the interpretation of the states of nuclei with closed shells (or near closed shells) of neutrons and protons. But the order of the particle levels in a spheroidal well is very different and is of great interest in the interpretation of the level structures in the deformed nuclei. The position of these levels has been calculated by NILLSON[3] using a modified oscillator well, by MOSZKOWSKI[4] using a square well of infinite depth, by GOTTFRIED[5] using a square well with spin-orbit coupling, and by URETSKY[6] using a square well of finite depth. The results have been applied by MOTTELSON and NILLSON[7] and by GOTTFRIED[5] to the interpretation of the spins and parities of the ground states and of excited states in the deformed nuclei. The deformation in part removes the degeneracy

[1] G. ALAGA, K. ALDER, A. BOHR and B. MOTTELSON: Dan. Mat. fys. Medd. **29**, No. 9 (1955).
[2] G. ALAGA: Phys. Rev. **100**, 432 (1955).
[3] S. G. NILLSON: Dan. Mat. fys. Medd. **29**, No. 16 (1955).
[4] S. A. MOSZKOWSKI: Phys. Rev. **99**, 803 (1955).
[5] K. GOTTFRIED: Phys. Rev. **103**, 1017 (1956).
[6] J. URETSKY: To be published.
[7] B. MOTTELSON and S. G. NILLSON: Phys. Rev. **99**, 1615 (1955). — Z. Physik **141**, 217 (1955).

of the levels; in order of increasing energy it distributes the levels corresponding to a given configuration in ascending order of spin for prolate and descending order of spin for oblate deformation. In Fig. 69, which is taken from MOTTELSON and NILLSON'S paper, the energy distribution of some of the states is plotted against the deformation (δ); in this figure, the δ is approximately equal to the parameter β (Sect. 76). The figures which are used to label the different components of the shell orbits refer to the projection of the angular momentum of the unpaired nucleon (Ω) on the axis of symmetry of the nucleus.

It will be seen that the order of appearance of levels of different spin and parity in deformed nuclei bears little resemblance to the shell structure of spherical nuclei. If the deformation is known, the level diagram (Fig. 69) indicates the order of appearance of levels with different values of Ω in odd mass number nuclei. It should be remembered that each of these states has a set of rotational levels based upon it. For the ground state, $K=\Omega=I$, the spin of the nucleus in its ground state, except when $\Omega=\frac{1}{2}$. In this case the spin of the ground state may not be equal to Ω, but it can be obtained from the empirical value of the decoupling parameter a which determines the order of levels in the "anomalous" system of levels. The range of values which a can take for a given value of the spin has been tabulated by NILLSON.

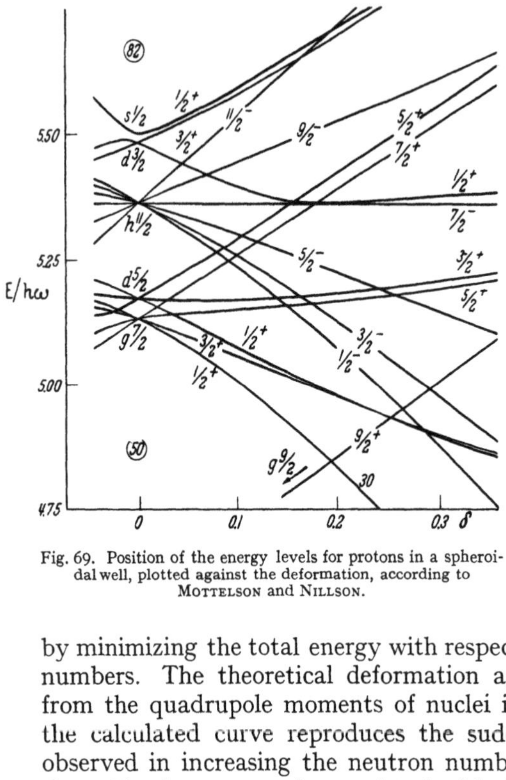

Fig. 69. Position of the energy levels for protons in a spheroidal well, plotted against the deformation, according to MOTTELSON and NILLSON.

MOTTELSON and NILLSON have calculated the equilibrium deformation by minimizing the total energy with respect to the deformation for different mass numbers. The theoretical deformation agrees rather well with those deduced from the quadrupole moments of nuclei in the range $150<A<190$. Moreover the calculated curve reproduces the sudden rise in the deformation which is observed in increasing the neutron number from 88 to 90 (Sect. 69). It seems that this phenomenon is associated with the filling of the $h_{\frac{11}{2}}$ proton shell at this point. In Eu^{153} ($N=90$) the ground state corresponds to $\Omega={}^5/_2{}^+$, and the excited state at 102 kev to $\Omega={}^3/_2{}^+$ which is what we should expect[1] from Fig. 69 for $\delta\approx 0.3$. For Eu^{151} ($N=88$), however, although the spin of the ground state is $\frac{5}{2}$, the magnetic moment requires odd parity and suggests that the unpaired proton in this nucleus is in one of the $h_{\frac{11}{2}}$ orbits.

It is too early to assess the validity and scope of NILLSON'S calculation but it already seems possible to explain qualitatively the order of appearance of the spins and parities of the ground states in the deformed nuclei and to account in some instances for the order of appearance of excited states. Some examples will be discussed in the next section.

---

[1] M. R. LEE and R. KATZ: Phys. Rev. **93**, 155 (1954). — M. MARTY: J. de Phys. **17**, 458 (1956); the transition between them is an allowed $M\,1$ γ-ray for which the half-life is 4.0×10^{-9} sec, which is consistent with Fig. 62 [R. L. GRAHAM and W. WALKER: Phys. Rev. **94**, 794 (1954)].

82. Examples of decay schemes in deformed nuclei. In this section we shall discuss some examples of decay schemes in nuclei in which more than one rotational family is excited. In some of these nuclei the γ-rays are quite numerous and the elucidation of the course of events in such nuclei is possible only if the energies of the γ-rays have been measured with some precision and if their multipole order has been determined from measurements of the absolute γ-ray intensities and of the intensities of the conversion electrons.

α) Lu^{175}. This nucleus is excited in the decay of Yb^{175} and of Hf^{175}. Several experimenters[1] have studied the radiations emitted by this nucleus; recently precision γ-ray measurements have been made by BOEHM *et al.*[2]. A decay scheme which comprises most of the γ-rays observed by the latter authors is shown in Fig. 70. It will be seen that the first level of the ground state rotational band is excited in both decays; in addition the decay of Hf^{175} excites two members of another band of even parity, while the decay of Yb^{175} excites an odd parity state. The theoretical interpretation of this level system has been given by CHASE and WILETS[3] on the basis of NILLSON's level system. These authors identify the ground state (spin $\frac{7}{2}$) with the $\frac{7}{2}$ member of the $g_{\frac{7}{2}}$ configuration; the level at 343 kev becomes a component of the $d_{\frac{5}{2}}$ level, while the odd state with spin $\frac{9}{2}$ is a component of $h_{\frac{11}{2}}$, in accord with NILLSON'S level order (see Fig. 69). Both the β- and the γ-transition probabilities are in general agreement[2] with theoretical predictions if suitable configurations are chosen for the ground states of Yb^{175} ($^7/_2{}^-$) and Hf^{175} ($^5/_2{}^-$). An interesting feature of the decay of $Yb^{175} \rightarrow Lu^{175}$ is the strong $M\,2$ content of two of the γ-rays emitted by the $^9/_2{}^-$ state. According to CHASE and WILETS the ratio of the reduced transition probabilities of these $M\,2$ components are in agreement with the ratios of the appropriate CLEBSCH-GORDAN coefficients (77.1); and the weakness of the $E\,1$

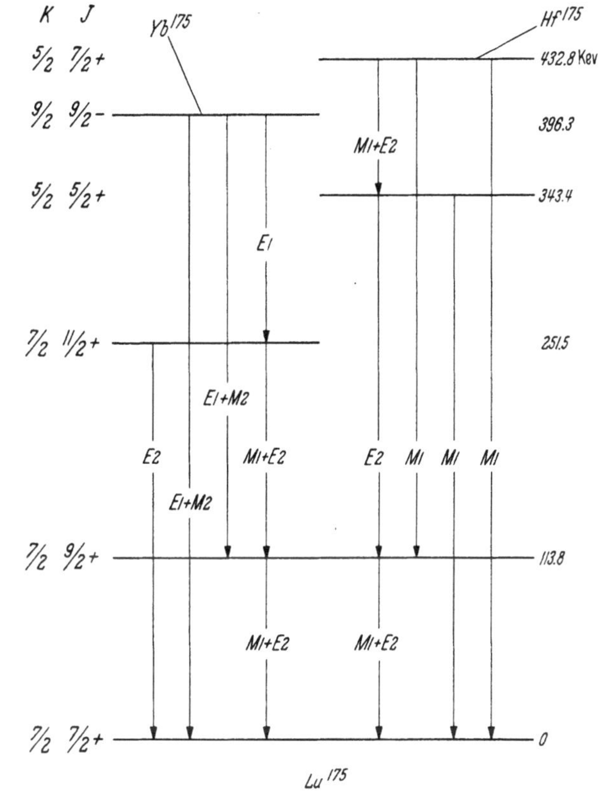

Fig. 70. Level scheme for Lu^{175} according to MIZE, BUNKER, and STARNER. The level energies are those measured by HATCH *et al.*

[1] H. DE WAARD: Phil. Mag. **46**, 445 (1955). — L. ÅKERLING, B. HARTMANN and T. WIEDLING: Phil. Mag. **46**, 448 (1955). — N. MARTY: C. R. Acad. Sci. Paris **240**, 963 (1955). — A. O. BURFORD, J. F. PERKINS and S. K. HAYNES: Phys. Rev. **99**, 3 (1955). — J. P. MIZE, M. E. BUNKER and J. W. STARNER: Phys. Rev. **100**, 1390 (1955); **103**, 182 (1956).

[2] E. BOEHM, E. N. HATCH, P. MARMIER and J. W. M. DUMOND: Bull. Amer. Phys. Soc. **1**, 170 (1956).

[3] D. M. CHASE and L. WILETS: Phys. Rev. **101**, 1038 (1956).

components is caused by a selection rule, valid for large deformations, determined by the quantum numbers representing the components of orbital angular momentum and spin in the direction of the symmetry axis.

β) W^{182} and W^{183}. The scheme devised by ALAGA *et al.*[1] to account for the levels of W^{182} is shown in Fig. 71; it is based on the experimental data of MURRAY, BOEHM, MARMIER, and DUMOND[2] obtained from mixed sources of Ta^{182} and Ta^{183}. In this remarkable work the energies of the γ-rays were determined with high accuracy by a diffraction method. The energies of the internal conversion electrons were also measured with precision and included many cases of separate

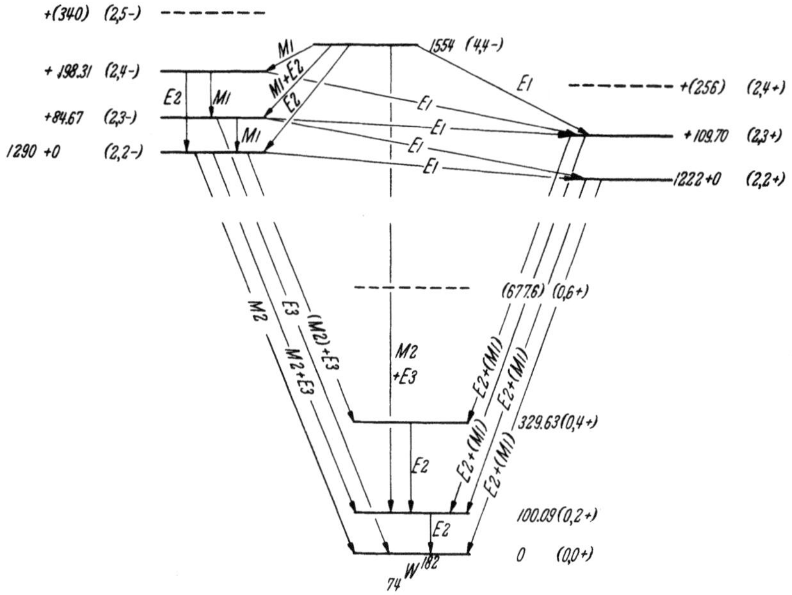

Fig. 71. Energy levels in W^{182} according to ALAGA *et al.* The figures represent the energy in kev, the value of K, the spin and the parity, in that order.

measurement of conversion in the L levels. There are at least three systems of levels: the ground state system ($K=0$); and another with odd parity ($K=2$) based on a level at 1290 kev and another with even parity based on a level as 1222 kev. There is also an odd level at 1554 kev which might be the starting point of another system with $K=4$. The level scheme shown in Fig. 71 is very similar to that deduced by MURRAY *et al.*[5] on the basis of a best fit for all the γ-ray energies and differs from it mainly in omitting an odd state at 1253 kev. The scheme satisfactorily accounts for the absence of $E1$ radiations in the higher energy transitions where their presence might be expected in MURRAY's scheme: they are forbidden by the K selection rule, and, in fact, only $M2$ and $E3$ components are observed. There are also low energy $E1$ γ-rays which connect one family to another.

The levels of W^{183} were also investigated by MURRAY *et al.*[2]. With the exception of one weak γ-ray (102 kev) which is of $M2$ type and does not appear to be a part of the main scheme, all the remaining γ-rays can be incorporated in three rotational families of odd parity, all apparently consistent with NILLSON's scheme. As already pointed out the energies of this system show irregularities

[1] See ref. 1, p. 347.
[2] J. J. MURRAY, F. BOEHM, P. MARMIER and J. W. M. DUMOND: Phys. Rev. **97**, 1007 (1955).

which have been explained by KERMAN in terms of perturbations of the rotation spectrum by particle interactions. The relative intensities are in good agreement with the results of these perturbation calculations.

γ) Pu^{239}. This nucleus can be excited by the α-decay[1] of Cm^{243} or by β-decay[2] of Np^{239}. The earliest results agreed in the existence of an excited state at 278 kev from which three alternative γ-rays are emitted. They lead to the ground state, an excited state at 49 kev, and another at 68 kev. However, in describing their recent measurements on the decay of Np^{239}, HOLLANDER, SMITH, and MIHELICH[3] pointed out that a strong 57 kev γ-ray does not fit into this scheme and have proposed instead the scheme shown in Fig. 72. (This is supported by their discovery of an additional weak γ-ray with an energy of 286 kev.) Since the spin of Pu^{239} is $\frac{1}{2}$, one may expect that the rotational level system based on the ground state to be of the "anomalous" type (76.4); in fact, this term scheme fits the formula well if the energy parameter $\hbar^2/2\mathfrak{J}$ is 6.25 kev, which is a little less than but consistent with the corresponding quantity for the even-even isotope Pu^{238} (7.27 kev). The choice of $K=\frac{5}{2}$ for the states E and F (Fig. 72) is determined by the necessity of accounting for the weakness of the M 1 components of the γ-rays emitted by the 286 kev state. This weakness is established by the fact, discovered by GRAHAM and BELL, that this state has a lifetime of 1.1×10^{-9} sec, which is longer than WEISSKOPF's estimates for M 1 radiation by a factor of the order of 10^4. The E 1 nature of the transitions from the 392 kev state was established by ENGELKEMEIR and MAGNUSSON[4]; the lifetime of this state was found to be 0.193 microsecond. Thus in the first half million electron volts in this nucleus, three rotational bands, or their respective "ground" states are identified. That this identification is substantially correct is borne out by comparison of the observed reduced transition probabilities for transitions within the ground state band and from the band above it. Within the band, the ratio of these quantities for the E 2 components of the 57 and 49 kev γ-rays should be 3.5 (Sect. 77); the experimental result was 3.2 ± 1.0. The observed reduced transition probabilities for the three forbidden M 1 transitions from the 286 kev state

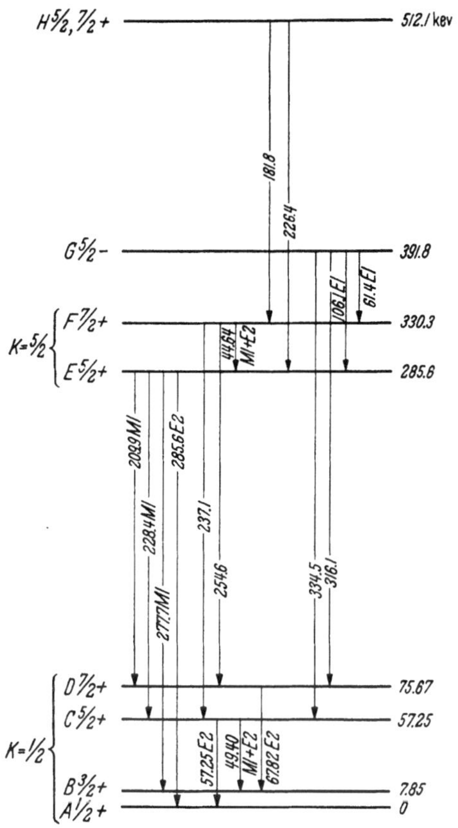

Fig. 72. Energy levels in Pu^{239} according to HOLLANDER SMITH, and MIHELICH.

---

[1] F. ASARO, S. G. THOMPSON and P. PERLMAN: Phys. Rev. **92**, 694 (1953).

[2] H. SLÄTIS: Ark. Mat., Astronom. Fys., Ser. A **35**, No. 3 (1948). — R. L. GRAHAM and R. E. BELL: Phys. Rev. **83**, 222 (1951). — E. P. TOMLINSON, H. W. FULBRIGHT and J. J. HOWLAND: Phys. Rev. **83**, 223 (1951).

[3] J. M. HOLLANDER, W. G. SMITH and J. W. MIHELICH: Phys. Rev. **102**, 740 (1956).

[4] D. ENGELKEMEIR and L. B. MAGNUSSON: Phys. Rev. **99**, 135 (1955).

are of interest, because if $K=\frac{5}{2}$ for this state, these transitions are strictly forbidden; in fact, they correspond more closely to the transition $K=\frac{3}{2}$ to $K=\frac{1}{2}$ which is understandable if it is remembered that their appearance may be caused by admixture of a $K=\frac{3}{2}$ configuration.

δ) Tm[169]. This nucleus is excited by the decay of Yb^{169}: the γ-rays emitted by it have been described by several authors[1] but a consistent decay scheme has only recently been constructed. JOHANSSON[2] has found that the first three levels fit an "anomalous" rotational level system for which the ground state has a spin of $\frac{1}{2}$. More accurate measurements by HATCH *et al.*[3], confirm JOHANSSON'S scheme; they find $\hbar^2/2\mathfrak{J}=11.97$ kev and the decoupling parameter $a=-0.77$. The scheme is shown in Fig. 73. Measurements of the rate of internal conversion of these γ-rays by HATCH *et al.* show that the transitions between levels in the

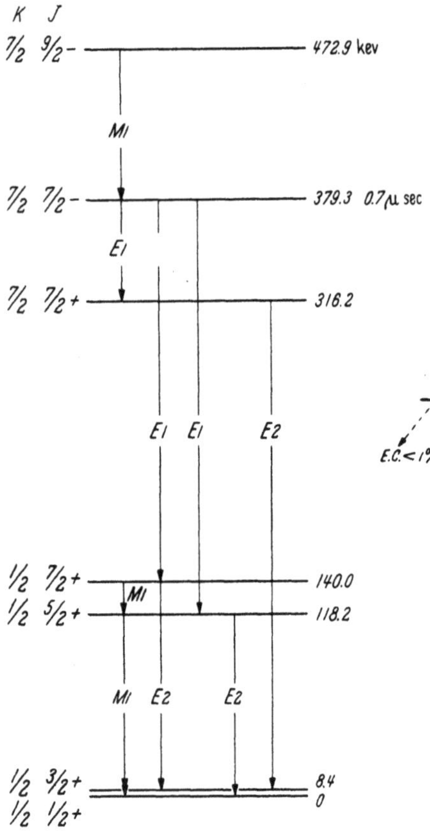

Fig. 73.

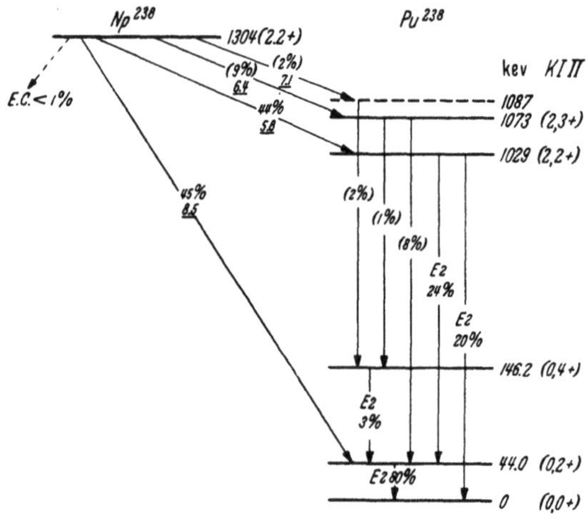

Fig. 74.

Fig. 73. Energy levels in Tm^{169} according to JOHANSSON and to HATCH *et al.*

Fig. 74. Energy levels in Pu^{238} (one of two possible schemes) according to RASMUSSEN and PASSELL. γ-ray intensities are given as percentages per disintegration; the underlined figures refer to log *ft* values.

rotational system based on the ground state are *M* 1, while the 130 kev transition from the 140 kev state to that at 8 kev is *E* 2, and that the γ-rays emitted by the 380 kev level are *E* 1. JOHANSSON showed that the 63 kev transition which is far stronger than the other two *E* 1 γ-rays (which have much higher energy) follows the 0.7 microsecond metastable state[4]. HATCH *et al.* suggested

[1] J. M. CORK, H. B. KELLER, W. C. RUTLEDGE and A. E. STODDARD: Phys. Rev. **78**, 95 (1950). — D. S. MARTIN, E. N. JENSEN, F. J. HUGHES and R. T. NICHOLS: Phys. Rev, **82**, 579 (1951).

[2] S. A. JOHANSSON: Phys. Rev. **100**, 835 (1955).

[3] E. N. HATCH, P. MARMIER, F. BOEHM and J. W. M. DUMOND: Bull. Amer. Phys. Soc. **1**, 170 (1956).

[4] S. DEBENEDETTI and F. K. MCGOWAN: Phys. Rev. **74**, 728 (1948).

that the 318 and 382 kev states both have spin $\frac{7}{2}$ and $K=\frac{7}{2}$, the latter having odd parity. This assignment would make the two higher energy $E\,1$ γ-rays K-forbidden. Thus we have the ground state band, based on $K={}^1/_2(^+)$ and the "ground" states of two others, with $K={}^7/_2(^+)$ and with $K=\frac{7}{2}$ and odd parity; the ground state is supposed to be a component of the $d_{\frac{3}{2}}$ level, the other two corresponding to $g_{\frac{7}{2}}$ and $h_{\frac{11}{2}}$, in qualitative agreement with the order predicted by MOTTELSON and NILLSON.

ε) Pu^{238}. This even-even nucleus is excited by the α-decay[1] of Cm^{242} or by the β-decay[2] of Np^{238}. The decay scheme is still far from complete and there are many inconsistencies. We mention this example here because it is a nucleus which, in addition to the usual rotational states based on the ground state, possesses another system of levels near 1 Mev. This phenomenon as pointed out in Sect. 75 is common to most heavy even-even nuclei; in this nucleus, the parity of these states seems to be even. The γ-rays and the conversion electrons have been studied recently by RASMUSSEN and his collaborators[3]; one of the two possible decay schemes is shown in Fig. 74. The assignment of $K=2$ to the states near 1 Mev is established by the ratio of the reduced transition probabilities for the two γ-rays leading to the ground state and to the first excited state, the experimental values being in agreement with the theoretical values only for $K=2$. This level, then forms the basis of another rotational system starting with spin 2 units. It seems that this latter system has its origin in particle excitation rather than vibrational excitation because it is excited by an allowed β-transition (it would be second-forbidden if it were a vibrational band of the γ type[4] because of a selection rule depending on the change in Ω).

83. Odd states in even-even nuclei. Experiments have revealed several examples of the presence of states in even-even nuclei with energies of the order of 1 Mev or more. The recent discovery of states with odd parity and much lower excitation energy in very heavy nuclei is of great interest because odd states with energies below that required to separate a pair of like nucleons presumably do not originate in particle excitation. This seems to be confirmed by intensity measurements on competing γ-rays emitted by the odd state; the data indicates that the odd state has $K=0$ like the ground state. Such states could originate in deformations of octupole (or pear) shape[5].

Low lying states apparently not related to rotational bands based on the ground states have been found by STEPHENS, ASARO, and PERLMAN[6] in Ra^{222}, Ra^{224}, and Th^{226}; their odd parity was established first by measurements of internal conversion coefficients (Ra^{224}) and later by measurements of the angular correlations in coincidences between the γ-rays emitted by these states and the α-particles producing them. The angular correlations, which are markedly different from those produced by $E\,2$ γ-rays, show that these states have spin 1; the odd parity follows immediately from considerations of parity conservation in the α-disintegrations of even-even nuclei. For the Ra isotopes the energies

---

[1] F. ASARO, S. G. THOMPSON and I. PERLMAN: Phys. Rev. **92**, 694 (1953).

[2] M. S. FREEDMAN, A. H. JAFFEY and F. WAGNER: Phys. Rev. **79**, 410 (1950). — H. SLÄTIS, J. O. RASMUSSEN and H. ATTERLING: Phys. Rev. **93**, 646 (1953).

[3] J. O. RASMUSSEN and T. O. PASSELL: Phys. Rev. **99**, 42 (1955). — J. O. RASMUSSEN, F. S. STEPHENS, D. STROMINGER and B. ASTROM: Phys. Rev. **99**, 47 (1955).

[4] See ref. 1, p. 347.

[5] According to reports attributed to A. BOHR. See also A. BOHR: U. N. Geneva Conf. **2**, 151 (1955).

[6] F. ASARO, F. S. STEPHENS and I. PERLMAN: Phys. Rev. **92**, 1495 (1953). — F. S. STEPHENS, F. ASARO and I. PERLMAN: Phys. Rev. **96**, 1568 (1954).

of these states lie between 200 and 300 kev (see Fig. 75). They decay by emission of two $E\,1$ γ-rays of comparable intensity to the ground state and to the first excited state[1]. In another communication[2] these authors report similar odd states in Ra^{226} and Th^{228}; the existence of such states seems well established both in[3] Ra^{226} and in[4] Th^{228}. In Pu^{238} two γ-rays differing in energy by the energy

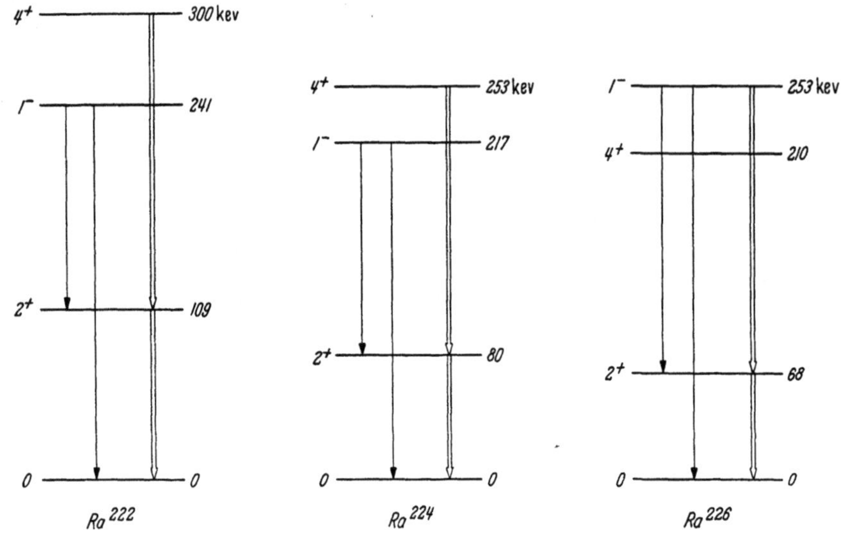

Fig. 75. Positions of the 1⁻ states in the Ra isotopes, according to ASARO, STEPHENS, and PERLMAN. Single lines, $E\,1$; double lines $E\,2$.

of the first excited state have been detected[2] when this nucleus is excited by the α-decay of Cm^{242}. Presumably these γ-rays originate in an odd state; they do not seem to have been detected yet in the decay of Np^{238} (see Sect. 82ε), nor do those excited in Th^{228} by α-decay appear to have been detected[5] in the β-decay of Ac^{228} (MsTh 2).

Table 8. *Ratios of reduced transition probabilities of the γ-rays emitted by 1^- levels in even-even nuclei.*

Nucleus	γ-ray energies in kev	Ratio of reduced transition probability
Pu^{238}	605 562	0.60 ± 0.15
Th^{228}	326 268	0.43 ± 0.08
Ra^{226}	253 184	0.49 ± 0.08
Th^{226}	232 159	0.51 ± 0.15
Ra^{224}	217 133	0.36 ± 0.15
Ra^{222}	242 130	0.48 ± 0.15

The ratio of the reduced transition probabilities of the higher energy γ-ray to that of the lower energy γ-ray are shown in Table 8. The average is about 0.5. Theoretically a state with spin 1 could have $K=1$ or $K=0$, and the ratio of the reduced transition probabilities should be given by the ratio of the

[1] In Ra^{226} the conversion electrons of a cascade transition from the 253 keV to the 70 keV and thence to the ground state seem to have been seen in a photographic emulsion by C. J. JARVIS and M. A. ROSS, Proc. Phys. Soc. Lond. A **64**, 535 (1951).

[2] F. S. STEPHENS, F. ASARO and I. PERLMAN: Phys. Rev. **100**, 1543 (1955).

[3] F. RASETTI and E. C. BOOTH: Phys. Rev. **91**, 315 (1953). — G. BOUISSIERES, P. FALK-VARIANT, M. RIOU, J. TEILLAC and C. VICTOR: C. R. Acad. Sci. Paris **236**, 1874 (1953). — G. M. TEMMER and J. M. WYCKOFF: Phys. Rev. **92**, 913 (1953). — P. FALK-VARIANT and G. PETIT: J. de Phys. **16**, 125 (1955). — E. C. BOOTH, L. MADANSKY and F. RASETTI: Phys. Rev. **102**, 800 (1956).

[4] F. ASARO and I. PERLMAN: Phys. Rev. **99**, 37 (1955). — G. SCHARFF-GOLDHABER, E. DER MATEOSIAN, G. HARBOTTLE and M. MCKEOWN: Phys. Rev. **99**, 180 (1955).

[5] H. C. BOX and G. S. KLAIBER: Phys. Rev. **95**, 1247 (1954). — J. KYLES, C. G. CAMPBELL and W. J. HENDERSON: Proc. Phys. Soc. Lond. A **66**, 519 (1953).

squares of two CLEBSCH-GORDAN coefficients (77.1). If $K=1$, this ratio should be 2, if $K=0$, the ratio should be 0.5 as observed.

The energy of the odd state, when plotted against the neutron number appears to have a minimum near $N=136$.

84. Lifetime of low energy $E1$ γ-rays. The lifetimes of a number of low energy $E1$ γ-rays have been measured in the heavy elements. Some of these lifetimes have been noted by GOLDHABER and SUNYAR[1] who pointed out that they are all much longer (by factors of 10^3 or more) than one would expect for $E1$ lifetimes calculated on the single particle model. While some of these γ-rays have been identified by conversion measurements, one, (e.g. the 27 kev γ-ray produced in Ac^{227} by the decay of Pa^{231}) has been identified merely by its intensity[2]. The measured lifetimes and the ratios of these lifetimes to those calculated from WEISSKOPF'S estimates are shown in Table 9.

Table 9. *Lifetimes if low energy $E1$ γ-rays.*

Nucleus	γ-ray energy in kev	Measured lifetime sec	Retardation
Pu^{239}	106	1.9×10^{-7} (a)	2×10^{6}
Np^{237}	60	1.6×10^{-7} (b)	3×10^{5}
Np^{237}	26	2.2×10^{-6} (b)	3×10^{5}
Pa^{231}	85	4.1×10^{-8} (e)	1×10^{4}
Ac^{227}	27	4.2×10^{-8} (c)	6×10^{5}
W^{182}	67 (?)	1.0×10^{-9} (d)	$\sim 2.5\times 10^{3}$
Tm^{169}	63	7.0×10^{-7} (f)	1.2×10^{6}

(a) D. ENGLEKEMEIR and L. B. MAGNUSSON: Phys. Rev. **99**, 135 (1955).
(b) J. K. BELING, J. O. NEWTON and B. ROSE: Phys. Rev. **87**, 670 (1952).
(c) J. TEILLAC, M. RIOU and P. DESNEIGES: C. R. Acad. Sci. Paris **237**, 41 (1953).
(d) Identification probable but not certain, see A. W. SUNYAR: Phys. Rev. **93**, 1122 (1954).
(e) D. STROMINGER and J. O. RASMUSSEN: Phys. Rev. **100**, 844 (1955).
(f) See Sect. 82 δ.

The low transition probabilities of $E1$ γ-rays may be attributed to the properties of deformed nuclei. In the limit of very strongly deformed nuclei, the emission of $E1$ radiation will be forbidden unless the selection rules pertaining to some five quantum numbers are satisfied (Sect. 80), of which the selection rule referring to changes in K is only one. In general, as inspection of NILSSON'S level scheme will show, the permissible changes in the quantum numbers which define the motion in the direction of the axis of symmetry and the orbital angular momentum in that direction, are rarely in accord with the actual changes in these quantities for transitions between the lowest levels differing in parity. Thus most low energy $E1$ transitions are forbidden[3]. We have already noted particular cases where $E1$ γ-rays are weak and where they could be forbidden by the K-selection rule. An interesting case of an $E1$ γ-ray which appears to be very highly forbidden by the K-selection rule[4] is the 57 kev γ-ray emitted by the isomer of Hf^{180}; the isomer has a spin of 9^- and decays to a final state with spin 8^+.

[1] See ref 1, p. 321.
[2] See ref. P. FALK-VARIANT and M. RIOU: J. de Phys. **14**, 65 (1953).
[3] J. O. RASMUSSEN and D. STROMINGER: Bull. Amer. Phys. Soc. **1**, 206 (1956).
[4] G. SCHARFF-GOLDHABER, M. MCKEOWN and J. W. MIHELICH: Bull. Amer. Phys. Soc. **1**, 206 (1956).

O. Energy levels in closed shell nuclei.

85. General features. It is necessary to distinguish here between the double closed shell nuclei, i.e. the few nuclei for which both neutron and proton shells are closed, and nuclei in which only one shell is closed. In both cases the energies of the first excited states are exceptionally high, those of double closed shell nuclei being the highest; the latter nuclei are also exceptions to the rule (Sect. 68) that the spins of the first excited states of even-even nuclei are 2 units.

α) Double closed shell nuclei. There are at least three double closed shell nuclei whose claims to this distinction will not be disputed. These are Ca^{40}, Ca^{48} and Pb^{208}. Although nothing is known about the spin of the first excited state in Ca^{48} (3.825 Mev), the first excited states of Ca^{40} (3.348 Mev) and Pb^{208} (2.615 Mev) are both exceptions to the rule that the spins of the first excited states of even-even nuclei are 2 units. The spin of the first state in Ca^{40} is zero with even parity, and in Pb^{208}, the spin of the first excited state seems to be 3 units with odd parity. These exceptions to the rule are to be expected if it is remembered that, for particle excitation, the first excited state is merely one state in the sequence of states which represent the configuration in which two like particles of spin j are coupled together $[0, 2, 4, \ldots (2j-1)]$. For doubly closed shell nuclei all these states are occupied; the lowest excited states, therefore, must represent the promotion of a nucleon from the closed shell to an empty one, or, as in Ca^{40}, the lowest state might possibly represent a pair of like particles coupled together in another configuration. The levels in the two Ca isotopes have been explored only very recently; as yet there is not sufficient knowledge of the level structure in these isotopes on which to base a theory. Rather more is known of the levels of Pb^{208} (Sect. 88).

There are a few nuclei for which the neutron shell is closed at $N=50$ or 82 but for which only a proton subshell is closed. Strictly these nuclei are not double closed shell nuclei; however, they are of interest in that the level structure of some of them is exceptional. For $N=50$, there are two such nuclei, e.g. Sr^{88} for which the $p_{\frac{3}{2}}$ proton subshell is closed, and Zr^{90}, for which both $p_{\frac{1}{2}}$ and $p_{\frac{3}{2}}$ subshells are closed. In Sr^{88} the only noteworthy feature is the second excited state which is odd and has spin 3[1]. Zr^{90} has some exceptional features which are discussed in Sect. 87. For $N=82$, the $g_{\frac{7}{2}}$ proton shell is full at Ce^{140} (which is excited by the decay of La^{140}). According to KELLY and WIEDENBECK[2] the first and second excited states in this nucleus follow the spin sequence 0—2—4 and, except for their spacing, are not exceptional. The existence of a 0^+ state in this nucleus at an energy below that of the 1.6 Mev state might well have escaped detection.

β) Single closed shell nuclei. Low lying levels in nuclei in which either the proton or the neutron shell is closed are likely to have their origin in the configuration of the open shell. Levels in nuclei with one nucleon above that required to close the shell, or one less (e.g. Bi^{209}, Pb^{209}, or Tl^{207}, Pb^{207}) should be the easiest to identify and the simplest to interpret theoretically. The positions of the levels in the spherical shell model drawn up by KLINKENBERG[3] are shown in Fig. 76. A nucleus with one extra nucleon would be expected to have as its ground state one of the levels lying above a dotted line in this figure, while a nucleus with one nucleon less than that required to fill a shell will occupy a level below the

---

[1] G. R. BISHOP and J. P. PEREZ Y JORBA: Phys. Rev. **98**, 89 (1955).

[2] W. H. KELLY and M. L. WIEDENBECK: Phys. Rev. **102**, 1130 (1956). See also H. H. BOLOTIN, C. H. PRUETT, P. L. ROGGENKAMP and R. G. WILKINSON: Phys. Rev. **99**, 62 (1955).

[3] P. E. A. KLINKENBERG: Rev. Mod. Phys. **24**, 63 (1952).

dotted line. The low lying excited states should correspond to the nucleon in excess (or to the vacancy) being transferred to other levels lying above (or below) the dotted line. A phenomenological theory of the level systems in nuclei in the Pb region has been developed by PRYCE[1]; he has extended it to include nuclei with as many as three vacancies.

86. Levels in the Ca isotopes. The levels in the Ca isotopes have been explored with great care by BRAAMS[2] with the aid of the (d, p) reaction and inelastic proton scattering; some of the results confirm data derived from β-decay.

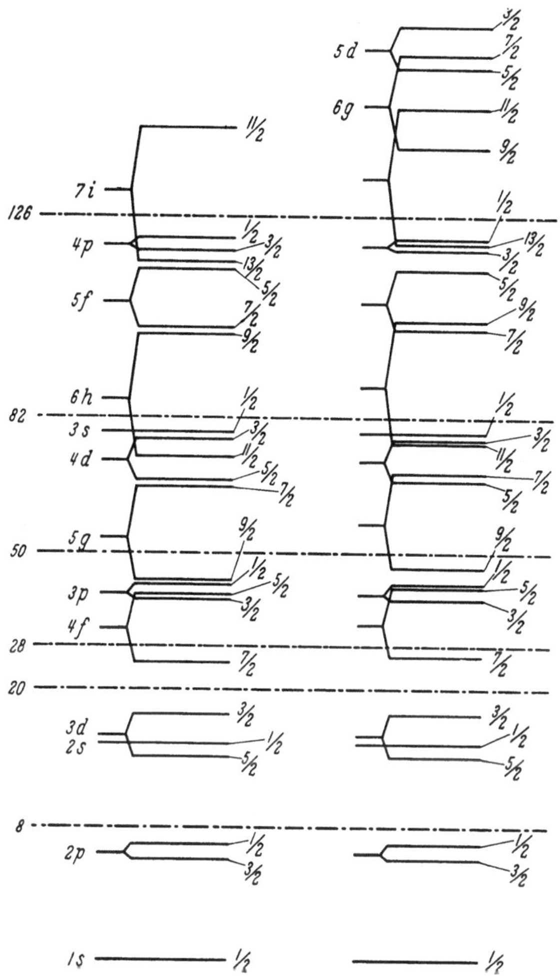

Fig. 76. Position of single particle levels according to the shell structure drawn up by KLINKENBERG.

α) Ca^{40}. The first excited state[2] in Ca^{40} lies at 3.348 Mev. This state seems to be 0^+ for pairs have been observed when Ca^{40} is excited by inelastic proton scattering[3] above a threshold near 3.35 Mev. The energy of the pairs, which were determined by a magnetic lens pair spectrometer, was 3.35 Mev in good agreement with the position of the level found by BRAAMS. This is confirmed by the detection of annihilation radiation by DAY[4] from inelastic neutron scattering in Ca, when the neutron energies were above 3.36 Mev. A state[5] at 3.73 Mev is strongly excited by the decay of Sc^{40}. The decay energy of K^{40} is too low to excite Ca^{40}. Little else is known about the higher states of this nucleus.

β) Ca^{48}. The states of Ca^{48} have been investigated by BRAAMS by inelastic proton scattering with proton energies up to 7 Mev. The first excited state is at 3.825 Mev which is even higher than that of Ca^{40}. Nothing is known about its spin; it might well be 0^+ like that of Ca^{40}.

---

[1] See ref. 5, p. 361.
[2] C. W. BRAAMS: Phys. Rev. **101**, 1764 (1956). — Thesis, Utrecht, 1956.
[3] R. D. BENT, T. W. BONNER and J. H. MCCRARY: Phys. Rev. **98**, 1325 (1955).
[4] R. B. DAY: Phys. Rev. **102**, 767 (1956).
[5] N. W. GLASS and J. R. RICHARDSON: Phys. Rev. **98**, 1251 (1955).

γ) Ca^{49} and Ca^{41}. These nuclei are of exceptional interest for they both have one neutron in addition to two closed shells. Three states in Ca^{49} have been detected by BRAAMS[1], but nothing is known about them but their energies; the energy of the first is 2.226 Mev.

The levels of Ca^{41} form a complex system starting at 1.947 Mev, and have been explored by BRAAMS with the (d, p) reaction and by various authors[2] by the γ-rays emitted by neutron capture in Ca^{40}. The low lying states appear to represent the position of the odd neutron in the shell scheme, the ground state being $f_{\frac{7}{2}}$, the first excited state probably $p_{\frac{3}{2}}$ and the $p_{\frac{1}{2}}$ state at 2.469 Mev; both p-states are strongly excited in the neutron capture reaction where they are produced by strong γ-rays by $E1$ emission from the capturing state, the target nucleus being Ca^{40}. The (d, p) measurements demonstrate the existence of a state just above the $p_{\frac{3}{2}}$ state, at 2.014 Mev; this may be the $f_{\frac{5}{2}}$ state. Above 2.5 Mev excitation the level system rapidly becomes very complex, presumably because of core excitation.

δ) Ca^{43} and Ca^{45}. The first four levels of Ca^{43} are excited in the decay[3] of K^{43} and Sc^{43}. It seems very likely that the first and second excited states have spins $\frac{5}{2}$ and $\frac{3}{2}$ and, like the ground state, have odd parity. The next two states appear to have even parity and spins of $\frac{1}{2}$, $\frac{3}{2}$ or $\frac{5}{2}$; however, there still remains some uncertainty about the nature and even the existence of some of these states.

LEVINSON and FORD[4], assuming that the ground state and the first three states of Ca^{43} form a $(f_{\frac{7}{2}})^3$ quadruplet, have calculated the position of these states, taking their data from the positions of the states of Ca^{41}, for which the first three excited states were assumed to be single-particle states in the expected sequence $p_{\frac{3}{2}}, f_{\frac{5}{2}}, p_{\frac{1}{2}}$. Their results, generally, are in very good agreement with experiment, if it is assumed that the third state lies at 0.8 Mev. [Although a state at this energy was called for by LINDQVIST and MITCHELL no corresponding state was found by BRAAMS in the (d, p) reaction]. An interesting feature of the γ-ray transitions in Ca^{43} is the strength of the $E2$ cross-over transition from the second state (spin $\frac{3}{2}$) to the ground state (spin $\frac{7}{2}$) compared with the transition between second and first excited states. As pointed out by FORD and LEVINSON, such an $E2$ transition would be highly forbidden if nothing but a neutron quadruplet contributed to transition probabilities between these states. They show that even a very weak surface coupling will account for the transition probabilities observed.

The states in Ca^{45} are less well studied; their distribution in energy does not look similar to those of Ca^{43} and the lower state lies much lower in energy. If pure jj coupling is assumed and if configuration mixing is neglected one would expect[5] that both Ca^{43} (with three extra neutrons) and Ca^{45} (with three vacancies) should have very similar level systems, the lowest states in ascending order having the spins $\frac{7}{2}, \frac{5}{2}, \frac{3}{2}, \frac{9}{2}, \frac{11}{2}$ and $\frac{15}{2}$, all with odd parity.

87. Levels in Zr^{90}. The position of Zr^{90} as a double closed shell nucleus is less secure than those already mentioned; it is of interest in that the energy gap between the $p_{\frac{1}{2}}$ proton subshell (which is full at this nucleus) and the $g_{\frac{9}{2}}$ shell which follows it, although smaller than the gap between a closed neutron

[1] See ref. 2, p. 357.
[2] L. V. GROSHEV *et al.*: Ref. 2, p. 308. Also B. B. KINSEY, G. A. BARTHOLOMEW and W. H WALKER: Phys. Rev. **85**, 1012 (1952). — T. H. BRAID: Phys. Rev. **102**, 1109 (1955).
[3] T. LINDQVIST and A. C. G. MITCHELL: Phys. Rev. **95**, 444 (1954).
[4] C. W. LEVINSON and K. W. FORD: Phys. Rev. **100**, 1, 13 (1955).
[5] I. TALMI: Helv. phys. Acta **25**, 185 (1952). — A. R. EDMONDS and B. H. FLOWERS: Proc. Roy. Soc. Lond., Ser. A **215**, 120 (1952).

shell and the next neutron state, is, nevertheless, sufficiently large to give this nucleus some of the properties of a double closed shell nucleus. This energy difference is revealed in the separation between the $p_{\frac{1}{2}}$ and $g_{\frac{9}{2}}$ states of Y^{89}, which is 0.91 Mev. FORD[1] has suggested that the 0^+ level of the excited $(g_{\frac{9}{2}})^2$ configuration might lie lower than the lowest member (spin 5) of the $(p_{\frac{1}{2}}\, g_{\frac{9}{2}})$ doublet, in which case the lowest state of Zr^{90} would be a 0^+ state and would decay to the ground state with pair emission. The existence of a strongly converting state (at 1.75 Mev) has been verified by JOHNSON, JOHNSON and LANGER[2]; they found a weak internal conversion line in the electron spectrum emitted by a strong source of Y^{90} and showed that γ-ray emission of this energy, if it occurs, must be weaker than the conversion line by a factor of 10. No other γ-radiation was observed. A weak positron emission, peaked at half the energy of the conversion line, might represent pair emission by this state[3]. These experiments suggest that the 1.75 Mev state is the lowest state in Zr^{90}; this conclusion is supported by the discovery of an isomeric state (0.8 sec) emitting a 2.30 Mev γ-ray by CAMPBELL *et. al.*[4], who prepared it (presumably by inelastic neutron scattering) by reactor activation in enriched Zr^{90}. This isomeric state may be the 5^- state referred to above; the lifetime is roughly that expected of an $E5$ transition. Both states have been observed by their conversion electrons in the decay of Nb^{90} by ONG PING HOK, KRAMER and MEIJER[5]; their results, however, do not appear to be consistent with those obtained from the decay of Y^{90} and more work needs to be done before the level scheme of Zr^{90} is established.

88. Levels in Pb^{208}. The study of the levels of Pb^{208} has drawn the attention of experimenters long before the discovery that this nucleus has the special properties associated with closed shells. The reason for this is that Pb^{208} is the final product of the family of related radioactive nuclei which start with Th^{232}; that the strong γ-ray (2.62 Mev) excited in this nucleus in the decay of Tl^{208} (ThC″) is a standard laboratory source of γ-rays; and that sources containing this latter nucleus are easily prepared.

The β-decay of Tl^{208} (ThC″) (3 min and 1.8 Mev) produces Pb^{208} at a high level of excitation. In the following we shall be concerned mainly with the γ-rays produced in the decay of ThC″, because the alternative method of exciting Pb^{208}, viz., the decay of Bi^{208}, a long-lived body decaying by electron capture, is not known to produce any γ-rays. Presumably the energy available is less than that of the first excited state of Pb^{208} (2.62 Mev).

In addition to the 2.62 Mev γ-ray, which has high intensity (usually assumed to be unity) there are γ rays with energies of 0.58, 0.51, 0.28 and 0.86 Mev in descending order of intensity. Although it has been clear for some time that the 0.86 Mev γ-ray is a cross-over γ-ray between two states one emitting the 0.28 Mev γ-ray, and the other being fed by the 0.58 Mev γ-ray, the sequence of events and the spins and parities of the levels have only recently been established. That the 2.62 Mev γ-ray has very nearly unit intensity and that it is in cascade with the 0.58 Mev γ-ray is clear from considerations of intensity and the energy balance in the alternative modes of decay of ThC (Bi^{212}), viz., ThC—C″—D, and ThC—C′—D. That its intensity is indeed very nearly unity now seems certain for

[1] K. W. FORD: Phys. Rev. **98**, 1516 (1955).
[2] O. E. JOHNSON, K. G. JOHNSON and L. M. LANGER: Phys. Rev. **98**, 1517 (1955).
[3] See ref. 2, this page. See also J. S. GREENBERG and M. DEUTSCH: Phys. Rev. **102**, 415 (1956).
[4] E. C. CAMPBELL, R. W. PEELLE, F. C. MAIENSCHEIN and P. H. STELSON: Phys. Rev. **98**, 1172 (1955).
[5] ONG PING HOK, P. KRAMER and G. MEIJER: Physica **20**, 1200 (1954).

recent work seems to have eliminated the possibility that other high energy γ-rays participate in the decay of ThC″. Conversion electrons of 1.6 and 1.8 Mev found by ELLIS[1] are now believed to be due to ThC′ which is always present when ThC″ is contained in a source of ThB or ThC; the existence of additional γ-rays at 1.35, 1.50 and 2.2 Mev, reported by ALICHANOW and DZELEPOV[2] and by LATYSHEV and KULCHITSKY[3] has not been confirmed by a recent very careful search for these γ-rays with a high resolution COMPTON spectrometer[4]. If present, the intensity of the 2.0 Mev γ-ray is less than 0.5% per disintegration[4].

The accumulated evidence, then, is in good accord with the assumption that the 2.62 Mev γ-ray is emitted by the first excited state with unit intensity. On the basis of measurements[5] of the pair conversion coefficient (which is in agreement with that expected of $E2$ radiation) and of angular correlations, the spin of the 2.62 Mev state was assumed to be 2 with even parity, although the measured photoelectric internal conversion coefficient lies between that of $M1$ and $E2$ radiation.

The first measurement of the internal conversion of this γ-ray which took account of the loss of the parent nucleus (ThC″) from the source by α-ray recoil, was that of FLAMMERSFELD[6] (0.23%). It is not clear if this result was corrected for the radioactive equilibrium of the genetically related activities. Recent accurate measurements by MARTIN, RICHARDSON and HSÜ[7] gave a somewhat lower value, viz. 0.18%, which, when repeated gave 0.178. One of the experimental difficulties of their method lies in the measurement of the ratio of the area under the peak to that under the β-ray spectrum (which may include the β-decay of ThB); this has been eliminated in recent measurements by ELLIOTT, GRAHAM and WOLFSON[8] who determined the absolute strength of the 2.62 Mev source by comparison against a known source of 2.75 Mev γ-rays from Na^{24} and found the effective aperture of the spectrometer by an independent measurement. Their result was (0.178 ± 0.012)% in agreement with above mentioned results. This result is close to the theoretical conversion coefficient of 0.186 for $E3$ radiation obtained by interpolation of ROSE's tables (cf. ref. [*19*]). (It lies roughly between the conversion coefficients of $E2$ and $M1$ radiation.) That the 2.62 Mev γ-ray is in fact an $E3$ γ-ray was confirmed by the same authors by angular correlation measurements between the 0.58 and the 2.62 Mev γ-rays using a liquid source; they found a much large anisotropy than that previously obtained by PETCH and JOHNS[9]. The results are in accord only with the spin sequence (0-3-5)[10].

[1] C. D. ELLIS: Proc. Roy. Soc. Lond. **143**, 350 (1934).

[2] A. I. ALICHANOW and V. I. DZELEPOV: C. R. Acad. Sci. USSR. **20**, 113 (1938).

[3] G. D. LATYSHEV and L. A. KULCHITSKY: J. Phys. USSR. **4**, 515 (1941).

[4] D. G. E. MARTIN and G. PARRY: Proc. Phys. Soc. Lond. A **68**, 1177 (1955).

[5] A. I. ALICHANOW and D. P. DZELEPOV: C. R. Acad. Sci. USSR. **20**, 113 (1938). — A. I. ALICHANOW: Bull. Acad. Sci. USSR. **1938**, No. 1/2, 33. — H. BRADT, J. HALTER, H.G. HEINE and P. SCHERRER: Helv. phys. Acta **19**, 431 (1946). — H. SLÄTIS and K. SIEGBAHN: Ark. Fysik **4**, 485 (1952).

[6] A. FLAMMERSFELD: Z. Physik **114**, 227 (1939).

[7] D. G. E. MARTIN, H. W. O. RICHARDSON and Y. K. HSÜ: Proc. Phys. Soc. Lond. A **60**, 466 (1948). — D. G. E. MARTIN and H. O. W. RICHARDSON: Proc. Roy. Soc. Lond., Ser. A **95**, 287 (1949).

[8] L. G. ELLIOTT, R. L. GRAHAM and J. L. WOLFSON: Phys. Rev. **93**, 356 (1954).

[9] H. E. PETCH and M. W. JOHNS: Phys. Rev. **80**, 478 (1950).

[10] It should be noted that although the E 3 character of this γ-ray seems well established, nevertheless it converts internally by pair formation as if it were E 2. This is not the only anomaly in a very anomalous nucleus: e.g. the close superposition discovered by ELLIS of the K-conversion line of the 0.58 Mev γ-ray on the L_I-conversion line of the 0.51 Mev γ-ray and the apparent equality of the energy of the latter γ-ray with that of annihilation radiation [A. HEDGRAN and D. LIND: Ark. Fysik **5**, 177 (1952)].

The lifetime was found to be less than 1×10^{-10} sec which is fast for an $E3$ transition at this energy. The 0.58 Mev γ-ray has a conversion coefficient in agreement with that of $E2$ radiation. The other γ-rays already mentioned fit the scheme shown in Fig. 77. It is probable[1] that even higher states are excited than those shown in this figure. Pair conversion of the transition between the 3.2 Mev level and the ground state has been detected although no γ-ray of this energy has been detected with ceráinty[2]. The absence of levels with low spins and energies up to 3 Mev is confirmed[3] by the unique $E1$ γ-ray produced by neutron capture in Pb^{207}.

The origin of the 2.62 Mev state is not clear. Its energy is lower than that to be expected from the promotion of a nucleon from a closed to an unfilled shell; for a neutron this should be roughly equal to the difference between the neutron binding energies of Pb^{209} and Pb^{208}, and for a proton, the difference between the proton binding energy of Bi^{209} and Pb^{208}, viz., 3.5 and 4.3 Mev respectively. The 2.62 Mev state does not appear in the levels found by HARVEY[4] in the Pb^{207} (d, p) Pb^{208} reaction. Since this reaction presumably involved the addition of a neutron, it has been suggested that the 2.62 Mev state concerns the displacement of a proton[5]. The lowest states found by HARVEY are at 3.4 and 3.6 Mev; it will be shown in Sect. 89 that these correspond to the neutron doublet $(p_{\frac{1}{2}}, g_{\frac{9}{2}})$.

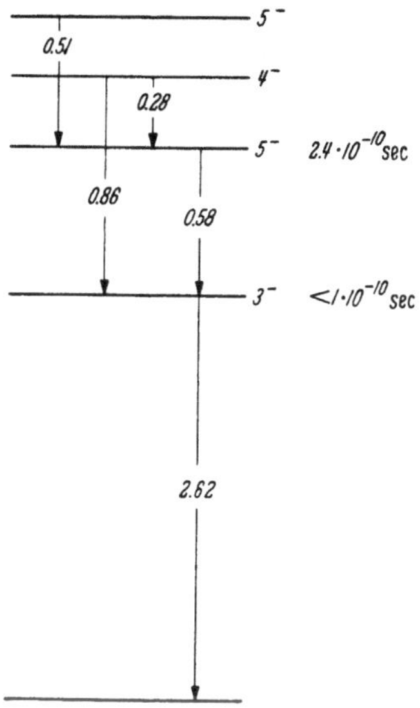

Fig. 77. Decay scheme for Pb^{208}.

89. Levels revealed by the (d, p) reaction. Measurements of the energies of the proton groups produced by (d, p) and (d, t) reactions in Bi and on separate isotopes of Pb have been made by HARVEY[4]. The energy of the deuteron beam was 14 Mev, very near the barrier height in Pb; the resolution was about 400 kev. The general appearance of the proton groups produced by Pb and Bi isotopes show remarkable similarities. When the Q values of the (d, p) reaction were plotted together (see Fig. 78a) the strong groups appear in the same position, the group corresponding to the ground state in Pb^{209} falling near the 3.5 Mev doublet in Pb^{208} and a level at 2.75 Mev in Pb^{207}.

HARVEY has pointed out that these regularities can be understood if there is little or no coupling between the neutron and proton shells and only weak coupling between the neutron shells and that there is a small interaction energy between unpaired nucleons which depends on their spin or resultant spin. The groups produced in Pb^{209} clearly represent the promotion of the 127-th neutron to the empty states above the closed shell in Pb^{208} (see Fig. 76). The three strong

---

[1] D. G. E. MARTIN and H. O. W. RICHARDSON: Proc. Phys. Soc. Lond. A **63**, 223 (1950).

[2] MARTIN and PARRY (see ref. 4, p. 360) put the intensity of this γ-ray at less than 0.1%; it is remarkable that the rate of internal pair conversion of this γ-ray is as high as it appears: viz. a few percent of that of the 2.62 Mev γ-ray. (G. D. LATYSHEV, ref. 3, p. 308).

[3] B. B. KINSEY, G. A. BARTHOLOMEW and W. H. WALKER: Phys. Rev. **82**, 380 (1951).

[4] J. A. HARVEY: Canad. J. Phys. **31**, 278 (1953).

[5] M. H. L. PRYCE: Proc. Phys. Soc. Lond. A **65**, 773 (1952). See also G. E. TAUBER: Phys. Rev. **99**, 176 (1955).

groups in Pb^{208} presumably are unresolved doublets and the corresponding groups in Bi^{210} mark the positions of groups of levels derived from the combination of the neutron state with the $h_{\frac{9}{2}}$ proton of Bi^{209}. In Pb^{207} the (d, p) reaction excites states at low energy in common with the (d, t) reaction. These states represent the displacement of the neutron vacancy in Pb^{207} from the $p_{\frac{1}{2}}$ ground state to other levels. It is noteworthy that these states do not appear in Pb^{208} where all vacancies are filled.

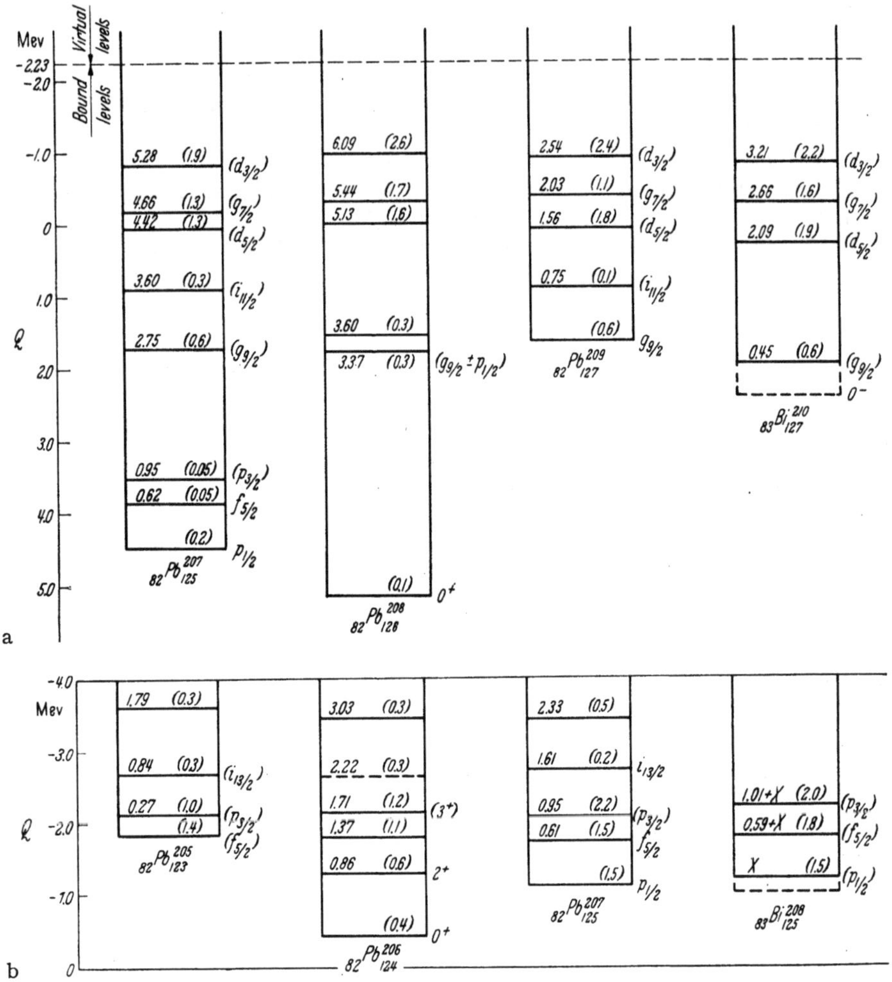

Fig. 78 a and b. Energy levels in Pb and Bi isotopes obtained by HARVEY, (a) from the (d, p) reaction; (b) from the (d, t) reaction. The figures in parentheses are proportional to the differential cross sections at 90° to the incident beam.

The intensities of the proton groups are a guide to the angular momentum carried away by the proton, for the deuteron energy being close to the COULOMB barrier, the centrifugal barrier must have a marked effect (see Sect. 7). This could account for the fact that the group of protons leading to the ground state of Bi^{210} was not detected (Fig. 78a) for the spin of Bi^{210} is low (Sect. 91) while that of the target (Bi^{209}) is high. For the same reason the weakly excited level at 0.75 Mev in Pb^{209} and 3.60 Mev in Pb^{207} is identified with the expected $i_{\frac{11}{2}}$ level (cf. Fig. 76); its partner the $i_{\frac{13}{2}}$ level appears as a vacancy in Pb^{207} at 1.6 Mev when a neutron is removed from Pb^{208} in the (d, t) reaction. This is the well-known isomeric state in Pb^{207} (Sect. 92). The same level appears as a doublet in Pb^{206} [of which the 7^- member is the isomeric state (Sect. 93)].

90. Levels in Pb^{209} and Bi^{209}. The nucleus Pb^{209} must have a spin of $\frac{9}{2}$ or $\frac{11}{2}$ since it decays to Bi^{209} by an allowed β-transition without emission of γ-radiation. Recently a short-lived isomeric state[1] has been found in Pb^{209} when produced by the decay of Tl^{209}. This state has a half-life of 3 millimicroseconds and decays to a state at 2.0 Mev by emission of an $E1$ γ-ray (120 kev). The latter state emits a 450 kev γ-ray to form the state at 1.56 Mev found by HARVEY[2], which then decays to the ground state without exciting the $i_{\frac{11}{2}}$ state found by HARVEY at 750 kev. The emission of the $E1$ γ-ray is strongly retarded (by 4×10^4).

Levels at 0.90, 1.61 and at 2.60 Mev in Bi^{209} have been observed[3] in studies of the inelastic scattering of neutrons and of the γ-rays produced by them. Presumably the first of these levels is the $f_{\frac{7}{2}}$ level.

91. Levels in Pb^{210}, Bi^{210}, Po^{210}. These three nuclei each have two nucleons above that required to close both neutron and proton shells.

α) Bi^{210}. This nucleus (as RaE) has been known for some fifty years; but little is yet known about its level system which is probably far more complicated[4] than that found in the (d, p) processes by HARVEY. A level at 46.7 kev has been known for many years; it is produced in 90% of the disintegrations[5] of Pb^{210} (RaD) and decays to the ground state with the emission of nearly pure $M1$ radiation. Many other low energy γ-rays have been reported but there remains some doubt that they exist[6]. In addition a long-lived α-emitting state has been found[7]; it is not yet clear whether this state or the β-emitter (RaE) lies the lower.

The spin of the ground state[8] is 1 and its parity seems to be odd. According to Fig. 76 we might expect that the lowest states of Bi^{210} are derived from the configuration $(g_{\frac{9}{2}}, h_{\frac{11}{2}})$ although they could well be combinations of several others. The energy distribution of these states has been discussed by PRYCE[9]. The distribution depends on whether the spin number is even or odd; states with the highest even spin must always lie lower than states with some intermediate spin. Since neutron capture by Bi^{209} must produce a compound nucleus with a spin of 4 or 5, i.e., a state with intermediate spin, it is clear that the subsequent γ-ray cascade could produce as readily states of very high spin as states of low spin. In fact, about a third of the neutron capture cross section of Bi appears to lead to the formation[10] of the isomeric state.

β) Pb^{210} (RaD). This nucleus is excited to a high energy by the decay of Tl^{210} (RaC″). It is difficult to study because the latter nucleus is only very

[1] D. STROMINGER, F. S. STEPHENS and J. O. RASMUSSEN: Phys. Rev. **103**, 748 (1956).
[2] See ref. 4, p. 361.
[3] M. J. POOLE: Phil. Mag. **44**, 1298 (1953). — E. A. ELIOT, D. HICKS, L. E. BEGHIAN and H. HALBAN: Phys. Rev. **94**, 144 (1954). — R. B. DAY: Phys. Rev. **102**, 767 (1956).
[4] D. M. BRINK: Proc. Phys. Soc. Lond. A **67**, 759 (1954).
[5] G. M. LEWIS: Proc. Phys. Soc. Lond. A **68**, 735 (1955).
[6] C. S. WU, F. BOEHM and E. NAGEL: Phys. Rev. **91**, 319 (1953). — R. W. FINK, G. W. WARREN, R. R. EDWARDS and P. E. DAMON: Phys. Rev. **103**, 651 (1956).
[7] H. M. NEWMAN, J. J. HOWLAND and I. PERLMAN: Phys. Rev. **77**, 720 (1950). — H. B. LEVY and I. PERLMAN: Phys. Rev. **85**, 758 (1952); **94**, 152 (1954).
[8] K. SMITH: To be published.
[9] M. H. L. PRYCE: See ref. 5, p. 361. See also ref. 4, this page.
[10] D. J. LITTLER and E. E. LOCKETT: Proc. Phys. Soc. Lond. A **66**, 700 (1953). — D. J. HUGHES and H. PALEVSKY: Phys. Rev. **92**, 1206 (1953).

infrequently produced by the α-decay of Bi^{214} (RaC). According to MAYER-KUCKUK[1] a succession of cascade γ-rays are produced by the decay of Tl^{210} starting at a level at 3.4 Mev. According to PRYCE the ground state is probably $(g_{\frac{9}{2}})^2$ and one may expect many levels corresponding to this configuration [and to others like $(g_{\frac{9}{2}}, j_{\frac{15}{2}})$].

γ) Po^{210}. This nucleus is not excited by the decay of RaE; it is, however, produced by the decay of At^{210}. A large number of γ-rays from this nucleus have been measured recently[2]. PRYCE has discussed the levels to be expected in such a nucleus but as yet no detailed theoretical analysis of Po^{210} has been published.

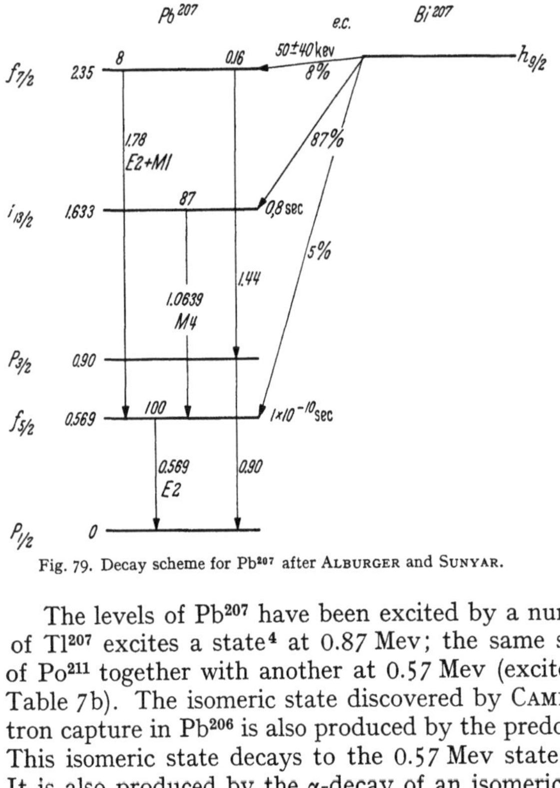

Fig. 79. Decay scheme for Pb^{207} after ALBURGER and SUNYAR.

92. Levels in Tl^{207} and Pb^{207}. These nuclei have each one vacancy in the shells closed at Pb^{208}. A level is known to exist in Tl^{207} (AcC″) at 0.35 Mev, excited by the α-decay[3] of Bi^{211} (AcC); the transition between this state and the ground state appears to be $M1$. This transition, and the β-decay between the ground state of Tl^{207} and Pb^{207} are consistent with shell model assignments of $s_{\frac{1}{2}}$ and $p_{\frac{1}{2}}$ respectively for the ground states.

The levels of Pb^{207} have been excited by a number of methods. The β-decay of Tl^{207} excites a state[4] at 0.87 Mev; the same state is excited by the α-decay of Po^{211} together with another at 0.57 Mev (excited by COULOMB excitation, see Table 7b). The isomeric state discovered by CAMPBELL and GOODRICH[5] by neutron capture in Pb^{206} is also produced by the predominant mode of decay of Bi^{207}. This isomeric state decays to the 0.57 Mev state by emission of $M4$ radiation. It is also produced by the α-decay of an isomeric state[6] of Po^{211}. The most detailed studies of the excited states and transitions in this nucleus have been made using the decay of Bi^{207}, by PRESCOTT[7], LAZAR and KLEMA[8] and by ALBURGER and SUNYAR[9]. Their results are shown in Fig. 79. The order of the levels is similar but not exactly given by the shell scheme in Fig. 76. The lifetime of the 0.57 Mev state is interesting (1.0×10^{-10} sec). It is of the order to be expected for a proton transition of $E2$ type; however, the transition here should

---

[1] T. MAYER-KUCKUK: Z. Naturforsch. **11**a, 627 (1956).

[2] J. W. MIHELICH, A. W. SCHARDT and E. SEGRÈ: Phys. Rev. **95** 1508 (1954).

[3] J. TEILLAC, P. FALK-VARIANT and C. VICTOR: J. de Phys. **13**, 313 (1952).

[4] J. SURUGUE: C. R. Acad. Sci. Paris **212**, 337 (1941). — J. de Phys. **7**, 145 (1946).

[5] E. C. CAMPBELL and M. GOODRICH: Phys. Rev. **78**, 640 (1950). — G. FRIEDLANDER, E. WILSON, A. GHIORSO and I. PERLMAN: Phys. Rev. **91**, 498 (1953). — E. C. CAMPBELL and F. NELSON: Phys. Rev. **91**, 499 (1953).

[6] W. JENTSCHKE, A. C. JUVELAND and G. H. KINSEY: Phys. Rev. **96**, 231 (1954).

[7] J. R. PRESCOTT: Proc. Phys. Soc. Lond. A **67**, 540 (1954).

[8] N. H. LAZAR and E. D. KLEMA: Phys. Rev. **98**, 710 (1955).

[9] D. E. ALBURGER and A. W. SUNYAR: Phys. Rev. **99**, 695 (1955).

be caused by the displacement of a neutron and should be much slower. This discrepancy indicates that collective effects must influence transition probabilities even in the simplified term scheme which obtains in this nucleus. TRUE[1] has calculated the effect to be expected from a weak coupling between the odd particle and the core and found fair agreement with experiment. The $p_{\frac{3}{2}}$ state at 0.90 Mev, which appears to pass to the ground state with an $M1$ transition, has not as yet been detected[2] in the γ-radiation produced by neutron capture in Pb^{206}. It has been suggested[3] that the weakness of this γ-ray (which should be $E1$) is connected with non-resonant capture of neutrons by Pb^{206}.

93. Levels in Pb^{206}. This nucleus has two neutron vacancies. There is no experimental data about the level system of the nucleus with two proton vacancies (Hg^{206}) nor is anything yet known about the level system of Tl^{206}—the analogue of Bi^{210}.

No γ-rays have been detected in the decay of Tl^{206}. One very weak γ-ray with an energy of 803 kev is produced by the decay of Po^{210}. Pb^{206} is most strongly excited in the decay of Bi^{206}. The first measurements of the γ-rays produced by the decay of this nucleus were made by ALBURGER and FRIEDLANDER[4]; their work has been repeated with greater precision, and has been interpreted theoretically, by ALBURGER and PRYCE[5]. The latter authors describe coincidence experiments and very accurate energy measurements which allow of the construction of a decay scheme in which many γ-rays participate (Fig. 80). They also discovered a 145 μsec isomeric state.

The basis of the interpretation of the level scheme of this nucleus is the assumption that the position of the levels can be calculated from the energies of the single-vacancy states of Pb^{207}, applying where necessary small corrections which can be roughly computed. The general validity of this assumption is clear from inspection of HARVEY's results. First it is assumed that the ground state of Pb^{206} consists of two vacancies in the $p_{\frac{1}{2}}$ shell. The energies of the two levels corresponding to some configuration, say $(p_{\frac{1}{2}} f_{\frac{5}{2}})$ is then given by the energy of the $f_{\frac{5}{2}}$ level in Pb^{207} (0.57 Mev) plus the interaction energy of the two $p_{\frac{1}{2}}$ neutrons (or vacancies) in Pb^{206} together with a small correction term which is zero for some states in the configuration and small and finite for others. The successful identification of levels in Pb^{206} with states for which this correction term is zero serves to fix the amount by which the $(p_{\frac{1}{2}})^2$ combination is depressed. If the 1.34 Mev (3^+) level is identified with onc of thc $(p_{\frac{1}{2}} f_{\frac{5}{2}})$ levels, the correction term is zero, and the ground state is depressed by $1.34-0.57=0.77$ Mev. A similar result is obtained if the 1.64 Mev (6^-) level is identified with the $(p_{\frac{1}{2}} i_{\frac{13}{2}})$ term. Knowing roughly the magnitude of the correction terms, when they appear, a level scheme was constructed which gives the positions of the levels and their spins with fair accuracy. The course of the decay of Bi^{206}, and the further identification of the predicted levels with those obtained empirically was obtained when it was noticed that the lowest odd level with spin 7 was theoretically lower than the lowest states of spin 5 and 6. Such a state, therefore, if excited, could decay only by $E3$ emission to a state with spin 4 and even parity. This led to the identification of the isomeric state with that at 2.20 Mev. These assignments are in general agreement with the positions of the undisplaced components of

[1] W. W. TRUE: Phys. Rev. **101**, 1342 (1956).
[2] See ref. 3, p. 361.
[3] L. K. PEKER: J. exp. theor. Phys. USSR. **29**, 865 (1956). — A. V. SHUT'KO and D. F. ZARETSKI: **29**, 866 (1956).
[4] D. E. ALBURGER and G. FRIEDLANDER: Phys. Rev. **81**, 523 (1951).
[5] D. E. ALBURGER and M. H. L. PRYCE: Phys. Rev. **92**, 514 (1953); **95**, 1482 (1954).

the $(p_{\frac{1}{2}} f_{\frac{5}{2}})$ and the $(p_{\frac{1}{2}} i_{\frac{13}{2}})$ doublets obtained in the (d, t) reaction with Pb^{207} (see Fig. 78b). In fact, since a neutron is removed in this reaction the levels most readily detected by that process must correspond to a $p_{\frac{1}{2}}$ vacancy, e.g. $(p_{\frac{1}{2}} p_{\frac{3}{2}})$

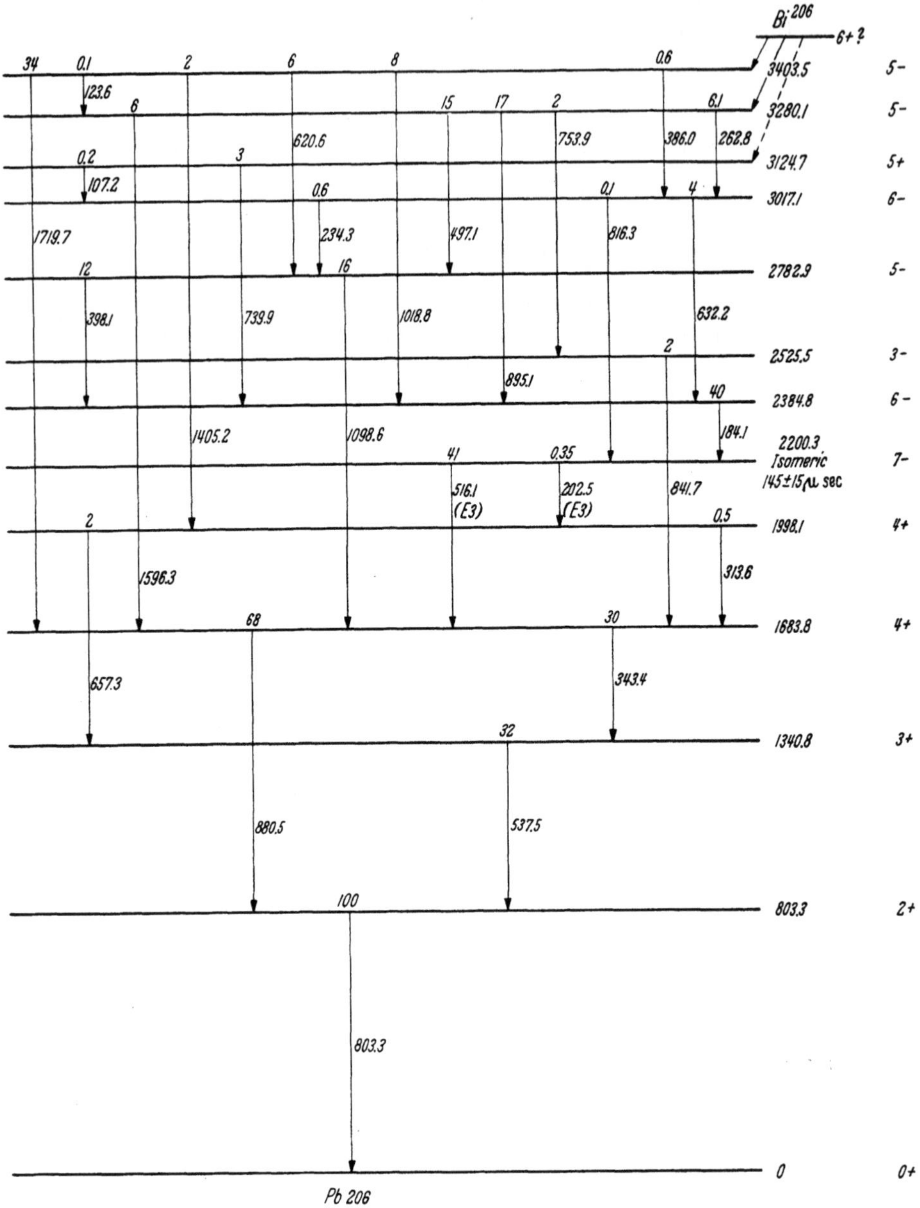

Fig. 80. Decay scheme for Pb^{206} according to Alburger and Pryce.

at 1.7 Mev, $(p_{\frac{1}{2}} h_{\frac{9}{2}})$ at 3.0 Mev and so on (neither of which are observed in Bi^{206} decay). The position of the observed and the predicted levels are shown in Fig. 81. This figure makes it very clear that the course of the γ-ray cascade which follows the electron capture into odd states with spin 5 selects only a small fraction of the total number of states which must exist.

As in Pb^{207}, the probabilities of electric transitions are much higher than would be expected if the electric moment of the transition was developed by the displacement of a neutron. This is shown by the isomeric $E3$ transition, which is only 10 times slower than would be expected for a proton, and by the strength of the 880 kev $E2$ cross-over γ-ray.

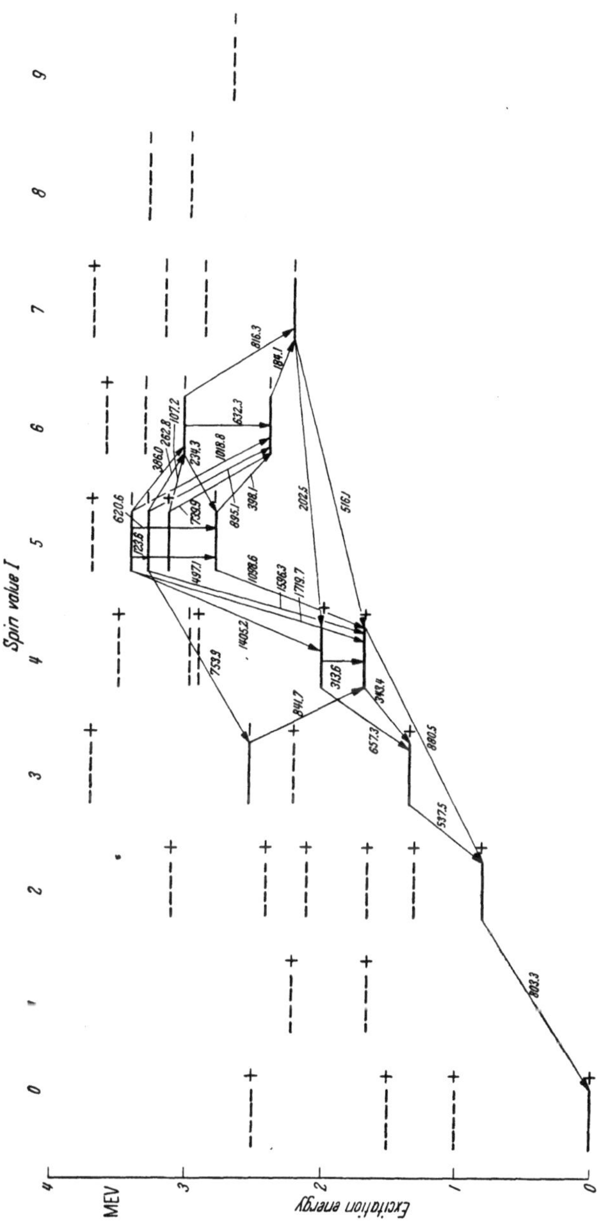

Fig. 81. Level scheme for Pb^{206} according to ALBURGER and PRYCE; broken lines: theoretical levels (not excited); full lines: levels found experimentally and theoretically predicted.

94. Levels in lighter Pb isotopes. *α) Pb^{205}.* The nucleus Pb^{205} is unstable[1]; to account for its lifetime and that of its neighbours[2], its spin must be $\frac{5}{2}$. Containing three neutrons less than that required to complete the closed shell of 126 neutrons, this spin is probably caused by a vacancy in the $f_{\frac{5}{2}}$ shell. Levels in Pb^{205} are

[1] J. R. HUIZENGA and J. WING: Phys. Rev. **102**, 926 (1956). See also Phys. Rev. **103**, 955 (1956).

[2] M. H. L. PRYCE: Nuclear Phys. **2**, 226 (1956).

excited in the α-decay of Po^{209} (570 and 840 kev) and by electron capture[1] by Bi^{205}. PRYCE has attempted to interpret the γ-rays observed in the latter disintegration with the method he has used so successfully for Pb^{206}. Owing to the greater complexity of a nucleus with three vacancies, Pb^{205} should contain more levels still; his calculations indicate that an isomeric state ($^{13}/_2{}^+$) may exist with an energy near 1 Mev.

β) Pb^{204}. Most of the levels known to exist in this nucleus are those disclosed by the decay of its isomeric state. The sequence of the γ-ray emission which follows this isomeric state has been disputed; the most recent results indicate that the decay of this isomeric state is produced by the emission of an $E5$ γ-ray (894 kev) followed by two $E2$ γ-rays in succession (375 and 899 kev respectively[2]), the final γ-ray having an energy very close to that of the first. The spin of the states are believed to be 0, 2^+, 4^+, and 9^-, in order of increasing energy. In addition, the emission of a 217 kev γ-ray in the decay of Bi^{205} has been reported[3]. The 375 kev transition, starting at the 4^+ level, is remarkable in that it is quite exceptionally slow[4] (2.6×10^{-7} sec), slow even by comparison with that expected of a single particle transition ($\sim10^{-9}$ sec). This is unusual among $E2$ γ-rays which, as we have seen, are usually accelerated by collective effects, even in closed shell nuclei. The long life of this 4^+ state is not yet understood. TRUE[5] has attempted to account for the level system in this nucleus; calculations of the lifetime of the 4^+ state using a weak coupling between the surface and the four vacancies in the closed shell gave much too fast a transition. Measurements of the gyromagnetic ratio of the 4^+ state by KROHN and RABOY[6] gives a result which is not consistent with that expected of this state if it is part of the ground state configuration $(f_{\frac{5}{2}})^2\,(p_{\frac{1}{2}})^2$.

γ) Pb^{203}, Pb^{202}, and Pb^{201}. Isomeric states in Pb^{203} and Pb^{201} have been found by FISCHER[7]. They decay in each case to the ground state with an $M4$ transition. Although the existence of these isomeric states is no doubt caused by a vacancy in the $i_{\frac{13}{2}}$ shell, they are not, as in Pb^{207}, obviously related to it, for many combinations of states can be derived from five or more vacancies in the uppermost shells in Pb nuclei.

A state at 440 kev has been excited[8] in Pb^{202} by the decay of Tl^{202}; MAEDER and WAPSTRA[9] have studied the decay of an isomeric state, said to have spin 9^-. According to their interpretation the level sequence excited in the decay of this isomeric state is 0, 2^+, 4^+, 5^-, 9^-.

P. Energy levels of nuclei with $N < 90$.

95. Levels in even-even nuclei. The systematics of even-even nuclei have been briefly discussed in Sects. 68 and 69. As we have already pointed out the spins of the first excited states of almost all even-even nuclei are 2 units, the only exceptions being the double closed shell nuclei and the single instance of Ge^{72}

[1] D. G. KARRAKER and D. H. TEMPLETON: Phys. Rev. **81**, 510 (1951).

[2] D. MAEDER, A. H. WAPSTRA, G. J. NIJGH and L. T. ORNSTEIN: Physica **20**, 521 (1954). See also ref. 6, this page.

[3] A. W. SUNYAR, D. E. ALBURGER, G. FRIEDLANDER, M. GOLDHABER and G. SCHARFF-GOLDHABER: Phys. Rev. **78**, 326 (1950).

[4] A. W. SUNYAR, D. ALBURGER, G. FRIEDLANDER, M. GOLDHABER and G. SCHARFF-GOLDHABER: Phys. Rev. **79**, 181 (1950).

[5] See ref. 1, p. 365.

[6] V. E. KROHN and S. RABOY: Phys. Rev. **97**, 1017 (1955); **102**, 1063 (1956).

[7] V. K. FISCHER: Phys. Rev. **99**, 764 (1955).

[8] P. KRAMER, H. C. HAMERS and G. MEIJER: Physica **22**, 208 (1955).

[9] D. MAEDER and A. H. WAPSTRA: Phys. Rev. **93**, 1433 (1954).

for which the first excited state (0.69 Mev) is 0^+ and decays to the ground state by internal conversion. In all other cases the first excited state decays to the ground state through the emission of pure $E2$ radiation, the transition probability being very much higher than that expected of single particle transitions. The second excited state, as we have pointed out in Sect. 69, usually has a spin of 2 or 4 units[1], and even parity. Those with spin 4 units cascade to the first excited state with emission of $E2$ radiation. But those with spin 2 decay preferentially to the first excited state (spin 2), the intensity of the higher energy cross-over γ-ray being comparable or lower than that of the cascade γ-ray (the KRAUSCHAAR-GOLDHABER rule). In the few instances where it has been possible to determine the $M1$ content of the cascade γ-ray, it is found that the intensity of this component is less than that of the $E2$ component. Assuming then that all of the intensity of the cascade γ-ray is derived from its $E2$ component, the calculated reduced $E2$ transition probabilities of the cross-over γ-rays are very much less than those of the cascade γ-rays. Finally the energies of the first and second excited states are usually roughly in the ratio of 2:1 regardless of the spin of the latter.

Data on the first and second excited states and on the γ-rays emitted by the latter are listed in Table 10, for nuclei where the spin sequence is 0-2-2[2]. In

Table 10. *Relative probabilities of cascade and cross-over transitions in even-even nuclei.*

Nucleus	E_1 Mev	E_2 Mev	E_2/E_1	n_{20}/n_{21}	B_{20}/B_{21} %	δ	Reference
Fe^{58}	0.805	1.62	2.01	0.31	1.0	2.2	(a)
Ni^{60}	1.33	2.13	1.60	0.38	0.3		(b)
Zn^{66}	1.04	2.40	2.30	0.43	0.7		(c)
Se^{72}	0.84	1.46	1.74	0.4	0.23		(d)
Se^{76}	0.55	1.20	2.16	1.0 (f) 2.8 (e)	5	~1 (o)	(e)
Kr^{82}	0.766	1.469	1.91	0.5	1.3		(g)
Te^{122}	0.566	1.258	2.26	0.19	1.0		(h)
Te^{126}	0.65	1.42	2.18	0.094	0.44		(i)
Xe^{126}	0.38	0.86	2.25	0.19	1.0		(i)
Xe^{128}	0.455	0.99	2.16	0.16	0.8		(j)
Pt^{192}	0.316	0.612	1.95	0.36	1.0		(k), (l)
Pt^{194}	0.327	0.618	1.89	0.31	0.7		(m)
Hg^{198}	0.411	1.089	2.66	0.19	1.8	1.7	(n)

(a) H. FRAUENFELDER, N. LEVINE, A. ROSSI and S. SINGER: Phys. Rev. **103**, 352 (1956).
(b) See ref. 6, p. 370.
(c) L. G. MANN, W. E. MEYERHOF and H. I. WEST: Phys. Rev. **92**, 1481 (1953).
(d) E. BRUN, J. J. KRAUSCHAAR and W. E. MEYERHOF: Phys. Rev. **102**, 808 (1956).
(e) J. D. KURBATOV, B. B. MURRAY and M. SAKAI: Phys. Rev. **98**, 674 (1955).
(f) P. HUBERT: J. de Phys. **12**, 823 (1951).
(g) R. C. WADDELL and E. N. JENSEN: Phys. Rev. **102**, 816 (1956).
(h) B. FARELLY, L. KOERTS, N. BENCZER, R. VAN LIESHOUT and C. S. WU: Phys. Rev. **99**, 1440 (1955).
(i) M. L. PERLMAN and J. P. WELKER: Phys. Rev. **95**, 133 (1954).
(j) N. BENCZER, B. FARELLY, L. KOERTS and C. S. WU: Phys. Rev. **101**, 1027 (1956).
(k) D. C. LU and M. L. WIEDENBECK: Phys. Rev. **94**, 501 (1954).
(l) L. L. BAGGERLY, P. MARMIER, F. BOEHM and J. W. M. DUMOND: Phys. Rev. **100**, 1364 (1955).
(m) M. T. THIEME and E. BLEULER: Phys. Rev. **102**, 195 (1956).
(n) L. G. ELLIOTT, M. A. PRESTON and J. L. WOLFSON: Canad. J. Phys. **32**, 153 (1954).
(o) J. J. KRAUSCHAAR and M. GOLDHABER: See ref. 3, p. 324.

[1] There are a few exceptions to this rule; one is Pd^{106} for which the second state seems to be 0^+, cf. KRAUSCHAAR and GOLDHABER ref. 3, p. 324.

[2] A more comprehensive but less exclusive table has been published recently by M. SAKAI: Progr. Theor. Phys., Japan **15**, 416 (1956).

this table the ratio n_{20}/n_{21} represents the ratio of the intensities of the cross-over to the cascade γ-rays; B_{20}/B_{21}, the ratio of the reduced transition probabilities calculated on the assumption that both γ-rays are entirely $E2$; and δ is the ratio of the amplitude of the $E2$ to that of the $M1$ component, where this quantity has been found from angular correlation measurements. It will be seen that the ratio of the reduced transition probabilities is always low, of the order of 1% or less. This is true of two adjacent 2^+ states in nuclei where the first excited state has some other spin. For example, it is true of the transition between the third and second states in[1] Ge^{72} and also[2] in Fe^{56} in which both states have spins of 2 units. For single particle transitions of 1 Mev, and for $A \approx 100$, the $M1$ transition probability is expected to be some 400 times greater than that of $E2$ transitions[3]. Thus, although the absolute values of the transition probabilities of the γ-rays produced by the second excited states are not known, the occurrence of a strong $E2$ component in the cascade γ-ray is a clear indication that either that component is greatly enhanced above that expected of single particle transitions or that the $M1$ component is far weaker than expected. The existence of the greatly accelerated $E2$ transitions from the first excited state argues strongly in favour of the former rather than the latter possibility.

The transition probabilities and the energy ratios of the first and second excited states suggest very strongly that these states originate in collective motions of the nucleus. The collective motion of a nucleus in its simplest form can be represented by simple harmonic oscillations of quadrupole type about a spherical equilibrium shape. The energy levels of such a system are $\hbar\omega$, $2\hbar\omega$, etc., where the fundamental frequency ω can be estimated from hydrodynamical arguments[4] and from surface and COULOMB energies. It is about 2 Mev for $A \approx 100$. The observed spacing of the first few excited states is usually less than this quantity. It seems that perturbations of the collective motion by coupling with the nucleons[5] and plausible arguments about the form of the potential which appears in the collective motion[6] all have the effect of reducing the level spacing below that expected from such elementary estimates. The forbidden nature of the cross-over transitions results very simply from the orthogonality of the wavefunctions. However, it would appear that the second excited state in general should have the form of a degenerate triplet with the angular momenta 0, 2, and 4. Although there is, as yet, very little evidence in support of such a grouping of states with these spins, most of the data is derived from studies of β-decay and the levels which have been observed, therefore, are excited by highly selective processes. Ni^{60} is an interesting example of a nucleus in which the spin sequence, originally thought to be 0-2-4 from measurements on the decay of Co^{60}, is now known to be 0-2-2, as a result of the discovery[7] of another 2^+ level excited in the decay of Cu^{60}. There may well be other examples; one can only conclude that the evidence obtained from β-decay in itself is insufficient to determine the level sequence. Data from different energy level studies are clearly required, e.g., from the (d, p) or (d, n) reaction, or inelastic neutron or proton scattering.

96. Levels in odd mass number nuclei. The energies of the lowest excited states of odd mass number nuclei can take very low values. In some nuclei

---

[1] E. BRUN, J. J. KRAUSCHAAR and W. E. MEYERHOF: Phys. Rev. **102**, 808 (1956).
[2] F. R. METZGER and W. B. TODD: Phys. Rev. **92**, 904 (1953).
[3] See ref. 1, p. 310.
[4] A. BOHR: Dan. Mat. fys. Med. **26**, No. 14 (1952). — See also ref. 2, p. 335.
[5] See ref. 2, p. 323.
[6] See ref. 5, p. 324.
[7] R. H. NUSSBAUM, R. VAN LIESHOUT, A. H. WAPSTRA, N. F. VERSTER, F. E. L. TEN HAAR, G. J. NIJGH and L. T. ORNSTEIN: Physica **20**, 555 (1953).

the lowest states have energies of only a few kilovolts (14 kev in Fe^{57} and Ge^{73}). Thus there appear to be no obvious restrictions on the energies of the lowest states such as seem to exist in even-even nuclei. We shall not discuss here the details of the level structure in any of these nuclei; the evidence, which for the most part is very scanty, is summarized elsewhere[1].

There exist however certain regularities in both the energies and the spins of the lowest states. These regularities are best seen in the properties of odd mass number isotopes of the same element or in corresponding isobars. In the Te, Xe, Ba region these regularities have been remarked by GOLDHABER and HILL[2]. For example the spins of the stable odd mass number Te isotopes are $\frac{1}{2}$ in agreement with the shell structure assignment of $s_{\frac{1}{2}}$ states, while the first and second states correspond to the expected $d_{\frac{3}{2}}$ and $h_{\frac{11}{2}}$ states. When the energy separation between these states is plotted against the proton number it is found to change continuously from one odd mass number nucleus to the next[3]. In this mass region these states are neutron states. MIHELICH and DE-SHALIT[4] have pointed out that the same $d_{\frac{3}{2}}$ and $h_{\frac{11}{2}}$ states are produced by protons and appear as ground states or as excited states in the Au isotopes; when plotted against the neutron number they move relative to each other in the same direction as those mentioned above. If there were no interaction between the nucleons one would expect that the difference in the energy of two states would remain constant (as pairs of like nucleons are added to a nucleus) and change sign when the configuration with the lower energy was full. The change in the observed energy differences presumably is a measure of these interactions or a measure of the shift in the position of the levels with changing deformation. The change over from one configuration to the next is actually seen in some elements (e.g., in Hg for which the lighter isotopes have a $p_{\frac{3}{2}}$ ground state, while the heavier isotopes are $p_{\frac{1}{2}}$). These levels however are probably well mixed; in fact, the mixing of states has been invoked by DE-SHALIT and GOLDHABER[5] to account for some anomalously slow β-transitions in nuclei with nearly closed shells of neutrons and protons.

It should be noted that if the lowest states in even-even nuclei have a collective origin, similar states might be expected to occur in odd mass number nuclei. None have yet been identified as such in nuclei with $N < 90$. Nevertheless the collective properties of the nucleus have a considerable effect on transition probabilities, for where the lowest states have been excited by COULOMB excitation they have usually been found to have a high $E2$ transition probability (e.g., the $s_{\frac{1}{2}} \to d_{\frac{3}{2}}$ transition in the Te isotopes, Table 6b). The effect persists, as we have seen, even into the closed shell nuclei[6].

The existence of rotational states in nuclei with $N < 90$ seems to be established only in some light nuclei. However, it is possible that rotational states exist in the two Ag isotopes, and in Rh^{103}, because the energies of the two states excited by COULOMB excitation are consistent with an "anomalous" rotational scheme

[1] Nuclear level schemes, $A = 40$ to $A = 92$, U.S. Atomic Energy Commission (1955). Another such publication covering the heavier elements is said to be in preparation.
[2] M. GOLDHABER and D. HILL: Rev. Mod. Phys. **24**, 179 (1952).
[3] For a study of such regularities in Sb, I, Cs, and La, see A. H. WAPSTRA: Physica, Haag **19**, 671 (1953).
[4] J. W. MIHELICH and A. DE-SHALIT: Phys. Rev. **93**, 135 (1954).
[5] See ref. 1, p. 323.
[6] Cf. Pb^{207} (Sect. 92); furthermore according to a recent measurement by METZGER [Phys. Rev. **103**, 983 (1956)] the lifetime of the 1.33 Mev state in Ni^{60} ($Z = 28$, $N = 32$) is five times shorter than the single particle estimate.

with a plausible value for the decoupling parameter[1], and the $E2$ components of the γ-rays in the transitions between these states and the ground state are roughly in the ratio expected of rotational states[2].

General references.

[1] PEASLEE, D. C.: Nuclear reactions of intermediate energy heavy particles. Ann. Rev. Nucl. Sci. **5**, 99 (1955).
[2] WIGNER, E. P.: On the development of the compound nucleus model. Amer. J. Phys. **23**, 371 (1955).
[3] FRIEDMAN, F. L., and V. F. WEISSKOPF: The compound nucleus; NIELS BOHR and the development of physics. London: Pergamon Press 1955.
[4] BLATT, J. M., and V. F. WEISSKOPF: Theoretical Nuclear Physics. New York: J. Wiley & Son 1953.
[5] DEVONS, S.: Excited States of Nuclei. Cambridge: University Press 1949.
[6] ADAIR, R. K.: Nuclear cross sections of the elements. Rev. Mod. Phys. **22**, 249 (1950).
[7] HUGHES, D. J.: Pile neutron research. Cambridge, Mass.: Addison-Wesley Publishing Co. 1953.
[8] BACON, G. E.: Neutron diffraction. Oxford: Clarendon Press 1955.
[9] SACHS, R. G.: Nuclear Theory, chap. 9. Cambridge, Mass.: Addison-Wesley Publishing Co. 1953.
[10] Lord RUTHERFORD, J. CHADWICK and C. ELLIS: Radiation from radioactive substances. Cambridge: University Press 1930.
[11] HANSON, A. O., R. F. TASCHEK and J. H. WILLIAMS: Monoergic neutrons from charged particle reactions. Rev. Mod. Phys. **21**, 635 (1949).
[12] BARSCHALL, H. H., L. ROSEN, R. F. TASCHEK and J. H. WILLIAMS: Measurement of fast neutron flux. Rev. Mod. Phys. **24**, 1 (1952).
[13] CRANBERG, L., R. B. DAY, L. ROSEN, R. F. TASCHEK and M. WALT: Techniques for measuring elastic and non-elastic neutron cross sections. Progr. Nucl. Energy **1**, 107 (1956).
[14] BARSCHALL, H. H.: Methods of measuring fast neutrons cross sections. Rev. Mod. Phys. **24**, 120 (1952).
[15] FREEMAN, J.: The inelastic scattering of fast neutrons. Progr. Nucl. Phys. **5**, 27 (1956).
[16] WHITEHOUSE, J.: Nuclear fission. Progr. Nucl. Phys. **2**, 120 (1952).
[17] HUBY, R.: Stripping reactions. Progr. Nucl. Phys. **3**, 177 (1953).
[18] LEVINGER, R.: Theories of photonuclear reactions. Ann. Rev. Nucl. Sci. **4**, 13 (1954).
[19] Beta- and gamma-ray spectroscopy, edited by K. SIEGBAHN. Amsterdam: North-Holland Publishing Co. 1955.
[20] BELL, R. E.: Nuclear particle detection: fast electronics. Ann. Rev. Nucl. Sci. **4**, 93 (1954).
[21] BOHR, A.: Rotational states of atomic nuclei. Copenhagen: Ejnar Munksgaards 1954.
[22] NEWTON, J. O.: Nuclear properties of the very heavy elements. Progr. Nucl. Phys. **4**, 234 (1955).
[23] PERLMAN, I., and F. ASARO: Alpha radioactivity. Ann. Rev. Nucl. Sci. **4**, 157 (1954).
[24] BLIN-STOYLE, R. J.: Theories of nuclear moments. Rev. Mod. Phys. **28**, 75 (1956).

[1] L. W. FAGG *et al.*: See Table 6b.
[2] See ref. 1, p. 347.

Resonance Processes by Neutrons.

By

JAMES RAINWATER.

With 63 Figures.

I. Introduction.

1. General considerations. Resonance processes of neutrons with nuclei have been studied with great interest from both the experimental and theoretical sides for over twenty years. Such processes are fundamentally similar, in principle, to those where protons or larger nuclear units are used in place of neutrons. The resonances in the various reaction processes are, in all cases, associated with quasi stationary energy states of the compound system of target nucleus plus incident particle. In comparing the use of neutrons with that of protons, deuterons, or heavier particles for such studies one notes the following features, among others:

1. For a given target nucleus the particular compound nucleus under investigation depends on the choice of bombarding particle. Thus the use of neutrons results in the investigation of the properties of different compound systems than for protons, etc.

Since neutrons alone of the above category of nuclear particles are without electric charge, there are important qualitative and quantitative differences in the experimental techniques and in the theoretical treatment.

2. With positive singly charged particles the COULOMB repulsion (V_c) of the nucleus at a radius $r = 1.4\,A^{\frac{1}{3}} \times 10^{-13}$ cm is equal to $Z A^{-\frac{1}{3}}$ Mev. Since this is the order of magnitude of the effective "nuclear radius" for reaction processes, this energy is approximately the height of the COULOMB barrier for charged particle reactions. For kinetic energies of relative motion $E \ll V_c$ it is still possible, in principle, to observe resonance processes. However, the "width" of the resonance is expected to be reduced by a barrier penetration factor B_c below the value in the absence of the barrier[1]. Since B_c approaches zero extremely rapidly as E/B_c decreases below unity, one cannot expect to observe resonances with charged particles for $E \ll B_c$. The limitation in the possible energy resolution for charged particle reaction studies is partly in obtaining a flux of monoenergetic particles, and partly in using sufficiently thin reaction targets that the variation of E as the particle undergoes ionization loss traversing the target is not excessive. In practice it is usually hard to have E monoenergetic to better than 1 kev, while $\sim$100 ev resolution is very unusual. If the mean level spacing of the compound system is $\gg$1 kev, as is the case for nuclei of low A, or where Z and A are nearly those for a closed shell configuration of nucleons (in Pb or Bi for example), charged particles can serve as excellent probe particles for studying compound state level structure. For most nuclei having $A \gtrsim 100$, however, the level spacing of the compound system is $<$ 1000 ev and studies of interactions for neutrons of $E <$ $\sim$10 kev represent the only method for a detailed resolution of level structure. For example, studies using the Columbia University Nevis Synchrocyclotron velocity selector system show eleven levels for E between 75 and 106 ev using

[1] Partial width if competing reactions are present.

Ta^{183} as the target nucleus. Levels for $T = 40.4$ ev and 41.6 ev in Ag appear as well separated peaks in σ vs. T.

3. In the range of energies, $E > \sim 10$ kev, mono-energetic neutrons for cross section measurements are generally made by charged particle reactions using mono-energetic charged particles in the first step. Thus uncertainty in T is usually less for charged particle reaction studies in this energy region than for neutrons.

4. In many cases the interpretation of neutron cross sections is simpler because of the lack of COULOMB scattering as such, and of interference effects between COULOMB and nuclear scattering. The formulation and use of the basic reaction theory is certainly simplified when one does not have to make extensive use of the regular and irregular COULOMB radial functions. As a modification of these remarks we must note that it is frequently possible to use the interference between COULOMB and nuclear scattering to obtain information which is not available using neutrons. For example p-p scattering $(d\sigma/d\Omega)$ shows that there could be no bound He^2 singlet $l=0$ state, even if the COULOMB repulsion were absent.

Techniques for studying neutron resonances divide into different categories according to method for obtaining neutron energy definition.

α) Pulsed accelerator time of flight slow neutron velocity spectrometers. In these systems a charged particle accelerator delivers a brief intense burst of electrons (betatron, synchrotron, or electron linear accelerator), protons, deuterons, etc. (pulse transformer, pulsed VAN DE GRAAFF, pulsed cyclotron, synchro-cyclotron, or linear accelerator) onto a target. When electrons are used, x-rays are produced which in turn can eject photoneutrons. The relative ease of obtaining very intense electron currents partly compensates for the relative inefficiency of the neutron photoproduction. When charged particles other than electrons are used, the production efficiency is also low if the range of the particles in the target is small compared to one mean free path for a nuclear encounter. A synchrocyclotron is a particularly effective source since $\gtrsim 400$ Mev protons have a stopping power range greater than a mean free path for nuclear encounters. When a 400 Mev proton strikes a nucleus and excites it to several hundred Mev, several "evaporation" neutrons are emitted on the average before the final nucleus reaches its ground state. Thus the number of evaporation neutrons is comparable to, or exceeds, the number of protons striking the target.

In systems of this type a hydrogeneous slab slows the fast neutrons to lower energies and a burst of neutrons is emitted containing all energies from the fast neutron E_0 to zero energy. The detection pulses at a detector at distance l from the source slab come at time $t = l/v$ after the source pulse. By sorting the detection pulses into a large number of time of flight intervals, each of width $\Delta_1 t$, the neutrons contributing to each time of flight detection interval are sorted according to velocity with a resolution dependent on $\Delta_1 t$, the duration of the fast neutron source burst, and uncertainties Δl in the path length l due to detector length, etc. In such a system the mean slowing down time for the emission of neutrons of final energy E is 3τ, with an r.m.s. spread $\sqrt{3}\,\tau$[1]. Here $\tau = v l_0$ is the mean time for collisons with hydrogen atoms in the slab and l_0 is the mean free path for such collisions at the final energy E. For paraffin $l_0 \sim 0.7$ cm so the slowing down process is best regarded as increasing the effective path length by $3 l_0$, with an r.m.s. smearing of $\sqrt{3}\,l_0$ (~ 1.2 cm in paraffin).

β) Pile fast choppers. Typical of such devices are the velocity selector systems for use with the Brookhaven, Argonne, and M.T.R. reactors. The high neutron

[1] This is shown by H. J. GROENWALD and H. GROENDIJK: Physica, Haag **13**, 141 (1947).

flux is obtained by using a suitably constructed and shielded "hole" to the interior of the reactor where moderator material is viewed. The collimated neutron beam is blocked by a rotating shutter which passes bursts of neutrons of $\lesssim 1\,\mu$sec duration. These are then "viewed" by a distant detector using time of flight techniques as for the pulsed accelerator systems. Resolutions of ~ 0.05 to $0.1\,\mu$sec per meter have been used extensively at higher energies where Δl effects are not too large.

The pile fast chopper systems so far have contributed most of the available information on slow neutron resonances. They have advantages of a steady source, a very small minimum slit size ($\ll 1$ cm^2) that permits measurements using separated isotopes, with relatively high intensity and resolution.

γ) Pile crystal spectrometer (cf. [*BNL—cs 12*]). The crystal spectrometers utilize coherent BRAGG reflections of neutrons from crystal planes. Recent improvements in techniques have given the crystal spectrometer the lead in effectiveness for measurements below ~ 10 ev. They are slow neutron monochrometers, and thus count continuously rather than in "gated intervals". Resolution widths $\sim 0.1\,\mu$sec/m have been obtained.

δ) VAN DER GRAAFF-*p-n sources without time of flight.* If the monoenergetic beam of protons (or deuterons) is incident on a thin (in terms of proton energy loss traversing the target) target of a suitable element, a source of monoenergetic intermediate energy neutrons (~ 1 kev to > 1 Mev) results. This occurs if only one reaction exit channel is present for neutron emission at that bombarding energy. Particular use has been made of $p + \mathrm{Li}^7 \to \mathrm{Be}^7 + n$. The energy resolution is limited by the uncertainties in initial proton energy, and in the target thickness energy loss. Most measurements of this type have used $\lesssim 1$ kev resolution. The group at the University of Wisconsin under H. H. BARSCHALL has made particularly effective use of this technique and has accumulated the large mass of experimental information that has been roughly matched by the predictions of the "cloudy crystal ball" model of FESHBACK, PORTER, and WEISSKOPF.

ε) Intermediate energy time of flight measurements[1]. A relatively recent method is that of pulsing a source of roughly monoenergetic neutrons, and then using a millimicrosecond time of flight method in addition to obtain superior resolution. Since only a limited energy band is present in the pulse, the repetition rate may be made relatively high without confusing neutrons from preceding cycles. This method is particularly promising in the region above 1 kev. The papers in the above reference describe the Oak Ridge millimicrosecond neutron spectrometer.

Results obtained using the various methods described are only partially published at present. Most of the unpublished results have been presented at meetings of the American Physical Society. The important experimental papers are listed by laboratory and method in the section on references.

The basic results which can be obtained from such studies can be roughly divided into matters mainly of engineering, or mainly pure physics interest. Thus the particular fact that Cd^{113} and Xe^{135} have resonances in the thermal energy region is of great interest for reactor engineering. Cd is used as a thermal neutron shielding material, while Xe^{135} is a fission product that tends to "poison" piles due to its $> 10^6$ barn cross section at thermal energies as shown in Fig. 1. On the physics side is information on nuclear level densities at excitation Q to $Q + \sim 1$ Mev above the ground state, where Q is the binding energy of a neutron in the compound nucleus, $Q \approx 8$ Mev on the average. It is of interest to note the variation of level density with nuclear size, to see how shell structure

[1] Cf. papers Y 8 to Y 11: Bull. Amer. Phys. Soc., Ser. II **1**, 178, 179 (1956).

or other effects perturb any otherwise smooth variation with A. It is also of interest to investigate the theory of the resonance process. Emphasis recently has focussed on the variation of Γ_n^0/D for a given nucleus, and on the variation of the average of this quantity from nucleus to nucleus. Here Γ_n^0 is the neutron width for the level divided by the square root of the level energy in ev, and D is the level spacing. Another point of interest is the distribution D/D_{av} of the spacing between adjacent levels for a given isotope and spin state in terms of the average level spacing. These topics are discussed in the section on theory, and in many of the figures.

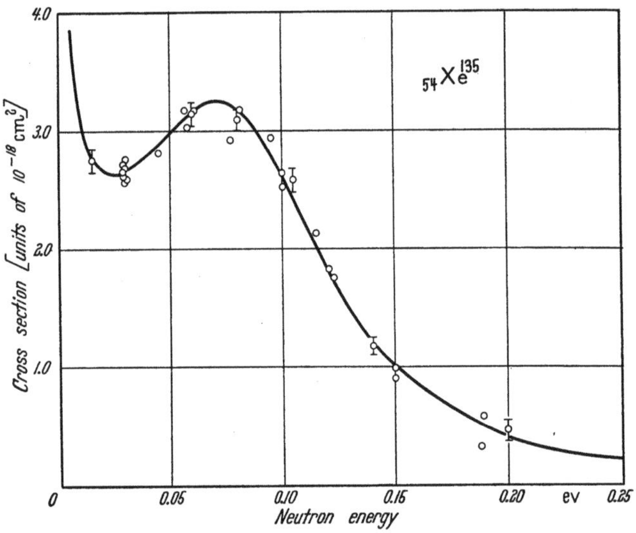

Fig. 1. The total cross section of Xe^{135} vs. energy in the region of the thermal energy resonance. This resonance has an unusually large σ_0 value. This is from [*ONL—cs 3*].

2. Remarks concerning the measurement of slow neutron levels. It has been customary for reviews of neutron cross sections to present a large collection of curves using logarithmic energy and cross section scales, showing "measured" σ vs. E. The "measured" cross sections are usually total cross sections measured by alternating a sample of "thickness" n atoms/cm² in and out of a neutron beam to measure the transmission

$$T = (e^{-n\sigma})_{av} \tag{2.1}$$

at energy E, averaged over the energy resolution function of the measuring system. The direct experimental quantities are the known sample thickness n, and a set of transmission values $T(E_j)$ at mean energies E_j. The $T(E_j)$ are known only to within some uncertainty δT due to statistical effects related to the finite number of neutrons detected, and possibly also due to effects of background counts and variation of counting efficiency with counting rate. In an ideal experiment one would measure T with infinitely good energy resolution, and zero statistical uncertainty for target nuclei at rest in the laboratory system. The $\sigma(E)$ thus obtained is the true σ vs. E which is considered in the basic theory. In practice, the nuclei are usually bound in a solid near room temperature, so they have thermal motion which gives a variation in the energy of relative motion even for monoenergetic neutrons in the laboratory frame. This DOPPLER broadening of the level can be calculated assuming a GAUSSian velocity distribution (MAXWELL-BOLTZMANN) for the atoms appropriate to a gas (see BETHE [*37*]), or attempts can be made to take account of the modifications which occur due to the binding of the atom in a chemical lattice. This effect was treated by LAMB[1] who showed that there are two effects. The first is a distortion of the shape from the symmetric form for a GAUSSian distribution of velocities (gas model). Vibration energy must be given up to, or absorbed from the lattice. There is an energy shift relative to the gas case corresponding to situations where the formation of the compound nucleus is elastic with respect to the lattice as a whole, with an inelastic tail. The energies kT, Γ, and E_0 must be treated in units of $k\Theta$, where Θ is the DEBYE temperature of the substance. To the extent that the effect for a crystal can be related to that for a gas using a higher effective temperature T', $\frac{3}{2}kT'$ should correspond to the average kinetic energy, including zero point vibration effects. LAMB gives a plot of T'/T vs. T/Θ where $T'/T = 1.30$, 1.13, 1.05, and 1.00 for $T/\Theta = 0.5$, 1.0, 1.5, and 2.0 respectively (to within the accuracy with which the curve could be read). He treats the case of silver where $\Theta = 210°$ K, so $T/\Theta \approx 1.4$ at ordinary temperature. For most calculations where $T/\Theta > 1$ and the level width is $> k\Theta$, the difference is usually ignored and the gas theory is applied.

Neglecting the $1/v$ factor in capture and effects of potential scattering, the BREIT-WIGNER one level formula can be written as

$$\sigma(x)/\sigma_0 = \frac{1}{1+x^2} \quad \text{(no DOPPLER broadening)} \tag{2.2}$$

where

$$x = 2(E - E_0)/\Gamma \tag{2.3}$$

and σ_0, E_0, E, and Γ are the peak cross sections at exact resonance, the resonance energy, the actual energy, and the full level width at half maximum respectively. The effect of DOPPLER broadening is to change the effective x for a given $E - E_0$ to a different value y that depends on the particular velocity of the atom at the time of the collision. Integrating over the distribution of atom velocities gives

$$\sigma(x) = \sigma_0\,\psi(x, t) = \frac{1}{2\sqrt{\pi t}} \int_{-\infty}^{\infty} \frac{e^{-\frac{(x-y)^2}{4t}}}{1+y^2}\,dy \tag{2.4}$$

where $t = \Delta^2/\Gamma^2$ and $\Delta = 2\,(E\,k\,t\,m/M)^{\frac{1}{2}}$, where m and M are the neutron and nucleus masses and Δ is the "DOPPLER width". Duc to the formal similarity of this formula to the problem of one-dimensional heat flow in an infinite slab having $(1+y^2)^{-1}$ initial temperature distribution, tables of this function have been generated by solving the partial differential equation for the heat flow problem using the UNIVAC computor at New York University. Tables are given in the Brookhaven National Laboratory report BNL—257 (T—40), issued in September 1953, by Dr. V. L. SAILOR. More complete tables are given by M. E. ROSE, W. MIRANKER, P. LEAK, L. ROSENTHAL and J. K. HENDRICKSON in Westinghouse Electric Corporation, Atomic Power Division, report WAPD—SR—506 dated October 1954. Earlier tables are given by M. BORN, "Optik" (Julius Springer, Berlin, 1933) p. 482.

The new peak cross section is $\sigma_\Delta = \sigma_0\,\psi(0, t)$. Fig. 2 shows σ_Δ/σ_0 vs. Δ/Γ. The function $\psi(x, t)$ for $t \neq 0$ no longer is of the form $(1+x^2)^{-1}$, but it is of interest to compare the half width Γ'' of the DOPPLER broadened σ to Γ. The following listing of numbers gives Γ''/Γ corresponding to the values of t in parentheses: 1.00 (0.0), 1.13 (0.05), 1.23 (0.10), 1.41 (0.2), 1.56 (0.3), 1.69 (0.4), 1.80 (0.5),

[1] W. E. LAMB jr.: Phys. Rev. **55**, 190 (1938).

2.05 (0.75), 2.26 (1.00), 2.63 (1.5), 2.94 (2.0), 3.46 (3.0) and 3.90 (4.0). Since one knows Δ a priori once E_0 and T are known, it is useful to express Γ' and Γ in

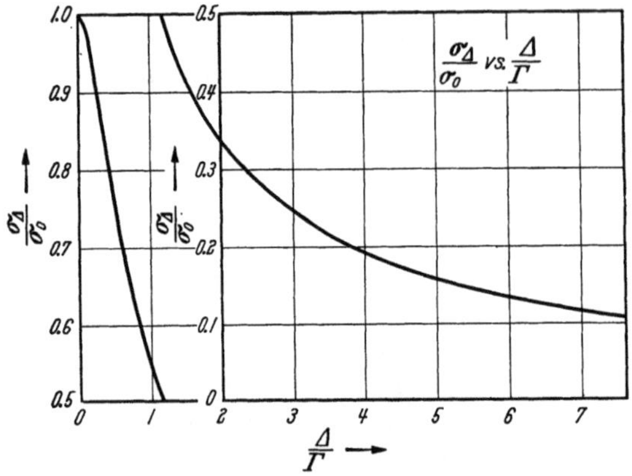

Fig. 2. The ratio of the DOPPLER broadened peak cross section σ_Δ to the true peak cross section σ_0 as a function of the ratio of the DOPPLER width Δ to the level width Γ.

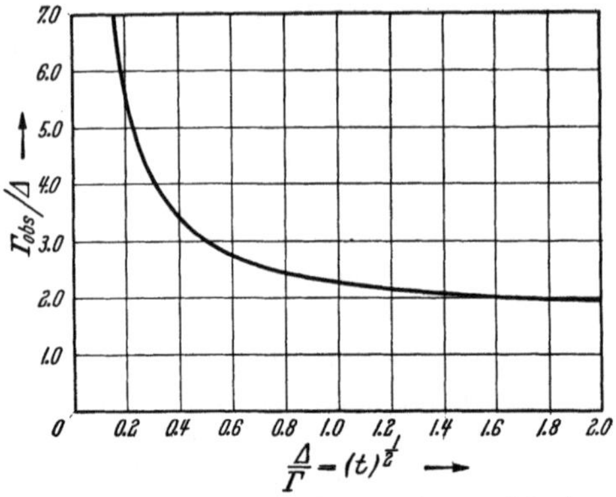

Fig. 3. The ratio of the DOPPLER broadened level width Γ' to the DOPPLER width Δ as a function of the ratio of Δ to the true level width Γ. From [*BNL—cs 9*].

units of Δ. Fig. 3 shows Γ'/Δ vs. Δ/Γ for $\Delta/\Gamma = 0$ to 2. For large Δ/Γ the value of $\sigma_\Delta \Delta$ is roughly 0.8 $\sigma_0 \Gamma$. The following relation holds to within $\sim 2\%$ or better.

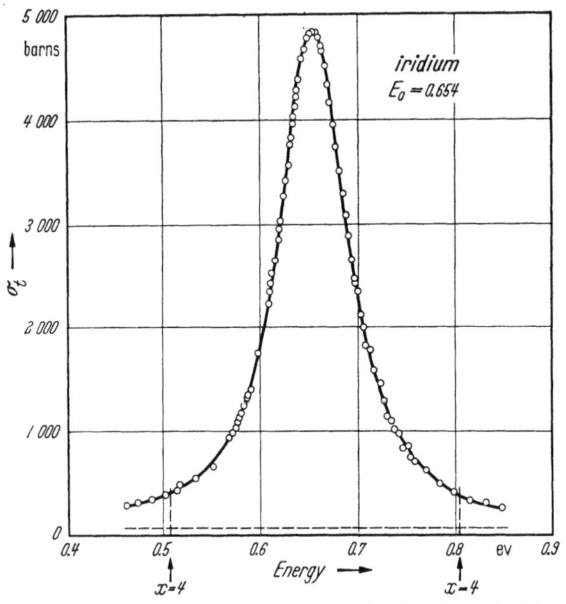

Fig. 4. The "measured" total cross section for the 0.654 ev level in Ir. In this and the subsequent curves showing "measured" cross sections, both DOPPLER and possible resolution effects are present. From [*BNL—cs 11*].

$$\sigma_0 \Gamma \approx [1.11 + 3\Gamma/4\Delta]\, \sigma_\Delta \Delta \qquad (2.5)$$

for $0.50 \leq \Delta/\Gamma \leq 10$. Since

$$\sigma_0 \Gamma = 4\pi \lambda\!\!\!^{-}{}_0^2\, g\, \Gamma_n \qquad (2.6)$$

where $E_0 = \hbar^2/2m\lambda\!\!\!^{-}{}_0^2$, and g is a spin statistical weight factor, and Γ_n is the "neutron partial width", the quantity $\sigma_0\, \Gamma$ is more useful for obtaining $g\Gamma_n$ than either σ_0 or Γ separately. For $\Delta/\Gamma > 1$, a knowledge of Δ and a measurement of σ_Δ permits $g\Gamma_n$ to be known with much better percentage accuracy than Γ is known.

Measurements where the values of σ_0 and Γ are directly obtained from measurements of σ vs. E with good resolution and good statistical accuracy are relatively few, and are limited to levels below ~ 5 to 10 ev. Most of these measurements were made using the high resolution Brookhaven crystal spectrometer (see papers under [*BNL—cs*] in the references). Figs. 4 to 9 illustrate some of the work of that group.

A more common method of analysis utilizes the area A under a transmission dip. If the transmission curve is re-normalized so $T \to 1$ at the sides of resonance (i.e. the potential scattering effects removed), then a plot of $(1 - T)$ vs. E bounds an area A in energy units. For a thin sample, where $n\sigma_\Delta \ll 1$, we can take $(1-T) \approx n\sigma$ so $A = n\int \sigma\, dE = \pi n \sigma_0 \Gamma/2$ independent of Δ/Γ. For a thick sample, where $n\sigma_\Delta \gg 1$, the value of $(1-T)$ saturates at unity as $n\sigma_\Delta \to \infty$. Thus $(1-T) \approx 1$ while $n\sigma > 1$ and $(1-T)$ recovers to zero as $n\sigma \to 0$ in the wings. If we let $W/2 = (E-E_0) = (n\sigma_0 \Gamma^2/2)^{\frac{1}{2}}$ for E such that $n\sigma = 1$, then A will be of the order of magnitude of W, or proportional to $(n\sigma_0\Gamma^2)^{\frac{1}{2}}$. Thus it is commonly noted that a thin sample area measurement tends to give $\sigma_0\Gamma$, while a thick sample area measurement tends to give $\sigma_0\Gamma^2$. When DOPPLER broadening is included the relation between A/Δ and $n\sigma_0\Gamma/\Delta$ is shown for various Δ/Γ in Fig. 10. It is seen that the curves tend to merge for small Δ/Γ, where the logarithmic slope of $\approx +1$ indicates an approximately linear relation. Fig. 11 gives a similar plot for $n\sigma_0\Gamma^2/\Delta^2$ vs. A/Δ for various Δ/Γ. The curves are seen to merge for large A/Δ and not too small Δ/Γ. It is important to note that the experimental resolution broadens the experimental transmission dip and decreases the peak magnitude of $(1-T)$, but does not alter the area A of the curve. Thus the area method of analysis using Fig. 10 has been used for many of the levels for which values of Γ_n are reported in Table 1. If two samples of thicknesses n_1 and n_2 give A_1 and A_2,

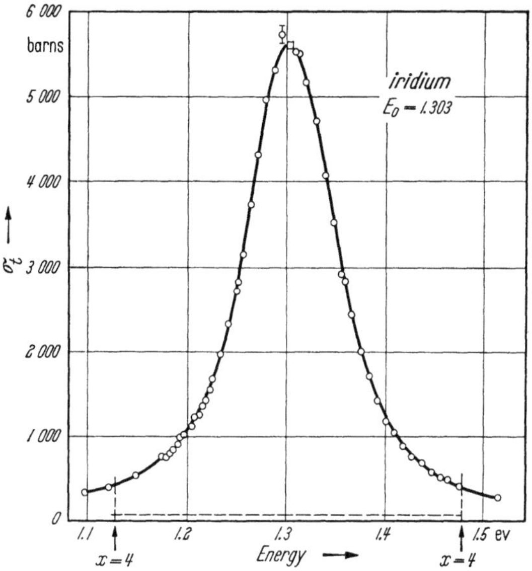

Fig. 5. The "measured" total cross section for the 1.303 ev level in Ir. From [BNL—cs 11].

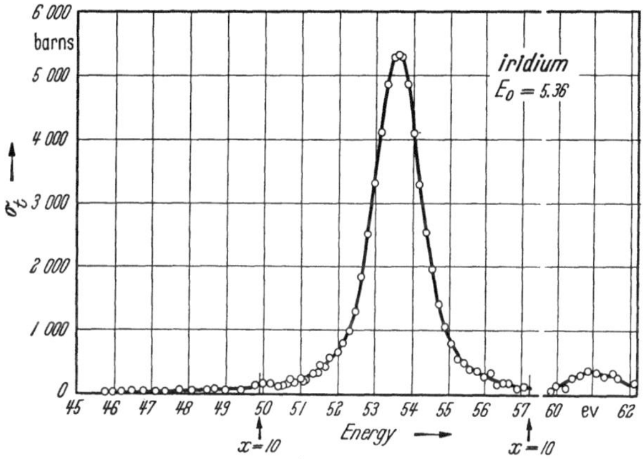

Fig. 6. The "measured" total cross section for the 5.36 and 6.1 ev levels in Ir. From [BNL—cs 11].

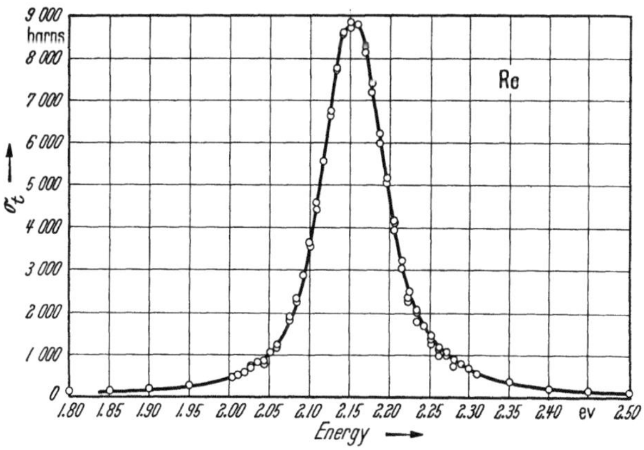

Fig. 7. The "measured" total cross section for the 2.156 ev level in Re^{185}. From [BNL—cs 10].

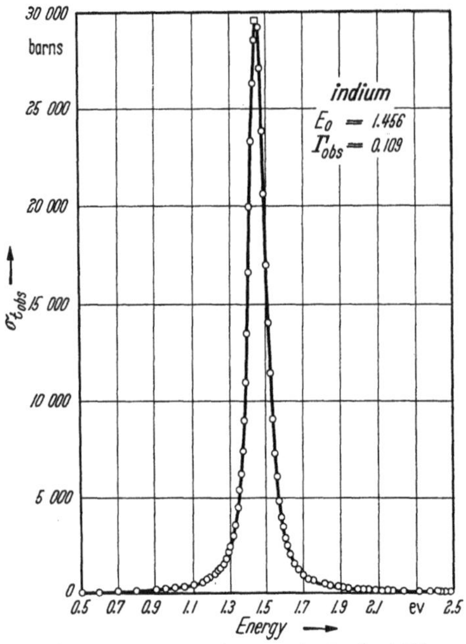

Fig. 8. The "measured" total cross section for the 1.456 ev level in indium. From [*BNL–cs 9*].

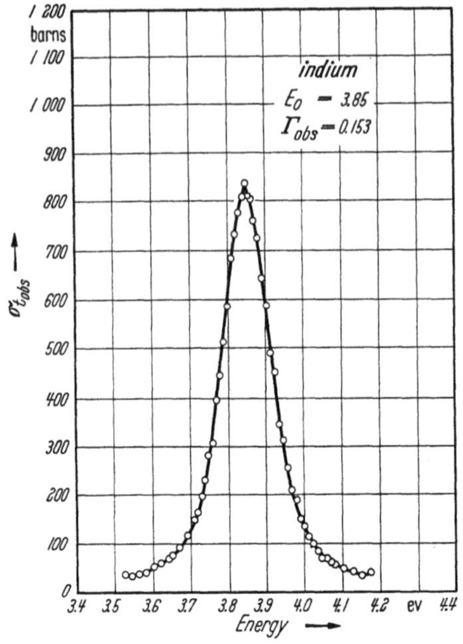

Fig. 9. The "measured" total cross section for the 3.85 ev level in indium. From [*BNL–cs 9*].

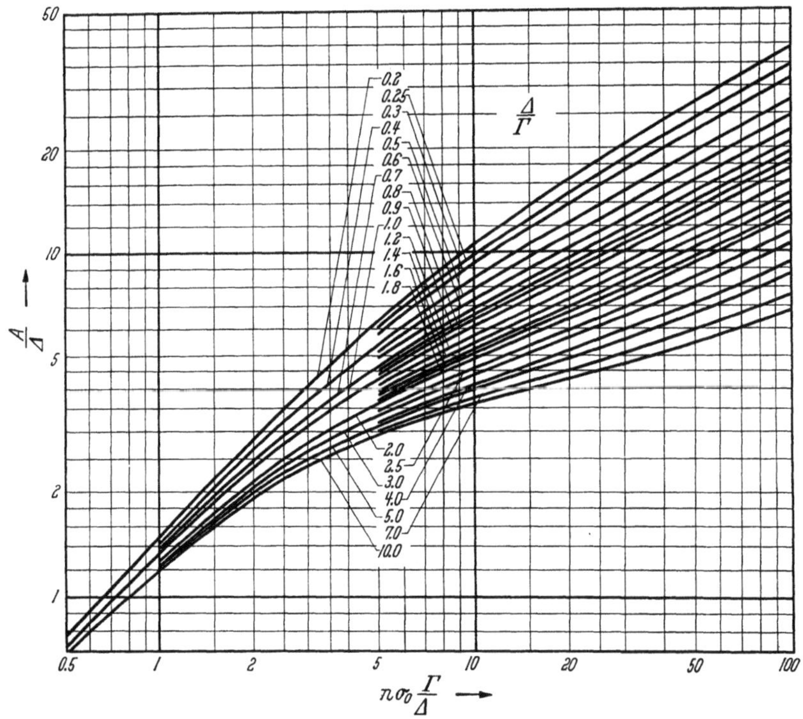

Fig. 10. A plot of the ratio A/Δ vs. $n\sigma_0\Gamma/\Delta$ for various Δ/Γ. Here A is the measured "area" of the resonance transmission dip in ev (see text of article), Δ is the DOPPLER width of the level, Γ is the level width, n is the sample "thickness" in atoms/cm², and σ_0 is the peak total cross section in cm²/atom. This extremely useful figure was prepared by the *BNL* slow neutron cross sections group.

then A_1/Δ and A_2/Δ and n_2/n_1 are known. If the predicted ratios of $n_2\sigma_0\Gamma/\Delta$ to $n_1\sigma_0\Gamma/\Delta$ for various Δ/Γ are compared with the known n_2/n_1, the values of

Δ/Γ that give agreement within the experimental uncertainty may cover a sufficiently small range that Δ/Γ, and thus Γ, may be said to have also been determined. This figure and method of analysis is due to the Brookhaven velocity selector group.

Returning now to a consideration of curves showing experimental cross section vs. energy, we note that these curves will nearly give the true DOPPLER broadened cross section curve if the levels are measured using proper sample

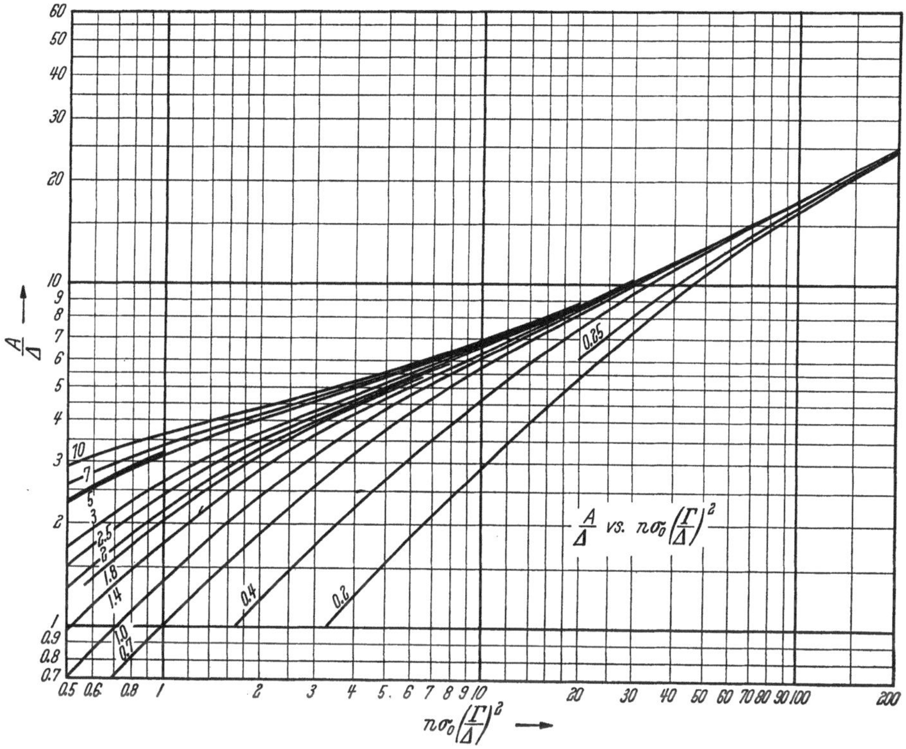

Fig. 11. A plot of A/Δ vs. $n\sigma_0\Gamma^2/\Delta^2$ for various Δ/Γ. This plot was prepared from Fig. 10 by a simple left-right displacement of the curves. It shows for $\Delta/\Gamma \gtrsim 1$ and $A/\Delta > 5$, that $n\sigma_0\Gamma^2/\Delta^2$ is roughly independent of Δ/Γ. For small A/Δ and $\Delta/\Gamma \gtrsim 1$, Fig. 10 shows that $n\sigma_0\Gamma/\Delta$ is roughly independent of Δ/Γ.

thicknesses and instrument resolution $\ll(\Gamma+\Delta)$. This has only been done for levels having $E_0 < \sim 10$ ev for nuclei having a level spacing $< \sim 100$ ev. For nuclei having level spacing $> \sim 10$ kev the structure seen using intermediate energy neutrons may give nearly the true $\sigma_0 = 4\pi g \lambda\!\!\!^{-2}$ for an $l=0$ resonance since $4\pi\lambda\!\!\!^{-2} = 2600000/E$ barns for neutrons decreases to a value nearer to the potential scattering amplitude and Γ_n, which is proportional to $E^{\frac{1}{2}}D$, becomes much larger. In the case of nuclei having $D < \sim 100$ ev the plot of σ vs. E will be highly misleading, since it may show a ratio of σ at resonance to σ between resonance less by a factor of 10 to 10^3 than the true ratio. The peak σ shown will tend to be of the order of magnitude of, or less than, the $(1/n)$ of the sample. The area under the σ vs. A curve also bears little resemblence to the true area. Furthermore, such curves usually extend to energies where the separate levels are not always resolved. As an example of this point, Fig. 12 shows a curve for silver prepared by the Brookhaven group [*BNL—fc 2*] to show the improvement in resolution achieved over the past ten years. Even the best curve of the [*BNL—fc*] group shows peak σ values orders of magnitude smaller than the true σ_0 values

Table 1. *Resonance parameters of nuclei.*

This table is mainly based on a similar table in BNL 325 with some additions. (See the comments preceding the references to the experimental work.) The basic theory is given in the text of this article. Γ_n, Γ_γ, etc. are the partial level width for decay by neutron, gamma ray, etc. emission. The factor $g\Gamma_n$ appears in the theory, where g is a spin weight factor $\approx \frac{1}{2}$ if two compound spin states are possible. In most cases g is not known, but is taken to be $\frac{1}{2}$. The quoted uncertainties in Γ_n and $\Gamma_n^0 = E^{-\frac{1}{2}}\Gamma_n$ include estimated uncertainties in Γ_γ, but not in g, in the analysis for Γ_n.

Isotope	I	E_0 kev	J	l	Γ_n kev	References
$_2He^4$	0					*Wis 17*
		1150 ± 50	$^3/_2$	1	1400 ± 200	Ground state of He^5
$_3Li^6$	1					*ANL—ac 4; ORNL—ac 2*
		248 ± 4	$^5/_2$	1	Γ 90 ± 10 Γ_n 60 ± 15 Γ_α 30 ± 10	
$_3Li^7$	$^3/_2$					*ANL—ac 4; MIT 1; Wis 11*
		258 ± 3	3	1	35 ± 5	
$_4Be^9$	$^3/_2$					*ORNL—ac 4; Wis 19*
		620 ± 10	3	1	25 ± 4	
		810 ± 10	2	1, 2	8 ± 3	
$_5B^{11}$	$^3/_2$					*ORNL—ac 4; Wis 19*
		430 ± 10	2	1	40 ± 5	
		1260 ± 20	3	2	140 ± 20	
		1780 ± 20	1	1, 2	60 ± 20	
		2450 ± 20	2	1, 2	120 ± 40	
		2580 ± 20	3	1, 2	60 ± 20	
$_6C^{12}$	0					*Wis 19*
		2080 ± 10	$\geq {}^3/_2$	2	≤ 10	
		2950 ± 100	$^3/_2$	2	60 ± 20	
		3650 ± 100	$^3/_2$	2	1200 ± 400	
$_7N^{14}$	1					*MIT 2; ORNL—ac 5; Wis 14*
		430 ± 5	$\geq {}^3/_2$	≥ 1	$\Gamma_n < 3$ $\Gamma_p < 0.01$	
		495 ± 5			$\Gamma_p < 10$ $\Gamma_n < 3$	
		639 ± 5	$^1/_2$	0	Γ_n 34 ± 4 Γ_p 9 ± 3	
		998 ± 5	$^3/_2$	0	Γ_p 45 ± 5 Γ_p 0.8 ± 0.3	
		1120 ± 6	$^5/_2$	1, 2	Γ_n 19 ± 3 Γ_p 0.20 ± 0.12	
		1188 ± 6	$\geq {}^3/_2$	≥ 1	$\Gamma_n < 2$ $\Gamma_p < 0.1$	
		1211 ± 7	$^1/_2$	1	Γ_n 12 ± 2 Γ_p 0.4 ± 0.2	
		1350 ± 7	$^5/_2$	1, 2	Γ_n 21 ± 4 Γ_p 1.0 ± 0.6	
		1401 ± 8	$^3/_2$	1	Γ_n 42 ± 10 Γ_p 10 ± 3 Γ_α 2 ± 1	
		1595 ± 8	$^5/_2$	1,2	Γ_n 21 ± 3 Γ_p 0.4 ± 0.2 Γ_α 0.20 ± 0.15	
		1779 ± 10	$^5/_2$	 1 2	Γ_n 18 ± 4 Γ_α 6 ± 2 Γ_p 0.20 ± 0.15	

Table 1. (Continued.)

Isotope	I	E_0 kev	J	l	Γ_n kev	References
$_8O^{16}$	0					*Wis 19*
		435 ± 5	$^3/_2$	1	40 ± 4	
		1000 ± 10	$^3/_2$	2	100 ± 10	
		1320 ± 10	$^3/_2$	1	35 ± 4	
		1660 ± 10	$\geqq {}^3/_2$	$\geqq 1$	< 7	
		1840 ± 10	$\geqq {}^3/_2$	$\geqq 1$	< 10	
		1910 ± 20	$^1/_2$	1	30 ± 6	
		2370 ± 20	$^1/_2$	0	140 ± 50	
		3330 ± 30	$^3/_2$	2	220 ± 40	
		3800 ± 80	$^3/_2$	1	800 ± 200	
		4400 ± 40	$^1/_2$	1	280 ± 80	
$_9F^{19}$	$^1/_2$					*ANL—ac 4; Duke 1; Wis 12*
		28 ± 1	$\geqq 1$	$\geqq 1$	< 0.4	
		50 ± 2	$\geqq 1$	$\geqq 1$	< 1.5	
		100 ± 2	1	1, 2	11 ± 2	
		275 ± 10			25 ± 10	
		340 ± 20			200 ± 100	
		420 ± 10			25 ± 15	
		510 ± 10			25 ± 15	
		590 ± 10			25 ± 10	
$_{11}Na^{23}$	$^3/_2$					*ANL—ac 3; BNL—fc 3; Duke 1; MIT 3*
		2.9 ± 0.2	2	0	0.24 ± 0.03	
		53 ± 2			< 5	
		204 ± 3	1	1	5 ± 2	
		217 ± 3	0	1	14 ± 10	
		243 ± 3	1, 2	1	7 ± 2	
		297 ± 3	1	0	4.0 ± 1.0	
		396 ± 4	0, 1	1	23 ± 3	
		451 ± 4	1	1, 2	9 ± 3	
		542 ± 5	1	0	39 ± 7	
		602 ± 4	$\geqq 1$		6 ± 4	
		710 ± 7	$\geqq 5$		72 ± 10	
		784 ± 5	$\geqq 2$		38 ± 6	
		914 ± 5	$\geqq 3$		36 ± 5	
		988 ± 5	$\geqq 1$		24 ± 10	
$_{12}Mg^{24}$	0					*Wis 17*
		85 ± 3	$^3/_2$	1	13 ± 3	
		275 ± 8	$^1/_2$	1	80 ± 20	
		430 ± 5	$^3/_2$	1	30 ± 10	
$_{13}Al^{27}$	$^5/_2$	5.6		0	0.45	*Duke 1*
		35 ± 2	3	0	1.2 ± 0.5	
		90 ± 4	3	0	7 ± 2	
$_{14}Si^{28}$	0					*Wis 17*
		195 ± 6	$^1/_2$	0	60 ± 10	
		570 ± 5	$^3/_2$	1	15 ± 5	
$_{16}S^{32}$	0					*Wis 9*
		111 ± 2	$^1/_2$	0	18 ± 3	
		203 ± 2		$\geqq 1$	< 2	
		274 ± 2		$\geqq 1$	< 3	
		290 ± 2		$\geqq 1$	< 3	
		375 ± 3	$^1/_2$	0	12 ± 2	
		585 ± 3	$^3/_2$	1	1.4 ± 0.5	
		700 ± 4	$^1/_2$	0	14 ± 3	

Table 1. (Continued.)

Isotope	I	E_0 ev	Γ_γ mv	Γ_n mv	Γ_n^0 mv	Reference
$_{25}Mn^{55}$	$^5/_2$					*ANL—fc 4; BNL—fc 1; KAPL 1*
		337 ± 6		22000 ± 4000	1200 ± 200	
		1080 ± 20		14200 ± 1500	400 ± 40	
		2360 ± 60		340000 ± 16000 $(J=3)$	7000 ± 300 $(J=3)$	
		7300		250000	2900	
		9000		710000	7500	
$_{27}Co^{59}$	$^7/_2$					*BNL—fc 1; KAPL 1*
		132 ± 2		4900 ± 700	430 ± 60	
$_{28}Ni^{58}$	0					*Duke 1*
		5000				
		17000		1300000	10000	
		65000		3600000	14000	
		109000		1200000	3600	
		124000		<500000	<1400	
		142000		2300000	6100	
		162000		4300000	10700	
		192000		2400000	5500	
		211000		5200000	11300	
		236000				
$_{28}Ni^{60}$	0					*Duke 1*
		15000		1800000	15000	
		30000		700000	4000	
		67000		700000	2700	
		88000		~400000	~1350	
		99000		~600000	~1900	
		109000		~500000	~1500	
		162000				
		188000				
		200000		3700000	8300	
		257000		~2900000	5700	
$_{28}Ni^{62}$	0					*ORNL—fc 1*
		4200 ± 1000		1300000 ± 400000	20000 ± 6000	
$_{30}Zn^{64}$	0					*ANL—fc 5*
		2750 ± 100		70000 ± 10000	1300 ± 200	
		4600 ± 300		60000 ± 30000	900 ± 400	
$_{30}Zn^{67}$	$^5/_2$					*ANL—fc 5*
		225 ± 3		1300 ± 200	87 ± 13	
		455 ± 10		13000 ± 3000	610 ± 140	
		1620 ± 70		19000 ± 9000	500 ± 200	
		2300 ± 100		29000 ± 15000	600 ± 300	
$_{30}Zn^{68}$	0					*ANL—fc 5*
		530 ± 11		10000 ± 2000	440 ± 90	
$_{5}Br^{79}$						*ANL—fc 6*
		35.6	400	45	7.5	
		54.2	345	30	4.1	
		188	360	86	6.3	
		240				
		317				
Nb^{93}						*BNL—fc 3*
		36.1		0.08	0.013	
		42.4		0.088	0.0135	
		95.0		0.40	0.041	
		106		0.44	0.043	
		119		3.0	0.275	
		195		60	4.3	
		246		1.8	0.115	

Table 1. (Continued.)

Isotope	I	E_0 ev	Γ_γ mv	Γ_n mv	Γ_n^0 mv	References
$_{42}Mo^{95}$	$^5/_2$		260 ± 80			*BNL—fc 2*
		45.0 ± 0.6	210 ± 60	174 ± 10	26.0 ± 1.6	
		162 ± 4		13.6 ± 1.9	1.1 ± 0.2	
		570 ± 30		120 ± 60	5 ± 3	
		700 ± 40		740 ± 130	28 ± 5	
$_{42}Mo^{96}$	0		260 ± 80			*BNL—fc 2*
		133 ± 2		200 ± 15	17.3 ± 1.3	
$_{42}Mo^{97}$	$^5/_2$		260 ± 80			*BNL—fc 2*
		71.5 ± 1.2	330 ± 80	16.6 ± 1.8	2.0 ± 0.2	
		292 ± 10		75 ± 15	4.4 ± 0.9	
		406 ± 17		80 ± 40	4 ± 2	
		580 ± 30		670 ± 130	28 ± 6	
$_{42}Mo^{98}$	0		260 ± 80			*BNL—fc 2*
		480 ± 20		740 ± 110	34 ± 5	
$_{42}Mo^{100}$	0		260 ± 80			*BNL—fc 2*
		367 ± 15		1000 ± 120	52 ± 6	
$_{44}Ru$ *			150 ± 50			*Col 12*
		9.8 ± 0.2		0.8 ± 0.3*	0.26 ± 0.09*	
		15.2 ± 0.3		1.1 ± 0.4*	0.28 ± 0.09*	
		24.1 ± 0.5		2.2 ± 0.8*	0.45 ± 0.15*	
		40.9 ± 0.8		7 ± 3*	1.1 ± 0.5*	
$_{45}Rh^{103}$	$^1/_2$					*BNL—cs 2*
		1.260 ± 0.004	155 ± 5	0.76 ± 0.04	0.68 ± 0.04	
$_{47}Ag^{107}$	$^1/_2$					*BNL—fc 1;*
		16.30	(130)	6.35	1.56	*MTR—fc 1;*
		41.6		13	2.0	*Col 13*
		45.0		1.55	0.23	
		51.6		37.4	5.2	
$_{47}Ag^{109}$	$^1/_2$					*BNL—fc 1;*
		5.120	136 ± 6	13.4 ± 0.6	5.9 ± 0.3	*MTR—fc 1;*
				$(J=3)$		*Col 13;*
		30.5		15.5	2.8	*BNL—cs 15*
		40.5		13	2.0	
		55.9		21.6	2.9	
		71.2		47	5.6	
		88.0		9.4	1.0	
$_{47}Ag$ **	$^1/_2$					*Col 13*
		133.4	(130)	127	11	
		139.0		4.5	0.38	
		143.5		13.1	1.1	
		172.4		92	7	
		201.4		28	2.0	
		208		40	2.8	
		251		206	13	
$_{48}Cd^{110}$	0					*BNL—fc 3*
		90	112	130	13.7	
$_{48}Cd^{111}$						
		27.7	90	3.6	0.68	
		67		1.32	0.16	
		101		58	5.8	
$_{48}Cd^{112}$	0					*BNL—fc 3*
		67	90	5.7	0.70	

* The resonances have not been identified with a particular isotope, and as a result the neutron widths listed are actually $a\Gamma_n$ and $a\Gamma_n^0$ where a is the abundance of the isotope.

** The resonances have not been identified by isotope, but both Ag^{107} and Ag^{109} have $\sim 50\%$ abundance so the calculations are made as if the isotope were known.

Table 1. (Continued.)

Isotope	I	E_0 ev	Γ_γ mv	Γ_n mv	Γ_n^0 mv	References
$_{48}Cd^{113}$	$^9/_2$					*BNL—fc 3; Col 3; CR 1*
		0.178 ± 0.002	113 ± 5	0.65 ± 0.02 ($J = 1$)	150 ± 0.05 ($J = 1$)	
		18.4		0.3	0.070	
		109		28	2.7	
$_{48}Cd^{114}$	0					*BNL—fc 3*
		121	121	55	5.0	
$_{49}In^{113}$	$^9/_2$		80 ± 20			*BNL—cs 1; BNL—fc 2*
		1.80 ± 0.03		< 0.1		
		4.71 ± 0.04		0.104 ± 0.016	0.048 ± 0.007	
		14.7 ± 0.1	60 ± 20	7.7 ± 1.0	2.0 ± 0.3	
		21.7 ± 0.2		4.4 ± 0.9	0.95 ± 0.20	
		25.2 ± 0.2	110 ± 40	9.7 ± 1.6	1.9 ± 0.3	
		32.5 ± 0.4		8.5 ± 1.0	1.49 ± 0.18	
		45.6 ± 0.6		4.7 ± 1.2	0.69 ± 0.018	
		63.6 ± 1.0		1.0 ± 0.4	0.13 ± 0.06	
		70.6 ± 1.2		4.7 ± 0.7	0.56 ± 0.08	
		82.5 ± 1.5		2.9 ± 1.0	0.32 ± 0.11	
		93.2 ± 1.8		40 ± 8	4.1 ± 0.8	
		105 ± 2		29 ± 9	2.8 ± 0.9	
$_{49}In^{115}$	$^9/_2$		77 ± 15			*BNL—cs 9; BNL—fc 2*
		1.458 ± 0.003	72 ± 2	3.36 ± 0.10	2.78 ± 0.08	
		3.86 ± 0.04	81 ± 4	0.318 ± 0.015	0.162 ± 0.008	
		9.10 ± 0.09	80 ± 40	1.73 ± 0.17	0.57 ± 0.06	
		12.1 ± 0.1	140 ± 60	0.106 ± 0.013	0.031 ± 0.004	
		23.0 ± 0.2		1.0 ± 0.2	0.21 ± 0.04	
		39.9 ± 0.5	140 ± 50	3.5 ± 0.4	0.55 ± 0.06	
		46.3 ± 0.6		0.43 ± 0.09	0.063 ± 0.013	
		48.6 ± 0.7		0.65 ± 0.08	0.093 ± 0.012	
		63.2 ± 1.0		1.1 ± 0.3	0.14 ± 0.04	
		83.3 ± 1.5		11 ± 2	1.2 ± 0.2	
		95 ± 2		2.8 ± 0.6	0.29 ± 0.06	
$_{50}Sn^{112}$	0		110 ± 30			*BNL—fc 1*
		96.5 ± 2.0		85 ± 10	8.7 ± 1.0	
		280 ± 9		36 ± 17	2.2 ± 1.0	
$_{50}Sn^{114}$	0		110 ± 30			*BNL—fc 2*
		280 ± 9		450 ± 90	27 ± 5	
$_{50}Sn^{115}$	$^1/_2$		110 ± 30			*BNL—fc 2*
		290 ± 10		260 ± 130	3.0 ± 1.5	
$_{50}Sn^{116}$	0		110 ± 30			*BNL—fc 2*
		112 ± 2		58 ± 6	5.5 ± 0.6	
		149 ± 4		4.5 ± 1.4	0.37 ± 0.11	
$_{50}Sn^{117}$	$^1/_2$		110 ± 30			*BNL—fc 2*
		39.4 ± 0.5	106 ± 25	5.4 ± 0.5	0.86 ± 0.09	
		122 ± 3		15 ± 3	1.4 ± 0.3	
		125 ± 3		2.8 ± 0.8	0.25 ± 0.08	
		197 ± 6		35 ± 9	2.5 ± 0.6	
		259 ± 8		12 ± 4	0.8 ± 0.3	
		343 ± 13		80 ± 60	4 ± 3	
		425 ± 18		190 ± 80	9 ± 4	
		460 ± 20		120 ± 70	6 ± 4	
$_{50}Sn^{118}$	0		110 ± 30			*BNL—fc 2*
		46.3 ± 0.6		0.7 ± 0.2	0.11 ± 0.03	
		368 ± 14		420 ± 30	22.0 ± 1.3	

Table 1. (Continued.)

Isotope	I	E_0 ev	Γ_γ mv	Γ_n mv	Γ_n^0 mv	References
$_{50}$Sn^{119}	$^1/_2$		110 ± 30			*BNL—fc 2*
		141 ± 3		30 ± 12	2.5 ± 1.0	
		222 ± 7		24 ± 6	1.6 ± 0.4	
		460 ± 20		200 ± 70	9 ± 3	
$_{50}$Sn^{120}	0		110 ± 30			*BNL—fc 2*
		425 ± 18		44 ± 18	2.1 ± 0.8	
$_{50}$Sn^{124}	0		110 ± 30			*BNL—fc 2*
$_{51}$Sb^{121}						*ANL—fc 6*
		15.6	(80)	9	2.3	
		29.8		7.2	1.3	
		53.5		3.2	0.44	
		62.5	63	3.2	0.40	
		64		1.0	0.125	
		74		10	1.16	
		90		38	4.0	
$_{51}$Sb^{123}						*ANL—fc 6*
		21.6	(80)	28	6.0	
		50.5		5.8	0.82	
		76.5		18	2.1	
		105		66	6.4	
$_{52}$Te^{123}	$^1/_2$					*BNL—cs 14*
		2.334 ± 0.008	104 ± 3	10.4 ± 0.5 ($J=1$)	6.8 ± 0.4 ($J=1$)	
$_{53}$I^{127}	$^5/_2$		100 ± 30			*BNL—fc 1*
		20.5 ± 0.3		1.6 ± 0.5	0.34 ± 0.10	
		31.4 ± 0.5		21 ± 7	3.8 ± 1.2	
		37.7 ± 0.7		40 ± 13	7 ± 2	
		46 ± 1		22 ± 10	3.3 ± 1.5	
		66 ± 2		3 ± 2	0.3 ± 0.2	
		78 ± 2		30 ± 20	3 ± 2	
		91 ± 3		30 ± 20	3 ± 2	
$_{54}$Xe^{135}	$^3/_2$					*ORNL—fc 1; ORNL—cs 3*
		0.0851 ± 0.001	86 ± 11	24 ± 4		
$_{55}$Cc^{133}	$^7/_2$		110 ± 30			*BNL—cs 4; BNL—fc 2*
		5.90 ± 0.04	115 ± 20	5.2 ± 0.7	2.1 ± 0.3	
		22.6 ± 0.3	120 ± 40	6.6 ± 1.1	1.4 ± 0.2	
		47.8 ± 0.6	140 ± 60	19 ± 3	2.8 ± 0.4	
		83.1 ± 1.5		9 ± 3	1.0 ± 0.3	
		94.8 ± 1.8		19 ± 6	2.0 ± 0.6	
		128 ± 2		112 ± 19	9.9 ± 1.7	
		143 ± 3		9 ± 4	0.8 ± 0.4	
		149 ± 4		45 ± 20	3.7 ± 1.6	
		182 ± 5		3.4 ± 1.1	0.25 ± 0.08	
		204 ± 6		57 ± 22	4.0 ± 1.5	
		224 ± 6		49 ± 22	3.3 ± 1.5	
		240 ± 7		480 ± 90	31 ± 6	
		301 ± 10		190 ± 40	11 ± 2	
		370 ± 14		160 ± 60	8 ± 3	
		407 ± 16		570 ± 200	28 ± 11	
		443 ± 19		290 ± 140	14 ± 6	
		530 ± 25		800 ± 300	35 ± 13	
$_{62}$Sm^{149}	$^7/_2$					*BNL—cs 16; JENER 1*
		0.096 ± 0.001	65 ± 2	0.56 ± 0.04	1.81 ± 0.13	

Table 1. (Continued.)

Isotope	I	E_0 ev	Γ_γ mv	Γ_n mv	Γ_n^0 mv	References
$_{63}Eu^{151}$	$^5/_2$		85 ± 15			*BNL—cs 6*; *BNL—cs 9*; *JENER 2*
		0.0006 ± 0.0010	67 ± 5		0.10 ± 0.02	
		0.327 ± 0.003	70 ± 10	0.088 ± 0.010	0.15 ± 0.02	
		0.461 ± 0.002	93 ± 3	0.80 ± 0.04	1.11 ± 0.06	
		1.055 ± 0.005	94 ± 4	0.27 ± 0.04	0.25 ± 0.04	
		2.73 ± 0.05		0.08 ± 0.03	0.05 ± 0.02	
		3.35 ± 0.02		3.4 ± 0.4	1.8 ± 0.2	
$_{63}Eu^{153}$	$^5/_2$		90 ± 20			*BNL—cs 6*
		1.76 ± 0.02		0.07 ± 0.02	0.06 ± 0.02	
		2.46 ± 0.01		1.8 ± 0.2	1.20 ± 0.15	
		3.84 ± 0.02		2.4 ± 0.6	1.3 ± 0.3	
$_{63}Eu^{151,3}$	$^5/_2$		90 ± 20			*BNL—fc 2*
		4.83 ± 0.04		0.048 ± 0.008	0.022 ± 0.004	
		5.47 ± 0.05		0.133 ± 0.018	0.057 ± 0.008	
		6.03 ± 0.08		0.35 ± 0.05	0.14 ± 0.02	
		6.25 ± 0.08		0.41 ± 0.07	0.16 ± 0.03	
		7.24 ± 0.10		2.4 ± 0.5	0.89 ± 0.18	
		7.47 ± 0.10		2.4 ± 0.5	0.88 ± 0.18	
		8.90 ± 0.09		4.2 ± 0.4	1.41 ± 0.14	
		10.5 ± 0.1		1.90 ± 0.13	0.59 ± 0.04	
		11.1 ± 0.2		0.5 ± 0.2	0.15 ± 0.06	
		11.7 ± 0.1		3.7 ± 0.3	1.08 ± 0.08	
		12.3 ± 0.2		1.2 ± 0.2	0.34 ± 0.07	
		12.7 ± 0.1		2.2 ± 0.4	0.62 ± 0.12	
		13.6 ± 0.2		0.4 ± 0.2	0.11 ± 0.06	
		14.8 ± 0.2		1.9 ± 0.3	0.49 ± 0.07	
		15.2 ± 0.2		2.2 ± 0.3	0.57 ± 0.08	
		16.8 ± 0.2		1.5 ± 0.2	0.37 ± 0.05	
		18.0 ± 0.2		5.0 ± 0.5	1.18 ± 0.13	
		18.8 ± 0.2		3.9 ± 0.8	0.90 ± 0.18	
		19.3 ± 0.2		4.3 ± 0.8	0.98 ± 0.19	
		20.1 ± 0.2		11.5 ± 1.2	2.6 ± 0.3	
		22.0 ± 0.3		2.9 ± 0.7	0.62 ± 0.15	
		22.5 ± 0.3		6.7 ± 1.3	1.4 ± 0.3	
		24.1 ± 0.3		3.0 ± 0.8	0.61 ± 0.15	
		25.0 ± 0.3		5.8 ± 1.2	1.2 ± 0.3	
		27.0 ± 0.4		6.7 ± 1.4	1.3 ± 0.3	
		29.5 ± 0.4		2.9 ± 1.5	0.5 ± 0.3	
		30.8 ± 0.4		3.5 ± 1.8	0.6 ± 0.3	
		31.8 ± 0.4		3.3 ± 1.7	0.6 ± 0.3	
		35.1 ± 0.5		6 ± 2	1.0 ± 0.3	
		36.9 ± 0.5		12 ± 4	2.0 ± 0.7	
		38.3 ± 0.6		13 ± 4	2.1 ± 0.7	
		42.2 ± 0.7		22 ± 7	3.4 ± 1.1	
		47.4 ± 0.8		41 ± 12	6 ± 2	
$_{64}Gd^{157}$	$^7/_2$					*ANL—cs 2*; *ANL—sc 1*
		0.030 ± 0.003	100 ± 30	0.65 ± 0.10	3.7 ± 0.5	
$_{64}Gd$						*ANL—fc 6*
		2.58 ± 0.05	70 ± 10			
$_{65}Tb^{159}$	$^3/_2$		90 ± 30			*BNL—fc 2*
		3.35 ± 0.03		0.43 ± 0.05	0.24 ± 0.03	
		4.99 ± 0.05		0.055 ± 0.008	0.025 ± 0.004	
		11.14 ± 0.10		9.2 ± 1.8	2.8 ± 0.6	
		14.4 ± 0.2		0.54 ± 0.11	0.14 ± 0.03	
		21.4 ± 0.2		2.1 ± 0.2	0.46 ± 0.05	
		24.7 ± 0.3		5.8 ± 0.7	1.17 ± 0.14	
		27.8 ± 0.3		0.90 ± 0.12	0.17 ± 0.02	
		34.1 ± 0.4		3.3 ± 0.7	0.57 ± 0.11	
		44.2 ± 0.6		5.6 ± 0.8	0.84 ± 0.13	

Table 1. (Continued.)

Isotope	I	E_0 ev	Γ_γ mv	Γ_n mv	Γ_n^0 mv	References
$_{65}Tb^{159}$ (cont.)	$^3/_2$					
		46.6 ± 0.6		15 ± 3	2.2 ± 0.4	
		51.1 ± 0.7		3.7 ± 0.7	0.52 ± 0.10	
		54.9 ± 0.7		1.6 ± 0.6	0.21 ± 0.08	
		58.7 ± 0.8		4.6 ± 1.1	0.60 ± 0.14	
		66.1 ± 1.1		15 ± 4	1.9 ± 0.5	
		74.6 ± 1.3		20 ± 5	2.3 ± 0.6	
		78.2 ± 1.4		15 ± 4	1.7 ± 0.4	
		91.9 ± 1.8		24 ± 6	2.5 ± 0.7	
		97.9 ± 1.9		45 ± 11	4.5 ± 1.1	
		112.5 ± 2.0		65 ± 25	6.0 ± 2.5	
		115 ± 2		70 ± 30	7 ± 3	
		122 ± 3		50 ± 30	4 ± 3	
		131 ± 3		16 ± 8	1.4 ± 0.7	
		144 ± 3		120 ± 30	10 ± 3	
		156 ± 4		160 ± 40	13 ± 3	
$_{67}Ho^{165}$	$^7/_2$		90 ± 20			*BNL—fc 2*
		3.92 ± 0.03		2.5 ± 0.5	1.3 ± 0.2	
		12.8 ± 0.1		13.1 ± 1.8	3.4 ± 0.5	
		18.2 ± 0.2	180 ± 90	0.92 ± 0.17	0.22 ± 0.04	
		21.3 ± 0.2		0.73 ± 0.12	0.16 ± 0.03	
		35.9 ± 0.4	60 ± 20	7.5 ± 1.1	1.20 ± 0.18	
		37.9 ± 0.5		0.36 ± 0.07	0.059 ± 0.012	
		40.3 ± 0.5	60 ± 30	21 ± 3	3.3 ± 0.5	
		48.5 ± 0.6	60 ± 30	25 ± 4	3.6 ± 0.5	
		52.2 ± 0.7		51 ± 6	7.0 ± 0.8	
		55.3 ± 0.8		7 ± 3	0.9 ± 0.4	
		66.3 ± 1.1		38 ± 6	4.6 ± 0.7	
		70.0 ± 1.2		1.1 ± 0.4	0.13 ± 0.04	
		73.1 ± 1.3	140 ± 70	35 ± 6	4.1 ± 0.7	
		83.4 ± 1.5		1.2 ± 0.6	0.13 ± 0.06	
		87.2 ± 1.6	260 ± 130	96 ± 10	10.3 ± 1.1	
		96 ± 2		125 ± 15	12.7 ± 1.5	
		104 ± 2		65 ± 13	6.4 ± 1.3	
		123 ± 3		34 ± 16	3.1 ± 1.5	
		130 ± 3		160 ± 25	14 ± 2	
$_{69}Tm^{169}$	$^1/_2$		70 ± 20			*BNL—fc 2*
		3.92 ± 0.03		12 ± 4	6 ± 2	
		14.4 ± 0.1	75 ± 20	5.7 ± 0.8	1.5 ± 0.2	
		17.6 ± 0.2	63 ± 16	3.2 ± 0.4	0.76 ± 0.10	
		29.1 ± 0.3		0.37 ± 0.13	0.07 ± 0.02	
		35.2 ± 0.4	110 ± 40	13 ± 2	2.2 ± 0.4	
		38.1 ± 0.5		0.61 ± 0.12	0.10 ± 0.02	
		45.6 ± 0.6		6.4 ± 1.3	0.9 ± 0.2	
		51.6 ± 0.7		7.7 ± 1.6	1.1 ± 0.2	
		59.8 ± 0.8		24 ± 5	3.1 ± 0.6	
		66.8 ± 1.1		100 ± 15	12 ± 2	
		84.4 ± 1.5		11 ± 3	1.2 ± 0.4	
		96 ± 2		74 ± 11	7.6 ± 1.1	
		104 ± 2		2.7 ± 1.4	0.26 ± 0.14	
		118 ± 2		41 ± 12	3.8 ± 1.1	
		128 ± 3		10 ± 5	0.9 ± 0.4	
		139 ± 3		45 ± 30	4 ± 2	
		158 ± 4		160 ± 40	13 ± 3	
$_{70}Yb^{168}$	0					*BNL—cs 7*
		0.597 ± 0.003	70 ± 10	3.3 ± 0.5	4.3 ± 0.7	

Table 1. (Continued.)

Isotope	I	E_0 ev	Γ_γ mv	Γ_n mv	Γ_n^0 mv	References
$_{71}\mathrm{Lu}^{176}$	≥ 7		70 ± 20			*BNL—cs 3, 11*
		0.143 ± 0.001	61 ± 2	0.092 ± 0.009	0.25 ± 0.03	
		1.57 ± 0.01		0.6 ± 0.2	0.5 ± 0.2	
		2.604 ± 0.010	50 ± 1	0.12 ± 0.02	0.07 ± 0.02	
$_{71}\mathrm{Lu}$ *						*BNL—cs 3; BNL—fc 2*
		4.39 ± 0.03		0.012 ± 0.006 *	0.006 ± 0.003 *	
		4.78 ± 0.03		0.28 ± 0.03 *	0.128 ± 0.014 *	
		5.22 ± 0.04		1.2 ± 0.2 *	0.52 ± 0.09[1]	
		6.17 ± 0.05		0.062 ± 0.012 *	0.025 ± 0.005 *	
		9.8 ± 0.1		0.044 ± 0.012 *	0.014 ± 0.004 *	
		11.3 ± 0.1	40 ± 20	3.2 ± 0.5 *	0.95 ± 0.15 *	
		14.1 ± 0.1		18 ± 3 *	4.8 ± 0.7 *	
		15.5 ± 0.2		1.40 ± 0.15 *	0.35 ± 0.04 *	
		20.0 ± 0.2		0.08 ± 0.02 *	0.018 ± 0.005 *	
		20.7 ± 0.2	160 ± 50	2.3 ± 0.3 *	0.51 ± 0.07 *	
		22.0 ± 0.2		0.07 ± 0.02 *	0.015 ± 0.004 *	
		23.7 ± 0.2	70 ± 20	4.9 ± 0.7 *	1.01 ± 0.14 *	
		24.8 ± 0.3		0.17 ± 0.06 *	0.034 ± 0.012 *	
		27.3 ± 0.3		0.27 ± 0.08 *	0.052 ± 0.016 *	
		28.2 ± 0.3		1.4 ± 0.3 *	0.26 ± 0.05 *	
		30.4 ± 0.3		9.2 ± 1.7 *	1.7 ± 0.3 *	
		31.2 ± 0.3		2.3 ± 0.6 *	0.41 ± 0.10 *	
		37.1 ± 0.4	90 ± 30	6.4 ± 0.9 *	1.05 ± 0.14 *	
		41.5 ± 0.5	80 ± 30	24 ± 3 *	3.7 ± 0.4 *	
		46.2 ± 0.6		0.39 ± 0.15 *	0.06 ± 0.02 *	
		50.7 ± 0.7		38 ± 4 *	5.3 ± 0.6 *	
		54.3 ± 0.7		0.8 ± 0.3 *	0.11 ± 0.04 *	
		57.5 ± 1.0		{ 3.7 ± 0.7 *	0.49 ± 0.09 *	
				3.7 ± 0.7 *	0.49 ± 0.09 *	
		2 equal resonances				
		63.0 ± 1.0		1.3 ± 0.6 *	0.16 ± 0.08 *	
		71.0 ± 1.2		10 ± 3 *	1.2 ± 0.4 *	
		86.0 ± 1.5		4 ± 2 *	0.4 ± 0.2 *	
		89.4 ± 1.6		4 ± 2 *	0.4 ± 0.2 *	
		98.9 ± 1.8		93 ± 11 *	9.3 ± 1.1 *	
		104 ± 2		36 ± 11 *	3.5 ± 1.1 *	
		110 ± 2		39 ± 10 *	3.7 ± 0.9 *	
		119 ± 3		83 ± 11 *	7.6 ± 1.0 *	
		132 ± 3		150 ± 15 *	13.1 ± 1.3 *	
$_{72}\mathrm{Hf}^{174}$	0		60 ± 20			*BNL—fc 2*
		30.5 ± 0.4		49 ± 6	8.9 ± 1.2	
$_{72}\mathrm{Hf}^{177}$	1/2		56 ± 15			*ANL—fc 3; BNL—fc 2, 3; BNL—cs 13*
	3/2	1.100 ± 0.005	67 ± 2	2.1 ± 0.05	2.0 ± 0.05	
		2.39 ± 0.01	60 ± 1	9.3 ± 0.2	6.1 ± 0.15	
		5.9 ± 0.1		5.1 ± 1.5	2.1 ± 0.6	
		6.6 ± 0.1	44 ± 20	11 ± 3	4.3 ± 1.3	
		8.8 ± 0.1		8 ± 3	2.7 ± 1.1	
		13.7 ± 0.2		0.67 ± 0.08	0.18 ± 0.02	
		14.1 ± 0.2		2.2 ± 0.3	0.59 ± 0.08	
		22.2 ± 0.2		2.7 ± 0.3	0.57 ± 0.06	
		23.5 ± 0.2		1.6 ± 0.4	0.33 ± 0.07	
		25.9 ± 0.3		0.41 ± 0.08	0.08 ± 0.02	
		27.2 ± 0.3		1.80 ± 0.18	0.35 ± 0.04	
		33.2 ± 0.4		1.2 ± 0.2	0.21 ± 0.04	
		37.2 ± 0.5		23 ± 5	3.8 ± 0.8	
		43.6 ± 0.6		4.3 ± 0.5	0.65 ± 0.08	

* The resonances have not been identified with a particular isotope, and as a result the neutron widths listed are actually $a\Gamma_n$ and $a\Gamma_n^0$ where a is the abundance of the isotope.

Table 1. (Continued.)

Isotope	I	E_0 ev	Γ_γ mv	Γ_n mv	Γ_n^0 mv	References
		45.7 ± 0.6		4.6 ± 0.6	0.68 ± 0.10	
$_{72}$Hf^{177} (cont.)	$^1/_2$					
	$^3/_2$	46.8 ± 0.6		4.9 ± 0.8	0.72 ± 0.12	
		49.4 ± 0.7		55 ± 5	7.8 ± 0.6	
		55.6 ± 0.7		18 ± 4	2.4 ± 0.5	
		57.2 ± 0.8		13 ± 3	1.7 ± 0.3	
		60.3 ± 1.0		2.8 ± 0.8	0.36 ± 0.11	
		64.4 ± 1.0		66 ± 8	8.2 ± 0.9	
		67.7 ± 1.1		36 ± 4	4.4 ± 0.5	
		72.3 ± 1.2		14 ± 4	1.6 ± 0.4	
		77.2 ± 1.3		16 ± 3	1.8 ± 0.3	
		84.3 ± 1.5		3.4 ± 0.9	0.37 ± 1.10 ?	
		86.2 ± 1.6		22 ± 4	2.4 ± 0.5	
		93.6 ± 1.8		6.4 ± 1.3	0.66 ± 0.13	
		98.5 ± 1.9		13 ± 3	1.3 ± 0.3	
		103 ± 2		20 ± 7	2.0 ± 0.7	
		105 ± 2		32 ± 10	3.1 ± 1.0	
$_{72}$Hf^{178}	0		60 ± 20			*BNL—fc 2*
		7.80 ± 0.10		49 ± 3	17.5 ± 1.0	
$_{72}$Hf^{179}	$^1/_2$		60 ± 20			*BNL—fc 2*
	$^3/_2$	5.69 ± 0.05		4.2 ± 1.3	1.8 ± 0.6	
		17.8 ± 0.2		2.0 ± 0.2	0.47 ± 0.05	
		24.0 ± 0.2		5.3 ± 1.4	1.1 ± 0.3	
		27.0 ± 0.3		1.25 ± 0.19	0.24 ± 0.04	
		31.5 ± 0.4		6.0 ± 1.2	1.1 ± 0.2	
		36.8 ± 0.5		17 ± 5	2.8 ± 0.8	
		40.6 ± 0.5		20 ± 4	3.1 ± 0.6	
		42.8 ± 0.6		11.9 ± 1.7	1.8 ± 0.2	
		44.7 ± 0.6		0.41 ± 0.12	0.06 ± 0.02	
		48.1 ± 0.7		0.8 ± 0.2	0.12 ± 0.03	
		51.1 ± 0.7		0.9 ± 0.2	0.13 ± 0.03	
		51.7 ± 0.7		0.51 ± 0.10	0.07 ± 0.02	
		52.4 ± 0.7		0.53 ± 0.10	0.07 ± 0.02	
		53.5 ± 0.7		0.56 ± 0.12	0.08 ± 0.02	
		55.4 ± 0.8		3.7 ± 0.5	0.50 ± 0.07	
		61.2 ± 1.0		0.63 ± 0.16	0.08 ± 0.02	
		63.0 ± 1.0		0.73 ± 0.15	0.09 ± 0.02	
		70.1 ± 1.2		8.1 ± 1.5	1.0 ± 0.2	
		78.2 ± 1.3		3.1 ± 0.5	0.35 ± 0.06	
		80.7 ± 1.4		1.7 ± 0.4	0.19 ± 0.04	
		84.3 ± 1.5		5.7 ± 1.2	0.62 ± 0.13	
		86.7 ± 1.6		3.9 ± 0.7	0.12 ± 0.08	
		93.6 ± 1.8		32 ± 8	3.3 ± 0.8	
		103 ± 2		99 ± 15	9.8 ± 1.4	
		106 ± 2		3.8 ± 1.1	0.37 ± 0.11	
		110 ± 2		12 ± 3	1.1 ± 0.3	
$_{72}$Hf^{180}	0		60 ± 20			*BNL—fc 2*
		73.9 ± 1.2		50 ± 6	5.8 ± 0.7	
$_{73}$Ta^{180}						*MTR—cs 1*
		0.433 ± 0.004	30 ± 5	0.12 ± 0.02	0.18 ± 0.03	
$_{73}$Ta^{181}	$^7/_2$					*BNL—cs 16; BNL—fc 2; MTR—fc 1*
		4.28 ± 0.02	49 ± 6	4.3 ± 0.7	2.1 ± 0.3	
		10.38 ± 0.10	49 ± 11	4.5 ± 0.5	1.40 ± 0.17	
		13.95 ± 0.12	50 ± 10	1.10 ± 0.05	0.29 ± 0.02	
		20.5 ± 0.2	51 ± 10	1.10 ± 0.06	0.24 ± 0.02	
		22.8 ± 0.3		0.25 ± 0.02	0.052 ± 0.008	
		24.1 ± 0.3	50 ± 15	7.0 ± 0.6	1.43 ± 0.13	
		29.8 ± 0.3		0.26 ± 0.03	0.048 ± 0.005	
		35.4 ± 0.4		15 ± 2	2.5 ± 0.3	

Table 1. (Continued.)

Isotope	I	E_0 ev	Γ_γ mv	Γ_n mv	Γ_n^0 mv	References
$_{73}Ta^{181}$ (cont.)	$^7/_2$	36.1 ± 0.4		17 ± 2	2.8 ± 0.3	
		39.3 ± 0.5	40 ± 15	51 ± 5	8.1 ± 0.8	
		49.4 ± 0.6		1.1 ± 0.3	0.16 ± 0.04	
		57.5 ± 0.7		0.5 ± 0.2	0.07 ± 0.03	
		63.2		8.0	1.0	*Col 13*
		76.7		14	1.6	
		77.5		6.5	0.74	
		78.6		3.6	0.41	
		82.7		16.3	1.8	
		84.9		6.2	0.67	
		89.4		4.5	0.48	
		91.1		3.6	0.38	
		96.7		4.2	0.43	
		99.0		120	12	
		103.2		9	0.09	
		105.4		31	3.0	
		114.8		44	4.1	
		118.0		2.4	0.22	
		126.0		48.4	4.3	
		135.9		23.3	2.0	
		138.0		18.8	1.6	
		143.6		2.4	0.20	
		148.0		4.8	0.39	
		149.2		8.1	0.66	
		165.5		10.5	0.82	
		175		185	14	
$_{74}W^{182}$	0	4.14 ± 0.03	46 ± 2	1.43 ± 0.03	0.70 ± 0.02	*ANL—fc 1; BNL—cs 11; BNL—fc 3*
		21.2		37.2	8.1	
		117		293	27	
		253		244	15.3	
$_{74}W^{183}$	$^1/_2$	7.62	52	1.64	0.59	*BNL 3; ANL—fc 2*
		27.1	57	30	5.8	
		40.8	(55)	0.5	0.078	
		46.6		127	18.6	
		48.1		13	1.9	
		65.7		1.06	0.13	
		102		85	8.3	
		136		2.8	0.24	
		146		48	4.0	
		157		215	17	
		177		70	5.3	
$_{74}W^{184}$	0	187		1042	76.5	*BNL—fc 3*
$_{74}W^{186}$	0	19.2 ± 0.3	(83)	140	32	*BNL—fc 3*
		175		23	1.7	
		221		485	33	
$_{75}Re^{185}$	$^5/_2$	2.156 ± 0.004	55.7 ± 0.6	6.6 ± 0.6	4.6 ± 0.5	*Col 12; BNL—cs 10*
$_{75}Re^{187}$	$^5/_2$	4.416 ± 0.008	45 ± 1	0.64 ± 0.1	0.31 ± 0.05	*Col 12*
$_{75}Re^{185,7}$	$^5/_2$	5.92 ± 0.08	90 ± 30	0.25 ± 0.15	0.10 ± 0.06	*Col 12; BNL—cs 10*
		7.18 ± 0.15		1.1 ± 0.5	0.4 ± 0.2	

Table 1. (Continued.)

Isotope	I	E_0 ev	Γ_γ mv	Γ_n mv	Γ_n^0 mv	References
$_{75}Re^{185,7}$ (cont.)	$^5/_2$					
		11.3 ± 0.2		4 ± 2	1.3 ± 0.6	
		13.1 ± 0.3		4 ± 2	1.0 ± 0.6	
		17.7 ± 0.4		12 ± 7	2.7 ± 1.5	
		21.1 ± 0.5		6 ± 4	1.3 ± 0.8	
$_{77}Ir^{191}$	$^1/_2$, $^3/_2$					*BNL—cs 16*
		0.654 ± 0.006	74 ± 3			
		5.36 ± 0.04	67 ± 5	7.2	3.1	
$_{77}Ir^{193}$	$^3/_2$					*BNL—cs 16*
		1.303 ± 0.010	87 ± 3			
$_{79}Au^{197}$	$^3/_2$		125 ± 30			*ANL fc 6; BNL—cs 8*
		4.906 ± 0.010	124 ± 3	15.6 ± 0.4 ($J = 2$)	7.1 ± 0.2 ($J = 2$)	
		58.1 ± 0.7		< 10		
		61.5 ± 1.0	170 ± 80	110 ± 20	14 ± 3	
		80.2 ± 1.5		15 ± 5	1.7 ± 0.5	
		110 ± 3		9 ± 4	0.9 ± 0.3	
		153 ± 4		50 ± 30	4 ± 2	
		168 ± 5		100 ± 50	8 ± 4	
		194 ± 6		50 ± 30	4 ± 2	
$_{80}Hg^{198}$	0					*BNL—fc 3*
		23.3 ± 0.2	145 ± 20	5.8 ± 0.5	1.20 ± 0.10	
$_{80}Hg^{199}$	$^1/_2$					*BNL—fc 3*
		34.0 ± 0.5	250 ± 70	78 ± 6	13.4 ± 1.0	
$_{82}Pb^{208}$	0					
		$(352 + 3) \times 10^3$		$(8 \pm 2) \times 10^6$ ($J = {}^1/_2$, $l = 1$)		*Wis 10*
		$(524 \pm 5) \times 10^3$		$(6 \pm 2) \times 10^6$ ($J = {}^3/_2$, $l = 1$)		
		$(718 \pm 7) \times 10^3$		$(6 \pm 2) \times 10^6$ ($J = {}^3/_2$, $l = 1$)		
$_{83}Bi^{209}$	$^9/_2$					*ANL—fc 4*
		810 ± 30		5300 ± 1000	190 ± 40	
		2370 ± 100		19000 ± 4000	390 ± 80	
$_{90}Th^{232}$	0		30 ± 10			*BNL—fc 3*
		22.0 ± 0.2	30 ± 10	2.0 ± 0.4	0.43 ± 0.08	
		23.7 ± 0.3	30 ± 10	3.7 ± 0.6	0.76 ± 0.13	
		59.6 ± 0.6		4.5 ± 0.9	0.59 ± 0.12	
		70.1 ± 0.8		40 ± 8	4.7 ± 0.9	
		114 ± 2		10 ± 4	0.9 ± 0.3	
		123 ± 2		27 ± 6	2.4 ± 0.6	
		131 ± 3		11 ± 5	0.9 ± 0.4	
		152 ± 3		15 ± 6	1.2 ± 0.5	
		174 ± 3		70 ± 13	5.3 ± 1.0	
		195 ± 5		30 ± 14	2.1 ± 1.0	
		202 ± 5		19 ± 10	1.3 ± 0.7	
		215 ± 5		2.3 ± 1.2	0.16 ± 0.08	
		226 ± 6		41 ± 16	2.7 ± 1.1	
		235 ± 6		1.3 ± 0.6	0.08 ± 0.04	
		256 ± 7		51 ± 19	3.2 ± 1.2	
		269 ± 7		22 ± 12	1.4 ± 0.7	
		290 ± 8		57 ± 18	3.3 ± 1.0	
		310 ± 9		105 ± 30	6.0 ± 1.6	

Table 1. (Continued.)

Isotope	I	E_0 ev	Γ_γ mv	Γ_F mv	Γ_n mv	Γ_n^0 mv	References
$_{92}U^{233}$							*ANL—fc 6; CR 3*
		neg. energy res.					
		1.76 ± 0.03	70 ± 3	380 ± 50	0.49 ± 0.03	0.37 ± 0.02	
$_{92}U^{234}$	0						*BNL—fc 3*
		5.10 ± 0.05	19 ± 9		4.5 ± 0.8	2.0 ± 0.4	
$_{92}U^{235}$	7/2		30 ± 6				*BNL—fc 3; Col 13*
		neg. energy res.					
		0.29 ± 0.01	31 ± 10	110 ± 30	0.0040 ± 0.0006	0.007 ± 0.001	
		1.12 ± 0.02	15 ± 15	130 ± 20	0.016 ± 0.003	0.015 ± 0.003	
		2.04 ± 0.01	27 ± 10	20 ± 12	0.008 ± 0.002	0.005 ± 0.001	
		2.86 ± 0.07		60 ± 40	0.004 ± 0.001	0.002 ± 0.001	
		3.17 ± 0.02	30 ± 15	130 ± 40	0.024 ± 0.003	0.014 ± 0.002	
		3.60 ± 0.02	40 ± 20	110 ± 20	0.050 ± 0.005	0.027 ± 0.003	
		4.85 ± 0.05		20 ± 10	0.052 ± 0.004	0.024 ± 0.002	
		5.4 ± 0.2 5.9 ± 0.2			0.04 ± 0.02	0.017 ± 0.009	
		6.2 ± 0.2			0.04 ± 0.02	0.016 ± 0.008	
		6.4 ± 0.1	30 ± 20	16 ± 6	0.34 ± 0.03	0.13 ± 0.01	
		7.1 ± 0.1	30 ± 20	30 ± 20	0.13 ± 0.01	0.49 ± 0.005	
		8.8 ± 0.1	34 ± 14	100 ± 30	1.08 ± 0.09	0.36 ± 0.03	
		9.3 ± 0.2			0.14 ± 0.03	0.046 ± 0.009	
		9.5 ± 0.2 9.8 ± 0.2			0.06 ± 0.03	0.02 ± 0.01	
		10.2 ± 0.1			0.09 ± 0.02	0.027 ± 0.005	
		11.7 ± 0.1		50 ± 20	0.63 ± 0.06	0.19 ± 0.02	
		12.5 ± 0.1		40 ± 20	1.54 ± 0.12	0.44 ± 0.03	
		13.3 ± 0.3			0.05 ± 0.03	0.014 ± 0.009	
		14.1 ± 0.2			0.51 ± 0.13	0.14 ± 0.04	
		14.6 ± 0.3			0.19 ± 0.06	0.05 ± 0.02	
		15.4 ± 0.2			0.25 ± 0.05	0.06 ± 0.02	
		16.2 ± 0.2			0.34 ± 0.08	0.08 ± 0.02	
		16.8 ± 0.2			0.25 ± 0.08	0.06 ± 0.02	
		18.2 ± 0.2			0.43 ± 0.11	0.10 ± 0.03	
		19.3 ± 0.2		45 ± 20	4.9 ± 0.5	1.11 ± 0.10	
		21.2 ± 0.2		90 ± 30	3.4 ± 0.3	0.74 ± 0.07	
		23.0 ± 0.4			0.3 ± 0.2	0.06 ± 0.04	
		23.7 ± 0.3		140 ± 70	2.0 ± 0.4	0.40 ± 0.08	
		24.6 ± 0.4			0.5 ± 0.1	0.11 ± 0.03	
		25.7 ± 0.3			0.8 ± 0.2	0.15 ± 0.04	
		26.8 ± 0.4			0.64 ± 0.13	0.12 ± 0.02	
		28.0 ± 0.4			0.9 ± 0.02	0.17 ± 0.04	
		31.2 ± 0.4			1.1 ± 0.4	0.20 ± 0.07	
		32.4 ± 0.5			1.9 ± 0.3	0.33 ± 0.05	
		33.9 ± 0.7			5 ± 2	0.8 ± 0.3	
		35.3 ± 0.6		110 ± 50	10 ± 2	1.6 ± 0.3	
$_{92}U^{238}$	0		25 ± 5				*BNL—fc 1; Har 4; MTR—fc 1*
		6.70 ± 0.06	24 ± 2		1.52 ± 0.07	0.59 ± 0.03	
		21.0 ± 0.2	25 ± 5		8.9 ± 0.4	1.94 ± 0.11	
		36.9 ± 0.2	29 ± 9		32.5 ± 1.9	5.3 ± 0.3	
		66.3 ± 0.5	27 ± 10		25 ± 2	3.1 ± 0.3	
		81.3 ± 0.5			2.1 ± 0.4	0.23 ± 0.04	
		90 ± 2			0.09 ± 0.03	0.009 ± 0.003	
		103.5 ± 0.7			67 ± 9	6.6 ± 0.9	
		117.5 ± 0.8			15 ± 2	1.4 ± 0.2	
		146 ± 3			0.9 ± 0.4	0.07 ± 0.03	
		166 ± 2			4 ± 2	0.31 ± 0.16	
		192 ± 2			140 ± 20	10.1 ± 1.5	

Table 1. (Continued.)

Isotope	I	E_0 ev	Γ_γ mv	Γ_F mv	Γ_n mv	Γ_n^0 mv	References
$_{92}U^{238}$ (cont.)	0	212 ± 5			60 ± 30	4 ± 2	
		242 ± 6			60 ± 30	4 ± 2	
		258 ± 6			1.3 ± 0.6	0.08 ± 0.04	
		278 ± 7			40 ± 20	2.4 ± 1.3	
		297 ± 8			40 ± 20	2.3 ± 1.3	
		368 ± 9			130 ± 50	7 ± 3	
		418 ± 10			80 ± 50	4 ± 2	
$_{93}Np^{237}$	5/2		32 ± 6				
		0.489 ± 0.005	32 ± 3		0.032 ± 0.002	0.046 ± 0.003	*MTR—cs 2*
		1.34 ± 0.02			0.030 ± 0.006	0.026 ± 0.005	
		1.49 ± 0.02			0.16 ± 0.04	0.13 ± 0.03	
$_{94}Pu_{239}$							*BNL—fc 3; Col 13; CR 3; LA 5*
		neg. energy res.					
		0.296 ± 0.004	41 ± 7	47 ± 7	0.12 ± 00.1	0.21 ± 0.02	
$^{94}Pu_{240}$							*Col 13*
		1.06 ± 0.02	42 ± 15		3.2 ± 1.4	3.1 ± 1.4	

Table 2. *Additional comments on σ vs. E.*

$_1H^1$ The cross section of 20.36 ± 0.1 barns just above the thermal region is mainly due to the near resonance of the singlet spin state (virtual level). For $E < \sim 10$ Mev the singlet and triplet cross sections can be expressed in terms of the effective range theory using the zero energy scattering lengths (cf. SACHS [*18*]).

$_1H^2$ The cross section of 3.4 barns near zero energy decreases smoothly for E up to 100 Mev and higher. Very roughly, the drop off is approximately given by a multiplying factor

$$\frac{5\text{ Mev}}{E+5\text{ Mev}}\cdot\frac{200\text{ Mev}}{E+200\text{ Mev}}.$$

$_1H^3$ This has a maximum in σ near 3.7 Mev where $\sigma \sim 2.5$ b. while $E\sigma$ increases slowly to the highest measured value of $\sigma \sim 1.0$ at 20 Mev.

$_3Li^6$ This has an (n, α) reaction for slow neutrons, which gives the main capture term $\sigma_c = 140 E^{-\frac{1}{2}}$ b (ev)$^{-\frac{1}{2}}$ at low energies. The first resonance for this process is at 248 kev with $\sigma_t \sim 116$ and $\sigma(n, \alpha) \sim 3.5$ b.

$_3Li^7$ This has its first resonance at 258 kev.

$_4Be^9$ The behavior of σ in the region of the two lowest levels at 620 and 810 kev is shown in Fig. 48.

$_{13}Al$ This has levels at 5.6, 35, 90, 145, 157, 210, 290, 315, 370, (etc.), kev.

$_{14}Si$ This has levels at 68, 195, 570, 830, and 1000 kev (and higher).

$_{15}P$ This shows ~ 6 levels per 100 kev in the region 120 to 830 kev.

$_{17}Cl$ This shows >30 peaks between 140 and 750 kev with many peaks probably not resolved. Thus there are an estimated 6 to 10 levels per 100 kev.

$_{18}A$ This shows 4 to 6 levels per 100 kev, but many may be missed.

$_{20}Ca$ This shows ~ 4 levels per 100 kev.

$_{21}Sc$ This has levels at 3.4, 4.4, and 6.8 kev with more peaks at higher energy. This represents a sudden decrease in level spacing from the above nuclei.

$_{22}Ti$ This has levels 3.2, 4.0, 9.0, and group near 20 kev.

$_{23}V$ This has levels at 170 ev (weak), 1450 ev (weak), 4250 ev (strong), and 6200 ev (strong), plus structure at higher energy.

$_{24}$Cr This has ~3 levels in the 4 to 7 kev region. However, even nucleus Cr^{52} has levels at 58 and 103 kev.

$_{26}$Fe This has a level at 8.2 kev and structure for higher E.

$_{27}$Co In addition to the 132 ev level, others have been seen at 4.8 and 8.0 kev.

$_{28}$Cu Levels have been seen at 235 ev (Cu^{65}), 595 ev (Cu^{63}), 2100 ev (Cu^{63}), 2600 ev (Cu^{65}), plus structure at higher energies.

$_{31}$Ga Levels have been seen at 96, 112, 295, 390, and 710 ev with structure at higher energy.

$_{32}$Ge Levels have been seen at 103, 205, 230, 335, 375, 490, and 550 ev.

$_{33}$As Levels have been seen at 27, 210, 270, 380, and 630 ev.

$_{38}$Sr Levels have been seen at 36 and 600 ev with weaker levels possibly present between these.

$_{39}Y^{89}$ No levels were seen to 200 ev, and σ in the 10 to 75 kev region shows structure, suggesting a few kev average level spacing. This increase in level spacing is for 50 neutrons (closed shell).

$_{40}$Zr Weak levels have been seen at 295, 690, and 1550 ev with the possibility that weak levels were missed. Zr^{90} is 51.5% abundant and may be responsible for the structure seen in the 10 to 80 kev region.

$_{41}$Nb Levels have been seen at 36, 43, 94, 106, 120, 195, and 245 kev. This represents a much larger spacing than for $_{31}$Ga, $_{32}$Ge, $_{33}$As, etc.

$_{43}Te^{99}$ Levels have been seen at 2.15, 5.6, 20.5, 40, and 53.5 ev. This is a return to a smaller level spacing after $Z=39$, 40, and 41.

$_{46}$Pd Levels have been seen at 13.2, 25, and 33.5 ev.

$_{47}$Ag This has Ag^{107} and Ag^{109} each in roughly 50% abundance. Each can form compound states of $I+{}^1/_2$ or $I-{}^1/_2$ for 4 total spin-isotope compound systems. Levels are observed at 5.12, 16.3, 30.5, 40.5, 41.6, 45, 51.6, 55.9, 71.2, 88.0, 133.4, 139.0, 143.5, 172.4, 201.4, 208, 251, 259, 264, 272, 290, 300, 310, 316, 326, 360, 386, 396, 403, 427, 443, 469, 466, 486, 499, 513, 523, 554, 563, 575, and 586 ev.

$_{52}$Te In addition to the 2.3 ev level in Te^{123}, levels have been seen at 7.4 and 25 ev for Te^{125}.

$_{54}$Xe In addition to the famous level at 0.082 ev in Xe^{135} (which causes pile poisoning) levels have been seen at 9.1 and 14.0 ev in ordinary Xe.

$_{56}$Ba Levels have been seen at 24.5, 81, and 88 ev for Ba^{135}.

$_{57}$La A level has been seen at 72 ev.

$_{58}$Ce The Duke group have seen structure in the 6 to 100 kev region.

$_{59}$Pr Levels were seen at 85, 220, 238, 360, 530, and 640 ev.

$_{60}$Nd Levels have been seen at 4.4 (Nd^{145}), 43 ev (Nd^{145}), 56 ev (Nd^{143}), and at 103, 128, 137, 156, 180, 190 ev, etc.

$_{62}$Sm In addition to the level at 0.096 ev, others have been seen at 3.46 ev (Sm^{147}), 4.8 ev (Sm^{149}), 11 ev, 15.5 ev (Sm^{152}), and at 12, 15, 17.5, 18.5, 20.5 ev, etc.

$_{64}$Gd Besides the levels at 0.03 and 2.58 ev, other levels have been seen at 2.02, 6.3, 7.7, 11.8, 14.5, 17 ev, etc.

$_{66}$Dy Levels have been seen at 1.7, 2.7, 3.7, 4.3, 5.4, 7.9, 10.5, 12.6, 14.2, 16.3, 18.6, 20.5 ev, etc.

$_{68}$Er Levels have been seen at 0.46, 0.58, 6.1, 8.0, 9.5 ev, etc.

$_{70}$Yb Besides the level at 0.597 ev (Yb^{168}) others have been seen at 4.6, 8.1, 13.5, and 18 ev.

$_{73}Ta^{181}$ Levels have been observed at 4.28, 10.38, 13.95, 20.5, 22.8, 24.1, 29.8, 35.4, 36.1, 39.3, 49.4, 57.5, 63.2, 76.7, 77.5, 78.6, 82.7, 84.9, 89.4, 91.1, 96.7, 99.0, 103.2, 105.4, 114.8, 118.0, 126.0, 135.9, 138.0, 143.6, 148.0, 149.2, 165.5, 175, 178, 181.5, 185, 188, 193.7, 199.0, 203.5, 207.3, 213.9, 215.3, 218.7, 221.1, 224.1, 229.5, 230.8, 236.4, 241.5, and 246.0 ev.

$_{76}$As Levels have been seen at 6.3, 8.9, 20 ev and higher.

$_{78}$Pt Levels have been seen at 11.8 ev (Pt^{195}), 19.5 ev (Pt^{195}). 66 ev (Pt^{195}), 96 ev (Pt^{198}), and at 110, 120, 140, and 152 ev.

$_{80}$Hg Levels have been seen at 23 ev (Hg^{198}), 34 ev (Hg^{199}), and at 43, 72, and 90 ev.

$_{81}$Tl A level has been seen at 235 ev (Tl^{203}) plus structure at higher energy. However, the Duke group has obtained complicated structure in the 10 to 80 kev region. This great increase in level spacing also holds for Pb and Bi and is associated with the double closed shell $Z=82$, $N=126$ being near.

for the levels. For contrast, Figs. 14 to 17 give some of the experimental curves for silver using the higher resolution Nevis synchrocyclotron velocity spectrometer and a self detection method as discussed below. Analyzed values of $\sigma_0 = 2600000\, g\Gamma_n/\Gamma E_0$ barns may be calculated from the values given in Table 1 (E_0 in ev).

Nevis synchrocyclotron velocity spectrometer. The highest resolution slow neutron measurements to date have been made using the 385 Mev Nevis synchrocyclotron of Columbia University as a source. The time average proton current of ~0.1 to 1.0 microamperes is concentrated into ~60 bursts/sec., each of <0.1 microseconds duration, by using a pulsed electrostatic deflector to deflect the proton "bunch" when they reach ~69 inch radius in their acceleration cycle. The protons strike a tungsten slab directly below a box containing nylon slabs for moderation. Since the range of protons of this energy is greater than the mean free path for nuclear encounters, most of the protons strike W nuclei to give several evaporation neutrons on the average. Thus the instantaneous production of evaporation neutrons during a burst is ~10^{17} to 10^{18}/sec. A flight path of ~35 meters has been used with a foil of the element being studied suspended perpendicular to the beam at the detector position (see Fig. 13). A bank of scintillation counters above and below the foil detect (n, γ) capture gamma rays from the foil. A similar foil can be alternated in and out of the beam at a transmission position to obtain self indication transmission measurements. This method has only been described briefly at meetings of the American Physical Society, and involves a somewhat different interpretation than ordinary transmission measurements. A discussion of the methods of analysis for such measurements is given below.

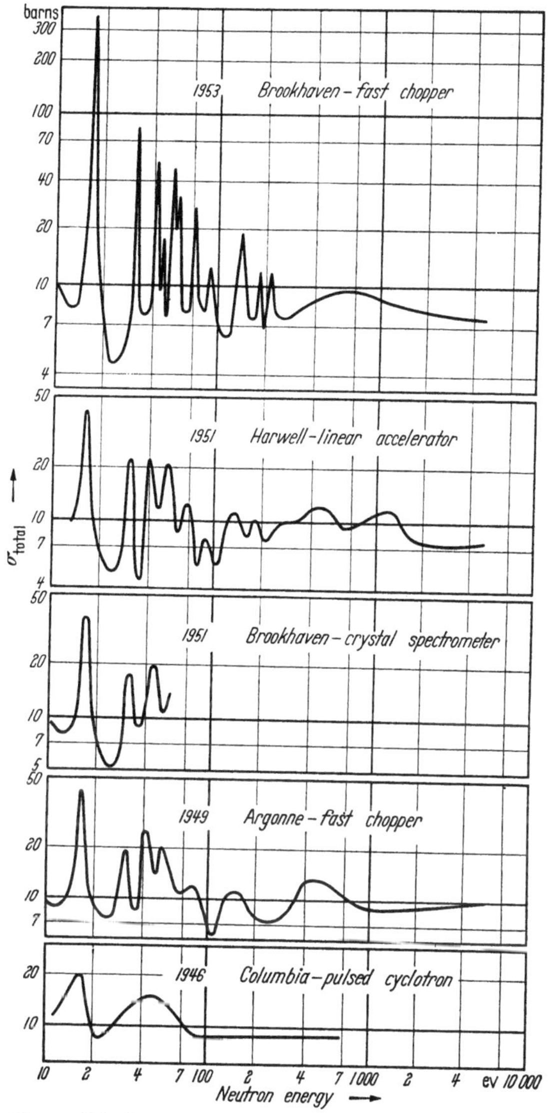

Fig. 12. This figure, given in [*BNL—fc 1*], shows dramatically how a plot of measured σ vs. E over a region containing resonances can be very misleading. The true σ_0 values are >1000 barns for most of the levels. Figs. 14 to 16 show the much greater resolution achieved with the new Columbia Nevis velocity spectrometer system. It is urged that Table 1 be used, rather than such curves as these, to determine the true behavior of σ vs. E near resonances.

The resolution of the instrument for measurements of levels in Ag and Ta is ~0.01 μsec/m. This is made up of a $\Delta t < 0.1$ μsec from the source "on time";

a $\Delta t = 0.25$ μsec from the detection interval gate width. There is an effective $\Delta l \sim 3$ cm equivalent due to the r.m.s. spread in the time delay of neutrons of

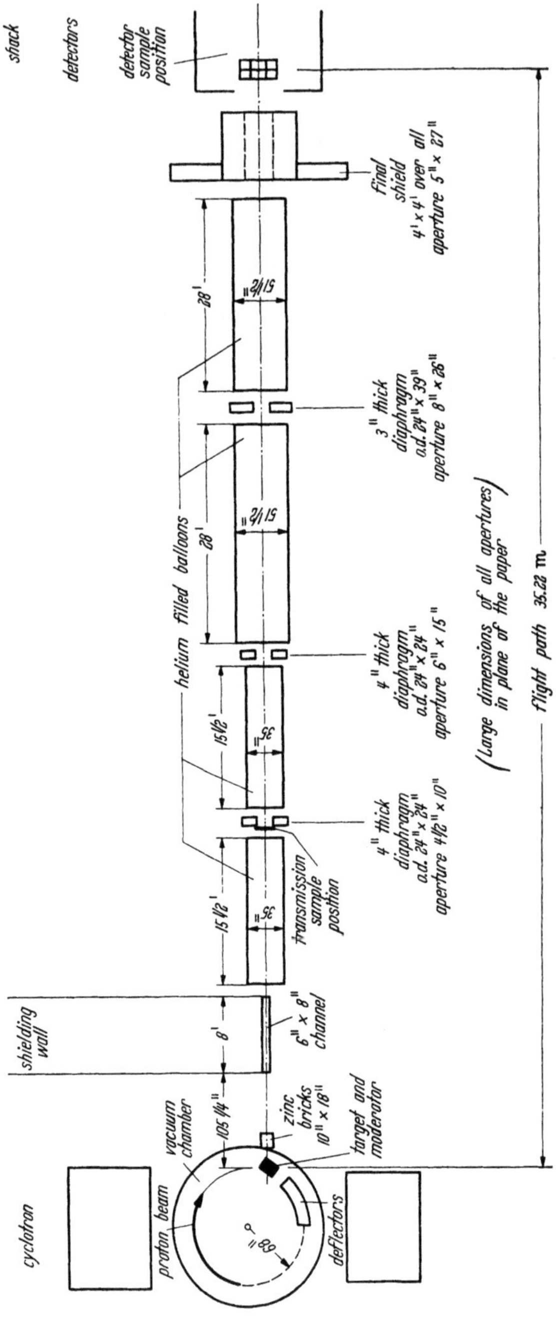

Fig. 13. Basic floor geometry for the new Columbia Nevis velocity spectrometer system. See the text of this article for a brief description.

energy E being emitted from the moderator slab (i.e. the time is inversely as the final neutron velocity and thus acts like a distance effect). The effective thickness of the detector foil is usually ~ 1 cm including the accuracy of positioning of

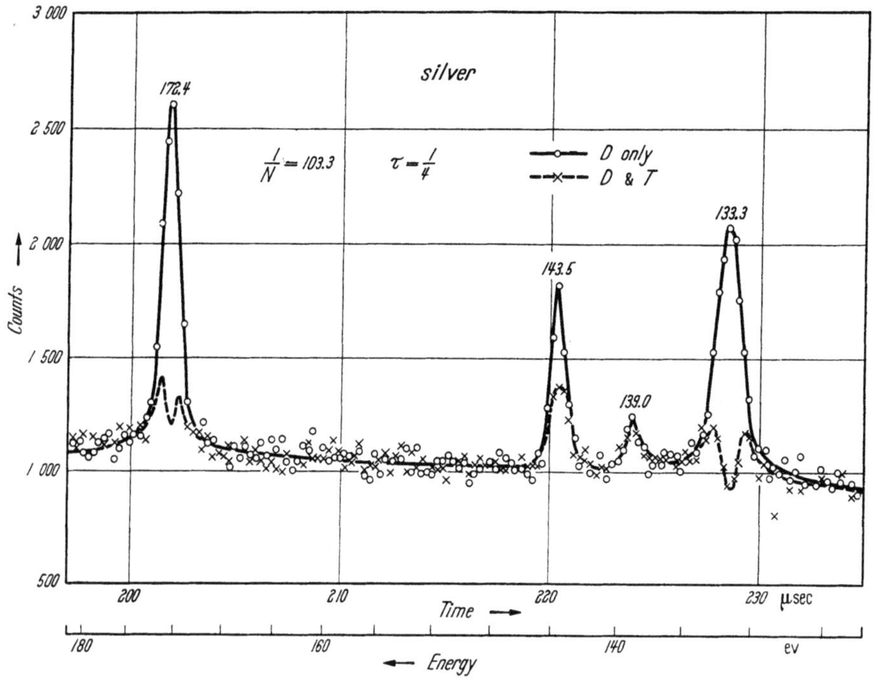

Fig. 14. Curves for detector gamma ray counting rate vs. neutron time of flight for 35.22 meters for a $1/n = 103.3$ barns/atom thickness sample of normal Ag at the detector position (*D* only) and for foils having $1/n = 103.3$ barns/atom thickness at both detector and transmission positions. The measurements were made using the Columbia Nevis velocity spectrometer system with 0,25 μsec detection interval gates for ~0.01 μsec/m resolution.

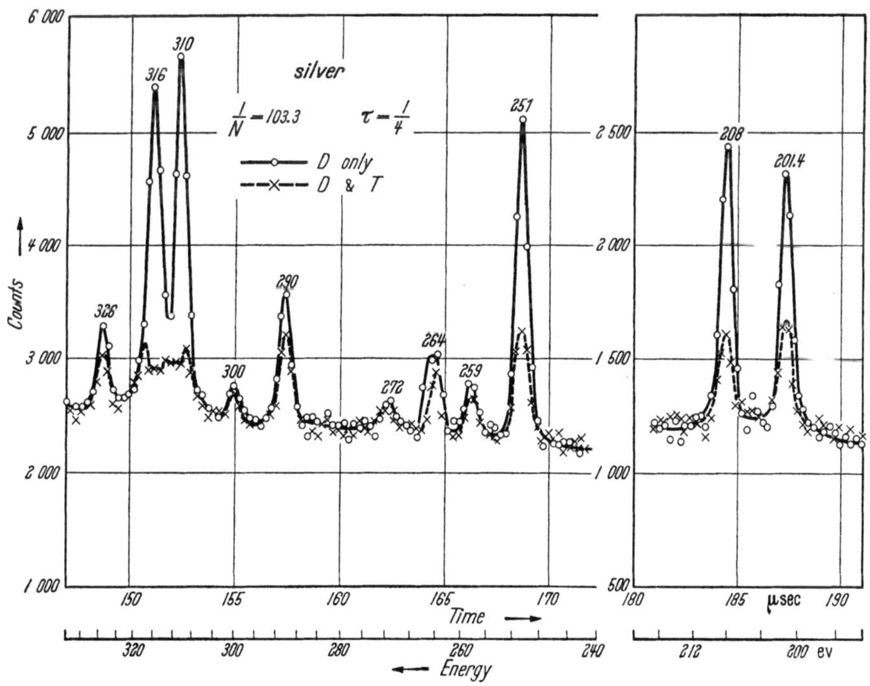

Fig. 15. Curves similar to those of Fig. 14, but for a higher energy region.

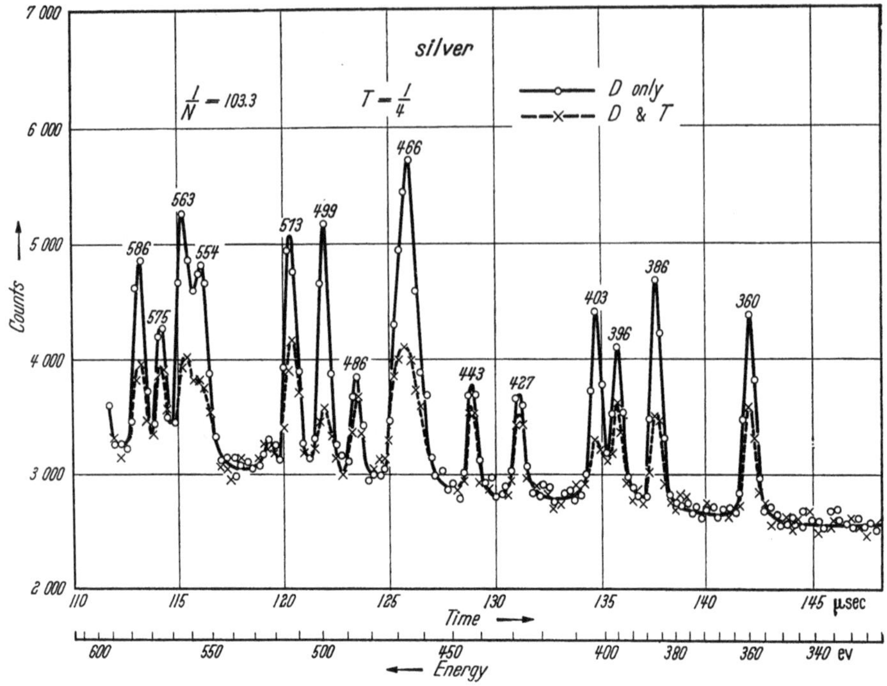

Fig. 16. Curves similar to those of Figs. 14 and 15, but for a higher energy region.

the 6 inch by 30 inch foil perpendicular to the source-detector path. The effect of $\Delta l/l \sim 0.001$ becomes equal to the timing Δt at $E \sim 50$ ev.

Figs. 14 to 21 show typical curves for silver and tantulum with and without the foil at the transmission position (D and $D+T$). For a strong level the $D+T$

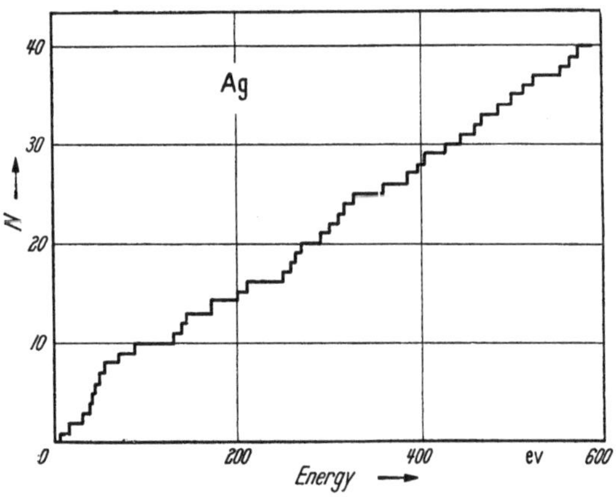

Fig. 17. The observed number of levels in silver below energy E. See also Table 2 for a listing of energy values.

curve shows a reversal dip to zero effect at the center of the resonance curve while the "D only" curve gives a "saturation effect", defined as the effect when all of the neutrons interact (capture + scattering) in the D sample. If only capture were present, the "D only" curve could be regarded as an inverted transmission curve and an area analysis used. For levels at ~ 100 ev or higher this analysis is also applicable if the scattering is present, but not $\gg$ capture. The energy loss

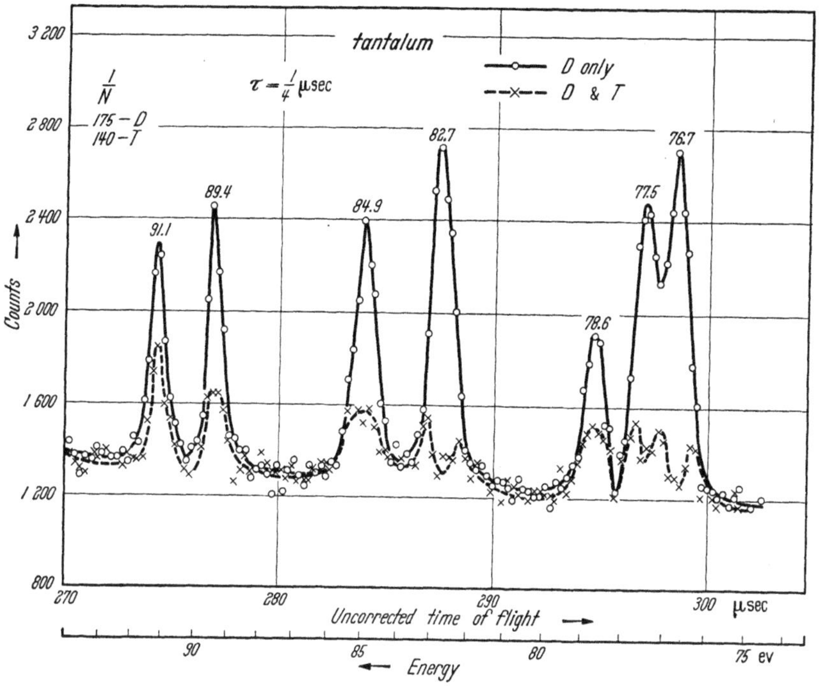

Fig. 18. Curves similar to those of Figs. 14, 15, and 16, but using Ta rather than Ag.

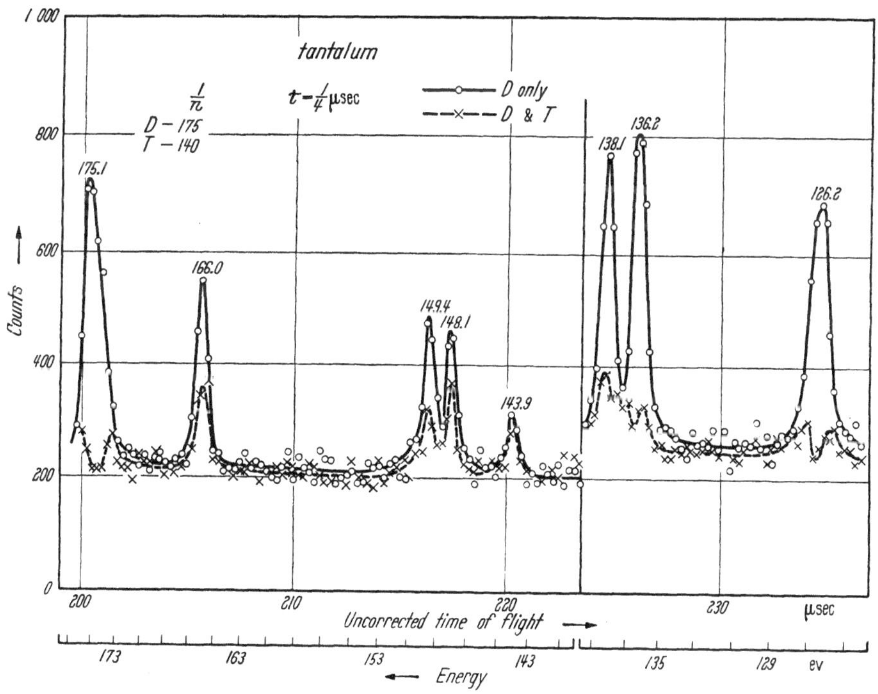

Fig. 19. A continuation of the curve of Fig. 18.

on scattering has equal probability per unit energy for zero to $\sim 4Em/M$ which is large compared to Γ. Thus several alternate analysis techniques have been

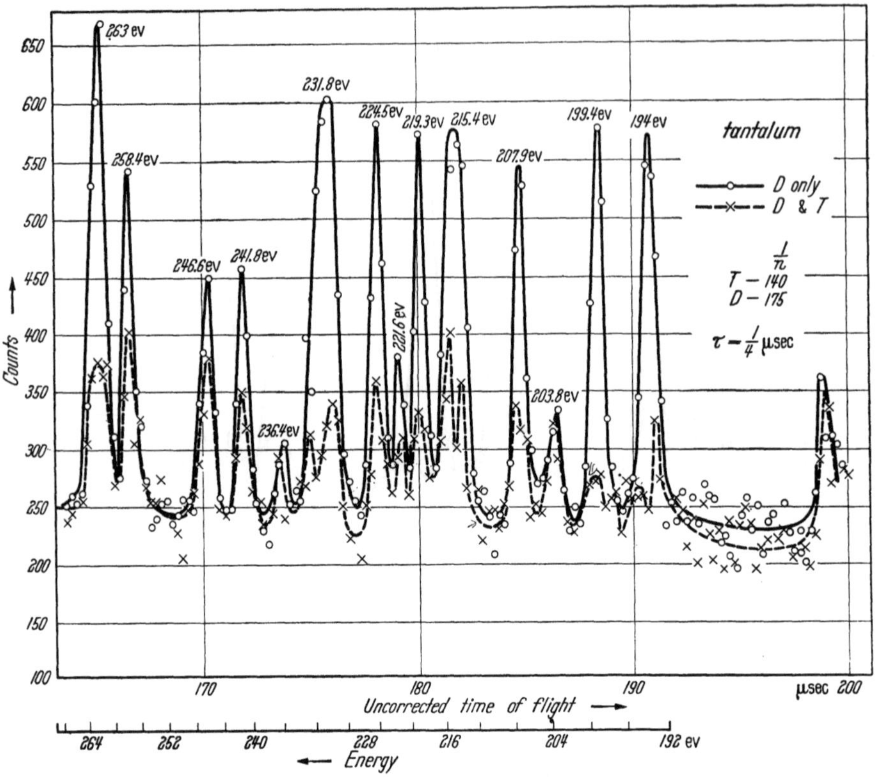

Fig. 20. A continuation of the curve of Fig. 19.

developed for these measurements including self indication transmission measurements. The net effect is that σ_Δ can often be obtained relatively independent

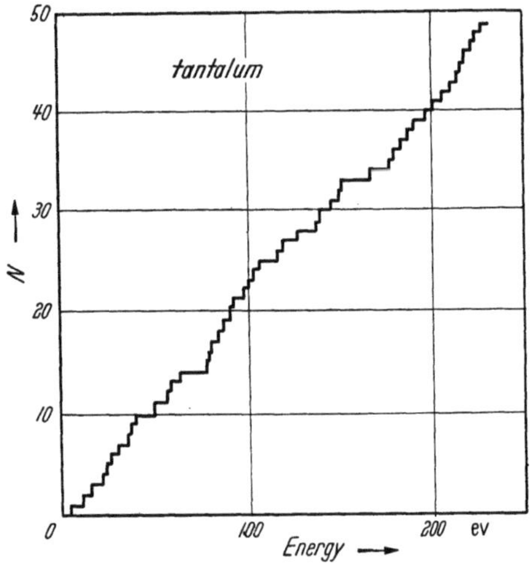

Fig. 21. The observed number of levels in Ta below energy E (see also Table 2).

of the assumed Γ, while $\sigma_0 \Gamma^2$ can often be obtained from D only area measurements Most of the results for Ag and Ta in Table 1 for the upper energy region were obtained in this manner.

II. Theory and experimental results.

3. Elementary derivation of the BREIT-WIGNER one level formula. In the following discussions we shall consider a few elementary derivations of the BREIT-WIGNER one level formula which use different types of approximations and assumptions, and thus, concern themselves differently in approach and emphasis with different phases of the process.

This derivation, which ignores all coherent interference between "potential" and resonance scattering, is the simplest for keeping spin factors in order. We consider a reaction where an incident particle a reacts with an initial nucleus A to form a compound nucleus C, which then decays by one of a number of possible decay modes symbolized by $b+B$ into an outgoing nuclear subunit, or photon, b, plus a final nucleus B. The final nucleus B and the fragment b may be in their ground states with respect to internal degrees of freedom, or in one of various possible excited states. The assumptions are:

1. The significant contribution to the build-up of the final state probability only occurs via the intermediate state C. Thus we omit the contribution from "potential" scattering.

2. The intermediate state C shall be a single quasistationary (long lived relative to characteristic nuclear transit times) state of the compound system. Thus, interference between levels is ignored.

3. The decay in probability of finding the excited state of the compound system is associated with an increase in probability for continuum final states in a manner independent of its mode of formation.

The rate balance formula is

$$\varrho_A R_{A\to C} = \varrho_A v_A \sigma_{A\to C} = \varrho_C R_{C\to A} \tag{3.1}$$

where v_A = velocity of relative motion of a and A and $\sigma_{A\to C}$ = total cross section for forming the compound system. ϱ_A is the density of relative motion states for the initial system times g_A, the spin statistical weight.

$$g_A = (2J_a + 1)(2J_A + 1) \tag{3.2}$$

where a and A have spin angular momenta J_a and J_A. It is convenient to use normalization per unit volume for continuum states in which case ϱ_A/g_A reflects only phase space contributions

$$\varrho_A = \frac{4\pi p_A^2 g_A}{(2\pi\hbar)^3 v_A} = \frac{g_A}{2\pi^2 \hbar \lambda\!\!\!^{-}{}_A^2 v_A}. \tag{3.3}$$

Here p_A = initial system relative motion momentum and $\lambda\!\!\!^{-}{}_A = \hbar/p_A$ is the wave length per radian.

$R_{C\to A}$ is the partial decay rate from the compound state to the initial channel A. If τ = mean life for decay, then

$$R_{C\to \text{all}} \equiv \frac{1}{\tau} = \sum_j \frac{1}{\tau_j} \equiv \sum_j \frac{\Gamma_j}{\hbar} = \frac{\Gamma}{\hbar}. \tag{3.4}$$

Here τ_j^{-1}, the partial decay rate by mode j, is set equal to $\Gamma_j/\hbar$ to define the partial width Γ_j for decay via channel j and Γ is the total level width

$$R_{C\to A} = R_{C\to \text{all}} \frac{\Gamma_A}{\Gamma} = \frac{\Gamma_A}{\hbar}. \tag{3.5}$$

The only remaining factor to consider is ϱ_C which contains the essential resonance part of the formula and requires the most clarification. If the quasi-stationary state C of the compound nucleus (C.N.) were a true eigenstate of the exact Hamiltonian H of the system, the state function (non relativistic) could be represented as the product of a portion ψ_0 referring to all internal degrees of freedom times $e^{-iE_0t/\hbar}$ which contains the total time dependence. In the actual situation where various decay channels are "open", ψ_0 can be regarded as the solution for a modified Hamiltonian H_0. Here H_0 is adjusted to give the proper energy levels by including the level shift contribution of all virtual processes not ending in a real decay, excluding only parts which give real decays. (Reflecting mirrors are assumed for particle channels.) If the compound state is established at $t=0$, one has a time dependence for the quasi-stationary state

$$\psi_1 = \psi_0\, e^{-iE_0t/\hbar}\, e^{-\Gamma t/2\hbar} \tag{3.6}$$

for the C.N. part of configuration space, with a corresponding build-up from zero for the amplitude in regions of configuration space corresponding to continuum final states. The time dependent state (3.6) may be regarded as a wave packet constructed of exact eigenstates $\psi(E)$ of the total Hamiltonian. These exact stationary state eigenfunctions can be extremely complicated and represent various possible "statistical equilibrium" situations for all possible reaction products. Those (continuum) normalized exact eigenstates of energy E ($E \approx E_0$ relative to the level spacing) which are to contribute appreciably to the expansion will have the form $A(E)\psi_0$ in the C.N. part of configuration space. Being continuum functions, they occupy, with significant amplitude, an infinite volume in total configuration space, so their normalization depends on their amplitude in regions of configuration space other than the small C.N. region. To construct a wave packet ψ_1 which at $t=0$ has the form $\psi_1=\psi_0$ in the C.N. region and zero elsewhere, the normalized continuum eigenstates will be multiplied by a function $B(E)=\int\psi(E)\psi_0$ which is proportional to $A(E)$. Thus[1] the contribution to the amplitude of ψ_1 in the C.N. portion of configuration space from $\psi(E)$, for E to $E+dE$, is proportional to $A^2(E)\,\psi_0$.

The packet ψ_1 thus has the following expansion for the C.N. region of configuration space:

$$\psi_1 = \psi_0\, e^{-i(E_0 - i\Gamma/2)t/\hbar} = C\psi_0 \int A^2(E)\, e^{-iEt/\hbar}\, dE. \tag{3.7}$$

So $A^2(E)$ is the Fourier transform of $e^{-i(E_0-i\Gamma/2)t/\hbar}$ for $t \geq 0$. We must specify a form for ψ_1 for $t<0$, however. The $\psi(E)$ all combine at $t=0$ to give zero amplitude through all of configuration space except for the C.N. region. For $t>0$ they do not cancel in outside regions, so ψ_1 builds up in the outside region but decreases in the C.N. region in such a way that $\int|\psi_1|^2$ over all configuration space remains unchanged. If t is replaced by $(-t)$ in (3.7) a similar effect would be

[1] This is easily demonstrated by the following simple example of the α decay calculation ($l=0$) where we can take $V=0$ for $0 \leq r \leq a$ and $b \leq r \leq R \to \infty$, while $V=V_0>E_0$ for $a<r<b$. The boundary conditions for the exact solution are $\varphi = r\psi = 0$ at $r=0$ and R, and energy E_0 is defined as the lowest for which φ has the form $M_0 \sin k_0 r$ for $r<a$, $N_0 e^{-\alpha_0(r-a)}$ for $a \leq r \leq b$, and $P_0 \sin(k_0 r + \beta_0)$ for $b<r<R$. For the case of $\alpha_0(b-a) \gg 1$, $(M_0/P_0) \sim e^{\alpha(b-a)}$. For $E=(E_0+X)$, the solutions have the form $M \sin kr$, $[Ne^{-\alpha(r-a)} + Qe^{\alpha(r-a)}]$, and $P_0 \sin(kr+\beta)$. As X changes from zero to some small value, (M/P_0) rapidly decreases to <1. Since $R\to\infty$, the net normalization is completely determined by the outside region and $P_0 \approx (4\pi)^{-\frac{1}{2}}(2/R)^{\frac{1}{2}}$ independent of E. Here the inside amplitude $A(E)$ is $M(E)$. The initial wave packet ψ_1 can be taken as proportional to $\psi(E_0)$ for $r \leq b$ and zero for $r>b$. The amplitude $B(E)$ given to each $\psi(E)$ to form ψ_1 is $\int\psi_0\psi$ which is proportional to $A(E)$ or $M(E)$. The amplitude contribution of $\psi(E)$ to ψ_1 in the important region $r<a$ is thus $B(E)A(E)$ which is proportional to $A^2(E)$, or $M^2(E)$. The decay in time as $e^{-\Gamma|t|/2\hbar}$ in the C. N. region is explicitly obtained in this case.

expected, so we choose $e^{-iE_0t/\hbar}\, e^{-\Gamma|t|/2\hbar}$ for all t. This gives

$$A^2(E) = |A(E)|^2 = \frac{C}{2\pi} \frac{\Gamma}{(E-E_0)^2 + (\Gamma/2)^2} = \varrho_C(E). \tag{3.8}$$

The total level weight is given by

$$\int |A(E)|^2 dE = g_C \tag{3.9}$$

which gives $C = g_C$. The total cross section is thus

$$\sigma_{A\to C} = \pi \lambda_A^2 \frac{g_C}{g_A} \frac{\Gamma_A \Gamma}{(E-E_0)^2 + (\Gamma/2)^2} \tag{3.10}$$

and a typical partial cross-section $A \to C \to B$ is

$$\sigma_{A\to C\to B} = \pi \lambda_A^2 \frac{g_C}{g_A} \frac{\Gamma_A \Gamma_B}{(E-E_0)^2 + (\Gamma/2)^2}. \tag{3.11}$$

The partial decay rates Γ_A and Γ_B may each be regarded as the product of a decay rate Γ_{A_j} or Γ_{B_j} to representative (continuum) eigenstates j of A or B times $\varrho_A(E)$ or $\varrho_B(E)$. It is plausible that the Γ_{A_j} or Γ_{B_j} should be roughly independent of $(E - E_0)$ for not too large $(E - E_0)$. (It would be independent of $(E - E_0)$ using perturbation theory.] Thus we take, (where ϱ_0 etc. refers to energy E_0)

$$\Gamma_A(E) = \Gamma_A(E_0) \frac{\varrho_A(E)}{\varrho_A(E_0)} = \Gamma_A(E_0) \frac{P^2 v_0}{P_0^2 v}. \tag{3.12}$$

For particle emission (non relativistic), this gives

$$\Gamma_A(E) = \Gamma_A(E_0) \frac{\lambda_0}{\lambda} = \Gamma_{A_0} \frac{\lambda_0}{\lambda}. \tag{3.13}$$

If channel B refers to photon emission (capture), then

$$\Gamma_B(E) \approx \Gamma_B(E_0) = \Gamma_{B_0} \tag{3.14}$$

since the photon energy is large compared to the range of $(E - E_0)$ to be considered. Thus, for slow neutron processes where only resonance scattering or capture are possible, $\Gamma_A \to \Gamma_N$, $\Gamma_N \to \Gamma_\gamma$, and

$$\sigma_{\text{scatt}} = \pi \lambda_0^2 \frac{g_C}{g_A} \frac{\Gamma_{N_0} \Gamma_{\gamma_0}}{(E-E_0)^2 + (\Gamma/2)^2}, \tag{3.15}$$

$$\sigma_{\text{capt}} = \pi \lambda_0 \lambda \frac{g_C}{g_A} \frac{\Gamma_{N_0}^2}{(E-E_0)^2 + (\Gamma/2)^2}, \tag{3.16}$$

If the variation of λ is not important for σ_{capt} then, for $|E - E_0| \ll E_0$,

$$\sigma_{\text{tot}} \approx \pi \lambda_0^2 \frac{g_C}{g_A} \frac{\Gamma_{N_0} \Gamma_0}{(E-E_0)^2 + (\Gamma_0/2)^2}. \tag{3.17}$$

Each process is proportional to the products of the Γ_j values for the initial and final channels. The factor λ in (3.16) is the famous $(1/v)$ factor in capture. The maximum peak cross section for elastic scattering is $4\pi\lambda_0^2(g_C/g_A)$, which occurs if $\Gamma_A = \Gamma_0$. The maximum total cross section for all incoherent (i.e. $B \neq A$) processes combined is obtained when $\Gamma_A = \Gamma - \Gamma_A = \Gamma/2$ to give $\pi\lambda_0^2(g_C/g_A)$. For slow neutrons, $l = 0$, $s = \frac{1}{2}$, so $J_C = J_A \pm \frac{1}{2}$ and

$$\frac{g_C}{g_A} = \frac{2J_A \pm 1 + 1}{4J_A + 2} \leq 1. \tag{3.18}$$

The equality sign in (3.18) applies for $J_A = 0$ when only one compound spin state, $J_C = \frac{1}{2}$, is allowed. For higher energy neutrons the compound state has a definite J_C and either the same or opposite parity from that of the initial nucleus. A change of parity implies that the scattering is for odd l, while no change in parity implies even l. For $l = 1$, $J_A = 0$, $J_C = \frac{3}{2}$ or $\frac{1}{2}$ so $(g_C/g_A) = 2$ or 1. For the situation $J_a = J_A = 0$, $g_A = 1$, then $J_C = l$ and (g_C/g_A) gives the full weighting $(2l+1)$. In all other cases the total weighting $(2l+1)$ is shared among several possible J_C states.

When interference between potential and resonance scattering is considered, resonance scattering of given l only interferes with the potential scattering of the same l in the total cross section [i.e. in $\int (d\sigma/d\Omega)$]. For low energy neutron resonances $(l=0)$ it may be shown that the interference is constructive for $E > E_0$ and destructive for $E < E_0$. In the region where $l = 1$ resonances first are evident, the $l = 1$ potential scattering is very small so $l = 0$ resonances show strong interference effects with potential scattering while $l = 1$ resonances do not. This offers one means of distinction. Other means include:

1. Measuring the peak cross section σ_0. At these higher energies $\Gamma_A \approx \Gamma$ usually so $\sigma_0 \approx 4\pi\lambdabar^2 (g_C/g_A)$. For neutrons

$$4\pi\lambdabar^2 = \frac{2\,600\,000 \text{ barns}}{E \text{ (in ev)}}. \tag{3.19}$$

Thus σ_0 gives (g_C/g_A) and g_C, and thus J_C, can be deduced.

2. Perhaps the most unambiguous method is the measurement of the angular distribution of the resonance elastic scattering. For $l = 0$ this is symmetric in the center of mass system. For $l = l_1 \neq 0$ but $J_a = J_A = 0$, the angular distribution is of the form

$$\frac{d\sigma}{d\Omega} = F(E)\,[P_{l_1}(\cos\vartheta)]^2. \tag{3.20}$$

For neutrons $J_a = \frac{1}{2}$. For $J_A = 0$ the resonance state has $J_C = l_1 \pm \frac{1}{2}$. For $j = l_1 + \frac{1}{2}$ the scattering amplitude factor is

$$f_+ \equiv \lambdabar\, e^{i\delta_+} \sin\delta_+ \tag{3.21}$$

(where δ_+ is the phase shift for $j = l_1 + \frac{1}{2}$). For $l_1 - \frac{1}{2}$ the corresponding value

$$f_- = \lambdabar\, e^{i\delta_-} \sin\delta_- \tag{3.22}$$

where f_+ and f_- can be quite different in magnitude.

There is then a net scattering amplitude $(l+1)\,f_+ + l f_-$ associated with $l = l_1$ scattering where the neutron spin is not changed, and an amplitude $(f_- - f_+)$ associated with spin flip. The non spin flip part has a factor $P_{l_1}(\cos\vartheta)$ (surface spherical harmonic for $l = l_1$, $m = 0$) while the spin flip amplitude is multiplied by the surface spherical harmonic for $l = l_1$, $m = \mp 1$ to give a factor $e^{\mp i\varphi} \sin\vartheta$ times a function of $\cos\vartheta$. For $J_A \neq 0$ it is necessary to state more fully how $\boldsymbol{J}_C$ is composed from $\boldsymbol{J}_a$, $\boldsymbol{J}_A$, and $\boldsymbol{l}$. The specification of the angular dependence of the scattering is much more complicated in this general case.

4. Second derivation of the Breit-Wigner one level formula. The preceding derivation was based on very general considerations and must be valid within the accuracy of the special assumptions. It suffers in that it does not predict level spacings, interference effects, or the values of the parameters such as Γ_N and Γ_γ which appear. It does, however, clearly give their physical significance. Two treatments will be briefly considered that consider the behavior of the logarithmic derivative of the radial part of the wave function near the nuclear surface since they offer some additional insight on the processes. The treatment of this section is a modification of that of FESHBACH, PEASLEE, and WEISSKOPF [11]. Only $l=0$ low energy neutron processes will be treated, and we neglect all angular momentum effects for simplicity. The $l=0$ part of $r\psi$ for the relative motion $r>R=$ nucleon radius (i.e. beyond the range of the interaction) has the form

$$\varphi = c\left[e^{-ikr} - \eta\, e^{ikr}\right] \tag{4.1}$$

where the amplitude η for the outgoing radial wave of momentum $\hbar k$ is unity for no interaction and $|\eta|\leq|$ is required for conservation of probability.

$$\sigma_{\text{scatt}} = \pi \lambda\!\!\!^{-2}\, |1-\eta|^2 = \pi \lambda\!\!\!^{-2}\left[1+\eta\bar{\eta} - (\eta+\bar{\eta})\right]. \tag{4.2a}$$

$$\sigma_{\text{abs}} = \pi \lambda\!\!\!^{-2}\left[1-|\eta|^2\right] = \pi \lambda\!\!\!^{-2}\left[1-\eta\bar{\eta}\right]. \tag{4.2b}$$

$$\sigma_{\text{tot}} = \pi \lambda\!\!\!^{-2}\left[2-(\eta+\bar{\eta})\right]. \tag{4.2c}$$

For some $r=a$ outside the nuclear surface let f be the complex logarithmic derivative of φ, and $x=ka$

$$f = f_0 - i f_i \equiv \left(\frac{r\,\varphi'}{\varphi}\right)_{r=a}. \tag{4.3}$$

Inserting Eq. (4.1) into Eq. (4.3) and solving for η gives:

$$\eta = e^{-2ix}\left[\frac{f_0 + i(x-f_i)}{f_0 - i(x+f_i)}\right]. \tag{4.4}$$

f_0 and f_i are real and f_i must be positive in order that

$$|\eta|^2 = \frac{f_0^2 + (x-f_i)^2}{f_0^2 + (x+f_i)^2} \leq 1.$$

If we group as "absorption" all processes not returning to the incident channel, then $|\eta|=1$ for no absorption and <1 when absorption occurs. Since f_i depends on absorption, we shall later associate it with Γ_B. Eq. (4.4) gives:

$$\sigma_{\text{abs}} = \pi \lambda\!\!\!^{-2}\left[\frac{4x f_i}{f_0^2 + (x+f_i)^2}\right], \tag{4.5a}$$

$$\sigma_{\text{scatt}} = \pi \lambda\!\!\!^{-2}\left|\sin x + \frac{x\,e^{-ix}}{i(x+f_i) - f_0}\right|^2. \tag{4.5b}$$

For $f_i=0$, a resonance is defined as the energy at which the phase shift δ is such that $|\sin\delta|=1$. For $ka=x\ll 1$, this is essentially equivalent to the condition $f_0=0$ and we shall, with these limitations in mind, so define resonance energy E_0. It will be shown later on general grounds (WIGNER-EISENBUD method) that:

$$f_0' \equiv \left(\frac{d f_0}{dE}\right)_{E=E_0} \tag{4.6}$$

is negative.

The usual one level formula is obtained by assuming that f_0 can be represented with sufficient accuracy by using only the linear term in a TAYLOR series expansion about $E=E_0$:

$$f_0 \to (E-E_0)\, f_0'. \tag{4.7}$$

In the case where absorption is not negligible, we note that the "resonance only" BREIT-WIGNER formulas for σ_{abs} and σ_{scatt} are obtained if the potential scattering factor $\sin x$ in Eq. (4.5b) is ignored, and $x f_0'^{-1}$ and $f_i f_0'^{-1}$ are identified as $\pm\Gamma_A/2$ and $\pm\Gamma_B/2$ respectively, with the same choice of sign in the two cases. The choice may be established using the following argument. Let a reflecting mirror be placed in the entrance channel at some suitable radius near $r=a$. The decay in time of the trapped packet is then similar to Eq. (3.6) except that the contribution of Γ_A to the exponential decay is removed, so we obtain the form

$$e^{-i(E_0-i\Gamma_B/2)t/\hbar} \equiv e^{-iE_R t/\hbar}. \tag{4.8}$$

This defines a complex "resonance energy"

$$E_R = E_0 - i\,\Gamma_B/2. \tag{4.9}$$

The assumption is now made that f can be treated as an analytic function of E in the complex domain such that $f_0=0$ at $E=E_0$, that $f=f_0-if_i=0$ at $E=E_R$, and that the complex f can be represented near resonance using only the linear term in a series expansion

$$f = f_0 - i f_i = (E-E_R)\,f_0' = (E-E_0)\,f_0' + \frac{i\,\Gamma_B f_0'}{2} \tag{4.10}$$

to give

$$f_0 = (E-E_0)\,f_0', \tag{4.7}$$

$$f_i = -f_0'\,\Gamma_B/2 \tag{4.11}$$

and

$$x = -f_0'\,\Gamma_A/2. \tag{4.12}$$

This shows that f_0' must be negative, since f_i, x, Γ_A, and Γ_B are positive. If we use $e^{-ix}\approx 1$ and $\sin x\approx x$ in Eq. (4.5b),

$$\sigma_{\text{abs}} = \pi\,\lambda\!\!\!^{-2}\left[\frac{\Gamma_A\,\Gamma_B}{(E-E_0)^2+(\Gamma/2)^2}\right] \tag{4.13a}$$

and

$$\sigma_{\text{scatt}} = 4\pi\,\lambda\!\!\!^{-2}\left|\frac{a}{\lambda\!\!\!^{-}} + \frac{\Gamma_A/2}{(E-E_0)+i\,\Gamma/2}\right|^2. \tag{4.13b}$$

Eq. (4.13b) is the usual single level expression showing the combined effects of resonance and potential scattering. Far from resonance the second term is small compared to $a/\lambda\!\!\!^{-}$ and $\sigma_{\text{scatt}}\to 4\pi a^2$. Near enough to resonance the second term dominates to give Eq. (3.15) without the factor (g_C/g_A) which should also be present here. For $(E-E_0)>\Gamma/2$ the second term has its real part positive and greater than the imaginary part to give constructive interference, while for $(E-E_0)<-\Gamma/2$ it is negative to give destructive interference with the potential part.

If df_0/dE is everywhere real and always decreases as it passes through zero at each resonance, it must behave similarly to the function $-\tan\int_{E_0}^{E} b(E)\,dE$, with $b(E)$ real positive. It is interesting to explore the possibility that there exists a region of configuration space for $r\approx R$ where ψ_0 can be represented with reasonable accuracy by

$$r\,\psi_0 \approx \chi_1(\text{int})\,\chi_2(\text{ang})\,\psi_3(r) \tag{4.14}$$

where χ_1 is a function of the internal degrees of freedom of the fragments, χ_2 depends on the angular part of the relative motion (including l and spins in general to keep J_C a good quantum number), while $\psi_3(r)$ depends only on the separation r of the fragments. If this separation of terms is approximately valid in a region where the fragments experience a strong interaction $V_0 \sim -20$ to -40 Mev, assumed constant, then ψ_3 in this region has the form (for the case $f_i = 0$) for $r < a$

$$\psi_3 = c \cos\left[K(r-a) + \int_{E_0}^{E} b(E)\, dE\right] \tag{4.15}$$

where $b(E)$ is real and positive if only scattering is present. $\hbar K$ is a momentum appropriate to an "inside kinetic energy of relative motion" of ~ 20 to 40 Mev. Applying (4.3) to (4.15) gives

$$f = -(K a) \tan\left[\int_{E_0}^{E} b(E)\, dE\right]. \tag{4.16}$$

An appropriate sign for the imaginary part of f is obtained if a (small) positive imaginary phase angle $i\mu(E)$ is added, so

$$f = f_0 - i f_i = -(K a) \tan\left[\int_{E_0}^{E} b(E)\, dE + i\mu(E)\right]. \tag{4.17}$$

General arguments have been advanced (cf. BLATT and WEISSKOPF [3]) showing that ,"other things being equal", one would expect $\mu(E)$ to be relatively slowly varying between resonances, and for $b(E)$ to have the same average value near resonance as between resonances. As a simplest test case to explore some of the features of this approach, we consider the consequence of keeping $b(E)$ and $\mu(E)$ constant with $b = \pi/D$. Here D is the level spacing (assumed constant). Setting $E - E_0 = \Delta E$, this gives

$$f_0 - i f_i = -(K a) \tan[\pi \Delta E/D + i\mu]. \tag{4.18}$$

If $\tan i\mu \approx i\mu$ this gives

$$f_0 = -(K a)\left[\frac{(1-\mu^2)\tan \pi \Delta E/D}{1 + \mu^2 \tan^2 \pi \Delta E/D}\right], \tag{4.19a}$$

$$f_i = (K a)\left[\frac{\mu \sec^2 \pi \Delta E/D}{1 + \mu^2 \tan^2 \pi \Delta E/D}\right], \tag{4.19b}$$

which gives an indication of the extent to which the linear approximation of Eq. (4.10) might be expected to fail when $|\Delta E/D|$ is not $\ll 1$. An approach similar to this has been explored by KOFOED-HANSEN [16] in fitting the theory to the experimental results for the lowest levels in Au and Ag. The important feature is that contributions from several levels become important when $|\Delta E/D|$ is not $\ll 1$. Eqs. (4.19a) and (4.19b) give $f_0' = -Ka(1-\mu^2)\pi/D \approx -Ka\pi/D$ for $\mu \ll 1$, and $f_i = Ka\mu$. Using Eqs. (4.11) and (4.12), this gives

$$\frac{\Gamma_A}{D} \approx \frac{2}{\pi}\frac{k}{K} \approx \frac{2}{\pi}\left(\frac{E}{30\ \text{Mev}}\right)^{\frac{1}{2}}, \tag{4.20a}$$

$$\frac{\Gamma_B}{D} \approx \frac{2\mu}{\pi} \tag{4.20b}$$

which shows that $\mu \sim \Gamma_B/D \ll 1$ as stated, and that Γ_A is proportional to $\sqrt{E}$. If E is taken as the numerical value of E in ev units, then Γ_A^0 is called the "reduced partial width" (reduced to a 1 ev basis to remove the effect of k). This implies that Γ_A^0 should be proportional to D with

$$\Gamma_A^0/D \sim 1.2 \times 10^{-4}. \tag{4.20c}$$

For the case of slow neutron resonances, this has the consequence that $\Gamma_N^0 \ll \Gamma_\gamma$ when D is relatively small (i.e. in cases where $D < \sim 100$ ev) and $\Gamma_N \gg \Gamma_\gamma$ for larger D when E is not too near zero energy.

Experimentally it is found that Γ_N^0 for a given isotope and spin state of the compound nucleus (for slow neutron resonances) does tend to follow Eq. (4.20c) on the average. However, Γ_N^0 values for individual levels show a very large fluctuation about the average value indicated. Here D is replaced by an average D since level spacings also show considerable fluctuations about the average values. There seems to be no clear indication that the large Γ_N^0 are accompanied by large local spacing of adjacent levels. This is best tested by nuclei like Th^{232} and U^{238} where only one isotope and spin state is involved (see Table 1). For U^{238} levels are seen at the following energies in ev having Γ_N^0 values in millivolts indicated in parentheses following the energy value. They are 6.7 (0.59), 21.0 (1.9), 36.9 (5.3), 66.3 (3.1), 81.3 (0.23), 90 (0.009), 103.5 (6.6), 117.5 (1.4), 146 (0.07), 166 (0.31), 192 (10.1), 212 (4). The average D is $\overline{D} = 18.6$ for this region, and, excluding the Γ_N^0 values of the first and last levels, the average Γ_N^0 is $\overline{\Gamma_N^0} = 2.90$ mv for $\overline{\Gamma_N^0}/\overline{D} = 1.56 \times 10^{-4}$. A plausible suggested theory for the distribution of Γ_N^0 values is that, if $\Gamma_N^0 = \gamma_N^2$, the γ_N values have a Gaussian distribution peaked at $\bar{\gamma}_N = 0$. To discuss this matter properly, however, we first consider an elementary simplified formulation of the Wigner-Eisenbud theory of nuclear reactions.

5. Elementary development of the Wigner-Eisenbud formalism for nuclear reactions [*9*]. Here we consider the various possible ways in which the compound nucleus can break up which are energetically allowed (open channels) or energetically forbidden. For simplicity we assume that prompt effects will involve only two fragments which are specified by the state χ_{1s} of the internal degrees of freedom of the fragments and a part χ_{2s} which involves the angular part of the relative motion wave function for the fragments when they are separated some minimum distance a_s or more. Specification of χ_{1s} and χ_{2s} (or χ_{12s} if a separation $\chi_{1s}\chi_{2s}$ is not allowed) together defines the channel s, while a_s is known as the "channel radius". As an example we might say that the fragments are a neutron and the final nucleus in its ground state with a relative angular momentum $l = 0$. For a compound nucleus of energy E, one must also specify the radial function of the relative motion coordinate r_s for each s. One can consider a closed volume in configuration space which corresponds to the C.N. and is bounded by the channel surfaces at each $r_s = a_s$. Within this bounded region of configuration space one can define a complete set of orthonormal states χ_λ which satisfy

$$H\chi_\lambda = E_\lambda \chi_\lambda \tag{5.1a}$$

and a suitable boundary condition at each channel surface $r_s = a_s$, near which χ_λ can with good accuracy be factored into

$$r_s \chi_\lambda = \chi_{1s}\chi_{2s}\chi_{3\lambda s}(r_s). \tag{5.1b}$$

Here the $\chi_{1s}\chi_{2s}$, or χ_{12s}, contain all but the radial part of the relative motion functions and are completely defined by the definition of the channel s. The radial functions $\chi_{3\lambda s}$ must conform to the boundary conditions for every s, and this can only happen at certain energies E_λ. For $E \neq E_\lambda$ the boundary conditions can be satisfied for all but one channel and we define the functions ψ_α such that

$$H\psi_\alpha = E\psi_\alpha \tag{5.2a}$$

and

$$r_s \psi_{\alpha s} = \chi_{1s}\chi_{2s}\varphi_{3\alpha s}(r) \tag{5.2b}$$

near the surface for channel s. There is a function ψ_α for each E and for each channel, since α designates the choice of single channel for which the boundary condition is to be relaxed.

The standard boundary condition which was selected is that

$$\left(\frac{\partial \chi_{3\lambda s}}{\partial r_s}\right)_{r_s=a_s} = 0, \tag{5.3a}$$

$$\left(\frac{\partial \varphi_{3\alpha s}}{\partial r_s}\right)_{r_s=a_s} = \delta_{\alpha s}\left(\frac{m_s}{\hbar}\right)^{\frac{1}{2}}. \tag{5.3b}$$

Here $\delta_{\alpha s}=0$ if $\alpha \neq s$ and $=1$ if $\alpha = s$ while m_s is the reduced mass of the relative motion for channel s. Thus the zero derivative boundary condition was chosen as in the fourier series using $\cos n x$ for $0 \leq x \leq \pi$. The amplitudes of the $\chi_{3\lambda s}$ at the a_s are completely defined by the requirement that the $\chi_{3\lambda}$ form a complete ortho-normal set and that $\chi_{1s}\chi_{2s}$ be separately normalized. Their amplitudes, however, define the γ matrix.

$$\chi_{3\lambda s}(a_s) \equiv \left(\frac{2m_s}{\hbar^2}\right)^{\frac{1}{2}} \gamma_{s\lambda}. \tag{5.4a}$$

The $\gamma_{s\lambda}$ are related to the partial widths for the various possible decay channels. The magnitude of the $\varphi_{3\alpha s}(a_s)$ are completely determined by Eq. (5.3b) and, even for $s=\alpha$, define the R matrix

$$\varphi_{3\alpha s}(a_s) \equiv \left(\frac{m_s}{\hbar}\right)^{\frac{1}{2}} R_{\alpha s}. \tag{5.4b}$$

Since the χ_λ form a complete set for the closed region of configuration space, the ψ_α can be expanded in terms of them.

$$\psi_a = \sum_\lambda A_{\alpha\lambda}\chi_\lambda \tag{5.5a}$$

where

$$A_{\alpha\lambda} = \int \chi_\lambda^* \psi_\alpha \, d\tau. \tag{5.5b}$$

At this point we by pass most of the technical complications faced in more thorough treatments by assuming that the ψ_α and χ_λ are real. When angular momentum states must be used, we assume that $+m_l$ and $-m_l$ can be combined to give forms like $\sin m\varphi$ or $\cos m\varphi$ in place of $e^{\pm i\varphi}$. In this case $\chi_\lambda^* = \chi_\lambda$ in Eq. (5.5b). The alternative is a careful consideration of such matters as the behavior of angular momentum states under such transformations as time reversal, etc. (This avoids most of the complications present for a more exact presentation.) It should be noted that this expansion is similar to expanding $f(x)$ in $b_0 + \Sigma b_n \cos n x$ for $0 \leq x \leq \pi$ when $f'(x) \neq 0$ at $x=\pi$. In each case the expansion is valid for ψ_α or $f(x)$, but cannot be differentiated without destroying the equality at a_s, since all of the χ_α (or $\cos n x$) have zero derivatives at the surfaces. Consider

$$\int [\chi_\lambda H \psi_\alpha - \psi_\alpha H \chi_\lambda]\, d\tau = (E - E_\lambda) \int \chi_\lambda \psi_\alpha \, d\tau \tag{5.6}$$

from Eqs. (5.1a) and (5.2a). The left side is zero for the potential energy part of H, so H can be replaced by the kinetic energy operator $-\hbar^2 \sum_j \vec{\nabla}_j \cdot \vec{\nabla}_j / 2m_j$ with j summed over the A nucleons of the compound nucleus. Here $m_j \sim$ mass m for atomic weight unity. This $3A$-dimensional integral becomes

$$-\frac{\hbar^2}{2m} \int \nabla \cdot [\chi_\lambda \nabla \psi_\alpha - \psi_\alpha \nabla \chi_\lambda]\, d\tau$$

where ∇ is the gradient or divergence operator in $3A$ dimensions. This is equal to the surface integral of the normal component of the quantity in the brackets at the surface of the region of configuration space. The surface includes all of the channel surfaces. These are inclined with respect to the original nucleon coordinate directions such that only the separations of the centers of mass of the nuclear subunits define the channel surfaces. This has the effect that m is replaced by the reduced mass m_s at each channel surface. The normalized $\chi_{1s}^2\,\chi_{2s}^2$ can be integrated over the internal coordinates and angular part of each s to leave only

$$-\sum_s \frac{\hbar^2}{2m_s}\left[(\chi_{3\lambda s}/r_s)\frac{\partial}{\partial r_s}(\varphi_{3\alpha s}/r_s)-(\varphi_{3\alpha s}/r_s)\frac{\partial}{\partial r_s}(\chi_{3\lambda s}/r_s)\right]_{r_s=a_s} a_s^2,$$

$$=-\sum_s \frac{\hbar^2}{2m_s}\left[\chi_{3\lambda s}\frac{\partial}{\partial r_s}\varphi_{3\alpha s}-\varphi_{3\alpha s}\frac{\partial}{\partial r_s}\chi_{3\lambda s}\right]_{r_s=a_s}$$

which, using Eqs. (5.3 a, b) and (5.6) gives

$$\frac{-\hbar^2}{2m_\alpha}\chi_{3\lambda\alpha}(a_\alpha)\left(\frac{m_\alpha}{\hbar}\right)^{\frac{1}{2}}=(E-E_\lambda)\int\psi_\alpha\chi_\lambda\,d\tau.$$

Using Eqs. (5.4a) and (5.5a, b) converts this to

$$A_{\alpha\lambda}=\left(\frac{\hbar}{2}\right)^{\frac{1}{2}}\frac{\gamma_{\alpha\lambda}}{(E_\lambda-E)}, \tag{5.7a}$$

which basic result shows that the A matrix coefficients depend only on the energy independent $\gamma_{\alpha\lambda}$ times a resonance denominator. Using Eqs. (5.4a), (5.4b), (5.5a), and (5.7a) gives

$$R_{\alpha s}=\sum_\lambda\frac{\gamma_{\alpha\lambda}\gamma_{s\lambda}}{(E_\lambda-E)}. \tag{5.7b}$$

This shows that the energy dependence of the R matrix is expressed in terms of the sum of contributions

$$\frac{\gamma_{s\lambda}\gamma_{\alpha\lambda}}{(E_\lambda-E)}$$

from each of the levels λ, where $\gamma_{s\lambda}\gamma_{\alpha\lambda}$ is independent of E, since the $\gamma_{s\lambda}$ depend only on the properties of the χ_λ. For $E=$ one of the E_λ, $R_{\alpha s}\to\infty$ for all s if the numerator is not zero. In the simple case of only one channel $s=\alpha$, $R_{\alpha\alpha}$ is the inverse of the logarithmic derivative f_0 at $R_\alpha=a_\alpha$.

The S (scattering) matrix is defined by considering the amplitudes of the incoming spherical waves in the incident channel and the outgoing spherical waves in all channels. Let $U_{3\alpha s}$ correspond to $r_s\psi$ for this case for $r_s \gtrsim a_s$ (if no Coulomb field is involved)

$$U_{3\alpha s}=\left(\frac{m_s}{\hbar k_s}\right)^{\frac{1}{2}}\left[\delta_{\alpha s}e^{-ik_s r_s}-S_{\alpha s}e^{ik_s r_s}\right], \tag{5.8}$$

and

$$\sigma_{\alpha s}=\pi\lambda_\alpha^2\,|\,\delta_{\alpha s}-S_{\alpha s}|^2 g \tag{5.9}$$

is the reaction cross section for incident channel α and outgoing channel s. Here $g=2l+1$ if the α channel fragment spins are zero; otherwise g contains more complicated statistical weight factors. Solving for $U_{3\alpha s}(a_s)$ and $U'_{3\alpha s}(a_s)\equiv(dU_{3\alpha s}/dr_s)_{r_s=a_s}$ gives

$$U_{3\alpha s}(a_s)=\left(\frac{m_s}{\hbar k_s}\right)^{\frac{1}{2}}\left[\delta_{\alpha s}e^{-ik_s a_s}-S_{\alpha s}e^{ik_s a_s}\right], \tag{5.10a}$$

$$U'_{3\alpha s}(a_s)=\left(\frac{m_s k_s}{\hbar}\right)^{\frac{1}{2}}(-i)\left[\delta_{\alpha s}e^{-ik_s a_s}+S_{\alpha s}e^{ik_s a_s}\right] \tag{5.10b}$$

which compares with Eq. (5.4b)

$$\varphi_{3\alpha s}(a_s) = \left(\frac{m_s}{\hbar}\right)^{\frac{1}{2}} R_{\alpha s}$$

and Eq. (5.3b)

$$\varphi'_{3\alpha s} = \left(\frac{m_s}{\hbar}\right)^{\frac{1}{2}} \delta_{\alpha s}.$$

We can express the $\varphi_{3\alpha s}$ in terms of the $U_{3\beta s}$

$$\varphi_{3\alpha s} = \sum_{\beta} c_{\alpha\beta} U_{3\beta s} \tag{5.11}$$

and use the above relations to solve for the S matrix in terms of the R matrix. We define matrices φ, U, φ', U', ω, B, C, R, S, and P, where $\varphi_{\alpha s} \equiv \varphi_{3\alpha s}(a_s)$, $U_{\alpha s} \equiv U_{3\alpha s}(a_s)$,

$$\omega_{\alpha\beta} = \delta_{\alpha\beta} e^{-i k_\alpha a_\alpha} = \delta_{\alpha\beta} e^{-i k_\beta a_\beta},$$

$$B_{\alpha\beta} \equiv \delta_{\alpha\beta} \sqrt{k_\alpha} = \delta_{\alpha\beta} \sqrt{k_\beta},$$

and

$$P_{\alpha\beta} \equiv \delta_{\alpha\beta} \left(\frac{m_\beta}{\hbar}\right)^{\frac{1}{2}} = \delta_{\alpha\beta} \left(\frac{m_\alpha}{\hbar}\right)^{\frac{1}{2}}.$$

Then

$$U = (\omega B^{-1} - S\omega^{-1} B^{-1}) P, \qquad i U' = (\omega B + S\omega^{-1} B) P,$$

$$B\varphi = B R P = B C U = (B C \omega - B C S \omega^{-1}) B^{-1} P,$$

$$i B \varphi' = i B P = i B C U' = (B C \omega + B C S \omega^{-1}) B P$$

giving

$$B R B = B C \omega - B C S \omega^{-1}$$

and

$$i I = B C \omega + B C S \omega^{-1}.$$

Using sums and differences this gives

$$-2i B C \omega = (I - i B R B),$$

$$-2i B C \omega \omega^{-1} S \omega^{-1} = (I + i B R B)$$

and

$$S = \omega \left[\frac{I + i B R B}{I - i B R B}\right] \omega. \tag{5.12}$$

This gives the customary form of the relation for S in terms of R. Usually BRB gives an effect small compared to unity if E is not near an E_λ. In that case

$$S \approx \omega (I + 2i B R B) \omega. \tag{5.13}$$

The behavior of (5.12) for slow neutrons, if only scattering is possible, is found be replacing the matrices by their single remaining component for $s = \alpha$. Thus $B = \sqrt{k}$, $\omega = e^{-ika}$, and

$$R = R_{\alpha\alpha} = \sum_{\lambda} \frac{(\gamma_{\alpha\lambda})^2}{(E_\lambda - E)} = \sum_{\lambda} \frac{\gamma_\lambda^2}{(E_\lambda - E)}. \tag{5.14}$$

$$|1 - S|^2 = \left| e^{ika} - e^{-ika} \left(\frac{1 + ikR}{1 - ikR}\right)\right|^2 = \frac{4}{1 + k^2 R^2} |\sin k a - k R \cos k a|^2.$$

For $ka \ll 1$ we approximate $\sin ka \to ka$, $\cos ka \to 1$ to give (for $l = 0$)

$$\sigma = \pi \lambda\!\!\!^{-2} |1 - S|^2 = \frac{4\pi}{1 + k^2 R^2} |a - R|^2. \tag{5.15}$$

Far from resonance $k^2R^2 \ll 1$ and

$$\sigma \approx 4\pi \left[a - \sum_{\lambda} \frac{\gamma_\lambda^2}{(E_\lambda - E)}\right]^2. \tag{5.16}$$

Note that $R = +\infty$ for $E = E_\lambda - 0$ and $R = -\infty$ for $E = E_\lambda + 0$. To the extent that a particular nearby resonance dominates R, $\sigma < 4\pi a^2$ for $E < E_\lambda$ and $\sigma > 4\pi a^2$ for $E > E_\lambda$. To some extent the generality of the WIGNER-EISENBUD method is also a weakness. For example, for no interaction we can still choose some a and obtain γ_λ and E_λ. In this case, however, $S = I =$ unit matrix and

$$R = i\,B^{-1}\left[\frac{1 - \omega^{-1} S\,\omega^{-1}}{1 + \omega^{-1} S\,\omega^{-1}}\right] B^{-1}$$

for this case becomes $R = \lambda\!\!\!^{-} \tan x$, where $x = ka$. This agrees with the result obtained starting with $\varphi = \text{const} \sin kr$ and $R = (\varphi/\varphi')_{r=a}$. The $k_\lambda a$ for $\varphi' = 0$ give $k_\lambda a = \pi(n + \tfrac{1}{2})$ for $n = 0, 1, \ldots$ Setting $\int\limits_0^a \chi_{3\lambda}^2\, dr = 1$ and $\gamma_\lambda = (\hbar^2/2m)^{\frac{1}{2}} \chi_{3\lambda}(a)$ gives $\gamma_\lambda = \text{const} = (\hbar^2/m a)^{\frac{1}{2}} = \gamma$, from Eq. (5.14)

$$\left.\begin{aligned} R &= \lambda\!\!\!^{-} \sum_{n=0}^{\infty} \frac{2x}{\pi^2 (n + \frac{1}{2})^2 - x^2} \\ \text{or} \qquad & \\ R &= \lambda\!\!\!^{-} \sum_{n=-\infty}^{\infty} \frac{1}{\pi(n + \frac{1}{2}) - x}. \end{aligned}\right\} \tag{5.17}$$

This series has all of its poles at the same places and of the same strength as $\lambda\!\!\!^{-} \tan x$, and it can be shown to be the same function. The energies E_λ here depend on a and the formal appearance of resonances for no interaction expresses a mathematical identity of no physical significance. Similarly the E_λ depend on the $k_s a_s$ chosen. Since the energies at which (physical) resonances occur cannot depend on the formal choices of the a_s, the ω matrix must neutralize this effect through the $e^{-i k_s a_s}$ factors. This is related to the remark following Eq. (4.6) that $f_0 = 0$ does not imply $|\sin\delta| = 1$ unless ka is negligible.

In the region between resonances the net contribution of distant resonances may not cancel for scattering, but may give a systematic modification of the effective scattering length. The consequences of this point are explored in greater detail in the following section in connection with the theory of FESHBACH, PORTER, and WEISSKOPF.

Returning to Eq. (5.15) for E near resonance, where $|R| \gg a$, we note that $|a - R|^2 \to \infty$ since the denominators in R have $E - E_\lambda$ instead of $E - E_\lambda + i\Gamma_\lambda/2$ as in Eq. (4.13b). It was the $i\Gamma/2$ that gave the proper σ when $E = E_\lambda$ in Eq. (4.13b). In Eq. (5.15) this role is performed by the factor $(1 + k^2 R^2)$ in the denominator that limits σ to $4\pi\lambda\!\!\!^{-2}$ at resonance. The formal comparison with the previous formalism may be made by defining

$$\Gamma_{s\lambda} \equiv 2k_s \gamma_{s\lambda}^2 \tag{5.18}$$

which becomes $\Gamma_\lambda = 2k\gamma_\lambda^2$ for the case of Eq. (5.15) so

$$R = \frac{\lambda\!\!\!^{-}}{2} \sum_\lambda \frac{\Gamma_\lambda}{E_\lambda - E},$$

$$\sigma = \frac{4\pi \left\{a + \dfrac{\lambda\!\!\!^{-}}{2} \sum\limits_\lambda \dfrac{\Gamma_\lambda}{E - E_\lambda}\right\}^2}{1 + \dfrac{1}{4}\left\{\sum\limits_\lambda \dfrac{\Gamma_\lambda}{E - E_\lambda}\right\}^2}. \tag{5.19}$$

If only one level $E_\lambda = E_0$ is important in giving the energy dependence, while the others effectively change the a to some modified a',

$$\sigma \approx 4\pi \lambda\!\!\!^{-2} \frac{\{k\, a'(E - E_0) + \Gamma/2\}^2}{(E - E_0)^2 + (\Gamma/2)^2}. \tag{5.20}$$

This formula would be the same as Eq. (4.13b) if $k^2 a'^2 \Gamma^2/4$ were added to the squared bracket in the numerator. This does not significantly alter the cross section.

The formulas in the case of many open channels bear similar resemblences to the customary BREIT-WIGNER formula when only one level is important. The stated proportionality of $\Gamma_{s\lambda}$ to the final state momentum is explicitly demonstrated in Eq. (5.18). In the case where the effects of many levels are included, the energy dependence of the contributions of the other levels may show up for E relatively near the E of the given level. In this connection see the discussions following Eq. (4.19b) and Eq. (5.17). Note that for the elastic scattering, all resonances having $E_\lambda > E$ contribute to R with a sign opposite to that of a, while levels having $E > E_\lambda$ contribute with the same sign as a.

For discussions of this formalism see [*9*], [*19*], [*3*] and [*18*].

6. Optical model for scattering.—FESHBACH, PORTER, WEISSKOPF model [*13*], Although individual particle nuclear models had been explored before 1947. the agreement with experiment had been discouraging. This situation changed greatly with the discovery of the MAYER-JENSEN shell model and the experimental developments since that time. It is now generally conceded that a good starting approximation to the nuclear wave function for many purposes is obtained by assigning the individual nucleons single particle wave functions in a common nuclear potential well. This well is roughly square, with rounded edges, and with a spin-orbit coupling dependence. If one starts with the level order for a three dimensional harmonic oscillator potential, the major shells, of energy $(n + \frac{3}{2})\hbar\omega$, are $1s$, $1p$, $2s + 1d$, $2p + 1f$, $3s + 2d + 1g$, $3p + 2f + 1h$, $4s + 3d + 2g + 1i$, etc. ($3p$ means the third $l = 1$ state, etc.). The states of higher l within each group are now to be lowered relative to those of low l, and individual particle states having $j = l + \frac{1}{2}$ are lowered relative to those of $j = l - \frac{1}{2}$ (for a given l). This picture definitely requires modifications which take account of correlations and of the non spherical shape of nuclei between major closed shell nuclear numbers 2, 8, 20, 50, 82, and 126 for Z or $N = A - Z$ (particularly between 50 and 82, and 82 and 126). The classification of nuclear excited states on this picture before considering interactions between nucleons in different orbits, is on the basis of the number of nucleons in excited state individual particle states (i.e. states not occupied for the nuclear ground state). For ~ 8 Mev excitation the observed level spacing is ~ 10 ev for $A \sim 150$. These levels correspond to the large number of ways by which excited states can be formed having many nucleons simultaneously excited, but where total j and parity are restricted by the experimental methods. In considering neutron total cross sections for $0 \leq E \lesssim$ a few Mev, a strict individual particle model would predict that only elastic scattering would occur and that this would just be the scattering expected from a potential well of ~ 40 Mev central depth and of effective radius $b \approx 1.4\, A^{\frac{1}{3}} \times 10^{-13}$ cm. For relative motion states of a given l and j the resonances would have extremely large spacings. Consider below the reasoning for the case of a nucleus having ~ 100 neutrons. The neutron $3p$, $2f$, $1h_{\frac{9}{2}}$, and $1i_{\frac{13}{2}}$ (bound) states have ~ 2 to 12 Mev binding ($E = -\varepsilon$) with two, one, zero, and zero internal nodes respectively, and with $|r\psi|$ at $r = b$ decreasing from its final maximum at $r < b$ (for $E = -\varepsilon$). For $E \approx 0$, the phase of the inside function at $r = b$ is only slightly

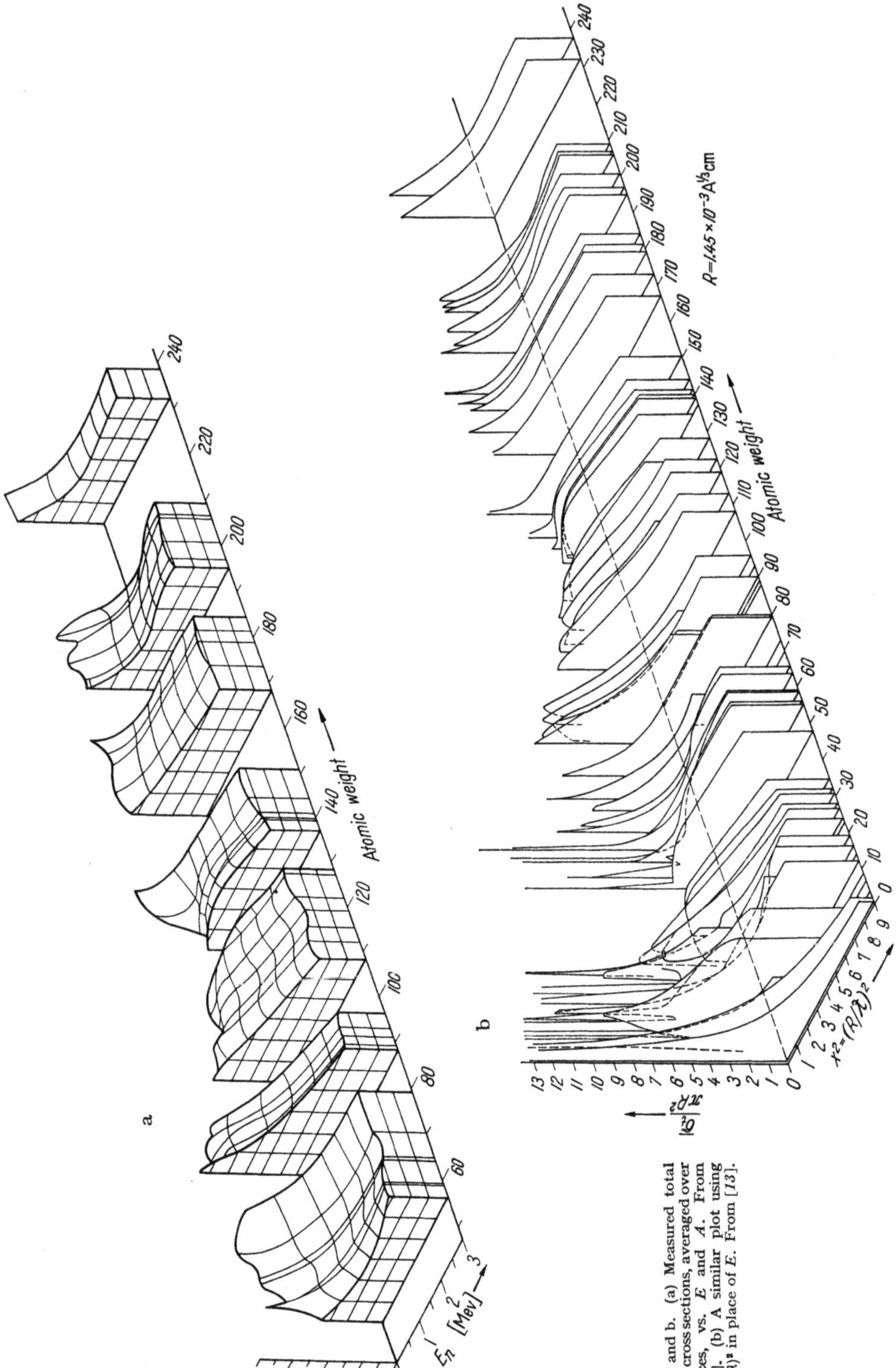

Fig. 22a and b. (a) Measured total neutron cross sections, averaged over resonances, vs. *E* and *A*. From [*Wis 23*]. (b) A similar plot using $x^2 \equiv (R/\lambda)^2$ in place of *E*. From [*13*].

greater than for $E = -\varepsilon$, which suggests the behavior of $f = (r\varphi'/\varphi)$ at $r = b$, and thus the behavior of σ [Eqs. (4.3) to (4.13)]. For the s, d, $g_{\frac{7}{2}}$, and $h_{\frac{11}{2}}$ single particle states, the corresponding value of φ'/φ at $r = b$ would occur for a total neutron number near the center of the preceding, or following, major shells Thus φ is apt to have an internal node just before $r = b$, with φ'/φ positive at $r = b$. This predicts that there will be only a few resonances for each l for $E = 0$ to ∞,

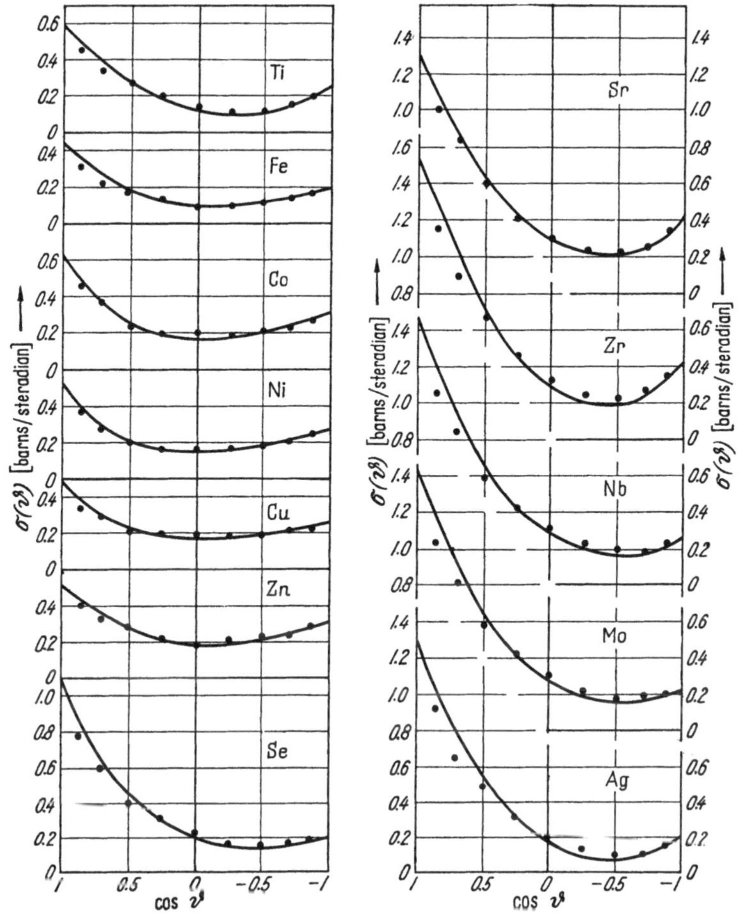

Fig. 23. Measured $d\sigma/d\Omega$ vs. $\cos\vartheta$ for 1 Mev neutrons for 12 elements. From WALT and BARSHALL: Phys. Rev. 93, 1062 (1954).

with extremely large level spacings. The cross section for $l = 0$ and $E \approx 0$ is expected to show nuclear size resonances. The resonances occur where $Kb = (n - \frac{1}{2})\,\pi$ where $n = 1, 2, 3, \ldots$ The following minima occur where $Kb \gtrsim n\pi$. The n-p singlet interaction has Kb near, but slightly under, $\pi/2$. The n-p triplet interaction already has $Kb > \pi/2$. The first minimum occurs for Kb about twice this value, or near $A = 4$ to 12. In general a maximum for $l = 0$ would occur for some neutron number near the beginning of each major closed shell having bound $l = 0$ states, while the minima for $l = 0$ occur near the maxima for $l = 1$.

The consequences of considering the interaction of the neutron with the nucleus as being able to be represented by a simple potential well was explored soon after the discovery of the neutron. The fact that $l = 0$ resonances occur every few ev for $A \lesssim 150$, however, seems to rule out this approach and Bohr's strong coupling approach was favored. Here the incoming neutron wave is considered to be attenuated in a distance $\sim K^{-1}$ by processes where it shares its energy with other nucleons. The absorption to give excited states of the nucleus as a whole can be represented by an imaginary component of the nuclear potential. Feshbach and Weisskopf [12] developed a theory of the expected scattering and

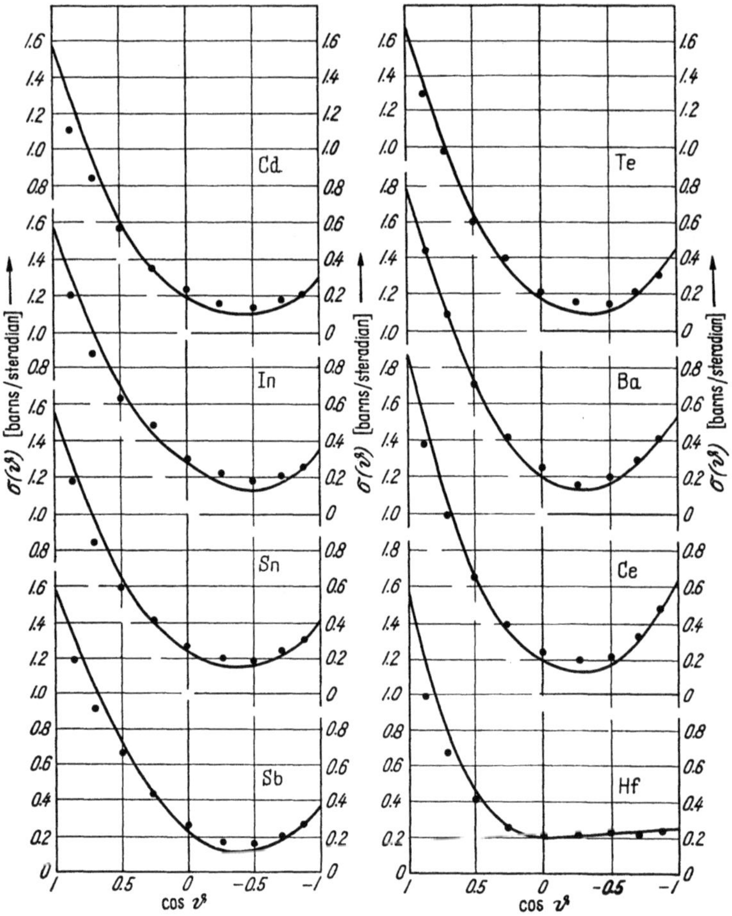

Fig. 24. A continuation of the curves of Fig. 23.

reaction cross sections averaged over many levels, as a function of nuclear size and energy. They used the surface boundary condition $\varphi'/\varphi = -iK$ at $r = b$. This predicted a smooth structureless dependence of the averaged cross sections on E and A. They compared the predicted relations with the experimental results for poor resolution measurements above ~ 100 kev and found only very rough agreement. As more and more experimental results collected, however, it became evident that the experimental curves (averaged over many resonances), when plotted in three dimensions as functions of E and A, or E and ϑ, showed structure which varied systematically with A as well as E. This is shown in Figs. 22 to 26.

Feshbach, Porter, and Weisskopf [13] found that they could obtain theoretical curves closely resembling the experimental curves in their gross

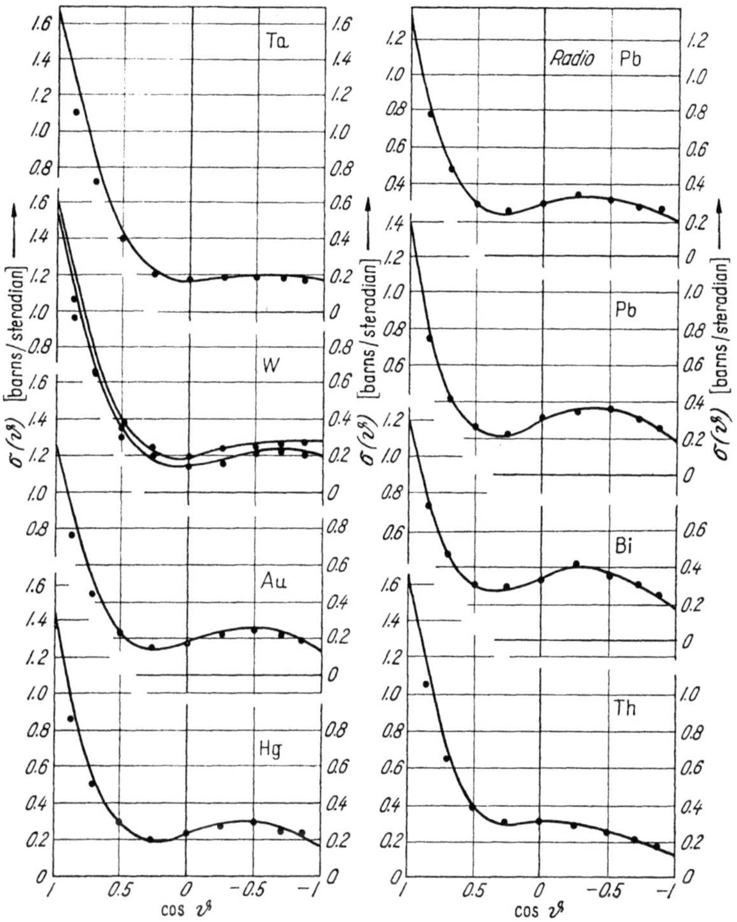

Fig. 25. A continuation of the curves of Figs. 23 and 24.

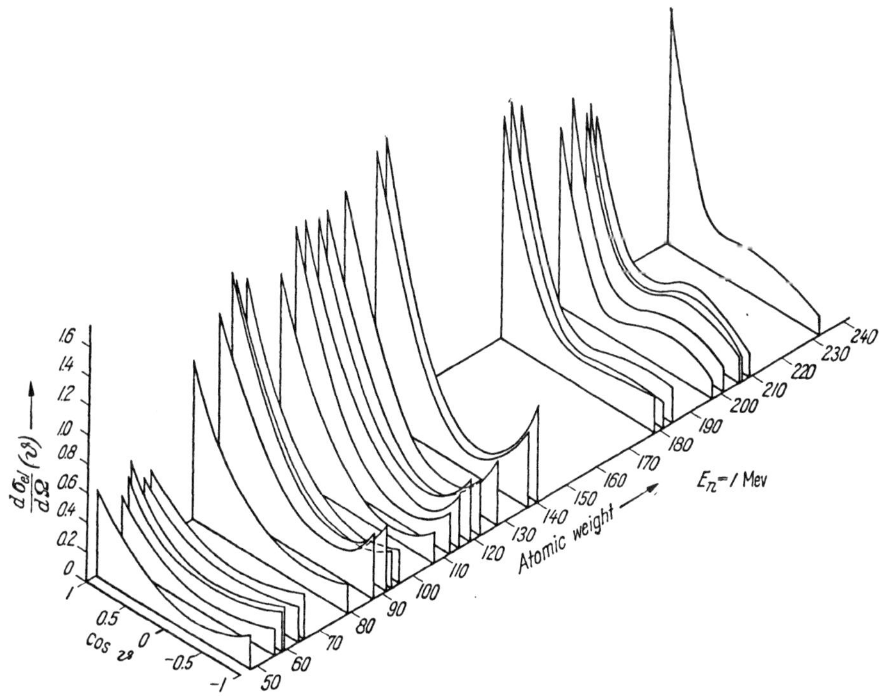

Fig. 26. Observed $d\sigma/d\Omega$ vs. $\cos\vartheta$ at 1 Mev, as in Figs. 23, 24, and 25, but plotted together do define a smooth surface vs. A.

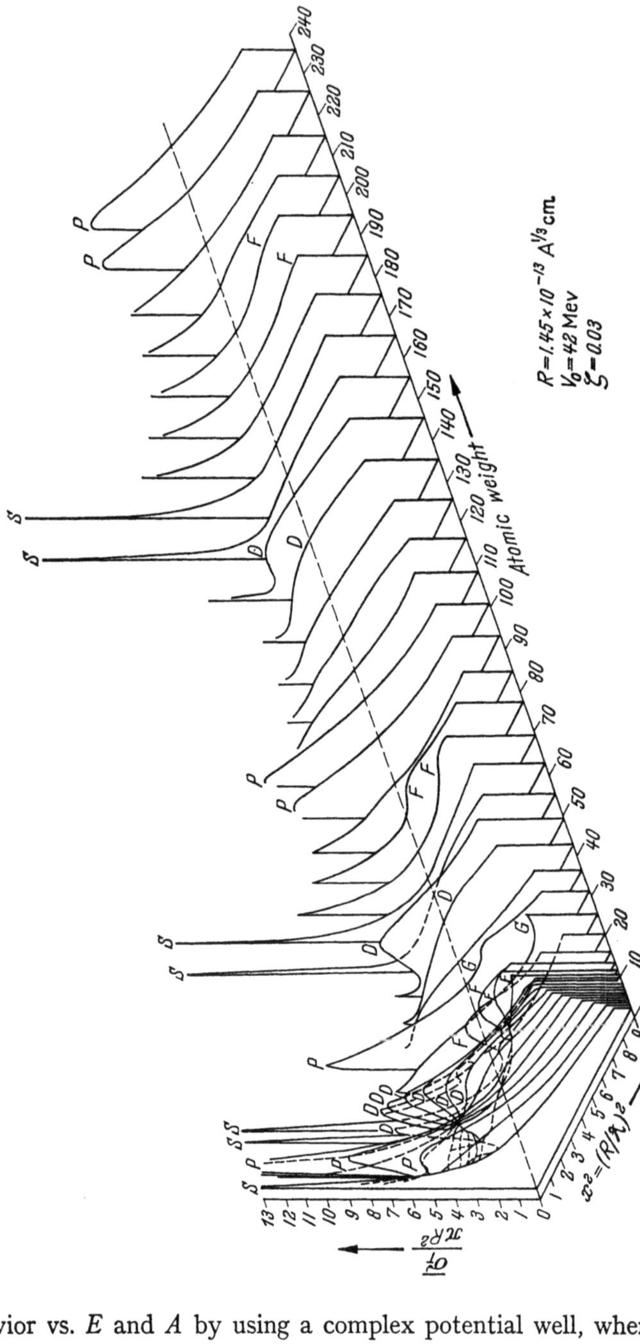

Fig. 27. Optical model calculated total neutron cross sections vs. $x^2 = (R/\lambda)^2$ and A for different choices of imaginary part of the potential. From [13]. This curve should be compared with Fig. 22b.

behavior vs. E and A by using a complex potential well, where

$$\left.\begin{aligned} &V = -42(1 + i\zeta)\ \text{Mev} \quad \text{for} \quad r < b,\\ &V = 0 \quad \text{for} \quad r > b \\ \text{where} \qquad & \\ &b = 1.45\, A^{\frac{1}{3}} \times 10^{-13}\ \text{cm},\\ &\zeta = 0.03 \pm \sim 0.02. \end{aligned}\right\} \tag{6.1}$$

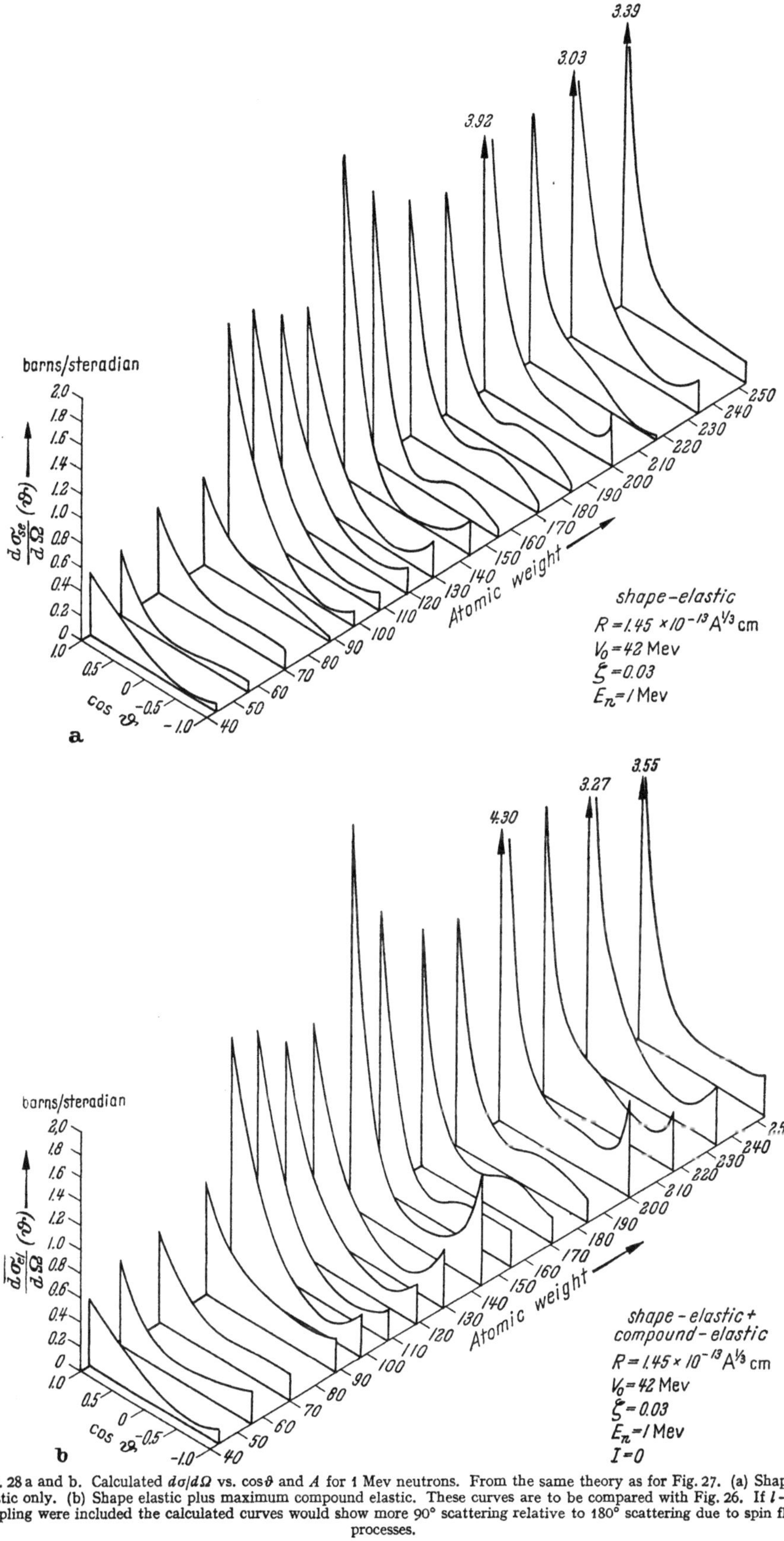

Fig. 28a and b. Calculated $d\sigma/d\Omega$ vs. $\cos\vartheta$ and A for 1 Mev neutrons. From the same theory as for Fig. 27. (a) Shape-elastic only. (b) Shape elastic plus maximum compound elastic. These curves are to be compared with Fig. 26. If $l-s$ coupling were included the calculated curves would show more 90° scattering relative to 180° scattering due to spin flip processes.

The calculated curves are shown in Figs. 27 and 28. These authors break the total cross section into an elastic part plus an absorption part. The absorption is due entirely to the compound nucleus formation, while the elastic is broken into a "shape elastic" part plus a compound elastic part. In the notation of Eqs. (4.1) and (4.2a) to (4.2c),

$$\sigma_{\text{tot}} \to \pi \lambda\!\!\!^{-2} [2 - (\eta + \bar{\eta})_{\text{av}}]\, g \tag{6.2a}$$

when averaged over many levels (here $\bar{\eta}$ is the conjugate complex of η). They define the shape elastic scattering as

$$\sigma_{\text{se}} = \pi \lambda\!\!\!^{-2} [1 + |\eta_{\text{av}}|^2 - (\eta + \bar{\eta})_{\text{av}}]\, g \tag{6.2b}$$

while the compound elastic scattering is

$$\sigma_{\text{ce}} = \pi \lambda\!\!\!^{-2} [|\eta^2|_{\text{av}} - |\eta_{\text{av}}|^2]\, g. \tag{6.2c}$$

The absorption and compound elastic cross sections are grouped together as the compound nucleus cross section σ_c so $\sigma_{\text{tot}} = \sigma_{\text{se}} + \sigma_c$. The authors give arguments to show that the absorption predicted by the optical model calculation should be identified with σ_c for compound nucleus formation, while the calculated scattering cross section should be identified with σ_{se}, the shape elastic cross section. The calculated shape $d\sigma_{\text{se}}/d\Omega$ for 1 Mev neutrons vs. A is shown in Fig. 28a. The experimental curves, shown in Fig. 26, also include σ_{ce}. Fig. 28b shows the result when the maximum amount of compound elastic scattering is added to the calculated shape elastic scattering. Thus the actual predicted elastic cross section should be intermediate between the two figures. If the common feature of a peaking in the forward direction for all of the curves is ignored, the feature of interest is the amount of peaking or drop off near 180°, and the behavior around 80 to 150°. In this respect the comparison is only fair. The experimental curves show a drop near 180° for $A \gtrsim 180$, while the calculated curves show a peak at 180° for $A \sim 200$ to 230.

As a summary of the above comparisons, we may note a general semi-quantitative similarity between the experimental and theoretical curves. The square well potential with sharp edges, however, does not give the same level order as the Mayer-Jensen scheme. In particular, the presence of spin-orbit coupling will separate and weaken the peaks in the scattering for a given l into the $j = l \pm \frac{1}{2}$ components. This will also give spin flip scattering which contains a $\sin\vartheta$ factor to increase 90° scattering relative to 180° scattering.

It seems probable that calculations using a potential more compatible with the shell model will give an improved agreement with experiment. As possible modifications, it is not unreasonable that the depths of the real and imaginary parts of the potential should depend on A and E. It is interesting to note that the authors originally used a potential of only half the depth (−19 Mev) to obtain reasonable agreement over a considerable range of A. However this potential was quite incompatible with the requirements of the shell model and the deeper potential was employed at the suggestion of Adair[1]. The authors relate the predictions of the optical model calculation to the parameters of the Wigner-Eisenbud formalism. In particular, the ratio Γ_N^0/D to $2/\pi K$ in Eq. (4.20a) is now expected to show peaks for nuclear size, A, corresponding to maxima for states of a given j. This also implies a systematic variation in

[1] R. K. Adair: Phys. Rev. **94**, 737 (1954).

the potential scattering length a' [see Eq. (5.20) and the associated discussion] since

$$a' = a + \sum_\lambda{}' \frac{\gamma^2_{\alpha\lambda}}{E - E_\lambda} \tag{6.3}$$

where a = channel radius $\approx$ nuclear radius, and the expression is a generalization of the bracket in Eq. (5.16). The sum $\sum'$ is to include only levels such that $|E - E_\lambda| > \Delta$, where Δ is large compared to the nuclear level spacing, but small compared to the width of the peaks in the Γ^0_N/D curve.

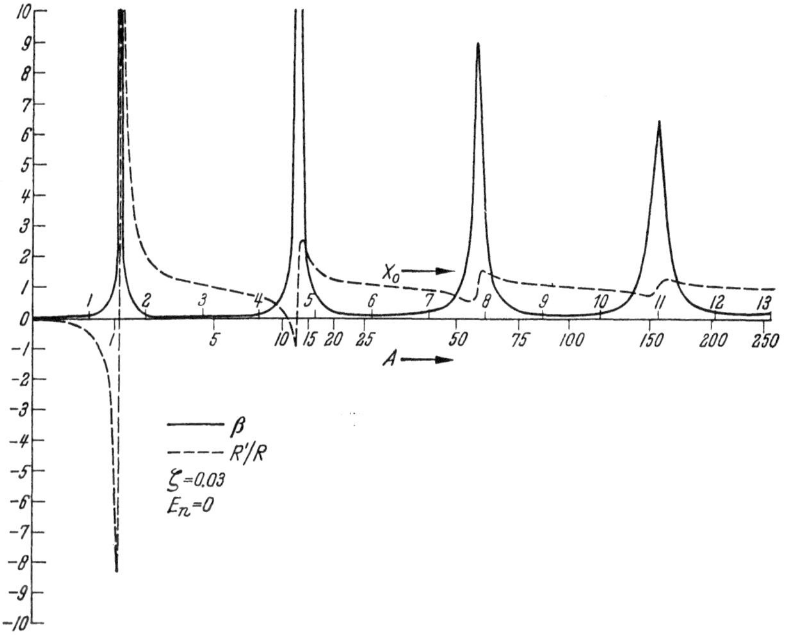

Fig. 29. Predicted variation of R'/R and $\beta = \pi K \Gamma_n / 2kD$ vs. A from the same theory as for Figs. 27 and 28. Here R' = scattering length, $R = 1.45 A^{\frac{1}{3}} \times 10^{-13}$ cm, $K/k = (V_0/E)^{\frac{1}{2}}$ = inside momentum/outside momentum of relative motion. Γ_n and D are the average neutron width and level spacing, averaged over many levels covering an energy region $\ll$ energy scale factor for changes predicted by the giant resonance theory.

If $\gamma^2_{\alpha\lambda}$ values show a resonance peak near energy E_1, then at $E \approx E_1$ contributions from $E < E_1$ and $E > E_1$ in the sum $\sum_\lambda{}' \gamma^2_{\alpha\lambda}/(E - E_\lambda)$ will roughly cancel to give $a' \approx a$. For $E > E_1$ by $\sim$ width of the peak in $\gamma^2_{\alpha\lambda}$ values, the strong contribution from the "giant resonance" will all add to make $a' > a$ while for E similarly $< E_1$, $a' < a$. The predicted low energy dependences of (a'/a), which they call R'/R, and of $\beta = (\pi/2)(K/k)(\Gamma_a/D)$ as functions of A, or of $x_0 \equiv Ka$, are shown in Fig. 29. The choice of complex potential is that previously given.

In concluding this section it should be mentioned that Bohr and Mottelson, in Appendix 5 of their comprehensive treatise on the collective model [4] give a somewhat similar treatment of resonance processes. They treat the coupling of the single particle excited state with the other excited configurations as occurring in a thin layer near the nuclear surface.

7. The giant resonance interpretation of the nucleon-nucleus interaction of Lane, Thomas, and Wigner [17]. Following the rough success of the Feshbach, Porter, and Weisskopf optical model match to the average cross sections over a region containing many levels, attempts were made to rephrase this concept in the R matrix formalism starting from a representation using nuclear states given by a modified independent particle model. On an extreme individual particle model, the excited states can be classified on the basis of 1, 2, 3, ... nucleons excited at a time. Only the one nucleon excitation states would be observed in neutron cross sections if there were no interaction between 1 particle excitation states and many particle excitation states. If some interaction is introduced to correspond better to the physical situation, the observation of the nuclear resonances corresponding to many nucleons excited occurs through

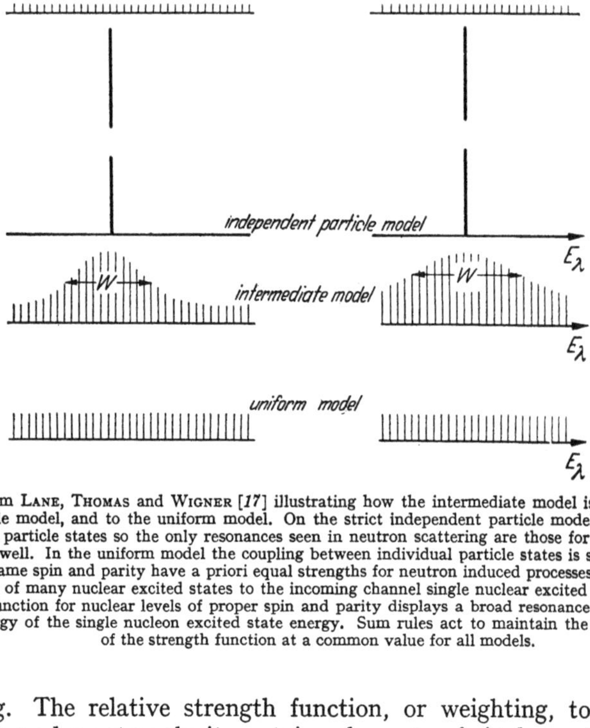

Fig. 30. Figure from LANE, THOMAS and WIGNER [17] illustrating how the intermediate model is related to the strict independent particle model, and to the uniform model. On the strict independent particle model there is no coupling between individual particle states so the only resonances seen in neutron scattering are those for a single particle in a spherical potential well. In the uniform model the coupling between individual particle states is so strong that all configurations of the same spin and parity have a priori equal strengths for neutron induced processes. In the intermediate model the coupling of many nuclear excited states to the incoming channel single nuclear excited state is such that the average strength function for nuclear levels of proper spin and parity displays a broad resonance behavior of width W peaked at the energy of the single nucleon excited state energy. Sum rules act to maintain the summed contribution of the strength function at a common value for all models.

this coupling. The relative strength function, or weighting, to a given level then depends on how strongly it contains elements of single particle excitation. Thus the strength function might be expected to reach a maximum near those energies $E_{1\lambda}$ of the single particle levels in the absence of coupling. The strength function would show a "giant resonance" behavior in that it would have a half width about each $E_{1\lambda}$ which would be very large compared to observed nuclear level spacings.

LANE, THOMAS, and WIGNER show, on the basis of closure sum rules, that the sum of the level strengths is the same as it would be in an extreme single particle model. As the width of the distribution is made narrower, the height at maximum increases in a compensating manner. This is illustrated in Fig. 30. They show that the width of these giant resonances in $\gamma_{\alpha\lambda}^2/D$ can, in principle, be calculated in terms of the force law between nucleons. In carrying out such a calculation for the second moment of the strength distribution, however, they obtain a result that is much too large to agree with experiment. This result appears to be quite sensitive to the assumptions made in the calculations, however, since VOGT [21] later obtained a result in much better agreement with experiment by using state functions that take account of nucleon-core polarization effects. It is likely that

Table 3. *Scattering cross sections.*

	σ_{coh} (sign)	$\sigma_{f.a.}\left(\frac{A+1}{A}\right)^2$	$\sigma_s(<1\text{ kev})$	$\sigma_s(10^3-10^4\text{ ev})$ or remarks
$_1H^1$	1.79 ± 0.02 (−)	81.5 ± 0.4	20.26	20
$_1H^2$	5.4 ± 0.3 (+)	7.6 ± 0.1	3.4	
$_2He^4$	1.1 ± 0.2 (+)	1.3 ± 0.2	0.8	
$_3Li^6$	6 ± 3 (+)			
$_3Li^7$	0.8 ± 0.05 (−)	1.4 ± 0.2	1.1	1.07
$_4Be^9$	7.53 ± 0.07 (+)	7.54 ± 0.07	6.0	6.0
$_5B^{10}$		4.0 ± 0.5		
$_5B^{11}$		4.4 ± 0.3		
$_6C^{12}$	5.50 ± 0.04	5.51 ± 0.03	4.7	4.7
$_6C^{13}$	4.5 ± 0.6 (+)	5.5 ± 1.0		
$_7N^{14}$	11.0 ± 0.5 (+)	11.4 ± 0.5	9	
$_8O^{16}$	4.2 ± 0.3 (+)	4.24 ± 0.02	3.7	3.7
$_9F^{19}$	3.8 ± 0.3 (+)	4.0 ± 0.1	3.5	3.7
$_{10}Ne$		2.9 ± 0.2		
$_{11}Na^{23}$	1.55 ± 0.05 (+)	3.4 ± 0.2	3.1	
$_{12}Mg$	3.60 ± 0.10 (+)	3.70 ± 0.10	3.4	
$_{13}Al^{27}$	1.5 ± 0.1 (+)	1.51 ± 0.03	1.42	1.4
$_{14}Si$	2.0 ± 0.2 (+)	2.4 ± 0.2	2.25	
$_{15}P$	3.1 ± 0.16 (+)	3.6 ± 0.3	3.4	
$_{16}S$	1.20 ± 0.08 (+)	1.2 ± 0.2	1.1	
$_{17}Cl$	12.1 ± 0.8 (+)		~14	but drops to ~2 at 10 kev
$_{18}A$	0.5 ± 0.1 (+)	0.9 ± 0.2	0.7	
$_{19}K$	1.5 ± 0.1 (+)	2.2 ± 0.1	2.0	
$_{20}Ca$	3.0 ± 0.1 (+)	3.2 ± 0.3		
$_{20}Ca^{40}$	3.0 ± 0.1 (+)	3.1 ± 0.3		
$_{21}Sc^{45}$	17.5 ± 1.5 (+)		2.3	dip at 1 kev to 0.4 and then resonances
$_{22}Ti$	1.4 ± 0.3 (−)	4.4 ± 0.2	4.3	(levels above 2 kev)
46		3.3 ± 1.0		
47		5.2 ± 1.0		
48		9 ± 4		
49		2.8 ± 1.0		
50		3.3 ± 1.0		
$_{23}V^{51}$	0.032 ± 0.008 (−)	5.1 ± 0.1	6.0	(resonances in 1 to 10 kev reg.)
$_{24}Cr$	1.56 ± 0.03 (+)	4.1 ± 0.3	4.1	
$_{25}Mn^{55}$	1.7 ± 0.1 (−)	2.0 ± 0.1	2.1	(below resonances) ~10 ev
$_{26}Fe$	11.36 ± 0.05 (+)	11.80 ± 0.04	11.5	
54	2.20 ± 0.13 (+)	2.5 ± 0.3		
56	12.8 ± 0.2 (+)	12.8 ± 0.2		
57	0.64 ± 0.04 (+)	2.0 ± 0.5		
$_{27}Co^{59}$	1.00 ± 0.06 (+)	6 ± 1	6.0	(then resonances)
$_{28}Ni$	13.2 ± 0.2 (+)	18.01 ± 0.05	17	(resonances 4500 ev)
58	25.9 ± 0.3 (+)	24.4 ± 0.5		
60	1.1 ± 0.1 (+)	1.0 ± 0.1		
62	9.5 ± 0.4 (−)	9 ± 1		
$_{29}Cu$	7.0 ± 0.4 (+)	8.0 ± 0.1	7.2	5.5
$_{30}Zn$	4.3 ± 0.3 (+)	4.1 ± 0.2	4.0	
$_{31}Ga$		7.5 ± 0.5	7 ± 1	
$_{32}Ge$	8.8 ± 0.5 (+)	9.0 ± 0.5	9.0	
$_{33}As$	5.0 ± 0.3 (+)	8 ± 1	8 ± 1	
$_{34}Se$	10.0 ± 0.6 (+)	8 ± 1	7 ± 1	
$_{35}Br$	5.7 ± 0.4 (+)	6.1 ± 0.2	5.5	
$_{37}Rb$	3.8 ± 0.3 (+)	5.5 ± 0.5		
$_{38}Sr$	4.1 ± 0.3 (+)	10 ± 2	5.4	
$_{40}Zr$	5.0 ± 0.3 (+)	6.3 ± 0.3	6.2	
$_{41}Nb$	6.0 ± 0.2 (+)	6.6 ± 0.3	6.3	
$_{42}Mo$	5.6 ± 0.2 (+)	6.1 ± 0.2	5.8 ± 0.5	
$_{44}Ru$		6.6 ± 0.5	6.6	
$_{45}Rh$	4.5 ± 0.5 (+)	5.6 ± 1.0		
$_{46}Pd$	5.0 ± 0.3 (+)	4.8 ± 0.3	4.5 ± 1	

Table 3. (Continued.)

	σ_{coh} (sign)	$\sigma_{f.a.}\left(\frac{A+1}{A}\right)^2$	$\sigma_s(<1$ kev)	σ_s (10^3-10^4 ev) or remarks
$_{47}Ag^{107}$	8.7 ± 0.5 (+)	10 ± 2		
$_{47}Ag^{109}$	2.3 ± 0.2 (+)	6 ± 1	6 ± 1	
$_{49}In$			5.5 ± 1	
$_{50}Sn$	4.6 ± 0.3 (+)	4.9 ± 0.5	5 ± 1	
$_{51}Sb(^{121}_{123})$	3.7 ± 0.3 (+)	4.2 ± 0.3	4.5	
$_{52}Te$	4.0 ± 0.3 (+)	4.5 ± 0.3	4.5	
$_{53}I^{127}$	3.4 ± 0.2 (+)	3.8 ± 0.4	4	
$_{55}Cs$	3.0 ± 0.2 (+)	7 ± 1	7	
$_{56}Ba$	3.5 ± 0.2 (+)	6 ± 1	7	
$_{57}La$	8.7 ± 0.3 (+)	9.3 ± 0.7	8 ± 1	
$_{58}Ce$	2.7 ± 0.2 (+)	2.8 ± 0.5	3	
140	2.8 ± 0.11 (+)	2.8 ± 0.5		
142	2.6 ± 0.2 (+)	2.6 ± 0.5		
$_{59}Pr^{141}$	2.4 ± 0.2 (+)	4.0 ± 0.4		
$_{60}Nd$	6.5 ± 0.4 (+)	16 ± 3	12	
142	7.5 ± 0.6 (+)			
144	1.0 ± 0.2 (+)			
146	9.5 ± 0.4 (+)			
$_{67}Ho$			8	
$_{68}Er$	7.8 ± 0.4 (+)			
$_{71}Lu$			~ 8	
$_{73}Ta$	6.1 ± 0.4 (+)	6 ± 1	6 ± 1	
$_{74}W$	2.74 ± 0.05 (+)	5.7 ± 0.6		
$_{76}Os$		15 ± 2	15	
$_{78}Pt$	11.2 ± 0.7 (+)	12 ± 1	12	
$_{79}Au$	7.3 ± 0.1 (+)		10	
$_{80}Hg$	22 ± 2 (+)		13 ± 1	
$_{81}Tl$	9 ± 2	10.0 ± 0.5	9	
$_{82}Pb$	11.5 ± 0.2 (+)	11.4 ± 0.1	11	
$_{83}Bi^{209}$	9.35 ± 0.04 (+)	9.37 ± 0.03	9	
$_{90}Th^{232}$	12.8 ± 0.5 (+)	12.6 ± 0.2	12	
$_{92}U^{238}$	9.0 ± 0.5 (+)		10.5	

this will remain an active field for theoretical investigation for some time before the matter will be regarded as being essentially solved.

Table 3 lists most of the known slow neutron scattering cross sections for comparison with the above concepts. The main difficulty in using these figures for such a comparison is in the problem of separating the effect of individual nearby resonances from the more long range effects of the giant resonances. The thermal coherent cross section is listed together with the sign of the coherent scattering length, where known. A cross section $4\pi a^2(+)$ implies that the node of the (extrapolated) outside $r\psi$ occurs at $r=+a$. Negative scattering lengths are thus the exception, and show that φ'/φ at the nuclear surface is small and positive. FORD and BOHM [*14*] first called attention to the systematic variation of the scattering lengths with A. Referring to the table, the following features may be noted.

1. For H^1 the value of $4\sigma_{f.a.}$ of 81.5 is very large compared with $\sigma_{coh}=1.79(-)$. This reflects the fact that the singlet and triplet scattering lengths are of opposite sign and give near cancellation for coherent scattering. The singlet scattering dominates in the total free atom cross section. The first size maximum is in this region.

2. The H^2 value of $\sigma_{coh}=5.4(+)$ is unusually large for such a small nucleus, but this is to be expected from the known binding energy of H^3.

3. The cross section of He^4 is near zero (+) while that of Li^7 is also small (−). The Li^6 value is uncertain and may be confused by the large $(n-\alpha)$ reaction effect. This region would be interpreted as the first size minimum.

4. Be^9, B, C, O, and F have relatively large scattering cross sections, with the Be^9 value largest. [The N^{14} value is omitted because of the (n, p) reaction.] This region would be interpreted as the second maximum.

5. Beyond ${}_9F^{19}$ to ${}_{20}Ca^{40}$ the values are systematically lower. In most cases σ_{coh} and $\sigma_{f.a.}$ are in reasonable agreement, indicating that both spin states (for initial nuclear spin $\neq 0$) agree. This decreases the probability that a single

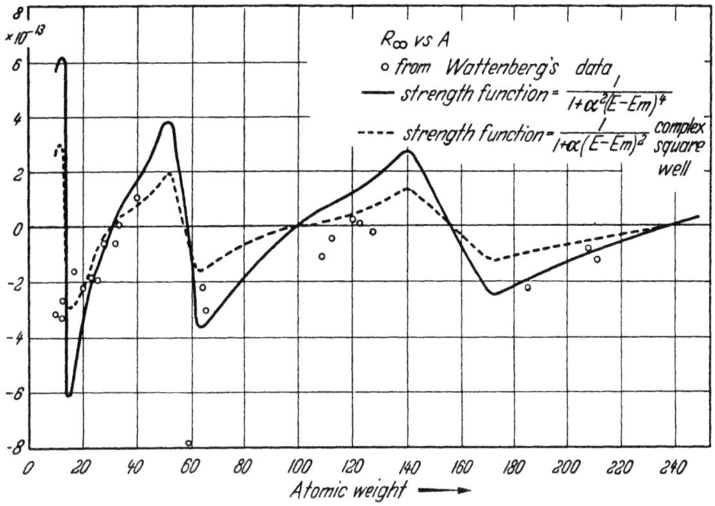

Fig. 31. Curve prepared by E. Vogt, report BNL 331 (c-21), comparing the variation of R_∞ vs. A predicted using the two indicated forms for the shape of the strength function in the intermediate coupling theory to the experimental values deduced from [*ANL*−γ*n* *1*]. Here R_∞ is related to the shift in the average value of the R matrix for neutron scattering from the value it would have for the uniform model. The average behavior near individual nuclear levels is $R \approx -R_0 \cot(\pi E/D) + R_\infty$, where E = distance from nearest level. The cross section averaged over levels is $\bar{\sigma} = 4\pi[(a-R_\infty)^2 + \lambda R_0]$, where a = nuclear radius.

resonance in one spin state dominates the behavior. The values for ${}_{13}Al^{27}$ (1.5+), ${}_{16}S^{32}$ (1.2+), and ${}_{18}A$ (0.5+) are particularly low. Thus this may be taken as the region of the second minimum.

6. In the region ${}_{21}Sc^{45}$, ${}_{22}Ti^{46\text{ to }50}$, ${}_{23}V^{51}$, ${}_{24}Cr$, and ${}_{25}Mn^{55}$ the cross sections fluctuate between large and small with σ_{coh}, $\sigma_{f.a.}$, and σ_s (< 1 kev) not agreeing in most cases. This suggests that the size is near a critical value where small changes in the effective Kb cause large changes in the scattering length.

7. The remarks of the preceding section also apply to some extent in the region from ${}_{26}Fe$ to ${}_{34}Se$. There is, however, an unusually large density of high values for the scattering cross section. There are nine cases of values $\gtrsim 8$ for $\sigma_{f.a.}\left(\frac{A+1}{A}\right)$. This region can be taken as the third maximum.

8. In the region from ${}_{37}Rb$ to ${}_{58}Ce$, with a few exceptions, the general trend with increasing A seems to be a slow decrease in the cross sections. Both ${}_{58}Ce^{140}$ and ${}_{58}Ce^{142}$ have $\sigma_{scatt} \approx 2.7$, for example.

9. The region around $A = 150$ to 170 has unusually large $\gamma^2_{\alpha\lambda}/D$ values associated with a giant resonance (see Fig. 33). The dense resonances in the low energy region make it relatively difficult to draw meaningful conclusions about the scattering in this region. The low values for ${}_{58}Ce^{140,142}$ indicate that the third minimum comes near this region of A.

10. The values of σ_s (<1 kev) for $_{76}$Os, $_{78}$Au, and $_{80}$Hg are systematically large compared to the region $_{51}$Sb to $_{58}$Ce, and there is a decreasing trend going from $_{76}$Os to $_{83}$Bi^{209}. Thus the fourth maximum might be identified with the beginning of this region. All slow neutron effects refer to $l=0$, of course.

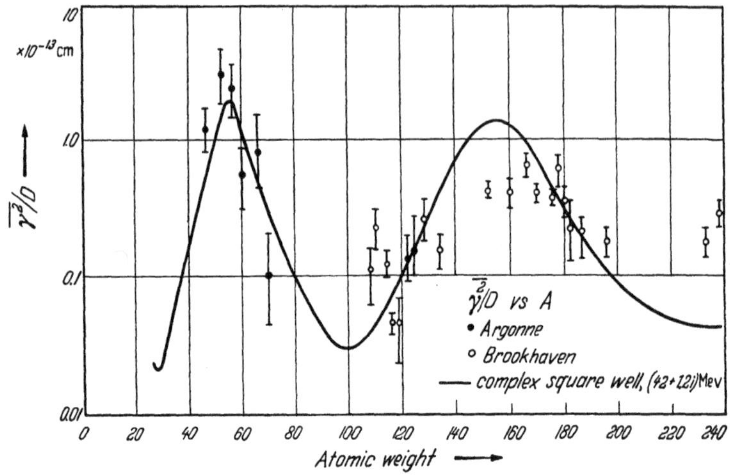

Fig. 32. Curve prepared by E. Vogt, BNL 331 (c-21) showing the behavior of the strength function expressed as $\overline{\gamma^2}/D$, which is proportional to $\overline{\Gamma_n^0}/\bar{D}$, as a function of A as indicated by the experimental results. This shows the giant (size) resonance peaking for $A\sim 60$ and 160. Fig. 33 gives a better presentation of the results near $A=160$.

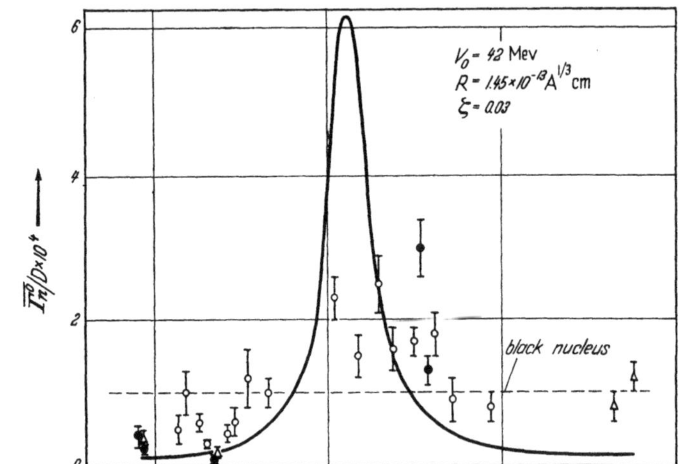

Fig. 33. Curve from [BNL—fc 2] showing the variation of Γ_n^0/D with A in the region of the $A\sim 160$ giant (size) resonance. This curve supplements Fig. 32.

Fig. 31 was presented by E. Vogt[1] who compares the predictions of the theory of Lane, Thomas, and Wigner with the photoneutron results of Wattenberg in 1947 [$ANL-\gamma n 1$].

Fig. 32 was also presented by Vogt at the same meeting to demonstrate the variation in the reduced neutron width divided by the level spacing, γ^2/D,

[1] At a conference on the Statistical Aspects of the Nucleus held at Brookhaven National Laboratory, January 24 to 26, 1955, and reported in BNL 331 (C-21).

showing resonances near $A=55$ and 60. Fig. 33 was given in [*BNL*—*fc* 2] and shows a more detailed comparison of theory and experiment. The experi-

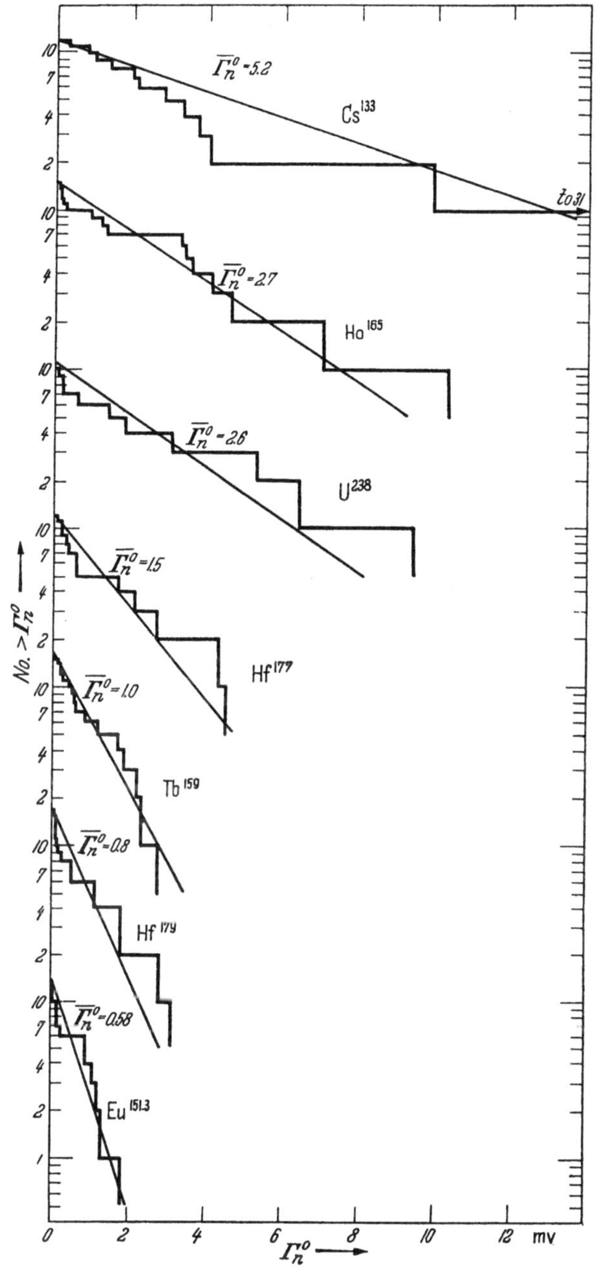

Fig. 34. Curves from [*BNL*—*fc* 2] showing the number of analyzed resonances having $\Gamma_n^0 >$ the value of the abscissa for each of several isotopes or elements.

mental peak is less sharp than the theoretical one. This could possible be due to the large quadrupole distortions of nuclei in this region.

8. Variation of $\Gamma_n^0/\bar{D}$ for a single isotope. Various semi-arbitrary "plausible" suggestions or "theories" have been advanced as to the expected distribution law for Γ_n^0/D for a given isotope and spin state. SCOTT[1] has suggested that Γ_n^0 should follow a simple exponential distribution. Alternate proposals are that

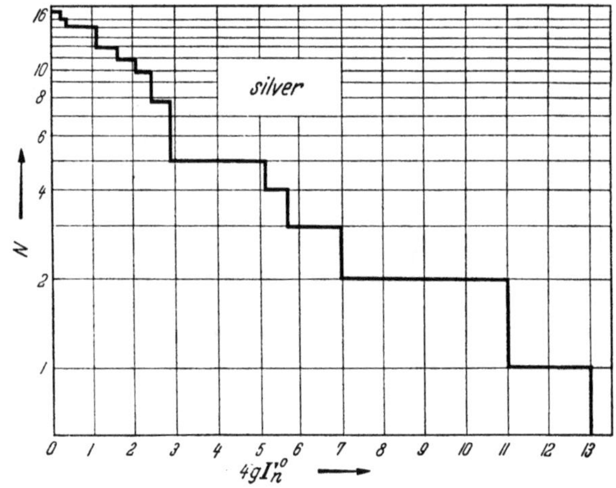

Fig. 35. Curves similar to those of Fig. 34 showing the number of levels having $4g\,\Gamma_n^0 >$ the abscissa value for Ag ($4g \approx 1$). The results are mainly those using the new Columbia Nevis velocity spectrometer.

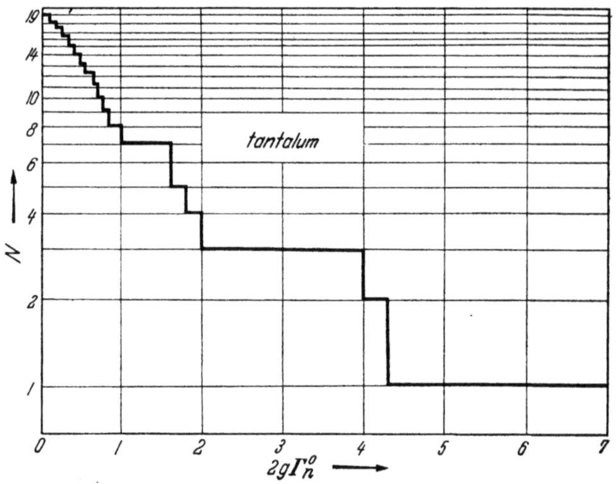

Fig. 36. Results similar to those of Fig. 35, but for Ta ($2g \approx 1$).

$\sqrt{\Gamma_n^0}$ should have an exponential distribution, or that $\gamma_n = \sqrt{\Gamma_n^0}$ should have a GAUSSian distribution. Fig. 34 shows the number of analyzed levels, for each of a number of isotopes, having Γ_n^0 greater than the value indicated. This was analyzed [*BNL—fc 2*] in a fashion intended to test the exponential distribution law for Γ_n^0. Figs. 35 and 36 give the Columbia results for silver and tantalum. Here $2g \approx 1$ for Ta and $4g \approx 1$ for Ag, since g was taken to include both spin and isotope factors. Fig. 37 shows results of the *MTR—fc* group for U^{235} [*MTR—fc 1*]. Curve A is the integral curve corresponding to a GAUSSian for $\sqrt{\Gamma_n^0}$, and curve B

[1] J. M. C. SCOTT: Phil. Mag. **45**, 1322 (1954).

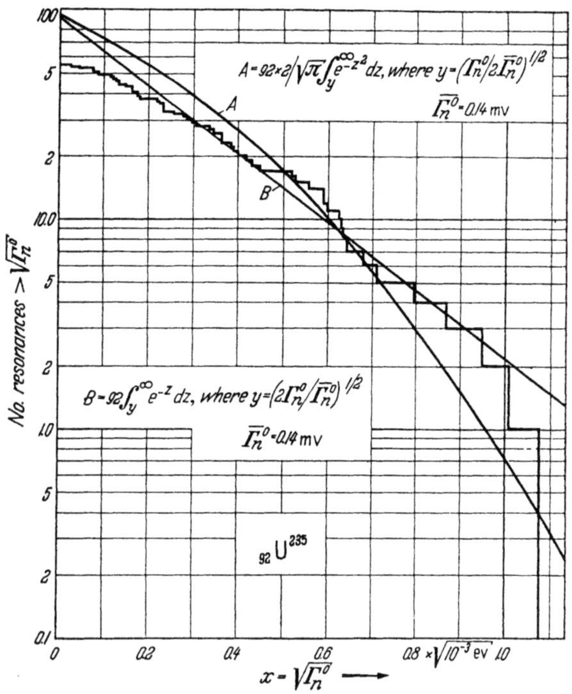

Fig 37. Results similar to those of Figs. 34 to 36 for U^{235} obtained by the [*MTR—fc*] group (O. D. Simpson, R. G. Fluharty and F. B. Simpson, submitted for publication in the Physical Review).

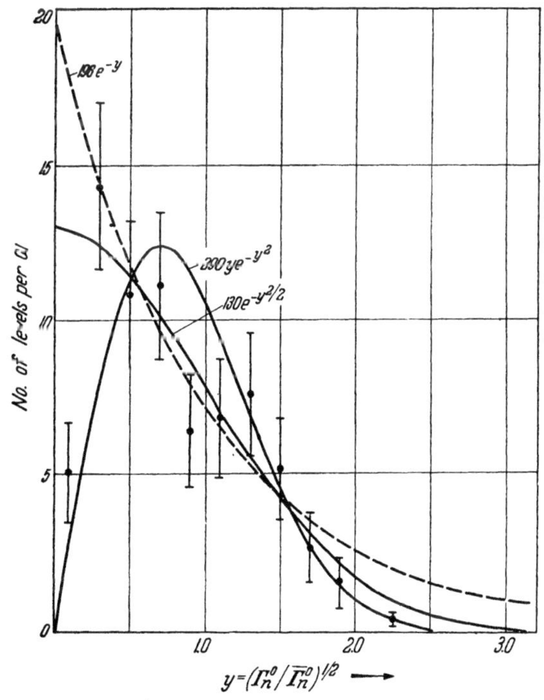

Fig. 38. Curve prepared by D. J. Hughes and J. A. Harvey (submitted for publication in the Physical Review) combining the results of the measured distribution of $(\Gamma_n/\bar{\Gamma}_n^0)^{\frac{1}{2}}$ values for many isotopes. Resonances having $\Gamma_n/\bar{\Gamma}_n^0$ very small may be missed experimentally, and this should be considered in interpreting the results.

is for an exponential in $\sqrt{\Gamma_n^0}$. Figs. 38 and 39 are from a paper by HUGHES and HARVEY[1]. They combined the results for $\Gamma_n^0/\overline{\Gamma_n^0}$ for many levels to increase the statistical accuracy. Fig. 38 gives a differential distribution, while Fig. 39 gives an integral distribution. In studying these various results, it should be noted that levels having small Γ_n^0 are much more likely to be missed than levels having

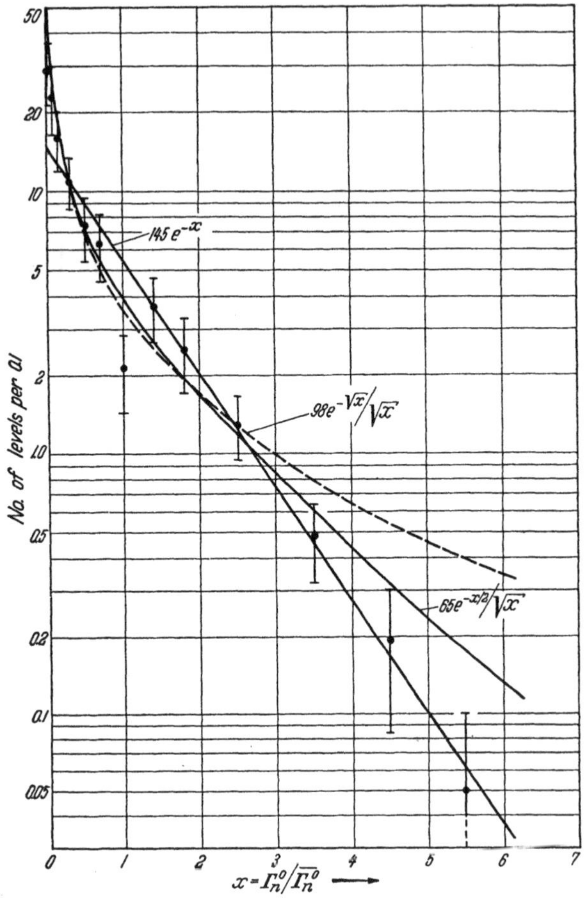

Fig. 39. A plot of the same results as in Fig. 38, but for $\Gamma_n^0/\overline{\Gamma}_n^0$, rather than the square root of this quantity.

large Γ_n^0. Also, if two or more levels are counted as one, they will contribute a larger Γ_n^0 than proper. When Γ_n is obtained from transmission measurements using a "thin sample" approximation, the effect of not resolving levels may affect the level count, but not the $\sum \Gamma_n^0$. Figs. 40 and 41 illustrate this for Tm^{169} [*BNL—fc 2*]. Figs. 17 and 21 show the Columbia results for Ag and Ta for the number of levels $< E$ vs. E. In such figures one looks for a systematic drooping at high E over and above any statistical fluctuations in level spacings.

9. Variation of Γ_γ with A. Careful measurements of level parameters for low energy resonances permit Γ_γ to be separately evaluated. This work has been mainly carried out by the BNL—cs group. Fig. 42 was given by H. H. LANDON [*BNL—cs 11*] to summarize the experimental results for the variation of

[1] D. J. HUGHES and J. A. HARVEY: Phys. Rev. (in press).

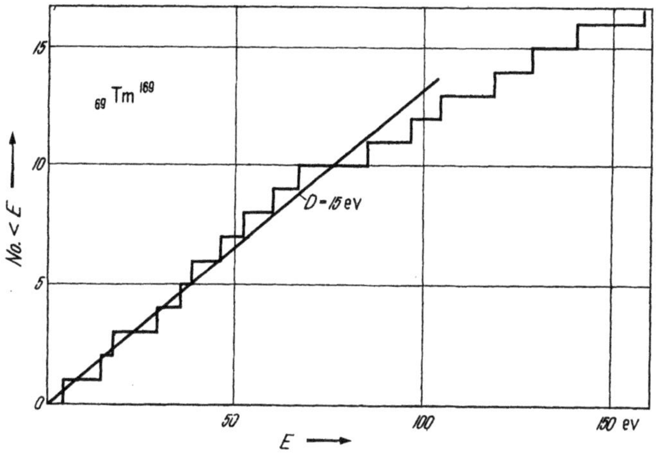

Fig. 40. Plot for Tm^{169} from [*BNL—fc 2*] showing the number of levels observed for $E <$ abscissa value to demonstrate that levels are probably not always observed or resolved above ~ 70 ev. This figure is to be compared with Fig. 41.

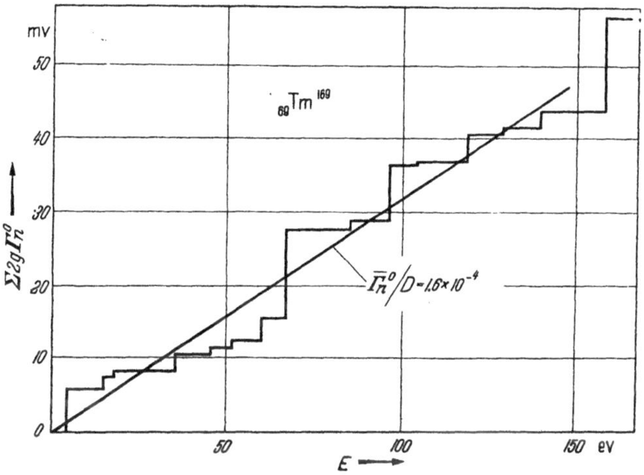

Fig. 41. Plot supplementing Fig. 40 showing the sum of the reduced neutron widths for levels having $E <$ abscissa value. This shows that this quantity is more reliable to ~ 150 ev. Although individual levels are not always resolved above ~ 70 ev, the Γ_n^0 values are roughly linearly additive in the analysis of the experimental results.

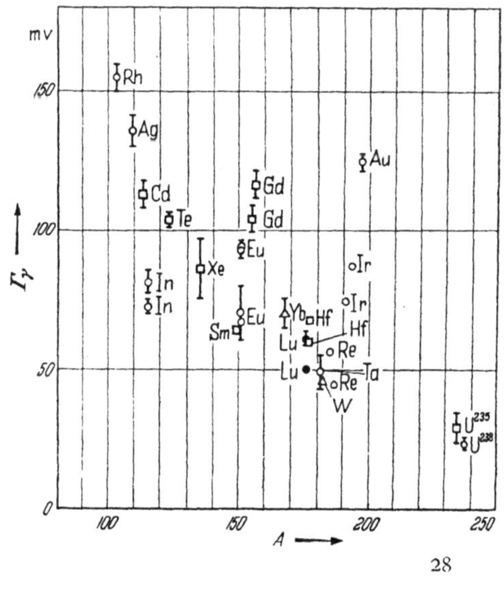

Fig. 42. Measured Γ_γ values for well resolved levels vs. A (from [*BNL—cs 11*]). There appears to be relatively little variation of the Γ_γ values for a given isotope, and Γ_γ shows a slow decrease with A.

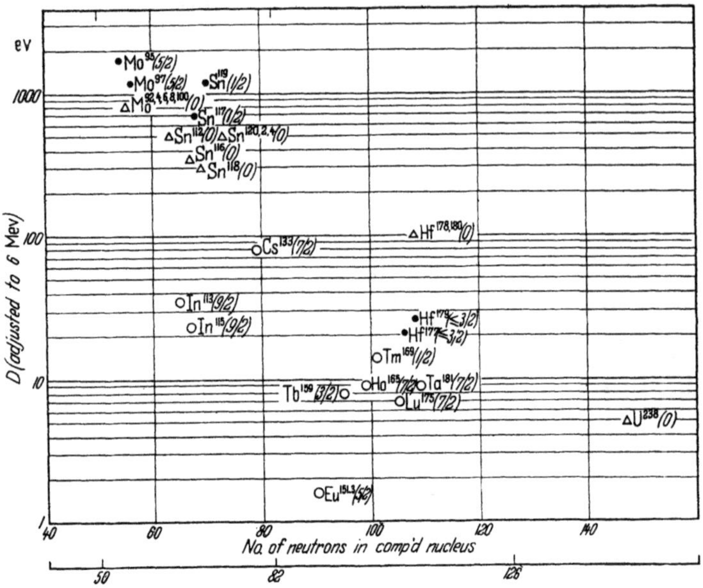

Fig. 43. The level spacings, D, adjusted to 6 Mev, as a function of neutron number in the compound nucleus. Each target nucleus is indicated and its spin is given in parentheses. The symbols ○, ●, and △ refer to odd Z—even N, even Z—odd N, and even Z—even N target nuclei respectively. Points labelled with several isotopic numbers are averages for the isotopes. (From [*BNL—fc 2*].)

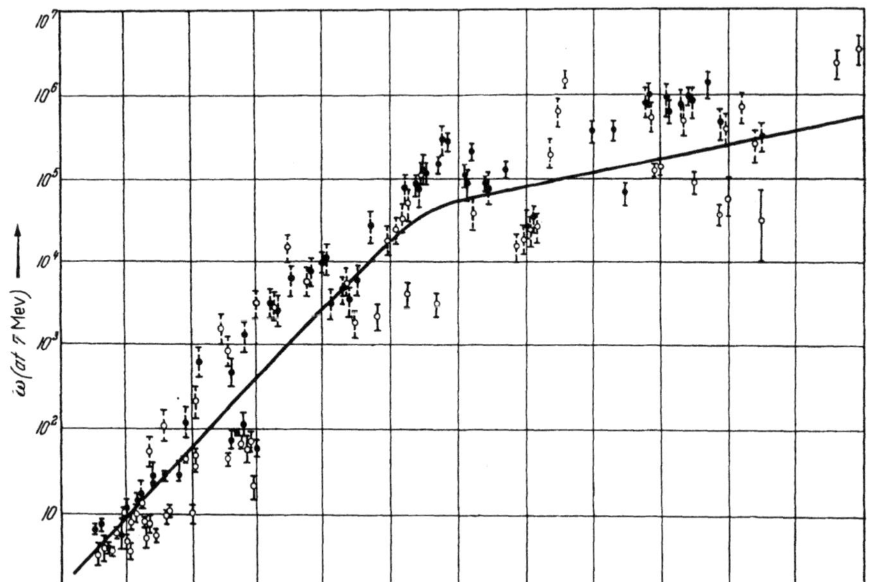

Fig. 44. The observed level densities (number of levels per Mev), extrapolated to 7 Mev, vs. atomic number. ϕ ϕ — obtained from the total level density of light nuclei ($A \leq 60$) and the spacings of slow neutron resonances ($A \geq 60$), ϕ ϕ obtained from the 1 Mev neutron cross sections, ϕ ϕ even—odd, odd—even or even—even compound nuclei, ϕ ϕ odd-odd compound nuclei. The full line in the density of levels at 7 Mev as given by WEISSKOPF's formula [3]. The vertical bars are a measure of the number of levels, N, from which each level density was derived; the probable error represented by these bars is $0.67\,(N)^{-\frac{1}{2}} \times 100\%$. The dashed bars for the 1 Mev data correspond to the 50% error quoted by HUGHES, GARTH, and LEVIN: Phys. Rev. **91**, 1423 (1953). This figure was prepared by ERICH VOGT, BNL 331 (c-21).

Γ_γ with A. The values of Γ_γ show relatively small fluctuation for levels of the same isotope, reflecting the fact that such transitions can occur to any of a large number

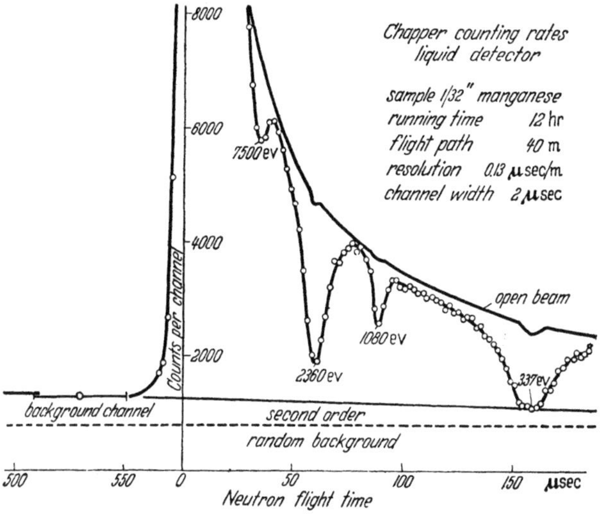

Fig. 45. This, and Figs. 46 and 47, are examples of the Argonne fast chopper results [*ANL—fc 4*]. This shows the directly measured counts vs. timing, with and without a Mn sample at the transmission position.

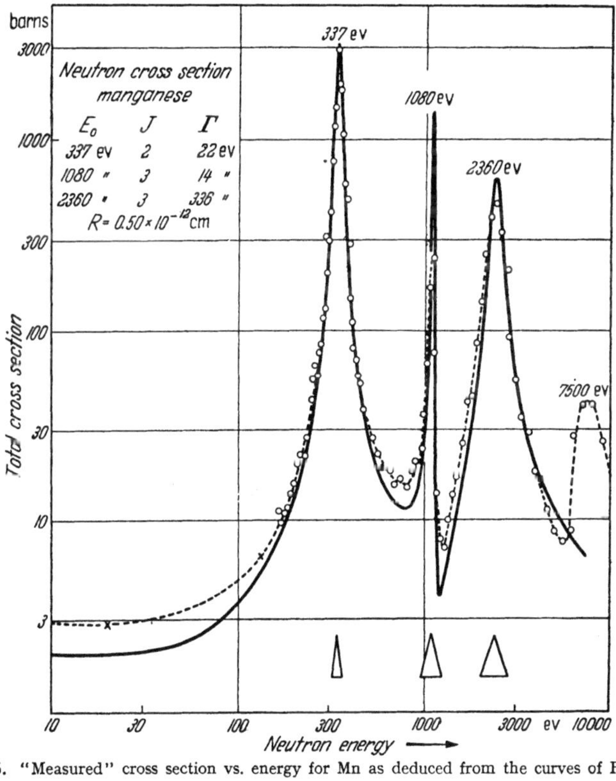

Fig. 46. "Measured" cross section vs. energy for Mn as deduced from the curves of Fig. 45.

of excited states above the ground state. The results can be made plausible by theoretical reasoning[1].

[1] See LEVIN and HUGHES: Phys. Rev. **101**, 1328 (1956) for a more detailed discussion.

10. Level densities. Fig. 43 shows a plot [*BNL—fc 2*] of level spacings vs. A, adjusted by the empirical WEISSKOPF formula to a common basis of 6 Mev excitation above the ground state. Fig. 44 is a plot presented by E. VOGT at the

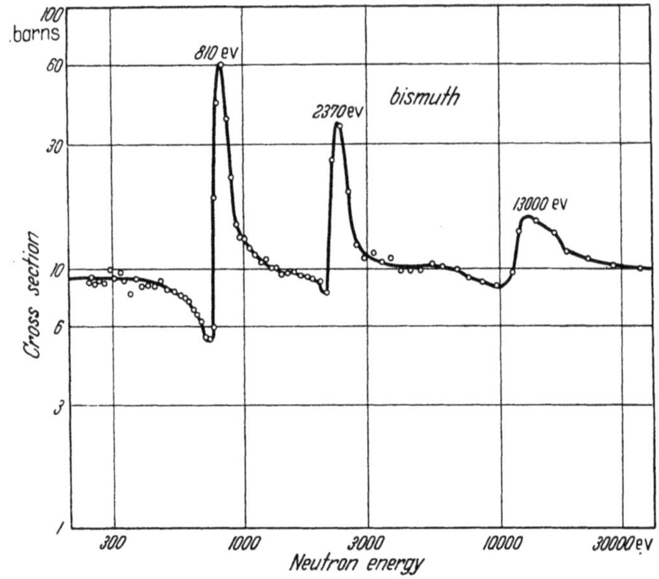

Fig. 47. Similar "measured" cross section vs. energy curve for bismuth from [*ANL—fc 4*].

Brookhaven Conference of January 1955, BNL 331 (c—21), on the basis of more data and corrected to give level density $\omega \equiv 1/D$ at 7 Mev excitation. Both sets of data show evidence of shell structure effects superimposed on a smooth trend.

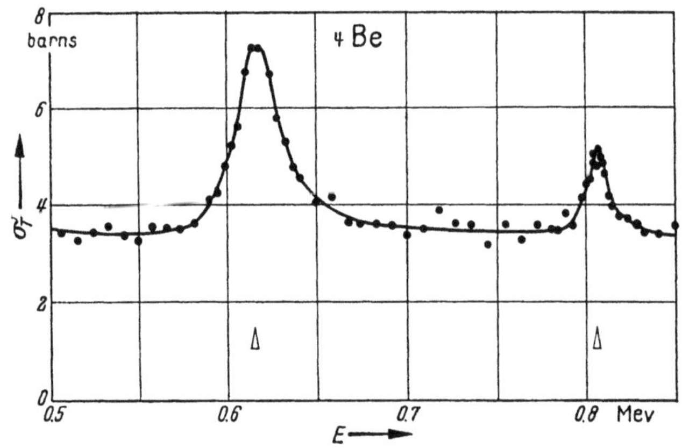

Fig. 48. The "measured" total cross section of $_4$Be from 0.5 to 0.85 Mev showing ~200 kev level spacing and large Γ_n from [*MIT 11*]). Figs. 48 to 63 are from the large compilation of neutron cross sections in BNL 325.

The reader can also refer to Tables 1 and 2 for a review of the experimental evidence. The change from large to small spacings (> 1 kev to < 100 ev) shows up in the type of data which shows resonance. For ≫1 kev level spacing, the levels are seen by intermediate energy neutron techniques, while for < 1 kev spacing they are seen using slow neutron techniques. Figs. 45 to 63 show selected "measured" cross sections vs. energy for a range of elements. Note the general comments earlier in this article concerning the vast difference between such "measured" cross sections near resonances, and true cross sections in all but very

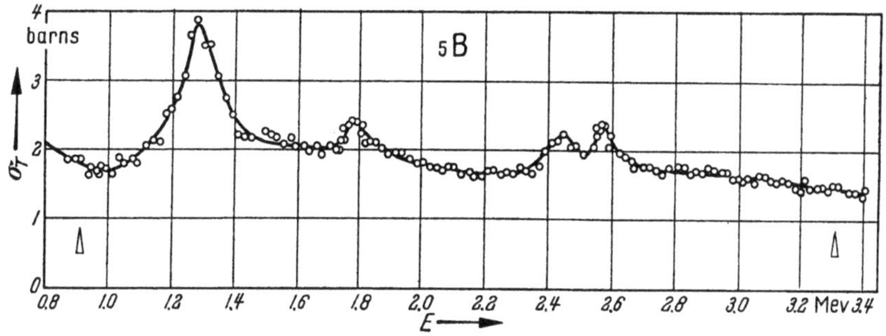

Fig. 49. The "measured" total cross section of $_5$B. (From [*Wis 12, 19*].)

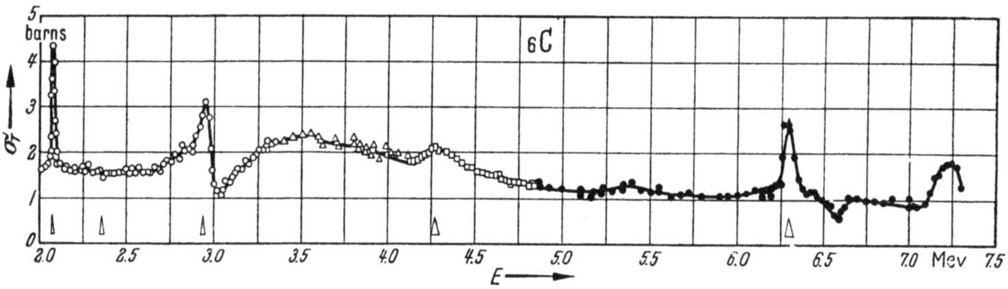

Fig. 50. The "measured" total cross section of $_6$C from 2 to 7.3 Mev. Results from [*Wis 19*], [*Minn 2*], [*BNL—ac 1, 2*] and [*LA 3, 5*].

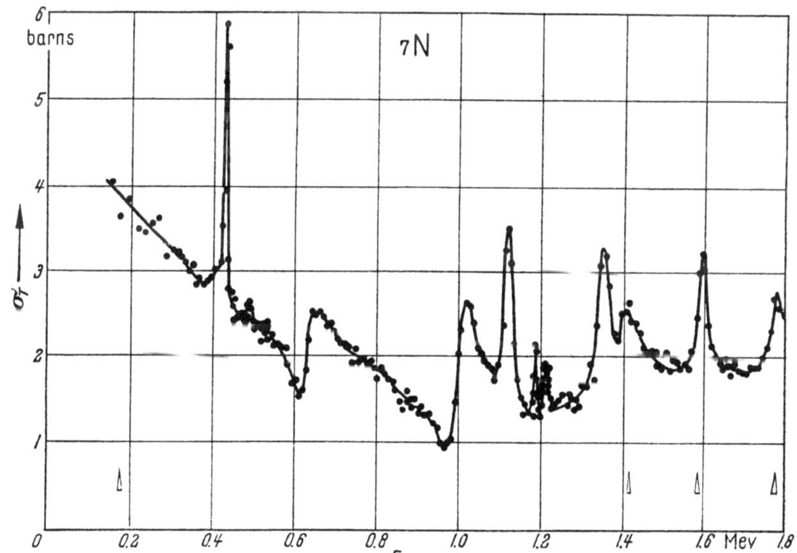

Fig. 51. The "measured" total cross section of $_7$N from 0.2 to 1.8 Mev. Results from [*MIT 2*].

favorable cases. Fig. 45 to 47 show the [*ANL—fc*] results for $_{25}$Mn and $_{83}$Bi. Note that, if anything, Bi shows a larger $\bar{D}$ than Mn.

Detailed treatments of the theory of level densities will be found in the references listed under this heading. A few remarks and emphasis may be in order,

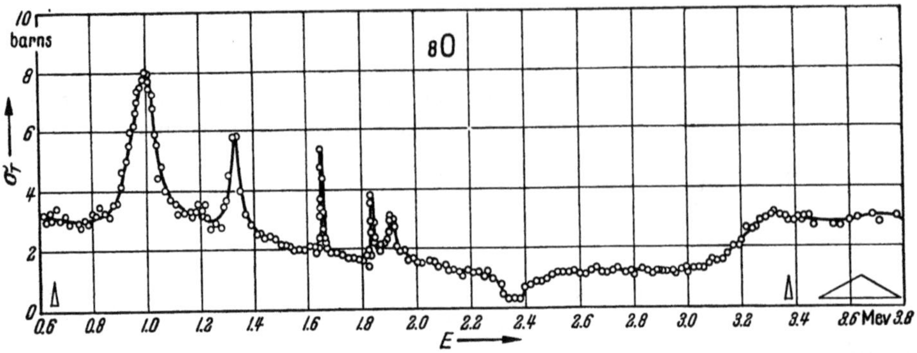

Fig. 52. The "measured" total cross section of $_8$O from 0.6 to 3.8 Mev. Results from [*Wis 3, 12, 19*], and [*Minn 2*].

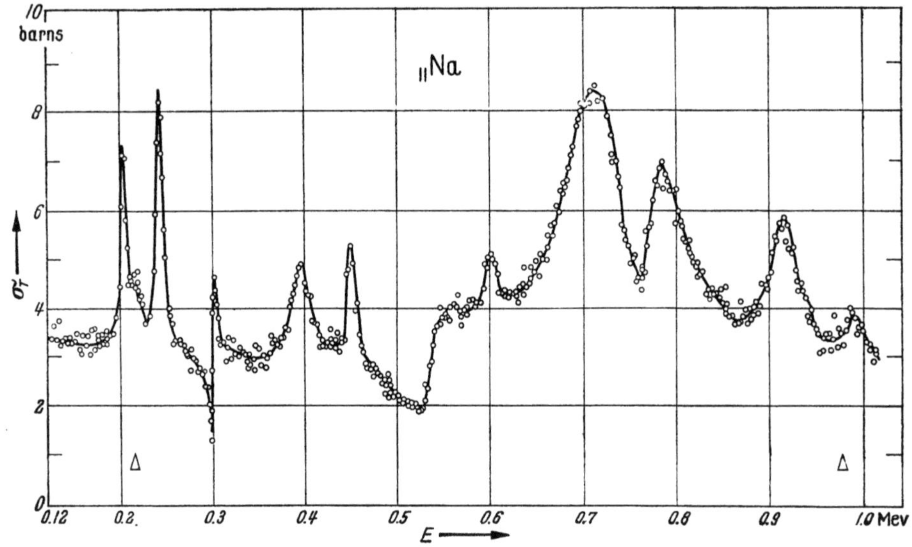

Fig. 53. The "measured" total cross section of $_{11}$Na from 0.12 to 1.0 Mev. Results from [*MIT 3*].

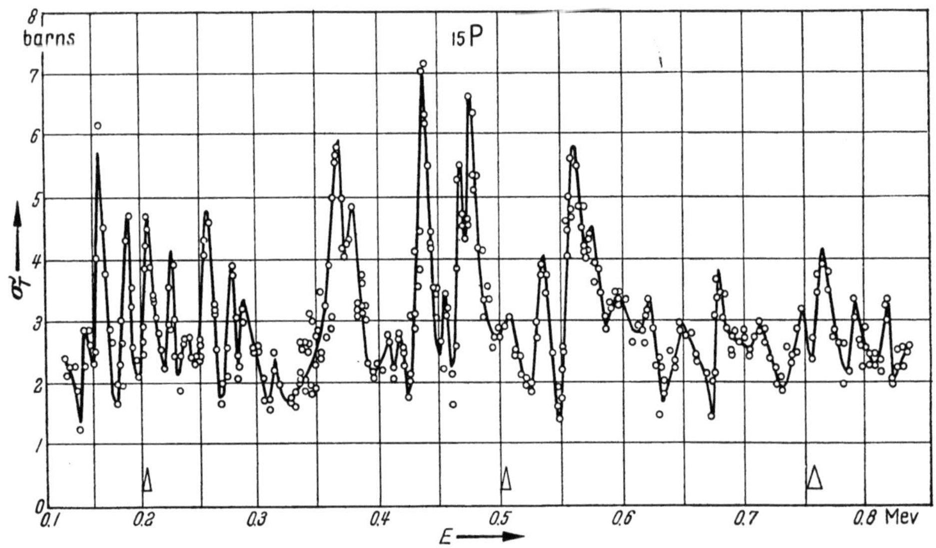

Fig. 54. The "measured" total cross section of $_{15}$P from 0.12 to 0.82 Mev. Results from [*MIT 7*].

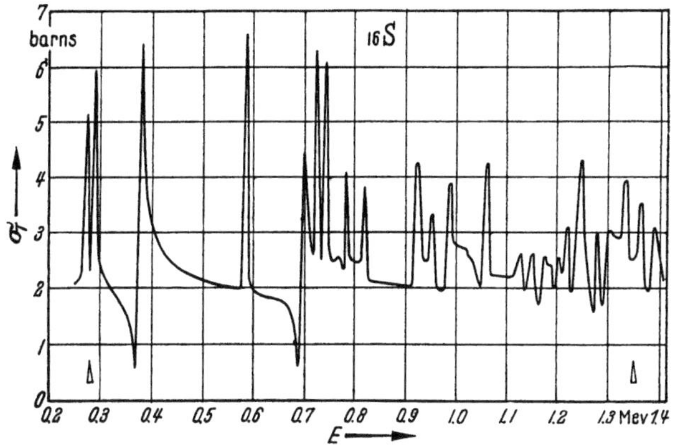

Fig. 55. The "measured" total cross section of $_{16}$S from 0.26 to 1.4 Mev. Results from [*Wis 5, 9*].

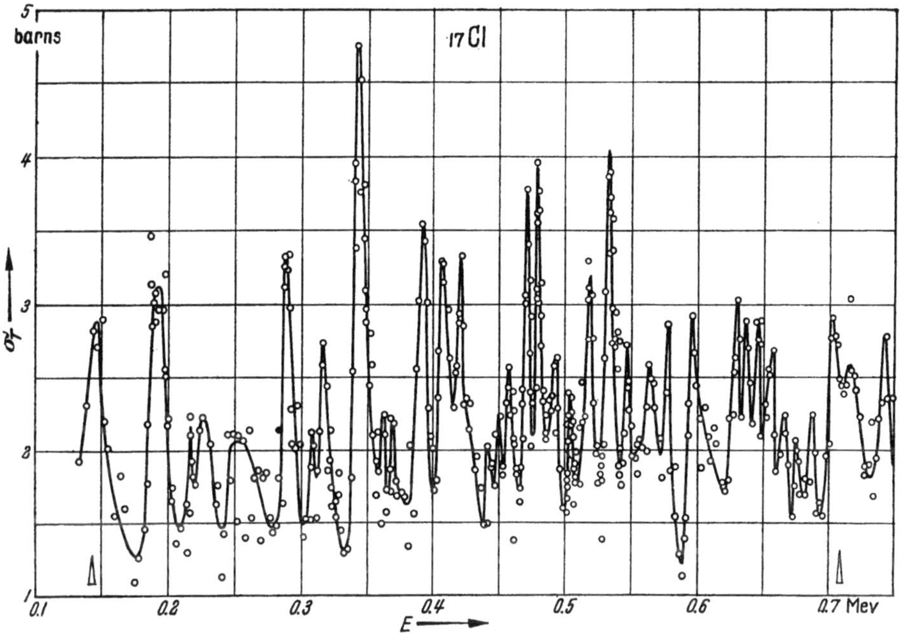

Fig. 56. The "measured" total cross section of $_{17}$Cl from 0.13 to 0.75 Mev. Results from [*MIT 5*].

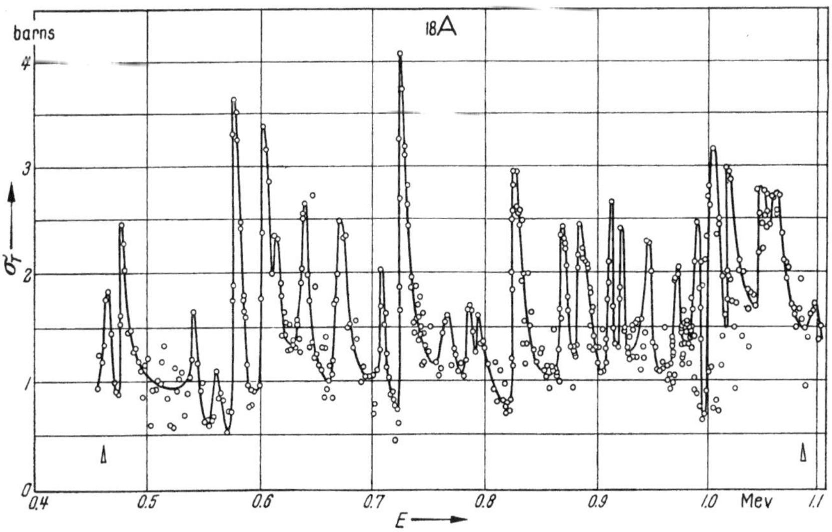

Fig. 57. The "measured" total cross section of $_{18}$A from 0.45 to 1.1 Mev. Results frcm [*MIT 6*].

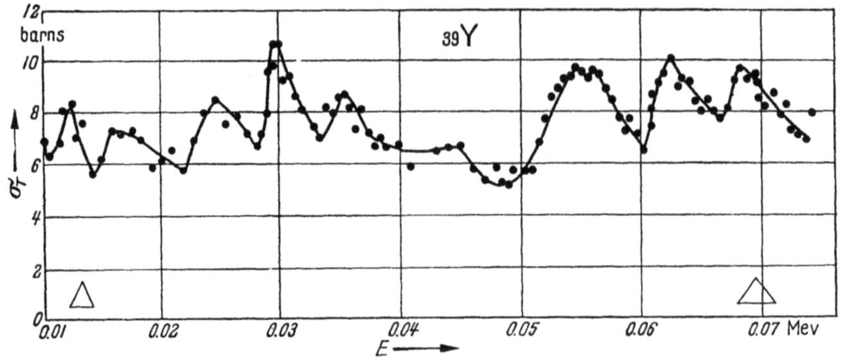

Fig. 58. The "measured" total cross section of $_{39}$Y from 10 to 75 kev. Results from [*Duke 1*].

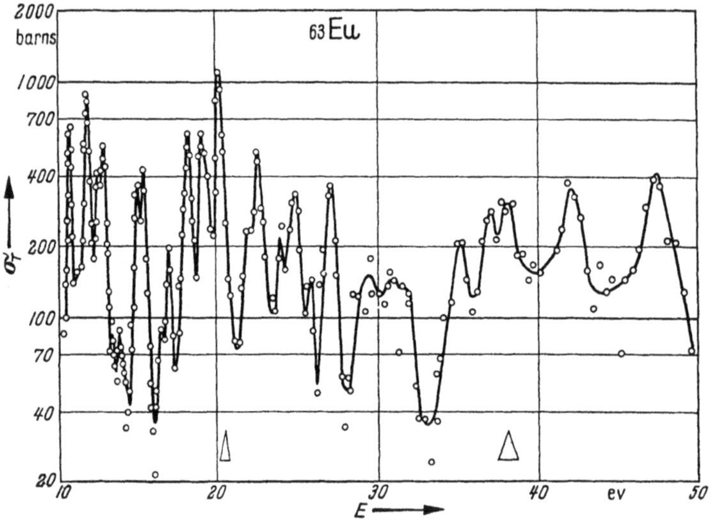

Fig. 59. The "measured" total cross section of $_{63}$Eu from 10 to 50 ev. Results from [*BNL—fc 2*].

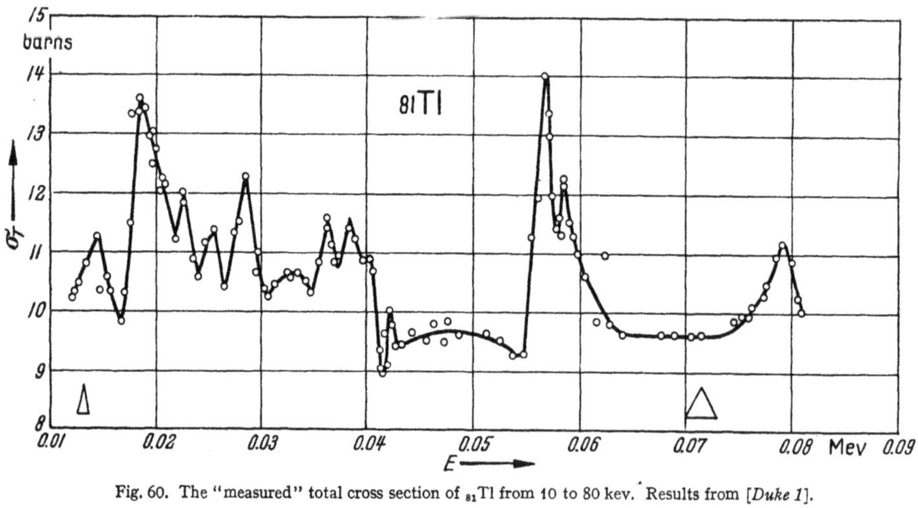

Fig. 60. The "measured" total cross section of $_{81}$Tl from 10 to 80 kev. Results from [*Duke 1*].

however, since most treatments use a full statistical treatment that ignores known relations from the shell model.

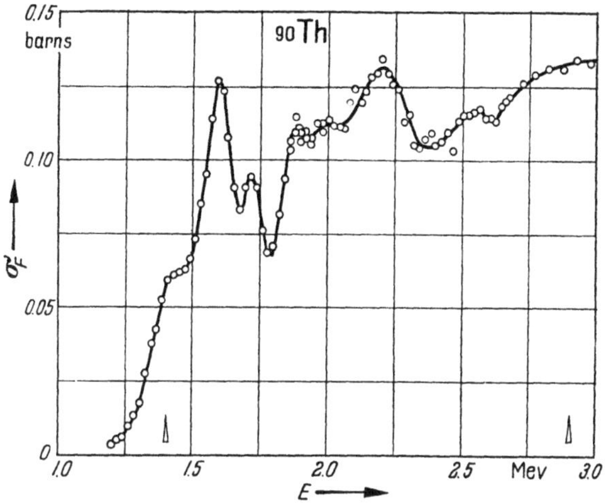

Fig. 61. The "measured" fission cross section of $_{90}$Th from 1.3 to 3.0 Mev. Results from [*LA-ac 20*].

As a simple model, let us consider an independent particle in a box approach where A nucleons are in a potential $\sim$40 Mev deep with the lowest A single particle states filled and the rest empty for the ground state configuration.

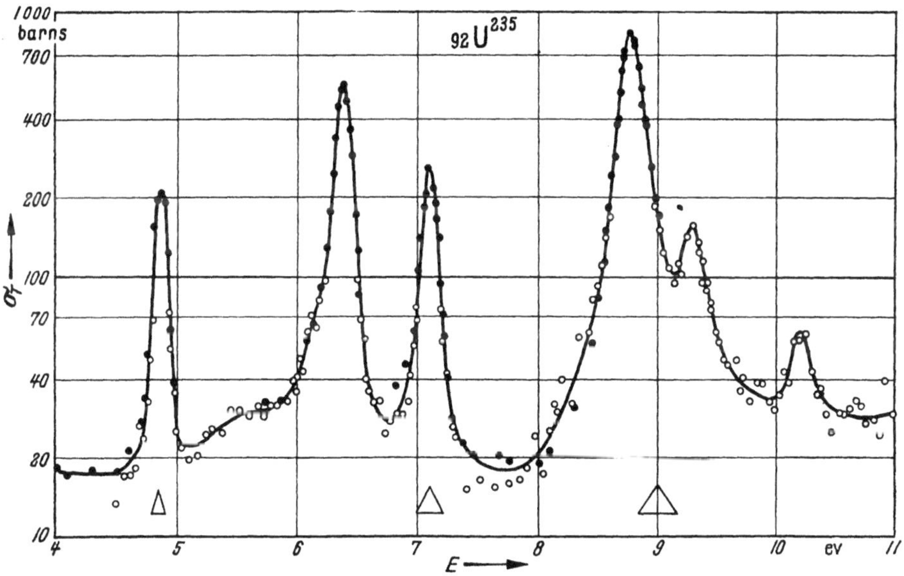

Fig. 62. The "measured" total cross section of $_{92}U^{235}$ from 4 to 11 ev. Results from [*BNL-fc 3*] and [*BNL-cs 10*].

Neglecting spin-orbit effects and j degeneracy, we assume that each space state can hold two neutrons and two protons (space + spin + isotopic spin define a single particle state). Since the last filled state has $\sim$8 Mev binding, the well is $\sim$40 Mev deep, and the total number of states varies roughly as $(E-V)^{\frac{3}{2}}$, the single particle space states are spaced $\sim$80 Mev/A for atomic number A. If we divide this by four we get $\varepsilon\sim$20 Mev/A as the spacing of a type of single particle state that would hold only one nucleon. The probability that a single particle state be occupied at temperature kT (in energy units) $=\tau$ is approximately

$$[1+e^{-E_j/\tau}]^{-1}$$

where E_j = energy of single particle state j above that of the last filled state at $\tau=0$. The entropy of the system may be calculated, see Bethe [1], and thus the density ϱ of levels of the complete system per unit energy. This will not be the observed level density for many reasons. Aside from the fact that the model

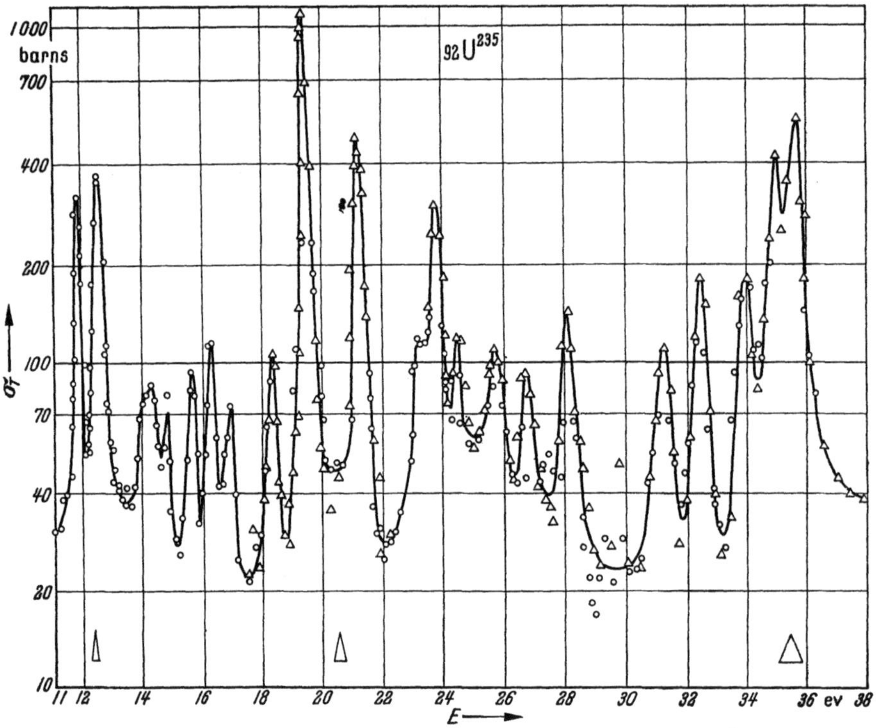

Fig. 63. The "measured" total cross section of $_{92}U^{235}$ from 11 to 38 ev. Results from [*MTR*–*fc*-*1*] and [*BNL*–*fc* *3*].

is very inexact, the density ϱ includes all possible total spin and parity for the nucleus, while the experiments using $l=0$ neutrons select $I\pm\frac{1}{2}$, no parity change. The net compromise effect is usually that Weisskopf's suggestion that the predicted exponential dependence of ϱ on $U^{\frac{1}{2}}$ (where U = nucleus energy above the ground state) be maintained, but that things otherwise be left somewhat flexible. The usual formula for the expected observed level density, ω, is

$$\omega=\frac{c_1}{\varepsilon_0}\,e^{(c_2\,U/\varepsilon_0)^{\frac{1}{2}}}.$$

On the simplest theory ε should vary as A^{-1}, but effects, such as those mentioned above, would be expected to greatly decrease the dependence on A. In Fig. 44 it is seen that there is relatively little change in ω from $A\sim 100$ to $A\sim 240$ if fluctuations are ignored. If one tries to take the Mayer-Jensen shell model seriously, it is noted that high j goes with individual particle states for large A, so the spacing per j state is increased by an average $(2j+1)$ as A^2 increases. The number of ways individual large j value can couple to give a net $J\sim 0$ are relatively small.

Another effect which has seemed to the present author to be of major importance is the shell structure fluctuations in the individual ε. This effect was considered by MARGENAU [24] using the shell pattern of a spherical box without l-s coupling. Briefly, the effect may be demonstrated in the following somewhat oversimplified fashion. Suppose the single particle level spacing is ε within each major shell, but there is a gap $\varepsilon' \gg \varepsilon$ at the close of a major shell before starting the next shell. The statistical model shows that the important factor for ω is U in units of a characteristic ε_0. For a partly closed shell this $\varepsilon_0 \approx \varepsilon$. For a just closed shell ε_0 is somewhere between ε and ε'. The value of ε_0 remains large compared to ε until the nucleon number differs appreciably from the closed shell value either above or below. It is instructive to let $\varepsilon' =$, say, 5ε and count the number of independent ways the system can be excited in one, two, three, etc. particle excitation states for $U = \varepsilon$, 2ε, 3ε, etc. for the last filled state at $\tau = 0$ chosen to be varying distances above or below the closed shell configuration. Such effects might also occur to a lesser degree for fluctuations in ε within a major shell. This approach essentially is one of modifying the purely statistical analysis to take account of the fact that one does have additional information about the system which can be used as input information in the estimates of how ω should vary with A.

Notes added in proof.

(Since the completion of the manuscript the following developments have occurred which seem of significance for this article.

1. Current theoretical interest is focussed on the work of BRUECKNER and his collaborators (BRUECKNER [8] and K. A. BRUECKNER, Phys. Rev. **103**, 172 (1956)) who have achieved some significant success in their attempt to develop nuclear theory starting from the basic two nucleon interaction. They reduce the problem in its later stages to that of LANE, THOMAS, and WIGNER [17]. See also developments by H. A. BETHE, Phys. Rev. **103**, 1353 (1956) who gives an extensive listing of earlier references.

2. The $MTR-fc$ and $BNL-fc$ results for U^{235} have now been published. SIMPSON, FLUHARTY, and SIMPSON, Phys. Rev. **103**, 971 (1956), and PILCHER, HARVEY, and HUGHES, Phys. Rev. **103**, 1342 (1956). The first of these papers gives the first detailed report of the excellent work of the $MTR-fc$ group. The $BNL-fc$ paper presents their analysis curve, Fig. 10, for the first time. It also gives a comparison with theory for the fluctuations in the ratio of the fission width, Γ_f, to the average level spacing, $\overline{D}$, and for $\overline{\Gamma_f}/\overline{D}$. Further reports on fission cross section, and other neutron cross section measurements will be found in Vol. 2 and 4 of the United Nations published report of the 1955 Geneva conference entitled "Proceedings of the International Conference on the Peaceful Uses of Atomic Energy". In Vol. 4, p. 155, WHEELER mentions his statistical theory that predicts that $\overline{\Gamma_f}/\overline{D}$ should equal $N/2\pi$. Here $N =$ effective number of fission channels. This formula is analogous to Eq. (4.20a) for the single channel neutron escape if the "barrier reflection coefficient" k/K is set equal to unity. The not too conclusive experimental results give $N \sim 0.5$ on this test, suggesting that some additional factor less than unity should appear as a factor multiplying $N/2\pi$. An alternate statistical analysis has been made of the distribution of $\Gamma_f/\overline{D}$ values to determine an effective number, N, of open fission channels. This interesting method of approach seems to have been initiated by R. G. THOMAS and A. M. LANE and may be briefly described as follows. In discussing the expected distribution for $\Gamma_N^0/\overline{D}$, a Gaussian distribution for $\gamma_{nn} = \sqrt{\Gamma_n^0}$ was favored

having zero as the most probable value. If N independent channels lead to the same final state f and their γ_{nf} are uncorrelated, one may assign the individual γ_{nf} of each channel orthogonal directions in an N dimensional space. The total $\Gamma_f =$ total γ_{nf}^2 then may be treated as the sum of the squares of the individual γ_{nf}^2 values and represents the square of the length of the vector to a given point in the N dimensional γ_{nf} space. If the individual $\overline{\gamma_{nf}}^2$ are equal, this gives a spherically symmetric Gaussian distribution for total γ_{nf} in an N dimensional γ_{nf} space. The size $\gamma_{nf}^{N-1} d\gamma_{nf}$ of a volume element then multiplies the Gaussian function for the distribution in *scalar* γ_{nf}. The expected distribution for Γ_f is then of the form $\Gamma_f^{\frac{N-2}{2}} \exp\{-\text{const}\ \Gamma_f\}$. For the neutron channel $N=1$, for γ emission N must be $\sim 10^4/a^2$ if Γ_γ shows fluctuations $\sim a\%$ about its mean value. The value $N \approx 4$ reported in the above reference should probably be decreased to 2 or less on the basis of discussions of more recent results reported at conferences at Amsterdam and at Harwell during July 1956.

3. The following papers give the first detailed report of the work of the Duke-acc group: JOHN H. GIBBONS, Phys. Rev. **102**, 1574 (1956) and NEWSON, GIBBONS, MARSHAK, WILLIAMSON, MOBLEY, TOLLER, and BLOCK, Phys. Rev. **102**, 1580 (1956).

4. The Columbia Nevis synchrocyclotron velocity selector group consists of J. RAINWATER, W. W. HAVENS jr., G. GRIMM, S. DESJARDINS, and J. ROSEN. E. MELKONIAN has assisted on some of the analysis of results.

References.

a) Theoretical.

The theory given in this article is not intended to represent an exhaustive review of the theoretical developments in this field, but to present the main concepts on an elementary level as a guide in interpreting the experimental material. The reader is referred to the article by G. BREIT in this series for a more complete theoretical treatment of the subject. The following partial listing of references includes most of the main developments. The later articles give references to most of the articles not included in this listing.

1. Resonance theory.

[1] BETHE, H. A.: Rev. Mod. Phys. **9**, 71 (1937).
[2] BETHE, H. A.: Phys. Rev. **57**, 1125 (1940).
[3] BLATT, J. M., and V. F. WEISSKOPF: Theoretical Nuclear Physics. New York: John Wiley & Sons, Inc. 1952. This is a general reference for all phases of the subject through 1951.
[4] BOHR, N., and F. KALCKAR: Kgl. danske Vid. Selsk. mat.-fys. Medd. **14**, 10 (1937).
[5] BOHR, A., and B. R. MOTTELSON: Kgl. danske Vid. Selsk. mat.-fys. Medd. **27**, 16 (1953), Appendix 5.
[6] BREIT, G., and E. P. WIGNER: Phys. Rev. **49**, 519, 642 (1936).
[7] BREIT, G.: Phys. Rev. **69**, 472 (1946).
[8] BRUECKER, K. A., R. J. EDEN and N. C. FRANCIS: Phys. Rev. **100**, 891 (1955).
[9] EISENBUD, L., and E. P. WIGNER: Phys. Rev. **72**, 29 (1947).
[10] EKSTEIN, H.: Phys. Rev. **101**, 880 (1956).
[11] FESHBACH, H., D. PEASLEE and V. F. WEISSKOPF: Phys. Rev. **71**, 145 (1947).
[12] FESHBACH, H., and V. F. WEISSKOPF: Phys. Rev. **71**, 145 (1947).
[13] FESHBACH, H., C. E. PORTER and V. F. WEISSKOPF: Phys. Rev. **96**, 448 (1954).
[14] FORD, K. W., and D. BOHM: Phys. Rev. **79**, 745 (1950).
[15] KAPUR, P. L., and R. PEIERLS: Proc. Roy. Soc. Lond., Ser. A **166**, 277 (1938).
[16] KOFOED-HANSEN, O.: Phys. Rev. **99**, 154 (1955).
[17] LANE, A. M., R. G. THOMAS and E. P. WIGNER: Phys. Rev. **99**, 693 (1955).
[18] SACHS, R. G.: Nuclear Theory. Cambridge: Addison-Wesley Publishing Company 1953. Chap. 10 and Appendix 3 cover the theory of nuclear reactions.
[19] TEICHMANN, T., and E. P. WIGNER: Phys. Rev. **87**, 123 (1952).
[20] THOMAS, R. G.: Phys. Rev. **97**, 224 (1955).
[21] VOGT, E.: Phys. Rev. **101**, 1792 (1956).

2. Statistical models of level density.

The listings below include a few references of particular interest on this subject. In addition, many of the references of the preceding group cover this subject, in particular, BETHE [*2*], and BLATT [*3*].

[*22*] BLOCH, C.: Phys. Rev. **93**, 1094 (1954).
[*23*] COHEN, B. L.: Phys. Rev. **92**, 1245 (1953).
[*24*] MARGENAU, H.: Phys. Rev. **59**, 627 (1941). The author considers the effect of shell structure on an independent particle model.
[*25*] WEISSKOPF, V., and D. H. EWING: Phys. Rev. **57**, 472, 935 (1940).

b) Experimental.

The tables in this paper and the reference listings in this section are largely taken, with some additions, from the comprehensive 328 page compilation by D. J. HUGHES and J. A. HARVEY: Neutron Cross Sections, Brookhaven National Laboratory Report BNL 325 (for sale by the Superintendent of Documents, U.S. Government Printing Office, Washington 25, D.C.). Their compilation is an outgrowth of a similar compilation, AECU-2040, of May 1952, prepared under the auspices of the United States Atomic Energy Commission's Neutron Cross Section Advisory Group. BNL 325 presents 261 pages, 10″×15″, showing cross sections vs. neutron energy using logarithmic cross section and energy scales. The curves give "experimental" cross sections as obtained from transmission measurements

$$\sigma = -(1/n) \ln T_E$$

where T_E is the measured transmission. Between resonances this gives a good indication of σ if the nearest level is at least one resolution width away. In the region of a resonance, however, the cross sections given are usually rather meaningless, except for a relatively small number of very low energy resonances that have been studied using a resolution width small compared to the level width. Usually, however, the peak cross section shown may be ten to a thousand times less than the true σ_0, or the doppler broadened σ_Δ. Even the area under the σ vs. E peak bears little relation to true value. For this reason, it is urged that the tables of resonance parameters be consulted for analyzed resonances, or that such curves otherwise merely be interpreted as indicating the presence of a single level, or two or more unresolved levels. (This subject is discussed in detail in the text.)

In preparing the table of resonance parameters, the author is greatly indebted to Dr. ROBERT CARTER, of Brookhaven National Laboratory, for making available unpublished cross section information which is being compiled for a supplement to BNL-325 to be issued soon. I also wish to thank Dr. REX FLUHARTY of the M.T.R. fast chopper group for making available some of the results of that group prior to publication.

The results for the angular distribution of the scattering of 1 Mev neutrons are from M. WALT and H. H. BARSCHALL, Phys. Rev. **93**, 1062 (1954). The references pertaining to cross section vs. energy values are listed as following according to laboratory and method. The listing is the same as the references for the cross section curves in BNL 325 with some addition.

ANL (Argonne National Laboratory).
1. FERMI, STURM and SACHS: Phys. Rev. **71**, 589 (1947).

ANL—ac (Argonne National Laboratory—accelerator).
1. BLAIR, J. M., and J. R. WALLACE: Phys. Rev. **79**, 28 (1950).
2. STUBBINS, W. F.: Phys. Rev. **84**, 902 (1951).
3. HIBDON, LANGSDORF jr. and HOLLAND: Phys. Rev. **85**, 595 (1952).
4. Unpublished.

ANL—cs (Argonne National Laboratory—crystal spectrometer).
1. STURM, W. J., and G. P. ARNOLD: Phys. Rev. **71**, 556 (1947).
2. STURM, W. J.: Phys. Rev. **71**, 757 (1947).

ANL—fc (Argonne National Laboratory—fast chopper).
1. SELOVE, W.: Phys. Rev. **84**, 869 (1951).
2. HARRIS, S. P.: Phys. Rev. **89**, 904 (1953).
3. BOLLINGER, HARRIS, HIBDON and MUEHLHAUSE: Phys. Rev. **92**, 1527 (1953).
4. BOLLINGER, DAHLBERG, PALMER and THOMAS: Phys. Rev. **100**, 126 (1955).
5. DAHLBERG and BOLLINGER: Phys. Rev. **95**, 645 (1954).
6. Unpublished.

ANL—fn (Argonne National Laboratory—fission neutrons).
1. HUGHES, GARTH and LEVIN: Phys. Rev. **91**, 1423 (1953).

ANL—γn (Argonne National Laboratory—photo-neutrons).
1. Fields, Russell, Sachs and Wattenberg: Phys. Rev. **71**, 508 (1947).
2. Hummel, V., and B. Hamermesh: Phys. Rev. **82**, 67 (1951).
3. Unpublished.

ANL—rd (Argonne National Laboratory—resonance detector).
1. Hibdon, C. T., and C. O. Muehlhause: Phys. Rev. **76**, 100 (1949).

ANL—sc (Argonne National Laboratory—slow chopper).
1. Brill, T., and H. V. Lichtenberger: Phys. Rev. **72**, 585 (1947).

Bar (Bartol Research Foundation).
1. Snowdon, S. C., and W. D. Whitehead: Phys. Rev. **90**, 615 (1953).

BNL—ac (Brookhaven National Laboratory—accelerator).
1. Poss, Salant, Snow and Yuan: Phys. Rev. **87**, 11 (1952).
2. Hafner, Hornyak, Falk, Snow and Coor: Phys. Rev. **89**, 204 (1953).
3. Toppel, B. J., and S. D. Bloom: Phys. Rev. **91**, 473 (1953).

BNL—cs (Brookhaven National Laboratory—crystal spectrometer).
1. Sailor, V. L., and L. B. Borst: Phys. Rev. **87**, 161 (1952).
2. Sailor, V. L.: Phys. Rev. **91**, 53 (1953).
3. Foote, Landon and Sailor: Phys. Rev. **92**, 656 (1953).
4. Landon, H. H., and V. L. Sailor: Phys. Rev. **93**, 1030 (1954).
5. Christensen, R. L.: Phys. Rev. **92**, 1509 (1953).
6. Sailor, Landon and Foote: Phys. Rev. **93**, 1292 (1954).
7. Sailor, Landon and Foote: Phys. Rev. **96**, 1014 (1954).
8. Wood, Landon and Sailor: Phys. Rev. **98**, 639 (1955).
9. Landon and Sailor: Phys. Rev. **98**, 1267 (1955).
10. Igo, G.: Phys. Rev. **100**, 1338 (1955).
11. Landon, H. H.: Phys. Rev. **100**, 1414 (1955).
12. Sailor, Foote, Landon and Wood: Rev. Sci. Instrum. **27**, 26 (1956).
13. Igo and Landon: Phys. Rev. **101**, 726 (1956).
14. Foote jr., H. L.: Phys. Rev. **94**, 790 (1954).
15. Wood, R. E.: Phys. Rev. **95**, 644 (1954).
16. Unpublished.

BNL—fc (Brookhaven National Laboratory—fast chopper).
1. Seidl, Hughes, Palevsky, Levin, Kato and Sjostrand: Phys. Rev. **95**, 476 (1954).
2. Harvey, Hughes, Carter and Pilcher: Phys. Rev. **99**, 10 (1955).
3. Unpublished.

BNL—sc (Brookhaven National Laboratory—slow chopper).
1. Carter, Palevsky, Myers and Hughes: Phys. Rev. **92**, 716 (1953).
2. Hughes, D. J., and H. Palevsky: Phys. Rev. **92**, 1206 (1953).
3. Unpublished.

Cal (California).
1. Cook, McMillan, Peterson and Sewell: Phys. Rev. **75**, 7 (1949).
2. Juren, J. de, and N. Knable: Phys. Rev. **77**, 606 (1950).
3. Hildebrand, R. H., and C. E. Leith: Phys. Rev. **80**, 842 (1950).
4. Ragent, B., and W. I. Linlor: Phys. Rev. **91**, 440 (1953).
5. Linlor, W. I., and B. Ragent: Phys. Rev. **92**, 835 (1953).
6. Unpublished.

Cav (Cavendish Laboratory).
1. Allen, Burchan and Wilkinson: Proc. Roy. Soc. Lond. A **192**, 114 (1947).
2. Cohen, S. G.: Nature, Lond. **161**, 475 (1948).
3. Bretscher, E., and E. B. Martin: Phys. Acta **23**, 15 (1950).
4. Stafford, G. H.: Proc. Phys. Soc. Lond. A **64**, 388 (1951).
5. Unpublished.

Col (Columbia University).
1. Zinn, Seely and Cohen: Phys. Rev. **56**, 260 (1939).
2. Havens jr., W. W., and L. J. Rainwater: Phys. Rev. **70**, 154 (1946).
3. Rainwater, Havens, Wu and Dunning: Phys. Rev. **71**, 65 (1947).
4. Havens, Wu, Rainwater and Meaker: Phys. Rev. **71**, 165 (1947).
5. Wu, Rainwater and Havens: Phys. Rev. **71**, 174 (1947).
6. Rainwater, Havens, Dunning and Wu: Phys. Rev. **73**, 733 (1948).
7. Havens, Rainwater, Wu and Dunning: Phys. Rev. **73**, 963 (1948).
8. Melkonian, E.: Phys. Rev. **76**, 1750 (1949).
9. Heindl, C., and I. W. Ruderman: Phys. Rev. **83**, 660 (1951).

10. Tittman, J., and C. Sheer: Phys. Rev. **83**, 746 (1951).
11. Havens jr., W. W., and L. J. Rainwater: Phys. Rev. **83**, 1123 (1951)
12. Melkonian, Havens and Rainwater: Phys. Rev. **92**, 702 (1953).
13. Unpublished.

Cor (Cornell University).
1. Jones jr., W. B.: Phys. Rev. **74**, 364 (1948).

CR (Chalk River Laboratory).
1. Paul, E. B., and R. L. Clark: Canad. J. Phys. **31**, 267 (1953).
2. Brockhouse, B. N.: Canad. J. Phys. **31**, 432 (1953).
3. Unpublished.

Duke (Duke University).
1. Unpublished.

Fr (France).
1. Halban, H. von, and L. Kowarski: Nature, Lond. **142**, 392 (1938).

Fr—cs (France—crystal spectrometer, Saclay Laboratory).
1. Unpublished.

Hrv (Harvad University).
1. Sherr, R.: Phys. Rev. **68**, 240 (1945).
2. Culler, V., and R. W. Waniek: Phys. Rev. **95**, 585 (1954).
3. Hillman, Stahl and Ramsey: Phys. Rev. **96**, 115 (1954).

Har (Harwell).
1. Egelstaff, P. A., and B. T. Taylor: Nature, Lond. **166**, 825 (1950).
2. Hodgson, Gallagher and Bowey: Proc. Phys. Soc. Lond. A **65**, 992 (1952).
3. Merrison, A. W., and E. R. Wiblin: Proc. Roy. Soc. Lond., Ser. A **215**, 278 (1952).
4. Unpublished.

Har—ac (Harwell—accelerator).
1. Tunnicliffe, P. R.: Phys. Rev. **89**, 1247 (1953).
2. Taylor, A. E., and E. Wood: Phil. Mag. **44**, 95 (1953).
3. Freeman, Lane and Rose: Phil. Mag. **46**, 17 (1955).
4. Freeman, J. M.: Phil. Mag. **46**, 17 (1955).
5. Unpublished.

Italy
1. Gatti, Germagnoli and Perona: Nuovo Cim. **11**, 262 (1954).

Japan
1. Aoki, H.: Proc. Phys.-Math. Soc. Japan **21**, 232 (1939).

JENER (Joint Establishment for Nuclear Energy Research, Norway).
1. McReynolds, A. W., and E. Andersen: Phys. Rev. **93**, 195 (1954).
2. Holt, N.: Phys. Rev. **98** (May 15, 1955).

KAPL (Knolls Atomic Power Laboratory).
1. Unpublished.

LA (Los Alamos).
1. Sutton, McDaniel, Anderson and Lavatelli: Phys. Rev. **71**, 272 (1947).
2. King, L. D. P., and L. Goldstein: Phys. Rev. **75**, 1366 (1949).
3. Nereson, N., and S. E. Darden: Phys. Rev. **89**, 775 (1953).
4. Nereson, N., and S. E. Darden: Phys. Rev. **94**, 1678 (1954).
5. Unpublished.

LA—ac (Los Alamos—accelerator).
1. Sutton, McDaniel, Anderson and Lavatelli: Phys. Rev. **71**, 272 (1947).
2. Fowler, J. L., and J. M. Slye jr.: Phys. Rev. **77**, 787 (1950).
3. Ribe, F. L.: Phys. Rev. **87**, 205 (1952).
4. Coon, Graves and Barschall: Phys. Rev. **88**, 562 (1952).
5. Phillips, Davis and Graves: Phys. Rev. **88**, 600 (1952).
6. Brolley, Fowler and Schlacks: Phys. Rev. **88**, 618 (1952).
7. Forbes, S. G.: Phys. Rev. **88**, 1309 (1952).
8. Battat, M. E., and F. L. Ribe: Phys. Rev. **89**, 80 (1953).
9. Martin, H. C., and R. F. Taschek: Phys. Rev. **89**, 1302 (1953).
10. Day, R. B., and R. L. Henkel: Phys. Rev. **92**, 358 (1953).
11. Martin, Diven and Taschek: Phys. Rev. **93**, 199 (1954).
12. Martin, H. C.: Phys. Rev. **93**, 498 (1954).
13. Frye jr., G. M.: Phys. Rev. **93**, 1086 (1954).
14. Henkel, Cranberg, Jarvis, Nobles and Perry: Phys. Rev. **94**, 141 (1954).

15. GRAVES, E. R., and R. W. DAVIS: Phys. Rev. **97**, 1205 (1955).
16. PRESTWOOD, R. J.: Phys. Rev. **98**, 47 (1955).
17. DAY, MILLS, PERRY and SCHERB: Phys. Rev. **98**, 279 (1955).
18. SEAGRAVE, J. D., and R. L. HENKEL: Phys. Rev. **98**, 666 (1955).
19. BEYSTER, HENKEL, NOBLES and KISTER: Phys. Rev. **98**, June 1 (1955).
20. Unpublished.

Mex (Mexico).

1. BONNER, ALBA, FERNANDEZ and MAZARI: Phys. Rev. **97**, 985 (1955).

Minn (University of Minnesota).

1. NUCKOLLS, BAILEY, BENNETT, BERGSTRAHL, RICHARDS and WILLIAMS: Phys. Rev. **70**, 805 (1946).
2. FREIER, FULK, LAMPI and WILLIAMS: Phys. Rev. **78**, 508 (1950).

MIT (Massachusetts Institute of Technology).

1. STELSON, P. H., and W. M. PRESTON: Phys. Rev. **84**, 162 (1951).
2. HINCHEY, STELSON and PRESTON: Phys. Rev. **86**, 483 (1952).
3. STELSON, P. H., and W. M. PRESTON: Phys. Rev. **88**, 1354 (1952).
4. ZIMMERMAN, COOPER and FRISCH: Phys. Rev. **90**, 339 (1953).
5. KIEHN, GOODMAN and HANSEN: Phys. Rev. **91**, 66 (1953).
6. GUERNSEY, J. B., and C. GOODMAN: Phys. Rev. **92**, 323 (1953).
7. HANSEN, KIEHN and GOODMAN: Phys. Rev. **92**, 652 (1953).
8. STORRS, C. L., and D. H. FRISCH: Phys. Rev. **95**, 1252 (1954).
9. KIEHN, R. M., and C. GOODMAN: Phys. Rev. **95**, 989 (1954).
10. GUERNSEY, J. B., and A. WATTENBERG: Phys. Rev. **98** (May 15, 1955).
11. Unpublished.

MTR—cs (Materials Testing Reactor—crystal spectrometer).

1. EVANS, JOKI and SMITH: Phys. Rev. **97**, 565 (1955).
2. Unpublished.

MTR—fc (Materials Testing Reactor—fast chopper).

1. Unpublished.

NWU (Northwestern University).

1. WEDDELL, J. B., and J. H. ROBERTS: Phys. Rev. **95**, 117 (1954).

ORNL (Oak Ridge National Laboratory).

1. RAYBURN, L. A., and E. O. WOLLAN: Phys. Rev. **87**, 174 (1952).

ORNL—ac (Oak Ridge National Laboratory—accelerator).

1. LAMPHERE, R. W.: Phys. Rev. **91**, 655 (1953).
2. JOHNSON, WILLARD and BAIR: Phys. Rev. **96**, 985 (1954).
3. STELSON, P. H., and E. C. CAMPBELL: Phys. Rev. **97**, 1222 (1955).
4. WILLARD, BLAIR and KINGSTON: Phys. Rev. **98**, 669 (1955).
5. FOWLER and JOHNSON: Phys. Rev. **98**, 728 (1955).
6. Unpublished.

ORNL—cs (Oak Ridge National Laboratory—crystal spectrometer).

1. BERNSTEIN, BORST, STANFORD, STEPHENSON and DIAL: Phys. Rev. **87**, 487 (1952).
2. ALLEN, STEPHENSON, STANFORD and BERNSTEIN: Phys. Rev. **96**, 1297 (1954).
3. BERNSTEIN, SHAPIRO, STANFORD, STEPHENSON, DIAL, FREED, PARKER, BROSI, HEBERT and DEWITT: Phys. Rev. **102**, 823 (1956).
4. Unpublished.

ORNL—fc (Oak Ridge National Laboratory—fast chopper).

1. Unpublished.

Oxf (Oxford University).

1. GRIFFITHS, J. H. E.: Proc. Roy. Soc. Lond., Ser. A **170**, 513 (1939).

Rice (Rice Institute).

1. LITTLE, A. B.: Phys. Rev. **87**, 716 (1952).
2. CONNER, J. P.: Phys. Rev. **89**, 712 (1953).
3. COOK, C. F., and T. W. BONNER: Phys. Rev. **94**, 651 (1954).
4. DUDLEY, J. D., and C. M. CLASS: Phys. Rev. **94**, 807 (1954).
5. Unpublished.

Russ (Russia).

1. DEMENTI, V. S., and D. V. TIMOSHUK: C. r. Acad. Sci. URSS. **27**, 929 (1940).
2. MESCHERYAKOV, M. G.: C. r. Acad. Sci. URSS. **48**, 555 (1945).

Switz (Switzerland).

1. BALDINGER, E., and P. HUBER: Helv. phys. Acta **12**, 330 (1939).
2. LUSCHER, RICAMO, SCHERRER and ZÜNTI: Helv. phys. Acta **23**, 561 (1950).

3. Ricamo, R.: Nuovo Cim. **8**, 383 (1951).
4. Adler, Huber and Halg: Helv. phys. Acta **26**, 349 (1953).
5. Meier, Ricamo, Scherrer and Üunti: Helv. phys. Acta **26**, 451 (1953).
6. Huber, P., and T. Hürlimann: Helv. phys. Acta **28**, 33 (1955).
7. Huber, P., and J. Seitz: Helv. phys. Acta **28** (1955).

Tex (University of Texas).

1. Dvorak, H. R., and R. N. Little jr.: Phys. Rev. **90**, 618 (1953).

Wis (University of Wisconsin).

1. Seagondollar, L. W., and H. H. Barschall: Phys. Rev. **72**, 439 (1947).
2. Barschall, Bockelman and Seagondollar: Phys. Rev. **73**, 659 (1948).
3. Adair, Barschall, Bockelman and Sala: Phys. Rev. **75**, 1124 (1949).
4. Bockelman, Peterson, Adair and Barschall: Phys. Rev. **76**, 277 (1949).
5. Adair, Bockelman and Peterson: Phys. Rev. **76**, 308 (1949).
6. Barschall, Bockelman, Peterson and Adair: Phys. Rev. **76**, 1146 (1949).
7. Peterson, R. E.: Phys. Rev. **77**, 747 (1950).
8. Adair, R. K.: Phys. Rev. **77**, 748 (1950).
9. Peterson, Barschall and Bockelman: Phys. Rev. **79**, 593 (1950).
10. Peterson, Adair and Barschall: Phys. Rev. **79**, 935 (1950).
11. Adair, R. K.: Phys. Rev. **79**, 1018 (1950).
12. Bockelman, C. K.: Phys. Rev. **80**, 1011 (1950).
13. Henkel, R. L., and H. H. Barschall: Phys. Rev. **80**, 145 (1950).
14. Johnson, C. H., and H. H. Barschall: Phys. Rev. **80**, 818 (1950).
15. Johnson, Bockelman and Barschall: Phys. Rev. **82**, 117 (1951).
16. Bashkin, Mooring and Petree: Phys. Rev. **82**, 378 (1951).
17. Fields, R. E., and M. Walt: Phys. Rev. **83**, 479 (1951).
18. Petree, Johnson and Miller: Phys. Rev. **83**, 1148 (1951).
19. Bockelman, Miller, Adair and Barschall: Phys. Rev. **84**, 69 (1951).
20. Miller, Adair, Bockelman and Darden: Phys. Rev. **88**, 83 (1952).
21. Adair, Okazaki and Walt: Phys. Rev. **89**, 1165 (1953).
22. Walt, Becker, Okazaki and Fields: Phys. Rev. **89**, 1271 (1953).
23. Okazaki, Darden and Walton: Phys. Rev. **93**, 461 (1954).
24. Fields, Becker and Adair: Phys. Rev. **94**, 389 (1954).
25. Unpublished.

Nuclear Reactions at High Energies.

By

ALBERT WATTENBERG.

With 34 Figures.

A. Resumé.

I. Introduction.

1. Prior to 1946 experiments on nuclear reactions above 50 Mev had to be performed with cosmic rays. Since then it has become much easier to perform measurements up to about 10^{10} ev due to the construction of high energy particle accelerators. With the aid of these machines, many aspects of high energy nuclear reactions have rapidly shifted from the stage of trying to understand gross features to the study of details. Other aspects of high energy nuclear physics are still in the very early stages of development.

In order to prevent the reader from becoming lost in fine points before he obtains an overall picture, this series of chapters proceeds from more general considerations to the more specific details. The remaining sections of this chapter attempt to summarize the current status of high energy nuclear physics from the point of view of what one might have hoped to learn about nuclei from high energy reactions. The following two chapters are devoted respectively to the general theoretical concepts and experimental techniques required to understand the experiments and results presented in the later chapters.

This article deals mainly with the nuclear reactions (above 50 Mev) of complex nuclei; the more fundamental two-body interactions are treated elsewhere in this Encyclopedia[1].

The author wishes to apologize to those whose contributions to high energy nuclear physics are not directly mentioned. It is hoped that it will be constantly remembered that the real sources of our understanding of high energy physics are the many articles listed in the bibliographies referred to in these chapters and not in just the few references given.

II. Information obtainable from high energy nuclear physics.

2. The spatial distribution of nuclear matter. During the 1920's and 1930's the spatial distribution of electrons in atoms was obtained from the interpretation of the X-ray scattering by the electrons of atoms. Analogously it is possible to obtain information on the spatial distribution of protons within a nucleus from the elastic scattering of high energy X-rays or electrons. The momentum of the X-rays or electrons must be sufficiently large so that the corresponding DE BROGLIE wavelength is smaller than the nucleus (see Sect. 7 on "form factors").

---

[1] Cf. L. HULTHÉN and M. SUGAWARA, Vol. XXXIX, this Eucyclopedia.

One of the most outstanding successes of high energy nuclear physics has been in providing more accurate information on the size of the nucleus and the distribution of charge within the nucleus. This information is the result of the interpretation of precise measurements of the angular distribution of the scattering of high energy electrons [8] (see Sect. 37).

If one naively pictures the nucleus as having a uniform (charge) density distribution up to a certain "radius", it now appears that the nuclear radius is somewhat smaller than earlier measurements indicated[1]. If the nuclear radius is taken to vary as $R = r_0 A^{\frac{1}{3}}$ for heavy nuclei, the constant r_0 is now believed to be about 1.2×10^{-13} cm whereas previously r_0 was thought to be 1.4×10^{-13} cm.

From the electron scattering results, it also now appears necessary to picture the nuclear density as tapering off slowly at the edges rather than having a sharp boundary at the "radius" of the nucleus.

The interpretation has been most readily performed for "spherical" nuclei. Scattering experiments have not yet been performed on aligned "ellipsoidal" nuclei.

Electrons and photons scatter through a COULOMB field interaction. Therefore, scattering experiments with these particles are studying the distribution of protons. The possibility exists that neutrons and protons do not have the same distribution in nuclei. Unfortunately, there are no convenient weak interactions of a neutron which would be useful for studying the spatial distributions of neutrons. As is discussed in the section on form factors (Sect. 7) it is not simple to interpret in terms of nuclear radii those measurements which are made with particles which interact strongly with neutrons and protons. Whether neutrons and protons have the same density distributions may be answered by theoretical developments rather than experimental advances.

Our present knowledge is in terms of average density distributions. Spatial fluctuations in density may be very important in understanding nuclei. For example, one can ask to what extent two or more nucleons have a spatial correlation. By this I mean two nucleons may have a higher probability of being closer together than the distance predicted from a "mean" density. Information on the question of correlation may come from studies of nuclear sub-units (see Sect. 43 on quasi-deuteron).

In principle, one might also hope to obtain such information on correlations from diffraction studies with particles which have very short wavelengths. However, there is some disagreement on the possibility of interpreting very short wavelength diffraction studies in terms of spatial correlations.

3. The momentum distribution of nucleons. It is perhaps easiest to understand how one determines momentum distributions by recalling what happens in the case of X-rays on atoms. The THOMPSON scattering dominates when the entire atom contributes to the scattering; whereas the COMPTON effect dominates where the momentum change of the photons is so high that a single electron is knocked out of the atom. At intermediate energies the COMPTON line, instead of being sharp, is modified and has a width that depends on the initial momentum of the electrons in the atom.

A broad class of phenomena in high energy nuclear physics have a somewhat analogous behavior to the modified COMPTON effect. For sufficiently high momentum changes, a single nucleon takes up the momentum rather than the entire nucleus. In such reactions, one finds a momentum region where the initial

[1] Cf. D. L. HILL's contribution to Vol. XXXIX of this Encyclopedia.

momentum of the nucleon modifies the kinematical relationships one would have obtained from a nucleon at rest.

The simplest example of this is high energy proton-proton scattering in a nucleus. If one scatters a proton off of hydrogen atoms (which are effectively protons at rest), the scattered and recoil protons are observed at 90° to each other. However, on bombarding, for example, Li with protons, a scattered and recoil proton are observed with some spread about 90°. The term quasi-elastic scattering is applied to this case (see Sect. 34). An analysis of the spread in angles (or energy) can be made in terms of the initial momentum of the protons in the nucleus.

More complex examples of the modification of the kinematics of single (or two) particle processes are observed in: meson production, "quasi-deuteron photodisintegration", and "pick-up", i.e. (p, d) reactions. Data from all of these reactions have been interpreted in terms of the momentum of nucleons inside of nuclei. The results obtained to date are listed in Table 1.

A (three dimensional) GAUSSian seems to fit most of the data for light nuclei fairly well. Almost all methods employed in momentum determinations observe only the surface nucleons from heavier nuclei (See Sect. 10 on mean free paths). Therefore, one of the outstanding problems in momentum distribution studies is to obtain information on heavier nuclei.

Table 1. *The momentum of nucleons in nuclei*

Element	Momentum distribution assumed or found	Experimental method
Li	FERMI distribution $E_{max} = 20$ Mev	$(p, 2p)$ Quasi-elastic scattering; CHAMBERLAIN and SEGRÈ[1]
C	GAUSSian distribution $1/e = 16$ Mev	$(p, 2p)$ Quasi-elastic scattering, CLADIS *et al.*[2]
Li, Be, B	GAUSSian distribution $1/e = 20$ Mev	$(p, 2p)$ Quasi-elastic scattering, WILCOX and MOYER[3]
C	CHEW-GOLDBERGER distribution[4] [6]	(n, d) Deuteron pickup, HADLEY and YORK[5]
He, Be, C	Empirical distributions $e^{-E/7} + 0.15\,e^{-E/50}$	(p, d) Deuteron pickup; SELOVE[6]
C	CHEW-GOLDBERGER [6]	(γ, π) Photo-meson production; STEINBERGER and BISHOP[7]
C	GAUSSian $E_{average} = 19$ Mev	(p, π) Proton-meson production; RICHMAN and WILCOX[8] and BLOCK *et al.*[9]
Li	GAUSSian $1/e = 6$ Mev	$(\gamma, p\,n)$ Quasi-deuteron disintegration; BARTON and SMITH[10]
Li	GAUSSian $1/e = 8$ Mev	(γ, pn) Quasi-deuteron disintegration; WATTENBERG *et al.*[11]
C	GAUSSian $1/e = 19$ Mev	
O	GAUSSian $1/e = 19$ Mev	

[1] O. CHAMBERLAIN and E. SEGRÈ: Phys. Rev. **87**, 181 (1952).
[2] J. B. CLADIS, W. N. HESS and B. J. MOYER: Phys. Rev. **87**, 425 (1952).
[3] J. M. WILCOX and B. J. MOYER: Phys. Rev. **99**, 875 (1955).
[4] G. F. CHEW and M. L. GOLDBERGER: Phys. Rev. **77**, 470 (1950). Their distribution is of the form $\frac{\alpha}{\pi(\alpha^2+p^2)^2}$ with $\frac{\alpha^2}{2M} = 18$ Mev.
[5] J. HADLEY and H. F. YORK: Phys. Rev. **80**, 345 (1950).
[6] W. SELOVE: Phys. Rev. **101**, 231 (1956).
[7] J. STEINBERGER and A. S. BISHOP: Phys. Rev. **78**, 494 (1950).
[8] C. RICHMAN and H. A. WILCOX: Phys. Rev. **78**, 496 (1950).
[9] M. M. BLOCK, S. PASSMAN and W. W. HAVENS: Phys. Rev. **88**, 1239 (1952).
[10] M. Q. BARTON and J. H. SMITH: Phys. Rev. (to be published).
[11] A. WATTENBERG, A. C. ODIAN, P. C. STEIN, R. WEINSTEIN and H. WILSON Phys. Rev. (to be published).

4. Nuclear forces. With high energy nucleons available in the laboratory, the possibility exists of obtaining experimental information which would provide a solution to some well known problems of nuclear physics. A partial list of such problems is: (1) What is the law of force between two nucleons and what is its dependence on energy? (2) Are there many-body forces? (3) What gives rise to the saturation of nuclear forces in complex nuclei? (4) How important are higher angular momenta? (5) What is the dependence on energy of the interaction of a nucleon with a complex nucleus? There has been a disappointing lack of success in obtaining insight into the first three questions.

The status of these problems are individually summarized below.

α) The two-body interaction. When 100 Mev protons and deuterons were produced in the laboratories, experiments were undertaken immediately to study the *p-p* and *n-p* reactions. At first there were discrepancies between the results obtained at different laboratories, and early theoretical efforts to explain the observed data were not very rewarding. With time, the experimental discrepancies have disappeared and the data have become more precise. Unfortunately, the theoretical situation has not successfully gone beyond a phenomenological description of the data.

One of the most recent theoretical advances has been to correlate almost all *n-p* and *p-p* data below 300 Mev in terms of boundary conditions on *S*, *P* and *D* states with only one energy dependent parameter (see BREIT[1] and FESHBACH[2]). However, there is still no answer to whether it is possible to write an analytical expression as a function of energy for the law of force between two nucleons. It may be that the need of a LORENTZ invariant expression for the interaction is the fundamental reason why potentials (even of the repulsive core type (JASTROW[3])] specifically have not been able to fit all of the data.

Although our insight on the purely theoretical side moves slowly, the amount of experimental information is currently increasing very rapidly, partly because of theoretical suggestions. One such suggestion was to perform multiple scattering experiments to measure the amount of polarization in nucleon-nucleon scattering (see Sect. 33). The polarization is very helpful in obtaining phase shifts. The experimental techniques for polarization measurements have been constantly improving. Some recent polarization experiments have been interpreted in terms of twelve phase shifts (see CHAMBERLAIN *et al.*[4]).

One can hope that further data from polarization experiments and higher energy machines may provide the clues needed by the theoretical physicists. However, it is not clear that the theoretical understanding we desire may not come from meson physics rather than from further nucleon experiments.

β) Many-body forces. One of the old questions of nuclear physics is whether there are many-body forces between three or more nucleons. The combination of a lack of knowledge about the two-body forces and the difficulties of theoretically treating many-body problems has left unanswered even the qualitative question of whether there are many-body forces.

One might hope to obtain at least an answer to the qualitative question from the interpretation of high energy data. However, there has been no success to date.

---

[1] G. BREIT: Proceedings of the Fifth Annual Rochester Conference, p. 143, 1955.

[2] H. FESHBACH: Phys. Rev. **102** (to be published).

[3] R. JASTROW: Phys. Rev. **81**, 105 (1951).

[4] O. CHAMBERLAIN, R. DONALDSON, E. SEGRÈ, R. TRIPP, C. WIEGAND and T. YPSILANTIS: Phys. Rev. **95**, 850 (1954).

γ) Saturation of nuclear forces. The problem of the saturation of nuclear forces arises in trying to understand the binding energy and stability of heavy nuclei. With high energy nucleons one can study what happens when nucleons are close together which bears on the collapsed state of a nucleus.

The early interpretation of high energy nucleon collision experiments led to the Serber mixture of nuclear forces. In this mixture, the exchange forces are equal to the non-exchange. However, such a mixture will not lead to saturation; the ratio should be four to one (q.v. "Wigner condition" Blatt and Weisskopf [3], Chapt III). The Wigner condition of 4 to 1 is on the assumption that nuclear forces are two-body attractive forces which are velocity independent.

The assumptions for the Wigner condition do not seem to be valid at the present moment. Jastrow interprets the high energy scattering data in terms of a repulsive core. Feshbach's boundary condition formulation of the two-body data implies velocity dependent forces (in that the boundary conditions are energy independent). Feshbach's solutions are also not in agreement with the "Serber mixture".

It appears at the present moment that several factors play a role in the saturation of nuclear forces. Lacking information on the existence of many-body forces, we cannot even be certain that this problem will be solved by better information on the two-body interaction.

δ) Higher angular momentum and spin dependence. In the case of complex nuclei, higher angular momenta have manifested themselves in elastic scattering diffraction patterns and in producing polarization in elastic scattering.

The diffraction effects are just those one expects to see from a particle with a fixed de Broglie wavelength impinging on a body of nuclear size (see Sect. 8).

Numerous measurements have already been made of the polarization in the elastic scattering of a nucleon by nuclei (see Sect. 33). A rather simple explanation of the observed data has been given by Fermi and others. The explanation is that the polarization arises from an LS coupling term. The coupling is the same as that required by the shell model to explain ground state phenomena of nuclei. In that one can explain the data in this manner, the experiments have not been informative on the basic origin of this LS coupling.

ε) The energy dependence of the interaction with complex nuclei. In that we lack a theory for the energy dependence of the two-body interaction, it is somewhat surprising and extremely gratifying to find that it is possible to understand the main features of the energy dependence of the interaction of a nucleon with complex nuclei. Serber [4] gave a semi-classical picture of high energy nuclear reactions in which one considers the interactions of an incoming nucleon with the individual nucleons of the target nucleus (see Sect. 6). The input data required for this approach are the differential cross sections for n-p and p-p scattering.

From the n-p and p-p scattering cross section, one can calculate a mean free path in nuclear matter (see Sect. 10). The concept of a mean free path and the associated attenuation has been successful in quantitatively explaining a broad group of phenomena.

5. New phenomena. It was anticipated that processes familiar at low energies might be appreciably modified at higher energies and that new reactions would occur. This section briefly describes such phenomena which have attracted appreciable interest.

α) Impulsive reactions. The term "impulsive reactions" covers those processes in which one or two nucleons act as if they were almost free. Essentially, they are reactions in which an appreciable transfer of momentum takes place.

The most outstanding case of this type is the quasi-elastic scattering of protons (Sect. 34).

Another such process is the so-called "*deuteron-pick-up*" process (Sect. 35). In this case, an incoming proton captures a neutron from the nucleus and to a first approximation leaves the rest of the nucleus unchanged.

Another example is the case of photo-meson production. A fraction of the meson production in complex nuclei is accompanied by the ejection of a single nucleon. The kinematical relationships are essentially those of meson production from a free nucleon.

Another process that might be put in this category is the photodisintegration of quasi-deuterons in a complex nucleus (Sect. 45). When high energy photons bombard a nucleus, protons and neutrons are observed in coincidence. They are ejected with approximately the kinematical relationships of a free deuteron.

β) Photon scattering by nuclei. At low energies, it was not possible, until a few years ago, to observe the scattering of photons by a nucleus because of the much more intense scattering from electrons. However, at high energies, photon scattering by nuclei is observable. To date, only the elastic scattering of photons by nuclei has been studied (Sect. 39).

γ) Inelastic electron scattering. Until recently the excitation of nuclei by electrons was known only through the radioactivity induced. At high energies with good energy resolution it has become possible to measure the shift in energy of the electrons which are inelastically scattered from specific levels in light nuclei. This is of particular interest in that the high energy electron experiments permit the study of some interesting transitions as for example, 0-0 transitions which are frequently difficult to observe by other techniques (Sect. 38).

δ) Spallation. When nuclei are bombarded by high energy particles, it becomes energetically possible to split off a rather large nuclear fragment and to give it sufficient energy to escape through the COULOMB barrier. Light nuclei have been observed as fragments in some reactions (Sect. 36).

ε) New radioactive species. When a nucleus is sufficiently highly excited, many particles are frequently observed to be emitted. The charged particles can be seen in photographic emulsions and are referred to as the "*prongs of stars*" (Sect. 46). From studies of the distribution of radioisotopes formed, it seems that such processes generally tend to favor leaving the nucleus close to the region of stable nuclei. However, these reactions sometimes leave the nucleus far from the region of stability and thus, new radioactive species have been formed.

In the case of a few light nuclei, the radioisotope formed is also unstable against neutron emission, and delayed neutron emitters have been found.

ζ) Mesic atoms[1]. High energy machines have made possible the production in the laboratory of beams of π and μ mesons. When negative π or μ mesons come to rest in matter, they are attracted by the positively charged nucleus in the same way as an electron.

The π^- and μ^- mesons will go from continuum states into "bound atom states" with the accompanying emission of X-rays. They thus temporarily form what are known as "mesic atoms". The X-rays from these transitions have been the observable manifestation of the existence of these mesic atoms. The μ-mesic atoms have been useful in providing data on the charge distribution of nuclei. The results agree with those obtained from electron scattering experiments.

[1] Mesic atoms are treated by S. FLÜGGE in Vol. XLIII of this Encyclopedia.

η) Hyperons and nuclear hyperfragments. Rather recently, it has been found that unstable particles (hyperons[1]) exist which are slightly heavier than protons and neutrons and which decay in about 10^{-10} seconds. The new machines with energies several times 10^9 ev produce these particles.

These hyperons appear to be bound inside some light nuclei to (temporarily) form a variety of new strange nuclei referred to as "hyperfragments". Studies of the binding energies of these hyperfragments are underway. The binding energies of different hyperfragments are felt to be intimately connected with the structure of light nuclei. It is too early to tell what information of interest to nuclear physics will come from studies of hyperfragments.

B. Theoretical concepts useful for high energy reactions.

6. Semi-classical picture. The first results of high energy machine experiments with nucleons were immediately interpreted by SERBER [*4*] on the basis of a very simple picture; it continues as the basis for our physical understanding of many high energy experiments. It is difficult to improve on SERBER's statements so several direct quotations from his paper follows:

"The general features of the high energy nuclear reactions which have been observed at the Radiation Laboratory (Univ. of California, Berkeley, California) can be understood in terms of a picture which is in its main outlines quite simple, though quite different from the description appropriate at lower energies. In trying to understand what happens when a nucleus is bombarded by a high energy neutron or proton, the first consideration that comes to mind is that the collision time between the incident particle and a particle in the nucleus is short compared to the time between collisions of the particles in the nucleus. This suggests that the first step in the process can be regarded in terms of collisions between the incident particle and the individual nuclear particles. We are thus led to ask the properties of the high energy scattering between free nucleons ..."

"... If nuclear matter is considered as a degenerate FERMI gas, it is clear that collisions with small momentum transfers will be discouraged, since these tend to lead from an occupied state to another already occupied. An estimate of this effect indicates that as a result the mean free path of a high energy particle ($\sim$100 Mev) traversing nuclear matter will be increased over what would be expected for collisions between free particles by a factor of $\frac{5}{3}$ ..."

"We estimate that the mean free path for a 100 Mev nucleon is about 4×10^{-13} cm ... Since the mean free path is comparable to nuclear radii, one cannot describe what goes on in terms of formation of a compound nucleus. In fact what happens will depend on the particular trajectory of the incident particle. If it happens to pass through the nucleus near its edge, it may make a single collision and emerge having lost only 25 Mev of its energy ... Or, if it strikes the center of the nucleus and has to pass through the full diameter, it may make several collisions, lose all its energy, and end its range still inside the nucleus. There are thus a variety of possibilities ranging from the bombarding particle emerging with most of its energy intact to the loss of the entire incident energy to the nucleus."

"Since the struck particles have much lower energy and shorter mean free paths than the incident one, they can escape from the nucleus without further collisions only if the collision occurs near the edge of the nucleus with the struck particle heading outward ... Otherwise it will collide with other nuclear particles,

[1] For hyperons see L. LEPRINCE-RINGUET *et al.* in Vol. XLIII of this Encyclopedia.

the energy will be distributed over the nucleus, and the subsequent events can be described in terms of the usual evaporation model, the nuclear excitation energy being dissipated by successive boiling off of particles each with a few million volts of kinetic energy."

"When the nucleus is bombarded with 200 Mev deuterons or 400 Mev α particles, the binding of the incident nucleons is important chiefly in causing a spatial correlation between them, and what goes on can be thought of in terms of a simultaneous bombardment by several nucleons."

"... Finally we should expect the transparency of nuclear matter to show up in measurements of the total cross-section for absorption or scattering of the incident particle. It should be mentioned that the attenuation of the wave representing the incident particle in passing through the nucleus will give rise to diffraction scattering at small angles ($\vartheta \sim \hbar/Rp$, where R is the nuclear radius, p the momentum of the bombarding particle). The cross-section for diffraction scattering is equal to the inelastic and absorption cross-section; for good geometry attenuation experiments it just doubles the cross-section ... One sees that for the heaviest elements one would expect a total cross-section still close to $2\pi R^2$, but for light elements the cross-section should drop considerably below this value."

At these high energies there are a sufficient number of energy and angular momentum states involved that one is approaching the "classical limit" of BOHR's Correspondence Principle and this is why one can anticipate the success of such a semi-classical picture.

The sections below (except for those on elastic scattering) are essentially the quantitative application of SERBER's picture. The fundamental principles apply also to mesonic processes and the high energy photoeffect with minor modifications.

The question arises where does this picture break down as one goes to lower energies. The question is not answered in terms of a specific energy. Some of the considerations involved in judging the validity of a semi-classical approach are given in Sects. 9 to 11 on the impulse approximation, the mean free path in nuclear matter, and the cascade theory. If one mutilates the simplicity of SERBER's picture by taking account of the correlations between nucleons and the concomitant interaction with groups of nucleons and the phase relationships involved, one can extend such a model to low energies. In this case one is dealing with a microscopic representation of low energy nuclear physics which must be equivalent to other successful models of nuclear reactions.

One of the most beautiful things in high energy nuclear physics is the success of SERBER's picture where it explains high energy nuclear phenomena directly in terms of the fundamental interactions between individual particles.

7. Form factors. Those familiar with the elastic scattering of X-rays by the electrons of an atom will realize that there must be an analogous behavior when high energy X-rays are scattered by the protons in a nucleus. The amplitude of scattering from all contributing electrons is added coherently. The resultant cross-section $\sigma_{\text{atom}}(\vartheta)$ for the coherent scattering from the entire atom is then conveniently expressed as $\sigma_{\text{atom}}(\vartheta) = \sigma_e(\vartheta)\, Z_e^2 F^2$ where $\sigma_e(\vartheta)$ is just the THOMPSON scattering from an individual electron which has an angular dependence $(1+\cos^2\vartheta)$; Z_e is the number of effective electrons and F is the "atomic structure factor" or "form factor"[1].

[1] See A. H. COMPTON and S. K. ALLISON: X-Rays in Theory and Experiment, p. 135—159. New York: D. van Nostrand Co. 1935.

For the nuclear scattering of X-rays by protons at sufficiently high energies, $\sigma_e(\vartheta)$ must be replaced by $\sigma_p(\vartheta)$, the appropriate cross-section for the X-ray scattering by a proton, and a form factor appropriate to the nucleus must be calculated or obtained from the observed data.

This formulation is not limited to the scattering of X-rays, but it should apply to any weak interactions in which plane waves constitute a reasonable approximation and the initial and final states of the nucleus are the same. $\sigma_p(\vartheta)$ is then changed to the cross-section appropriate for the process. Processes where it is convenient to think in terms of form factors include the elastic scattering of electrons (from low Z nuclei), elastic photon scattering, μ meson scattering, μ meson pair production, and the scattering of those K mesons which have a weak nuclear interaction. The nuclear form factor is also a convenient first approximation in some strong interactions such as the photoproduction of neutral π mesons.

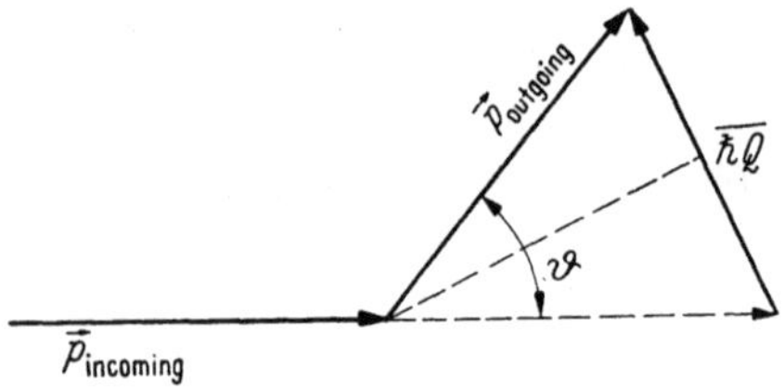

Fig. 1. Momentum diagram for an elastic scattering. By simple geometry

$$p \sin\left(\frac{\vartheta}{2}\right) = \frac{\hbar Q}{2}. \qquad \text{In text } p = \frac{\hbar}{\lambda\!\!\!^{-}}.$$

The form factor is essentially just the summation of the phase factors from the various parts of the nucleus, i.e. the differences in relative phases of the incoming and outgoing waves. For a density distribution of scattering centers $\varrho(r, \psi)$, the form factor is

$$F(Q R) = \tfrac{1}{2} \int_0^\infty \int_0^\pi \varrho(r, \psi)\, e^{i Q r \cos\psi}\, r^2\, dr\, d(\cos\psi), \tag{7.1}$$

where r and ψ are coordinates within the nucleus and Q is the change in wave number of the incoming and outgoing waves. $\hbar Q$ is the "collisive momentum", namely the momentum change (see Fig. 1). If the momentum of the incoming and outgoing wave are $\hbar/\lambda\!\!\!^{-}$, one gets immediately

$$Q = \frac{2}{\lambda\!\!\!^{-}} \sin(\vartheta/2),$$

where $\lambda\!\!\!^{-}$ is the DE BROGLIE wavelength divided by 2π.

Expression (7.1) is for a cylindrically symmetric nucleus. If the nucleus is spherically symmetric the form factor reduces to

$$F(Q R) = \int_0^\infty \varrho(r) \frac{\sin|Q r|}{|Q r|} r^2\, dr \quad \text{with} \quad R^2 = 3 \int_0^\infty \varrho(r)\, r^2\, dr. \tag{7.2}$$

The above expression is readily evaluated for certain simple shapes such as a constant density or a GAUSSian density; F^2 goes to unity as QR goes to zero, and it drops off very rapidly for values of QR greater than 3.

The density $\varrho(r)$ in Eqs. (7.1) and (7.2) came from a semi-classical treatment employing scattering centers. Quantum mechanically one frequently obtains an identical expression. In the quantum mechanical treatment one starts off with the matrix element between identical initial and final states of all the nucleons involved in the process. In those cases where the matrix element can be reduced to a sum of individual nucleon matrix elements and where plane waves can be employed for the scattered particle (BORN approximation) one obtains expression (7.1) and (7.2)[1]. If the calculation are carried through considering all

[1] The definition of a form factor is somewhat arbitrary. In this article the term is restricted to those cases where $\varrho(r)$ contains no phase factor; that is $\varrho(r)$ is a positive definite quantity.

available final states and not just identical initial and final states, one obtains other terms which include inealstic scattering and/or incoherent scattering. These terms must be considered in analyzing the data from experiments in which it is not possible to establish that the nucleus remains unchanged.

In those cases where the interaction is sufficiently strong that plane waves are a poor approximation the form factor calculation must be modified. Such is the case for the scattering of charged particles from high Z nuclei. It is necessary to use COULOMB functions, and the factor $e^{iQr\cos\psi}$ becomes a more complicated expression.

In the case of the photoproduction of neutral π mesons, some of the mesons are absorbed before they leave the nucleus. One could try to put a complex (real and imaginary) phase factor into the outgoing wave to take care of attenuation. However, it is also necessary to consider multiple collisions within the nucleus and the associated change in direction. The calculation is obviously no longer that of a simple form factor but a quite formidable undertaking. It becomes a many-body problem. Similar considerations are involved in the elastic scattering of nucleons from complex nuclei.

8. Potential well representation and many-body reactions. It has been very popular to change the many-body problem into an equivalent two-body problem by introducing a nuclear potential to represent the multiple scatterings and phase relationships that are present when a nucleon interacts with a complex nucleus. The interaction of the incoming (and/or outgoing) nucleon with the entire nucleus is treated phenomenologically by using the experimental data to determine the parameters of the potential. As one tries to fit more data, one has to introduce more parameters.

If one wants to fit total cross-sections for a series of nuclei one needs a complex potential i.e. a potential with a real and an imaginary part. The complex square well must then be modified to a well with tapered edges if it is desired to adapt the potential to treat the angular distributions observed in nucleon scattering. If the potential is to explain the angular distribution of the polarization in nucleon scattering, an additional LS coupling term must be added to the tapered edge complex potential.

The term "Optical Model" is sometimes loosely used for potential well representations because one can immediately obtain an index of refraction from a potential. Specifically if n is the index of refraction

$$n = \left[\frac{(E+V)\left(1+\frac{E+V}{2MC^2}\right)}{E\left(1+\frac{E}{2MC^2}\right)}\right]^{\frac{1}{2}} = \frac{k_0+k_1}{k_0}, \tag{8.1}$$

where E is the kinetic energy of the free particle and V is the depth of the potential where k_0 is the wave number of the impinging nucleon outside the nucleus (i.e. $\hbar k_0$ = momentum of free nucleon), and where k_0+k_1 is the wave number of the nucleon inside the nucleus. At high energies the optical model appears to be a good approximation; many theories express their results in terms of optical parameters such as the index of refraction or the directly related quantity

$$\frac{k_1}{k_0} = n - 1. \tag{8.2}$$

The "Optical Model" was first applied to high energy neutron interactions by FERNBACH, SERBER, and TAYLOR [7]. They used a real potential to get the real part of the index of refraction and an attenuation (mean free path) inside the nucleus to obtain the imaginary part of the index of refraction. The mean free path is in agreement with the observed high energy two-nucleon cross-sections (see Sect. 10).

TAYLOR[1] and MANDL and SKYRME[2] found that the parameters of the optical model are a function of energy. They used the optical model in the energy range 50 to 400 Mev and found that, to fit the data, it was necessary to have the depth of the square well decrease rapidly from 30 Mev at 50 Mev to about 15 Mev at 130 Mev after which it falls off slowly with increasing energy. The mean free path increases from about 2.2×10^{-13} cm at 50 Mev to about 4.0×10^{-13} cm at 150 Mev. Quite a few of the experimental papers contain analyses of their data in terms of these optical model parameters with essentially similar conclusions (see Sect. 29).

The aforementioned optical model calculations assumed square well potentials in order to interpret total cross-section data. In attempting to fit the observed angular distribution of elastically scattered protons, WOODS and SAXON[3] introduced a tapered edge to the complex potential of FESHBACH, PORTER, and WEISSKOPF[4]. The form WOODS and SAXON employ is

$$V(r) = \frac{V + i W}{1 + \exp\left(\frac{r - r_0}{a}\right)}. \tag{8.3}$$

V is the depth of the real part of the potential and W is the depth of the imaginary part, r_0 is a measure of the nuclear size and a determines the diffuseness of the nuclear surface. They claim that this exponentially decreasing surface leads to better agreement with the data than would a square well potential; specifically the cross-sections at large angles are smaller.

STERNHEIMER[5] has employed the same type of tapered edge potential in attempting to fit the data on the polarization of elastically scattered protons. The polarization arises from an LS coupling term. He employs the shape of the WOODS-SAXON potential [Eq. (8.2)] combined with LS coupling, but instead of evaluating the imaginary potential he employs a (reciprocal) mean free path. STERNHEIMER finds this model gives results in reasonable agreement with the observed angular dependence of the polarization. He used an a of about 0.5×10^{-13} cm and an r_0 of $1.23 \times 10^{-13} A^{\frac{1}{3}}$ cm which corresponds to an effective square well radius of $1.4 \times 10^{-13} A^{\frac{1}{3}}$ cm. The depth of his well was 55 Mev at 50 Mev and decreased to 15 Mev at 130 Mev.

It is possible to calculate the imaginary part of the potential from a mean free path obtained from p-p and n-p scattering cross-sections. Such a calculation has been performed by LANE and WANDL[6]. They find that W, the depth of the imaginary part of the potential increases from the order of 2 Mev at low energies to about 15 Mev at 100 Mev where it seems to go through a maximum. The PAULI exclusion principle is responsible for the low value at low energies.

The relationship of the potential model parameters to the fundamental interaction of nucleons is not generally a simple matter. The problem of obtaining

[1] T. B. TAYLOR: Phys. Rev. **92**, 831 (1953).
[2] F. MANDL and T. H. R. SKYRME: Phil. Mag. **44**, 1028 (1953).
[3] R. D. WOODS and D. S. SAXON: Phys. Rev. **95**, 577 (1954).
[4] H. FESHBACH, C. E. PORTER and V. F. WEISSKOPF: Phys. Rev. **90**, 166 (1953). — Phys. Rev. **96**, 448 (1954).
[5] R. M. STERNHEIMER: Phys. Rev. **100**, 886 (1955).
[6] A. M. LANE and C. F. WANDL: Phys. Rev. **98**, 1524 (1955).

the depth of the real potential from two-body interactions is being worked on by a series of authors: specifically BRUECKNER, EDEN, FRANCIS, and WATSON[1].

These groups have been developing mathematical formalisms for solving the many-body problem which they claim are applicable to both low and high energy phenomena. At this time only a crude descriptive statement of their starting point can be given without going into the details of the mathematics. WATSON *et al.* employ a multiple scattering formalism; BRUECKNER *et al.* start from the SCHRÖDINGER equation for many nucleons in strong interaction with each other.

These theoretical investigations have been very provocative to the entire field of nuclear theory and have already been helpful in producing a clarification of the equivalence of various theoretical representations of nuclei and their reactions. They are also providing a better understanding of the validity of some of the simple models employed in high energy physics.

9. The impulse approximation. The SERBER picture [*4*] (Sect. 6) was applied by CHEW[2] to calculate the inelastic scattering of 90 Mev neutrons by deuterons. He assumed that the scattering process takes place in a time which is short compared to the period of the nucleons in the deuteron and that the scattering is therefore just the sum of the amplitudes for *n-n* scattering and *n-p* scattering. The name "impulse approximation" was given to those calculations in which one can assume the interaction between the incoming particle and a single nucleon occurs in a time that is short compared to the period of the nucleons in a nucleus. Impulse approximation calculations have been applied to many reactions including: quasi-elastic proton proton scattering, meson scattering, meson photoproduction, and meson production by nucleon collisions (this latter case is a questionable application). The high energy photoeffect (the photo-disintegration of the quasi-deuteron) can also be thought of as a process in which one may be able to apply the impulse approximation if one uses the term descriptively rather than in a strict sense.

CHEW and WICK[3] have given a strict definition of the impulse approximation. They limit the application to those cases where all of the following assumptions hold: (1) The incident particle interacts only with one single nucleon at a time; (2) the amplitude of the incident wave is not appreciably diminished in crossing the nucleus, and (3) the binding forces between the constituents of the system are negligible during the decisive phase of the collision when the incident particle interacts strongly.

If these conditions apply, then one finds that the matrix element is the sum of the amplitudes of the interaction between two nucleons of momentum p, weighted by the probability of their initially having momentum p. Where one has experimental data on the interaction between two nucleons as a function of energy (momentum) the empirical data can be put directly into the impulse approximation formulation. In the impulse approximation the presence of the other nucleons and their mutual interactions manifest themselves in providing the initial momentum distribution. Therefore one is essentially carrying out a momentum calculation; as is typical in such calculations, conservation of energy in the individual interactions is omitted. There are various ways of taking care of the conservation of energy which one knows must be there after the process is over. One way is by use of a closure argument.

[1] For a list of the earlier papers of BRUECKNER, EDEN and FRANCIS see K. A. BRUECKNER, R. J. EDEN and N. C. FRANCIS: Phys. Rev. **100**, 891 (1955). For references to WATSON's work see N. C. FRANCIS and K. M. WATSON: Phys. Rev. **92**, 291 (1953).

[2] G. F. CHEW: Phys. Rev. **80**, 196 (1950).

[3] G. F. CHEW and G. C. WICK: Phys. Rev. **85**, 636 (1952).

CHEW and WICK investigated condition (3) quantitatively. Their criterion is that if the target nucleons are bound by a potential U and if τ is the time during which the "sudden" interaction takes place, then $\tau \ll \hbar/U$. If this is the case one can neglect conservation of energy during the "sudden" interaction. Unfortunately it is not always obvious how to estimate τ. They do not discuss the apparently alternative though more limited experimentally observable criterion that the collisive momentum be large compared with the momenta normally present in a complex nucleus.

CHEW and GOLDBERGER[1] have investigated all three assumptions of the impulse approximation in a more rigorous way for the case of the scattering of particles by a complex nucleus. They have extended the usefulness of the impulse approximation in showing that it is possible to make internal scattering corrections for a nucleus in which condition (2) does not hold rigorously.

Where one is employing the impulse approximation and does not have empirical data for the individual nucleon interactions but must calculate the matrix element, it is important to investigate that all degrees of freedom are included in the matrix element. Specifically, one must inquire as to whether all virtual and real mesonic processes are included.

10. The mean free path in nuclear matter. The mean free path has already been seen to be an important parameter in employing SERBER's [*4*] picture and in the Optical Model. It also enters in cascade theory (Sect. 11), in meson reactions, and in the photo-ejection of nucleons (Sect. 45).

One can obtain approximate values of the mean free path which are good to 20% quickly and naively; however it will be seen that more reliable values require more sophistication.

The mean free path Λ is defined as

$$\Lambda \equiv \frac{1}{\varrho\sigma} \tag{10.1}$$

where ϱ is the number of nucleons per cm³ and σ is the cross-section. The cross-sections come from the experimental nucleon-nucleon data which are summarized in Fig. 2; obviously the mean free path is a function of energy. To obtain a value of ϱ it is convenient to assume that the nucleus is a sphere with a uniform density up to radius R. Then the number of protons per cm³ is

$$\varrho_P = \frac{Z}{\frac{4}{3}\pi R^3} \tag{10.2a}$$

and the number of neutrons per cm³ is

$$\varrho_N = \frac{A-Z}{\frac{4}{3}\pi R^3}\,. \tag{10.2b}$$

It will seen from Eqs. (10.1) and (10.2) that the value of the mean free path depends on the value of the radius of the nucleus.

The reciprocal of the mean free path, K, the attenuation coefficient, (sometimes referred to in the literature as an absorption coefficient or an opacity coefficient) for a proton inside a nucleus is

$$K_P \equiv \frac{1}{\Lambda_P} = \varrho_N \sigma_{NP} + \varrho_P \sigma_{PP}, \tag{10.3a}$$

and for a neutron inside a nucleus

$$K_N \equiv \frac{1}{\Lambda_N} = \varrho_N \sigma_{PP} + \varrho_P \sigma_{NP}\,. \tag{10.3b}$$

[1] G. F. CHEW and M. L. GOLDBERGER: Phys. Rev. **87**, 778 (1952).

It is assumed that the neutron-neutron cross-section, σ_{NN}, is the same as the proton-proton cross-section, σ_{PP}, at high energies.

For nuclei where there are equal number of neutrons and protons

$$K = K_N = K_P = \frac{3A}{8\pi R^3}\left[\sigma_{NP} + \sigma_{PP}\right]. \tag{10.4}$$

If R is assumed to vary as $R = r_0 A^{\frac{1}{3}}$, then the mean free path becomes independent of the atomic weight of the nucleus

$$K = \frac{1}{\Lambda} = \frac{3}{8\pi r_0^3}\left[\sigma_{NP} + \sigma_{PP}\right]. \tag{10.5}$$

For an accurate value for the mean free path it is necessary to correct the cross-sections (given in Fig. 2) for the initial motion of the nucleons in the nucleus and for the effect of the PAULI exclusion principle. SERBER [4] pointed out that one is dealing with collisions within a degenerate FERMI gas and the PAULI exclusion principle will forbid those collisions in which one particle ends up in a state that is already occupied. The cross-sections in Eq. (10.5) should be reduced to an *effective* cross-section. The reduction factor has been estimated approximately by GOLDBERGER [5] to be given by

$$\sigma_{\text{effective}} = \sigma_{\text{free}}\left(1 - \frac{7}{5}\frac{p_F^2}{p_i^2}\right) \tag{10.6}$$

where p_F is the maximum momentum of the FERMI gas $\left(\frac{p_F^2}{2M} \sim 20 \text{ Mev}\right)$, for a nucleus of radius R, and p_i is the momentum of the incident nucleon. Eq. (10.6) is limited to $p_i > \sqrt{2}\, p_F$.

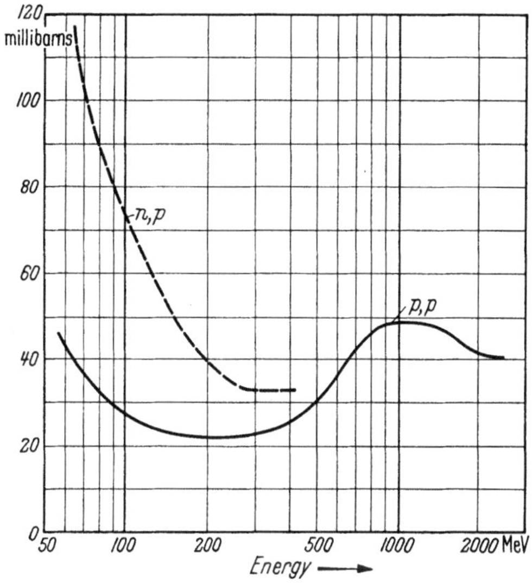

Fig. 2. Total cross-sections for the neutron-proton and proton-proton reactions in the laboratory as a function of energy. The data for the curves were obtained from the references of tables 4 and 6 and from A. SHAPIRO (Private Communication). For information on angular distribution see: G. BREIT and R. L. GLUCKSTERN, Ann. Rev. Nucl. Sci. **2**, 365 (1953); N. P. BOGACHEV and I. K. VZOROV, Dokl. Akad. Nauk USSR. **99**, 931 (1954); V. P. DZHELEPOV and Y. M. KAZARINOV. Dokl. Akad. Nauk USSR. **99**, 939 (1954).

A simple geometrical construction for understanding the correction factor is given by MORRISON [2].

The cross-sections in Fig. 2 were obtained from collisions between a particle at rest and one in motion. In the case of collisions inside a nucleus, the target nucleon is also in motion. Furthermore at low energies one must also augment the energy of the impinging particle for the assumed depth of the potential well it experiences on entering the nucleus. The initial motion of the target nucleons smears out the kinematical angular relationships and this becomes important in estimating angular distributions.

The cross-section to be employed for σ_{free} [in Eq. (10.6)] should thus be a weighted average over a broad energy range. The breadth of the energy is rather large; specifically an (initial) 100 Mev neutron will undergo collisions inside a nucleus which would correspond to energies from about 40 to 250 Mev for a neutron approaching a nucleon at rest. Due to the target nucleons having a fixed range of momenta and the energy dependence of the cross-section shown in Fig. 2, the effect of the initial motion is most serious at lower energies. GOLDBERGER [5] provides a convenient coordinate system for estimating the average cross-section; he states that the calculation is rather tedious.

Results of total cross-section measurements and absorption[1] cross-section measurements are frequently expressed in terms of the reciprocal of the mean free path, K. The relationship between K and the absorption cross-section, σ_a, has been given by FERNBACH, SERBER, and TAYLOR [7] for a uniform density sphere of radius R as

$$\sigma = \pi R^2 \left[1 - \frac{1-(1+2KR)\,e^{-2KR}}{2K^2 R^2}\right]. \tag{10.6}$$

For a non-uniform density distribution such as a nucleus with a tapered edge, K is no longer a constant across the nucleus but instead becomes a function of r and ψ, the coordinates of points inside the nucleus. Obviously expression (10.6) does not hold and one must reformulate the problem as a function of r and ψ. For the case of absorption cross-sections a convenient integral formulation has been given by WILLIAMS[2] for the case where K is a function of r only. However for realistic density distributions (as obtained from recent electron scattering measurements (see Sect. 37) these integrals have to be solved numerically.

From the above it should be evident that the mean free path concept has become one in which people are concerned about details that will allow them to make estimates to better than 20%. Where the mean free path is desired for an approximate calculation, expression (10.5) and the cross-section from (10.6) should be adequate if one can not employ the values of Table 4 (p. 483).

11. Cascade theory and the Monte Carlo method. The general principles of high energy nuclear reactions enumerated by SERBER [4] suggests a separation into two phases for the non-elastic reactions of nucleons with a nucleus. The first phase is the rapid development of a nucleonic cascade, and the second phase is the de-excitation of a thermally excited nucleus by the relatively slower process of evaporation.

The cascade develops through the incident nucleon colliding with one of the nucleons of the target nucleus and the recoil nucleons either escaping from the nucleus or individually colliding with other nucleons (in the same nucleus); the recoil nucleons from these secondary collisions either escape from the nucleus or collide with other nucleons, and so on. The cascade process is visualized as continuing until the energy of each of the individual nucleons in the collisions becomes less than the energy required by a nucleon to directly escape from the nucleus. The entire cascade process is estimated to take place in the order of 10^{-22} sec.

Those nucleons which do not escape from the nucleus are assumed to share their energy with the entire residual nucleus. This leads to the second phase, namely the de-excitation of an excited nucleus. The dissipation of this energy is conveniently treated by the statistical model of the nucleus and the evaporation theory of WEISSKOPF[3].

The cascade process is ideally suited to the Monte Carlo method of calculation, and the method was first applied by GOLDBERGER [5] to investigate the interactions of 90 Mev neutrons with a heavy nucleus.

The Monte Carlo method is applicable for an ordered sequence of processes each of which can be characterized by its own probability distribution. A single

[1] The absorption cross-section is the total cross-section minus the elastic scattering cross-section. It includes all non-elastic processes.

[2] R. W. WILLIAMS: Phys. Rev. **98**, 1387 (1955).

[3] V. WEISSKOPF: Phys. Rev. **52**, 295 (1937).

particle is followed through the sequence of events determined by chance, and each of the collision partners of that particle are followed through their life histories. One then starts with a new particle and follows its chance experiences and so on. The final results one obtains from following an individual particle have no significance. However by combining the results found for N particles (N individual Monte Carlo calculations), one obtains a set of final distributions with a statistical fluctuation appropriate to N particles.

If N is a sufficiently large number and if the assumptions and the input information are correct, one should obtain distributions from this calculational experiment which are in agreement with the distributions obtained in laboratory experiments with N particles. The input information consists of the nucleon-nucleon cross-sections and their energy and angular dependence. The assumptions made by GOLDBERGER include the following: the particles have classical trajectories in between collisions; only two nucleons are involved in each collision; the nucleus is a sphere of uniform density; an average cross-section can be used for both neutrons and protons, the angular distribution can be assumed isotropic in the center of mass, and the cross-section varies roughly as the inverse of the momentum in the center of mass. BERNARDINI, BOOTH, and LINDENBAUM[1] used essentially the same assumption in applying the Monte Carlo method to the interaction of 400 Mev protons with nuclei in a photographic emulsion.

The above assumptions are not fundamental to the Monte Carlo method. The assumptions were made in order to reduce the amount of calculations that had to be performed in applying the Monte Carlo method. The method could be readily (but more tediously) carried through with more realistic assumptions than the above, except for the first two assumptions.

GOLDBERGER [5], BERNARDINI *et al.*, and MORRISON [2] all give a clear explanation of how to apply the Monte Carlo method. A brief prescription for the application of the Monte Carlo method follows: (step two is given in more detail for illustrative purposes).

Step 1. Determine by chance where nucleon number 1 enters the nucleus. This can be done by subdividing the surface into rings of equal areas normal to the axis of the incoming particle, and then by some process of chance choose one of these area intervals.

Step 2. Determine by a method of chance how far nucleon number 1 travels in the nucleus before making a collision. The path interval from zero to infinity is divided into small path intervals of equal probability according to the law

$$p(x) = e^{-x}/\Lambda' \tag{11.1}$$

where $p(x)$ is the probability for a nucleon to penetrate a distance x and Λ' is the mean free path in nuclear matter *not* taking account of the PAULI principle (see Sect. 10).

In equal probability divisions (11.1) becomes

$$x_n = \Lambda' / \ln p_n. \tag{11.2}$$

For example if one divides p_n into one hundred equal intervals corresponding to 0.00 to 0.01, 0.01 to 0.02, 0.02 to 0.03, ..., 0.99 to 1.00 one can associate the set of two-digit numbers from 00 to 99 with each of these intervals of p_n. Then by some method of chance one picks two digits. Specifically one can choose from a table of random numbers the second two digits of the first entry in the

[1] G. BERNARDINI, E. T. BOOTH and S. L. LINDENBAUM: Phys. Rev. **88**, 1017 (1952).

table. (This assumes the first two digits were used in step 1.) Particle number 1 moves to x_n determined from the chosen value of p_n, and it is assumed to have a collision at that point.

Step 3. The next problem is to choose (by chance) the momentum of the nucleon (in the target) which has been struck. The uniform density model of the nucleus has a concomitant FERMI momentum sphere whose size is fixed by the number of particles in the nucleus and the radius of the nucleus. The momentum intervals must be equally subdivided taking into account the probability (i.e. cross-section) for a nucleon-nucleon collision for particles with various momenta with the incoming particle.

Step 4. The next decision to be made is that of the scattering angle. Here one has to weight the solid angle in the center of gravity system by the differential cross-section. Kinematics then determines the final momenta of the collision partners. At this point one may find the final momentum of one of the collision particles lies within the already occupied FERMI momentum sphere. If this is the case the collision is forbidden, and the trajectory of particle number 1 is continued by choosing a new path length (step 2) from the point of the forbidden collision. One then continues to follow particle number 1 through steps 3, 4 etc.

Step 5. If both of the partners of the collision have allowed momenta, each separately starts off on its new trajectory and step 2 is repeated, then step 3, etc. for each particle. Some of the particles will escape from the nucleus at step 2, and the angle and energy of emission of the particle are recorded.

One keeps repeating steps 2 and so forth for each particle in subsequent collisions until every particle involved in a collision either escapes from the nucleus or is captured by the nucleus. A nucleon which is captured is one whose energy after a collision is found to be below the energy required to escape from the potential well. GOLDBERGER used a 33 Mev deep well with a FERMI energy of 18 Mev.

After the original nucleon number 1 and all its cascade products have either escaped or been captured one starts off fresh at step number 1 with nucleon number 2 and follows it and all its cascade products. GOLDBERGER repeated this series of steps for 100 incident (90 Mev) neutrons. He could follow about three particles in a day. It is possible to perform such calculations with electronic computing machines. BERNARDINI *et al.* followed 90 incident protons (of 400 Mev); 30 of these went through the nucleus without interacting and 60 resulted in inelastic events. Two of the cases followed by BERNARDINI *et al.* are shown in Fig. 3 and 4. Fig. 4 shows a case in which both nucleons of the first collision escape without further interactions; this case corresponds to a quasi-elastic scattering event. Obviously quasi-elastic scattering events are more probable in light nuclei where $R \ll \Lambda$.

GOLDBERGER was interested in interpreting data first obtained with counting equipment. BERNARDINI *et al.* were trying to understand star prong data obtained in photographic emulsions. They both found that cascade particles have a peaking at forward angles. Almost all of the very energetic nucleons that escape are in the forward hemisphere.

The sum of the energies of the captured nucleons from one incident nucleon gives the thermal excitation of the residual nucleus which is dissipated through evaporation. GOLDBERGER found the mean thermal excitation due to 90 Mev neutrons on lead was 42 Mev; however the distribution of thermal excitation energies was very broad extending from about 5 to 90 Mev with no pronounced peaking. BERNARDINI *et al.* for 400 Mev protons on a nucleus of mass 100 found a mean thermal excitation of 50 Mev and the maximum thermal excitation was 200 Mev. Considering the difference in the energy of the incident particles, this is a rather small rise in the thermal excitation; however it can be readily accounted for on the basis that high energy nucleons make high energy secondaries for which the nucleus is relatively more transparent.

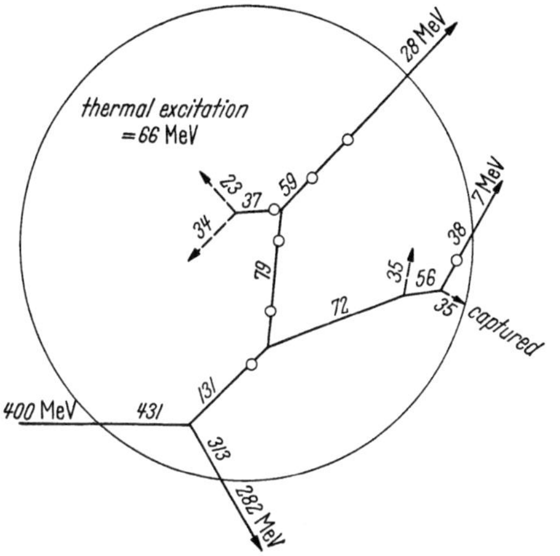

Fig. 3.

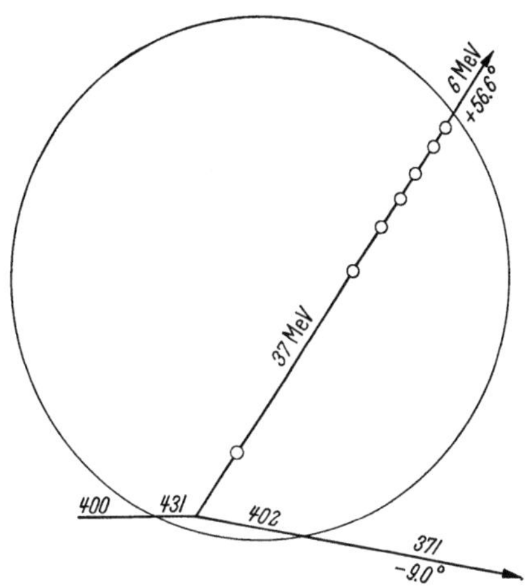

Fig. 4.

Fig. 3. A two-dimensional diagram for the Monte Carlo method showing the development of a cascade. The entering nucleon has 400 Mev of kinetic energy; the numbers indicate the energy of the nucleons involved. The path of each nucleon is shown until it leaves the nucleus or its energy becomes less than 35 Mev, indicated by a broken line and arrow. An open circle indicates the position at which a collision would have occurred but was forbidden by the Pauli exclusion principle. In this collision three cascade particles emerge in the forward hemisphere and the nucleus is left with a thermal excitation of 66 Mev. (Figure is redrawn from the work of Booth *et al.*)

Fig. 4. A two-dimensional diagram for the Monte Carlo method which illustrates a case where a quasi-elastic scattering interaction developed. For further explanation see Fig. 3. [From Booth *et al.*: Phys. Rev. **88**, 1013 (1952).]

12. High energy evaporation theory. The dissipation of this thermal excitation by means of the evaporation process is similar to the de-excitation of the familiar compound nucleus. Most of the short black prongs observed in stars in photographic emulsions are a manifestation of the evaporation process. The true evaporation prongs have a modified Maxwellian energy distribution and an isotropic angular distribution.

The spread in the cascade and evaporation process leads to the many radioactive species observed by the radiochemists when high energy nucleons bombarded nuclei. Extensive thermodynamic and statistical mechanics treatments of the high energy evaporation process are given by LeCouteur[1], Morrison [2] and Fujimoto and Yamaguchi[2]. The latter paper gives references to a set of earlier papers by these authors.

There are several factors that appear to come in at high energy that are not normally considered in the evaporation theory treatment of lower energy phenomena. These factors essentially result from the thermal expansion of the nucleus which LeCouteur claims is required by thermodynamic arguments. One result of a thermal expansion is the lowering of the Coulomb barrier for charged particles. A shift in size also affects the Fermi (kinetic) energy; the evaluation of this gets one into the usual difficulty of choosing a nuclear model in order to evaluate the partition function for a Fermi-Dirac distribution at

[1] K. J. LeCouteur: Proc. Phys. Soc. Lond. A **63**, 259 (1950).

[2] Y. Fujimoto and Y. Yamaguchi: Progr. Theor. Phys. **5**, 787 (1950). — Y. Yamaguchi: Progr. Theor. Phys. **6**, 529 (1951).

temperatures above absolute zero. LECOUTEUR has evaluated the entropy term due to this effect and finds it adds to the probability of proton emission. A third effect of a thermal expansion is that the surface tension decreases. This decrease in surface tension will affect the thresholds for fission and the probability for spallation (see Sect. 36). However quantitative predictions on spallation depend on the particular nuclear model chosen to predict the surface vibrations.

LECOUTEUR is rather definite in predicting the energy dependence of the ratio of singly to doubly charged evaporation fragments and the energy distribution of the protons. He also expects that about 28% of the singly charged particles should be deuterons. These predictions need to be subjected to further investigation.

There are several things that should be pointed out. One is, that it is not clear that the nucleus is ever left at thermal excitations as high as are used in these thermodynamic treatments. Another is, that the thermal expansion of the nucleus may have been overestimated in order to obtain a desired reduction in the COULOMB barrier (see LECOUTEUR). MORRISON [2] discusses other possible sources for a reduction of the COULOMB barrier.

The thermodynamic treatment is very appealing in that it tries to handle a very complex set of phenomena in terms of familiar classical parameters. However the relation of the thermodynamic parameters to the parameter of other nuclear models is frequently obscure.

The thermodynamic treatment has been applied mainly to understanding stars in photographic emulsions. The end products are also the radioactive nuclei observed in radiochemical studies. The application of evaporation theory to prediction of the yields of radioisotopes is rather limited (see MORRISON [2]). The form that a more general application might take may be indicated by HALPERN'S treatment of the yield of radioisotopes from photo-nuclear reactions (see Sect. 47).

C. Experimental techniques of high energy physics.

I. High energy machines.

13. High energy machines provide particles in the laboratory at energies up to 6×10^9 ev as of the present time. Machines now in the design stage will increase this by a factor of 10. These energies are still many orders of magnitude less than those available in cosmic rays. Cosmic ray physicists have measured nuclear cross-sections up to energies of 10^{11} ev.

For obtaining high energy particles or photons in the laboratory four types of machines are now employed: betatrons, linear accelerators, synchrocyclotrons, and synchrotrons. The betatron was developed by KERST[1]. All betatrons in existence accelerate electrons. The electrons travel in a circular orbit of effectively constant radius. In a betatron the magnetic flux inside the orbit increasing with time accelerates the electrons.

The original idea for linear accelerators came from WIDEROE[2] and since then there have been many variations. The modern high energy linear accelerators were made possible by the development during the 1940's of radiofrequency techniques and equipment. The particles are accelerated by a sequence of oscillating electric fields along a straight line, the oscillators spaced so that the particles stay in phase as their energy increases. Linear accelerators are used to produce

[1] D. W. KERST: Phys. Rev. **60**, 47 (1941). — See also KERST'S article in Vol. XLIV of this Encyclopedia.

[2] R. WIDEROE: Arch. Elektrotechn. **21**, 387 (1929).

both high energy electrons and protons, with a simpler design for electrons because they approach the velocity of light at a lower energy.

Synchrocyclotrons are a modification (by McMillan[1] and Veksler[2]) of the original cyclotron of Lawrence and Livingston[3]. Synchrocyclotrons (also called frequency modulated cyclotrons) compensate for the relativistic increase in mass of heavy particles by appropriately modulating the radio-frequency on the dees of the cyclotron. In a cyclotron, in contrast to other high energy circular accelerators, the magnetic field is kept constant in time and the particles increase their orbits (i.e. spiral paths). Synchrocyclotrons are used to accelerate protons, deuterons, or helium ions.

Synchrotrons stem from the ideas of McMillan and Veksler. They are generally circular machines in which the orbit is effectively constant. The acceleration is by means of an alternating electric field (radio-frequency) across one or more gaps. For electrons the frequency is almost constant; for protons the frequency has to be modulated to correspond to the magnetic field (the velocity of the particles).

A very valuable principle that enables synchrotrons to be built more economically was independently suggested by Courant, Livingston, and Snyder[4] and by Christofilos[5].

Table 2. *Pulse lengths of various machines.*

Type of machine	Approximate pulse length in μ sec
Betatrons	1 or as long as 300
Linear accelerator . .	1
Synchrocyclotron . .	500
Electron synchrotrons	1 500 or as short as 1
Proton synchrotron .	20000

By means of alternating the gradient of the magnetic field in adjacent sections of the magnet, one obtains an effective focusing action which reduces the radial and vertical spread of the particle orbits. Several synchrotrons are under construction or in the design stage which employ this alternating gradient (sometimes called "strong focussing") principle. The new Cornell electron synchrotron is the first machine to operate with alternating gradients.

The above is only a brief statement of the differences between the machines. Much more information on the principles of these machines can be found in Vol. XLIV of this Encyclopedia, dedicated especially to them, or in Livingston's[6] concise book.

The various types of machines are limited in the time duration of the pulse which in an important way affects the experiments. Table 2 gives approximate pulse durations for various machines.

II. Sources of particles.

14. Protons, deuterons, and helium ions. The most common sources of high energy protons up to 800 Mev are synchrocyclotrons. At the present time the three proton synchrotrons which produce protons of higher energy are the

[1] E. M. McMillan: Phys. Rev. **68**, 143 (1945).
[2] V. Veksler: J. Phys. USSR. **9**, 153 (1945).
[3] E. O. Lawrence and M. S. Livingston: Phys. Rev. **38**, 834 (1931).
[4] E. Courant, M. S. Livingston and H. Snyder: Phys. Rev. **86**, 1190 (1952).
[5] Nicholas Christofilos: Communication to the Univ. of California Radiation Laboratory, 1954.
[6] M. S. Livingston: High Energy Accelerator. New York, N.Y.: Interscience Publishers, Inc. 1950.

Birmingham machine[1] (1×10^9 ev), the Brookhaven Cosmotron[2] (3×10^9 ev), and the Berkeley Bevatron[3] (6×10^9 ev).

These high energy proton synchrotrons give one pulse every few seconds. In the Brookhaven Cosmotron the pulse rate is one every five seconds, and there are the order of 10^{10} protons in a pulse. The pulse rate of synchrocyclotrons is greater than that of the higher energy sychrotrons, namely 60 to 300 pulses per second, and the average beam current is the order of a microampère.

If one employs an internal target in a synchrocyclotron with no deflecting system the energy of the protons varies from 3 to 10% about the mean energy depending on the size of the radial oscillations present in the machine. Better energy resolution can be obtained with external beams which can be produced by a variety of techniques. CHAMBERLAIN *et al.*[4] obtained an energy spread of 2% in an external beam produced by scattering the internal protons into a magnetic shielding tube. With the scattering technique the external beam is only a small fraction of the internal beam.

LECOUTEUR[5] has proposed a more efficient technique for obtaining external beams from synchrocyclotrons which has been quite successful.

An external beam from proton synchrotrons has been obtained by PICCIONI *et al.*[6] at the Brookhaven Cosmotron. The intensity of the external beam is approximately half that of the internal beam. The technique takes advantage of the magnetic field free straight sections of the machine. A decrease in radius is produced by passing the beam through solid matter, which causes an energy loss by ionization. Due to the energy lost the beam shrinks to a smaller radius and at the smaller radius the beam enters a magnetic deflecting system which then curves the beam outward. Their method is similar to one suggested for the Berkeley Bevatron by WRIGHT[7].

15. Neutron sources. The most prolific sources of useful high energy neutrons are the proton (or deuteron) accelerators. The two reactions which are employed to produce high energy neutrons are the *deuteron stripping reaction* and the *proton charge exchange scattering reaction*. The deuteron stripping reaction can be crudely pictured as a process in which the proton in the deuteron strikes the edge of the nucleus and is stripped off with the neutron continuing on its way. The theory for the process has been given by SERBER[8]. The neutrons emitted in a narrow cone at small angles have approximately one-half the energy of the deuteron with a calculated spread in energy of about ± 20% (for 200 Mev deuterons). With a one-half inch thick beryllium target HELMHOLZ, MCMILLAN and SEWELL[9] obtained neutron yields that were about 2% of the original deuteron beam.

With a synchrocyclotron, proton charge exchange scattering is capable of producing higher energy neutron beams than the deuteron stripping reaction. The term charge exchange scattering denotes those proton-neutron scatterings which leave the outgoing neutron with almost the same energy and momentum as the original incoming proton.

---

[1] Staff: Proton Synchrotron of the Univ. of Birmingham. Nature, Lond. **172**, 704 (1953).
[2] Cosmotron Staff: Rev. Sci. Instrum. **24**, 723 (1953).
[3] E. J. LOFGREN: Invited paper at the June 1954 meeting of the Amer. Phys. Soc. Phys. Rev. **96**, 838 (1954).
[4] O. CHAMBERLAIN, E. SEGRÈ and C. WIEGAND: Phys. Rev. **83**, 923 (1951).
[5] K. LECOUTEUR: Proc. Roy. Soc. Lond., Ser. A **232**, 242 (1955).
[6] O. PICCIONI, D. CLARK, G. FRIEDLANDER and D. KASSNER: Phys. Rev. **98**, 275 (1955).
[7] B. T. WRIGHT: Rev. Sci. Instrum. **24**, 290 (1955).
[8] R. SERBER: Phys. Rev. **72**, 1008 (1947).
[9] A. C. HELMHOLZ, E. M. MCMILLAN and D. C. SEWELL: Phys. Rev. **73**, 1003 (1947).

The energy spread in neutron beams arises from the internal momentum distribution of the neutrons in the target nucleus and from two experimental sources. The latter come from the initial spread in energy of the charged particles and from multiple traverses of the target by the protons (in circular accelerators). The problem has been studied by KNOX[1] who interpreted his data mainly in terms of the multiple traversals of his target. However, an energy spread of about 10% can arise from the initial spread in proton energies associated with the radial oscillations of the protons in the machine.

With an external proton beam both of these experimental sources of energy spread can be avoided. The type of neutron energy spectrum one obtains are illustrated by the measurements of CASSELS *et al.*[2] at 145 Mev, of NELSON *et al.*[3] at 220 Mev, of GOODELL *et al.*[4] at 350 Mev and of NEDZEL[5] at 410 Mev.

Knowledge of the spectrum of higher energy neutrons ($\sim 1.4 \times 10^3$ Mev) from proton synchrotrons is still rather preliminary, and neutrons associated with meson production probably introduce an appreciable low energy tail. Experiments with neutrons at these energies resemble those with cosmic rays in that the detection technique determines the effective energy spread as in the experiments of COOR *et al.*[6]. Their neutron beam was the order of 10^3 neutrons per cm^2 per pulse of 10^9 protons.

16. Mesons. At the present time useful external beams of charged π mesons are available at proton accelerators. As an example[7], the (122 ± 3) Mev π^- beam from the Chicago synchrocyclotron has a flux of about 1000 π^- mesons per cm^2 per second over an area of 50 cm^2. This is one of their best beams and at 200 Mev they only obtain about $^1/_{50}$-th of this flux. ASHKIN *et al.*[8] find that a π^+ beam has about $^1/_{30}$-th the intensity of a π^- beam. The contamination in π beams consists of about 7% of μ mesons and 1% of electrons of the same charge; however for beams of lower energy π mesons, the contamination increases.

External beams of mesons with energies greater than 10^9 ev are available from the high energy proton synchrotrons. The beams have the order of 10π mesons from a pulse of about 10^{10} protons (on an internal target). The development of intense external proton beams will improve this situation.

Beams of μ mesons have been obtained from synchrocyclotrons; the μ mesons arise from π mesons which decay in flight. WINSBERG[9] obtained approximately 10^3 μ^- mesons per second with an energy of (40 ± 15) Mev.

At the Berkeley Bevatron and the Brookhaven Cosmotron "beams" of charged K mesons are obtained (by K mesons we mean particles with a mass about 966 electron masses and a lifetime of about 10^{-8} sec). Unfortunately in the so-called "K^+ meson beam" there are actually about 20 times as many π mesons present as K^+ mesons. At Berkeley in the arrangement of BIRGE *et al.*[10], for every

[1] W. J. KNOX: Phys. Rev. **81**, 693 (1951).

[2] J. M. CASSELLS, T. C. RANDLE and T. G. PICKAVANCE: Phil. Mag. **42**, 215 (1951).

[3] B. K. NELSON, G. GUERNSEY and G. MOTT: Phys. Rev. **88**, 1 (1952).

[4] W. F. GOODELL, H. H. LOAR, R. R. DURDIN and W. W. HAVENS: Phys. Rev. **89**, 724 (1953).

[5] V. A. NEDZEL: Phys. Rev. **94**, 174 (1954).

[6] T. COOR, D. A. HILL, W. F. HORNIAK, L. W. SMITH and G. SNOW: Phys. Rev. **98**, 1369 (1955).

[7] H. L. ANDERSON, J. MARSHALL, L. J. KORNBLITH, L. SCHWARCA and R. MILLER: Rev. Sci. Instrum. **23**, 707 (1952).

[8] J. ASHKIN, J. P. BLASER, E. FEINER, J. G. GORMAN and M. O. STERN: Phys. Rev. **96**, 1105 (1954).

[9] L. WINSBERG: Rev. Sci. Instrum. **25**, 1079 (1954).

[10] R. W. BIRGE, D. H. STORK, R. P. HADDOCK, L. T. KERTH, J. R. PETERSON, J. SANDWEISS and M. N. WHITEHEAD: Phys. Rev. **99**, 335 (1955).

proton hitting the target they obtain about $2\times10^{-9}\,K^+$ mesons and $2\times10^{-12}\,K^-$ mesons.

Hyperfragments referred to in Sect. 5 are produced when K^- mesons are absorbed in nuclei.

17. Electrons. The circular electron synchrotrons and betatrons have been used more for the X-rays they produce rather than for studies of the interaction of electrons with nuclei. Experiments with electrons have been more conveniently performed at machines with straight sections (e.g. University of Michigan race-track) or with linear accelerators.

The most outstanding research with electrons is that performed at the University of Stanford. A detailed description of the Stanford machine is given in a set of articles by the staff[1]. At the present time the maximum energy of the Stanford machine is about 630 Mev; lower energies are available at stations along the way. The machine operates at 60 pulses per second with about 6×10^{11} electrons in each pulse. The pulse duration is one microsecond, however the main portion of the intensity occurs in 0.3 microseconds and this creates many experimental difficulties. The energy spread is narrow having a full width at half maximum of 2%. In many experiments a double magnetic deflecting system is employed which gives extremely good energy resolution and frees the beam of the contamination of X-rays and neutrons.

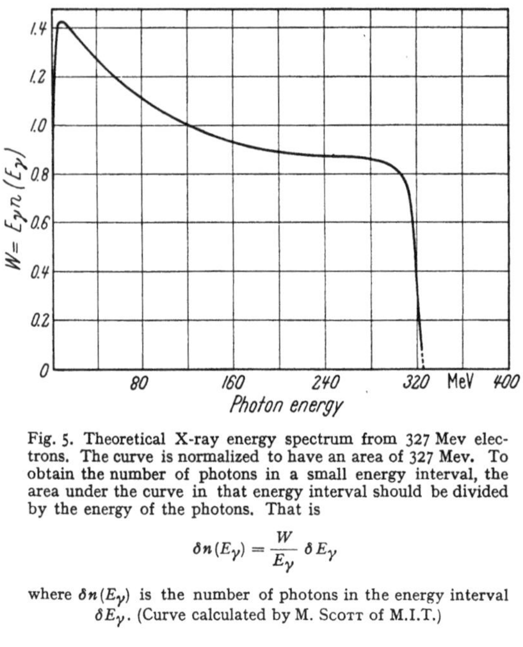

Fig. 5. Theoretical X-ray energy spectrum from 327 Mev electrons. The curve is normalized to have an area of 327 Mev. To obtain the number of photons in a small energy interval, the area under the curve in that energy interval should be divided by the energy of the photons. That is

$$\delta n(E_\gamma) = \frac{W}{E_\gamma}\,\delta E_\gamma$$

where $\delta n(E_\gamma)$ is the number of photons in the energy interval δE_γ. (Curve calculated by M. SCOTT of M.I.T.)

The electron current is monitored or measured in several different ways. One way is the direct catching and retaining of the entire beam and the secondary charged particles produced. At 300 Mev their beam catcher is designed to contain 99.9% of the charged components of the electron shower.

18. Photons. By allowing the electrons to strike a metal target, the high energy electron accelerators can be conveniently used to produce a bremsstrahlung spectrum (i.e. a continuous X-ray spectrum). Copper, wolfram, and platinum are employed as targets generally in the form of a metal foil less than a millimeter thick.

In circular machines the targets are displaced from the stable electron orbits and a perturbation of the orbit causes the electrons to hit the target. The type of perturbation introduced determines the time duration of the beam. The various techniques employed are described by THOMAS *et al.*[2], by SCHIFF[3], and by DAZEY *et al.*[4].

[1] M. CHODOROW, E. L. GINZTON, W. W. HANSEN, R. L. KYHL, R. B. NEAL, W. K. H. PANOFSKY and staff: Rev. Sci. Instrum. **26**, 134 (1955).

[2] J. E. THOMAS jr., W. L. KRAUSHAAR and I. HALPERN: Ann. Rev. Nucl. Sci. **1**, 175 (1952).

[3] L. I. SCHIFF: Ann. Rev. Nucl. Sci. **2**, 79 (1953).

[4] DAZEY, FRANK, HELMHOLZ, NUNAN and PETERSON: Rev. Sci. Instrum. **21**, 436 (1950).

The theory for the continuous bremsstrahlung spectrum has been given by BETHE and HEITLER[1]. The theory for the angular distribution and spectrum has to be modified because of the small angle scatterings of the electrons which occur in targets of finite thickness. The spectrum to be expected from finite thin targets has been worked out by SCHIFF[2] and is shown in Fig. 5. Experimental investigations of the spectrum have been made by LEISS, YAMAGATA and HANSON[3] and by POWELL, HARTROUGH and HILL[4]. The experimental data fit the SCHIFF spectrum in the energy regions investigated.

The X-rays[5] are produced in a forward cone, the angular spread of which decreases with increasing electron energy as $\frac{mc^2}{E_e}\ln\left(\frac{E_e}{mc^2}\right)$ where E_e is the energy of the electron, m the mass of the electron, and c the velocity of light. In the usual targets, the small angle multiple scattering of the electrons is commensurate with the spread predicted by the BETHE-HEITLER theory. In most experiments the angular spread is limited by a collimating system.

In the X-ray beam there are electrons and neutrons which result from the beam passing through any matter or colliding with the walls of the collimating system. Most of the electrons can be removed by passing the X-rays through a strong magnetic field. The general background of neutrons around photon machines is a source of difficulty and a physiological hazard.

III. Detection techniques.

19. Introductory remarks. Many of the detectors employed by cosmic ray physicists and low energy nuclear physics have proved useful in high energy nuclear physics. In that there are an abundant number of texts on most of these techniques[6], the sections below deal with the problems involved in employing only a few of the most useful of these detectors, namely: scintillation counters, CERENKOV counters, photographic plates, bubble chambers and cloud chambers.

The range of high energy particles may be the order of 50 gm/cm² (or more) in contrast to ranges of milligrams/cm² at low energies. This is very convenient in that one does not have to worry about window thicknesses, vacuum systems, or target thicknesses in many cases. However it becomes too much of a good thing when the range of particles becomes larger than a mean free path for nuclear reactions (namely above 300 Mev for protons). Then one finds it improbable to bring the particles to rest and runs into many experimental difficulties.

Also as one goes to increasingly higher energies a great variety of reactions become possible. With the exception of scattering, the products of a reaction generally depend more upon the energy of the initiating particle than upon the nature of the particle. Therefore it is difficult to find a single reaction which can be used as a means of identifying a particle or determining its energy. This leads to high energy measurements requiring detecting systems that employ several simultaneous measurements in order to restrict the observations to the specific particles and energy group that it is desired to investigate. Detecting systems are very likely to include time-of-flight apparatus and magnetic deflection or magnetic lens systems.

---

[1] H. A. BETHE and W. HEITLER: Proc. Roy. Soc. Lond., Ser. A **146**, 83 (1934); also see J. M. JAUCH and F. ROHRLICH: The Theory of Photons and Electrons, p. 36. Cambridge, Mass.: Addison-Wesley Publishing Co., 1955.

[2] L. I. SCHIFF: Phys. Rev. **70**, 87 (1946); **83**, 252 (1951).

[3] J. E. LEISS: Doctoral Thesis, Univ. of Illinois 1953.

[4] W. M. POWELL, W. HARTROUGH and M. HILL: Phys. Rev. **81**, 213 (1951).

[5] M. STEARNS: Phys. Rev. **76**, 836 (1949).

[6] For details see Vol. XLV of this Encyclopedia.

20. Scintillation counters for charged particles. The general subject of scintillation counters has been reviewed recently by Swank[1] and luminescent materials have been reviewed by Garlick[2]. The organic scintillators are used in all forms: crystals, organic solutions (hereafter called "liquid scintillators") and solid solutions in plastics (referred to as "plastic scintillators").

In that coincidence circuits are generally employed and the beams are pulsed for a short time, the organic scintillators are more frequently used than NaI; NaI crystals have a luminescent decay time which is longer than that of organic scintillators. However for particles which come to rest in a detector, the light output of organic scintillators is not proportional to the energy lost (see Garlick); therefore when one wants to measure the energy lost in the detector special calibration measurements are required. The organic scintillators are very valuable if one wants solely to record the passage of a charged particle through a detector. The liquid scintillators and plastic scintillators are wonderful in being comparatively inexpensive detectors of any desired shape and size.

For mesons, protons, deuterons, tritons, etc. detecting telescopes are used which consist of several detectors through which the particles pass or come to rest. The problem of identifying the type of particle requires at least two measurements. If no magnet is employed, the two measurements are the rate of energy lost, (dE/dx), and the total energy of the particle; the latter most frequently from a range measurement. The range technique is limited to energies below about 300 Mev because of the range becoming commensurate with the mean free path for a nuclear interaction for all heavy particles except μ mesons. There are a variety of ways of using detecting telescopes; a two-crystal telescope in which the pulse heights in both crystals are measured is described by Feld *et al.*[3]. They discuss such telescopes and provide curves which may be helpful in understanding the response and problems of two- and three-counter telescopes.

Telescopes containing many detectors are very convenient for measuring several energy groups simultaneously. The group at the University of Illinois employ a five-detector telescope. The pulse heights are displayed on an oscilloscope and a photographic record is made of the scope display[4].

Such techniques are not limited to scintillation detectors; Cassels *et al.*[5] and Dzhelepov *et al.*[6] have employed proportional counters in such telescopes to obtain excellent energy resolution.

Positive mesons can be identified by observing their delayed decay products (again within the limit of the range being shorter than the mean free path for nuclear interactions). The delay can be measured directly by photographing a fast oscilloscope trace, or one can do it electronically by a series of delayed coincidence circuits.

For particles above several hundred Mev where the range has become too great, the determination of the energy and mass of the particle requires both magnetic analysis to determine the particle's momentum and some method of determining the particle's velocity. For the velocity measurement either a

[1] R. K. Swank: Ann. Rev. Nucl. Sci. **4**, 111 (1954). — Cf. also R. B. Sutton in Vol. XLV of this Encyclopedia.

[2] G. F. L. Garlick: Progr. in Nucl. Phys. **2**, 51 (1952) or in Vol. XXVI of this Encyclopedia.

[3] B. T. Feld, R. D. Godbole, A. Odian, F. Scherb, P. C. Stein and A. Wattenberg: Phys. Rev. **94**, 1000 (1954).

[4] T. Yamagata, M. Q. Barton, A. O. Hanson and J. H. Smith: Phys. Rev. **95**, 574 (1954).

[5] L. M. Cassels, T. C. Randle and T. G. Pickavance: Phil. Mag. **42**, 215 (1951).

[6] V. P. Dzhelepov and Yu. M. Kazarinov: Dokl. Akad. Nauk **99**, 939 (1954).

CERENKOV detector or a time-of-flight measurement is employed and sometimes both. The important requirement on the scintillation detector is that it have a fast response and good pulse height. The current limitation on the speed of time-of-flight equipment is in the photomultiplier tubes. The spread in time produced in photomultipliers is greater than 3×10^{-9} sec. This combined with length of the flight path determines the velocity resolution of such apparatus.

21. Scintillation counters for neutron detection. High energy neutron detectors *with good energy resolution* have measured the energy of proton recoils from the scattering of neutrons by a hydrogeneous target. The arrangement generally consists of a thin hydrocarbon target and a proton telescope (see previous section) at a well defined angle. The hydrocarbon target requires a carbon subtraction measurement. Such neutron detectors and the problems in using them have been reviewed elsewhere by the author[1].

Neutron detectors with good energy resolution generally have a low efficiency because of two requirements. One is, that the scattering target be kept thin so as to minimize the uncertainty in the energy lost in the target by the protons. The second requirement is, that the angle of the recoil proton be well defined relative to the direction of the initial neutron; therefore the recoil proton telescope must subtend a small solid angle. The efficiency of this type of neutron detector usually comes out to be less than 10^{-4}.

Neutron detectors (with poor energy resolution) with efficiencies as high as 10% can be made by employing appreciable volumes of liquid or plastic scintillators. For neutrons with energies greater than 100 Mev such detectors are generally at least 20 cm long and at least 10 cm in diameter. The events which produce scintillations are the recoil protons produced by neutrons scattering on the hydrogen or the charged particles produced by neutrons interacting with the carbon in the scintillating material. At high energies these proton recoils have an appreciable range and therefore the dimensions of such detectors must be large. If a background of charged particles or photons is present the neutron detector is surrounded by lead. If lead is employed in front of the detector, low energy secondary events are produced by neutron interactions in this lead. Due to these low energy secondaries and the charged particle reaction products from the carbon and the non-linear response of organic scintillators, the low part of the pulse height spectrum of this type of neutron detector differs drastically from the usual $1/E$ response (of low energy neutron recoil counters). A study of the efficiency and pulse height response of such detectors has been made by CHRISTIE *et al.*[2].

If one desires the maximum efficiency from this type of neutron detector, it is necessary to accept a broad spectrum of pulse heights and forego any energy discrimination. However, with an appreciable sacrifice in the efficiency, one can obtain some energy resolution. VOSS and WILSON[3] have employed liquid scintillator neutron detectors with an energy resolution of about 10% by limiting their observations to the recoil protons which had more than 80% of the maximum pulse height.

Such neutron detectors are simple to construct. For purposes of pulse height calibration, it is desirable to avoid leaving the photomultiplier surface in contact with the scintillating liquid.

---

[1] A. WATTENBERG: Ann. Rev. Nucl. Sci. **3**, 119 (1953).

[2] E. R. CHRISTIE, B. T. FELD, A. C. ODIAN, P. C. STEIN and A. WATTENBERG: Rev. Sci. Instrum. **27**, 127—131 (1956).

[3] L. L. THRESHER, R. G. P. VOSS and R. WILSON: Proc. Roy. Soc. Lond. **229**, 492 (1955); also R. G. P. VOSS and R. WILSON: Proc. Roy Soc. Lond. (to be published).

22. Scintillation counter for electrons and photons. High energy photons are generally detected by converting them into electron-positron pairs in a high Z material and then observing the electron-positron pairs.

The most common type of scintillation detector system for electrons or photons is a so-called range telescope; such are useful from about 25 to 150 Mev. A typical range telescope consists of a converter (if photons are being studied), an organic scintillator, an appreciable thickness (range) of a light element such as Be, C, or Al, and another (back) electron detector. It is becoming popular to employ a Cerenkov detector (see below) as the back electron detector. The range straggling for electrons is large, and there is an appreciable probability of radiative losses (bremsstrahlung) even in low Z elements as the energy increases. Consequently electron range telescopes only set a crude lower limit on the energy of the electrons. However, if a sufficient thickness of Be, C, or Al is used, they have the advantage of discriminating against heavier charged particles of the same energy or smaller range. A crude rule of thumb for choosing the thickness of the Be, C, or Al is that the total thickness of the telescope in gm/cm^2 should be equal to 0.4 times the energy of the electrons in Mev or $\frac{1}{2}$ that thickness for photons of the same energy. For example 100 Mev photons should produce at least one particle that has a high probability of going through a total thickness of 20 gm/cm^2.

The efficiency of such a range telescope for 50 Mev electrons can be over 90% if the diameter of the back detector is sufficiently large and if all components are kept as closely packed as possible. For 70 Mev photons, the overall efficiency was found to be about 30% for a system consisting of a 3 mm thick by 7.5 cm diameter lead converter and a back detector 15 cm in diameter.

For purposes of obtaining better energy resolution several large volume scintillation detectors have been studied by Pugh, Frisch, and Gomez[1]. A liquid scintillation detector they designed measured the total energy lost by ionization in the liquid. The detector was 16 inches in diameter and 30 inches long based on the maximum range and probable scattering of 150 Mev electrons. The energy resolution of such a detector is limited by the radiative losses of the fast electrons (or positrons). Despite this limitation, this detector has been extremely useful in a variety of experiments, and it is a vast improvement over range telescopes in regard to energy resolution. The large liquid scintillator is probably impractical for photon energies over 200 Mev because of the increase of range and increasing probability of radiative losses.

Pugh, Frisch, and Gomez describe another γ-ray detector in the same article. It is a multiplate shower detector composed of alternate thin sheets of scintillating plastic and lead. The idea is that an electron shower which develops in the lead plates will lose a fixed fraction of its energy in the scintillating material. The fraction lost in the scintillators should be a sufficient sample to give a good estimate of the total energy of the shower.

23. Cerenkov detectors. A relativistic charged particle moving through an optically transparent medium may have a velocity v, which is greater than the speed of light in the material. Specifically if the refractive index of the material is n, then the speed of light in the material is c/n, and it is possible to have $v/c \equiv \beta > 1/n$. There is a virtual photon field moving with the charged particle and one can make a Huygens construction as shown in Fig. 6. From the construction one sees that an electromagnetic (shock) wave results which

[1] G. E. Pugh, D. H. Frisch and R. Gomez: Rev. Sci. Instrum. **25**, 124 (1954).

moves at an angle ϑ to the direction of motion of the particle. The relationship is

$$v\cos\vartheta = c/n$$

or

$$\cos\vartheta = \frac{1}{\beta n} \quad \text{for} \quad \beta > \frac{1}{n}. \tag{23.1}$$

This electromagnetic (shock) wave is the radiation that was first observed by CERENKOV[1]. It is a classical electromagnetic effect that depends solely on the velocity of the particle and the index of refraction of the material. The refractive index depends upon the wavelength (or frequency) of the light, and the effect occurs only for those wavelengths that give real values of the angle ϑ in Eq. (23.1)

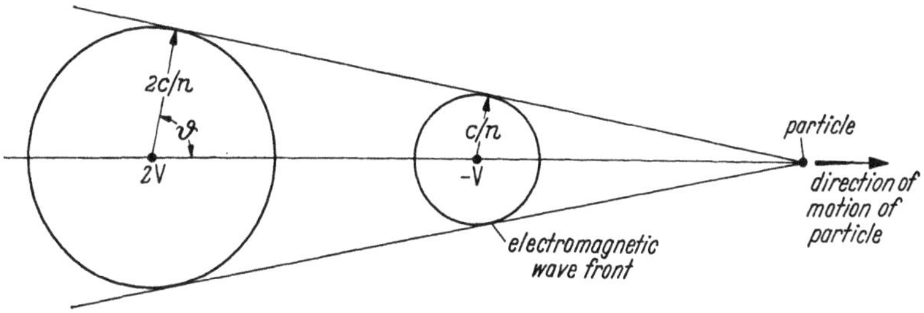

Fig. 6. HUYGENS construction for CERENKOV radiation showing geometric relationship of the electromagnetic wave and the motion of the particle

$$v\cos\vartheta = \frac{c}{n}.$$

above. The number of photons emitted per unit path length, I, has been calculated by FRANK and TAMM[2] to be

$$I = \frac{2\pi e^2 Z^2}{\hbar c^2} \Delta\nu \left(1 - \frac{1}{\beta^2 \eta^2}\right) \tag{23.2a}$$

or

$$I = \frac{2\pi Z^2}{137c} \Delta\nu \sin^2\vartheta. \tag{23.2b}$$

In practice the interval $\Delta\nu$ is determined by the frequency to which the light detector, the photomultiplier, is sensitive. An excellent review of CERENKOV detectors up to 1954 has been given by MARSHALL[3]. He estimates that with currently available photomultiplier tubes the useful frequency range is about 3×10^{14} cycles per second and with a plastic ($n=1.50$) about 250 photons per centimeter of path should be emitted. A CERENKOV detector several centimeters thick gives a signal well above the noise level of a good photomultiplier tube and is a comparatively simple detector of relativistic charged particles.

The only spread in time associated with CERENKOV detectors arises from the different path lengths the rays may have to travel; therefore they can be made extremely fast and are excellent for use in very fast coincidence work.

The most promising types of CERENKOV detectors are those which are designed with an optical system that restricts the response of the detector to a narrow band of velocities. (See the section in MARSHALL's review on "focusing CERENKOV counters".) If a CERENKOV detector with good velocity resolution is used in combination with a magnetic (momentum) analyzing system, it is possible to identify the mass of the particles being detected. An excellent example of the importance of velocity selective CERENKOV detectors is in the identification of

[1] P. A. CERENKOV: C. r. Acad. Sci. URSS 2, 451 (1934).
[2] I. FRANK and I. TAMM: C. r. Acad. Sci. URSS 14, 109 (1937).
[3] J. MARSHALL: Ann. Rev. Nucl. Sci. 4, 141 (1954).

the anti-proton by CHAMBERLAIN *et al.*[1]. Their optical system limits the CERENKOV counter to detecting particles in the narrow band $0.75 < v/c < 0.78$.

CERENKOV detectors are specifically high energy detectors and therefore are becoming increasingly important for counting experiments with higher energy machines.

One rather specialized type of CERENKOV detector should be mentioned as it is being used and studied currently at about half a dozen laboratories. It is a CERENKOV detector for high energy γ-rays and electrons (300 Mev or higher) which should have fairly good energy resolution. It consists of a large volume of lead glass, namely about ten radiation lengths in both the diameter and thickness. The energy is determined from the principle that the energy of an incoming event is roughly proportional (at high energies) to the integrated path length of all the electrons developed in a shower. The total CERENKOV radiation is proportional to the path lengths except for the last centimeters of track. At high energy, the total CERENKOV radiation should be proportional to the total energy, providing the fraction of energy which escapes is kept very small. This is why the detector must be so large.

In order to gather all light, independent of the direction of the shower particles, the counter should be coated with a highly (diffuse) reflecting substance such as titanium dioxide. Such detectors have been tested by SWARTZ[2] and by FRISCH and associates[3].

24. Track recording devices: General remarks. Under the title of "track recording devices" are included photographic emulsions, cloud chambers, bubble chambers, and solid devices in which the tracks of ionizing particles can be observed. In that these techniques generally afford several simultaneous measurements, they can often give the information needed to identify particles and determine the energy of a particle. There are many combinations of simultaneous measurements employed depending on the technique; most frequently the density of ionization is combined with a range measurement, amount of scattering, or curvature in a magnetic field. The fundamentals underlying all of these techniques are given by ROSSI [*1*].

At higher energies the track recording devices can be extremely useful in those cases where there are present a variety of reaction products or background particles. They are also very convenient in studying the energy spectrum of reaction products. To do the same job counting devices either become very elaborate or require using the machines over a long period of time.

The variety of track recording devices and variations in technique constitute an enormous subject which cannot even be reviewed here. For those who need details there are excellent reviews and entire books especially on photographic emulsion techniques[4] and cloud chamber techniques[5]. The remainder of this

[1] O. CHAMBERLAIN, E. SEGRÈ, C. WIEGAND and T. YPSILANTIS: Phys. Rev. **100**, 947 (1955).

[2] C. SWARTZ: Proceedings of the Scintillation Counter Conference 1956.

[3] D. H. FRISCH and B. RICHTER: Private Communication.

[4] *References for photographic emulsion technique:* M. M. SHAPIRO in Vol. XLV of this Encyclopedia. — M. TEUCHER: Ergebn. exakt. Naturw. **28**, 407 (1955). — Y. GOLDSCHMIDT-CLERMONT: Ann. Rev. Nucl. Sci. **3**, 141 (1953). — L. ROTBLAT: Progr. in Nucl. Phys. **1** (1950). — E. W. TITTERTON: Progr. in Nucl. Phys. **4**, 1 (1955). — H. YAGODA: Radiaoctive Measurements with Nuclear Emulsion. New York: John Wiley & Sons, Inc. 1949.

[5] *Cloud chamber techniques:* C. M. YORK jr. in Vol. XLV of this Encyclopedia. — W. B. FRITTER: Ann. Rev. Nucl. Sci. **5**, 145 (1955). — H. STAUB: Experimental Nuclear Physics, Vol. I, part I. — E. SEGRÈ editor: New York, N.Y.: John Wiley & Sons, Inc. 1951. — J. G. WILSON: The Principles of Cloud Chamber Techniques. Cambridge, England: Cambridge Univ. Press 1951. — W. GENTER, H. MAIER-LEIBNIZ and W. BOTHE: Atlas of Typical Expansion Chamber Photographs. London: Pergamon Press 1954.

chapter is subdivided into several very short subsections dealing with: recent advances in emulsion techniques not covered in the review articles, bubble chambers, a table for comparison of the uses of various track recording devices, and solid devices.

25. Recent advances in emulsion techniques. One of the most recent advances is the careful experimental determination of the range-energy relationship for emulsions by BARKAS and co-workers. It is somewhat ironic that the best known range-energy curve should be in other than a pure substance. At high energies range-energy relationships in pure substances are appreciably less certain at this time than in emulsions (see CALDWELL[1]).

Another development is the increase in the volume of an emulsion stack that can be reliably processed and analyzed. Volumes of emulsions as large as 4 cm by 30 cm by 50 cm are being employed by stacking several hundred 30×50 cm pellicles each 200 microns thick. This means that in an emulsion one can study reactions which may simultaneously emit a particle with a range of 100 microns and another particle with a range of 20 cm.

The possibility of following a single track through so many plates has been made easier by using a so called "Berkeley grid system". [Photographic Plate Groups at the University of California, Berkeley (private communication from G. GOLDHABER).] An exposure in ordinary light is made of the emulsion held in precise contact with a numbered grid work. The same numbered coordinate system appears on one surface of each emulsion pellicle after it is developed.

Several laboratories are experimenting with pulsed magnetic fields of sufficient strength (10^5—10^6 gauss) to provide useful track curvature measurements in emulsions.

26. Bubble chambers. The development of bubble chambers should be credited to GLASER[2]. He saw the need for track recording devices containing materials with appreciable density and in which the composition of the material could be varied. He patiently undertook their development and has successfully provided high energy physics with a new tool.

The basic principle underlying bubble chambers is that a liquid in a thermodynamic condition where boiling should occur remains temporarily in a superheated state until centers at which bubbles will grow are introduced. The passage of an ionizing particle through a liquid in the proper state of instability gives rise to local ionization centers at which bubbles form and grow. The path of the particle thus appears as a track of bubbles if one photographs the liquid immediately. For best results the photograph should be within the order of magnitude of 10 microseconds after the passage of the particle. The first photographs of ionizing tracks were obtained in 1953 in clean, smooth walled, small glass chambers containing ether. Since then larger chambers have been operated successfully with liquid hydrogen, nitrogen, carbon tetrachloride, and a variety of organic liquids. The larger chambers are made of metal and glass with gasket seals. Chambers containing the order of a liter of liquid have been constructed. It is not yet clear as to what will set an upper limit on the size of bubble chambers.

The superheated condition of the liquid is produced by a sudden reduction in pressure. Movable diaphragms directly on the chamber are successful as well

[1] D. O. CALDWELL: Phys. Rev. **100**, 291 (1955).

[2] D. A. GLASER: Phys. Rev. **97**, 474 (1955). This contains references to his earlier work. An excellent elementary article on the bubble chamber is D. A. GLASER, Sci. Amer. **192**, 46 (1955). FRETTER has reviewed the subject of bubble chambers: W. B. FRETTER: Ann. Rev. Nucl. Sci. **5**, 145 (1955); see also D. A. GLASER's contribution to Vol. XLV of this Encyclopedia.

as solenoid pop valves at the end of a tube leading from the chamber The reduction of the pressure must be produced in a very short time if there is boiling of the liquid at gasketed junctions and uncoated metal surfaces. The photographic techniques required with bubble chambers are similar to those used with cloud chambers.

The measurement of the relative density of ionization of a particle from the density of bubbles is in a promising state of development. Evidence of this can be seen in some of the lovely photographs collected by HILDEBRAND[1].

27. A comparison of emulsions and various chambers. In planning experiments with track recording devices some of their properties and characteristics are of importance. Table 3 is an attempt to provide a rapid basis for comparing emulsions, various cloud chambers and bubble chambers. The particular measurement that one is trying to perform determines which properties and characteristics are desirable and undesirable; it is a rare experiment which does not demand some compromise.

Table 3. *Properties and characteristics of photographic emulsions and various chambers.*

Property or characteristic	Photographic emulsions	Bubble chamber	Cloud chambers			
			Conventional	High pressure	Diffusion	Large multi-plate
The fraction[2] of the time they are sensitive	Continuous	10^{-4}	10^{-3}	10^{-3}	Continuous	10^{-3}
Can they be pulsed with a machine?	No	Yes	Yes	Yes	by flashing light	Yes
Can they be triggered by a counter system?	No	No	Yes	Yes	by flashing light	Yes
Possibility of varying chemical composition	Poor	Excellent	Good	Fair	Good	Good
Possibility of observing photons	No	Fair	No	No	No	Excellent
Possibility of observing low energy events	Excellent	No	Excellent	No	Good	No
Maximum range[2] that is observable in gm/cm²	10^2	10^2 †	10^{-2}	10	10^{-1}	10^2
Quantities that can be measured: Range	Yes	Yes	Yes	Yes	Yes	Yes
Quantities that can be measured: Density of ionization	Yes	? †	Yes	Yes	Yes	Yes
Quantities that can be measured: Magnetic curvature	No †	Yes	Yes	Yes	Yes	No †
Quantities that can be measured: Multiple scattering	Excellent	Fair	No	No	No	Fair
Quantities that can be measured: δ-rays	Excellent	Poor	Good	Fair	Good	Good
The maximum flux[2] of particles that can be present in a single observation in particles/cm²	10^5	20	1	1	1	1
Repetition rate in seconds with above flux of particles	—	10^{-1}	10	10^2	10	10^3
The possibility of observing a time delay in associated events	No	Yes 10^{-6} to 10^{-5} sec	Poor	Poor	Yes[3] 10^{-6} to 10^{-3} sec	Poor

[1] R. H. HILDEBRNAD: Proceedings of the 1954 Glasgow Conference on Nuclear and Meson Physics, London: Pergamon Press 1955.

[2] Indicates only an order of magnitude.

† Indicates future developments may improve the situation

[3] Personal communication from R. A. SCHLUTER.

28. Solid chambers. ZAVOISKII *et al.*[1] have reported observing the tracks from the passage of ionizing particles in a luminescent crystal of NaI (Th). A solid track detector could be very useful as it can be counter triggered, is very fast, and can be constructed of other luminescent materials. A lens is used to focus the light pulses from the crystal onto an optical (electronic) image amplifier. The output of the image amplifier is then photographed.

The main problems are in the image amplifier and the use of a lens. The sharp focus required on the image amplifier and the present demand that the lens subtend a large solid angle have been shown by WEINSTEIN[2] to require the crystal being rather thin. With their present equipment ZAVOISKII *et al.* use a crystal which is the order of a millimeter thick. The development of better light amplifiers might remove the requirement that the lens subtend a large solid angle. Astronomers and other groups interested in light amplifiers with better signal-to-noise ratios and more sensitive film may be of assistance in developing a new tool for nuclear physics.

D. Reactions of nucleons.

29. Total cross-section for neutrons. Throughout this chapter names will be applied to different types of cross-section measurements which are defined by the type of experiment being performed. *Total cross-section for neutrons are the cross-sections obtained from an attenuation measurement employing "good geometry".* A good geometry is one in which the detector subtends a very small solid angle at the sample. The angle has to be very small at high energies as most of the elastic scattering of high energy particles is in a diffraction peak at forward angles (see Sect. 31). The root mean square angle of the diffraction peak is the order of $(kR)^{-1}$ where $\hbar k$ is the momentum of the particle and R a measure of the nuclear radius. This angle is the order of 5° even for light nuclei for 300 Mev neutrons and protons. Where the intensity of neutrons and the efficiency of the neutron detector do not provide a sufficient counting rate for a really good geometry measurement, a series of measurements are made with the detector subtending different solid angles, i.e. with the detector at various distances from the sample. The data are then extrapolated to an ideal measurement at zero degrees with the aid of a theoretical expression for the diffraction scattering. To avoid these difficulties some of the experimentalists have employed solid angles of the order of 5×10^{-5} steradians.

As explained in the section on neutron sources (Sect. 15) the spectrum of neutrons from high energy machines is generally quite broad. In order to perform measurements with good energy resolution, the detector must accept a narrow energy band. Those detectors which consist of a hydrogenous target from which one observes recoil protons in a proton telescope provide the best energy resolution (see Sect. 21). Good energy resolution is always at the expense of the efficiency of the detecting system.

Table 4 contains a partial list of the measurement of total cross-sections for neutrons. The experiments are listed in order of increasing neutron energy as far as possible. The fifth column lists the detection techniques employed.

The results of some of these experiments are shown in Fig. 7 where the total cross-sections for several typical elements are plotted as a function of energy.

[1] E. K. ZAVOISKII, G. E. SMOLKIN, A. G. PLAKOV and M. M. BUTSLOV: Dokl. Akad. Nauk SSSR. **100**, 241 (1955).

[2] R. WEINSTEIN: Ann. Progr. Report of Laboratory of Nuclear Science of M.I.T., Cambridge, Mass., May 15, 1955.

(The energy is on a logarithmic scale.) The gross features are that with increasing energy the cross-sections all show an appreciable decrease up to about 200 Mev. Between 200 and 400 Mev they are fairly constant. At 1400 Mev the cross-sections have values larger than at 410 Mev.

The details of the variation with energy of the cross-sections has been of great interest. HILDEBRAND and LEITH (see Table 4) first noted that the cross-sections at 42 Mev did not increase smoothly as $A^{\frac{2}{3}}$. TAYLOR and WOOD (Table 4) established by measurements with good energy resolution that there were maxima and minima in the cross-sections below 100 Mev. Specifically, lead had a minimum at about 60 Mev, cadmium and copper had minima in the 30 to 40 Mev range. The light nuclei do not have any minima or maxima but just decrease monotonically.

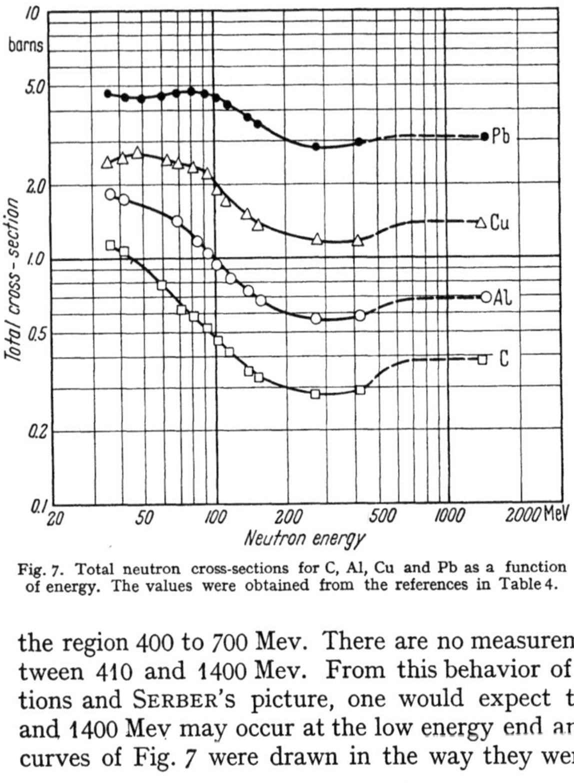

Fig. 7. Total neutron cross-sections for C, Al, Cu and Pb as a function of energy. The values were obtained from the references in Table 4.

The lighter nuclei have a larger fractional decrease with energy than the heavy nuclei, and they also have a larger fractional rise at 1400 Mev. This is in qualitative agreement with SERBER'S [4] semiclassical picture (see Sect. 6) where the cross-sections in complex nuclei are directly related to the fundamental nucleon-nucleon collision cross-sections. The fundamental nucleon-nucleon collision cross-sections decrease rapidly and go through a minimum in the 200 to 400 Mev region. They rise rapidly again in the region 400 to 700 Mev. There are no measurements on complex nuclei in between 410 and 1400 Mev. From this behavior of the nucleon-nucleon cross-sections and SERBER'S picture, one would expect the observed rise between 410 and 1400 Mev may occur at the low energy end and this is the reason the dotted curves of Fig. 7 were drawn in the way they were.

Attempts have been made to interpret many of the experiments according to the optical model of FERNBACH, SERBER and TAYLOR [7], which was first proposed to explain the data of COOK *et al.* (see Table 4). They employed a model which is only a first approximation even optically and neglected refraction and reflection at the nuclear surface. The parameters of their model are r_0, K, and k_1. r_0 is the radius parameter in the relationship $R = r_0 A^{\frac{1}{3}}$ for a sphere of radius R with uniform density. K is the reciprocal mean free path, and $\hbar k_1$ is the *increase* in the momentum of the neutron which occurs when it enters the nucleus (see Sect. 8). The values of these parameters which were assumed or determined in interpreting the experimental data are listed in the last three columns of Table 4. The values are the best values for the simple optical model; the model of FERNBACH *et al.* [7] is very appealing as a parametric form for representing the data; however some of the authors object to the values of the parameters they obtain. For example FOX *et al.* (see Table 4) obtain a value of K which is above that

Table 4. *Measurements of total cross-sections for neutrons.*

Energy in Mev	Energy reduction in Mev	No. of elements measured	Work performed by	Technique employed for neutron detection	r_0 (10^{-13} cm)	K (10^{12} cm^{-1})	k_1 (10^{12} cm^{-1})
42	Broad	33	R. H. HILDEBRAND and C. E. LEITH: Phys. Rev. **80**, 842 (1950)	Measurement of induced radio-activity in C^{12} (n, $2n$) C^{11} reaction			
12 different energies between 60 and 107	Mainly ± 0.5 or ± 1.0	12	V. CULLER and R. WANIAK: Phys. Rev. **99**, 740 (1955)	Proton range telescope employing 6 scintillators			
A series of energies between 30 and 153	± 1 at lower ± 3 at higher energies	6	A. E. TAYLOR and E. WOOD: Phil. Mag. **44**, 95 (1953)	Proton range telescope employing 3 proportional counters			
85	Broad	15	L. J. COOK, E. M. MCMILLAN, L. M. PETERSON and D. C. SEWELL: Phys. Rev. **75**, 7 (1949)	C^{12} (n, $2n$) C^{11} radioactivity	1.37	~2.3	3.22
					Interpreted by FERNBACH, SERBER an TAYLOR		
95	Broad	12	J. J. DEJUREN and N. KNABLE: Phys. Rev. **77**, 606 (1950)	Bismuth fission chamber detector	1.38	3.0	2.85
156	± 3	10	A. E. TAYLOR, T. G. PICKAVANCE, J. M. CASSELLS and T. C. RANDLE: Phil. Mag. **42**, 751 (1950)	Recoil protons by 3 proportional counters	1.37	3.75	0 to 0.7
270	Broad	11	J. J. DEJUREN: Phys. Rev. **80**, 27 (1950)	Bismuth fission chamber			
280	± 15	12	R. FOX, C. E. LEITH, L. WOUTENS and K. R. MACKENZIE: Phys. Rev. **80**, 23 (1950)	Proton telescope of 2 scintillation counters accepted all energies above 250 Mev			
410	± 20	14	V. A. NEDZEL: Phys. Rev. **94**, 174 (1954)	Proton telescope of 3 scintillation counters	1.23	5.1	0
1400	Broad	8	T. COOR, D. A. HILL, W. F. HORNYAK, L. W. SMITH and G. A. SNOW: Phys. Rev. **98**, 1369 (1955)	Proton telescope of 4 counters	1.28	4.9	1.5

calculated from the free proton-proton and neutron-proton cross-section (see Sect. 10).

TAYLOR[1] has tried to fit the observations from 65 to 410 Mev holding the radius of the nuclei constant. He abandons $R=r_0 A^{\frac{1}{3}}$ and obtains the values shown in Table 5. He finds K decreases rapidly and has a broad minimum in the 150 to 300 Mev region with a value of about $1.6\times 10^{12}\,\text{cm}^{-1}$. k_1 decreases monotonically due to V leveling off at a value of 13 Mev at energies above 150 Mev. It is somewhat disconcerting to find the same theory applied to the same data giving entirely different values for the parameters of the theory depending upon the person who is applying the theory.

The minima and maxima observed in the heavier elements below 100 Mev would require the potential of the optical model to go up and down as a function of energy and various nuclei. The values required for lead are given by MANDL and SKYRME[2]. They also compare a variational calculation to the predictions of the optical model.

Table 5. TAYLOR'S *values for nuclear radii as determined by a best fit to total neutron cross-sections.*

Nucleus	R (10^{-13} cm)	$RA^{-\frac{1}{3}}$ (10^{-13} cm)
C	3.52	1.54
Al	4.49	1.50
Cu	5.80	1.45
Cd	6.86	1.42
Pb	8.12	1.37

In that FERNBACH *et al.* [7] used only a first approximation optical theory and neglected refraction and reflection at the nuclear surface they do not predict the minima and maxima observed in the heavier elements below 100 Mev. A more exact optical theory taking account of the effect of the surface would give such minima and maxima according to LAWSON[3]. A more exact treatment of the optical model has been given by MEMMERT[4], who formulates the refractive index, n, in the same manner as JASTROW[5], namely

$$n = 1 + \left(\frac{\pi\varrho}{k^2}\right)[f_{nn}(0) + f_{np}(0)]$$

where ϱ is the density of nucleons and $f_{nn}(0)$ and $f_{np}(0)$ are the real parts of the forward scattering amplitudes in the laboratory.

JASTROW points out that the real parts of the forward scattering amplitude are not necessarily the square root of the scattering cross-sections at zero degrees. He tries to obtain information about the nucleon-nucleon potential from the values of n obtained from interpreting total neutron cross-sections. He interprets the value of $n\sim 1.0$ (i.e. $k_1\sim 0$) at about 300 Mev as evidence favoring a repulsive core for the two-body interaction.

30. Total cross-sections for protons. Despite the fact that proton detectors are much more efficient than neutron detectors and that the energy spread in a proton beam can be made very narrow, comparatively few total cross-section measurements have been performed with protons. To measure total nuclear cross-sections requires an experiment with a good geometry in order to include the elastic scattering diffraction peak. However in a good geometry experiment with charged particles, one encounters the complication of the COULOMB (or RUTHERFORD) scattering. The COULOMB scattering cross-section at small angles is much larger than the total nuclear cross-section. Therefore it is not possible

[1] T. B. TAYLOR: Phys. Rev. **92**, 831 (1953).
[2] F. MANDL and T. H. R. SKYRME: Phil. Mag. **44**, 1030 (1953).
[3] J. D. LAWSON: Phil. Mag. **44**, 102 (1953).
[4] G. MEMMERT: Z. Physik **134**, 42 (1953).
[5] R. JASTROW: Phys. Rev. **82**, 261 (1951).

to actually measure the total nuclear cross-section for protons. The total nuclear cross-section is obtained by extrapolating a series of measurements approaching good geometry.

To extrapolate the nuclear part to zero degrees it is desirable to have a theoretical expression for how the COULOMB scattering varies with angle. There are several difficulties in obtaining such a theoretical extrapolation. The scattering cross-section at forward angles is the square of the sum of the amplitudes for nuclear elastic scattering, $f_n(\vartheta)$, and COULOMB scattering, $f_c(\vartheta)$. Thus the elastic scattering cross-section contains an interference term, namely

$$\sigma(\vartheta)_{\text{elastic}} = f_n^2(\vartheta) + 2\,|f_n(\vartheta)\, f_c(\vartheta)| + f_c^2(\vartheta). \tag{30.1}$$

At high energies the COULOMB scattering is not the simple RUTHERFORD cross-section from a point charge. The finite extension of the charge density of the nucleus must be taken into account. In BORN approximation the COULOMB scattering amplitude from a finite sized charged distribution becomes

$$f_c(\vartheta) = \frac{Z e^2}{2 p v} \frac{1}{\sin^2(\vartheta/2)} F(Q R) \tag{30.2}$$

where $F(QR)$ is the form factor for a nucleus of radius R (see Sect. 7; the BORN approximation is not valid for high Z nuclei). In extrapolating to the total nuclear cross-section, CHEN, LEAVITT, and SHAPIRO[1] used

$$F(Q R) = \frac{1}{1 + [k R \sin(\vartheta/2)]^2}. \tag{30.3}$$

This is the form factor for a YUKAWA shaped density distribution, namely

$$\varrho(r) = \frac{e^{-r/R}}{R^2 r}. \tag{30.4}$$

Further details on the interference between COULOMB and nuclear scattering are given in the next section on elastic scattering cross-sections (Sect. 31).

There are three sets of measurements on total cross-sections for protons that have been reported and they are listed in Table 6. CHEN, LEAVITT and SHAPIRO at 860 Mev defined their beam by three scintillation counters in coincidence; the sample was then placed between the last of these three counters and a fourth counter. All four counters were in coincidence. The transmission of the sample was measured as a function of the distance of the fourth counter

Table 6. *Measurements of total cross-sections for protons.*

Energy Mev	Energy spread Mev	Elements measured	Work performed by	Technique employed
208	± 4	H. D, Be C. O, Al Cl, S	H. G. DECARVALHO: Phys. Rev. **96**, 398 (1954),	Anti-coincidence measurement varying geometry
315	± 8(?)			
408	± 10	H. D, Li Be, C, O	J. MARSHALL, L. MARSHALL and V. A. NEDZEL: Phys. Rev. **91**, 767 (1953)	Anti-coincidence measurement varying geometry
860	± 50	Be, C, Al, Cu, Sn, Pb	F. P. CHEN, C. P. LEAVITT and A. M. SHAPIRO: Phys. Rev. **99**, 857 (1955)	Attenuation measurement in both poor and good geometry

[1] F. P. CHEN, C. P. LEAVITT and A. M. SHAPIRO: Phys. Rev. **99**, 857 (1955).

behind the sample. The half angle subtended by the fourth detector was varied from about 20 to about 2°. They also give a detailed discussion of the corrections required in such measurements. Their samples had transmissions between 0.6 and 0.8; therefore there was a correction for COULOMB multiple scattering. They kept this correction (to the transmission) smaller than 10%.

The multiple COULOMB scattering can be treated as a random walk problem with a mean square scattering angle (in radians) which can be approximately estimated from the relationship

$$(\vartheta_s^2)_{\text{average}} = \left(\frac{2l}{p\,v}\right)^2 \frac{\Delta x}{l_r} \tag{30.5}$$

where $p\,v$ is to be expressed in Mev. Δx is the thickness of the sample, and l_r is the radiation length (see ROSSI [1]). From Eq. (30.5) it will be seen that the multiple scattering increases appreciably at lower energies. To keep the multiple scattering correction small, MARSHALL, MARSHALL, and NEDZEL[1] and DE CARVALHO[2] at the lower energies used very thin samples. This meant that the transmission of the samples was close to unity. To avoid the difficulties of exactly measuring almost equal counting rates, the measurements at the lower energies were made by an anti-coincidence technique. Specifically the counter behind the sample was in anti-coincidence with those in front of the sample and the number of protons removed from the beam was measured. They also varied the solid angle of the detector behind the sample; the angles subtended varied from 2 to 15°.

Table 7. *Extrapolated total cross-sections for protons (in millibarns).*

Element	208 Mev	315 Mev	408 Mev	860 Mev
D	62 ± 4		55.6 ± 2.2	
Be	247 ± 6	229 ± 6	242 ± 6	316 ± 16
C	296 ± 4	292 ± 6	285 ± 14	405 ± 23
O	395 ± 8	379 ± 10	406 ± 3	
Al	592 ± 10	580 ± 16		750 ± 50
S	680 ± 20	686 ± 18		
Cl	740 ± 20	740 ± 20		

The values of the extrapolated total cross-section for seven elements obtained from these three measurements are listed in Table 7 (see Table 6 for references). (It is to be noted that there are unexplained discrepancies in the tables of results given by DECAVALHO; only these values are quoted which seem to be free of discrepancies.) Cross-sections for the heavier elements are not given by the authors because of the uncertainties in the COULOMB scattering corrections. The values of the total proton cross-sections are almost all in agreement within the experimental error with the values obtained for neutron total cross-sections at similar energies. They should be the same on the basis of the assumption of the charge independence of nuclear forces at these energies.

The dependence of the proton cross-sections on energy is the same as that pointed out previously for neutrons. The cross-sections are almost independent of energy, between 200 and 400 Mev and then rise slightly beyond 400 Mev. In that the values of the cross-sections for 560 Mev protons are in agreement with the values for 1400 Mev neutrons, there is some experimental basis for the manner in which the dotted curves were drawn in Fig. 7 for the neutron cross-sections.

The values for the parameters of the optical model for the protons seem to be $r_0 = 1.24 \times 10^{-13}$ cm and $K = 5.0 \times 10^{12}$ cm^{-1} at lower energies and $K = 5.6 \times 10^{12}$ cm^{-1} at 860 Mev. However the authors felt that they would find a better fit to their data if they abandoned the simplicity of a uniform density sphere of

[1] J. MARSHALL, L. MARSHALL and V. A. NEDZEL: Phys. Rev. **91**, 767 (1953).
[2] H. G. DE CARVALHO: Phys. Rev. **96**, 398 (1954).

radius $R = r_0 A^{\frac{1}{3}}$. In the case of CHEN *et al.* the basis for this dissatisfaction is in the shapes of the extrapolation curves. Such curves are intimately connected with the angular distribution of elastic scattering where others also find it necessary to abandon the uniform density model of the nucleon. The details are given in the next section.

31. Angular distribution of scattered nucleons. From the previous section, it will be realized that it is easier to understand the scattering of neutrons than protons because the neutrons are free of the complication of COULOMB scattering. For this reason neutrons are treated first.

The angular distributions of 84 Mev neutrons scattered from Al, Cu, and Pb were measured by BRATENAHL, FERNBACH, HILDEBRAND, LEITH and MOYER[1]. The distributions were studied from 2°.5 to about 25°. They used the broad spectrum of neutrons from the stripping of 190 Mev deuterons on a Be target (see Sect. 35). The scattered neutrons were detected by measuring the radioactivity induced in the C^{12} $(n, 2n)$ C^{11} reaction. Therefore inelastically scattered neutrons were being measured as well as the elastic. Fig. 8 shows the results they obtained. As mentioned previously the scattering is mostly in a diffraction peak at forward angles.

The curves drawn are theoretical curves for elastic scattering using the optical model of a transparent nucleus with a uniform density. The uniform density model fails to give the correct angular distribution.

RICHARDSON, BALL, LEITH and MOYER[2] measured the angular distribution of 340 Mev protons scattered by 12 elements ranging from C to Bi.

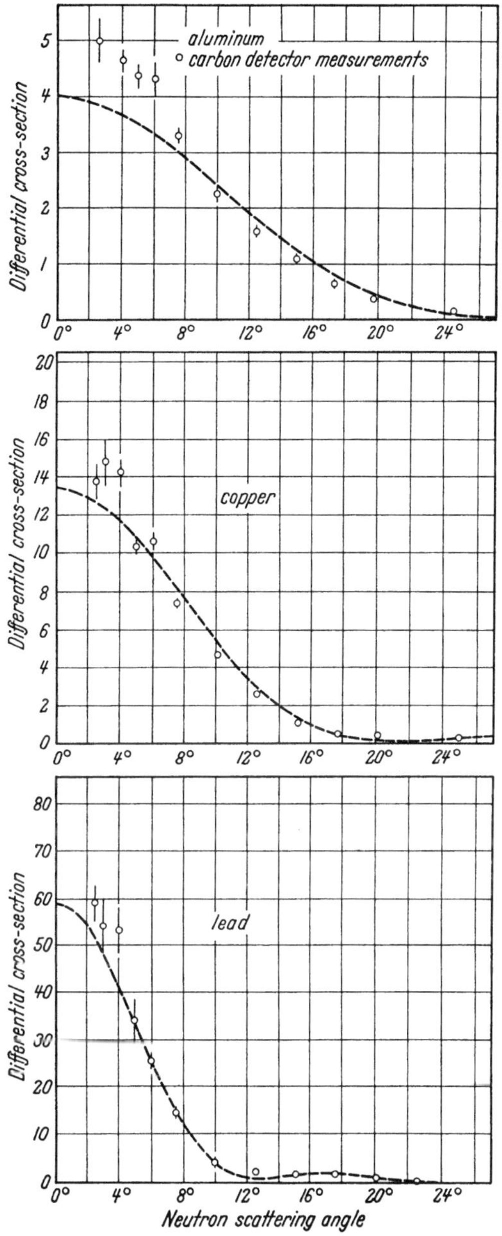

Fig. 8. The angular distributions of 84 Mev neutrons scattered from Al, Cu, and Pb. The dotted curves are those predicted by the optical model [7] for a square well potential. [Figure from BRATENAHL *et al.*: Phys. Rev. 77, 597 (1950).]

[1] A. BRATENAHL, S. FERNBACH, R. H. HILDEBRAND, C. E. LEITH and B. J. MOYER: Phys. Rev. **77**, 597 (1950).

[2] R. E. RICHARDSON, W. P. BALL, C. E. LEITH jr. and B. J. MOYER: Phys. Rev. **83**, 859 (1951); **86**, 29 (1952).

They employed a proton range telescope with a threshold at about 330 Mev in most of their measurements. The type of results they obtained are shown in Figs. 9 and 10 for the case of carbon and lead. At small angles the COULOMB scattering interferes with the nuclear scattering. The heavier nuclei with larger radii have much narrower diffraction peaks. However, even in a light nucleus like carbon at 340 Mev, the differential cross-section has dropped by a factor of about 10^3 by 30°.

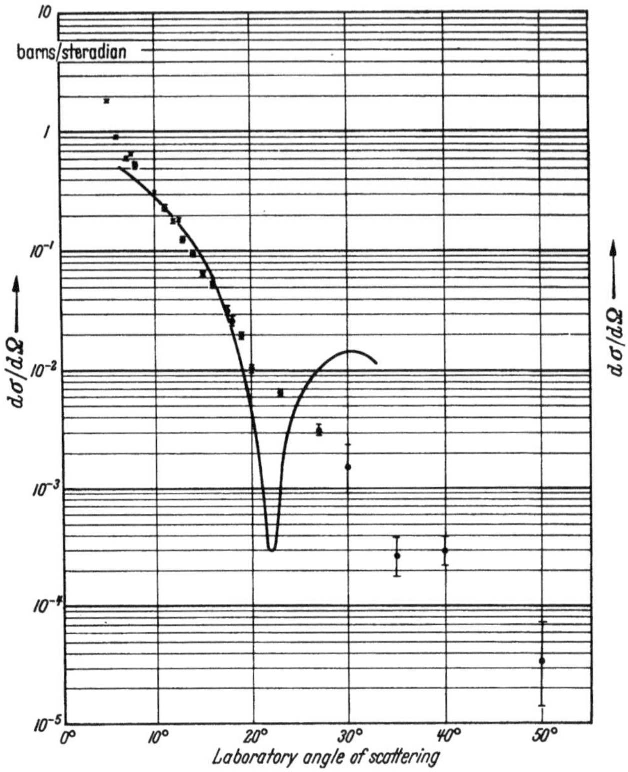

Fig. 9. Differential elastic scattering cross-section as a function of angle for 340 Mev protons on carbon. The curve is that predicted by the optical model [7] for a uniform potential with a correction applied for the effect of the finite angular resolution (~ ½ degree) of the experiment. [Figure from RICHARDSON *et al.*: Phys. Rev. **86**, 29 (1952).]

An indication of the amount of inelastic scattering can be obtained from the work of STRAUCH and TITUS[1]. They employed a range telescope with six counters which had an energy resolution of 2 Mev, and they studied both the energy and angular distribution of the scattered protons. The initial proton energy was 95 Mev; the overall energy resolution was 3 Mev, including the spread in beam energy, the energy loss in the targets, and the resolution of the detector. The energy distribution of the protons scattered from C at 44°, is shown in the accompanying Fig. 11. It will be seen that the inelastic scatterings which leave the nucleus in excited states are readily distinguishable in this light nucleus. In heavier nuclei where there are several excited states within 3 Mev of the ground state, the inelastic scattering runs into the elastic. The highest energy peak mostly due to elastic scattering. The elastic and inelastic scattering from carbon

[1] K. STRAUCH and W. F. TITUS: Phys. Rev. **95**, 854 (1954). Further work will be published in Phys. Rev.

are commensurate at 40°. At smaller angles the elastic rises more rapidly than the inelastic. STRAUCH's later work indicates that the differential cross-section for elastic and inelastic scattering from a single level fall off at approximately the same rate at larger angles.

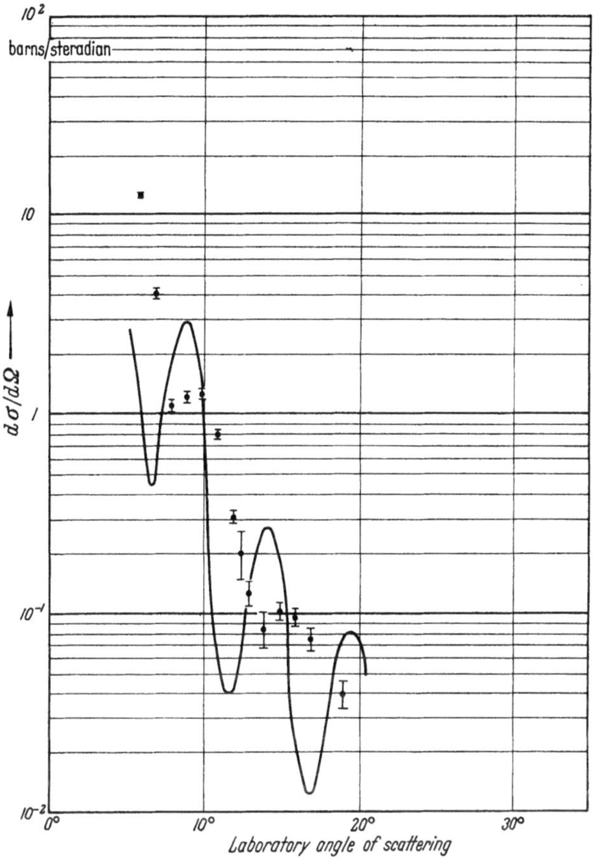

Fig. 10. The differential elastic scattering cross-section as a function of angle for 340 Mev protons on Pb. The curve is that predicted by the optical model [7] for a uniform potential with a correction applied for the effect of the finite angular resolution ($\sim \frac{1}{6}$ degree) of the experiment. [Figure from RICHARDSON *et al.*: Phys. Rev. **86**, 29 (1952).]

STRAUCH and TITUS and RICHARDSON *et al.* observe slight maxima and minima (see Figs. 9—11). From optics one gets that the diffraction pattern from a disc should have deep minima at the values of ϑ given by

$$J_1\left(2kR' \sin\frac{\vartheta}{2}\right) = 0 \qquad (31.1)$$

where J_1 is the first order BESSEL function. These minima would go to zero for neutron scattering if the nucleus could be considered an opaque disc of radius R'. For a semi-transparent nucleus, R' is slightly smaller than the radius of the nucleus. The depth of the minima anticipated for protons on the optical model of FERNBACH, SERBER, and TAYLOR [7] are shown in Figs. 9 and 10 with the finite instrumental angular resolution taken into account. It is seen that the experimental minima are not as deep as predicted by the theory. GATHA and RIDDELL[1] have carried out calculations using the W.K.B. method and have

[1] K. M. GATHA and R. J. RIDDELL jr.: Phys. Rev. **86**, 1035 (1952).

evaluated the phase shifts approximately. With a uniform density model (a sharp edge) for the nucleus they also get much more pronounced maxima and minima than are observed. They also took account of the interference of COULOMB and nuclear elastic scattering [see Eq. (30.1)].

The maxima and minima become shallow as observed experimentally if one uses a nuclear potential which is a function of the radial displacement. STERNHEIMER[1] showed that the maxima and minima are not as deep as those of a

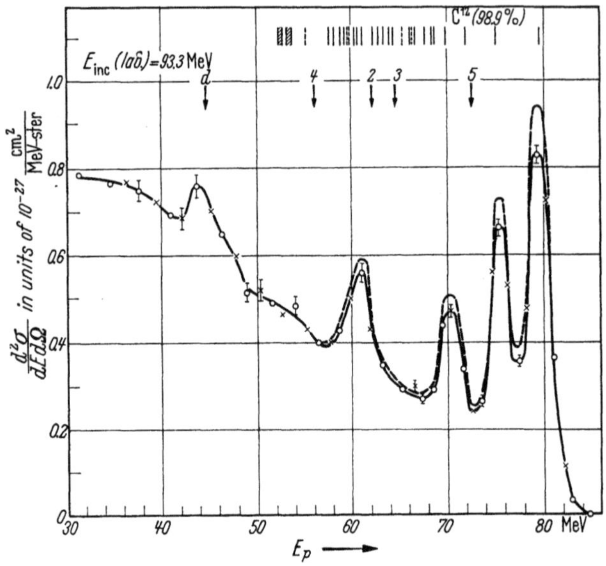

Fig. 11. The energy distribution of protons scattered from a carbon target at 43.9° in the center of mass system. The incident protons have an energy of 93.3 Mev in the laboratory. The vertical lines at the top correspond to the energy levels of the C^{12} nucleus and the scattering corresponding to the first ex tced state appears to be resolved. The numbered arrows indicate the threshold for the following reactions; 2(p, pn); 3(p, $2p$); 4(p, dp); 5(p, p). [Figure from STRAUCH and TITUS: Phys. Rev. 95, 854 (1955).]

uniform density (or square well potential) if one employs a radial dependence of the type first used by SAXON and WOODS[2], namely

$$V = \frac{V_r + i\,V_i}{1 + \exp\left[(r - r_0)/a\right]} \tag{31.2}$$

with a value of a of about 0.5×10^{-13} cm (see Sect. 8). However, even this model apparently gives maxima and minima that are more pronounced than those observed experimentally. A GAUSSian shaped potential would fit the experimental data for light nuclei better than Eq. (31.2). The theoretical situation is being appreciably aided by the measurement of the polarization of the scattered nucleons (see Sect. 33).

It might appear that the situation is not completely satisfactory from a theoretical point of view; however it should be noted that except for the most recent work of STRAUCH, it is really unsatisfactory on the experimental side as well. The inelastic scattering observed by STRAUCH does not have maxima and minima. Therefore in experiments with poor energy resolution or for those nuclei with low lying excited levels, the inelastic scattering will tend to make the maxima and minima appear less pronounced.

[1] G. A. SNOW, R. M. STERNHEIMER and C. N. YANG: Phys. Rev. 94, 1073 (1954). — R. M. STERNHEIMER: Phys. Rev. 97, 1314 (1955); 100, 886 (1955).

[2] R. D. WOODS and D. S. SAXON: Phys. Rev. 95, 577 (1954).

Also it should be pointed out that the minima and maxima are observed to be more pronounced in lead than in a nucleus like Ta with a large quadrupole moment, i.e. in nuclei which are known to have an ellipsoidal shape. More data are needed on nuclei which are believed to have a spherical shape; unfortunately these generally have many isotopes.

Despite the lack of quantitative agreement between experiments and theory on the depth and height of the maxima and minima, the angles at which the maxima and minima occur seem to establish the basic diffraction character of the elastic scattering. If one calculates R' in Eq. (31.1) from the position of the second and higher minima observed by STRAUCH for various nuclei, one finds $R' = (1.28 \pm 0.08)\, A^{\frac{1}{3}} \times 10^{-13}$ cm. (The position of the first minimum may be affected by the interference with the COULOMB field.)

The agreement with Eq. (31.1) may be fortuitous; however, independent of the nuclear model chosen, the maxima and minima should still occur in different nuclei for the same values of QR, that is for the same values of the collisive momentum, times a radial parameter. (This holds only beyond where COULOMB interference affects the position of the maxima and minima in proton measurements.)

The decrease at larger angles of the differential cross-section for inelastic scattering can be understood qualitatively in terms of a constructive interference from the many high angular momentum components which are present in a collision of a fast nucleon with a nucleus. Even in a nucleus as light as Al and at an energy as low as 100 Mev $kR \approx 10$, and 10 phase shifts need to be considered.

32. Absorption cross-section. From the measurement of the angular distribution of scattered particles (previous Sect. 31) it was established that the elastic scattering occurred in a narrow cone at forward angles. Therefore, in principle, *if one measures the attenuation in a "poor geometry" experiment one should obtain a measure of the absorption cross-section.* The "poor geometry" needs to be one in which the detector subtends a solid angle at the attenuator sufficient to include all the elastically scattered particles. At high energies this is a reasonably small solid angle so that inelastically scattered particles and reaction products should have a low probability of being observed in an attenuation measurement.

Many of the experimentalists who were studying total cross-sections by good geometry measurements (see references of Tables 4 and 5) also changed to poor geometry to get an indication of the absorption cross-section. BRATENAHL *et al.*[1] and DE JUREN and KNABLE[2] were amongst the first to do this. The main result of their poor-geometry measurements was to show that the absorption cross-section is slightly less than half of the total cross-section; they find the ratio of σ_a/σ_T (absorption cross-section/total cross-section) for various elements is between 0.39 and 0.50 for 90 Mev neutrons. More recent experiments at higher energies by CHEN *et al.*[3] (860 Mev proton) and COOR *et al.*[4] (1400 Mev neutrons) show an increase in the ratio σ_a/σ_T. They find σ_a/σ_T is between 0.50 and 0.65.

Three groups not referred to in Tables 4 and 5 have undertaken a series of experiments to determine absorption cross-sections. They are listed in the accompanying Table 8.

Some of the results of the measurements of absorption cross-sections are summarized in the accompanying Table 9. Except for the two low energy values,

[1] A. BRATENAHL, S. FERNBACH, R. H. HILDEBRAND, C. E. LEITH and B. J. MOYER: Phys. Rev. **77**, 579 (1950).
[2] J. DE JUREN and N. KNABLE: Phys. Rev. **77**, 606 (1950).
[3] F. P. CHEN, C. P. LEAVITT and A. M. SHAPIRO: Phys. Rev. **99**, 857 (1955).
[4] T. COOR, D. A. HILL, W. F. HORNYAK and L. W. SMITH: Phys. Rev. **98**, 1369 (1955).

the cross-section for each element is essentially constant up to 1400 Mev. There is also very little difference if any between proton absorption cross-sections and neutron absorption cross-sections.

The cross-sections for the complex charged particles are similar for the different charged particles and except for lead are more than double the single nucleon cross-sections. An interpretation of the deuteron cross-sections has been given by HECKROTTE[1]. He assumes that the most frequent process for a complex particle is the breaking up of the unit, e.g. deuteron stripping. The deuteron stripping process is discussed in Sect. 35 below.

Table 8. *Absorption cross-section experiments.*

Energy (Mev)	Resolution	Particle	Elements	Work performed by	Technique
55 81 105 140	10%	Neutrons	C, Al, Cu, Cd and Pb	R. P. G. VOSS and R. WILSON: Proc. Roy. Soc. Lond. Ser. A **236**, 41 (1956)	Large liquid scintillation neutron counter with pulse height analysis. Attenuation measurement
134	± 4 Mev	Protons	C, Al, Cu, Cd and Pb	L. M. CASSELLS and J. D. LAWSON: Proc. Phys. Soc. Lond., A **67**, 125 (1954)	Anticoincidence measurement of protons removed from beam. Gas counters and scintillation counters
160	± 15 Mev	Deuterons	Be, C, Al, Cu, Ta, Pb, Bi, U	G. P. MILLBURN, W. BIRNBAUM, W. E. CRANDALL and L. SCHECTER: Phys. Rev. **95**, 1268 (1954)	Attenuation measurement. Detector was a FARADAY cup which observed charged particles which penetrated a (range) absorber
290	± 15 Mev	Protons	Be, C, Al, Cu, Pb, and U		
240	± 15 Mev	He^4	C, Cu, Ta		
315	± 15 Mev	He^3	C, Al, Cu, Ta, and U		

The absorption cross-sections of VOSS and WILSON at low energies decrease monotonically with increasing energy whereas the total cross-sections for the heavy elements (see Sect. 29) show maxima and minima. If one assumes that $\sigma_{\text{total}} = \sigma_{\text{absorption}} + \sigma_{\text{elastic scattering}}$ then the maxima and minima must be associated with the elastic scattering (or diffraction scattering).

If one employs the simple optical model assumptions of FERNBACH, SERBER, and TAYLOR [7] of a uniform density nucleus of radius $R = r_0 A^{\frac{1}{3}}$, then the absorption cross-section is given by the relationship

$$\sigma_a = \pi R^2 \left[1 - \frac{1 - (1 + 2KR)\,e^{-2KR}}{2K^2 R^2}\right]. \tag{32.1}$$

The index of refraction, or k_1, is not in the expression for the absorption cross-section.

If one chooses a value of $K = 5 \times 10^{12}\ \text{cm}^{-1}$, then $r_0 = (1.28 \pm 0.3) \times 10^{-13}$ cm fits the values of the cross-sections fairly well in the region where they are essentially constant. A smaller value of K gives a larger value of r_0 and also provides a reasonable fit to the data.

[1] W. HECKROTTE: Phys. Rev. **95**, 1279 (1954).

The constancy of the absorption cross-sections seems to imply that K and r_0 are constant. This would be in disagreement with the variation of K with energy found by TAYLOR[1] to be required in order to explain the total cross-sections unless one wanted to let r_0 vary as a function of energy. Actually, from the nucleon-nucleon cross-section K is expected to vary, and therefore r_0 must vary. To some extent this illustrates that r_0 is just a convenient parameter with the properties of a unit of length.

It appears at this time that one should abandon the assumptions of a uniform density distribution and $R = r_0 A^{\frac{1}{3}}$ (see VOSS and WILSON). The absorption cross-section data can be satisfactorily interpreted at high energies using the density distributions required to explain other experimental data. WILLIAMS[2] has employed SERBER'S picture (see Sect. 6) and an approximate formula for the tapered

Table 9. *Absorption cross-sections for* C, Al, Cu, *and* Pb.

Particle	Energy (Mev)	C (barns)	Al (barns)	Cu (barns)	Pb (barns)
Neutrons	55	0.276	0.499	1.122	2.237
	81	0.202	0.412	0.840	1.977
	105	0.234	0.428	0.773	1.715
	140	0.221	0.418	0.741	1.704
	300[3]	0.203	0.390	0.755	1.72
	1400	0.200	0.410	0.670	1.73
Protons	134	0.220	0.373	0.752	1.78
	240[3]	0.202	0.383	0.667	1.57
	290	0.199	0.416	0.717	
	860	0.209	0.394	0.728	1.68
Cosmic ray neutrons and Protons	30000[4]	0.256			2.15
Deuterons	160	0.67	1.0	1.76	3.44
He^3	315	0.59	0.91	1.8	
He^4	240	0.64		1.8	

nuclear density distribution obtained from interpreting electron scattering experiments (see Sect. 37). K is then a function of the radial distance inside the nucleus. For interpreting the 1400 Mev neutrons data he takes $\sigma_{\text{effective}}$ inside nuclear matter to be equal to the observed σ_{np} cross-section (i.e. 43 millibarns). He then calculates the equivalent radii, and finds they are in good agreement with those obtained for the charge distribution calculations (see Sect. 37) from electron scattering. In terms of the size of the equivalent square well radius $(r^2)_{\text{av}}$ he gets

$$r_0 = 1.19 \times A^{\frac{1}{3}} \times 10^{-13} \text{ cm}.$$

WILLIAMS has also summarized and interpreted the equivalent of absorption cross-sections obtained in cosmic ray experiments. The experiments consist in general of observing the production of at least two charged particle events from the collision of a high energy incoming particle. He obtains average cross-sections from the data for both neutrons and protons, and these are listed in Table 9. He judges the mean energy to be about 30000 Mev.

It is to be noted these cross-sections at 30000 Mev are greater than those at 1400 Mev. Using the same density distribution employed in interpreting the

[1] T. B. TAYLOR: Phys. Rev. **92**, 831 (1953).
[2] R. W. WILLIAMS: Phys. Rev. **98**, 1387 (1955).
[3] From unpublished work of BALL listed by MILLBURN *et al.*
[4] Reaction cross-sections compiled by R. W. WILLIAMS (see text).

1400 Mev data, WILLIAMS finds that $\sigma_{\text{effective}}$ must be 120^{+30}_{-20} mb (millibarns) at about 30000 Mev. He interprets this as the value of the fundamental nucleon-nucleon cross-sections at these high energies; $\sigma_{np} = \sigma_{pp} = 120$ mb is an appreciable increase over the value of $\sigma_{np} = 43$ mb at 1400 Mev.

The absorption cross-sections are supposed to be the total cross-sections for all reactions except elastic scattering. The absorption cross-section should include the following processes: the quasi-elastic scattering, the deuteron pick-up process, the formation of stars in photographic emulsion, the radioactive products studied by radio-chemical techniques. All of these are treated separately in later sections. The absorption cross-section also includes most of the inelastic scattering (i.e. the scattering of the original particle with the nucleus left in an excited state). Information on such inelastic scattering is scanty (see Sect. 31). As one goes above 200 Mev π meson production sets in; this process would also contribute to the absorption cross-section. At very high energies, meson production is believed to be the dominant reaction.

None of the people who have studied absorption cross-sections have tried to account for the values they obtained in terms of the various reactions involved. The absorption cross-section can be related to the individual reactions quantitatively by means of a GOLDBERGER [5] type Monte Carlo calculation (see Sect. 11). For very light nuclei at not too high energies, one would anticipate that the quasi-elastic scattering is the most important single reaction. There is some indication of this in the type of experiments where the energy and angular distribution of ejected particles are studied. For example, HOFFMAN and STRAUCH[1] obtained the energy and angular distributions of the neutrons ejected when 95 Mev protons bombarded a series of elements. Related experiments are described in the section on quasi-elastic scattering (Sect. 34).

33. Polarization experiments. SCHWINGER[2] and WOLFENSTEIN[3] pointed out that if there are spin dependent forces that one will obtain a partially polarized beam of neutrons or protons after scattering the particle. The polarization is along an axis perpendicular to the plane of the scattering. The original calculations of WOLFENSTEIN were motivated as a check on nuclear models for low energy phenomena. However, WOLFENSTEIN[4] and ASHKIN have provided a derivation of the general relationships between the polarization produced in a single scattering and the asymmetry that should be observed in a double scattering.

The degree of polarization is conveniently defined by

$$P(\vartheta_1) \equiv \frac{N_+(\vartheta_1) - N_-(\vartheta_1)}{N_+(\vartheta_1) + N_-(\vartheta_1)} \tag{33.1}$$

where $N_+(\vartheta_1)$ and $N_-(\vartheta_1)$ are the number of particles scattered at an angle ϑ_1 with spin up and down (or vice versa). If the initial beam is unpolarized then

$$N_\pm(\vartheta_1) = C\,\sigma(\vartheta_1)\,[1 \pm \delta(\vartheta_1)] \tag{33.2}$$

where C is a constant independent of polarization and comes from the flux of particles and the target thickness; $\sigma(\vartheta_1)$ is the scattering cross-section and $\delta(\vartheta_1)$ is the fraction of the cross-section that leads to polarization. Combining Eqs. (33.1) and (33.2), one obtains

$$P(\vartheta_1) = \delta(\vartheta_1).$$

---

[1] J. A. HOFFMAN and K. STRAUCH: Phys. Rev. **90**, 449 (1953).
[2] J. SCHWINGER: Phys. Rev. **69**, 681 (1946); **73**, 407 (1948).
[3] L. WOLFENSTEIN: Phys. Rev. **75**, 1664 (1949); **76**, 541 (1949).
[4] L. WOLFENSTEIN and J. ASHKIN: Phys. Rev. **93**, 806 (1954).

Normally detectors are not sensitive to the polarization and a measurement of the scattering gives the total intensity

$$N_{\text{total}} = N_+ + N_- = 2C\,\sigma(\vartheta_1)\,. \tag{33.3}$$

However if one scatters the partially polarized beam again at angle ϑ_2 as shown in Fig. 12, one will observe an asymmetry in the number of particles scattered. The asymmetry manifests itself as a difference in the counting rates observed at angles ϑ_2 and $-\vartheta_2$ if the second scattering is in the same plane as the first scattering.

With a detector which is independent of polarization one measures the total intensity at $(+\vartheta_2)$ and $(-\vartheta_2)$

$$N(\vartheta_1, \pm\vartheta_2) = N_+(\vartheta_1, \pm\vartheta_2) + N_-(\vartheta_1, \pm\vartheta_2)\,. \tag{33.4}$$

The fractional asymmetry in the observed intensities is frequently expressed in

$$\varepsilon(\vartheta_1, \vartheta_2) = \frac{N(\vartheta_1, +\vartheta_2) - N(\vartheta_1, -\vartheta_2)}{N(\vartheta_1, +\vartheta_2) + N(\vartheta_1, -\vartheta_2)} = P(\vartheta_1)\,P(\vartheta_2)\,. \qquad (33.5)^{1}$$

If the second scattering is not in the same plane as the first scattering, but is in a plane that makes an angle φ (azimuthal angle) to the original plane, then

$$\left.\begin{aligned} &N(\vartheta_1, \vartheta_2, \varphi) = \\ &N(\vartheta_1, \vartheta_2, 0)\,(1+\varepsilon\cos\varphi)\,. \end{aligned}\right\} \tag{33.6}$$

Therefore the asymmetry disappears if the second scattering is in a plane perpendicular to the plane of the original scattering. Generally this is used as a check on the geometry and proper functioning of the experimental equipment.

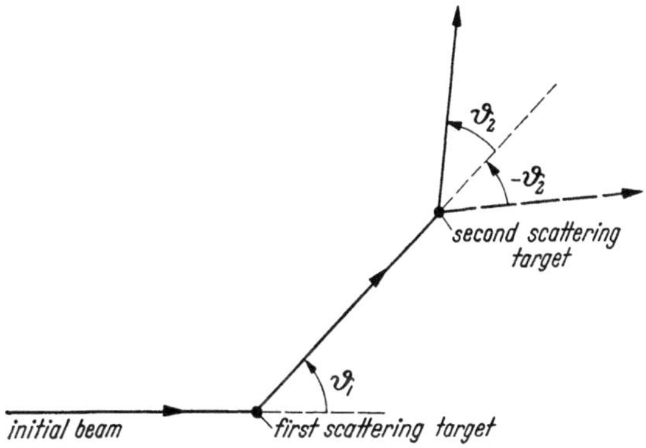

Fig. 12. The angles involved in a polarization experiment for the case where both planes of scatterings are in the plane of the paper.

Since the first experiments of OXLEY *et al.*[2] in the U.S. and DICKSON and SALTER[3] in Great Britain, almost every laboratory with high energy protons has undertaken polarization experiments. The number of experiments reported is extremely large and the amount of activity going on in this field make a summary obsolete in a short time. Reviews of polarization experiments can be found in the proceedings of the Rochester Conferences[4] and Glasgow Conference[5].

Frequently the angle of the first scattering is kept reasonably small in order to obtain mainly elastically scattered events. The detectors are placed so as to observe the scattering from the second target which may or may not be of the same element as the first target. Generally two sets of detectors are used and are placed symmetrically relative to the second target. In some experiments the accuracy of the angles relative to the second target is the order of two minutes of arc, although the detectors may subtend several degrees. As a check, the

[1] Some of the experimentalists have put in a factor of 2 because of (33.3).

[2] C. L. OXLEY, W. F. CARTWRIGHT and LEROUVINA: Phys. Rev. **93**, 806 (1954).

[3] J. M. DICKSON and D. C. SALTER: Proc. Phys. Soc. Lond. A **66**, 721 (1953).

[4] Proceedings of the fourth and fifth Annual Rochester Conference, Interscience Publishers, Inc., New York 1954, 1955.

[5] Proceedings of the 1954 Glasgow Conference on Nuclear and Meson Physics, Pergamon Press, London (1955).

detecting system is rotated into a plane perpendicular to the original plane of scattering. There should be no asymmetry when it is perpendicular [see Eq. (33.7)] The asymmetry is in the plane of the first scattering. In order to differentiate between elastic scattering and quasi-elastic scattering (see next Sect. 34) good energy resolution is required. The magnitude of the polarization can be determined if the same target material is used for first and second scattering targets and the two angles are the same. Then in theory $\varepsilon = P_1^2$; in practice there are complications due to the first scattering frequently being performed inside the cyclotrons where there is an energy spread in the initial beam so that some of the scattered particles observed are really from inelastic scatterings. Corrections need to be made for such effects and the energy loss in the targets especially if one is working at energies below 100 Mev.

Once one has determined $|P_1|$, then it is possible from asymmetry measurements with the second scattering angle changed, or the target changed, to measure $|P_2|$ as a function of angle and target material.

The most extensive studies of this type are those of DICKSON, ROSE, and SALTER[1] from which many of the following conclusions are drawn:

(1) The sign of the asymmetry remains the same for all targets measured, including hydrogen.

(2) The angular variations are not smooth functions of ϑ except for the lightest elements at low energies.

(3) Neighboring medium weight and heavy elements produce very similar variations of the asymmetry as a function of angle.

(4) The maxima and minima in the angular variation of the polarization lie close to the maxima and minima of the differential elastic scattering diffraction patterns.

(6) The asymmetry decreases in general with increasing atomic weight of the target material.

(7) As the energy is reduced from 130 to 66 Mev there is a reduction in the polarization by about a factor of five (for Be and C).

(8) At 130 Mev the maximum polarization is the order of 0.6 to 0.8 in the lighter nuclei.

(9) The maximum polarization remains fairly constant from 130 to 400 Mev (see CHAMBERLAIN *et al.*[2], DE CARVALHO *et al.*[3], and KANE *et al.*[4]).

(10) The elastically scattered protons are much more strongly polarized than the quasi-elastically scattered protons.

(11) By slowing down the protons and scattering them on He, MARSHALL and MARSHALL[5] established that, when the scattering is to the right, the direction of polarization of the proton is down.

FERMI pointed out that an LS coupling term added to a complex potential would give rise to the polarization. He estimated on a simple model the strength of the LS coupling term needed to account for the observed polarization, and he found that the amount of LS coupling was the same as that required by the nuclear shell model calculations. One gains no insight from this as to the basic origin of the LS coupling.

[1] J. M. DICKSON, B. ROSE and D. C. SALTER: Proc. Phys. Soc. Lond. A **68**, 361 (1955).
[2] O. CHAMBERLAIN, E. SEGRÈ, R. TRIPP, C. WIEGAND and T. YPSILANTIS: Phys. Rev. **95**, 1105 (1954).
[3] H. G. DE CARVALHO, J. MARSHALL and L. MARSHALL: Phys. Rev. **96**, 1081 (1954).
[4] J. A. KANE, R. A. STALLWOOD, R. B. SUTTON, T. H. FIELDS and J. G. FOX: Phys. Rev. **95**, 1694 (1954).
[5] L. MARSHALL and J. MARSHALL: Phys. Rev. **98**, 1398 (1955).

STERNHEIMER[1], HECKROTTE[2], and others have used the LS coupling added to a tapered potential model to try to account for the magnitude and angular dependence of the observed polarization. They are able to account fairly well for the behavior as a function of angle, and STERNHEIMER shows that the polarization need not change sign in passing through a minimum in the elastic diffraction scattering.

WOLFENSTEIN[3] has pointed out that one can obtain additional useful information from triple scattering experiments. The feasibility of such experiments has been demonstrated by the measurements of YPSILANTIS *et al.*[4].

It is to be noted that the knowledge of the polarization of proton beams and the scattering techniques for measuring the degree of polarization make possible an enormous variety of new experiments. One can obtain added information on most processes by investigating their dependence on polarization; this includes processes where a proton is the impinging particle or the product particle, and where mesons, neutrons, deuterons etc. are involved. (An asymmetry may be present in any experiment involving the measurement of three momenta where at least one of the particles has a spin.) Polarization experiments are not limited to the elastic scattering of protons. It is not possible in this brief report to cover polarization experiments already performed with particles other than protons.

34. Quasi-elastic scattering. In the course of making proton-proton scattering measurements, it occurred to CHAMBERLAIN and SEGRÈ[5] that light nuclei should be sufficiently transparent at high energies to allow observation of proton-proton collisions within the nucleus. Thus one might explore the model in which nucleons within nuclear mater act very much as do free nucleons.

The name "quasi-elastic" scattering was proposed by CLADIS, HESS, and MOYER[6] in their concurrent studies of this process. CHAMBERLAIN *et al.*[7] established that quasi-elastic scattering occurs by looking for the characteristic feature of a collision of two bodies of equal mass, namely that the particles after the collision come off at 90° to each other in the laboratory. They used a lithium target and the equipment from their proton-proton scattering experiment. They kept one counter fixed and moved the other counter. They observed coincidences at the expected angles (corrected for the binding energy in Li). Due to the protons in lithium being in motion the coincidences were observed over a 30° spread in angles which was consistent with the protons in lithium having a FERMI momentum distribution with a maximum energy of about 20 Mev. These experiments were of a preliminary nature mainly to establish the quasi-elastic scattering process and were not intended as a detailed study of the momentum distribution.

CLADIS *et al.* measured the energy spectrum of protons at 40° from the bombardment of carbon by 340 Mev protons. They obtained a bell shaped spectrum centered about 130 Mev which extended up to about 300 Mev. They used a 35 channel magnetic spectrometer.

WOLFF[8] analyzed the spectrum observed by CLADIS *et al.* in terms of various assumed momentum distributions. A FERMI momentum distribution (with

[1] R. M. STERNHEIMER: Phys. Rev. **97**, 1314 (1955); **100**, 886 (1955).
[2] W. HECKROTTE: Phys. Rev. **101**, 1406 (1956).
[3] L. WOLFENSTEIN: Phys. Rev. **96**, 1654 (1954).
[4] T. YPSILANTIS, C. WIEGAND, R. TRIPP, E. SEGRÈ and O. CHAMBERLAIN: Phys. Rev. **98**, 840 (1955).
[5] O. CHAMBERLAIN and E. SEGRÈ: Phys. Rev. **87**, 81 (1952).
[6] J. B. CLADIS, W. N. HESS and B. J. MOYER: Phys. Rev. **87**, 425 (1952).
[7] O. CHAMBERLAIN, E. SEGRÈ and C. WIEGAND: Phys. Rev. **83**, 923 (1951).
[8] P. A. WOLFF: Phys. Rev. **87**, 434 (1952).

$T=0°$) does not fit the data. A GAUSSian momentum distribution namely

$$N(p)\,dp \propto e^{-\frac{p^2}{2ME_G}}\,p^2\,dp \tag{34.1}$$

fits the data fairly well with E_G having values between 14 and 19 Mev. The best fit is with $E_G=16$ Mev. The CHEW-GOLDBERGER distribution

$$N(p)\,dp \propto \frac{\alpha\,p^2\,dp}{(\alpha^2+p^2)^2} \qquad \text{with} \qquad \frac{\alpha^2}{2M}=18 \text{ Mev}$$

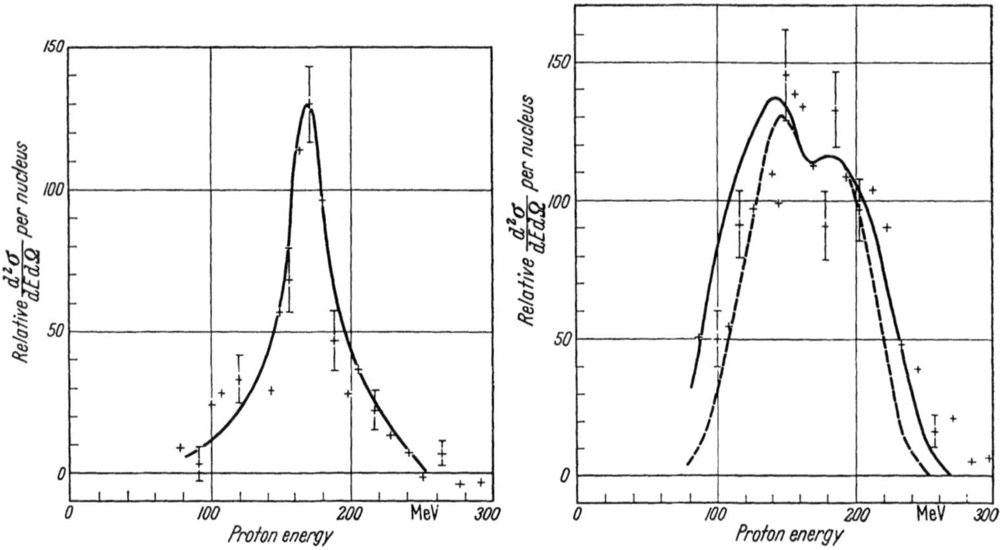

Fig. 13. Fig. 14.

Fig. 13. The energy spectrum of the protons emitted at 40° from the quasi elastic scattering of 340 Mev protons by Li. These protons are in coincidence with protons at −44° which can have any energy above that required to be detected. The line is an attempt to fit the data by the assumption of two momentum groups being present in Li. [Figure from WILCOX and MOYER: Phys. Rev. 99, 875 (1955).]

Fig. 14. The energy spectrum of protons quasi-elastically scattered from Be. The experimental arrangement is the same as that described in Fig. 13. The curves represent the spectrum calculated GAUSSian momentum distributions with a 1/e value of 25 Mev for the solid line, and a 1/e value of 16 Mev for the dashed line.

deviates in the region where CHEW and GOLDBERGER pointed out that it should not be applied.

It is not clear whether the data are sufficiently accurate to rule out the CHEW-GOLDBERGER distribution in the region where it was originally found to be useful.

More recent measurements of the quasi-elastic scattering have been reported by WILCOX and MOYER[1]. They demanded a coincidence at fixed angles and measured the energy spectrum of one of the protons (employing their 35 channel magnetic spectrometer). They bombarded Li, Be, and B with 340 Mev protons. There were pronounced qualitative differences between the curves obtained from the three targets. Their data are shown Figs. 13—15. They employ conservation of momentum (i.e. the impulse approximations) in interpreting their data. The beryllium data is fitted fairly well by a GAUSSian momentum distribution with $E_G=20$ Mev. Obviously the Li, and B curves are narrower than the Be curve indicating lower momentum for the protons in Li and B. They qualitatively

[1] J. M. WILCOX and B. J. MOYER: Phys. Rev. 99, 875 (1955).

attribute the shapes for Li and B to two groups of protons namely core protons and peripheral protons.

There is no doubt about the reality of the quasi-elastic process. As more data accumulate on the energy spectrum as a function of angle and incident energy, it should be possible to obtain a better knowledge of the momentum distribution in various nuclei. In analyzing more exact data more careful consideration should be given by the theorists to such questions as the scattering and attenuation of the outgoing particles, the effect of the binding energy, and the modifications to the impulse approximation required by nuclear correlations.

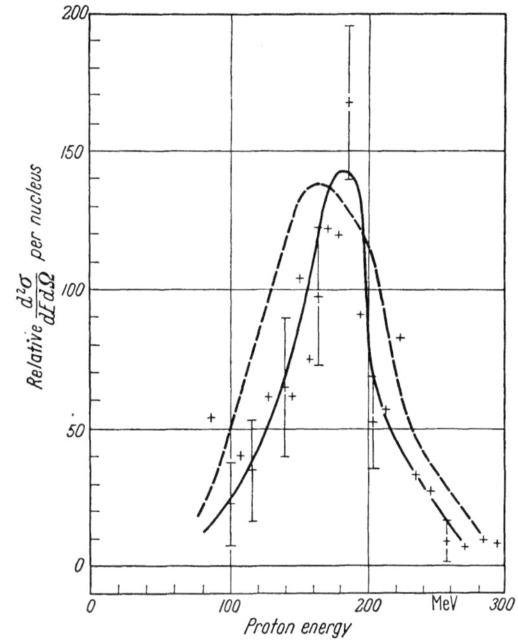

Fig. 15. The energy spectrum of protons quasi-elastically scattered from B and Be. The experimental arrangement is the same as that described in Fig. 13. The solid line is drawn through the experimental points for B; the dashed line indicates the experimental data for Be.

35. Deuteron pick-up and stripping and related reactions. HADLEY and YORK[1] undertook to investigate the angle and energy distribution of the charged particles emitted when 90 Mev neutrons bombarded targets of C, Cu, and Pb. They employed a magnet to determine the momentum of the charged particles and a proportional counter to measure the ionization loss of the particles. Combining these two pieces of information one obtains the mass of the particle and its energy. They observed protons, deuterons, and tritons were being emitted from the target. The number of high energy deuterons being emitted was much greater than the number predicted by the Monte Carlo [5] calculation of GOLDBERGER (see Sect. 11). They found that the deuterons emitted from carbon at 12° had an energy distribution with a peak at about 60 Mev and decreased for lower energy deuterons. Obviously, this is not the energy distribution of evaporated deuterons; it is also much too high an energy for a deuteron subunit which may have recoiled from a neutron collision. At larger angles the peak disappears.

The total cross-section for the emission of deuterons with energies greater than 27 Mev at all angles was estimated to be about 26 millibarns for carbon. This is greater than 10% of the cross-section (see Table 9) for all reactions other than elastic scattering. About half of these deuterons are emitted between 0 and 25°, and about 80% of the deuterons are emitted between 0 and 45°.

The data on tritons are sparse as the yield of tritons is, at most, the order of 10% of the yield of deuterons.

CHEW and GOLDBERGER [6] proposed that the high energy deuterons could be understood in terms of a "pick-up" process. A neutron passing by the nucleus can form a deuteron with one of the protons of the nucleus if the proton had the

[1] J. HADLEY and H. YORK: Phys. Rev. 80, 345 (1950).

appropriate initial momentum. There is really no exactly equivalent process for a free proton (neutron-proton capture without the emission of a γ-ray) due to the need for a third body to conserve energy and momentum.

CHEW and GOLDBERGER [6] gave a theoretical treatment of the "pick-up" process in BORN approximation; that is, they assumed the incoming neutron and the outgoing deuteron were plane waves. The theory requires the ground state wave function of the deuteron which is known to a fair approximation from lower energy experiments. The matrix element for the pick-up process contains an average binding energy for the proton in the original target and the initial momentum distribution for the protons in the target nucleus. They employed this theory and the data of HADLEY and YORK to obtain a momentum distribution for protons in a carbon nucleus. A reasonable fit to HADLEY and YORK's data was obtained from a momentum distribution of the form (in terms of k_p, the wave number)

$$\frac{8\pi\alpha_p}{(\alpha_p^2 + k_p^2)^2} \tag{35.1}$$

where α_p^2 corresponds to a proton of 18 Mev. Expression (35.1) is known as the CHEW-GOLDBERGER momentum distribution and has been applied to explain meson production spectra and other phenomena. It is an approximation which they point out should not be applicable at high momenta.

The basic assumption is that there is a sudden rearrangement of the entire nucleus and that effectively one can neglect conservation of energy. The justification of this lies in the impulse approximation (see Sect. 9). However the calculation for the pick-up process was a BORN approximation calculation, it differs in a very important way from most applications of the impulse approximation. In the "pick-up" process one does not have an equivalent process on a free nucleus; therefore one can not put into the matrix element the empirically observed cross-section for the process on a single nucleon.

A series of precise experiments on the pick-up process have been performed by SELOVE[1]. He used 95 Mev protons as the impinging particle in order to have their energy well determined; the deuterons were detected in a range $-dE/dx$ telescope of eight counters; the effective resolution of the experiment was about 3 Mev. The elements he investigated were He, Be, C, Al, Si, Cu, and Pb; however the most extensive measurements were on Be and C.

The energy distribution of the deuterons emitted at 18° from C and Si consists mainly of a fairly narrow energy group near the maximum energy; these deuterons are probably associated with the residual nucleus being left mainly in its ground state. In Al the high energy group is appreciably broader than C and Si and is probably due to contributions from several excited levels. In Be there are two pronounced peaks, one corresponding to the essentially free (peripheral) neutron, and the other peak is due to the neutrons associated with the alpha particle core of Be^9.

At larger angles the peaks at maximum energy decrease, and there is a relatively larger fraction of deuterons spread over all energies. These latter deuterons are attributed to more complicated processes than the direct pick-up.

SELOVE subtracted the contribution of the general background from the peak at maximum energy to obtain the angular dependence of the direct pick-up process. The type of angular distribution he obtains is shown in Fig. 16. For Be the two peaks have quite different angular distributions; the tightly bound peak has an angular distribution like that of C; however the loosely bound neutron

[1] W. SELOVE: Phys. Rev. **98**, 208 (1955); **101**, 231 (1956).

peak falls off much more rapidly. All of this is in agreement with the ideas of GOLDBERGER and CHEW.

SELOVE employed his data and the formulation of CHEW and GOLDBERGER to obtain momentum distributions for the neutrons in the lightest elements. For C he finds that he can fit his data very well with an empirical momentum density distribution of the form

$$e^{-\frac{E}{7}} + 0.15\, e^{-\frac{E}{50}} \tag{35.2}$$

where E (in Mev) is the kinetic energy of the neutron inside a carbon nucleus. For the lower momentum points a GAUSSian with a $1/e$ energy of 14 Mev fits SELOVE's data approximately.

HESS and MOYER[1] have studied the deuterons emitted from seven elements when bombarded by 300 Mev neutrons and protons. The energy of the neutrons was broad. They interpreted their data according to a theory of BRANSDEN[2] in which at such high energies indirect pick-up processes are considered the main source of deuterons. These would correspond to the background (energy continuum) deuterons in SELOVE's experiment. In the indirect process some of the nucleons in the early part of the cascade (see Sect. 11) escape from the nucleus and pick up a partner to form a deuterons as they leave the nucleus. They observed an appreciable number of deuterons at angles as large as 60° which would not be anticipated from the original formulation of the CHEW-GOLDBERGER theory of the direct pick-up. It is not clear that at these higher energies the production and reabsorption of real and virtual mesons may not be affecting the pick-up process. One sees this effect in the photo-disintegration of the deuteron at high energies (see Sect. 40), and the pick-up process must include contributions from the inverse of the photodisintegration of the deuteron. Theories for the pick-up process which include meson contributions explicitly require further work.

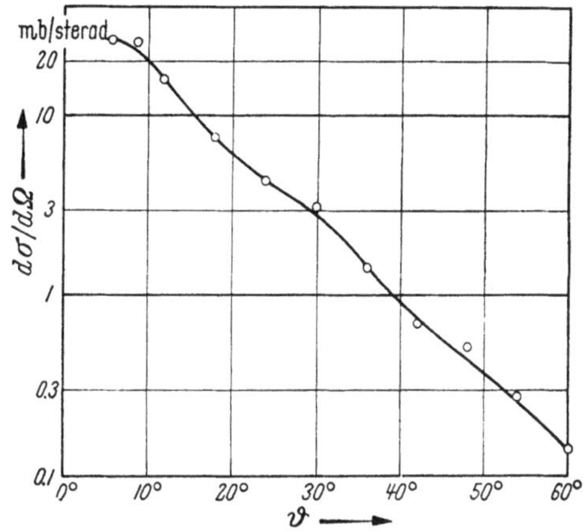

Fig. 16. The angular distribution of "pick-up" deuterons from 95 Mev protons on C^{12}. The distribution is for those deuterons whose energy corresponded to the C^{11} nucleus being left in its ground state. [Figure from SELOVE: Phys. Rev. **98**, 208 (1955).]

HESS and MOYER also observed tritons and find that at 10° they have the same dependence on the atomic weight of the target as the deuterons. They consider this evidence for the formation of tritons by BRANSDEN's multistep pick-up process.

The deuteron stripping reaction (d, p) or (d, n) would appear to be related to the pick-up process (p, d) or (n, d) through detailed reversibility. The stripping reaction was predicted by SERBER[3], and the neutrons from such reactions were one of the initial manifestations of the successful performance of the 184 inch cyclotron at Berkeley, California, as reported by BROBECK *et al.*[4].

[1] W. N. HESS and B. J. MOYER: Phys. Rev. **101**, 337 (1956).
[2] B. H. BRANSDEN: Proc. Phys. Soc. Lond. A **65**, 738 (1952).
[3] R. SERBER: Phys. Rev. **72**, 1008 (1947).
[4] W. M. BROBECK, E. O. LAWRENCE, K. R. MACKENZIE, E. M. MCMILLAN, R. SERBER, D. C. SEWELL and R. L. THORNTON: Phys. Rev. **71**, 449 (1947).

The stripping process has been a very important source of neutrons, and many of those employing neutrons have investigated various aspects of the reaction. The first extensive study of the angular distribution of the neutrons was made by HELMHOLTZ *et al.* [1]. More extensive studies of the yields of the neutrons were later made by KNOX [2] (who lists the work of others). HELMHOLTZ *et al.* observed that the neutrons were emitted in a rather narrow cone in the forward direction; for 190 Mev deuterons the half-width of the cone is about 0.16 radians for light nuclei and about 0.20 for heavy nuclei.

SERBER had anticipated this process on the basis of a simple mechanical picture. The proton in the deuteron strikes the edge of the nucleus and is stripped off. The neutron essentially continues its original trajectory which includes its internal momentum in the deuteron at the instant of the collision. This initial momentum of the neutron is what is mainly responsible for the angular spread in the neutrons. SERBER'S estimate of the cross-section was

$$\sigma = \tfrac{1}{2}\pi R R_D \tag{35.3}$$

where R is the nuclear radius and R_n the deuteron radius. Expression (35.3) becomes approximately

$$\sigma \approx 5\,A^{\frac{1}{3}} \times 10^{-26}\ \text{cm}^2, \tag{35.4}$$

somewhat depending on the choice of constants. SERBER obtained the proper value of the angular width of the neutron distribution from his simple picture. For heavier nuclei the COULOMB field has to be considered in estimating the half-width.

Actually for heavier nuclei the COULOMB field of high Z nuclei contributes to the cross-section. DANCOFF [3] evaluated the point first noted by OPPENHEIMER [4], namely that a high speed deuteron passing through a COULOMB field would be photodisintegrated by the virtual photons of the field. DANCOFF obtained the cross-section for this photodisintegration process to be $\sim 2 \times 10^{-29} Z^2$ cm² for 200 Mev deuterons. This becomes commensurate with expression (35.4) only for the highest Z nuclei.

MILLBURN *et al.* [5] in their studies on the reaction cross-sections for protons, deuterons, He^3, and He^4 particles measured the protons produced from bombardment of various targets by 160 Mev deuterons. They found the cross-section for carbon is 0.43 barns which is about three times the value predicted by Eq. (35.4). They also found that the A dependence was closer to $A^{\frac{1}{3}}$ then $A^{\frac{1}{3}}$. HECKROTTE [6] attempted to explain this discrepancy by modifying SERBER'S simple assumptions so that the nucleus had a potential well with a parabolic shape and an effective radius of $R = 1.6 A^{\frac{1}{3}} \times 10^{-13}$. This is an empirical model that is specific to the reactions of D, He^3, and He^4 and has no bearing on other nuclear phenomena.

A more appealing explanation of the discrepancy between the experimental values and Eq. (35.4) has been given by GLAUBER [7]. He noted that the diffraction of the neutron and proton in a deuteron leads to an effective increase of their kinetic energy and the dissociation of the deuteron. This occurs without either

---

[1] A. C. HELMHOLTZ, E. M. MCMILLAN and D. C. SEWELL: Phys. Rev. **72**, 1003 (1947).
[2] W. J. KNOX: Phys. Rev. **81**, 687 (1951).
[3] S. M. DANCOFF: Phys. Rev. **72**, 1017 (1947).
[4] J. R. OPPENHEIMER: Phys. Rev. **67**, 845 (1945).
[5] G. P. MILLBURN, W. BIRNBAUM, W. E. CRANDALL and L. SCHECTER: Phys. Rev. **95**, 1268 (1954).
[6] W. HECKROTTE: Phys. Rev. **95**, 1279 (1954).
[7] R. J. GLAUBER: Phys. Rev. **99**, 1515 (1955).

particle striking the nucleus. Both particles then continue in a fairly forward direction; neither the neutron nor the proton is absorbed. They should both be detectable in a coincidence experiment; such an experiment has not yet been reported. GLAUBER only worked out a preliminary estimate of the contribution of this diffraction stripping, but the value of the cross-section was commensurate with that of SERBER'S and should be added to (35.4). Essentially a complete optical model calculation still has to be performed including the partial transparency of the nucleus and the non-uniformity of the density distribution.

On SERBER'S picture, if one of the nucleons is absorbed into the ground state, then the cross-section for that part of the stripping reactions $[(d, p)$ or $(d, n)]$ should be directly calculable by detailed reversibility from the cross-section for the deuteron pick-up reactions $[(p, d)$, or $(n, d)]$. However those parts of the cross-section associated with GLAUBER'S picture or DANCOFF'S correspond to a (d, pn) reaction, and the pick-up process is not the inverse of these reactions.

36. Complex nuclear reactions. Although SERBER'S [4] picture of high energy nuclear reactions is a quite simple one, complex nuclear reactions are frequently the end products because of the cascade process and the subsequent evaporation process (see Sects. 6, 11, and 12). The multiprong stars observed, when photographic emulsions are bombarded by high energy particles, are manifestations of the cascade and evaporation processes. As pointed out in Sect. 11, a Monte Carlo calculation gave results in agreement with the experimental ones for the energetic prongs observed by BERNARDINI *et al.*[1]. For the very high energy prongs a Monte Carlo calculation [5] has to be performed for the particular experimental conditions being investigated. As regards the lower energy evaporation prongs, the theoretical situation is not completely satisfactory.

MORRISON [2] provides a list of experimental references up to 1952; more recent work can be found in the article of MORRISON *et al.*[2]. From the experiments it appears that for the low energy prongs the energy and charge distribution are independent of the nature of the bombarding particle. In Sect. 47 there is a discussion of the difficulty in correlating the photographic emulsion stars with the radionuclides obtained from bombardments with high energy particles.

A great many studies have been made of the radioactive species produced by high energy reactions, and in the eight years since these studies were started, over a hundred experiments have been reported. In this brief section it is not possible to present more than a few general conclusions most of which are elaborated upon in the review article of TEMPLETON[3].

From the bombardment of a single isotope of a medium weight element by a monoenergetic beam of high energy particles, as many as 30 or 40 different radionuclides have been observed (e.g. see BELMONT and MILLER[4]).

In order to compare the results of such studies for different experimental conditions, it is convenient to use a contour diagram in the Z-N plane as shown in Fig. 17. In such a diagram, the circles represent the various radionuclides; the star represents the target nucleus; the contour lines connect regions of approximately equal yield and represent a change in yield by a factor of ten; the dashed line indicates the center of the stable isotopes.

The radiochemists have studied these yields as a function of the element being bombarded, the energy of the bombarding particle, and the type of bombarding particle. The more general conclusions are:

[1] G. BERNARDINI, E. T. BOOTH and S. J. LINDENBAUM: Phys. Rev. **85**, 826 (1952).
[2] C. G. MORRISON, H. MUIRHEAD and W. G. V. ROSSER: Phil. Mag. **44**, 1326 (1953).
[3] D. H. TEMPLETON: Ann. Rev. Nucl. Sci. **2**, 93 (1952).
[4] E. BELMONT and L. M. MILLER: Phys. Rev. **95**, 1554 (1954).

(1) The maximum yields are close to the stable nuclei; an explanation of this is given in Sect. 47.

(2) The highest yields correspond to the ejection of three to seven nucleons. Reactions corresponding[1] to the ejection of 20 or more particles have yields which are smaller by a factor of 1000 or more.

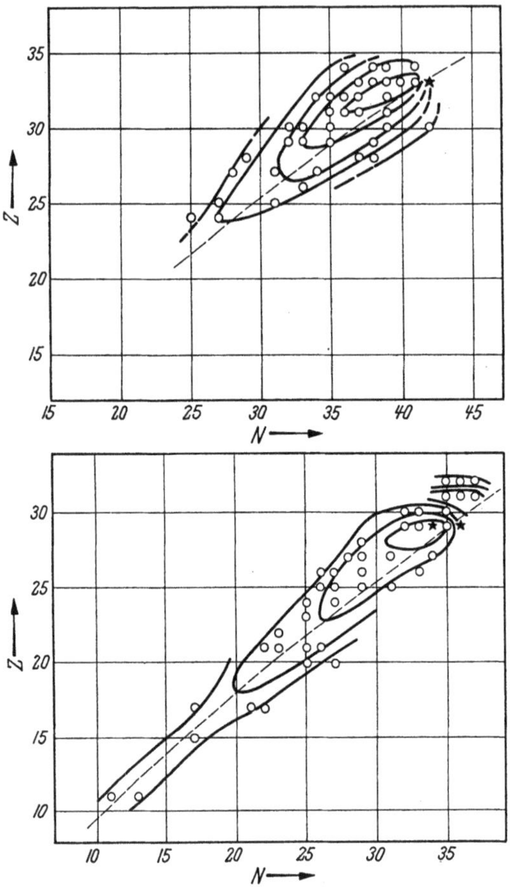

Fig. 17. Contour diagrams showing the relative yield of radionuclides plotted in the proton-neutron plane. The circles represent nuclides whose yields were measured; the star indicates the target nucleus (arsenic bombarded by 190 Mev deuterons in the upper, and copper bombarded by 340 Mev protons in the lower figure); the lines connect regions with approximately equal yields, each line representing a factor of ten in yield on an arbitrary scale. [Figure from TEMPLETON: Ann. Rev. Nucl. Sci. **2**, 93 (1953).]

(3) An increase in the energy of the bombarding particles makes it energetically possible for more reactions to occur, and at higher energies the contour lines become elongated along the line of stable nuclei.

(4) In the energy range from 400 to 3000 Mev, FRIEDLANDER *et al.*[2] and others find that the cross-section for the formation of particular radionuclides is almost a constant independent of energy.

(5) If one omits the yields of radionuclides which may arise from the emission of one or two nucleons, then the contour diagrams become independent of the type of bombarding particle.

(6) For elements heavier than calcium the contour diagrams are also qualitatively independent of the target nuclei except for long range trends with respect to atomic number.

(7) One of these trends is that the higher Z elements show a shift of the contour lines toward the neutron deficient side of the line of stability.

Some of the above points are discussed in more detail in Sect. 47.

For elements with atomic weights greater than about 180 one also observes fission from the bombardment of high energy particles. The separation of fission products from spallation products is somewhat arbitrary, and it may be a questionable distinction in the region where the radioactive products overlap.

The subject of high energy fission has been reviewed by SPENCE and FORD[3]. High energy fission is pictured as following a cascade process which has left

[1] It must be realized that from a radiochemical yield measurement in most cases one does not know whether there were ejected a large number of nucleons, or one or more alpha particles and a few nucleons, or a fragment like Li^8 possibly accompanied by a few nucleons.

[2] G. FRIEDLANDER, J. HUDIS and R. L. WOLFGANG: Phys. Rev. **99**, 263 (1955). They give the references to the work of others.

[3] R. W. SPENCE and G. P. FORD: Ann. Rev. Nucl. Sci. **2**, 399 (1953).

a heavy nucleus with a great deal of excess energy. The fission process can be considered as an alternative for the evaporation process. For the heaviest nuclei, it has been estimate that 10% of all the cascades are followed by fission. The asymmetry in fission product yields which is present at low energies disappears at high energies for most nuclei. However there is an appreciable tail of fission products with low atomic weights.

The technique employed by DEUTSCH[1] may be helpful in resolving some of the ambiguities that exist in the production of radionuclides. By using a photographic emulsion as a catcher, he is able to identify the nature of some of the secondary particles resulting from high energy nuclear bombardments. He has good results on most of the nuclear fragments up to Be^8.

E. Electron reactions.

37. Elastic scattering. The two reactions of high energy electrons which have been studied up to the present are the elastic and inelastic scattering. This section deals with the elastic scattering.

The outstanding experiments of HOFSTADTER and associates[2] at STANFORD have provided extensive data on the elastic scattering of high energy electrons by nuclei. Their data are at a sufficiently high energy so that it is possible to determine not only a nuclear charge radius but also to obtain information on the radial distribution of charges in nuclei. LYMAN *et al.*[3] and PIDD and HAMMER[4] have made measurements at lower energies which can only be interpreted in terms of a nuclear charge radius.

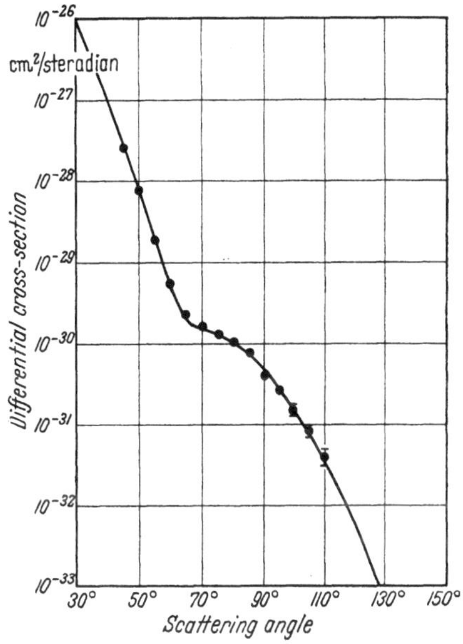

Fig. 18. Experimental and theoretical cross-sections for the scattering of 183 Mev electrons by Co^{59}. [Figure from HAHN *et al.*: Phys. Rev. **101**, 1127 (1956).]

HOFSTADTER *et al.* employ the electron beam of the Stanford Electron Accelerator. Measurements on Au^{197} and Pb^{208} have been made at 84, 126, 154, and 183 Mev. Measurements on Ca, V, Co, In, Sb, Ta, W, Au, Bi, Th, and U have been made at 183 Mev. The electron beam spot on the target is about 1 mm high and 3 mm wide. This allows good angular resolution, and they employ a spread of about $\pm 1.5°$. The energy resolution appears from their curves to be about 1 Mev at 183 Mev. This allows for the separation of elastic and inelastic scattering in nuclei where the first excited state of a nucleus is $\frac{1}{2}$ Mev or higher. This excellent energy and angular resolution is provided by an initial magnetic double deflection system which focuses the electrons on the target. The scattered electrons are then analyzed in a double focusing magnetic spectrometer (see HOFSTADTER,

[1] R. W. DEUTSCH: Phys. Rev. **97**, 110 (1955).

[2] R. HOFSTADTER, H. R. FECHTER and J. A. MCINTYRE: Phys. Rev. **92**, 978 (1953). — R. HOFSTADTER, B. HAHN, A. W. KNUDSEN and J. A. MCINTYRE: Phys. Rev. **95**, 512 (1954). — B. HAHN, D. G. RAVENHALL and R. HOFSTADTER: Phys. Rev. (to be published).

[3] E. M. LYMAN, A. O. HANSON and M. B. SCOTT: Phys. Rev. **84**, 626 (1951).

[4] R. W. PIDD, C. L. HAMMER and E. C. RAKA: Phys. Rev. **92**, 436 (1953). — R. W. PIDD and C. L. HAMMER: Phys. Rev. **99**, 1396 (1955).

FECHTER, and MCINTYRE). A CERENKOV counter is employed as a detector. The magnets are held constant to about 0.1%.

The type of results obtained by HOFSTADTER *et al.* are shown in the accompanying Figs. 18—20.

In interpreting the scattering of electrons by nuclei it is assumed that the interaction is purely electromagnetic, and that there are no deviations at small

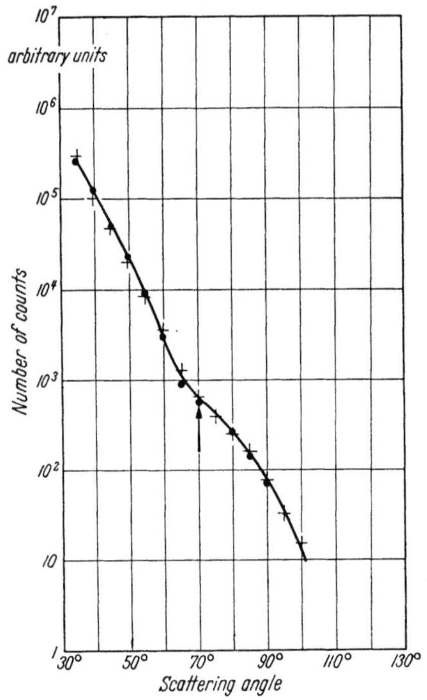

Fig. 19.

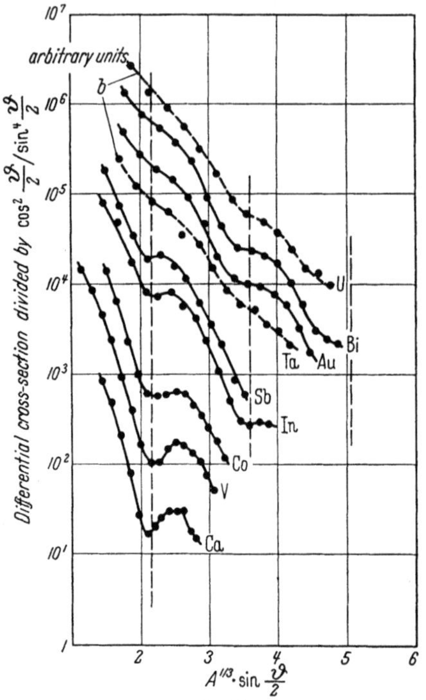

Fig. 20.

Fig. 19. The elastic scattering in Au^{197} and Pb^{208} of 183 and 186 Mev electrons respectively. The arrow indicates the angular position estimated for the diffraction minimum. The curves are normalized arbitrarily. Crosses: gold target of 2 mils, dots: lead target of 2.5 mils thickness. [Figure from HOFSTADTER *et al.*: Phys. Rev. **95**, 512 (1954).]

Fig. 20. Experimental results on 183 Mev electron scattering for several nuclei. The differential cross-section is divided by $\cos^2(\vartheta/2)/\sin^4(\vartheta/2)$ in order to display the diffraction structure. The data are plotted versus $A^{\frac{1}{3}}\sin(\vartheta/2)$. The dashed vertical lines indicate approximately the location of the first, second, and third diffraction minimum. The curves are shifted vertically for convenience of comparison. The nuclei labeled *b* are believed to have an ellipsoidal shape. [Figure from HAHN *et al.*: Phys. Rev. **101**, 1127 (1956).]

distances from the known electromagnetic laws. The COULOMB scattering of 183 Mev electrons from a point charge, should show a decrease by a factor of about 10^{-2} between 30° and 130°. The curves in the accompanying figures show a decrease of about 10^{-4}. The larger decrease is readily understood in terms of a diffraction effect if one has a finite sized nucleus. This pronounced effect of finite size is readily measurable (though not as pronounced) at lower energies. PIDD and HAMMER have employed the effect at 30, 40, and 60 Mev to determine the radius of the wolfram nucleus.

At higher energies a simple form factor calculation (see Sect. 7) of the diffraction will show that one should obtain minima and maxima if the charge distribution is uniform and cuts off sharply. With such a nuclear model, the simple form factor calculation (BORN approximation) leads to minima which go to zero at

$$\frac{R}{\lambda}\sin\frac{\vartheta}{2} = \pi \qquad (37.1)$$

where R is the nuclear radius, $\lambda\!\!\!^-$ the DE BROGLIE wavelength of the electron divided by 2π, and ϑ the angle of scattering.

HOFSTADTER'S early results showing that these maxima and minima were washed out was first interpreted as a smooth decrease of the charge density from the center of the nucleus. However, YENNIE, RAVENHALL, and WILSON[1] showed that the first BORN approximation is not sufficiently accurate for high Z elements. A more accurate phase shift calculation (up to 10 phase shifts are employed) showed that minima and maxima are not pronounced for high Z elements even from a uniform charge density with sharp boundaries. For low Z elements the minima and maxima should be there and the BORN approximation is more pertinent though not completely reliable. These theoretical and experimental results stimulated a desire for more exact experimental and theoretical information.

The theory for electron scattering has been worked on by YENNIE, RAVENHALL, and WILSON[2], HILL and FORD[3], and BROWN and ELTON[4]. They attempted to fit the data with various assumed shapes of charge distributions.

On the experimental side careful measurements were carried out on Au^{197} and Pb^{208} at a series of energies.

The best fit to the experimental data is provided by a charge distribution with tapered edges and a uniform central region. A simple GAUSSian or exponential shape will not fit either the 126 Mev or the 183 Mev data on Au and Pb.

The most recent data of HOFSTADTER *et al.* on a large number of nuclei have been interpreted in terms of two parameters. These two parameters are a mean radius and a surface thickness. They find the radial parameter (i.e. the distance to the point at which the charge density has dropped to one half its maximum value) varies as $A^{\frac{1}{3}}$ for spherical nuclei, namely $(1.07\times 0.02)A^{\frac{1}{3}}\times 10^{-13}$ cm. The surface thickness is defined as the thickness in going from 0.1 to 0.9 the maximum density. The surface thickness in nuclei above Ca can be taken as constant for the spherical nuclei, Ca, V, Co, In, Sb, Pb, namely $(2.4\pm 0.3)\times 10^{-13}$ cm. These conclusions of the variation with $A^{\frac{1}{3}}$ can be nicely seen in Fig. 20, where the data are plotted versus $A^{\frac{1}{3}}\sin\frac{\vartheta}{2}$ [q.v. Eq. (37.1)].

The nuclei labeled (b) in this figure are considered to be non-spherical nuclei according to the BOHR-MOTTELSON[5] theory. Inelastic scattering from low lying energy levels and a non-spherical charge distribution should flatten out the minima and maxima (see DOWNS *et al.*[6]). This effect is what is observed for the elements labeled b in Fig. 20.

The interpretation of the results on the spherical nuclei is in agreement with the interpretation of the X-rays from μ-atoms which give a charge radius of $R=1.2\times 10^{-13}A^{\frac{1}{3}}$ for heavier nuclei. It is to be noted that the results of charge radius measurements disagree with the radii of the potential wells employed to interpret nucleon scattering experiments (see Sects. 29 to 32). The subject of nuclear charge distributions has recently been reviewed by FORD and HILL[7].

---

[1] D. R. YENNIE, D. G. RAVENHALL and R. N. WILSON: Phys. Rev. **95**, 500 (1954).

[2] D. G. RAVENHALL, D. R. YENNIE and R. WILSON: Phys. Rev. **96**, 239 (1954).

[3] D. L. HILL and K. W. FORD: Phys. Rev. **94**, 1657 (1954); see also D. L. HILL's article in Vol. XXXIX of this Encyclopedia.

[4] G. E. BROWN and L. R. B. ELTON: Phil. Mag. **46**, 164 (1955).

[5] A. BOHR and B. R. MOTTELSON: Kgl. danske Vid. Selsk. mat-fys. Medd. **27**, No. 16 (1953).

[6] B. W. DOWNS, D. G. RAVENHALL and D. R. YENNIE: Phys. Rev. **98**, 227 (1955).

[7] D. L. HILL and K. W. FORD: Ann. Rev. Nucl. Sci. **5**, 25 (1955); see also D. L. HILL in Vol. XXXIX of this Encyclopedia.

The high energy electron scattering work of HOFSTADTER *et al.* is probably one of the most outstanding contributions of high energy physics to our knowledge of nuclear parameters.

HOFSTADTER and MCALLISTER[1] have been studying the fundamental process of the scattering of electrons by protons, which is of great interest. In this case the experimental data lie above those predicted for a point charge proton, as shown in Fig. 21. However the magnetic moment of the proton contributes to the scattering (see ROSENBLUTH[2]). The observed data lie below the theoretical curve predicted for a point charge and point (anamalous) magnetic moment. Also the data do not fit ROSENBLUTH'S curve with a DIRAC magnetic moment and a point charge.

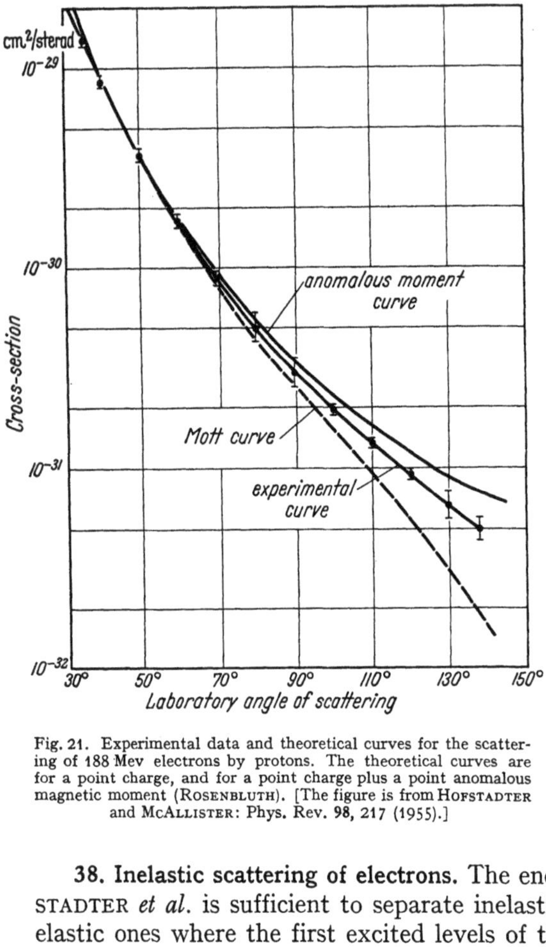

Fig. 21. Experimental data and theoretical curves for the scattering of 188 Mev electrons by protons. The theoretical curves are for a point charge, and for a point charge plus a point anomalous magnetic moment (ROSENBLUTH). [The figure is from HOFSTADTER and MCALLISTER: Phys. Rev. 98, 217 (1955).]

The variation of the data from the point charge theory can be accounted for on the assumption that both the charge and magnetic moment of the proton are spread out over a finite distance. On the assumption that the magnetic moment and electric charge have the same distribution, they find that a root mean square radius of about 7×10^{-14} cm fits the data they have obtained at several energies up to 236 Mev.

From precise data on the differential scattering as a function of angle at even higher energies, it is hoped to obtain information on the shape of the proton charge and magnetic moment distribution. Preliminary results of such measurements have been reported verbally[3].

38. Inelastic scattering of electrons. The energy resolution employed by HOFSTADTER *et al.* is sufficient to separate inelastically scattered electrons from the elastic ones where the first excited levels of the nucleus are the order of 1 Mev above the ground state. They have observed inelastic scattering in Be[4], C[5], Mg[6], Si[6], S[6], and Se[6].

In the case of Be they obtained some data on the angular distribution of the electrons scattered from the 2.5 and 6.8 Mev levels.

In the case of carbon, FREGEAU and HOFSTADTER have a series of good measurements on the inelastic scattering from the 4.43, 7.68, and 9.61 levels.

[1] R. HOFSTADTER and R. W. MCALLISTER: Phys. Rev. **98**, 217 (1955).
[2] M. N. ROSENBLUTH: Phys. Rev. **79**, 615 (1950).
[3] R. HOFSTADTER: Invited paper at the Feb. 1956 Meeting of the Amer. Phys. Soc.
[4] L. A. MCINTYRE, B. HAHN and R. HOFSTADTER: Phys. Rev. **94**, 1084 (1954).
[5] L. H. FREGEAU and R. HOFSTADTER: Phys. Rev. **99**, 1503 (1955).
[6] R. H. HELM and R. HOFSTADTER: To be published.

They have measured the angular distribution of the inelastic scattering from these levels with 150 and 187 Mev electrons. Fig. 22 illustrates a particularly favorable case, 187 Mev electrons at 90°. With lower energy electrons the inelastic scattering does not appear as pronounced because the elastic scattering cross-section is larger. There is a low energy tail to the elastic scattering (arising from energy loss by bremsstrahlung) which effectively produces a background in the region of the inelastic peaks.

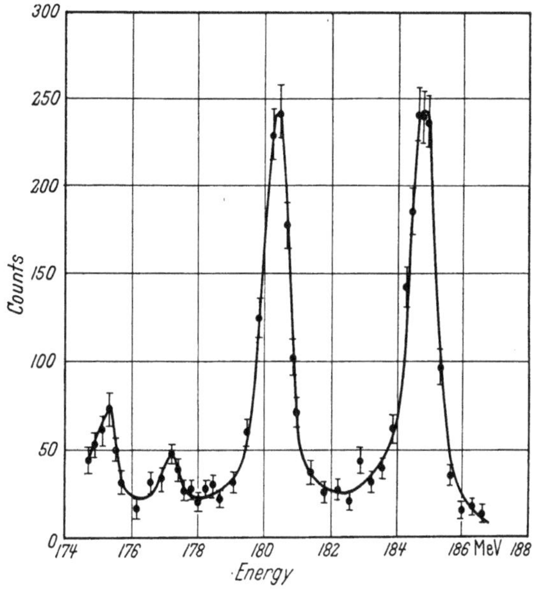

Fig. 22. The elastic and inelastic peaks from the scattering of 187 Mev electrons on carbon at 90° in the laboratory. The peak at the highest energy corresponds to scatterings which left carbon in its ground state. The three other peaks correspond to carbon having been left in the 4.43, 7.6 and 9.61 Mev excited levels. [Figure is from FREGEAU and HOFSTADTER: Phys. Rev. 99, 1503 (1955).]

The peaks due to inelastic scattering are also relatively more difficult to observe at smaller angles. Fig. 23 shows the dependence of the elastic and inelastic scattering as a function of angle. It will be observed that the inelastic scattering falls off less rapidly with angle than the elastic scattering. FREGEAU and HOFSTADTER refer to calculations of RAVENHALL[1] which give a so-called "form factor" for the inelastic scattering.

The inelastic scattering from the 7.68 Mev level falls off more rapidly with angle than the scattering from the 4.43 and 9.61 Mev levels. It is known that the 4.43 is a 2^+ level and the 7.68 is a 0^+ level, and the angular variation of the scattering depends on the nature of the final state.

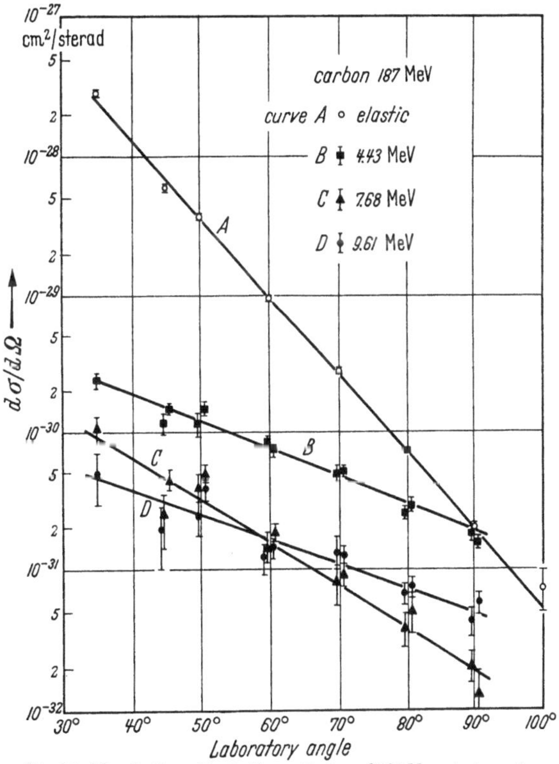

Fig. 23. The elastic and inelastic scattering of 187 Mev electrons by carbon as a function of laboratory angle. The points are obtained from integrating the area under the peaks in Fig. 22. [Figure from FREGEAU and HOFSTADTER: Phys. Rev. 99, 1503 (1955).]

SCHIFF[2] has pointed out that the C^{12} 0^+ to 0^+ is an electric monopole transition. When such levels are not the first excited levels it is difficult to

[1] D. G. RAVENHALL: To be published.
[2] L. I. SCHIFF: Science, Lancaster, Pa. 121, (3156) 881 (1955).

study them by the usual techniques of lower energy nuclear physics; the high energy electron inelastic scattering thus provides a very fine tool for investigating such transition and the 0^+ levels in even-even nuclei.

F. Nuclear reactions of X-rays.

39. X-ray scattering by nuclei. The scattering of light or X-rays by the electrons of an atom is arbitrarily classified into three main types: dispersive scattering (or RAYLEIGH scattering), THOMSON scattering, and COMPTON scattering. The analogues of these three types have their counterparts in the scattering of X-rays by nuclei. However, in general, the cross-sections are reduced by the ratio of the mass of the electron to the mass of the proton raised to the power that the electron mass enters the electron scattering cross-section.

One can think of the dispersive scattering of X-rays by nuclei as the resonance scattering associated with the giant (GOLDHABER-TELLER) resonance absorption that occurs in most nuclei between 20 and 30 Mev. This scattering has been studied by FULLER and HAYWARD[1]. The dispersive or resonance scattering has its main effects below 50 Mev.

The COMPTON scattering of X-rays by protons should have a cross-section of the order of magnitude

$$\left(\frac{e^2}{M c^2}\right)^2 \approx 10^{-32}\,\text{cm}^2$$

where M is the mass of the proton. This low cross-section makes it difficult to study, and there are experimental complications above 140 Mev. For energies less than 130 Mev the COMPTON scattering by protons has been observed very recently by OXLEY and TELEGDI[2] and by FRISCH, GOMEZ, PUGH, and JANES[3].

OXLEY and TELEGDI employed a bremsstrahlung beam and a γ-ray converter in front of an electron range telescope. Their observations were therefore integrals over the upper end of the 94 Mev bremsstrahlung spectrum. FRISCH *et al.* employed an energy selective γ-ray detector and a 130 Mev bremsstrahlung beam. Both groups have preliminary results which agree fairly well with the theoretical KLEIN-NISHINA cross-section modified for the mass of the proton. FRISCH *et al.* find that there is some indication of a reduced cross-section in the forward direction above 100 Mev. This would be in agreement with the predictions of GELL-MANN, GOLDBERGER, and THIRRING[4] which are based on the dispersion relationship (and photo-meson cross-sections).

The THOMSON scattering of high energy X-rays by the protons in a nucleus should be coherent and therefore modified by a form factor analogous to the atomic structure factor for the scattering of X-rays by the electrons in atoms. PUGH *et al.*[5] have measured the scattering of 35 to 130 Mev X-rays by a series of nuclei: Be, C, Al, Cu, In, Pb, and Bi. They employed a large volume of liquid scintillator to measure the energy of the (scattered) converted γ-rays. Mesurements made at 90° and 135° show a marked fall-off with increasing energy (larger collisive momentum) as predicted by a form factor calculation.

Their results are in agreement with a nuclear radius of $R = 1.1 \times 10^{-13} A^{\frac{1}{3}}$ cm. In principle such measurements should yield the same information as those of

[1] E. G. FULLER and E. HAYWARD: Phys. Rev. **95**, 1106 (1954); **101**, 692 (1956).
[2] C. L. OXLEY and V. L. TELEGDI: Phys. Rev. **100**, 435 (1955).
[3] D. H. FRISCH, R. GOMEZ, G. PUGH and G. S. JANES: Unpublished.
[4] M. GELL-MANN, M. L. GOLDBERG and W. E. THIRRING: Phys. Rev. **95**, 1612 (1954).
[5] G. PUGH, D. H. FRISCH and R. GOMEZ: Phys. Rev. **95**, 590 (1954).

HOFSTADTER *et al.* [8] with electron scattering (see Chap. E). However, experimentally the X-ray scattering experiments do not have the good energy resolution of the electron scattering experiments.

PUGH *et al.* find some indication that at higher energies (and larger angles) there is a deviation from the classical THOMSON scattering of $(1+\cos^2\vartheta)$. Such an effect can be thought of in terms of a frequency dependent magnetic polarizability of the nucleons and could arise from approaching the energy of the "resonance" in the photoproduction of mesons.

At higher energies both the neutrons and protons should contribute to the scattering. If neutrons do contribute appreciably to the coherent scattering of X-rays by nuclei, it would be worthwhile to study the X-ray scattering at such energies. Specifically a comparison of the results of such measurements with the results of HOFSTADTER might answer the question as to whether neutrons and protons have the same spatial distribution in nuclei.

In the earlier discussion of the scattering of X-rays by nuclei it was assumed that the scattering was elastic. Actually there should also be present some inelastic scattering. However, the techniques currently employed can not separate the inelastic from the elastic scattering.

40. Photo-disintegration of the deuteron. The photodisintegration of the deuteron has been much advertised as a source of information on the forces between two bound nucleons. A review of the low energy literature will show that the photodisintegration of the deuteron has been mainly a source of information on the interaction of the electromagnetic field with bound nucleons. The low energy measurements have confirmed what one expects for the electric dipole interaction. (The ground state wave function of the deuteron is better known from other data.) However, even at low energies, the contribution of exchange currents to the matrix elements for the magnetic-dipole interaction is not well known.

Based on the low energy electric interaction MARSHALL and GUTH[1] and SCHIFF[2] calculated the cross-section for the photodisintegration of the deuteron up to energies of about 150 Mev. At higher energies they anticipated that mesonic effects should become important.

After much experimental effort the story at high energies is even more extreme than that at low energies. The results are providing very valuable information on the interaction of the electromagnetic field with nucleon systems but not on nuclear forces.

The early experimental results of BENEDICT and WOODWARD[3] and GILBERT and ROSENGREN[4] showed that the cross-section is appreciably higher than that predicted by the phenomenological calculation of SCHIFF and MARSHALL and GUTH.

Since then, there have been many measurements covering the energy region from 22 to 450 Mev. ALLEN[5], using photographic plates, studied the energy region 22 to 65 Mev. HANSON[6] and associates in a series of experiments using both plates and counters have measured the cross-section from 65 Mev to about 300 Mev using the University of Illinois 300 Mev betatron. Measurements using

[1] J. F. MARSHALL and E. GUTH: Phys. Rev. **78**, 738 (1950).

[2] L. I. SCHIFF: Phys. Rev. **78**, 733 (1950).

[3] T. S. BENEDICT and W. M. WOODWARD: Phys. Rev. **85**, 924 (1952).

[4] W. L. GILBERT and J. W. ROSENGREN: Phys. Rev. **88**, 901 (1952).

[5] S. ALLEN: Phys. Rev. **98**, 705 (1955).

[6] E. A. WHALIN, B. DWIGHT-SCHRIEVER and A. O. HANSON: Phys. Rev. **101**, 377 (1956). References to HANSON'S *et al.* earlier work and that of others is given in this reference.

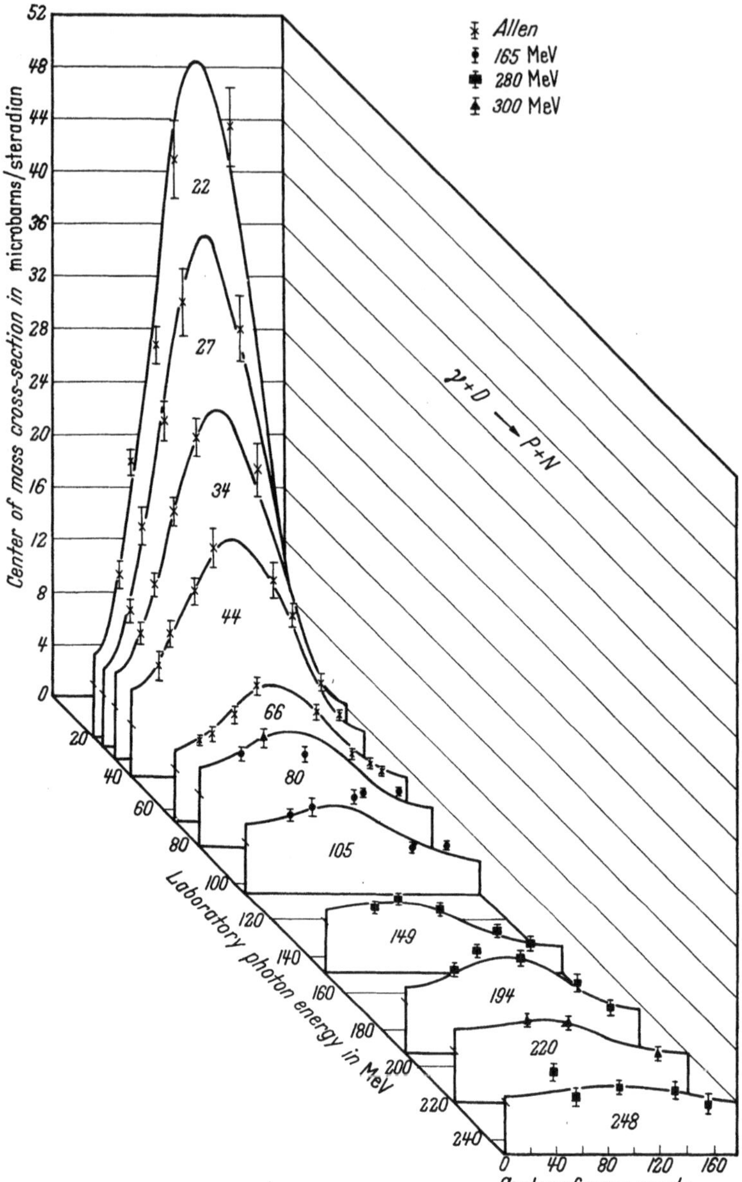

Fig. 24. The differential cross-section for the photodisintegration of the deuteron as a function of the center-of-mass angle and the photon energy in the laboratory. The curves are an attempt to fit the data to an angular dependence of the form

$$(A + B\sin^2\vartheta)(1 + 2\beta\cos\vartheta).$$

[Figure from WHALIN *et al.*: Phys. Rev. **101**, 377 (1956).]

counters have been made from 150 to 450 Mev by KECK and TOLLESTRUP[1] at the California Institute of Technology 500 Mev synchrotron. The above (recent) experiments at Illinois and Cal. Tech. (in the region where they overlap) are in

[1] L. C. KECK and, A. V. TOLLESTRUP: Phys. Rev. **101**, 360 (1956).

good agreement with each other and with the previous results of KECK *et al.*[1] obtained at Cornell with the 300 Mev synchrotron.

The most recent measurements were made by bombarding liquid or gas targets of pure deuterium. Only the protons were detected; their energy being determined from a measurement of their range. The maximum energy of the photon spectrum was adjusted to avoid observing protons from the two reactions $\gamma + D \rightarrow P + N + \pi^0$ and $\gamma + D \rightarrow P + P + \pi^-$.

The results of measurements of the differential cross-section (center of momentum angles) are shown in Fig. 24 as a three-dimensional drawing for energies up to 250 Mev. The differential cross-sections obtained by KECK *et al.* at higher energies are shown in Fig. 25. The curves are empirical fits to the data. The principle feature of the angular distributions are the following: at very low energies the distribution is essentially $\sin^2\vartheta$ plus an isotropic term; the $\sin^2\vartheta$ part decreases with energy; above 100 Mev the angular distribution becomes asymmetric about 90° with a strong forward maximum; above 250 Mev the isotropic component persists but the asymmetry diminishes, and from 350 Mev till 450 Mev the distributions are almost symmetric about 90°. There is no theoretical calculation that gives the observed asymmetric angular distributions above 100 Mev. Independent of any model, such an asymmetry must arise from the interference of transitions to states of different parities. Also, in that it is a forward asymmetry for the proton, the interaction must still be partly with the electric charge even at energies of 300 Mev.

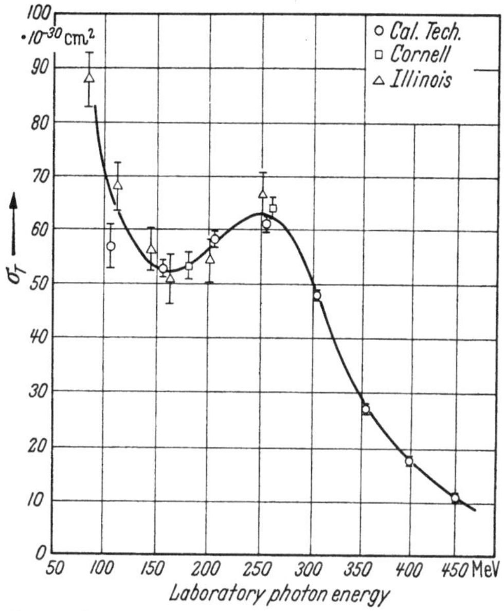

Fig. 25. Total cross-sections for the photodisintegration of the deuteron as a function of laboratory energy. The curve is drawn through the experimental points of several observers. [Figure from KECK and TOLLESTRUP: Phys. Rev. **101**, 368 (1956).]

From the empirical differential cross-section curves the total cross-section can be estimated. Fig. 25 shows the results of these estimates of the total cross-section as a function of photon energy. The more important conclusions are: that at about 100 Mev the total cross-section is already more than twice the value calculated by SCHIFF and by MARSHALL and GUTH, that the total cross-section goes through a slight minimum at 150 Mev and rises to a maximum of 60×10^{-30} cm² at 250 Mev at energies above 250 Mev, the total cross-section decreases to a value of 10×10^{-30} cm² at 450 Mev.

A semiquantitative explanation of the behavior of the total cross-section for the photodisintegration of the deuteron has been given by WILSON[2]. The explanation is based on the assumption that the photodisintegration of the deuteron at high energies is a process connected with meson production. The process can be qualitatively understood as follows: In the region above the threshold

[1] L. C. KECK, R. LITTAUER, G. K. O'NEILL, A. M. PERRY and W. M. WOODWARD: Phys. Rev. **93**, 827 (1954).

[2] R. R. WILSON: Phys. Rev. **86**, 125 (1952).

(~150 Mev) for photomeson production, the cross-section for photodisintegration is considerably higher than that calculated by SCHIFF and MARSHALL and GUTH. If a meson is produced from a nucleon in the deuteron, it may either be emitted or reabsorbed by the two-nucleon system; in the latter case its energy is shared as the kinetic energy of recoil of the nucleons. If the photomeson production process occurs when the two nucleons are relatively far apart (beyond a meson COMPTON wavelength of each other), the meson will escape. If, on the other hand, the two nucleons are close together, the phase space factor in the transition probabilities favors the process with the higher momentum, namely the reabsorption of the meson and recoil of the nucleons. The phase space factor favors the reabsorption, providing the photon energy is not too far above the meson threshold. Using the HULTHÉN wave functions for the ground state of the deuteron, to estimate the fraction of the time the two nucleons are within a pion COMPTON wavelength of each other, WILSON obtains that the photodisintegration cross-section should be about $\frac{1}{4}$ of the photomeson production cross-section at 250 Mev. The experimental value is about $\frac{1}{7}$.

A quantitative phenomenological treatment of the photodisintegration of the deuteron, based on WILSON's model, together with a specific ($I={}^3/_2{}^+$, $T={}^3/_2$ resonance) model for the meson production process has been given by FELD[1] and AUSTERN[2]. The most logically complete calculation to date has been made by ZACHARIASEN using a meson formalism which has been fairly successful at treating photomeson production. He finds that terms which the treatment of AUSTERN and FELD neglect lead to a pronounced minimum in the cross-section at 125 Mev. ZACHARIASEN'S calculated values seem to fit the data above 300 Mev fairly well. However, his calculated angular distributions do not show the strong forward maximum observed in the angular distribution.

A great deal of progress has been made toward understanding the high energy photo-disintegration of the deuteron; however much remains to be done theoretically and experimentally especially on the angular distributions. In the latter regard, it is to be noted that there are no measurements at less than 30° or above 150°.

As pointed out at the beginning of this section, an understanding of the photodisintegration of the deuteron is more involved with the nature of the electromagnetic interaction than with nuclear forces.

41. The photoproduction of nucleons, deuterons and alpha particles: General remarks. For γ-rays of energy between threshold and 50 Mev, most of the photoreactions arise from the well known "dipole resonance" reaction. The product particles result mainly from evaporation of the excited compound nucleus. However, even in the dipole resonance energy range, some of the nucleons ejected appear to result from a direct interaction between the γ-ray and the nucleon ejected. These "directly emitted" photonucleons exhibit two characteristic features, (1) they carry away a large fraction of the incident γ-ray energy and (2) their angular distribution unlike that of the evaporation products, is not isotropic, but tends to exhibit the $\sin^2\vartheta$ distribution characteristic of a photoelectric ejection process (see COURANT[3] and JENSEN[4]).

At higher energies, photon absorption usually results in multiple particle emission (see following sections). The energy dependence for photon absorption in the energy range 50 to 150 Mev is not yet well established, but once the dipole

[1] B. T. FELD: Nouvo Cim. **2**, Suppl. 1, 145 (1955).
[2] N. AUSTERN: Phys. Rev. **100**, 1522 (1955).
[3] E. D. COURANT: Phys. Rev. **82**, 703 (1951).
[4] P. JENSEN: Naturwiss. **35**, 190 (1948).

resonance has been passed, the cross-section appears to be small and either slowly decreasing or approximately constant with increasing energy. In the region of 140 Mev, which corresponds to the onset of the possibility of π meson production, the cross-section for photon absorption appears to increase, and the increase is believed to be due to processes connected with mesons. As in the case of the deuteron (see section above) an appreciable fraction of the mesons produced are reabsorbed in the same nucleus and give rise to the ejection or the evaporation of neutrons, protons, deuterons, α-particles, and possibly heavier nuclear fragments.

The following three sections deal with studies involving the observations of a single particle emitted from targets bombarded by high energy X-rays. The interpretation of such events as the ejection of a single particle is ambiguous in many cases as will be seen in Sect. 45 on the photodisintegration of the "Quasi-Deuteron". The ambiguity arises from the only available sources of appreciable intensity of photons being bremsstrahlung sources. The bremsstrahlung spectrum falls off approximately as $1/E$, so that the resulting observed particle spectrum frequently represents an integral in which the effects of the lower energy photons are strongly weighted. The group at Cornell[1] tried to avoid the difficulty of working with a bremsstrahlung spectrum; they developed a method of making observations of ejected particles in coincidence with the scattered electron that gave rise to the photon; this permitted them to work with monochromatic photons. However, owing to the very small intensities involved, the technique has been of rather limited applicability.

Despite the difficulties of interpreting the data in many cases, a great deal of experimental information has been accumulated.

42. Photo-protons. LEVENTHAL and SILVERMAN[2], bombarded C, Cu, and Pb with a 322 Mev X-ray beam and observed the protons in a proportional counter telescope. They studied the angle and energy distribution of protons with energies between 10 and 70 Mev. They obtained a proton energy spectrum at 90° that fell off with (proton) energy roughly as E^{-2}. The angular distribution of the protons with energies between 8 to 12 Mev was isotropic and were interpreted as being the result of evaporation processes. The protons with energies between 30 and 40 Mev, showed a distinct peaking at forward angles.

WALKER studied the protons emitted with energies of 70 Mev and above at a number of angles from a carbon target bombarded by a 195 Mev beam.

KECK[3] at Cornell, ROSENGREN and DUDLEY[4] and FELD *et al.*[5] using (approximately) 320 Mev bremsstrahlung beams extended these photo-proton studies to higher energies, a larger number of elements, and a greater range of angles. They all employed proton counter telescopes. Similar measurements have been made by HAXBY and associates[6] using a cloud chamber.

All of these workers obtained essentially similar results. The angular distributions are peaked at forward angles as shown in Fig. 26. The energy spectra falls as E^{-n} with $1.7 < n < 2.2$ up to an energy of 140 to 190 Mev where there is an apparent break in the spectrum's energy dependence. Beyond this break,

---

[1] J. W. WEIL and B. D. MCDANIEL: Phys. Rev. **92**, 391 (1953).
[2] C. LEVINTHAL and A. SILVERMAN: Phys. Rev. **82**, 822 (1951).
[3] J. C. KECK: Phys. Rev. **85**, 410 (1952).
[4] J. W. ROSENGREN and J. M. DUDLEY: Phys. Rev. **89**, 603 (1953).
[5] B. T. FELD, R. D. GODBOLE, A. ODIAN, F. SCHERB, P. C. STEIN and A. WATTENBERG: Phys. Rev. **94**, 1000 (1954).
[6] W. E. MATHESON, C. R. CARLSON, F. S. MATHEWS, R. O. HAXBY and R. M. WHALEY: Phys. Rev. **96**, 826(A) (1954).

the spectrum falls off with a much higher power of the energy. This apparent break led LEVINGER[1] to propose a two-nucleon model for the high energy photo-effect. Subsequent measurements (discussed in the section on "Photodisintegration of the Quasi-Deuteron") have established that the majority if not all of these high energy protons are initially associated with a neutron.

Measurements of photoprotons due to intermediate energy photons have been made by several groups. HENDEL[2] used a 150 Mev bremsstrahlung beam to bombard Be, C, Cu, and Pb; he observed the ejected photoprotons in photographic plates. HENDEL's results on the angular distribution and yields of protons with energies between 20 and 60 Mev are very similar to the results obtained with 300 Mev bremsstrahlung.

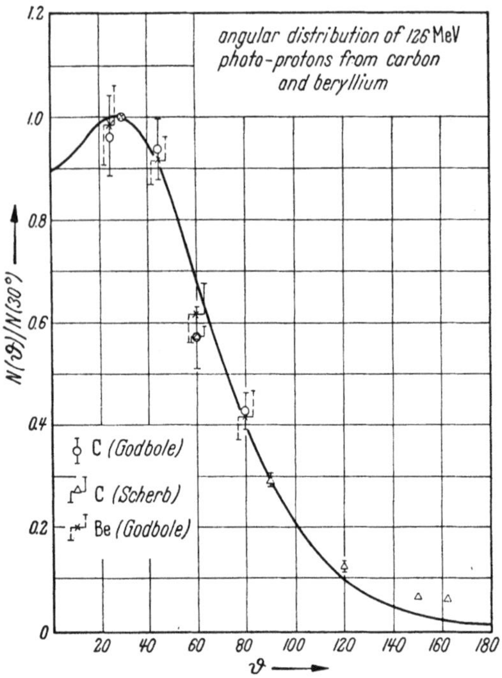

Fig. 26. The relative angular distribution of protons with 126 Mev emitted by Be and C when bombarded by a 330 Mev bremsstrahlung beam. The curve is an attempt to fit the data to a formula of the type $\frac{A + B\sin^2\vartheta}{(1-\beta\cos\vartheta)^4}$. [Figure from FELD *et al.*: Phys. Rev. **94**, 1000 (1954).]

JOHANSSON[3] used a 65 Mev bremsstrahlung beam to bombard C, Al, Ni, and Mo; his detector was sensitive to all protons above 14 Mev. He obtained angular distributions which peaked at about 60°; the yield of protons was approximately proportional to the Z of the target. He also presents excitation curves of the yield of proton versus maximum bremsstrahlung energy.

JOHANSSON states that his results are in reasonable agreement with COURANT'S[4] theory of the direct photo-effect (after making the correction required for retardation). DEDRICK, using LEVINGER'S model, has performed a numerical calculation for photon energies between 50 and 100 Mev. The results he obtains from this model also fit JOHANSSON'S data. Further work in this energy region is obviously necessary to decide between these models.

It is the author's personal belief that it is not fortuitous that the predictions of the quasi-deuteron model are in agreement with some aspects of COURANT'S theory at low energies.

43. Photoneutrons. The measurement of the energy of neutrons is rather difficult and because of this the early experiments on neutrons emitted in photo-reactions were integrations over all neutron energies. KERST and PRICE[5] and JONES and TERWILLIGER[6] detected the neutrons after slowing them down in a large block of paraffin. They bombarded a large number of elements in 320

[1] J. S. LEVINGER: Phys. Rev. **84**, 43 (1951).
[2] H. HENDEL: Z. Physik **135**, 168 (1953).
[3] S. A. E. JOHANSSON: Phys. Rev. **97**, 434 (1955).
[4] E. D. COURANT: Phys. Rev. **82**, 703 (1951).
[5] I. W. KERST and G. A. PRICE: Phys. Rev. **79**, 725 (1950).
[6] L. W. JONES and K. M. TERWILLIGER: Phys. Rev. **91**, 699 (1955).

and 330 Mev bremsstrahlung beams. The results of their measurements are shown in Fig. 27 and are compared to the results obtained in a 22 Mev bremsstrahlung beam at Illinois. In the light elements there is an even-odd fluctuation in yield. For heavier elements the yield goes as $A^{1.7}$ except for U and Th which have yields 30% above the extrapolated yield curve. These extra neutrons are attributed to the photofission reactions in the giant resonance. The cross-section for the emission of neutrons has been determined by JONES and TERWILLIGER for photon energies from 13.5 to 320 Mev for eleven elements. The results for heavier elements are shown in Fig. 28. One sees the "resonance" shape in the

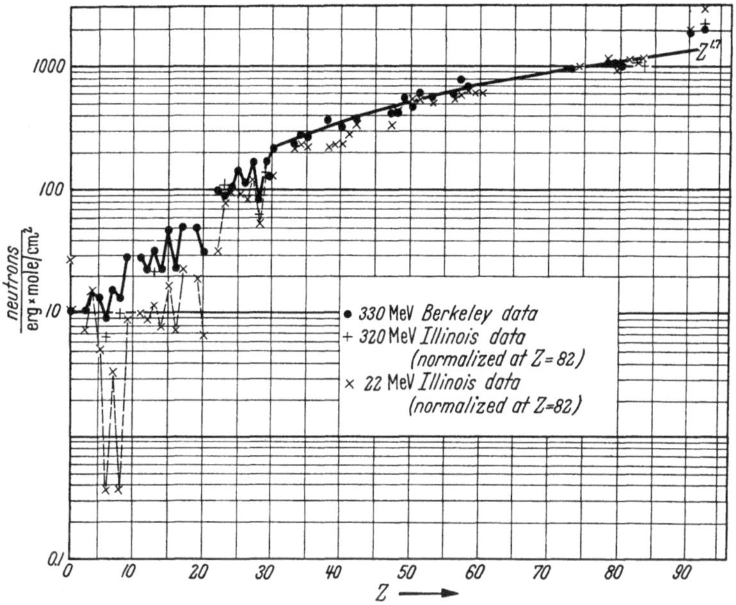

Fig. 27. The yield of photoneutrons as a function of the atomic number of the target for 330 Mev photons. The yields are expressed in terms of neutrons/(erg × moles/cm²). The data of KERST and PRICE at 22 and 320 Mev are included and are normalized at the value for lead. [Figure from TERWILLIGER *et al.*: Phys. Rev. **82**, 820 (1951).]

20 Mev region followed by a region extending to 80 Mev in which the cross-section is down by a factor of five to ten; there is an appreciable rise in cross-section as one approaches 300 Mev.

44. The photoproduction of deuterons, tritons and alpha-particles. SMITH and LAZLITT[1] and RING[2] have used a cloud chamber to study the deuterons produced by bombarding Cu and S with 65 Mev X-rays. RING reports a corrected ratio of deuterons to protons of 0.15 from S.

SILVERMAN and associates[3] have measured the ratio of photo-deuterons to photoprotons from Be, C, Al, Cu, Mo, Ag, Pt, and Pb targets placed in a 320 Mev X-ray beam.

The angular distribution of deuterons of 40 Mev average energy are roughly the same as those for photoprotons of the same energy. The ratio of photodeuterons to photoprotons emitted at 90° with energies of about 40 Mev increase from 0.12 for C to about 0.25 for the heavier elements. For higher energy deuterons and protons the ratio of deuterons to protons decreases steadily; at about 110 Mev the deuteron-to-proton ratio from C is about 0.04 and that from Pt 0.08.

[1] W. H. SMITH and L. J. LAZLITT: Phys. Rev. **86**, 523 (1952).
[2] L. S. RING: Phys. Rev. **99**, 137 (1955).
[3] J. W. DE WIRE, A. SILVERMAN and B. WOLGE: Phys. Rev. **92**, 519 (1953).

These ratios are amazingly high for any direct production process involving only the deuteron and the nucleus. As the charge-to-mass ratio is almost the same for a deuteron and a light nucleus like carbon, there is no electric dipole moment produced by separating a deuteron from carbon, therefore the electric

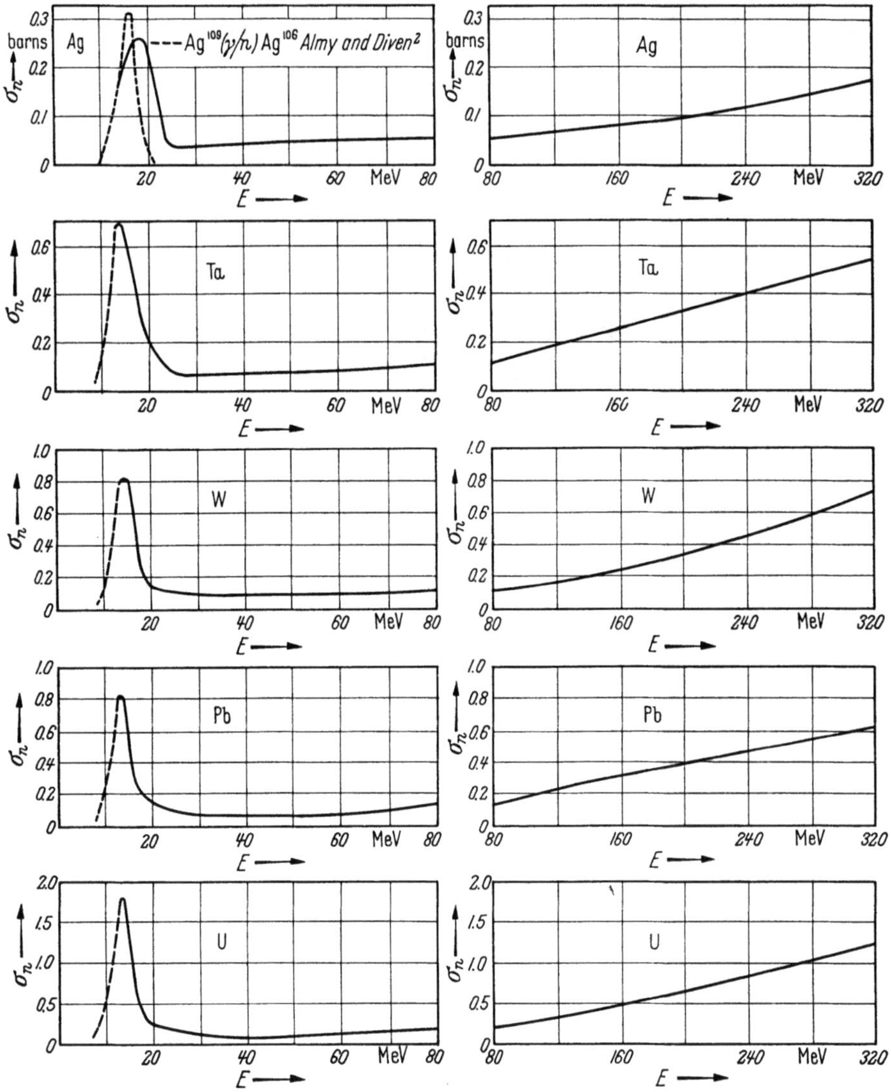

Fig. 28. The cross-section for the production of photoneutrons in Ag, Ta, W, Pb, and U as a function of photon energy. Up to 13.5 Mev, the curves are estimated. [Figure from JONES and TERWILLIGER: Phys. Rev. 91, 699 (1953).]

dipole matrix element is zero for carbon, and must be extremely small for all light nuclei. The quasi-deuteron model does not predict the direct production of photodeuterons. As pointed out by DE WIRE *et al.*, the large number of photodeuterons may be due to a photoproduced neutron or proton which in the cascade process undergoes an (internal) deuteron pick up process. An alternative explanation would be that of YOSHIDA[1], in which intermediate energy photons interact

[1] I. YOSHIDA: Progr. Theor. Phys. 6, 1032 (1951).

with subunits of the nucleus consisting of two protons and a neutron, or two neutrons and a proton. If this model is correct then some of the time a neutron or proton should be emitted simultaneously with the deuterons and at kinematically predictable angles.

The photoproduction of alpha particles has also been studied in photographic plates by MILLAR[1] and in cloud chambers by GAERTTNER and YEATER[2]. MILLAR finds that the energy spectrum of the alphas is what one would expect if they were the result of an evaporation process. The photoproduction of tritons has been observed[3] from the bombardment of lead by 320 Mev bremsstrahlung.

45. The "quasi-deuteron" photodisintegration process. The "breaks" observed in the high energy photoproton spectra (see Sect. 42) led to the proposal by LEVINGER[4] of a "quasi-deuteron" model for the high energy nuclear photoeffect. According to LEVINGER's model where the wavelength of the photon is short, compared to the nucleon diameter, the photon can be pictured as interacting with two nucleons in the nucleus; the ejection of high energy nucleons is, at least in part, by the same processes as lead to the photodisintegration of the deuteron. It should be made clear the LEVINGER's model does not require the actual existence of deuterons in the nucleus. The model only requires that the photons interact with a neutron-proton pair in reasonable proximity to each other. In fact DEDRICK[5] has performed calculations for intermediate energies employing the LEVINGER model and the assumption that the nucleons are in completely random motion. However it is convenient to employ the name "quasi-deuteron", and we will continue to do so without any implications regarding the correlation of nucleons.

Convincing evidence of the validity of the quasi-deuteron model has been provided by the observations of neutrons in coincidence with photoprotons by BARTON and SMITH[6] at the University of Illinois and by a group at M.I.T.[7].

There is an angular correlation of the neutron and proton due solely to kinematics. The kinematics differ slightly from those of the photodisintegration of a free deuteron due to the motion of the center of momentum of the neutron and proton pair in a complex nucleus. The neutron-proton coincidences as a function of the angle of the neutron detector are shown in Figs. 29—31 for targets of D, Li, and C. The measurements were all made with the proton angle fixed at about 76° in the laboratory and the proton detector set to observe protons with a mean energy of 140 Mev. The width of the deuterium curve is the instrumental angular resolution. As this is a two-body reaction, the energy and angle of the proton determines the energy and angle of the emitted neutron. The broadening of the curves from other targets is due to the initial motion of the nucleons in the nucleus. A simple calculation based on conservation of momentum can be readily applied to the data. It is found that the observed angular distribution cannot be fitted assuming a FERMI momentum distribution with $T=0°$. The data agree very well with the assumption of (three dimensional)

[1] C. H. MILLAR: Amer. J. Phys. **31**, 262 (1953).

[2] E. R. GAERTTNER and M. L. YEATER: Phys. Rev. **83**, 146 (1951) contains a list of earlier references.

[3] J. W. ROSENGREN and N. BARON: Phys. Rev. **101**, 410 (1956).

[4] J. S. LEVINGER: Phys. Rev. **84**, 43 (1951).

[5] K. G. DEDRICK: Phys. Rev. **100**, 58 (1955).

[6] M. Q. BARTON and J. H. SMITH: Phys. Rev. **95**, 573 (1954). Preliminary report.

[7] H. MEYERS, A. C. ODIAN, P. C. STEIN and A. WATTENBERG: Phys. Rev. **95**, 576 (1954). Preliminary report. A series of papers will cover the more extensive experiments and appear in the Phys. Rev. q.v. A. C. ODIAN, P. C. STEIN, A. WATTENBERG, B. T. FELD and R. WEINSTEIN: Phys. Rev. **102**, 837 (1956).

GAUSSIAN distributions with 1/e values of about 8, 18, and 18 Mev respectively for Li, C, and O. Due to approximations in the theory, the relative differences are more significant than the absolute values. BARTON and SMITH have compared the widths from He and Li, and find that the mean momentum of nucleons in He is higher than that in Li. It thus seems definite that the average momentum of the nucleons in Li is less than that of other light nuclei measured to date.

The interpretation of the widths of the curves in terms of the momentum of the initial nucleons is predicated on the assumption that a two body process is involved. In the range 140 to 300 Mev BARTON and SMITH find that the energy dependence of the cross-section of the quasi-deuteron process in Li is the same as that observed for the photodisintegration of a free deuteron, and ODIAN *et al.* find that the angular distribution is essentially the same as that for the free deuteron. These observations indicate that at the energies investigated the quasi-deuteron photodisintegration proceeds through the same mechanisms as are responsible for the photodisintegration of a (free) deuteron despite the presence of other nucleons in a nucleus.

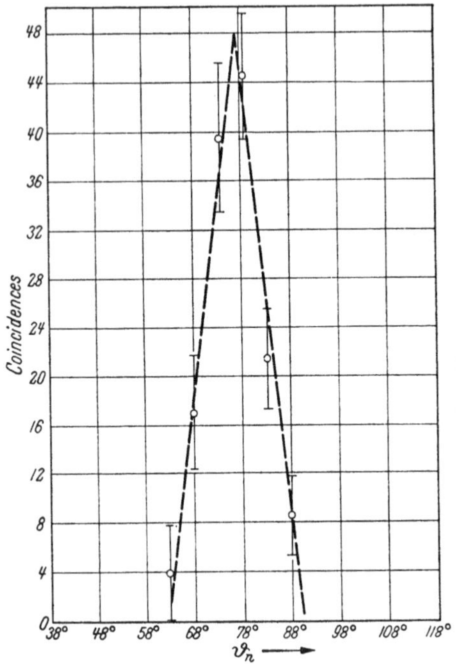

Fig. 29.

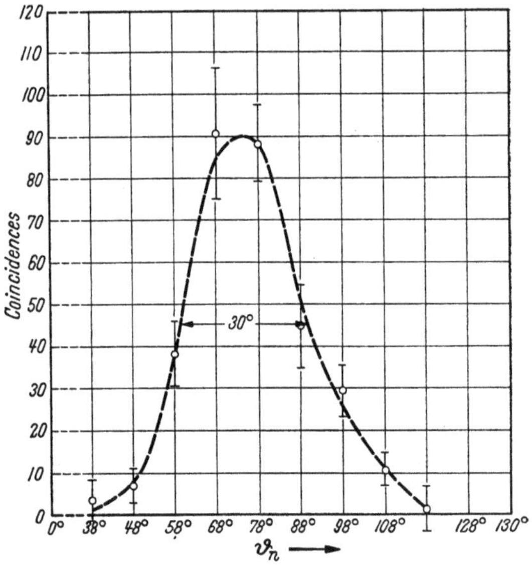

Fig. 30.

Fig 29. The number of neutron-proton coincidence as a function of neutron angle from the photodisintegration of deuterium by 280 Mev photons. In these measurements the proton counter was kept fixed at 76° (on the side opposite to the neutron counter) and the energy of the protons accepted by the detector was 129 Mev. The width of the curve arises from the finite angular resolution of the equipment. [Figure from WATTENBERG *et al.:* (to be published).]

Fig. 30. The number of neutron-proton coincidences as a function of neutron angle from Li for the same experimental conditions as in Fig. 29.

The M.I.T. group studied the integrated number of neutron-proton coincidences as a function of atomic weight between Li and Pb. The data quantitatively fit a formula of the type

$$\sigma(Z^A) = C\frac{NZ}{A}\sigma_D f\left(\frac{2R}{A}\right) \tag{45.1}$$

where C is a constant with a value of 4 ± 1, N is the number of neutrons, Z the number of protons, A the atomic weight, $\sigma(Z^A)$ the cross-section for the observation of a neutron and proton in coincidence, σ_D the observed cross-section for the photodisintegration of the deuteron (see Fig. 25 for numerical values), $f\left(\frac{2R}{\Lambda}\right)$ is a calculated attenuation factor which gives the probability of escape of correlated neutrons and protons as a function of the nuclear radius, R, and a mean free path Λ in nuclear matter. The major uncertainty in the constant C arises from the uncertainty in R/Λ.

The factor $f(2R/\Lambda)$ is to correct for those nucleons which suffer collisions with other nucleons before they escape from the nucleus. $f(2R/\Lambda)$ is about 40% for carbon and about 7% for lead and depends upon the values of R/Λ chosen.

In the light of the experimental results some revision of LEVINGER's original calculations are required. One should employ experimental values of σ_D instead of calculating the matrix element. WILSON's[1] model of the photodisintegration of the deuteron gives a value of 3.8 for the coefficient C, in Eq. (45.1) which is in excellent agreement with the experimental value of 4 ± 1, considering the simplicity of the assumptions.

The high energy nuclear photoeffect appears to be dominated by the interaction of photons with a pair of nucleons, and quantitative predictions can be made on the basis of a very simple picture.

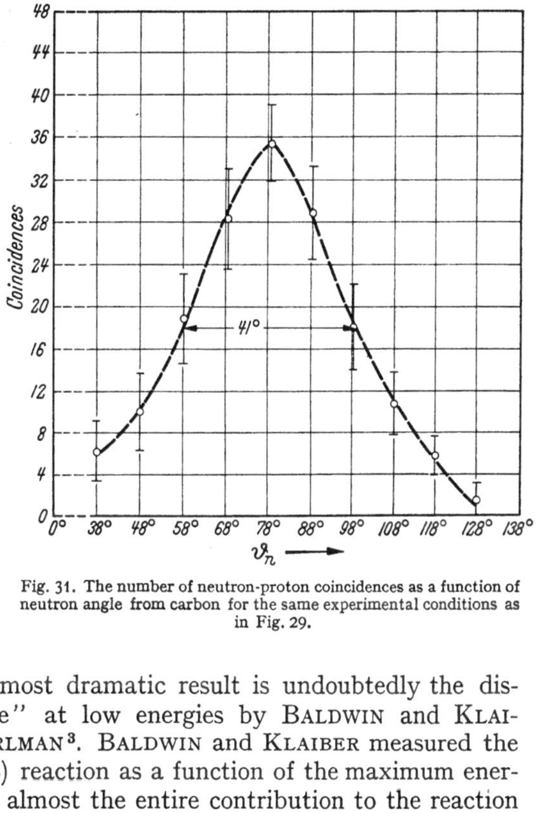

Fig. 31. The number of neutron-proton coincidences as a function of neutron angle from carbon for the same experimental conditions as in Fig. 29.

46. "Stars" and radioactivity induced by photons: Energy dependence. From studies with high energy X-ray machines the most dramatic result is undoubtedly the discovery of the "Dipole-Resonance" at low energies by BALDWIN and KLAIBER[2] and FRIEDLANDER and PERLMAN[3]. BALDWIN and KLAIBER measured the radioactivity induced by the (γ, n) reaction as a function of the maximum energy of the photons and found that almost the entire contribution to the reaction is due to photons of approximately 20 Mev. PERLMAN and FRIEDLANDER, studying other processes like (γ, p) $(\gamma, 2n)$, $(\gamma, 2p)$ by induced radioactivity, showed that the effect must be duc to the nature of the photon absorption. GOLDHABER and TELLER[4] explained these results as being due to a "giant dipole resonance" and suggested a nuclear model to give such a resonance.

A great deal of work at energies above 50 Mev measuring the induced radioactivity have confirmed the early result that the photon absorption is comparatively small once one is above the giant resonance region until one reaches the region of meson production. Studies up to 1952 are summarized by STRAUCH[5].

---

[1] R. R. WILSON: Phys. Rev. **86**, 125 (1952).
[2] G. C. BALDWIN and G. I. KLAIBER: Phys. Rev. **73**, 1156 (1948).
[3] M. L. PERLMAN and G. FRIEDLANDER: Phys. Rev. **74**, 442 (1948).
[4] M. GOLDHABER and E. TELLER: Phys. Rev. **74**, 1046 (1948).
[5] K. STRAUCH: Ann. Rev. Nucl. Sci. **2**, 105 (1953).

A complete list of references of both theoretical and experimental papers will be found in a bibliography of all photonuclear reactions compiled by TOMS[1].

STRAUCH describes his own experiments in which he obtained the energy dependence for reactions by determining an effective depth of a reaction in a stack of foils irradiated simultaneously ("transition curve" technique). The effective depth is related to a mean energy for a reaction by electromagnetic shower theory. Data on the energy dependence have also been obtained from radioactive yields as a function of the maximum energy of the X-rays. In the region below meson threshold, the radioactivity data are in agreement with those obtained in photoneutron experiments, namely that the cross-sections are small compared to the cross-section of the giant resonance. Similar results are observed in star data from cloud chambers; GAERTTNER and YEATER[2] find only a slight increase in the number of multiprong stars from 50 to 100 Mev.

Above about 100 Mev there is an increase in the cross-section for the production of radioactivity and "stars" in emulsions. For example from the X-ray bombardment of Mg and Al REAGAN[3] obtains yields of N^{17} (a delayed neutron emitter) which he interprets as a photoproduction cross-section of about 2 microbarns at 100 Mev and 30 microbarns at 400 Mev. The cross-sections at various photon energies for the production of stars with more than two prongs are listed in the accompanying table, taken from KIKUCHI[4]. The values given are commensurate with the entire cross-section for the quasi-deuteron process [see Eq. (45.1)]. The explanation for the high cross-section probably lies in meson production. Photomesons produced in complex nuclei have a very low probability of escaping without undergoing collisions and absorption. The initial production of a meson is thus followed by the cascade process (described in Chap. B, Sect. 11) with the possibility that the meson will be captured and give up its total energy (including a rest energy of about 140 Mev) to recoiling nucleons. REFF[5] has performed a phenomenological computation of photo-star production on the above model and finds that the process accounts for a large part of the observed photostars from heavy elements. However, he finds that for light elements the meson process will account for only a small fraction of the stars observed by MILLER[6], especially those found with 161 Mev X-rays.

Table 10. *Average cross-sections for the photoproduction of multi-prong stars in photographic emulsions.*

Energy (Mev)	150	200	250	300
Cross-section 10^{-27} cm^2/per equivalent quanta	1.95 ± 0.18	2.55 ± 0.19	5.6 ± 0.6	6.0 ± 0.4

47. "Stars" and radioactivity induced by photons: Types of reactions. The events observed by KIKUCHI and MILLER include some of the events which result in radioactive products where a charged particle is emitted; however they fail to observe those events where only neutrons are emitted. Some of the prongs observed by KIKUCHI were mesons and not nucleons. This result combined with

[1] M. E. TOMS: Bibliography of "Photonuclear Reaction". Naval Research Laboratory, June 1955 (545 references).

[2] E. R. GAERTTNER and M. L. YEATER: Phys. Rev. 77, 570 (1950).

[3] D. REAGAN: Phys. Rev. **100**, 113 (1955).

[4] S. KIKUCHI: Phys. Rev. **86**, 41 (1952); see also Phys. Rev. **81**, 1060 (1951).

[5] I. REFF: Phys. Rev. **91**, 150 (1953).

[6] R. D. MILLER: Ph. D. Thesis, University of California 1952.

the complication that the radioactive technique fails to observe the stable products produced, makes it almost impossible to quantitatively check the star data against the radioactive yield data. Nor is it clear as to how to combine two sets of data to obtain a total cross-section. For example what appears to a radiochemist as a $(\gamma, 2n)$ reaction might actually have been a $(\gamma, p n \pi^-)$ and would have appeared to KIKUCHI as an event in which two charged particles were emitted.

The stars and radioactive product distribution are very similar to those observed in reactions with high energy particles except for those events in which only one or two particles are emitted. One finds a great variety of end products,

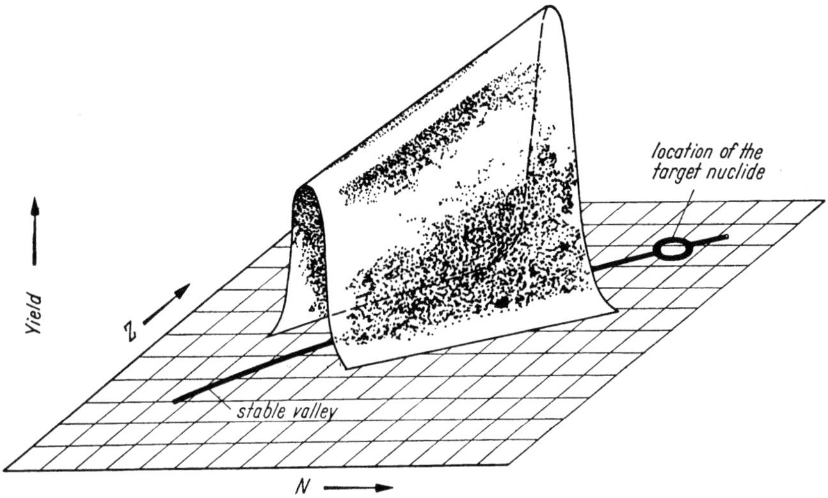

Fig. 32. A schematic representation of the relative yield of various nuclides from photonuclear reactions with 320 Mev X-rays. The yield is indicated by the height above the proton-neutron plane. The dark line in the proton-neutron plane indicates the center of the valley of stability. [Figure from HALPERN *et al.*: Phys. Rev. **97**, 1327 (1955).]

and at higher energies it is not possible to tell whether the process was through the emission of several alpha particles or the individual nucleons that would constitute such particles, and spallation also occurs. The relative yield of radionuclides can be understood by consideration of the nuclear energy surface just as in the case of the yield of charged particle reaction. The importance of these considerations and the details involved have been given by HALPERN *et al.*[1]. For nuclides of the same atomic weight produced from the same isotope, they showed that the yield falls off very rapidly with the distance from the valley of stability. For a variety of initial targets, the yield pattern is independent of the target. This result comes from the evaporation process compensating for fluctuations in emission that take place during the initial cascade process. The evaporation process becomes correlated with the initial cascade process because of the steepness of the rise of the mass energy surface on either side of the valley of stability.

If one plots the yields as a surface above the neutron proton plane, one observes it is fairly narrow as is seen in Fig. 32. The walls of the yield surface drop by a factor of 100 in three atomic mass units; the ridge of the yield surface is roughly parallel to the stable valley but is displaced about one mass unit to the neutron deficient side. These features of the yield surface seem to be rather independent of the nature of the bombarding agent or its energy, provided that it is more than 100 Mev.

[1] I. HALPERN, R. J. DEBS, J. T. EISINGER, A. W. FAUHALL and H. G. RICHTER: Phys. Rev. **97**, 1327 (1955); for data see Phys. Rev. **97**, 1325 (1955).

Below 100 Mev the relative yield of radionuclides is affected by the high thresholds for some complex reactions. Compilation of yield data for 50 and 87 Mev X-rays are given by HOLTZMAN and SUGARMAN[1], and for 70 Mev X-rays by EDWARDS and MCMILLAN[2]. In general one notices that the single particle reaction yields (γ, n), (γ, p), and $\begin{pmatrix}\gamma, d \\ \gamma, n\, p\end{pmatrix}$ are appreciably larger than the yields for more complex reactions.

From very light nuclei one finds some variation from this pattern probably due to the characteristics of the particular nucleus and its nearest neighbors. For example, COOPER and ZAFFARNO[3] find that the reactions C^{12} $(\gamma, 3\alpha)$, and O^{16} $(\gamma, 4\alpha)$, and N^{14} $(\gamma, np\, 3\alpha)$ occur frequently in photographic emulsions exposed to a 70 Mev bremsstrahlung beam. Similar peculiarities of individual light nuclei were observed by GAERTTNER and YEATER[4] in their cloud chamber exposures to a 100 Mev X-ray beam. In an exposure of a photographic emulsion to 300 Mev bremsstrahlung, SOFTKY[5] found that the C^{12} $(\gamma, 3\alpha)$ reaction was actually being produced by the low energy photons, namely those photons with energies less than 50 Mev.

The first exposure of photographic plates to 300 Mev bremsstrahlung was made by KIKUCHI[6] who found that the relative number of 2:3:4:5:6:7 pronged stars are respectively 160:100:89:38:64:1.6 (probably from silver and bromine). The most likely energy of the prongs are those expected from the evaporation of protons; higher energy events seem to constitute less than 10% of all events. This energy distribution probably arises from the $1/E$ nature of the bremsstrahlung spectrum combined with the high cross-section of the dipole resonance. An explanation of evaporation prong distributions (from cosmic ray stars) has been given by LECOUTEUR[7] (see Sect. 12).

For those physicists who like to study complex phenomena, the high energy photoproduction of stars and radionuclides with a continuous X-ray spectrum is probably one of the most complex situations that they can find in nuclear physics.

If the reader feels slightly confused by details at this point, it is hoped that he will re-read Sects. 2 through 5 (Chap. A) to some advantage.

G. Reactions of mesons.

48. The properties and interactions of mesons. This chapter is concerned with the contributions to our knowledge of nuclei which can be obtained from the reactions of mesons with nuclei. The characteristics of the various mesons and hyperons are listed in Table 11; the values are from a summary by SHAPIRO[8] dated April 1956. More detailed discussions of the properties of mesons and their interactions are contained in books by ROSSI [*1*], MARSHAK[9], BETHE and DE HOFFMANN[10], in a review article by TURKEVITCH[11] and in Vol. XLIII of this Encyclopedia.

[1] R. B. HOLTZMAN and N. SUGARMAN: Phys. Rev. **87**, 633 (1952).
[2] L. S. EDWARDS and F. A. MCMILLAN: Phys. Rev. **87**, 377 (1952).
[3] R. D. COOPER and D. L. ZAFFARNO: Phys. Rev. **98**, 1199A (1955).
[4] E. R. GAERTTNER and M. L. YEATER: Phys. Rev. **77**, 570 (1950); **79**, 401 (1950).
[5] S. D. SOFTKY: Phys. Rev. **98**, 173 (1955).
[6] S. KIKUCHI: Phys. Rev. **80**, 492 (1950).
[7] K. J. LECOUTEUR: Proc. Phys. Soc. Lond. A **63**, 259 (1950).
[8] A. M. SHAPIRO: Rev. Mod. Phys. **28**, 164 (1956).
[9] R. MARSHAK: Meson Physics. New York: McGraw-Hill Co. 1952.
[10] H. BETHE and F. DE HOFFMANN: Mesons and Fields. Evanston, Ill.: Row, Peterson & Co. 1955.
[11] A. TURKEVITCH: Ann. Rev. Phys. Chem. **4**, 119 (1953).

The mesons originally proposed by YUKAWA[1] were intimately related to nuclear forces with the consequence that such particles should have strong interactions with nucleons and with nuclei, and one should find cross-sections commensurate with the size of the nucleus. Also negative mesons brought to rest in matter should disappear by nuclear capture rather than undergo free decay. π mesons and K^- mesons have been found to behave in the anticipated fashion indicating that they interact strongly with nuclei.

Table 11. *Table of properties of mesons and hyperons.*

Particle symbol and charge	Mass		Decay products	Q	Mean life	Spin
	(m_e)	(Mev)		Mev	(sec)	
L mesons $\mu^\pm$	206.7	105.6	$e^\pm + \nu + \nu$	105.1	2.22×10^{-6}	$(\frac{1}{2})$
π^+	273.3	139.7	$\mu^+ + \nu$	34.1	2.53×10^{-8}	0
π^-	272.8	139.4	$\mu^- + \nu$	33.8	2.55×10^{-8}	0
π^0	264.3	135.0	$\gamma + \gamma$	135.0	$(1-5) \times 10^{-15}$	0
K mesons τ^+	966.1	493.7	$\pi^+ + \pi^+ + \pi^-$ $(\pi^+ + \pi^0 + \pi^0)$	75.0	1.27×10^{-8}	(0)
τ^-	(966.1)	(493.7)	$\pi^- + \pi^+ + \pi^-$ $(\pi^- + \pi^0 + \pi^0)$	(75.0)	(1.27×10^{-8})	(0)
ϑ^0	965	493	$\pi^+ + \pi^-$ $(\pi^0 + \pi^0)$	214	1.3×10^{-10}	Int. $\neq$ 1
$K\pi 2^{(\pm)}$ (a)	966 ± 2 Av		$\pi^{(\pm)} + \pi^0$	219.1	$12.1^{+1.1}_{-1.0} \times 10^{-9}$	
$K\pi 3^{(\pm)}$ (a)	965 ± 7 Av		$\pi^{(\pm)} + \pi^0 + \pi^0$	84.1	10×10^{-9}	
$K\mu 2^{(\pm)}$ (a)	966 ± 2 Av		$\mu^{(\pm)} + \nu$	378.0	$11.7^{+0.8}_{-0.7} \times 10^{-9}$ 10×10^{-9}	
$K\mu 2^{(\pm)}$ (a)	964 ± 5 Av		$\mu^{(\pm)} + \pi^0 +$ $(\mu^{(\pm)} + \gamma +)$	253.0	10×10^{-9}	
$Ke 3^{(\pm)}$ (a)	967 ± 9 Av		$e^{(\pm)} + ? + ?$ $(e^{(\pm)} + \pi^0 +)$ $(e^{(\pm)} + \gamma +)$		10×10^{-9}	
Hyperons Λ^0	2181.5	1114.7	$p^+ + \pi^-$ $(n^0 + \pi^0)$	36.9	3.7×10^{-10}	$\frac{1}{2}$ Int.
Σ^+	2326.9	1189.0	$p^+ + \pi^0$ $n^0 + \pi^+$	115.8 109.9	3.4×10^{-11}	$\frac{1}{2}$ Int.
Σ^-	(2326.9)	(1189.0)	$n^0 + \pi^-$	(109.9)	3.4×10^{-11}	$\frac{1}{2}$ Int.
Ξ	2586	1321	$\Lambda^0 + \pi^-$	67	1×10^{-10}	$\frac{1}{2}$ Int.

(a) The various K mesons listed may be alternate decay modes and different charge states of the same particle.

However it has developed that μ mesons have weak interactions with nuclei (see the following sections). Specifically μ-mesons interact with nuclei through their electric charge and through the FERMI interactions (reactions involving the emission or absorption of neutrinos). Due to our having some familiarity with these two types of interactions and that they are weak ones, μ-mesons are currently useful for probing nuclei. However, it should be kept in mind that the more fundamental dynamics of nuclei can only be comprehended from an understanding of the interactions with nucleons of at least π mesons and perhaps K mesons also.

[1] H. YUKAWA: Proc. Phys. Mat. Soc. Japan 17, 48 (1935).

49. Mesic atoms and X-rays. Charged mesons are brought to rest in matter by collisions with atoms according to the conventional theory of ionizing (and radiative) collisions (see FERMI, TELLER, and WEISSKOPF[1]). A meson in solid matter will slow down from 2000 ev to rest in less than 10^{-12} sec. If it is a positive meson, it will then undergo free decay via the decay modes (and with the lifetimes) listed in Table 11. However, if it is a negative meson, it will be attracted by the positive electric charge of the nucleus, and by a series of radiative (and AUGER) transitions, the meson will fall into lower and lower states (BOHR orbits). In that the COULOMB force between a nucleus and a negative meson is the same as that between a nucleus and an electron, mesonic atoms are formed. To a first approximation the energy of such atomic states varies as the mass of the particles; for a μ mesonic atom the energy of the 1 S state is thus of the order of $2.7\times 10^3 Z^2$ ev. The radiations accompanying the jumps between such BOHR orbits are observed as very high energy X-rays.

WHEELER[2] pointed out that the finite size of the nucleus appreciably reduces the energy of these states. One must use the COULOMB field from a finite sized charge distribution and not that from a point source.

In more detailed papers, WHEELER[3] stressed the possibility of obtaining the size of the charge distribution of the nucleus from measurements of the energy of these X-rays from μ mesonic atoms.

μ-mesonic X-rays were first observed by CHANG[4] in a cloud chamber containing lead plates. In seven cases where μ mesons stopped in lead, they observed the materialization of photons with energies in the range predicted by WHEELER.

Table 12. *Energies of μ-mesic X-ray.*
(All energies are in Mev; the calculated values are on the assumption that the mass of the μ meson is 210 m_e.)

Z	Calculated values				Observed values
	$R=0$		$R=1.3\times A^{\frac{1}{3}}\times 10^{-13}$ cm		
	$2p_{\frac{3}{2}}-1s_{\frac{1}{2}}$	$2p_{\frac{1}{2}}-1s_{\frac{1}{2}}$	$2p_{\frac{3}{2}}-1s_{\frac{1}{2}}$	$2p_{\frac{1}{2}}-1s_{\frac{1}{2}}$	
13	0.363	0.363			0.35
14	0.421	0.421			0.41
22	1.046	1.043	0.935	0.933	0.955
29	1.828	1.820	1.52	1.51	1.55
30	1.958	1.950			1.60
51	5.833	5.763	3.41	3.37	3.50
80	15.51	15.01			5.80
82	16.41	15.86	5.45	5.30	6.02
83	16.88	16.29			6.02

Further theoretical and experimental investigations were undertaken by quite a few groups; references to their work can be found in the paper of BURBRIDGE and DE BORDE[5] on the mesonic AUGER effect.

The most precise values of the energies of the X-rays from μ-mesic atoms are those obtained by FITCH and RAINWATER[6]. They used negative μ-mesons from

---

[1] E. FERMI, E. TELLER and V. F. WEISSKOPF: Phys. Rev. **71**, 314 (1947). — E. FERMI and E. TELLER: Phys. Rev. **72**, 399 (1947).

[2] J. A. WHEELER: Phys. Rev. **71**, 320 (1956).

[3] J. A. WHEELER: Rev. Mod. Phys. **21**, 133 (1949). — J. A. WHEELER: Phys. Rev. **92**, 812 (1953).

[4] W. Y. CHANG: Rev. Mod. Phys. **21**, 166 (1949).

[5] G. R. BURBRIDGE and A. H. DE BORDE: Phys. Rev. **89**, 189 (1953).

[6] V. L. FITCH and J. RAINWATER: Phys. Rev. **92**, 789 (1953).

the Columbia synchrocyclotron and had them come to rest in nine different targets ranging from Al to Bi. The arrival of a μ-meson and the subsequent emission of an X-ray were detected in a coincidence-anticoincidence-coincidence telescope. The energy of the X-ray was measured in a NaI scintillation spectrometer to better than 1 % for $Z>22$. Figs. 33 and 34 show the type of results they obtained. The energy of the X-rays found for the $2p_{\frac{3}{2}}-1s_{\frac{1}{2}}$ and $2p_{\frac{1}{2}}-1s_{\frac{1}{2}}$ transitions are listed in Table 12. Also included in the table are the calculated values one would expect for point source nuclei and for nuclei with radii of $R=1.3A^{\frac{1}{3}}\times 10^{-13}$ cm. From the values in the table one sees that the charge distribution in nuclei have radii smaller than $1.3\times A^{\frac{1}{3}}\times 10^{-13}$ cm.

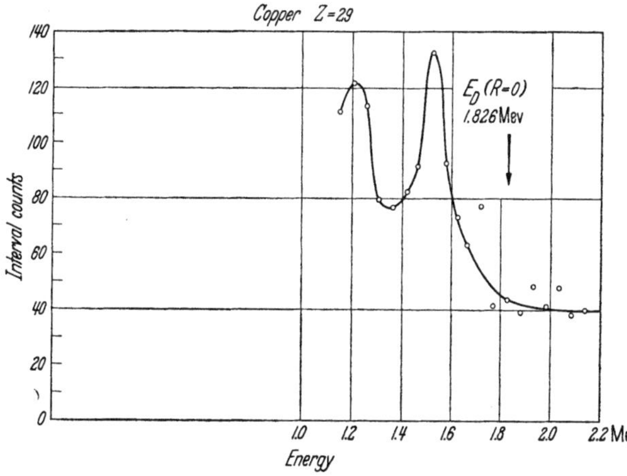

Fig. 33. The upper end of the pulse height distribution of μ-mesic X-rays from a Cu target. The upper peak represents the full energy of the X-rays emitted in the $2p-1s$ transition. The expected energy for a point nucleus is indicated by the arrow. The energy shift due to the finite size of the nucleus is seen to be about 18% of the measured value.

The curves for lighter nuclei (for example Fig. 33) were simple to interpret, because the COMPTON process gives the highest energy and dominant peak in the region of 1 Mev; it was not possible to resolve any fine structure. The curve for Pb, Fig. 34, and the other heavy elements were more difficult to interpret because at higher energies pair production is the dominant process in the NaI crystal. In pair production 1 Mev is lost; however when the positron annihilates one or both of these photons may materialize in the crystal and give up their energy of 0.51 Mev and one may observe several peaks separated by 0.51 Mev. The two peaks they observed were separated by about 0.2 Mev; they interpret this separation as most probably arising from the difference of the $2p_{\frac{3}{2}}$ and $2p_{\frac{1}{2}}$ transitions. However, further measurements would be desirable as such a splitting would definitely establish that the spin of the μ-meson is $\frac{1}{2}$.

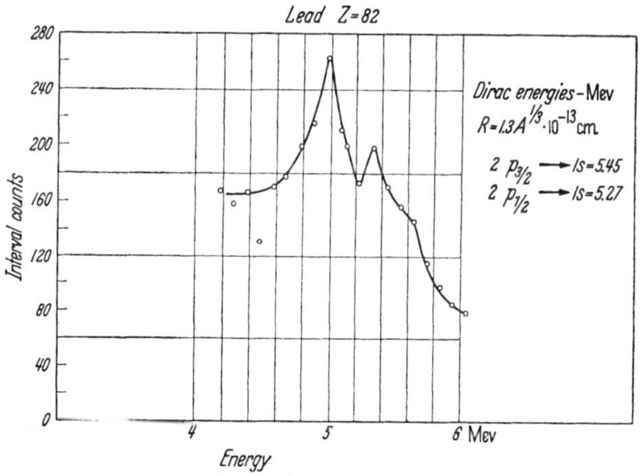

Fig. 34. The pulse height distribution for Pb with 100 kev interval width. The main peak at 5.00 Mev is interpreted as the pair production peak due to the μ-mesic X-rays from the $2p_{\frac{3}{2}}-1s_{\frac{1}{2}}$ transition. Further discussion of this curve is in the text. The energy shift due to the finite extension of the nucleus is 172% of the measured value.

FITCH and RAINWATER and COOPER and HENLEY[1] have carefully analyzed the data and found that agreement with the data could best be obtained with a

[1] L. N. COOPER and E. M. HENLEY: Phys. Rev. **92**, 801 (1953).

charge distribution of $R=1.2\times10^{-13}A^{\frac{1}{3}}$. This is the equivalent radius of a uniformly charged sphere; for purposes of comparison with non-uniform charge distributions this measurement depends on the mean square radius of the charge distribution. Such comparisons have been made by several authors and have been reviewed by FORD and HILL[1]. These comparisons show that the μ-mesic X-ray results are in agreement with the work of HOFSTADTER *et al.* [8] on the scattering of electrons.

WHEELER[2] has pointed out that further information about the nucleus can be obtained from more precise measurements of the X-ray spectrum especially of the fine structure that should be present. The things that one can study include nuclear quadrupole moments, the non-uniformity of the nuclear charge distribution, and nuclear polarizability.

The X-rays from π mesonic atoms have been extensively investigated. The work is covered by BETHE and DE HOFFMANN[3] and in the proceedings of the Annual Rochester Conference on High Energy Nuclear Physics[4]. The interaction between π mesons and nucleons is so strong that the energy levels of the π mesic atom are perturbed from those of a pure COULOMB field; the energy of the X-rays are used to study these perturbations in order to obtain information on the pionnucleon interaction. For example it is found that the energies of the K X-rays ($2p-1s$ transition) are less than those calculated including all known electromagnetic corrections (and the size of the nucleus). In that the L X-rays ($3d-2p$ transition) appear to have the correct energy, the discrepancy in the energy of the K X-rays is interpreted as a repulsive nucleon interaction for π mesons in an $1s$ state.

K X-rays are observed only for light nuclei, the yield of such X-rays (per pion stopped) decreases rapidly with Z. This is attributed to an increasing probability for nuclear capture from the $2p$ state and is in agreement with a known attractive interaction between a nucleon and a meson in a p state. The yield of K X-rays for $Z=8$ is down to about 3% of all pions that have stopped; L X-rays have a yield of about 8% for $Z=26$.

Most laboratories with π meson beams have such investigations underway, and more effort is being devoted presently to studies of π-mesic X-rays than to studies of μ-mesic X-rays.

50. Interactions of μ-mesons with nuclei. α) *The capture probability for slow μ-meons.* Negative μ^--mesons in the innermost BOHR orbit can be captured by a process that is analogous to the K electron capture process. As the innermost BOHR orbit of an atom has a radius that varies as $1/Z$, the total volume occupied by a particle in such a state varies as $1/Z^3$. The volume of a nucleus varies as A; therefore to a first approximation, a capture process has a transition probability which varies as the ratio (Volume of the nucleus)/(total volume occupied by the particle) or as Z^3A (see BLATT and WEISSKOPF [3]).

The surprise was that the analogy between the K capture of a μ^--meson and an electron was too good. The same interaction with the nucleus appears to be involved in the capture of both particles. PONTECORVO[5] first pointed out that

[1] K. W. FORD and D. L. HILL: Ann. Rev. Nucl. Phys. **5**, 25 (1955). Cf. also HILL's contribution to Vol. XXXIX of this Encyclopedia.

[2] J. A. WHEELER: Phys. Rev. **92**, 812 (1953).

[3] H. A. BETHE and F. DE HOFFMANN: Mesons and Fields, Vol. II. Evanston, Ill.: Row, Peterson & Co. 1955.

[4] Proceedings of the Annual Rochester Conference, Vols. 1—6. New York, N. Y.: Interscience Publishers, Inc. 1951—1956.

[5] B. PONTECORVO: Phys. Rev. **72**, 246 (1947).

the lifetime observed for μ^-capture was consistent with the two reactions being (with ν = neutrino):

$$e^- + P \rightarrow N + \nu, \tag{50.1a}$$

$$\mu^- + P \rightarrow N + \nu. \tag{50.1b}$$

This conclusion came from the results of the famous experiment of CONVERSI, PANCINI, and PICCIONI[1] when combined with the theory of FERMI, TELLER and WEISSKOPF[2] for the slowing down of mesons in matter.

CONVERSI *et al.* used a magnetic lens to separate (cosmic ray) μ^--mesons from μ^+-mesons. They measured the number of electrons emitted when μ^--mesons stopped in Fe and C. They found that μ^--mesons stopping in iron predominantly underwent nuclear capture. However, those mesons stopping in carbon mainly underwent free decay instead of capture. If they were the mesons proposed by YUKAWA, they should have also been captured in the carbon (see TOMONAGA and ARAKI[3]). The experiment of CONVERSI, PANCINI, and PICCIONI thus established that μ-mesons have a weak interaction with nucleons (like the beta decay interaction) instead of the strong interaction that one now finds existing between π-mesons and nucleons.

These early results were confirmed in further experiments which measured the lifetime and electron emitting power of μ-mesons captured in different elements; the early experiments are listed by SIGURGIERSSON and YAMAKAWA[4]. The quantitative interpretation of these experiments has been given by TIOMNO and WHEELER[5]. They estimate that the coupling constants for the beta decay process and μ-meson capture process, Eq. (50.1b) agree with one another within the limits of error of experiment and theory.

Experiments up to 1954 on the capture of μ-mesons and on the general subject of the interaction of slow μ-mesons are reviewed by SARD and CROUCH[6]. In that μ^--mesons interact only with protons in a nucleus, the capture probability is expected to vary roughly as Z^4 instead of $Z^3 A$. In light elements the free decay probability is commensurate with the capture probability. Therefore to a first approximation the resultant lifetime for a μ^--meson brought to rest in a low Z material should be[7]

$$\frac{1}{\tau} = \frac{1}{\tau_0}\left[1 + \left(\frac{Z}{Z_0}\right)^4\right] \tag{50.2}$$

where τ_0 is the mean life, 2.22×10^{-6} sec (Table 11). Z_0 has been found to be about 10.

More exact calculations of the mean life for capture have been made by TIOMNO and WHEELER, BUDINI[8], KENNEDY[9] and PRIMAKOFF[10]. KENNEDY has shown that the deviations from the Z^4 law found by KEUFFEL *et al.*[11] in heavy elements can be related to nuclear shell structure. PRIMAKOFF has shown that

---

[1] M. CONVERSI, E. PANCINI and O. PICCIONI: Phys. Rev. **71**, 209 (1947).
[2] E. FERMI, E. TELLER and V. WEISSKOPF: Phys. Rev. **71**, 314 (1947).
[3] S. TOMONAGA and S. ARAKI: Phys. Rev. **58**, 90 (1940).
[4] T. SIGURGIERSSON and K. A. YAMAKAWA: Rev. Mod. Phys. **21**, 124 (1949).
[5] J. TIOMNO and J. A. WHEELER: Rev. Mod. Phys. **21**, 153 (1949).
[6] R. D. SARD and M. F. CROUCH: Prog. Cosmic. Ray Phys. **2**, 3 (1954).
[7] J. A. WHEELER: Phys. Rev. **71**, 320 (1947).
[8] P. BUDINI: Nuovo Cim. **8**, 901 (1951).
[9] J. M. KENNEDY: Phys. Rev. **87**, 953 (1954).
[10] H. PRIMAKOFF: Proceedings of the 5th Annual Rochester Conference, p. 174. New York, N. Y.: Interscience Publishers, Inc. 1955.
[11] J. W. KEUFFEL, F. B. HARRISON, T. N. K. GODFREY and G. T. REYNOLDS: Phys. Rev. **87**, 942 (1952).

the capture rate is dependent upon the proton density inside a nucleus and on a nucleon-nucleon correlation function. PRIMAKOFF compared his theory with some unpublished data of KEUFFEL. More exact data are being obtained by LEDERMAN and coworkers[1]. It is obvious that further studies of this sort offer the possibility of giving information similar to that obtained from electron K capture. However, one is now able to look at normally stable nuclei.

β) The products of the capture of slow μ^--mesons. When a μ^--meson is captured, its rest energy of 105 Mev is released in the reaction

$$\mu^- + P \rightarrow N + \nu. \tag{50.3}$$

If the reaction were to occur in a proton at rest, conservation of energy and momentum would give the neutron only about 5.5 Mev, the majority of the energy being carried off by the neutrino.

The extensive experiments of SARD and co-workers showed that the emission of low energy (evaporation) neutrons was the dominant reaction product of the capture of μ^--mesons in complex nuclei (see SARD and CROUCH[2]).

These results were important in establishing the validity of the assumption of the reaction being (50.3). In studying the stopping of cosmic ray μ-mesons they employed cloud chambers in some experiments, a large neutron detector in other experiments, and the combination of the two techniques in one case.

The energy of the neutrons emitted and the number emitted depend on the wave functions of the nucleons inside a nucleus. One representation of this is that familiar to β-decay theory, namely that the transition probabilities depend upon the overlap of the initial configuration state with that of the various final states. Another representation is that the energy acquired by the neutron depends upon the initial momentum of the proton which captures the μ-meson.

From the latter point of view a neutron is likely to acquire a good deal more energy than 5.5 Mev because of the initial motion of the proton and the PAULI principle inhibiting small momentum transfers in a degenerate FERMI gas. These neutrons in general lead to excitation of the entire nucleus and eventually an evaporation process. TIOMNO and WHEELER have calculated the excitation energy to be expected for various nucleuar models; they find the excitation energies are very dependent on the nuclear model.

SARD and co-workers have found that on the average about 2.0 neutrons are emitted when a μ-meson is captured in lead. When one takes into consideration the binding energy of neutrons in lead, this number of neutrons indicates that the average excitation is probably in the range of 14 to 22 Mev instead of the 5.5 Mev one would get for capture by a proton at rest.

Similar conclusions can be drawn from WINSBERG'S[3] studies of the yields of radionuclides resulting from μ^--meson capture by I^{127}. WINSBERG employed radiochemical techniques and a μ^--meson beam from the Chicago Synchrocyclotron. The yields he obtained are given in Table 13.

If the neutron had been left with 5.5 Mev, it would not escape as this energy is less than the binding energy of the neutron in Te^{127}. Therefore the combined yield of 8.2% (for the reactions $I^{127}(\mu^-, \nu)$ Te^{127} and $I^{127}(\mu^-, \nu)$ Te^{127m} indicates that the probability of such a low momentum transfer is very much

[1] L. LEDERMAN and M. WEINRICH: Unpublished. Report at the CERN Symposium on High Energy Physics. CERN Service d'Information, Vol. II, p. 427. Geneva 1956.

[2] R. D. SARD and M. F. CROUCH: Prog. Cosmic Ray Phys. 2, 3 (1954). This review contains references to the work of others.

[3] L. WINSBERG: Phys. Rev. 95, 205 (1954).

inhibited. There are similar results and conclusions from the observation by GODRFEY[1] that there is only a 10% yield of the reaction C^{12} (μ^-, ν) B^{12}.

WINSBERG'S yield of 2.8 day Sb^{122} requires the nucleus to have been left with an excitation energy well above 50 Mev a small percentage of the time. This energy is above the excitation energy which could be acquired if the protons in I^{127} really had a FERMI momentum distribution. The same conclusion can be drawn from the observations of NIKLAS and LAUTERJUNG[2]; they found that some of the neutrons had energies of 45 Mev.

From the work reported in these last three sections, it seems clear that μ^--mesons and electrons interact with nucleons through the same two mechanisms. *The result of these measurements with low energy μ-mesons is that the μ-meson acts as if it were just a heavy (unstable) electron.* In that the interactions are understood and are weak ones, slow μ-mesons become a wonderful tool for probing the nucleus.

Table 13. *Yields of radioactive species from the capture of μ-mesons stopped in I^{127}.*

Nucleons emitted (n = neutron, p = proton)	Nuclide		Yield (in %)
0	9.3 hr.	Te^{127}	5.2 (standard)
0	110 day	Te^{127m}	3.0 ± 0.2
$2n$	58 day	Te^{125m}	12.6 ± 2.5
$4n$	104 day	Te^{123m}	5
$9n$	6 day	Te^{118}	0.03
$p, 2n$	60 day	Sb^{124}	0.23 ± 0.07
$p, 4n$	2.8 day	Sb^{122}	0.16 ± 0.02
$2p, xn$		Sn	none found

51. Reactions of fast μ-mesons. Although at low energies a μ-meson only interacts with nucleons through its electric charge and the FERMI interaction (coupling to the neutrino field), there is the possibility that other interactions may exist at high energies. Most of the studies of the reactions of fast μ-mesons with nuclei have been involved with investigating whether there is another interaction at high energies. Up to 1956 these studies of the interaction of μ-mesons in flight have been performed with μ-mesons from cosmic rays.

AMALDI and FIDECARO[3] looked for the scattering of μ-mesons into angles larger than 20°. They defined their incoming direction by a cosmic ray telescope of coincidence-anticoincidence counters; the telescope was combined with an elaborate array of counters to look for the scattered mesons. They set upper limits on the cross-section for the large angle scattering of μ-mesons- specifically, for μ-mesons with kinetic energy between 200 and 320 Mev the cross-section is less than 4.5×10^{-29} cm²/nucleon; for mesons with energies greater than 320 Mev the cross-section is less than 2.3×10^{-30} cm²/nucleon.

A variety of experiments employing photographic emulsions and cloud chambers and counters have been used to study the scattering of μ-mesons at small angles. These experiments are summarized by ROCHESTER and WOLFENDALE[4] and associates, by KANANGARA and SHRIKANTIA[5] and by GEORGE[6].

At small angles there is a large amount of COULOMB and multiple COULOMB scattering. The interpretation of the results requires an exact knowledge of this COULOMB scattering. Some of the authors found that their data did not agree

[1] T. N. K. GODFREY: Phys. Rev. **92**, 512 (1953).

[2] V. L. NIKLAS and K. H. LAUTERJUNG: Z. Naturforsch. A **8**, 214 (1953).

[3] E. AMALDI and G. FIDECARO: Phys. Rev. **81**, 339 (1951). They list references to earlier work.

[4] V. APPAPILAI, A. W. MAILVAGANAM and A. W. WOLFENDALE: Phil. Mag. **44**, 1059 (1954). — G. D. ROCHESTER and A. W. WOLFENDALE: Phil. Mag. **45**, 980 (1954).

[5] M. L. T. KANANGARA and G. S. SHRIKANTIA: Phil. Mag. **44**, 1091 (1953).

[6] E. P. GEORGE: Prog. Cosmic Ray Phys. **1**, 436 (1952).

with what they estimated for the COULOMB scattering. They attributed the disagreement to an anomalous scattering of the μ-mesons by the nucleus. However COOPER and RAINWATER[1] have calculated the multiple COULOMB scattering from finite size nuclei, and they question whether any anomalous μ-meson scattering has been observed.

The nuclear reaction of fast μ-mesons have been studied by GEORGE[2] in photographic plates exposed undergound, by counter techniques (see ANIS *et al.*[3]) and by the cloud chamber techniques (see ALTHAUS and SARD[4]).

In these experiments a weak but non-vanishing interaction is found, and the cross-section obtained depended on the type of particles being observed by the detection equipment. Where charged particles are being detected in photographic plates, as in the work of GEORGE and EVANS, the cross-section appeared to be as low as 4×10^{-30} cm²/nucleon. However in the work of ANNIS *et al.* where neutrons were being observed, the cross-section is the order of ten to thirty times larger. (The uncertainty is related to a lack of knowledge of the number of neutrons emitted in a reaction.)

It appears possible to explain these reactions as arising from the COULOMB excitation of the nucleus due to the rapidly moving charged meson passing near the nucleus. ANNIS *et al.* have calculated a cross-section for this process. They combined the virtual photon spectrum with the known cross-section for photoneutron production. The value they calculated was in agreement with the value they obtained experimentally. GEORGE and HIGASHI *et al.*[5] have made similar estimates for π meson production by the virtual photons of a fast μ-meson. It may be possible to explain the present observations on the interaction of μ-mesons in flight as arising from the electric charge of the μ-mesons; however the interpretation of the anomalous scattering experiments is still a subject of controversy.

Monochromatic beams of μ-mesons may be useful in extending to higher energies the current studies of the COULOMB excitation of nuclei. They have the advantage over protons of being free from the complication of other nuclear reactions that set in with high energy protons.

52. Meson production. μ-mesons in general are produced as the decay products of π and K mesons (see Table 11). However when high energy photons pass through a COULOMB field μ-meson pairs ($\mu^+\mu^-$) should be produced by a process completely analogous to electron-positron pair production. The cross-section for this process is very small and is not a useful source of μ-mesons.

As first pointed out by FELD *et al*[6], the collisive momenta involved in μ-pair production are so large that the finite extension of the COULOMB charge of the nucleus requires that the cross-section be modified by a form factor (see Sect. 7). MASEK and PANOFSKY[7] have observed μ-mesons being produced in approximate agreement with the theoretical cross-section modified by a form calculated by RAWITSCHER[8]. Further experiments on μ-meson pair production are in progress.

π-mesons are produced directly by the bombardment of nuclei by high energy nucleons, photons, or electrons. As mentioned in Sect. 3, the energy distribution

---

1 L. W. COOPER and J. RAINWATER: Phys. Rev. **97**, 492 (1955).

2 E. P. GEORGE: Prog. Cosmic Ray Phys. **1**, 428 (1952).

3 M. ANNIS, H. C. WILKINS and J. D. MILLER: Phys. Rev. **94**, 1038 (1954).

4 E. J. ALTHAUS and R. D. SARD: Phys. Rev. **91**, 382 (1953).

5 S. HIGASHI, M. ODA, T. OSHIO, H. SHIBATA, K. WATANABE and Y. WATASE: Progr. Theor. Phys. **16**, 250 (1956).

6 B. T. FELD, A. JULIAN, A. C. ODIAN, L. S. OSBORNE and A. WATTENBERG: Phys. Rev. **96**, 1386 (1954).

7 G. E. MASEK and W. K. H. PANOFSKY: Phys. Rev. **101**, 1094 (1956).

8 G. H. RAWITSCHER: Phys. Rev. **101**, 423 (1956).

of the π-mesons is intimately connected with the momentum of the nucleons in the nucleus. One can understand this with the aid of SERBER'S [4] model of high energy reactions; the bombarding nucleon or photon is assumed to create the meson in an interaction with a single nucleon in the target nucleus. The calculations are carried out in BORN approximation or by a somewhat questionable application of the "Impulse Approximation" (see Sect. 9). The cross-section for meson production from a single nucleon is used taking into account that the struck nucleon is in motion inside the nucleus; thus monoenergetic photons or nucleons hitting nucleons in motion result in a spread in energies for the mesons which are ejected at a fixed angle. In the case of production of π mesons by photons, the energy distribution of the pions from monoenergetic photons has to be integrated over the photon bremsstrahlung spectrum. Such a calculation has been carried through by LAX and FESHBACH[1], using the CHEW-GOLDBERGER momentum distribution [6]; they find a meson energy spectrum at 90° which is in agreement with the experimental results obtained from a carbon target by STEINBERGER and BISHOP[2].

In that a double integration is involved and that a theoretical instead of experimental single particle cross-section is employed, this agreement between theory and experiment should not be construed as a sensitive check on the momentum distribution of nucleons. Their calculation should be considered as indicating the feasibility of interpreting photomeson production from complex nuclei in terms of purely kinematical factors. The possibility of obtaining information on nucleon momentum distributions from more elaborate photomeson production experiments is under investigation by RANKIN *et al.*[3].

In that proton beams are monoenergetic, π-meson production experiments with protons are more sensitive checks on the momentum distribution. The early measurements of RICHMAN and WILCOX[4] and BLOCK *et al.* on π-mesons produced by the proton bombardment of carbon have been theoretically investigated by HENLEY and HUDDLESTONE[5] and by HENLEY. In the paper by HENLEY[6], the scattering and reabsorption of the mesons after their production is investigated. He also takes account of the PAULI principle which forbids those events where the final nucleon is left in an already occupied state. HENLEY compared the experimental data with the predictions of Gaussian, FERMI, and modified CHEW-GOLDBERGER momentum distributions. The observed π-meson spectrum from carbon is in closest agreement with the assumption of a Gaussian momentum distribution with an average energy of about 19 Mev (i.e. a $1/e$ energy = 13 Mev). BLOCK, PASSMAN, and HAVENS[7] also find their own data on C are fitted best by the assumption of a Gaussian momentum distribution with an average energy of about 19 Mev.

The reabsorption of mesons produced in a nucleus is a very important effect and even manifests itself in the photomeson production from deuterium (see Sect. 40); in larger nuclei one sees meson production varying as $A^{\frac{2}{3}}$ (see, for example, LITTAUER and WALKER[8]). The oversimplified picture is that those mesons produced in the central regions of a large nucleus do not escape. In the

[1] M. LAX and H. FESHBACH: Phys. Rev. **81**, 189 (1951).
[2] J. STEINBERGER and A. S. BISHOP: Phys. Rev. **78**, 494 (1950).
[3] W. RANKIN, P. C. STEIN and A. WATTENBERG: Unpublished.
[4] C. RICHMAN and H. A. WILCOX: Phys. Rev. **78**, 496 (1950).
[5] E. M. HENLEY and R. HUDDLESTONE: Phys. Rev. **82**, 754 (1951).
[6] E. M. HENLEY: Phys. Rev. **85**, 204 (1952).
[7] M. M. BLOCK, S. PASSMAN and W. W. HAVENS: Phys. Rev. **88**, 1239 (1950).
[8] R. M. LITTAUER and D. WALKER: Phys. Rev. **86**, 838 (1952).

case of the photoproduction of π^0-mesons there is constructive interference in the forward direction. BARRINGER and OSBORNE[1] observe that the photoproduction of π^0-mesons in the forward direction varies as $A^{\frac{4}{3}}$. One can crudely understand this on the basis of the number of surface nucleons varying as $A^{\frac{2}{3}}$ and the coherent production being dependent upon the square of the number of effective nucleons. However the meson production process is very much involved with the pion-nucleon forces, and one can not hope to obtain the values of cross-sections from purely kinematical relationship or only nuclear parameters.

53. Reactions of π-mesons with nuclei. The remark at the end of the last section holds true as regards the reactions of π-mesons with nuclei. The values of the cross-sections and the angular distributions in scattering are a result of pion-nucleon interactions and nucleon-nucleon interactions. Most of the experimental and theoretical investigations were motivated by a desire to study the pion-nucleon interaction; however it is frequently found that the interpretation of data involves nucleon-nucleon interactions.

As in most nuclear physics, it is possible to develop phenomenological theories that provide convenient formalisms for expressing the results of the experiments. For example the results of experiments on the scattering and absorption of fast mesons can be expressed in terms of potentials or the parameters of the optical model of FERNBACH, SERBER, and TAYLOR [7] (see Sects. 8 and 10) developed for nuclear reactions.

A good bibliography and a brief review of the nuclear reactions of fast π-mesons can be found in the article by KESSLER and LEDERMAN[2]; they present the following summary. At low energies the dominant process is absorption of the meson followed by star formation; at 40 Mev star formation has a cross-section which is roughly half the geometrical cross-section of a nucleus. Only a small elastic scattering cross-section is observed. As the energy of the π-mesons increases the cross-sections for star formation and elastic scattering rise. At 60 Mev inelastic scattering begins to occur with an appreciable probability which increases rapidly with energy. At 80 Mev the total reaction cross-section (elastic scattering is omitted) is approximately the geometrical size of the nucleus for all complex nuclei. Experiments performed with higher energy mesons appear to indicate a slow rise in the cross-section for some nuclei. KESSLER and LEDERMAN find that 125 Mev negative π-mesons have a geometric cross-section in carbon but a slightly higher cross-section in lead.

By use of a cloud chamber, they found the cross-sections listed in Table 14. They find that the optical model parameters for π-mesons which fit the 125 Mev data are a 30 Mev deep (attractive) potential well and a mean free path of 3×10^{-13} cm. At 60 Mev the corresponding parameters are a 20 Mev deep well and a mean free path 8×10^{-13} cm. The mean free path in nuclei of high energy pions is shorter than that of nucleons (see Sect. 32). However it is not as short as one would calculate from the pion-single nucleon cross-section (see Sect. 10). It is not clear whether this discrepancy arises from nucleon-nucleon interactions or the assumption of a uniform density distribution of nucleons.

There have been theoretical investigations which attempt to relate the optical model parameters to what is known about pion-nucleon interactions. STERNHEIMER'S[3] calculations of the optical parameters show that they are very energy dependent. The mean free path goes through a minimum at 180 Mev of about

[1] R. BARRINGER and L. S. OSBORNE: Unpublished.
[2] J. O. KESSLER and L. M. LEDERMAN: Phys. Rev. **94**, 689 (1954).
[3] R. M. STERNHEIMER: Phys. Rev. **101**, 384 (1956).

8×10^{-14} cm and then increases sharply; it goes through a broad maximum at about 500 Mev and then levels off. The cross-sections he calculates agree within experimental error with the results of KESSLER and LEDERMAN.

FRANK, GAMMEL, and WATSON[1] have investigated theoretically the basis for the parameters of the optical model for pion-nucleus scattering. They also find that the real part of the potential changes at about 160 Mev from being attractive at low energies to being repulsive at higher energies. They believe that studies of pion-nucleus scattering may provide evidence, for or against, the basic assumption of two-body forces in nuclei.

PEVSNER, RAINWATER, and associates[2], have measured the scattering of π-mesons in a series of experiments. In the case of Al, they attempted to fit the data by an optical model calculation and found poor agreement with the observed angular distribution. For Li they did a detailed calculation assuming the wave functions of the neutrons and protons were those given by an harmonic oscillator

Table 14. *Observed cross-section[3] for 125 Mev π^--mesons on C and Pb.*

	Geometric size[4]	Elastic scattering ($\vartheta > 20°$)	Inelastic scattering $\Delta E > 40$ Mev	Charge exchange scattering[5] (π^-, π^0)	Stars and stops	Total reaction
Carbon	325	179 ± 18	70 ± 11	20^{+20}_{-10}	220 ± 40	308 ± 43
Lead	2150	700 ± 84	530 ± 130	100^{+80}_{-40}	1840 ± 350	2477 ± 385

potential, and that the fundamental angular distribution of the scattered π-mesons was the same as that observed for the scattering of a pion by a single nucleon. The calculated angular distribution for Li agreed quite well the observed data except at small angles where the COULOMB scattering was neglected.

Only brief mention is made here of the large number of radiochemical and photographic emulsion studies of the capture of π^--mesons. An excellent review of this work up to 1953 has been given by TURKEVITCH[6] and more recent work can be found in the bibliography of WINSBERG's article[7]. The main conclusion is that the π^--meson when it is captured gives up its entire rest mass to the nucleus and a cascade and nuclear evaporation ensue (see Sects. 11 and 12). The end products have distribution similar to those described for nucleon reactions (Sect. 36) and photon reactions (Sect. 47). Meson induced fission has been studied by AL-SALAM[8], SUGARMAN[9], and JOHN and FRY[10].

There is the possibility that the initial reaction involved in π^--meson capture may more closely resemble high energy photon reactions than high energy nucleon reactions in that two particles may initiate the cascade. BRUECKNER,

[1] R. M. FRANK, J. L. GAMMEL and K. M. WATSON: Phys. Rev. **101**, 891 (1956).

[2] A. PEVSNER, J. RAINWATER, R. E. WILLIAMS and S. J. LINDENBAUM: Phys. Rev. **100**, 1419 (1955). — A. PEVSNER and J. RAINWATER: Phys. Rev. **100**, 1431 (1955). — R. WILLIAMS, J. RAINWATER and A. PEVSNER: Phys. Rev. **101**, 412 (1956).

[3] Cross-sections are in millibarns.

[4] The nuclear radius is supposed to be $1.40 A^{1/3} \times 10^{-13}$ cm.

[5] The charge exchange scattering reaction is a negative π-meson going in and neutral π-meson coming out with a proton in the nucleus changed into a neutron. The π^0-meson immediately decays ($\sim 10^{-16}$ sec) into two γ-rays.

[6] A. TURKEVITCH: Ann. Rev. Phys. Chem. **4**, 119 (1953).

[7] L. WINSBERG: Phys. Rev. **95**, 198 (1954).

[8] S. G. AL-SALAM: Phys. Rev. **84**, 254 (1951).

[9] N. SUGARMAN: Phys. Rev. **86**, 604 (1952).

[10] W. JOHN and W. F. FRY: Phys. Rev. **91**, 1234 (1953).

SERBER, and WATSON[1] have proposed that the basic mechanism for the absorption of a π-meson involves two nucleons. If this is the case one may use meson absorption experiments to obtain a measure of the correlation amongst nucleons in nuclei. From preliminary evidency BRUECKNER *et al.* find an indication that a high degree of correlation between nucleons is present in complex nuclei.

54. *K*-mesons and hyperfragments. At the time this article is being written there are available only early semiquantitative results on the interaction of K-mesons[2] and hyperons with nuclei. Most of the experiments with such particles have been reported at the Pisa conference[3] or at the recent Rochester conferences[4].

The main reaction of K^+-mesons with nuclei is elastic scattering; a few events have been observed which can be interpreted as charge exchange scatterings (K^+, ϑ^0); nuclear absorption of K^+-mesons does not appear to take place.

However, the main reaction for K^--mesons is nuclear absorption resulting in stars or hyperons; they also elastically and charge exchange scatter. The total cross-section for K^--meson interactions is commensurate with the geometrical size of the nucleus. The cross-section for scattering of K^+-mesons seems to be about one order of magnitude smaller than the K^- total cross-section.

K^--meson brought to rest in matter are captured by a nucleus instead of undergoing free decay (see π^--mesic atoms in Sect. 49). The capture manifests itself as stars in a photographic emulsion or bubble chamber containing complex nuclei. In emulsion it is observed that a hyperfragment comes out of about 5% of these stars. These hyperfragments are a new species of nuclei in which one of the constituents is a Λ° hyperon. These hyperfragments or Λ° containing nuclei decay in about 10^{-10} seconds and are characterized by a release of energy that is within a few Mev of the decay energy of a Λ° hyperon. An example of a hyperfragment decay is

$$_{\Lambda}\mathrm{He}^4 \rightarrow \mathrm{He}^3 + p + \pi^- .$$

Hyperfragments also come out of stars made by: high energy cosmic rays, protons with energies of 3×10^9 ev or higher, and high energy π^--mesons. However, the frequency of emission of hyperfragments from stars formed by such collisions is between 10^{-3} and 10^{-4} (hyperfragments per star).

The first hyperfragment was observed by DANYSZ and PNIEWSKI[5]. Since then many such events have been recognized; references to these observations can be found in articles by DALETZ[6], JONES and KNIPP[7], and FRY, SCHNEPS, and SWAMI[8]. FRY *et al.* have studied about 100 such hyperfragments. $_{\Lambda}\mathrm{H}^3$ seems to be the lightest nucleus which contains a bound Λ°, the heaviest observed by them is $_{\Lambda}\mathrm{C}^{11}$. The binding energy of the Λ° in $_{\Lambda}\mathrm{H}^3$ is about 0.2 Mev; for heavier hyperfragments the binding energy of the Λ° increases with the mass at a rate of about 1 Mev per nucleonic mass unit. Such binding energies are consistent with a potential well depth for the Λ° of about 20 Mev.

---

[1] K. A. BRUECKNER, R. SERBER and K. M. WATSON: Phys. Rev. **84**, 258 (1951).

[2] By K-mesons in this section are meant all the various mesons (of mass about 965 m_e) and their charge states including τ-mesons and ϑ°-mesons (see table 11).

[3] Report on the Pisa conference of 1955 (to be published in Nuovo Cimento).

[4] Proceedings of the Fifth and Sixth Annual Rochester Conferences. New York, N.Y.: Interscience Publishers, Inc. 1955 and 1956.

[5] M. DANYSZ and L. PNIEWSKI: Phil. Mag. **44**, 348 (1953).

[6] R. H. DALETZ: Phys. Rev. **99**, 1475 (1955).

[7] J. T. JONES and J. KNIPP: Nuovo Cim. **2**, 857 (1955).

[8] W. F. FRY, J. SCHNEPS and M. S. SWAMI: Phys. Rev. **99**, 1561 (1955); **101**, 1526 (1956).

Hyperfragments with a mass greater than 5 are more likely to decay by the ejection of several fast nucleons rather than by the emission of a low energy pion because of phase space arguments. The nucleonic decay mode can be looked upon as analogous to an internal conversion of the pion. It is hoped to obtain more information about the structure of such hyperfragments from studies of: their binding energies, the angular correlations between the emitted π^--meson and the proton, and the ratio of mesonic to nucleonic decays.

It should be noted that the Λ° is found to be bound in ${}_\Lambda He^5$ (specifically with about 5 Mev of binding). The existence of such a bound hyperfragment shows that the Λ° can naïvely be considered to maintain its identity inside a nucleus and is not affected by those considerations of the PAULI principle which lead to normal nuclei of mass 5 being unstable.

It is too early to foretell what the studies of these peculiar new nuclear species will contribute to our understanding of normal nuclei. It is to be hoped that a new dimension has been added to nuclear physics.

Acknowledgment.

The author wishes to express his gratitude to the U.S. Office of Naval Research and the U.S. Atomic Energy Commission for that part of the support he has received from their joint program.

General references.

[1] ROSSI, B.: High Energy Particles. New York: Prentice Hall, Inc. 1952.
[2] MORRISON, P.: Experimental Nuclear Physics, Vol. II, E. SEGRÈ, editor. New York: John Wiley & Sons, Inc. 1953.
[3] BLATT, J. M., and V. F. WEISSKOPF: Theoretical Nuclear Physics. New York: John Wiley & Sons (1952).
[4] SERBER, R.: Phys. Rev. **72**, 1144 (1947).
[5] GOLDBERGER, M. L.: Phys. Rev. **74**, 1269 (1948).
[6] CHEW, G. F., and M. L. GOLDBERGER: Phys. Rev. **77**, 470 (1950).
[7] FERNBACH. I., R. SERBER and T. B. TAYLOR: Phys. Rev. **75**, 1352 (1949).
[8] HOFSTADTER, R., and associates, see references in B. HAHN, D. G. RAVENHALL and R. HOFSTADTER: Phys. Rev. **101**, 1127 (1956).

Sachverzeichnis.

(Deutsch-Englisch.)

Bei gleicher Schreibweise in beiden Sprachen sind die Stichwörter nur einmal aufgeführt.

Subject Index.

(English-German.)

Where English and German spelling of a word is identical the German version is omitted.